Spon's
External Works
and Landscape
Price Book

2013

Spon's External Works and Landscape Price Book

Edited by
Davis Langdon, An AECOM Company

in association with
LandPro Ltd
Landscape Surveyors

2013

Thirty-Second edition

Spon Press
an imprint of Taylor & Francis
LONDON AND NEW YORK

184 548

692.5 SPO
NTBTA

First edition 1978
Thirty-Second edition published 2013
by Spon Press
2 Park Square, Milton Park, Abingdon, Oxon, OX14 4RN

Simultaneously published in the USA and Canada
by Spon Press
711 Third Avenue, New York, NY10017

Spon Press is an imprint of the Taylor & Francis Group, an informa business

British Library Cataloguing in Publication Data
A catalogue record for this book is available from the British Library

ISBN 13: 978-0-415-69079-9 (hardback)
ISBN 13: 978-0-203-07706-1 (ebook)

ISSN: 0267-4181

Typeset in Arial by Taylor & Francis Books

MIX
Paper from
responsible sources

FSC
www.fsc.org   **FSC® C004839**

Printed and bound in Great Britain by
TJ International Ltd, Padstow, Cornwall

Contents

PART 5: APPROXIMATE ESTIMATING RATES – MINOR WORKS   415

PART 6: PRICES FOR MEASURED WORKS – MINOR WORKS   437

Preface to the Thirty-Second Edition

Market conditions

The Landscape construction sector of the construction industry is a mixed bag this year. Whilst some contractors are reporting that they are enjoying profitable conditions, we perceive that there is a general trend of a harder and keener market. Costs of materials are up and there is pressure from the labour force to increase wages in line with inflation. Most employers however are managing to retain their labour at minimal increased rates or even at rates frozen to last year's scales. Tender prices are keener and margins on tender awards are lower. The constructed cost index for this year's book as published in the general section of this book shows an index of + 4%.

Key influences

- Construction inflation measures all falling
- Consumer price inflation higher than construction price inflation and forecast fall being stubbornly resisted
- Higher oil prices have impacted trades such as road building
- Manufacturing industries' input costs sharply lower

Adjustment formulae for construction contracts

Price Adjustment Formulae indices, compiled by the Building Cost Information Service (previously by the Department for Business Innovation & Skills), are designed for the calculation of increased costs on fluctuating or Variation of Price contracts. They provide useful guidance on cost changes in various trades and industry sectors and on the differential movement of work sections in Spon's Price Books. Over the last twelve months between April 2011 and April 2012, the 60 Building Work categories recorded an average rise of 3.3%, well down from 4.4% three months previously and 5.2% six months ago.

Those works categories showing the highest increases over the last twelve months that are relevant to this book are Pavings: coated macadam and asphalt – 10.7%

Supplier prices

Similarly to last year, there have been many suppliers who have only reflected minimal increases or even no increase in their prices especially for material supplies; The larger suppliers have increased their list costs by standard 5% – 7.5% but we perceive that there is room to negotiate on material cost rates.

Labour rates

This year we have reduced the price of the base cost of labour by £0.25p (Approximately £440.00 per annum). This is due to our perception of an increased pressure on labour during the coming 12 months and to ensure validity of the rates in this book. Readers may make adjustments to these rates in accordance with the instructions and the indices tables published in the general section following this preface.

The standard rate for this years' calculation's for Major works is £19.25 per hour a reduction of 1.3%. The labour rate was increased in last years' publication and this decrease finely tunes that adjustment.

Plant rates

The hired in heavy plant suppliers referenced in this publication have advised us during our update to keep their prices the same. The price of red diesel has increased and this would have some effect on the hourly rates of the plant used. Our supplier of prices to this publication comments that the large national cross-hiring companies are driving hire prices down.

Tables and memoranda

We remind readers of this section at the back of the book. It is often overlooked but contains a wealth of information useful to external works consultants and contractors.

Profit and overhead

Spon's External Works and Landscape prices do not allow for profit or site overhead. Readers should evaluate the market conditions in the sector or environment in which they operate and apply a percentage to these rates. Company overhead is allowed for within the labour rates used. Please refer to 'Part 1 General' for an explanation of the build-up of this year's labour rates.

Prices for suppliers and services

We acknowledge the support afforded to us by the suppliers of products and services who issue us with the base information used. Their contact details are published in the directory section at the front of this book. We advise that readers wishing to evaluate the cost of a product or service should approach these suppliers directly should they wish to confirm prices prior to submission of tenders or quotations.

Whilst every effort is made to ensure the accuracy of the information given in this publication, neither the editors nor the publishers in any way accept liability for loss of any kind resulting from the use made by any person of such information.

We remind readers that we would be grateful to hear from them during the course of the year with suggestions or comments on any aspect of the contents of this year's book and suggestions for improvements. Our contact details are shown below.

DAVIS LANGDON, An AECOM COMPANY
MidCity Place
71 High Holborn
London
WC1V 6QS

SAM HASSALL
LANDPRO LTD
Landscape Surveyors
14 Upper Bourne Lane
Farnham
Surrey
GU10 4RQ

Tel: 02070 617000
E-mail: spons@davislangdon.com

Tel: 01252 795030
E-mail: info@landpro.co.uk

Acknowledgements

This list has been compiled from the latest information available but as firms frequently change their names, addresses and telephone numbers due to reorganization, users are advised to check this information before placing orders.

ACO Technologies Plc
ACO Business Park
Hitchin Road
Shefford
Bedfordshire SG17 5TE
Tel: 01462 816666
Website: www.aco.co.uk
E-mail: technologies@aco.co.uk
Linear drainage

Addagrip Surface Treatments UK Ltd
Addagrip House
Bell Lane Industrial Estate
Uckfield
East Sussex TN22 1QL
Tel: 01825 761333
Website: www.addagrip.co.uk
E-mail: sales@addagrip.co.uk
Resin bound & bonded surfacing

Agripower Ltd
Broomfield Farm
Rignall Road
Great Missenden
Bucks HP16 9PE
Tel: 01494 866776
Website: www.agripower.co.uk
E-mail: info@agripower.co.uk
Land drainage and sports grounds

Amenity Land Solutions Ltd (ALS)
Allscott Park
Allscott
Telford
Shropshire TF6 5DY
Tel: 01952 641949
Website: www.amenity.co.uk
E-mail: sales@amenity.co.uk
Horticultural products supplier

Architectural Heritage Ltd
Taddington Manor
Cutsdean
Cheltenham
Gloucestershire GL54 5RY
Tel: 01386 584414
Website: www.architectural-heritage.co.uk
E-mail: puddy@architectural-heritage.co.uk
Ornamental stone buildings and water features

Autopa Ltd
Cottage Leap
Rugby
Warwickshire CV21 3XP
Tel: 01788 550556
Website: www.autopa.co.uk
E-mail: info@autopa.co.uk
Street furniture

AVS Fencing Supplies Ltd
Unit 1 AVS Trading Park
Chapel Lane
Milford
Surrey GU8 5HU
Tel: 01483 410960
Website: www.avsfencing.co.uk
E-mail: sales@avsfencing.co.uk
Fencing

Bauder Ltd
70 Landseer Road
Ipswich
Suffolk IP3 0DH
Tel: 01473 257671
Website: www.bauder.co.uk
E-mail: info@bauder.co.uk
Green roof systems

BBS Green Roofing Ltd
Vitis House
50 Dickens Street
Battersea
London SW8 3EQ
Tel: 0207 622 6225
Website: http://www.green-roofing.co.uk/
Email: bbs@green-roofing.co.uk
Specialist Green Roof Consultants / Installers

Bituchem Group
Forest Vale Industrial Estate
Cinderford
Gloucestershire GL14 2YH
Tel: 01594 826768
Website: www.bituchem.com
E-mail: info@bituchem.com
Coloured hard landscaping materials

Boddingtons Ltd
Fiberweb Geosynthetics Ltd
Blackwater Trading Estate
The Causeway
Maldon
Essex CM9 4GG
Tel: 01621 874200
Website: www.boddintons-ltd.com
E-mail: sales@boddingtons-ltd.com
Erosion control, soil stabilization, plant protection

Breedon Special Aggregates
(A division of Breedon Aggregates)
Breedon Quarry
Breedon-on-the-Hill
Derby
Derbyshire DE73 8 AP
Tel: 01332 694001
Website: www.breedon-special-aggregates.co.uk
E-mail: specialaggs@breedonaggregates.com
Specialist gravels

British Seed Houses Ltd
Camp Road
Witham St Hughs
Lincoln LN6 9QJ
Tel: 01522 868714
Website: www.bshamenity.com
E-mail: seeds@bshlincoln.co.uk
Grass and wildflower seed

British Sugar Topsoil
Sugar Way
Peterborough PE2 9 AY
Tel: 0870 240 2314
Website: www.bstopsoil.co.uk
E-mail: topsoil@britishsugar.com
Topsoil

Broxap Street Furniture
Rowhurst Industrial Estate
Chesterton
Newcastle-under-Lyme
Staffordshire ST5 6BD
Tel: 0844 800 4085
Website: www.broxap.com
E-mail: sales@broxap.com
Street furniture

Capital Garden Products Ltd
Gibbs Reed Barn
Pashley Road
Ticehurst
East Sussex TN5 7HE
Tel: 01580 201092
Website: www.capital-garden.com
E-mail: sales@capital-garden.com
Plant containers

CED Ltd
728 London Road
West Thurrock
Grays
Essex RM20 3LU
Tel: 01708 867237
Website: www.ced.ltd.uk
E-mail: sales@ced.ltd.uk
Natural stone

Cedar Nursery
Horsley Road
Cobham
Surrey KT11 3JX
Tel: 01932 862473
Website: www.landscaping.co.uk
E-mail: sales@landscaping.co.uk
Aggregate containment grids

Charcon Hard Landscaping
Hulland Ward
Ashbourne
Derbyshire DE6 3ET
Tel: 01335 372222
Website: www.aggregate.com
E-mail: ukenquiries@aggregate.com
Paving and street furniture

Cleartrack (EvL) Ltd
Ploughmans Barn
Station Farm
Denton Road
Horton
Northants NN7 2BG
Tel: 01604 871360
Website: www.cleartrack.co.uk
E-mail: info@cleartrack.co.uk
Ground clearance

Combined Harvesters Ltd
Britannia House
Dock Road
Birkenhead
Wirral CH41 1DF
Tel: 0151 639 0880
Website: www.combinedharvesters.co.uk
E-mail: info@combinedharvesters.co.uk
Rainwater harvesting

Cooper Clarke Civils and Lintels
Bloomfield Road
Farnworth
Bolton BL4 9LP
Tel: 01204 862222
Website: www.civilsandlintels.co.uk
E-mail: farnworthspecials@civilsandlintels.co.uk
Wall drainage and erosion control

Craft Pegg
34 Riverside Building
Trinity Buoy Wharf
London E14 0JY
Tel: 0207 538 9010
Website: www.craftpegg.com
E-mail: mail@craftpegg.com
Landscape architects

CU Phosco Ltd
Charles House
Lower Road
Great Amwell
Ware
Hertfordshire SG12 9TA
Tel: 01920 860600
Website: www.cuphosco.co.uk
E-mail: sales@cuphosco.co.uk
Lighting

D H Loveday & Son Ltd
96 Cotterells
Hemel Hempstead
Hertfordshire HP1 1JG
Tel: 01442 256254
Aggregates

Deepdale Trees Ltd
Tithe Farm
Hatley Road
Sandy
Bedfordshire SG19 2DX
Tel: 01767 262636
Website: www.deepdale-trees.co.uk
E-mail: mail@deepdale-trees.co.uk
Trees, multi-stem and hedging

DLF Trifolium Ltd
Thorn Farm
Evesham Road
Inkberrow
Worcestershire WR7 4LJ
Tel: 01386 791102
Website: www.dlf.co.uk
E-mail: amenity@dlf.co.uk
Grass and wildflower seed

Duracourt (Spadeoak) Ltd
Town Lane
Wooburn Green
High Wycombe
Buckinghamshire HP10 0PD
Tel: 01628 529421
Website: www.duracourt.co.uk
E-mail: info@duracourt.co.uk
Tennis courts

Earth Anchors Ltd
15 Campbell Road
Croydon
Surrey CR0 2SQ
Tel: 020 8684 9601
Website: www.earth-anchors.com
E-mail: sales@earth-anchors.com
Anchors for site furniture

Elliott
Manor Drive
Peterborough PE4 7 AP
Tel: 01733 298700
Website: www.elliottuk.com
Email: hirediv@elliottuk.com
Site offices

English Woodlands
Burrow Nursery
Cross in Hand
Heathfield
East Sussex TN21 0UG
Tel: 01435 862992
Website: www.ewburrownursery.co.uk
E-mail: sales@ewburrownursery.co.uk
Plant protection

Eura Conservation Ltd
Unit H10, Halesfield 19
Telford
Shropshire TF7 4QT
Tel: 01952 680 218
Website: www.eura.co.uk
E-mail: enquiries@eura.co.uk
Metalwork restoration

Everris Limited
Epsilon House
West Road
Ipswich
Suffolk IP3 9FJ
Tel: 01473 237100
Website: www.everris.com
E-mail: prof.sales@everris.com
Landscape chemicals

Eve Trakway
Bramley Vale
Chesterfield
Derbyshire S44 5GA
Tel: 08700 767676
Website: www.evetrakway.co.uk
E-mail: marketing@evetrakway.co.uk
Portable roads

Exterior Decking
Exterior Solutions Ltd
Unit 5b
Reed Industrial Estate
28 Plantation Road
Amersham
Buckinghamshire HP6 6HJ
Tel: 01494 722204
Website: www.exteriordecking.co.uk
E-mail: office@exteriordecking.co.uk
Decking

Fairwater Ltd
Lodge Farm
Malthouse Lane
Ashington
West Sussex RH20 3BU
Tel: 01903 892228
Website: www.fairwater.co.uk
E-mail: info@fairwater.co.uk
Water garden specialists

Farmura Environmental Products Ltd
Stone Hill
Egerton
Ashford
Kent TN27 9DU
Tel: 01233 756241
Website: www.farmura.com
E-mail: info@farmura.com
Organic fertilizer suppliers

Forticrete Ltd
Anstone Works
Kiverton Park Station
Kiverton Park
Sheffield S26 6NP
Tel: 0870 9034015
Website: www.forticrete.co.uk
E-mail: info@forticrete.com
Concrete retaining wall systems

FP McCann Ltd
Whitehill Road
Coalville
Leicester LE67 1ET
Tel: 01530 240000
Website: www.fpmccann.co.uk
E-mail: info@fpmccann.co.uk
Precast concrete pipes

Furnitubes Street Furniture
Meridian House
Royal Hill
Greenwich
London SE10 8RT
Tel: 020 8378 3200
Website: www.furnitubes.com
E-mail: spons@furnitubes.com
Street furniture

Gardenlink Ltd
27 Lesbourne Road
Reigate
Surrey RH2 7JS
Tel: 01737 243 224
Website: www.gardenlink.co.uk
E-mail: enquiries@gardenlink.co.uk
Steel edgings

The Garden Trellis Company
Unit 1 Brunel Road
Gorse Lane Industrial Estate
Clacton-On-Sea
Essex CO15 4LU
Tel: 01255 688361
Website: www.gardentrellis.co.uk
E-mail: info@gardentrellis.co.uk
Garden joinery specialists

Geosynthetics Ltd
Fleming Road
Harrowbrook Ind Est
Hinckley
Leicestershire LE10 3DU
Tel: 01455 617139
Website: www.geosyn.co.uk
E-mail: sales@geosyn.co.uk
Cellular containment systems

Goroots Ltd
37 The Chase
Eastcote HA5 1SH
Tel: 0208 429 8049
Website: www.goroots.co.uk
Hydroseeding

Grass Concrete Ltd
Duncan House
142 Thornes Lane
Thornes
West Yorkshire WF2 7RE
Tel: 01924 379443
Website: www.grasscrete.com
E-mail: info@grasscrete.com
Grass block paving

Greenfix Soil Stabilization and Erosion Control Ltd
Allens West
Durham Lane
Eaglescliffe
Stockton on Tees TS16 ORW
Tel: 01642 888693
Website: www.greenfix.co.uk
E-mail: stockton@greenfix.co.uk
Seeded erosion control mats

Greenleaf Horticulture
Ivyhouse Industrial Estate
Haywood Way
Hastings TN35 4PL
Tel: 01424 717797
E-mail: enquiries@greenleaftrees.co.uk
Root directors

Griffin Nurseries
New barn farm
Rake road
Milland
Liphook
Hants GU30 7JU
Tel: 01428 741655
Website: www.griffinnurseries.co.uk
E-mail: enquiries@griffinnurseries.co.uk
Hedging

Grundon Waste Management Ltd
Goulds Grove
Ewelme
Wallingford OX10 6PJ
Tel: 0870 4438278
Website: www.grundon.com
E-mail: sales@grundon.com
Gravel

Guncast Swimming Pools Ltd
Unit 4
Hampers Common Industrial Estate
Petworth
West Sussex GU28 9NR
Tel: 01798 343221
Website: www.guncast.com
E-mail: info@guncast.com
Swimming pools

Haddonstone Ltd
The Forge House
Church Lane
East Haddon
Northampton NN6 8DB
Tel: 01604 770711
Website: www.haddonstone.com
E-mail: info@haddonstone.co.uk
Landscape ornaments and architectural cast stone

Harrison External Display Systems
Borough Road
Darlington
Co Durham DL1 1SW
Tel: 01325 355433
Website: www.harrisoneds.com
E-mail: sales@harrisoneds.com
Flagpoles

Havells Sylvania Fixtures UK Ltd
Avis Way
Newhaven
East Sussex BN9 0ED
Tel: 0870 6062030
Website: www.havells-sylvania.com
E-mail: info.concord@havells-sylvania.com
Lighting

Headland Amenity Ltd
1010 Cambourne Business Park
Cambourne
Cambridgeshire CB23 6DP
Tel: 01223 597834
Website: www.headlandamenity.com
E-mail: info@headlandamenity.com
Chemicals

Heicom UK
4 Frog Lane
Tunbridge Wells
Kent TN1 1YT
Tel: 01892 522360
Website: www.treesand.co.uk
E-mail: mike.q@treesand.co.uk
Tree sand

Hepworth Wavin Plc
Hazelhead
Crow Edge
Sheffield S36 4HG
Tel: 0870 4436000
Website: www.hepworthdrainage.co.uk
E-mail: info@hepworthdrainage.co.uk
Drainage

HSS Hire
Unit 3
14 Wates Way
Mitcham
Surrey CR4 4HR
Tel: 020 8685 9500
Website: www.hss.com
E-mail: hire@hss.com
Tool and plant hire

Inturf
The Chestnuts
Wilberfoss
York YO41 5NT
Tel: 01759 321000
Website: www.inturf.com
E-mail: info@inturf.co.uk
Turf

J Toms Ltd
7 Marley Farm
Headcorn Road
Smarden
Kent TN27 8PJ
Tel: 01233 770066
Website: www.jtoms.co.uk
E-mail: jtoms@btopenworld.com
Plant protection

Jacksons Fencing
Stowting Common
Ashford
Kent TN25 6BN
Tel: 01233 750393
Website: www.jacksons-fencing.co.uk
E-mail: sales@jacksons-fencing.co.uk
Fencing

James Coles & Sons (Nurseries) Ltd
The Nurseries
Uppingham Road
Thurnby
Leicester LE7 9QB
Tel: 01162 412115
Website: www.colesnurseries.co.uk
E-mail: sales@colesnurseries.co.uk
Nurseries

Johnsons Wellfield Quarries Ltd
Crosland Hill
Huddersfield
West Yorkshire HD4 7 AB
Tel: 01484 652311
Website: www.johnsons-wellfield.co.uk
E-mail: sales@johnsons-wellfield.co.uk
Natural Yorkstone pavings

Jones of Oswestry
Whittington Road
Oswestry
Shropshire SY11 1HZ
Tel: 01691 653251
Website: www.jonesofoswestry.com
E-mail: sales@jonesofoswestry.com
Channels, gulleys, manhole covers

Kinley Systems
Haywood Way
Hastings
East Sussex TN35 4PL
Tel: 01424 201111
Website: www.kinleysystems.com
E-mail: sales@exceledge.co.uk
Aluminium edgings

Kompan Ltd
20 Denbigh Hall
Bletchley
Milton Keynes
Bucks MK3 7QT
Tel: 01908 642466
Website: www.kompan.com
E-mail: kompan.uk@kompan.com
Play equipment

Landline Ltd
1 Bluebridge Industrial Estate
Halstead
Essex CO9 2EX
Tel: 01787 476699
Website: www.landline.co.uk
E-mail: sales@landline.co.uk
Pond and lake installation

Lappset UK Ltd
Lappset House
Henson Way
Kettering
Northants NN16 8PX
Tel: 01536 412612
Website: www.lappset.com
E-mail: customerserviceuk@lappset.com
Play equipment

LDC Limited
The Charcoal House
Blacksmith Lane
Guildford
Surrey GU4 8NQ
Tel: 01483 573817
Website: www.ldclandscape.co.uk
E-mail: info@ldc.co.uk
Willow walling

Leaky Pipe Systems Ltd
Frith Farm
Dean Street
East Farleigh
Maidstone
Kent ME15 0PR
Tel: 01622 746495
Website: www.leakypipe.co.uk
E-mail: sales@leakypipe.co.uk
Irrigation systems

Lister Lutyens Co Ltd
6 Alder Close
Eastbourne
East Sussex BN23 6QF
Tel: 01323 431177
Website: www.listerteak.com
E-mail: sales@listerteak.com
Street furniture

Lorenz von Ehren
Maldfeldstrasse 4
D-21077 Hamburg
Germany
Tel: 0049 40 761080
Website: www.lve.de
E-mail: sales@lve.de
Topiary

Maccaferri Ltd
7400 The Quorum
Oxford Business Park North
Garsington Road
Oxford OX4 2JZ
Tel: 01865 770555
Website: www.maccaferri.co.uk
E-mail: oxford@maccaferri.co.uk
Gabions

Marshalls Plc
Landscape House
Premier Way
Lowfields Business Park
Elland HX5 9HT
Tel: 01422 312000
Website: www.marshalls.co.uk
Hard landscape materials, street furniture, linear drainage

Matta Products
19 Triumph Way
Woburn Road Industrial Estate
Kempston
Bedford MK42 7QB
Tel: 01234 848484
Website: www.matta-products.com
E- mail: mattasales@phs.co.uk
Safety surfacing

Maxit Building Products Ltd
Heath Business Park
Runcorn
Cheshire WA7 4QX
Tel: 01928 565656
Website: www.maxit-uk.co.uk/2367
E- mail: sales@maxit-uk.co.uk
Expanded clay aggregate

McArthur Group Ltd
Foundry Lane
Bristol BS5 7UE
Tel: 0117 943 0500
Website: www.mcarthur-group.com
E-mail: marketing@mcarthur-group.com
Security fencing

Melcourt Industries Ltd
Boldridge Brake
Long Newnton
Tetbury
Gloucestershire GL8 8RT
Tel: 01666 502711
Website: www.melcourt.co.uk
E-mail: mail@melcourt.co.uk
Mulch and compost

Milton Precast
Cooks Lane
Milton Regis
Sittingbourne
Kent ME10 2QF
Tel: 01795 425191
Website: www.miltonprecast.com
E-mail: sales@miltonprecast.com
Soakaway rings

Mobilane UK
PO Box 449
Stoke-on-Trent
Staffordshire ST6 9 AE
Tel: 07711 895261
Website: www.mobilane.co.uk
E-mail: sales@mobilane.co.uk
Green screens

Neptune Outdoor Furniture Ltd
Thompsons Lane
Marwell
Winchester
Hampshire SO21 1JH
Tel: 01962 777799
Website: www.nofl.co.uk
E-mail: info@nofl.co.uk
Street furniture

Nomix Enviro
A Division of Frontier Agriculture Ltd
The Grain Silos
Weyhill Road
Andover
Hants SP10 3NT
Tel: 01264 388050
Website: www.nomixenviro.co.uk
E-mail: nomixenviro@frontierag.co.uk
Herbicides

Norris and Gardiner Ltd
Lime Croft Road
Knaphill
Woking
Surrey GU21 2TH
Tel: 01483 289111
Website: www.norrisandgardiner.co.uk
E-mail: rich@norg.co.uk
Grounds maintenance

Oakover Nurseries Ltd
Maidstone Road
Hothfield
Ashford
Kent TN26 1 AR
Tel: 01233 713016
Website: www.oakovernurseries.co.uk
E-mail: enquiries@oakovernurseries.co.uk
Native plant nurseries

Orchard Street Furniture Ltd
Whistler House
51 The Green North
Warborough
Oxfordshire OX10 7DW
Tel: 01491 642123
Website: www.orchardstreet.co.uk
E-mail: sales@orchardstreet.co.uk
Street furniture

PBA Solutions
Bryn
Rake Road
Liss
Hants GU33 7HB
Tel: 01730 893460
Website: www.pba-solutions.com
E-mail: info@pba-solutions.com
Japanese Knotweed consultants

PC Landscapes Ltd
Abbott House
Hale Road
Farnham
Surrey GU9 9QH
Tel: 01252 891150
Website: www.pclandscapes.co.uk
Email: info@pclandscapes.co.uk
Landscaping services

Phi Group
Harcourt House
13 Royal Crescent
Cheltenham
Gloucestershire GL50 3DA
Tel: 01242 707600
Website: www.phigroup.co.uk
E-mail: southern@phigroup.co.uk
Retaining wall specialists

Plastech Southern
Home Park
Lyon Way
Frimley
Surrey GU16 7ER
Tel: 01276 24765
Website: www.plastechtitan.com
E-mail: sales@plastechsouthern.co.uk
Ducting

Platipus Anchors Ltd
Kingsfield Business Centre
Philanthropic Road
Redhill
Surrey RH1 4DP
Tel: 01737 762300
Website: www.platipus-anchors.com
E-mail: info@platipus-anchors.com
Tree anchors

Practicality Brown Ltd
Iver Stud Nursery
Iver
Bucks SLO 9LA
Tel: 01753 652022
Website: www.pracbrown.co.uk
E-mail: sales@pracbrown.co.uk
Hedging

Rigby Taylor Ltd
The Riverway Estate
Portsmouth Road
Peasmarsh
Guildford
Surrey GU3 1LZ
Tel: 01483 446900
Website: www.rigbytaylor.com
E-mail: sales@rigbytaylor.com
Horticultural supply

RIW Ltd
Arc House
Terrace Road South
Binfield
Bracknell
Berkshire RG42 4PZ
Tel: 01344 397777
Website: www.riw.co.uk
E-mail: enquiries@riw.co.uk
Waterproofing products

Road Equipment Ltd
28-34 Feltham Road
Ashford
Middlesex TW15 1DL
Tel: 01784 256565
E-mail: roadequipment@aol.com
Plant hire

Rolawn Ltd
York Road
Elvington
York YO41 4XR
Tel: 01904 608661
Website: www.rolawn.co.uk
E-mail: info@rolawn.co.uk
Industrial turf

RTS Ltd
UK Sales
Daisy Dene
Inglewhite Road
Goosnargh
Preston PR3 2EB
Tel: 01772 780234
Website: www.rtslimited.uk.com
E-mail: rtslimited@hotmail.co.uk
Permaloc edging

Sawhorse Ltd
Unit 4 Willows Gate
Stratton Audley
Bicester
Oxon OX27 9 AU
Tel: 07989 578661
Website: www.3rdspace.co.uk
E-mail: info@sawhorse-ltd.co.uk
Garden buildings

Scotscape Ltd
Unit 1
Wandle Technology Park
Mill Green Road
Mitcham Junction
Surrey CR4 4HZ
Tel: 02082 545000
Website: www.scotscape.net
E-mail: sales@scotscape.net
Living walls

Sleeper Supplies Ltd
PO Box 1377
Kirk Sandall
Doncaster DN3 1XT
Tel: 0845 230 8866
Website: www.sleeper-supplies.co.uk
E-mail: sales@sleeper-supplies.co.uk
Sleeper supplier

SMP Playgrounds Ltd
Ten Acre Lane
Thorpe
Egham
Surrey TW20 8RJ
Tel: 01784 489100
Website: www.smp.co.uk
E-mail: sales@smp.co.uk
Playground equipment

Southern Conveyors
Unit 2, Denton Slipways Site
Wharf Road
Gravesend DAI2 2RU
Tel: 01474 564145
Website: www.southernconveyors.co.uk
E-mail: sales@southernconveyors.co.uk
Conveyors

Spadeoak Construction Co Ltd
Town Lane
Wooburn Green
High Wycombe
Bucks HP10 0PD
Tel: 01628 529421
Website: www.spadeoak.co.uk
E-mail: email@spadeoak.co.uk
Macadam contractors

Steelway Brickhouse
Brickhouse Lane
West Bromwich
West Midlands B70 0DY
Tel: 01215 214500
Website: www.steelway.co.uk
E-mail: sales@steelwaybrickhouse.co.uk
Access covers

Steelway Fensecure
Queensgate Works
Bilston Road
Wolverhampton
West Midlands WV2 2NJ
Tel: 01902 490919
Website: www.steelway.co.uk
E-mail: sales@steelway.co.uk
Fencing

SteinTec
728 London Road
West Thurrock
Essex RM20 3LU
Tel: 01708 860049
Website: www.steintec.co.uk
E-mail: info@steintec.co.uk
Specialized mortars

Stonebank Ironcraft Ltd
Parkway Farm
Fossbridg
Northleach
Cheltenham
Gloucestershire GL54 3JL
Tel: 01285 720 737
Website: www.stonebank-ironcraft.co.uk
E-mail: info@stonebank-ironcraft.co.uk
Solid steel products

Sugg Lighting Ltd
Sussex Manor Business Park
Gatwick Road
Crawley
West Sussex RH10 9GD
Tel: 01293 540111
Website: www.sugglighting.co.uk
E-mail: sales@sugglighting.co.uk
Lighting

SureSet UK Limited
32 Deverill Road
Trading Estate
Sutton Veny
Warminster BA12 7BZ
Tel: 01985 841873
Website: www.sureset.co.uk
E-mail: mail@sureset.co.uk
Solid steel products

Targetti Poulsen UK Ltd
Unit C 44 Barwell Business Park
Leatherhead Road
Chessington KT9 2NY
Tel: 020 8397 4400
Website: www.targetti.com
Or: www.louispoulsen.com
E-mail: nak-uk@lpmail.com
Outdoor lighting

Tensar International
Cunningham Court
Shadsworth Business Park
Blackburn
Lancashire BB2 4PJ
Tel: 01254 262431
Website: www.tensar-international.com
E-mail: sales@tensar.co.uk
Erosion control, soil stabilization

Terranova Lifting Ltd
Terranova House
Bennett Road
Reading
Berks RG2 OQX
Tel: 0118 931 2345
Website: www.terranova-lifting.co.uk
E-mail: cranes@terranovagroup.co.uk
Crane hire

Townscape Products Ltd
Fulwood Road South
Sutton-in-Ashfield
Nottinghamshire NG17 2JZ
Tel: 01623 513355
Website: www.townscape-products.co.uk
E-mail: sales@townscape-products.co.uk
Hard landscaping

Trulawn Ltd
Unit 7E Vulcan Way
Sandhurst
Berkshire GU47 9DB
Tel: 01252 819695
Website: www.trulawn.co.uk
E-mail: sales@trulawn.co.uk
Artificial grass

Tubex Ltd
Aberaman Park
Aberaman
South Wales CF44 6DA
Tel: 01685 888000
Website: www.tubex.com
E-mail: plantcare@tubex.com
Plant protection, tree guards

Turf Management Systems
Seven Hills Road
Iver
Buckinghamshire SL0 0PA
Tel: 01895 834411
Liquid sod

Wade International Ltd
Third Avenue
Halstead
Essex CO9 2SX
Tel: 01787 475151
Website: www.wadedrainage.co.uk
E-mail: sales@wade.eu
Stainless steel drainage

Wavin Plastics Ltd
Parsonage Way
Chippenham
Wiltshire SN15 5PN
Tel: 01249 766600
Website: www.wavin.co.uk
E-mail: info@wavin.co.uk
Drainage products

Wicksteed Leisure Ltd
Digby Street
Kettering
Northamptonshire NN16 8YJ
Tel: 01536 517028
Website: www.wicksteed.co.uk
E-mail: sales@wicksteed.co.uk
Play equipment

Willowbank Services
Curload
Stoke St Gregory
Taunton Somerset TA3 6JD
Tel: 01823 690113
Website: www.willowbankservices.co.uk
E-mail: info@willowbankservices.co.uk
Waterside bank stabilization

Woodscape Ltd
Church Works
Church Street
Church
Lancashire BB5 4JT
Tel: 01254 383322
Website: www.woodscape.co.uk
E-mail: sales@woodscape.co.uk
Street furniture

Wybone Ltd
Mason Way
Platts Common Industrial Estate
Barnsley
South Yorkshire S74 9TF
Tel: 01226 744010
Website: www.wybone.co.uk
E-mail: sales@wybone.co.uk
Street furniture

Yeoman Aggregates Ltd
Stone Terminal
Horn Lane
Acton
London W3 9EH
Tel: 020 8896 6820
Website: www.foster-yeoman.co.uk
E-mail: sales@yeoman-aggregates.co.uk
Aggregates

Spon's Asia-Pacific Construction Costs Handbook

4th Edition

Edited by **Davis Langdon & Seah**

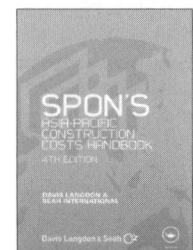

Spon's Asia Pacific Construction Costs Handbook includes construction cost data for 20 countries. This new edition has been extended to include Pakistan and Cambodia. Australia, UK and America are also included, to facilitate comparison with construction costs elsewhere. Information is presented for each country in the same way, as follows:

• key data on the main economic and construction indicators.

• an outline of the national construction industry, covering structure, tendering and contract procedures, materials cost data, regulations and standards

• labour and materials cost data

• Measured rates for a range of standard construction work items

• Approximate estimating costs per unit area for a range of building types

• price index data and exchange rate movements against £ sterling, $US and Japanese Yen

The book also includes a Comparative Data section to facilitate country-to-country comparisons. Figures from the national sections are grouped in tables according to national indicators, construction output, input costs and costs per square metre for factories, offices, warehouses, hospitals, schools, theatres, sports halls, hotels and housing.

This unique handbook will be an essential reference for all construction professionals involved in work outside their own country and for all developers or multinational companies assessing comparative development costs.

April 2010: 234x156: 480pp
Hb: 978-0-415-46565-6: **£120.00**

To Order: Tel: +44 (0) 1235 400524 **Fax:** +44 (0) 1235 400525
or Post: Taylor and Francis Customer Services,
Bookpoint Ltd, Unit T1, 200 Milton Park, Abingdon, Oxon, OX14 4TA UK
Email: book.orders@tandf.co.uk

For a complete listing of all our titles visit:
www.tandf.co.uk

Taylor & Francis
Taylor & Francis Group

How this Book is Compiled

INTRODUCTION

First-time users of *Spon's External Works and Landscape Price Book* and others who may not be familiar with the way in which prices are compiled may find it helpful to read this section before starting to calculate the costs of landscape works.

Please also refer to the 'How to use this book' notes on page 32.
The cost of an item of landscape construction or planting is made up of many components:

- the cost of the product
- the labour and additional materials needed to carry out the job
- the cost of running the contractor's business

These are described more fully below.

IMPORTANT NOTES ON THE PROFIT ELEMENT OF RATES IN THIS BOOK

The rates shown in the Approximate Estimates and Measured Works sections of this book do not contain a profit element unless the rate has been provided by a subcontractor. Prices are shown at cost. This is the cost including the costs of company overhead required to perform this task or project. For reference please see the tables on pages 6–9.

Analysed rates versus subcontractor rates

As a general rule if a rate is shown as an analysed rate in the Measured Works section, i.e. it has figures shown in the columns other than the 'Unit' and 'Total Rate' column, it can be assumed that this has no profit or overhead element and that this calculation is shown as a direct labour/material supply/plant task being performed by a contractor.

On the other hand, if a rate is shown as a Total Rate only, this would normally be a subcontractor's rate and would contain the profit and overhead for the subcontractor.

The foregoing applies for the most part to the Approximate Estimates section, however in some items there may be an element of subcontractor rates within direct works build-ups.

As an example of this, to excavate, lay a base and place a macadam surface, a general landscape contractor would normally perform the earthworks and base installation. The macadam surfacing would be performed by a specialist subcontractor to the landscape contractor.

The Approximate Estimate for this item uses rates from the Measured Works section to combine the excavation, disposal, base material supply and installation with all associated plant. There is no profit included on these elements. The cost of the surfacing, however, as supplied to us in the course of our annual price enquiry from the macadam subcontractor, would include the subcontractor's profit element.

The landscape contractor would add this to his section of the work but would normally apply a mark up at a lower percentage to the subcontract element of the composite task. Users of this book should therefore allow for the profit element at the prevailing rates. Please see the worked examples below and the notes on overheads for further clarification.

INTRODUCTORY NOTES ON PRICING CALCULATIONS USED IN THIS BOOK

There are two pricing sections to this book:

Major Works – Average contract value £100,000.00–£300,000.00

Minor Works – Average contract value £10,000.00–£70,000.00

Typical Project Profiles Used in this Book

	Major Works	Minor Works
Contract value	£100,000.00–£300,000.00	£10,000.00–£70,000.00
Labour rate (see page 3)	£19.25 per hour	£20.75 per hour
Labour rate for maintenance contracts	£16.50 per hour	–
Number of site staff	35	6–9
Project area	6000 m^2	1200 m^2
Project location	Outer London	Outer London
Project components	50% hard landscape	20% hard landscape
	50% soft landscape and planting	80% soft landscape and planting
Access to works areas	Very good	Very good
Contract	Main contract	Main contract
Delivery of materials	Full loads	Part loads

As explained in more detail later the prices are generally based on wage rates and material costs current at June 2011. They do not allow for preliminary items, which are dealt with in the Preliminaries section of this book, or for any Value Added Tax (VAT) which may be payable.

Adjustments should be made to standard rates for time, location, local conditions, site constraints and any other factors likely to affect the costs of a specific scheme.

Term contracts for general maintenance of large areas should be executed at rates somewhat lower than those given in this section.

There is now a facility available to readers that enables a comparison to be made between the level of prices given and those for projects carried out in regions other than outer London; this is dealt with on page 23.

Units of measurement

The units of measurement have been varied to suit the type of work and care should be taken when using any prices to ascertain the basis of measurement adopted.

The prices per unit of area for executing various mechanical operations are for work in the following areas and under the following conditions:

Prices per m^2 relate to areas not exceeding $100 m^2$
(any plan configuration)

Prices per $100 m^2$ relate to areas exceeding $100 m^2$ but not exceeding ¼ ha
(generally clear areas but with some subdivision)

Prices per ha relate to areas over ¼ ha
(clear areas suitable for the use of tractors and tractor-operated equipment)

The prices per unit area for executing various operations by hand generally vary in direct proportion to the change in unit area.

Approximate estimating

These are combined Measured Work prices which give an approximate cost for a complete section of landscape work. For example, the construction of a car park comprises excavation, levelling, road-base, surfacing and marking parking bays. Each of these jobs is priced separately in the Measured Works section, but a comprehensive price is given in the Approximate Estimates section, which is intended to provide a quick guide to the cost of the job. It will be seen that the more items that go to make up an approximate estimate, the more possibilities there are for variations in the PC prices and the user should ensure that any PC price included in the estimate corresponds to his specification.

The figures given in *Spon's External Works and Landscape Price Book* are intended for general guidance, and if a significant variation appears in any one of the cost groups, the Price for Measured Work should be recalculated.

Measured works

Prime Cost: Commonly known as the 'PC'. Prime Cost is the actual price of the material item being addressed such as paving, shrubs, bollards or turves, as sold by the supplier. Prime Cost is given 'per square metre', 'per 100 bags' or 'each' according to the way the supplier sells his product. In researching the material prices for the book we requested that the suppliers price for full loads of their product delivered to a site close to the M25 in London. Spon's rates do not include VAT. Some companies may be able to obtain greater discounts on the list prices than those shown in the book. Prime Cost prices for those products and plants which have a wide cost range will be found under the heading of Market Prices in the main sections of this book, so that the user may select the product most closely related to his specification.

Materials: The PC material plus the additional materials required to fix the PC material. Every job needs materials for its completion besides the product bought from the supplier. Paving needs sand for bedding, expansion joint strips and cement pointing; fencing needs concrete for post setting and nails or bolts; tree planting needs manure or fertilizer in the pit, tree stakes, guards and ties. If these items were to be priced out separately, Spon's External Works and Landscape Price Book (and the Bill of Quantities) would be impossibly unwieldy, so they are put together under the heading of Materials.

Labour: This figure covers the cost of planting shrubs or trees, laying paving, erecting fencing etc. and is calculated on the wage rate (skilled or unskilled) and the time needed for the job. Extras such as highly skilled craft work, difficult access, intermittent working and the need for labourers to back up the craftsman all add to the cost. Large regular areas of planting or paving are cheaper to install than smaller intricate areas, since less labour time is wasted moving from one area to another.

Plant, consumable stores and services: This rather impressive heading covers all the work required to carry out the job which cannot be attributed exactly to any one item. It covers the use of machinery ranging from JCBs to shovels and compactors, fuel, static plant, water supply (which is metered on a construction site), electricity and rubbish disposal. The cost of transport to site is deemed to be included elsewhere and should be allowed for elsewhere or as a preliminary item. Hired plant is calculated on an average of 36 hours working time per week.

Subcontract rates: Where there is no analysis against an item, this is deemed to be a rate supplied by a subcontractor. In most cases these are specialist items where most external works contractors would not have the expertise or the equipment to carry out the task described. It should be assumed that subcontractor rates include for the subcontractor's profit. An example of this may be found for the tennis court rates in section Q26.

Labour rates used in this edition

The rates for labour used in this edition have been based on surveys carried out on a cross section of external works contractors. These rates include for company overheads such as employee administration, transport insurance, and oncosts such as National Insurance. Please see the calculation of these rates shown on page 6. The rates do not include for profit.

Overheads: An allowance for this is included in the labour rates which are described above. The general overheads of the contract such as insurance, site huts, security, temporary roads and the statutory health and welfare of the labour force are not directly assignable to each item, so they are distributed as a percentage on each, or as a separate preliminary cost item. The contractor's and subcontractor's profits are not included in this group of costs. Site overheads, which will vary from contract to contract according to the difficulties of the site, labour shortages, inclement weather or involvement with other contractors, have not been taken into account in the build up of these rates, while overhead (or profit) may have to take into account losses on other jobs and the cost to the contractor of unsuccessful tendering.

ADJUSTMENT AND VARIATION OF THE RATES IN THIS BOOK

It will be appreciated that a variation in any one item in any group will affect the final Measured Works price. Any cost variation must be weighed against the total cost of the contract. A small variation in Prime Cost where the items are ordered in thousands may have more effect on the total cost than a large variation on a few items. A change in design that necessitates the use of earth moving equipment which must be brought to the site for that one job will cause a dramatic rise in the contract cost. Similarly, a small saving on multiple items will provide a useful reserve to cover unforeseen extras.

Worked examples

These are shown on page 34

HOW THIS BOOK IS UPDATED EACH YEAR

The basis for this book is a database of Material, Labour, Plant and Subcontractor resources each with its own area of the database.

Material, Plant and Subcontractor Resources

Each year the suppliers of each material, plant or subcontract item are approached and asked to update their prices to those that will prevail in September of that year.

These resource prices are individually updated in the database. Each resource is then linked to one or many tasks. The tasks in the task library section of the database is automatically updated by changes in the resource library. A quantity of the resource is calculated against the task. The calculation is generally performed once and the links remain in the database.

On occasions where new information or method or technology are discovered or suggested, these calculations would be revisited. A further source of information is simple time and production observations made during the course of the last year.

Labour resource update

Most tasks, except those shown as subcontractor rates (see above), employ an element of labour. The Data Department at Davis Langdon conducts ongoing research into the costs of labour in various parts of the country.

Tasks or entire sections are then re-examined and recalculated. Comments on the rates published in this book are welcomed and may be submitted to the contact addresses shown in the Preface.

Davis Langdon,
An AECOM Company

Program, Cost, Consultancy
www.davislangdon.com
www.aecom.com

Davis Langdon
An AECOM Company

A leading light in our industry

PART 1

General

This part of the book contains the following sections:

Labour Rates Used in this Edition

Based on surveys carried out on a cross section of external works contractors, the following rates for labour have been used in this edition.

These rates include for company overheads such as employee administration, transport insurance, and oncosts such as National Insurance.

The rates used do not include for profit

The rates used may include a subcontractor's profit where subcontractor's rates are used to build up the costs of an item in the book.

The rates for all labour used in this edition are as follows:

Major Works

General contracting	£19.25/hour
Maintenance contracting	£16.50/hour

Minor Works

General contracting	£20.75/hour

Notes on the Minor Works rates used in this edition

The figure shown above and used in the minor works calculations of this book does not reflect the rate shown in the cost model for wages of £22.21 per hour.

Whilst we feel that the correct cost of labour as per the tables on the following pages is £22.21 per hour we believe that prevailing market conditions for the year 2012/2013 are more in line with a labour cost rate of £20.75 per hour in the Minor Works sector and we have therefore used this rate.

Users of these data can adjust the rates shown in the data sections to suit their requirements and market conditions.

European Gardens
History, Philosophy and Design

Tom Turner

EUROPEAN GARDENS
HISTORY, PHILOSOPHY AND DESIGN

Tom Turner

Garden design and usage has been a feature of human civilisation as far back as Neolithic times, when the first gardens began to be used for residential, horticultural and sacred tasks. Tom Turner follows the entire history of the European garden from its prehistoric rootsright up to the present day in this beautifully illustrated book.

European Gardens is divided into ten periods of history and garden development, detailing the advancement of land usage for over 10,000 years. Some of the topics covered in this comprehensive book include the Egyptian gardens of the Pharaohs, the castle gardens of medieval times, eclectic gardens of the nineteenth century and abstract gardens of the last 100 years. The geographical scope of this book covers the whole of the European continent, and touches the garden designs of North Africa and the Middle East.

Tom Turner is a skilled landscape architect and garden historian, who supports his engaging writing with his own detailed plans and diagrams. *European Gardens* also features almost 1,000 colour photographs from across the continent allowing the reader to see for themselves how the design and structure of gardens has developed over time.

A companion to the *Asian Gardens* book, published by Routledge in 2010, *European Gardens* is a development of the original Garden History book from 2004

December 2010: 250x250: 424pp
Hb: 978-0-415-49684-1: **£35.00**

To Order: Tel: +44 (0) 1235 400524 **Fax:** +44 (0) 1235 400525
or Post: Taylor and Francis Customer Services,
Bookpoint Ltd, Unit T1, 200 Milton Park, Abingdon, Oxon, OX14 4TA UK
Email: book.orders@tandf.co.uk

For a complete listing of all our titles visit:
www.tandf.co.uk

Taylor & Francis
Taylor & Francis Group

Computation of Labour Rates Used in this Edition

Different organizations will have varying views on rates and costs, which will in any event be affected by the type of job, availability of labour and the extent to which mechanical plant can be used. However this information should assist the reader to:

(1)   compare the prices to those used in his own organization
(2)   calculate the effect of changes in wage rates or prices of materials
(3)   calculate prices for work similar to the examples given

From September 2011 – assumed basic weekly rates of pay for craft and general operatives are £407.94 and £306.93 respectively; to these rates have been added allowances for the items below in accordance with the recommended procedure of the Chartered Institute of Building in its 'Code of Estimating Practice'. The resultant hourly rates are £13.69 and £10.20 for craft operatives and general operatives respectively.

The items for which allowances have been made are:

- Lost time
- Construction Industry Training Board Levy
- Holidays with pay
- Accidental injury, retirement and death benefits scheme
- Sick pay
- National Insurance
- Severance pay and sundry costs
- Employer's liability and third party insurance

The tables that follow illustrate how the above hourly rates have been calculated. Productive time has been based on a total of 1802 hours worked per year for daywork calculations above and for 1953.54 hours worked per year for the all-in labour rates (including 5 hours per week average overtime) for the all-in labour rates below.

How the labour rate has been calculated

A survey of typical landscape/external works companies indicates that they are currently paying above average wages for multi-skilled operatives regardless of specialist or supervisory capability. In our current overhaul of labour constants and costs we have departed from our previous policy of labour rates based on national awards and used instead a gross rate per hour for all rate calculations. Estimators can readily adjust the rate if they feel it is inappropriate for their work.

The productive labour of any organization must return their salary plus the cost of the administration which supports the labour force.

The labour rate used in this edition is calculated as follows:

Team size

A three man team with annual salaries as shown and allowances for National Insurance, uniforms, site tools and a company vehicle, returns a basic labour rate of £15.05 per hour to which company overhead is added as per the following tables.

Basic labour rate calculation for Spon's 2013 Major and Minor Works

Standard Working Hours per year (2012/13)
=   1794
Less allowance for sick leave and down time
=   −39
Actual Working Hours per year (2012/13)
=   1755

Working Hours per year 2012/2013				*1755 Allows sick and down time of 39 hours*			
	Number of	Labour team	NI	Nett Cost	Clothing	Site Tools	TOTAL 3 Man Team
Foreman	1	25000.00	3200.00	28200.00	200.00	150.00	
Craftsman	1	20500.00	2624.00	23124.00	200.00	150.00	
Labourer	1	16000.00	2048.00	18048.00	200.00	150.00	
	3.00	61500.00	7872.00	69372.00	600.00	450.00	
						Subtotal	**£ 70,422.00**
Overtime Hours							
Nominal hours/annum at 1.5 times normal standard rate							
Foreman	40.00	21.37	2.74	964.10			
Craftsman	40.00	17.52	2.24	790.56			
Labourer	40.00	13.68	1.75	617.03			
Total staff /team	3.00			71743.69	600.00	450.00	**£ 72793.69**
Vehicle Costs	**Working**	**£/Day**					
Inclusive of Fuel	**Days**						
Insurances etc.							
	240.00	45.00					**£ 10,800.00**
						TOTAL	**£ 83,593.69**

The basic average labour rate excluding overhead costs (below) is

$$\frac{£83,539.69}{\text{Total staff in team (3)} \times \text{Working hours per year (1755)}} = £15.88$$

Add to the above the overhead costs as per the table on the next page.

Overhead costs

Add to the above basic rate the company overhead costs for a small company as per the table below. These costs are absorbed by the number of working men multiplied by the number of working hours supplied in the table below. This then generates an hourly overhead rate which is added to the Nett cost rate above.

Illustrative overhead for company employing 12–50 landscape operatives

Cost Centre	Number of	Cost	Total
MD: Salary only: excludes profits + NI + Vehicle	1	53100.00	53100.00
Senior contracts managers + NI + Vehicle	1	40000.00	40000.00
Other contracts managers + NI + Vehicle	1	30000.00	30000.00
Business administrator/secretary	1	20000.00	20000.00
Bookkeeper	1	19000.00	19000.00
Rental	12	1000.00	12000.00
Insurances	1	5000.00	5000.00
Telephone and mobiles	12	250.00	3000.00
Office equipment	1	1000.00	1000.00
Stationery	12	50.00	600.00
Advertising	12	200.00	2400.00
Other vehicles not allocated to contract teams	1	9000.00	9000.00
Other consultants	1	6000.00	6000.00
Accountancy	1	2500.00	2500.00
Lights heating water	12	200.00	2400.00
Other expenses	1	8000.00	10000.00
Total office overhead			216000.00

The final labour rate used is then generated as follows:

$$\text{Nett labour rate} \quad + \quad \frac{\text{Total Office Overhead (£216000.00)}}{\text{Working hours per year (1755)} \times \text{Labour resources employed}}$$

Total nr of Site Staff	Admin Cost per hour	Total Rate per Man Hour
12	10.26	26.13
15	8.21	24.08
18	6.84	22.71
20	6.15	22.03
25	4.92	20.80
35	3.52	19.39
40	3.08	18.95
45	2.74	18.61
50	2.46	18.34

Hourly labour rates used for *Spon's External Works and Landscape 2013* (Major Works)
This year's rates for Major Works are calculated on the rounded value of a 35 man working team rounded to the nearest £0.50

General contracting	£19.25
Maintenance contracting	£16.50

Minor works

- The minor works labour calculation is based on the above but the administration cost of the organization is £107,440.00 per annum
- The owner of the business does not carry out site works and is assumed to be paid a salary of £40,000.00 plus National insurance plus a motor vehicle per annum; this figure constitutes part of the £107440.00 shown
- There are between 6 and 15 site operatives

Illustrative overhead for small company employing 6–15 landscape operatives

Cost Centre	Number of	Cost	Total
MD: Salary only: excludes profits + NI + Vehicle	1	53100.00	53100.00
Secretary	1	17000.00	17000.00
Bookkeeper	1	8000.00	8000.00
Rental	12	300.00	3600.00
Insurances	1	4000.00	4000.00
Telephone and mobiles	12	200.00	2400.00
Office equipment	1	1000.00	1000.00
Stationery	12	20.00	240.00
Advertising	12	200.00	2400.00
Other vehicles not allocated to contract teams	1	4000.00	4000.00
Other consultants	1	1000.00	1000.00
Accountancy	1	1500.00	1500.00
Lights heating water	12	100.00	1200.00
Other expenses	1	8000.00	8000.00
Total office overhead			£107,440.00

The final labour rate used is then generated as follows:

$$\text{Nett labour rate} \quad + \quad \frac{\text{Total Office Overhead (£107440.00)}}{\text{Working hours per year (1755)} \times \text{Labour resources employed}}$$

Total nr of Site Staff	Admin Cost per hour	Total Rate per Man Hour
3	20.41	36.28
6	10.20	26.08
9	6.80	22.68
12	5.10	20.98
15	4.08	19.96

Hourly labour rates used for Spon's External Works and Landscape 2013 (Minor Works)

This year's rates for Minor Works are calculated on the rounded value of a 9 man working team rounded (£22.68)
We have used a rate of 20.75 due to indications on labour rates which we predict for the following 12 month period

General and Maintenance	£20.75

Calculation of annual hours worked

Normal hours worked		
Normal hours worked per week		39 hours
Normal hours worked per year	39 hrs × 52 wks	2028 hours
Less holidays		
Annual holidays	39 hrs × 4.2 wks	−163.8 hours
Public holidays (includes diamond jubilee day)	9 days @7.8 hours	−70.2 hours
Less sick and downtime	39 hrs × 1 wk	−39 hours
Actual hours worked per annum		1755 hours

Environmental Noise Barriers

2nd Edition

B. Kotzen et al.

Noise pollution generated by transport is widely recognised as a major environmental problem. The public is increasingly aware of the problems of excessive noise levels and compensation costs will continue to grow unless and until the impact of traffic noise is reduced. The use of environmental noise barriers, already common in Europe and the USA is now becoming increasingly important, changing the face of our road and railway networks. *Environmental Noise Barriers* examines both the acoustical and landscaping issues affecting the design of barriers. It addresses the needs of the whole design team, and provides expert guidance on good practice and highlights the pitfalls of poor design. Detailed consideration is given to materials, engineering, legal and environmental issues as well as the health and social impacts of noise barriers. And it illustrates the wide variety of design solutions for different acoustic and landscape situations in several European countries with a generous range of full colour photographs and illustrations.

This extensively revised new edition is updated in line with UK and EU legislation and international provision of barriers, the use of new materials and new designs and design thinking. It includes new methods for improving the acoustic performance of barriers and extended sections on road surfacing, tunnels and railways and aircraft/airport acoustic amelioration. The sections on vegetative barriers, the use of photovoltaics and integrated solutions have been expanded. *Environmental Noise Barriers* is a unique one-stop reference for practitioners, whether acoustical engineers, landscape architects or manufacturers, and for highways departments in local and central authorities.

April 2009: 288pp
Hb:978-0-415-43708-0: **£90.00**

To Order: Tel: +44 (0) 1235 400524 **Fax:** +44 (0) 1235 400525
or Post: Taylor and Francis Customer Services,
Bookpoint Ltd, Unit T1, 200 Milton Park, Abingdon, Oxon, OX14 4TA UK
Email: book.orders@tandf.co.uk

For a complete listing of all our titles visit:
www.tandf.co.uk

Computation of the Cost of Materials

Percentages of default waste are placed against material resources within the supplier database. These range from 2.5% (for bricks) to 20% (for topsoil). An allowance for the cost of unloading, stacking etc. should be added to the cost of materials.

The following are typical hours of labour for unloading and stacking some of the more common building materials.

Material	Unit	Labourer hours
Cement	tonne	0.67
Lime	tonne	0.67
Common bricks	1000	1.70
Light engineering bricks	1000	2.00
Heavy engineering bricks	1000	2.40

Construction Contracts Questions and Answers

2nd Edition

David Chappell

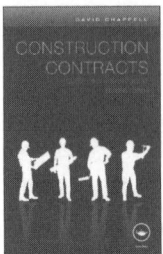

What they said about the first edition: "A fascinating concept, full of knowledgeable gems put in the most frank of styles... A book to sample when the time is right and to come back to when another time is right, maybe again and again."
– *David A Simmonds, Building Engineer magazine*

• Is there a difference between inspecting and supervizing?

• What does 'time-barred' mean?

• Is the contractor entitled to take possession of a section of the work even though it is the contractor's fault that possession is not practicable?

Construction law can be a minefield. Professionals need answers which are pithy and straightforward, as well as legally rigorous. The two hundred questions in the book are real questions, picked from the thousands of telephone enquiries David Chappell has received as a Specialist Adviser to the Royal Institute of British Architects. Although the enquiries were originally from architects, the answers to most of them are of interest to project managers, contractors, QSs, employers and others involved in construction.

The material is considerably updated from the first edition – weeded, extended and almost doubled in coverage. The questions range in content from extensions of time, liquidated damages and loss and/or expense to issues of warranties, bonds, novation, practical completion, defects, valuation, certificates and payment, architects' instructions, adjudication and fees. Brief footnotes and a table of cases will be retained for those who may wish to investigate further.

August 2010: 216x138: 344pp
Pb: 978-0-415-56650-6: **£34.99**

To Order: Tel: +44 (0) 1235 400524 **Fax:** +44 (0) 1235 400525
or Post: Taylor and Francis Customer Services,
Bookpoint Ltd, Unit T1, 200 Milton Park, Abingdon, Oxon, OX14 4TA UK
Email: book.orders@tandf.co.uk

For a complete listing of all our titles visit:
www.tandf.co.uk

Landfill Tax

The tax

The Landfill Tax came into operation on 1 October 1996. It is levied on operators of licensed landfill sites at the following rates with effect from 1 April 2012:

- Inactive or inert wastes   £2.50 per tonne   Included are soil, stones, brick, plain and reinforced concrete, plaster and glass
- All other taxable wastes   £64 per tonne   Included are timber, paint and other organic wastes generally found in demolition work and builders skips

The rate for all other taxable wastes will be increased by £8 per tonne each year at least until 2014 when the rate will be £80 per tonne. The lower rate for inactive or inert wastes will be frozen at £2.50 per tonne to 2012/13.

Mixtures containing wastes not classified as inactive or inert will not qualify for the lower rate of tax unless the amount of non-qualifying material is small and there is no potential for pollution. Water can be ignored and the weight discounted.

Calculating the weight of waste

There are two options:

- If licensed sites have a weighbridge, tax will be levied on the actual weight of waste.
- If licensed sites do not have a weighbridge, tax will be levied on the permitted weight of the lorry based on an alternative method of calculation based on volume to weight factors for various categories of waste.

Effect on prices

The tax is paid by Landfill site operators only. Tipping charges reflect this additional cost.

Active waste will normally be disposed of by skip and will probably be mixed with inactive waste. The tax levied will depend on the weight of materials in the skip which can vary significantly.

Exemptions

The following disposals are exempt from Landfill Tax subject to meeting certain conditions:

- dredgings which arise from the maintenance of inland waterways and harbours
- naturally occurring materials arising from mining or quarrying operations
- reclamation of contaminated land
- inert waste used to restore landfill sites and to fill working and old quarries where a planning condition or obligation is in existence

The exemption for waste from contaminated land will be phased out completely by 1 April 2012 and no new applications for landfill tax exemption are now accepted.

For further information contact the National Advisory Service, Telephone: 0845 010 9000.

Design for Outdoor Recreation

2nd Edition

S. Bell

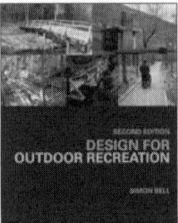

Design for Outdoor Recreation takes a detailed look at all aspects of design of facilities needed by visitors to outdoor recreation destinations. The book is a comprehensive manual for planners, designers and managers of recreation taking them through the processes of design and enabling them to find the most appropriate balance between visitor needs and the capacity of the landscape. A range of different aspects are covered including car parking, information signing, hiking, waterside activities, wildlife watching and camping.

This second edition incorporates new examples from overseas, including Australia, New Zealand, Japan and Eastern Europe as well as focusing on more current issues such as accessibility and the changing demands for recreational use.

July 2008: 276x219: 240pp
Pb: 978-0-415-44172-8: **£55.00**

To Order: Tel: +44 (0) 1235 400524 **Fax:** +44 (0) 1235 400525
or Post: Taylor and Francis Customer Services,
Bookpoint Ltd, Unit T1, 200 Milton Park, Abingdon, Oxon, OX14 4TA UK
Email: book.orders@tandf.co.uk

For a complete listing of all our titles visit:
www.tandf.co.uk

The Aggregates Levy

The Aggregates Levy came into operation on 1 April 2002 in the UK, except for Northern Ireland where it has been phased in over five years from 2003.

It was introduced to ensure that the external costs associated with the exploitation of aggregates are reflected in the price of aggregate, and to encourage the use of recycled aggregate. There continues to be strong evidence that the levy is achieving its environmental objectives, with sales of primary aggregate down and production of recycled aggregate up. The Government expects that the rates of the levy will at least keep pace with inflation over time, although it accepts that the levy is still bedding in.

The rate of the levy increased to £2.10 per tonne from 1 April 2012 and is levied on anyone considered to be responsible for commercially exploiting virgin aggregates in the UK and should naturally be passed by price increase to the ultimate user.

All materials falling within the definition of Aggregates are subject to the levy unless specifically exempted.

It does not apply to clay, soil, vegetable or other organic matter.

The intention is that it will:

- Encourage the use of alternative materials that would otherwise be disposed of to landfill sites
- Promote development of new recycling processes, such as using waste tyres and glass
- Promote greater efficiency in the use of virgin aggregates
- Reduce noise and vibration, dust and other emissions to air, visual intrusion, loss of amenity and damage to wildlife habitats

Definitions

'Aggregates' means any rock, gravel or sand which is extracted or dredged in the UK for aggregates use. It includes whatever substances are for the time being incorporated in it or naturally occur mixed with it.

'Exploitation' is defined as involving any one or a combination of any of the following:

- Being removed from its original site
- Becoming subject to a contract or other agreement to supply to any person
- Being used for construction purposes
- Being mixed with any material or substance other than water, except in permitted circumstances

Incidence

It is a tax on primary aggregates production – i.e. virgin aggregates won from a source and used in a location within the UK territorial boundaries (land or sea). The tax is not levied on aggregates which are exported or on aggregates imported from outside the UK territorial boundaries.

It is levied at the point of sale.

Exemption from tax

An aggregate is exempt from the levy if it is:

- Material which has previously been used for construction purposes
- Aggregate that has already been subject to a charge to the Aggregates Levy
- Aggregate which was previously removed from its originating site before the start date of the levy
- Aggregate which is being returned to the land from which it was won
- Aggregate won from a farm land or forest where used on that farm or forest
- Rock which has not been subjected to an industrial crushing process
- Aggregate won by being removed from the ground on the site of any building or proposed building in the course of excavations carried out in connection with the modification or erection of the building and exclusively for the purpose of laying foundations or of laying any pipe or cable
- Aggregate won by being removed from the bed of any river, canal or watercourse or channel in or approach to any port or harbour (natural or artificial), in the course of carrying out any dredging exclusively for the purpose of creating, restoring, improving or maintaining that body of water
- Aggregate won by being removed from the ground along the line of any highway or proposed highway in the course of excavations for improving, maintaining or constructing the highway otherwise than purely to extract the aggregate
- Drill cuttings from petroleum operations on land and on the seabed
- Aggregate resulting from works carried out in exercise of powers under the New Road and Street Works Act 1991, the Roads (Northern Ireland) Order 1993 or the Street Works (Northern Ireland) Order 1995
- Aggregate removed for the purpose of cutting of rock to produce dimension stone, or the production of lime or cement from limestone
- Aggregate arising as a waste material during the processing of the following industrial minerals:
  - ball clay
  - barytes
  - calcite
  - china clay
  - coal, lignite, slate or shale
  - feldspar
  - flint
  - fluorspar
  - fuller's earth
  - gems and semi-precious stones
  - gypsum
  - metal or the ore from any metal
  - muscovite
  - perlite
  - potash
  - pumice
  - rock phosphates
  - sodium chloride
  - talc
  - vermiculite

However, the levy is still chargeable on any aggregates arising as the spoil or waste from or the by-products of the above exempt processes. This includes quarry overburden.

Anything that consists 'wholly or mainly' of the following is exempt from the levy (note that 'wholly' is defined as 100% but 'mainly' as more than 50%, thus exempting any contained aggregates amounting to less than 50% of the original volumes:

- clay, soil, vegetable or other organic matter
- coal, slate or shale
- china clay waste and ball clay waste

Relief from the levy either in the form of credit or repayment is obtainable where:

- it is subsequently exported from the UK in the form of aggregate
- it is used in an exempt process
- where it is used in a prescribed industrial or agricultural process
- it is waste aggregate disposed of by dumping or otherwise, e.g. sent to landfill or returned to the originating site

The Aggregates Levy Credit Scheme (ALCS) for Northern Ireland was suspended with effect from 1 December 2010 following a ruling by the European General Court.

A new exemption for aggregate obtained as a by-product of railway, tramway and monorail improvement, maintenance and construction was introduced in 2007.

Discounts

From 1 July 2005 the standard added water percentage discounts listed below can be used. Alternatively a more exact percentage can be agreed and this must be done for dust dampening of aggregates.

- washed sand = 7%
- washed gravel = 3.5%
- washed rock/aggregate = 4%

Impact

The British Aggregates Association suggests that the additional cost imposed by quarries is more likely to be in the order of £3.40 per tonne on mainstream products, applying an above average rate on these in order that by-products and low grade waste products can be held at competitive rates, as well as making some allowance for administration and increased finance charges.

With many gravel aggregates costing in the region of £16.00 to £18.00 per tonne, there is a significant impact on construction costs.

Avoidance

An alternative to using new aggregates in filling operations is to crush and screen rubble which may become available during the process of demolition and site clearance as well as removal of obstacles during the excavation processes.

Example:

Assuming that the material would be suitable for fill material under buildings or roads, a simple cost comparison would be as follows (note that for the purpose of the exercise, the material is taken to be 1.80 tonne per m³ and the total quantity involved less than 1000 m³):

	£/m³	£/tonne
Importing fill material:		
Cost of 'new' aggregates delivered to site	31.23	17.35
Addition for Aggregates Tax	3.78	2.10
Total cost of importing fill materials	35.01	19.45
Disposing of site material:		
Cost of removing materials from site materials	21.52	11.95
Crushing site materials:		
Transportation of material from excavations or demolition to stockpiles	3.00	1.67
Transportation of material from temporary stockpiles to the crushing plant	4.00	2.22
Establishing plant and equipment on site; removing on completion	2.00	1.11
Maintain and operate plant	9.00	5.00
Crushing hard materials on site	13.00	7.22
Screening material on site	2.00	1.11
Total cost of crushing site materials	33.00	18.33

From the above it can be seen that potentially there is a great benefit in crushing site materials for filling rather than importing fill materials.

Setting the cost of crushing against the import price would produce a saving of £2.10 per m³. If the site materials were otherwise intended to be removed from the site, then the cost benefit increases by the saved disposal cost to £23.53 per m³.

Even if there is no call for any or all of the crushed material on site, it ought to be regarded as a useful asset and either sold on in crushed form or else sold with the prospects of crushing elsewhere.

Specimen Unit rates	unit³	£
Establishing plant and equipment on site; removing on completion		
Crushing plant	trip	1,200.00
Screening plant	trip	600.00
Maintain and operate plant		
Crushing plant	week	7,200.00
Screening plant	week	1,800.00
Transportation of material from excavations or demolition places to temporary stockpiles	m³	3.00
Transportation of material from temporary stockpiles to the crushing plant	m³	2.40
Breaking up material on site using impact breakers		
mass concrete	m³	14.00
reinforced concrete	m³	16.00
Brickwork	m³	6.00
Crushing material on site		
mass concrete not exceeding 1000 m³	m³	13.00
mass concrete 1000–5000 m³	m³	12.00
mass concrete over 5000 m³	m³	11.00
reinforced concrete not exceeding 1000 m³	m³	15.00
reinforced concrete 1000–5000 m³	m³	14.00
reinforced concrete over 5000 m³	m³	13.00
brickwork not exceeding 1000 m³	m³	12.00
brickwork 1000–5000 m³	m³	11.00
brickwork over 5000 m³	m³	10.00
Screening material on site	m³	2.00

More detailed information can be found on the HMRC website (www.hmrc.gov.uk) in Notice AGL 1 Aggregates Levy published in May 2009.

The Planning Game

A Lord

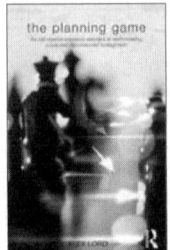

The sharing of information and knowledge is essential in the processes and negotiations involved in planning. In this book, Alex Lord sets out a new way of looking at the transfer of information and the cooperation of groups in planning by exploring the strand of economics known as information economics, including game theory. He starts by discussing theories of information economics, then moves into actual accounts of bargaining in planning practice.

April 2012: 216x138: 208pp
Hb: 978-0-415-59905-4: **£100.00**
Pb: 978-0-415-59906-1: **£27.99**

Cost Indices

The purpose of this section is to show changes in the cost of carrying out landscape work (hard surfacing and planting) since 1990. It is important to distinguish between costs and tender prices: the following table reflects the change in cost to contractors but does not necessarily reflect changes in tender prices. In addition to changes in labour and material costs, which are reflected in the indices given below, tender prices are also affected by factors such as the degree of competition at the time of tender and in the particular area where the work is to be carried out, the availability of labour and materials, and the general economic situation. This can mean that in a period when work is scarce, tender prices may fall despite the fact that costs are rising, and when there is plenty of work available, tender prices may increase at a faster rate than costs.

The Constructed Cost Index

A Constructed Cost Index based on PSA Price Adjustment Formulae for Construction Contracts (Series 2). Cost indices for the various trades employed in a building contract are published monthly by HMSO and are reproduced in the technical press.

The indices comprise 49 Building Work indices plus seven 'Appendices' and other specialist indices. The Building Work indices are compiled by monitoring the cost of labour and materials for each category and applying a weighting to these to calculate a single index.

Although the PSA indices are prepared for use with price-adjustment formulae for calculating reimbursement of increased costs during the course of a contract, they also present a time series of cost indices for the main components of landscaping projects. They can therefore be used as the basis of an index for landscaping costs.

The method used here is to construct a composite index by allocating weightings to the indices representing the usual work categories found in a landscaping contract, the weightings being established from an analysis of actual projects. These weightings totalled 100 in 1976 and the composite index is calculated by applying the appropriate weightings to the appropriate PSA indices on a monthly basis, which is then compiled into a quarterly index and rebased to 1976 = 100.

Constructed Landscaping (Hard Surfacing and Planting) Cost Index

Based on approxim

Year	First Quarter	Second Quarter	Third Quarter	Fourth Quarter	Annual Average
2000	493	496	513	514	504
2001	512	514	530	530	522
2002	531	540	573	574	555
2003	577	582	603	602	591
2004	602	610	639	639	623
2005	641	648	684	683	664
2006	688	692	710	710	710
2007	714	718	737	742	728
2008	752	771	803	794	780
2009	789	792	796	802	795
2010	804	810	814	817	811
2011	825	835	845	852	839
2012	858*	860*			

* *Provisional*

This index is updated every quarter in Spon's Price Book Update. The updating service is available, free of charge, to all purchasers of Spon's Price Books (complete the reply-paid card enclosed).

Regional Variations

Prices in *Spon's External Works and Landscape Price Book* are based upon conditions prevailing for a competitive tender in the outer London area. For the benefit of readers, this edition includes regional variation adjustment factors which can be used for an assessment of price levels in other regions.

Special further adjustment may be necessary when considering city centre or very isolated locations.

Region	Adjustment Factor
Outer London	1.00
Inner London	1.09
South East	1.00
South West	0.95
East Midlands	0.91
West Midlands	0.92
East Anglia	0.98
Yorkshire & Humberside	0.92
Northern	0.87
North West	0.84
Scotland	0.91
Wales	0.87
Northern Ireland	0.54
Channel Islands	1.69

The following example illustrates the adjustment of prices for regions other than Outer London, by use of regional variation adjustment factors.

A. Value of items priced using *Spon's External Works and Landscape Price Book* for Outer London   £100,000

B. Adjustment to value of A. to reflect Northern Region price level 100,000 × 0.87   £ 87,000

Introduction to Environmental Impact Assessment

4th Edition

J Glasson et al.

This is a comprehensive, clearly structured and readable overview of the subject, and a fourth edition of the book that has established itself as the leading introduction to EIA.

This edition has comprehensive appendices, with a wealth of important reference material, including key websites. It is also presented in a new and innovative format, including use of colour illustrations, and chapter questions for discussion. Written by three authors with extensive research, training and practical experience of EIA, this book brings together the most up-to-date information from many sources

December 2011: 246x174: 416pp
Hb: 978-0-415-66468-4: **£95.00**
Pb: 978-0-415-66470-7: **£32.99**

To Order: Tel: +44 (0) 1235 400524 **Fax:** +44 (0) 1235 400525
or Post: Taylor and Francis Customer Services,
Bookpoint Ltd, Unit T1, 200 Milton Park, Abingdon, Oxon, OX14 4TA UK
Email: book.orders@tandf.co.uk

For a complete listing of all our titles visit:
www.tandf.co.uk

Rates of Wages – Building Industry

Building Regulations in Brief

7th Edition

R Tricker & S Alford

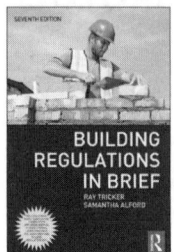

The seventh edition of the most popular and trusted guide to the building regulations includes the latest on all the significant amendments to Building Regulations, Planning Permission and the Approved Documents that occurred in October 2010 and includes changes to Parts F and L, as well as Approved Documents A, C, and J. There are also changes reflecting the consolidation of the building regulations included.

A ready reference giving practical information, it enables compliance in the simplest and most cost-effective manner possible. *Building Regulations in Brief* cuts through the confusion to explain the meaning of the regulations, their history, current status, requirements, associated documentation and how local authorities view their importance, as well as emphasizing the benefits and requirements of each regulation.

February 2012: 234x156: 1056pp
Pb: 978-0-415-80969-8: **£28.99**

To Order: Tel: +44 (0) 1235 400524 **Fax:** +44 (0) 1235 400525
or Post: Taylor and Francis Customer Services,
Bookpoint Ltd, Unit T1, 200 Milton Park, Abingdon, Oxon, OX14 4TA UK
Email: book.orders@tandf.co.uk

For a complete listing of all our titles visit:
www.tandf.co.uk

BUILDING INDUSTRY WAGES

Note
The following data is published for information only and is not used for within the calculation of rates for this publication. The exception is the calculation shown for the working hours per year.

The Working Rule Agreement includes a pay structure with general operative and additional skilled rates of pay as well as craft rate. Plus rates and additional payments will be consolidated into basic pay to provide the following rates (for a normal 39 hour week) which will come into effect from the following dates:

Effective from 5 September 2011

The following basic rates of pay are:

	Rate per 39-hour week (£)	Rate per hour (£)
Craft Rate	407.94	10.46
Skill Rate 1	388.83	9.97
Skill Rate 2	374.40	9.60
Skill Rate 3	350.22	8.98
Skill Rate 4	330.72	8.48
General operative	306.93	7.87

Example of Computation using National Wage Awards – Building Craft and General Operatives (not used for rate calculations in this book)

			Craft operatives		General operatives	
			£	£	£	£
Wages at standard basic rate						
Productive time	44.1	weeks	407.94	17,990.15	306.93	13,535.61
Lost time allowance	0.9	weeks	407.94	367.15	306.93	276.24
Overtime	0	weeks	0.00	0.00	0.00	0.00
				£ 18,357.30		£ 13,811.85
Extra payments under National Working Rules	45	weeks				
Sick Pay	1	week				
CITB Allowance (0.50% of payroll)	1	year		104.02		78.27
Holiday pay	4.2	weeks	407.94	1713.35	306.93	1289.11
Public Holiday pay (includes diamond jubilee holiday in the first week of June 2012)	1.8	weeks	407.94	734.29	306.93	552.47
Employer's contribution to:						
EasyBuild Stakeholder Pension	52	weeks	5.00	260.00	5.00	260.00
National Insurance (average weekly payment)	48	weeks	32.91	1579.68	19.84	952.32
				22748.64		16944.02
Severance pay and sundry costs	Plus	%	1.5	341.23	1.50	254.16
				23089.87		17198.18
Employer's Liability and Third Party Insurance	Plus	%	2.00	461.80	2.00	343.96
Total cost per annum				**£ 23551.67**		**£ 17,542.14**
Total cost per hour				**£ 13.69**		**£ 10.20**

DAYWORK RATES

Example Calculations of Prime Cost of Labour in Daywork

Effective from 5 September 2011

Example 1

		Rate (£)	Craft Operative	Rate (£)	General Operative
Basic Wages:	46 weeks	407.94	£18,765.24	306.93	£14,118.78
Extra Payments:	Where applicable		0.00		0.00
Subtotal:			£18,765.24		£14,118.78
Employer's National Insurance contribution (13.8% after the first £144 per week):			£1,675.49		£1,034.28
Holidays with Pay:	234 hours	10.46	£2,447.64	7.87	£1,841.58
Welfare Benefit:	52 weeks stamps	11.39	£592.28	11.39	£592.28
CITB Levy:	0.5% of payroll		£106.06		£79.80
Annual labour cost:			**£23,586.71**		**£17,666.72**
Hourly base rates as defined			**£13.15**		**£9.85**

Example 2

		Rate (£)	Craft Operative	Rate (£)	General Operative
Basic Wages:	46.0 weeks	407.94	£18,765.24	306.93	£14,118.78
Extra Payments:	Where applicable		0.00		0.00
Subtotal:			£18,765.24		£14,118.78
Employer's National Insurance contribution (13.8% after the first £144 per week):			£1,675.49		£1,034.28
CITB Levy:	0.5% of payroll		£93.83		£70.59
Annual labour cost:			**£20,534.56**		**£15,223.65**
Hourly Base Rate:			**£11.45**		**£8.49**

NOTES:

1. Calculated following Definition of Prime Cost of Daywork carried out under a Building Contract, published by the Royal Institution of Chartered Surveyors and the Construction Confederation.
2. Standard basic rates effective from 5 September 2011.
3. Standard working hours per annum calculated as follows:

52 weeks @ 39 hours	=	2028
Less		
4.2 weeks holiday @ 39 hours	=	163.8
9 days public holidays @ 7.8 hours	=	70.2
		234
		1794.0

4. All labour costs incurred by the contractor in his capacity as an employer, other than those contained in the hourly base rate, are to be taken into account under Section 6.
5. The above example is for guidance only and does not form part of the Definition; all the basic costs are subject to re-examination according to the time when and in the area where the daywork is executed.
6. N.I. payments are at not-contracted out rates applicable from April 2012.
7. Basic rate and GMB number of weeks = 52 weeks – 4.2 weeks annual holiday – 1.8 weeks public holiday = 46.0 weeks.

PART 3

Approximate Estimating Rates – Major Works

Prices in this section are based upon the Prices for Measured Works, but allow for incidentals which would normally be measured separately in a Bill of Quantities. They do not include for Preliminaries which are priced elsewhere in this book.

Items shown as subcontract or specialist rates would normally include the specialist's overhead and profit. All other items which could fall within the scope of works of general landscape and external works contractors would not include profit.

Based on current commercial rates, profits of 10% to 35% may be added to these rates to indicate the likely 'with profit' values of the tasks below. The variation quoted above is dependent on the sector in which the works are taking place – domestic, public or commercial.

HOW TO USE THIS BOOK

Item Excluding site overheads and profit	Unit	Total rate £
HOW THIS BOOK IS PRESENTED		
All prices in this book exclude overhead and profit.		

There are two pricing sections to this book which are split into two subsections as follows:
White pages
    These pages are the 'Major Works' section of the book. Please see page no 115 for the definitions of a Major Works project used in this book
Yellow pages
    These pages are the 'Minor Works' section of the book. Please see page no 437 for the definitions of a Minor Works project used in this book
Each of these sections is further subdivided into the following two sections:
    Approximate Estimates – This is presented in a 3 column format. This section combines prices from the Measured Works items to form a composite rate for the work described.
    Measured Works – This is presented in an 8 column format and is a detailed analysis of single tasks commonly employed throughout the external works and landscape sector.		

COLUMN FORMATS		

Description of the column formats used in this book.
Approximate Estimates
    Only the 3 column format is used; users of the book cannot see the components which make up the prices in the book format. An electronic version is supplied free with this book. Users of the electronic system can see more detail in the electronic format.
Measured Works
    Where an 8 column analysis is shown these are tasks carried out directly by the external works or landscape contractor and are shown at cost.
    Where prices are shown in the 8 column section without analysis, prices have been supplied by specialist subcontractors and are assumed to have the specialist subcontractor's selling price of the item included.		

Explanation of the columns used in the Measured Works analysis Prime Cost Commonly known as the 'PC'. Prime Cost is the actual price of the material item being addressed such as paving, shrubs, bollards or turves, as sold by the supplier. Prime Cost is given 'per square metre', 'per 100 bags' or 'each' according to the way the supplier sells his product. In researching the material prices for the book we requested that the suppliers price for full loads of their product delivered to a site close to the M25 in London. Spons rates do not include VAT. Some companies may be able to obtain greater discounts on the list prices than those shown in the book. Prime Cost prices for those products and plants which have a wide cost range will be found under the heading of Market Prices in the main sections of this book, so that the user may select the product most closely related to their specification.		

HOW TO USE THIS BOOK

Item Excluding site overheads and profit	Unit	**Total rate £**
Labour Hours This is the total amount of man hours required to carry out the quantity and unit of the measured works item which is being analysed. **Labour £** This is the value of the labour used. Please refer to the labour rates used in this edition which are detailed at the preceding the Approximate estimates of both the Major and Minor Works sections. **Plant** This covers the cost of machinery used to work out the cost of the item addressed. Plant can range from heavy excavators to light electrical breakers. All plant is assumed as hired in except for concrete mixers and normal site tools which a contractor would be expected to carry in his or her company vehicle (hand drills, grinders etc.). Plant includes for fuel and down time. The rates used are shown in the section of this book 'Computation of Mechanical Plant Costs' which follows this section. Delivery and removal to site costs are not included in any rates. **Materials** The PC material plus the additional materials required to fix the PC material. Every job needs materials for its completion besides the product bought from the supplier. Paving needs sand for bedding, expansion joint strips and cement pointing; fencing needs concrete for post setting and nails or bolts; tree planting needs manure or fertilizer in the pit, tree stakes, guards and ties. If these items were to be priced out separately, *Spon's External Works and Landscape Price Book* (and the Bill of Quantities) would be impossibly unwieldy, so they are put together under the heading of Materials. **Delivery Prices** These are not generally included. Exceptions would be aggregates or loose materials which are delivered by the load. Our specification to suppliers of information of the Major Works section of this book is that all materials are delivered in full loads to a site close to the M25 in North London. The Minor Works assumes part loads or smaller load sizes for aggregates. **Unit** The unit being used to measure the item. **Total Price £** The total price 'At Cost' – without profit of the item being addressed. **Where only a figure is shown in the columns** This is either a subcontract rate which has been received from a specialist subcontractor. This rate would generally include the subcontractor's overhead and profit but not the main contractors oncosts. **Where a figure is shown in the Material column only** This is generally the market price of a single material only.		

HOW TO USE THIS BOOK

Item Excluding site overheads and profit	Unit	Total rate £
WORKED EXAMPLES USING DATA IN THIS EDITION		
The following operations describe the trench excavation, pipe laying and backfilling operations as contained in section R12 of this edition.		
Example 1		
Pipe laying 100 m		
Remove topsoil 150 × 300 mm wide	100 m	**48.00**
Excavate for drain 150 × 450 mm; excavated material to spoil heaps on site by machine	100 m	**710.00**
Lay flexible plastic pipe 100 mm diameter wide	100 m	**155.00**
Backfilling with gravel rejects, blinding with sand and topping with 150 mm topsoil	100 m	**335.00**
TOTAL FOR ALL OF THE ABOVE COMBINED	100 m	**1250.00**
This total above item is a composite of all of the above 'Measured Works' items combined and is typical of an entry in the Approximate Estimates section of this book.		
Example 2		
The following individual items are shown as components of a tree planting specification for one tree. These individual items are all extracted from the paving section of this book.		
The PC of 900 × 600 × 50 mm precast concrete flags is shown in the PC (Prime cost) column as:	nr	**4.35**
This equates to the rate for the PC material of:	m²	**8.05**
Costs of bedding and pointing are added to give a total Materials cost of:	m²	**4.80**
Further costs for labour and mechanical plant give a resultant price of:	m²	**20.50**
If a quotation of PC £8.00/m² was received from a supplier of the flags the resultant price would be calculated as:		
Original price of the flags:	m²	**8.05**
Gives an extra over of:	m²	**−0.04**
Less the original cost PC (£7.48) plus the revised cost (£8.00) to give:	m²	**20.50**
Example 2		
The following samples are extracted from the tree planting section of this edition.		
The prime cost of an Acer platanoides 8–10 cm bare root tree is given as:	nr	**12.00**
To which the cost of handling and planting only to treepits excavated separately is added:	nr	**7.70**
Giving an overall cost of:	nr	**19.70**
The following items are also shown as component prices in the Measured Works section		
labour and mechanical plant for mechanical excavation of a 600 × 600 × 600 mm pit	nr	**3.50**
a single tree stake	nr	**5.35**
importing Topgrow compost in 75 litre bags	nr	**2.15**
backfilling with imported topsoil	nr	**8.70**
transporting the excavated material to spoil heaps on site 25 m distant	nr	**1.00**
and disposing of the excavated material	nr	**4.65**

HOW TO USE THIS BOOK

Item Excluding site overheads and profit	Unit	Total rate £
The following price which is a composite of all the above might be shown in the Approximate Estimates section: Tree planting; supply and plant Acer platanoides 8–10 cm in 600 × 600 mm tree pit excavated mechanically; allow for a single tree stake and Topgrow compost and backfilling with imported topsoil; all excavated material removed from site	nr	46.50
ADJUSTMENT AND VARIATION OF THE RATES IN THIS BOOK		
It will be appreciated that a variation in any one item in any group will affect the final Measured Works price. Any cost variation must be weighed against the total cost of the contract. A small variation in Prime Cost where the items are ordered in thousands may have more effect on the total cost than a large variation on a few items. A change in design that necessitates the use of earth moving equipment which must be brought to the site for that one job will cause a dramatic rise in the contract cost. Similarly, a small saving on multiple items will provide a useful reserve to cover unforeseen extras. Using the tree planting example above, if the tree size was to be increased from a 8–10 cm tree to a 10–12 cm tree of the same variety, this might necessitate an increase in tree pit size, more compost, a larger tree stake and more excavated material being disposed of off site. The resultant variation would again be referenced from the tree planting section as follows:		
The revised PC (prime cost) of an Acer platanoides 10–12 cm bare root tree is given as:	nr	19.50
To which the cost of handling and planting only to treepits excavated separately is added:	nr	11.20
Giving an overall cost of:	nr	30.50
Add to this the following variations as described:		
Tree pit increased in size from 600 × 600 × 600 mm to 900 × 900 × 600 mm	nr	7.85
Additional compost	nr	7.10
Increased staking requirements to two stakes	nr	7.40
Increased volume of imported topsoil	nr	19.60
increased disposal volume	nr	11.50
Giving a resultant price for the increased tree size as	nr	77.00
The variation therefore from the original specification for the 8–10 cm tree to the revised 10–12 cm tree is calculated as an additional sum per tree of	nr	30.00
LABOUR RATES USED IN THIS EDITION		
Major Works generally		
General labour rate for all works	hr	19.25
Major Works maintenance		
General labour rate for all works	hr	16.50
Minor Works generally		
General labour rate for all works	hr	20.75

HOW TO USE THIS BOOK

Item Excluding site overheads and profit	Unit	Total rate £
COMPUTATION OF MECHANICAL PLANT COSTS		
Hired Plant; Road Equipment Ltd		
Figures in Brackets reflect the actual working hours per week of the hired machinery		
Dumper 3 tonne and operator composite	hr	27.00
Dumper 3 tonne Thwaites self drive (32 hours/week)	hr	46.00
Dumper 5 tonne Thwaites self drive (32 hours/week)	hr	29.00
Dumper 6 tonne Thwaites self drive (32 hours/week)	hr	4.40
Excavator 360 Tracked 21 ton operated (32 hours/week)	hr	84.00
Excavator 360 Tracked 7 ton self drive (32 hours/week)	hr	69.00
Excavator Tracked 5 ton self drive (32 hours/week)	hr	56.00
Fork lift telehandler self drive (24 hours/week)	hr	32.00
JCB 3CX 4 × 4 Sitemaster + breaker operated (32 hours/week)	hr	39.00
JCB 3CX 4 × 4 Sitemaster operated (32 hours/week)	hr	37.50
Mini Excavator 1.5 tonne self drive (32 hours/week)	hr	4.05
Mini Excavator JCB 803 Rubber tracks self drive (32 hours/week)	hr	7.00
Mini Excavator JCB 803 Steel tracks self drive (32 hours/week)	hr	7.50
Skip loader 1 tonne (32 hours/week)	hr	3.45
Fuel charge Red diesel (32 hours/week)	l	1.15
HSS Ltd		
Access tower; alloy; 5.2 m	day	34.50
Cultivator; 110 kg	hr	8.85
Diamond blade consumable; 450 mm; concrete	mm	29.00
Heavy-duty breaker; 110 v; 2200 w	hr	4.70
Heavy-duty breaker; petrol; 5 hrs/day; single tool	hr	7.05
Mesh fence; temporary security fencing; 2.85 × 2.0 m high	1 m/ wk	2.35
Oxy-acetylene cutting kit	hr	4.70
Petrol masonry saw bench; 350 mm	hr	5.70
Petrol poker vibrator + 50 mm head	hr	2.20
Post hole borer; 1 man; weekly rate	hr	4.40
Vibrating plate compactor	hr	3.10
Vibrating roller; 136 kg; 10.1 kN; 4 hrs/day	hr	6.20
Vibration damped breaker (light); 1600 w; 110 v	hr	2.00

PRELIMINARIES

Item Excluding site overheads and profit	Unit	Total rate £
CONTRACT ADMINISTRATION AND MANAGEMENT		
Prepare tender bid for external works or landscape project; measured works contract; inclusive of bid preparation, provisional program and method statements		
Measured works bid; measured bills provided by the employer; project value		
£30,000.00	nr	310.00
£50,000.00	nr	510.00
£100,000.00	nr	1150.00
£200,000.00	nr	1800.00
£500,000.00	nr	2325.00
£1,000,000.00	nr	3300.00
Lump sum bid; scope document, drawings and specification provided by the employer; project value		
£30,000.00	nr	670.00
£50,000.00	nr	870.00
£100,000.00	nr	1725.00
£200,000.00	nr	2950.00
£500,000.00	nr	4050.00
£1,000,000.00	nr	6200.00
Prepare and maintain Health and Safety file; prepare risk and COSHH assessments, method statements, works programmes before the start of works on site; project value		
£35,000.00	nr	150.00
£75,000.00	nr	230.00
£100,000.00	nr	310.00
£200,000.00–£500,000.00	nr	770.00
Contract management		
Site management by site surveyor carrying out supervision and administration duties only; inclusive of oncosts		
full time management	week	1200.00
managing one other contract	week	600.00
Parking		
Parking expenses where vehicles do not park on the site area; per vehicle		
metropolitan area; city centre	week	200.00
Metropolitan area; outer areas	week	160.00
suburban restricted parking areas	week	40.00
Congestion charging		
London only	week	40.00
SITE SETUP AND SITE ESTABLISHMENT		
Site fencing; supply and erect temporary protective fencing and remove at completion of works		
Cleft chestnut paling		
1.20 m high 75 mm larch posts at 3 m centres	100 m	840.00
1.50 m high 75 mm larch posts at 3 m centres	100 m	1025.00

PRELIMINARIES

Item Excluding site overheads and profit	Unit	Total rate £
SITE SETUP AND SITE ESTABLISHMENT – cont		
Site fencing – cont		
Heras fencing		
2 week hire period	100 m	930.00
4 week hire period	100 m	1400.00
8 week hire period	100 m	2350.00
12 week hire period	100 m	3300.00
16 week hire period	100 m	4200.00
24 week hire period	100 m	6100.00
Site compounds		
Establish site administration area on reduced compacted Type 1 base; 150 mm thick; allow for initial setup of site office and welfare facilities for site manager and maximum 8 site operatives		
temporary base or hard standing; 20 × 10 m	200 m²	2325.00
To hardstanding area above; erect office and welfare accommodation; allow for furniture, telephone connection, and power; restrict access using temporary safety fencing to the perimeter of the site administration area		
Establishment; delivery and collectioncosts only		
site office and chemical toilet; 20 m of site fencing	nr	800.00
site office and chemical toilet; 40 m of site fencing	nr	860.00
site office and chemical toilet; 60 m of site fencing	nr	930.00
Weekly hire rates of site compound equipment		
site office 2.4 × 3.6 m; chemical toilet; 20 m of site fencing	week	100.00
site office 4.8 × 2.4 m; chemical toilet; 20 m of site fencing; armoured store 2.4 × 3.6 m	week	115.00
site office 2.4 × 3.6 m; chemical toilet; 40 m of site fencing	week	150.00
site office 2.4 × 3.6 m; chemical toilet; 60 m of site fencing	week	195.00
Removal of site compound		
Remove base; fill with topsoil and reinstate to turf on completion of site works	200 m²	1625.00
Surveying and setting out		
Note: In all instances a quotation should be obtained from a surveyor. The following cost projections are based on the costs of a surveyor using EDM (electronic distance measuring equipment) where an average of 400 different points on site are recorded in a day. The surveyed points are drawn in a CAD application and the assumption used that there is one day of drawing up for each day of on-site surveying.		
Survey existing site using laser survey equipment and produce plans of existing site conditions		
site 1,000 to 10,000 m² with sparse detail and features; boundary and level information only; average five surveying stations	nr	750.00
site up to 1,000 m² with dense detail such as buildings, paths, walls and existing vegetation; average five survey stations	nr	1500.00
site up to 4000 m² with dense detail such as buildings, paths, walls and existing vegetation; average 10 survey stations	nr	2250.00

C DEMOLITION AND SITE CLEARANCE

Item Excluding site overheads and profit	Unit	Total rate £
VEGETATION CLEARANCE		
Clear existing scrub vegetation including shrubs and hedges and stack on site		
By machine		
light scrub and grasses	100 m²	21.00
undergrowth brambles and heavy weed growth	100 m²	51.00
small shrubs	100 m²	51.00
As above but remove vegetation to licensed tip	100 m²	40.00
By hand	100 m²	85.00
As above but remove vegetation to licensed tip	100 m²	60.00
Fell trees on site; grub up roots; all by machine; remove debris to licensed tip		
trees 600 mm girth	each	285.00
trees 1.5–3.00 m girth	each	870.00
trees over 3.00 m girth	each	1375.00
Chip existing vegetation on site; remove to stockpile 100 m		
light scrub and bramble; average height 500 mm	100 m²	23.50
dense scrub and bramble; average height 1.00 m	100 m²	47.00
mixed scrub and shrubs average 1.50 m high	100 m²	75.00
dense overgrown shrubs and bramble; average 2.00 m high	100 m²	100.00
Turf stripping for preservation		
Cut and strip existing turf area by machine; turves 50 mm thick; turf to be lifted for preservation		
load to dumper or palette by hand; stack on site not exceeding 100 m travel to stack	100 m²	21.00
as above but all by hand; including barrowing	100 m²	450.00
Turf stripping for disposal; spray turf with glyphosate and return to site after 3 weeks; strip dead grass; load to stockpile on site; rotavate to prepare area for new surface treatments to within 50 mm of final levels		
By knapsack sprayer; excavator and dumper; pedestrian rotavator		
5 tonne excavator and 6 tonne dumper; stock pile distance 25 m	100 m²	64.00
5 tonne excavator and 3 tonne dumper; stock pile distance 50 m	100 m²	66.00
5 tonne excavator and 3 tonne dumper; stock pile distance 100 m	100 m²	70.00
By tractor drawn equipment		
spraying with glyphosate; 2 passes	100 m²	770.00
Clear site; demolish existing closeboard fence; clear shrubs 50% in shrub beds and 50% free-standing; spray turf and return to clear ground of sprayed vegetation; chip vegetation on site and transport to spoil heaps 50 m; works by mechanical excavator		
Rectangular site with 70% turf and 30% shrub area		
1000 m²	m²	1875.00
2000 m²	m²	3400.00
4000 m²	m²	6400.00
Cleartrack (Evl) Ltd; as above but clearance by mechanical spraying and in situ mulcher; vegetation only to spoil heap; turf mulched in		
5000 m²	m²	2175.00

C DEMOLITION AND SITE CLEARANCE

Item Excluding site overheads and profit	Unit	Total rate £
VEGETATION CLEARANCE – cont		
Clear site; demolish existing chain-link fence; clear shrubs 50% in shrub beds and 50% free-standing; spray turf and return to clear ground of sprayed vegetation; chip vegetation on site and transport to spoil heaps 50 m; works by high capacity mechanical mulcher		
Cleartrack (Evl) Ltd		
overgrown areas of weeds and brambles with maximum stem diameter of 100 mm	m²	3500.00
overgrown areas of weeds and brambles with maximum stem diameter of 150 mm	m²	6900.00
Rectangular site with 60% turf grass/scrub area and 40% shrubs and overgrowth		
1000 m²	m²	2075.00
2000 m²	m²	3900.00
4000 m²	m²	7300.00
CLEARANCE OF SURFACES		
Demolish existing surfaces		
Break up plain concrete slab; remove to licensed tip		
150 mm thick	m²	9.10
200 mm thick	m²	12.20
300 mm thick	m²	18.30
Break up plain concrete slab and preserve arisings for hardcore; break down to maximum size of 200 × 200 mm and transport to location; maximum distance 50 m		
150 mm thick	m²	6.00
200 mm thick	m²	8.05
300 mm thick	m²	12.10
Break up reinforced concrete slab and remove to licensed tip		
150 mm thick	m²	13.40
200 mm thick	m²	17.90
300 mm thick	m²	27.00
Break out existing surface and associated 150 mm thick granular base load to 20 tonne muck away vehicle for removal off site		
macadam surface; 70 mm thick	m²	8.10
macadam surface; 150 mm thick	m²	11.60
block paving; 50 mm thick	m²	8.00
block paving; 80 mm thick	m²	8.65
CLEARANCE OF STRUCTURES		
Demolish existing free-standing walls; grub out foundations; remove arisings to tip		
Brick wall; 112 mm thick		
300 mm high	m	11.70
500 mm high	m	17.50
Brick wall; 225 mm thick		
300 mm high	m	14.70
500 mm high	m	19.00
1.00 m high	m	30.50
1.20 m high	m	32.50
1.50 m high	m	40.00
1.80 m high	m	45.00

C DEMOLITION AND SITE CLEARANCE

Item Excluding site overheads and profit	Unit	Total rate £
Demolish existing structures		
Timber sheds on concrete base 150 mm thick inclusive of disposal and backfilling with excavated material; plan area of building		
10 m²	nr	**200.00**
15 m²	nr	**285.00**
Building of cavity wall construction; insulated with tiled roof; inclusive of foundations; allow for disconnection of electrical water and other services; grub out and remove 10 m of electrical cable, water and waste pipes		
10 m²	nr	**1550.00**
20 m²	nr	**2050.00**
CLEARANCE OF FENCING AND GATES		
Demolish fencing and gates		
Demolish fencing; break out posts, remove fencing stack and load mechanically for removal to tip		
Chain link fencing; vegetation not present on angle iron stakes driven in	100 m	**220.00**
Chain link fencing; overgrown with vegetation on angle iron stakes driven in	100 m	**590.00**

D GROUNDWORK

Item Excluding site overheads and profit	Unit	Total rate £
GROUNDWORK		
Prices for excavating and reducing to levels are for work in light or medium soils; multiplying factors for other soils are as follows		
Clay	1.5	–
Compact gravel	1.2	–
Soft chalk	2.0	–
Hard rock	3.0	–
Excavating topsoil for preservation		
Remove topsoil average depth 300 mm; deposit in spoil heaps not exceeding 100 m travel to heap; turn heap once during storage		
by machine	m³	13.50
by hand	m³	120.00
Excavate to reduce levels; remove spoil to stockpile not exceeding 100 m travel; all by machine		
Average 0.25 m deep	m²	2.10
Average 0.25 m deep	m³	8.40
Average 0.30 m deep	m²	2.55
Average 0.30 m deep	m³	7.70
Average 1.0 m deep; using 21 tonne 360 tracked excavator	m³	4.90
Average 1.0 m deep; using JCB	m³	7.70
Excavate to reduce levels; remove spoil to stockpile not exceeding 100 m travel; all by hand		
Average 0.10 m deep	m²	10.30
Average 0.20 m deep	m²	21.00
Average 0.30 m deep	m²	31.00
Average 100–300 mm deep	m³	100.00
Average 1.0 m deep	m³	120.00
Extra for carting spoil to licensed tip off site (20 tonnes)	m³	22.00
Extra for carting spoil to licensed tip off site (by skip; machine loaded)	m³	43.00
Extra for carting spoil to licensed tip off site (by skip; hand loaded)	m³	97.00
Spread excavated material to levels in layers not exceeding 150 mm; using scraper blade or bucket		
Average thickness 100 mm	m²	0.40
Average thickness 100 mm but with imported topsoil	m²	3.75
Average thickness 200 mm	m²	0.80
Average thickness 200 mm but with imported topsoil	m²	7.50
Average thickness 250 mm	m²	1.00
Average thickness 200–250 mm	m³	4.20
Average thickness 250 mm but with imported topsoil	m²	9.35
Average thickness 250 mm but with imported topsoil	m³	37.50
Extra for work to banks exceeding 30° slope	30%	–

D GROUNDWORK

Item Excluding site overheads and profit	Unit	Total rate £
Rip subsoil using approved subsoiling machine to a depth of 600 mm below topsoil at 600 mm centres; all tree roots and debris over 150 × 150 mm to be removed; cultivate ground where shown on drawings to depths as shown		
100 mm	100 m²	8.90
200 mm	100 m²	9.10
300 mm	100 m²	9.40
400 mm	100 m²	12.20
Form landform from imported material; mounds to falls and grades to receive turf or seeding treatments; material dumped by 20 tonne vehicle to location; volume 100 m³ spread over area; thickness varies from 150–500 mm; prices to not include for surface preparations		
Imported standard grade topsoil		
by large excavator	m³	34.00
by 5 tonne excavator	m³	36.50
by 3 tonne excavator	m³	38.00
Form landform from excavated material; mounds to falls and grades to receive turf or seeding treatments; material moved by dumper to location maximum 50 m; volume 100 m³ spread over a 200 m² area; thickness varies from 150–500 mm; prices to not include for surface preparations		
Imported recycled topsoil		
by large excavator	m³	3.70
by 5 tonne excavator	m³	6.30
by 3 tonne excavator	m³	7.70
TRENCHES		
Excavate trenches for foundations; trenches 225 mm deeper than specified foundation thickness; pour plain concrete foundations GEN 1 10 N/mm and to thickness as described; disposal of excavated material off site; dimensions of concrete foundations		
250 mm wide × 250 mm deep	m	10.90
300 mm wide × 300 mm deep	m	14.60
400 mm wide × 600 mm deep	m	35.50
600 mm wide × 400 mm deep	m	37.00
Excavate trenches for foundations; trenches 225 mm deeper than specified foundation thickness; pour plain concrete foundations GEN 1 10 N/mm and to thickness as described; disposal of excavated material off site; dimensions of concrete foundations		
250 mm wide × 250 mm deep	m³	205.00
300 mm wide × 300 mm deep	m³	160.00
600 mm wide × 400 mm deep	m³	155.00
400 mm wide × 600 mm deep	m³	140.00

D GROUNDWORK

Item Excluding site overheads and profit	Unit	Total rate £
TRENCHES – cont		
Excavate trenches for services; grade bottoms of excavations to required falls; remove 20% of excavated material off site		
Trenches 300 mm wide		
600 mm deep	m	6.00
750 mm deep	m	7.70
900 mm deep	m	9.20
1.00 m deep	m	10.10
Ha-ha		
Excavate ditch 1200 mm deep × 900 mm wide at bottom battered one side 45° slope; excavate for foundation to wall and place GEN 1 concrete foundation 150 × 500 mm; construct one and a half brick wall (brick PC £300.00/1000) battered 10° from vertical; laid in cement: lime: sand (1:1:6) mortar in English garden wall bond; precast concrete coping weathered and throated set 150 mm above ground on high side; rake bottom and sides of ditch and seed with low maintenance grass at 35 g/m²		
wall 600 mm high	m	255.00
wall 900 mm high	m	320.00
wall 1200 mm high	m	395.00
Excavate ditch 1200 mm deep × 900 mm wide at bottom battered one side 45° slope; place deer or cattle fence 1.20 m high at centre of excavated trench; rake bottom and sides of ditch and seed with low maintenance grass at 35 g/m²		
fence 1200 mm high	m	20.00

D GROUND STABILIZATION

Item Excluding site overheads and profit	Unit	Total rate £
GRADING AND PREPARATION OF BANKS		
Excavate and grade banks to grade to receive stabilization treatments below;		
removal of excavated material not included		
By 21 tonne excavator	m³	0.35
By 7 tonne excavator	m³	2.55
By 5 tonne excavator	m³	2.95
By 3 tonne excavator	m³	4.35
Excavate sloped bank to vertical to receive retaining or stabilization treatment		
priced below; allow 1.00 m working space at top of bank; backfill working space		
and remove balance of arisings off site; measured as length along the top of the		
bank		
Bank height 1.00 m; excavation by 5 tonne excavator		
10°	m	69.00
30°	m	21.50
45°	m	12.30
60°	m	7.15
Bank height 1.00 m; excavation by 5 tonne excavator; arisings moved to stockpile on		
site		
10°	m	12.70
30°	m	6.50
45°	m	3.75
60°	m	2.20
Bank height 2.00 m; excavation by 5 tonne excavator		
10°	m	280.00
30°	m	84.00
45°	m	49.00
60°	m	28.00
Bank height 2.00 m; excavation by 5 tonne excavator but arisings moved to stockpile		
on site		
10°	m	85.00
30°	m	25.50
45°	m	15.00
60°	m	8.60
Bank height 3.00 m; excavation by 21 tonne excavator		
20°	m	250.00
30°	m	160.00
45°	m	87.00
60°	m	47.50
As above but arisings moved to stockpile on site		
20°	m	40.50
30°	m	25.50
45°	m	10.30
60°	m	5.60

D GROUND STABILIZATION

Item	Unit	Total rate £
Excluding site overheads and profit		
EROSION CONTROL MATS		
Bank stabilization; erosion control mats		
Excavate fixing trench for erosion control mat 300 × 300 mm; backfill after placing mat selected below	m	6.10
To anchor trench above; place mat to slope to be retained; allow extra over to the slope for the area of matting required to the anchor trench		
By machine; final grade only of excavated bank by excavator; lay erosion control solution; sow with low maintenance grass at 35 g/m²; spread imported topsoil 25 mm thick incorporating medium grade sedge peat at 3 kg/m² and fertilizer at 30 g/m²; water lightly		
unseeded Eromat Light	m²	2.75
seeded Covamat Standard	m²	3.40
lay Greenfix biodegradable pre-seeded erosion control mat; fix with 6 × 300 mm steel pegs at 1.0 m centres	m²	3.90
Tensar open textured erosion mat	m²	5.40
By hand; final grade only of excavated bank by hand; lay erosion control solution; sow with low maintenance grass at 35 g/m²; spread imported topsoil 25 mm thick incorporating medium grade sedge peat at 3 kg/m² and fertilizer at 30 g/m²; water lightly		
unseeded Eromat Light	m²	3.20
seeded Covamat Standard	m²	3.80
lay Greenfix biodegradable pre-seeded erosion control mat; fix with 6 × 300 mm steel pegs at 1.0 m centres	m²	4.30
Tensar open textured erosion mat	m²	5.85
Cellular retaining systems; excavate vertical bank to slope; grade to even grade; lay imported topsoil 200 mm thick		
Slope angle 20°		
excavation by 21 tonne excavator	m	310.00
SOIL REINFORCEMENT		
Geogrid soil reinforcement (Note: measured per metre run at the top of the bank)		
Backfill excavated bank; lay Tensar geogrid; cover with excavated material as work proceeds		
2.00 m high bank; geogrid at 1.00 m vertical lifts		
10° slope	m	590.00
20° slope	m	330.00
30° slope	m	240.00
2.00 m high bank; geogrid at 0.50 m vertical lifts		
45° slope	m	60.00
60° slope	m	48.00
3.00 m high bank		
10° slope	m	590.00
20° slope	m	330.00
30° slope	m	240.00
45° slope	m	510.00

D GROUND STABILIZATION

Item Excluding site overheads and profit	Unit	Total rate £
RETAINING WALL SYSTEMS		
Note: The following models do not allow for excavation to the existing bank in preparation of the retaining system. Please amalgamate with the 'Grading and preparation' section above.		
Excavate trench and lay foundation concrete 1:3:6; 600 × 300 mm deep		
By machine	m	30.50
By hand	m	41.00
Concrete block retaining walls; Stepoc		
Excavate trench 750 mm deep and lay concrete foundation 600 mm wide × 600 mm deep; construct Forticrete precast hollow concrete block wall with 450 mm below ground laid all in accordance with manufacturer's instructions; fix reinforcing bar 12 mm as work proceeds; fill blocks with concrete 1:3:6 as work proceeds		
Walls 1.00 m high		
type 190; 400 × 225 × 200 mm	m	185.00
type 256; 400 × 225 × 256 mm	m	210.00
As above but 1.50 m high		
type 190; 400 × 225 × 200 mm	m	240.00
type 256; 400 × 225 × 256 mm	m	270.00
Walls 1.50 m high as above but with foundation 1.80 m wide × 600 mm deep		
type 190; 400 × 225 × 200 mm	m	290.00
type 256; 400 × 225 × 256 mm	m	330.00
Walls 1.80 m high		
type 190; 400 × 225 × 200 mm	m	315.00
type 256; 400 × 225 × 256 mm	m	355.00
On foundation measured above, supply and install Milton Precast Concrete precast concrete L shaped units, constructed all in accordance with manufacturer's instructions; backfill with approved excavated material compacted as the work proceeds		
1500 mm high × 1000 mm wide	m	215.00
2500 mm high × 1000 mm wide	m	450.00
3000 mm high × 1000 mm wide	m	580.00
Gabion walls; excavation costs excluded; see 'Grading and preparation of banks' above		
Gabions 500 mm high; allowance of 1.00 m working space at top of bank; backfill working space and remove balance of arisings off site; lay concrete footing; 200 mm deep × 1.50 m wide		
Height retained		
500 mm	m	91.00
1.00 m	m	140.00
1.50 m	m	195.00
2.00 m	m	240.00

D GROUND STABILIZATION

Item Excluding site overheads and profit	Unit	Total rate £
RETAINING WALL SYSTEMS – cont		
Timber log retaining walls		
Excavate trench 300 mm wide to one third of the finished height of the retaining walls below; lay 100 mm hardcore; fix machine rounded logs set in concrete 1:3:6; remove excavated material from site; fix geofabric to rear of timber logs; backfill with previously excavated material set aside in position; all works by machine		
100 mm diameter logs		
500 mm high (constructed from 1.80 m lengths)	m	70.00
1.20 mm high (constructed from 1.80 m lengths)	m	110.00
1.60 mm high (constructed from 2.40 m lengths)	m	160.00
2.00 mm high (constructed from 3.00 m lengths)	m	210.00
150 mm diameter logs		
1.20 mm high (constructed from 1.80 m lengths)	m	200.00
1.60 mm high (constructed from 2.40 m lengths)	m	230.00
200 mm diameter logs		
2.00 mm high (constructed from 3.00 m lengths)	m	300.00
Timber crib wall		
Excavate trench to receive foundation 300 mm deep; place plain concrete foundation 150 mm thick in 11.50 N/mm² concrete (sulphate-resisting cement); construct timber crib retaining wall and backfill with excavated spoil behind units		
timber crib wall system; average 5.0 m high	m	1000.00
timber crib wall system; average 4.0 m high	m	730.00
timber crib wall system; average 2.0 m high	m	320.00
timber crib wall system; average 1.00 m high	m	160.00
REVETMENTS		
Gabion mattress revetments		
Construct revetment of Maccaferri Ltd Reno mattress gabions laid on firm level ground; tightly packed with broken stone or concrete and securely wired; all in accordance with manufacturer's instructions		
gabions 6 × 2 × 0.17 m; one course	m²	34.00
gabions 6 × 2 × 0.17 m; two courses	m²	56.00
Grass concrete; bank revetments; excluding bulk earthworks		
Bring bank to final grade by machine; lay regulating layer of Type 1; lay filter membrane; lay 100 mm drainage layer of broken stone or approved hardcore 28–10 mm size; blind with 25 mm sharp sand; lay grass concrete surface (price does not include edgings or toe beams); fill with approved topsoil and fertilizer at 35 g/m²; seed with dwarf rye based grass seed mix		
Grasscrete in situ reinforced concrete surfacing GC 1; 100 mm thick	m²	57.00
Grasscrete in situ reinforced concrete surfacing GC 2; 150 mm thick	m²	70.00
Grassblock 103 mm open matrix blocks; 406 × 406 × 103 mm	m²	54.00

D GROUND STABILIZATION

Item Excluding site overheads and profit	Unit	Total rate £
Neoweb; cellular slope retention; clear existing bank of light scrub, shrubs and grasses; spray with glyphosate; grade to even grade; lay soil retaining system		
Backfilled with excavated material		
100 mm deep	m^2	**13.60**
200 mm deep	m^2	**23.00**
Backfilled with ballast		
100 mm deep	m^2	**15.70**
200 mm deep	m^2	**27.00**
Backfilled with concrete 10 Nmm		
100 mm deep	m^2	**19.80**
200 mm deep	m^2	**36.00**
Backfilled with imported topsoil and seeded		
100 mm deep	m^2	**18.70**

E IN SITU CONCRETE

Item Excluding site overheads and profit	Unit	Total rate £
SITE MIXED CONCRETE		
Mix concrete on site; aggregates delivered in 20 tonne loads; deliver mixed concrete to location by mechanical dumper distance 25 m		
1:3:6	m³	88.00
1:2:4	m³	100.00
As above but ready mixed concrete		
10 N/mm²	m³	105.00
15 N/mm²	m³	110.00
Mix concrete on site; aggregates delivered in 20 tonne loads; deliver mixed concrete to location by barrow distance 25 m		
1:3:6	m³	130.00
1:2:4	m³	145.00
As above but aggregates delivered in 1 tonne bags		
1:3:6	m³	175.00
1:2:4	m³	185.00
As above but ready mixed concrete		
10 N/mm²	m³	150.00
15 N/mm²	m³	150.00
As above but concrete discharged directly from ready mix vehicle to required location		
10 N/mm²	m³	100.00
15 N/mm²	m³	105.00
IN SITU CONCRETE FOUNDATIONS		
Excavate foundation trench mechanically; remove spoil off site; lay 1:3:6 site mixed concrete foundations; distance from mixer 25 m; depth of trench to be 225 mm deeper than foundation to allow for three underground brick courses priced separately		
Concrete poured to blinded exposed ground		
200 mm deep × 400 mm wide	m	21.00
300 mm deep × 500 mm wide	m	34.50
400 mm deep × 400 mm wide	m	34.00
400 mm deep × 600 mm wide	m	51.00
600 mm deep × 600 mm wide	m	71.00
Concrete poured to formwork		
200 mm deep × 400 mm wide	m	45.00
300 mm deep × 500 mm wide	m	59.00
300 mm deep × 600 mm wide	m	65.00
300 mm deep × 800 mm wide	m	80.00
400 mm deep × 400 mm wide	m	58.00
400 mm deep × 600 mm wide	m	89.00
600 mm deep × 600 mm wide	m	110.00

E IN SITU CONCRETE

Item Excluding site overheads and profit	Unit	Total rate £
Excavate foundation trench by hand; remove spoil off site; lay 1:3:6 site mixed concrete foundations; distance from mixer 25 m; depth of trench to be 225 mm deeper than foundation to allow for three underground brick courses priced separately		
Disposal to spoil heap 25 m by barrow and off site by grab vehicle		
200 mm deep × 400 mm wide	m	60.00
300 mm deep × 500 mm wide	m	86.00
400 mm deep × 400 mm wide	m	86.00
400 mm deep × 600 mm wide	m	125.00
600 mm deep × 600 mm wide	m	170.00
Excavate foundation trench mechanically; remove spoil off site; lay ready mixed concrete GEN1 discharged directly from delivery vehicle to location; depth of trench to be 225 mm deeper than foundation to allow for three underground brick courses priced separately; foundation size		
200 mm deep × 400 mm wide	m	16.80
300 mm deep × 500 mm wide	m	27.00
400 mm deep × 400 mm wide	m	26.00
400 mm deep × 600 mm wide	m	39.00
600 mm deep × 600 mm wide	m	56.00
Reinforced concrete wall to foundations above (site mixed concrete)		
Up to 1.00 m high × 200 mm thick	m²	64.00
Up to 1.00 m high × 300 mm thick	m²	64.00
Reinforced concrete wall to foundations above (ready mix concrete RC35)		
1.00 m high × 200 mm thick	m²	94.00
1.00 m high × 300 mm thick	m²	105.00
Grading; excavate to reduce levels for concrete slab; lay waterproof membrane, 100 mm hardcore and form concrete slab reinforced with A142 mesh in 1:2:4 site mixed concrete to thickness; remove excavated material off site		
Concrete 1:2:4 site mixed		
100 mm thick	m²	45.50
150 mm thick	m²	53.00
250 mm thick	m²	68.00
300 mm thick	m²	75.00
Concrete ready mixed GEN 2		
100 mm thick	m²	43.50
150 mm thick	m²	50.00
250 mm thick	m²	63.00
300 mm thick	m²	69.00

E IN SITU CONCRETE

Item Excluding site overheads and profit	Unit	Total rate £
IN SITU CONCRETE WALLS		
Excavate foundation trench mechanically; remove spoil off site; fix reinforcement starter bars 12 mm at 200 mm centres; lay 1:3:6 site mixed concrete foundations; distance from mixer 25 m; depth of trench to be 225 mm deeper than foundation; cast in situ concrete walls inclusive of bar reinforcement 12 mm; footings cast to blinded exposed ground		
Wall height 500 mm above ground; on foundation 400 mm × 200 mm thick		
150 thick wall; site mixed concrete 20 tonne aggregate loads	m	82.00
150 thick wall; ready mixed concrete ST2	m	81.00
250 thick wall; site mixed concrete; 20 tonne aggregate loads	m	120.00
250 thick wall; ready mixed concrete ST2	m	125.00
Wall 1.00 mm high on foundation 500 mm wide × 300 mm deep		
150 thick wall; site mixed concrete; 20 tonne aggregate loads	m	135.00
150 thick wall; ready mixed concrete ST2	m	130.00
250 thick wall; site mixed concrete; 20 tonne aggregate loads	m	150.00
250 thick wall; ready mixed concrete ST2	m	140.00
Wall 1.50 mm high on foundation 500 mm wide × 300 mm deep		
250 thick wall; site mixed concrete; 20 tonne aggregate loads	m	285.00
250 thick wall; ready mixed concrete ST2	m	270.00
Wall 1.80 mm high on foundation 600 mm wide × 300 mm deep		
250 thick wall; site mixed concrete; 20 tonne aggregate loads	m	360.00
250 thick wall; ready mixed concrete ST2	m	340.00
Wall 1.80 mm high on foundation 800 mm wide × 300 mm deep		
250 thick wall; site mixed concrete; 20 tonne aggregate loads	m	370.00
250 thick wall; ready mixed concrete ST2	m	350.00

F BRICK/BLOCK WALLING

Item Excluding site overheads and profit	Unit	Total rate £
BRICK WALLING		
One brick thick wall (225 mm thick); excavation 400 mm deep; remove arisings off site; lay GEN 1 concrete foundations 450 mm wide × 250 mm thick; all in English Garden Wall bond; laid in cement: lime: sand (1:1:6) mortar with flush joints, fair face one side; DPC two courses engineering brick in cement: sand (1:3) mortar; precast concrete coping 152 × 75 mm		
Wall 900 mm high above DPC		
engineering brick (class B)	m	145.00
brick PC £300.00/1000	m	220.00
brick PC £800.00/1000	m	205.00
Wall 1200 mm high above DPC		
engineering brick (class B)	m	185.00
brick PC £300.00/1000	m	275.00
brick PC £800.00/1000	m	390.00
Wall 1800 mm high above DPC		
engineering brick (class B)	m	250.00
brick PC £300.00/1000	m	405.00
brick PC £800.00/1000	m	4700.00
One and a half brick wall; excavate 450 mm deep; remove arisings off site; lay GEN 1 concrete foundations 600 × 300 mm thick; two thick brick piers at 3.0 m centres; all in English Garden Wall bond; laid in cement: lime: sand (1:1:6) mortar with flush joints; fair face one side; DPC two courses engineering brick in cement: sand (1:3) mortar; coping of headers on edge		
Wall 900 mm high above DPC		
engineering brick (class B)	m	570.00
brick PC £300.00/1000	m	225.00
Wall 1200 mm high above DPC		
engineering brick (class B)	m	325.00
brick PC £300.00/1000	m	285.00
Wall 1800 mm high above DPC		
engineering brick (class B)	m	400.00
brick PC £300.00/1000	m	440.00
brick PC £800.00/1000	m	970.00
BLOCK WALLING		
Notes: Measurements allow for works above ground only		
Concrete block walls; including excavation of foundation trench 450 mm deep; remove spoil off site; lay GEN 1 concrete foundations 600 × 300 mm thick; 1 block below ground		
Solid blocks 7 N/mm^2; 100 mm thick		
500 mm high	m^2	52.00
750 mm high	m^2	60.00
1.00 m high	m^2	69.00
1.25 m high	m^2	77.00
1.50 m high	m^2	85.00
1.80 m high	m^2	95.00

F BRICK/BLOCK WALLING

Item Excluding site overheads and profit	Unit	Total rate £
BLOCK WALLING – cont		
Concrete block walls – cont		
140 mm thick		
140 mm thick	m²	**70.00**
100 mm blocks; laid on flat		
500 mm high	m²	**77.00**
750 mm high	m²	**98.00**
100 mm blocks laid flat; 2 courses underground		
1.00 m high	m²	**125.00**
1.20 m high	m²	**140.00**
1.50 m high	m²	**165.00**
1.80 m high	m²	**190.00**
Hollow blocks filled with concrete; 440 × 215 × 215 mm		
500 mm high	m²	**79.00**
750 mm high	m²	**97.00**
1.00 m high	m²	**115.00**
1.25 m high	m²	**125.00**
1.50 m high	m²	**150.00**
1.80 m high	m²	**170.00**
Concrete block retaining walls; 2 courses underground; including excavation of foundation trench 450 mm deep; remove spoil off site; lay GEN 1 concrete foundations 1200 × 400 mm thick		
Solid blocks 7 N/mm²; 100 mm thick		
1.00 m high	m	**130.00**
1.50 m high	m	**155.00**
1.50 m high	m	**170.00**
100 mm blocks; laid on flat		
1.00 m high	m²	**195.00**
1.50 m high	m²	**235.00**
1.80 m high	m²	**260.00**
Hollow blocks filled with concrete; 215 mm thick		
1.00 m high	m²	**220.00**
1.50 m high	m²	**265.00**
1.80 m high	m²	**285.00**
Hollow blocks with steel bar cast into the foundation		
1.00 m high	m²	**240.00**
1.50 m high	m²	**295.00**
1.80 m high	m²	**315.00**
Concrete block wall with brick face 112.5 mm thick to stretcher bond; including excavation of foundation trench 450 mm deep; remove spoil off site; lay GEN 1 concrete foundations 600 × 300 mm thick; place stainless steel ties at 4 nr/m² of wall face		
Solid blocks 7 N/mm² with stretcher bond brick face; reclaimed bricks PC £800.00/1000		
100 mm thick	m²	**2050.00**
140 mm thick	m²	**2050.00**

F BRICK/BLOCK WALLING

Item Excluding site overheads and profit	Unit	Total rate £
BRICK/BLOCK WALLING WITH PIERS		
Half brick thick wall; 102.5 mm thick; with one brick piers at 2.00 m centres; excavate foundation trench 500 mm deep; remove spoil off site; lay site mixed concrete foundations 1:3:6 350 × 150 mm thick; laid in cement: lime: sand (1:1:6) mortar with flush joints; fair face one side; DPC two courses underground; engineering brick in cement: sand (1:3) mortar; coping of headers on end		
Wall 900 mm high above DPC		
engineering brick (class B)	m	170.00
brick PC £300.00/1000	m	100.00
brick PC £800.00/1000	m	160.00
One brick thick wall; 225 mm thick; with one and a half brick piers at 3.0 m centres; excavation 400 mm deep; remove arisings off site; lay GEN 1 concrete foundations 450 mm wide × 250 mm thick; all in English Garden Wall bond; laid in cement: lime: sand (1:1:6) mortar with flush joints; fair face one side; DPC two courses engineering brick in cement: sand (1:3) mortar; precast concrete coping 152 × 75 mm		
Wall 900 mm high above DPC		
engineering brick (class B)	m	280.00
brick PC £300.00/1000	m	220.00
brick PC £800.00/1000	m	430.00
Wall 1200 mm high above DPC		
engineering brick (class B)	m	315.00
brick PC £300.00/1000	m	275.00
brick PC £800.00/1000	m	3250.00
Wall 1800 mm high above DPC		
engineering brick (class B)	m	460.00
brick PC £300.00/1000	m	405.00
brick PC £800.00/1000	m	4700.00
BRICK PIERS		
Brick piers on concrete footings 500 × 500 × 250 mm thick; inclusive of excavations and disposal off site; coping of engineering brick on edge; 2 courses underground		
Pier 215 × 215 mm; one brick thick; brick PC £300/1000		
500 mm high	nr	44.50
750 mm high	nr	55.00
1.00 m high	nr	66.00
1.25 m high	nr	76.00
1.50 m high	nr	87.00
1.80 m high	nr	99.00
Pier 215 × 215 mm; one brick thick; brick PC £800/1000		
500 mm high	nr	54.00
750 mm high	nr	68.00
1.00 m high	nr	82.00
1.25 m high	nr	97.00
1.50 m high	nr	110.00
1.80 m high	nr	130.00

F BRICK/BLOCK WALLING

Item Excluding site overheads and profit	Unit	Total rate £
BRICK PIERS – cont		
Brick piers on concrete footings 500 × 500 × 250 mm thick – cont		
Pier 337.5 × 337.5 mm; one and a half brick thick; brick PC £300/1000		
500 mm high	nr	77.00
750 mm high	nr	99.00
1.00 m high	nr	120.00
1.25 m high	nr	140.00
1.50 m high	nr	165.00
1.80 m high	nr	190.00
Pier 337.5 × 337.5 mm; one and a half brick thick; brick PC £800/1000		
500 mm high	nr	240.00
750 mm high	nr	320.00
1.00 m high	nr	410.00
1.25 m high	nr	490.00
1.50 m high	nr	580.00
1.80 m high	nr	680.00
Brick piers on concrete footings 800 × 800 × 600 mm thick; inclusive of excavations and disposal off site; coping of Haddonstone S150C classically moulded corbelled piercap PC £114.00; 120 mm thick overall		
Pier 450 × 450 mm; two bricks thick; brick PC £300/1000; height of brickwork above ground		
500 mm high	nr	290.00
750 mm high	nr	320.00
1.00 m high	nr	360.00
1.25 m high	nr	390.00
1.50 m high	nr	425.00
1.80 m high	nr	470.00
2.00 m high	nr	495.00
Pier 450 × 450 mm; two bricks thick; brick PC £800/1000; height of brickwork above ground		
500 mm high	nr	325.00
750 mm high	nr	370.00
1.00 m high	nr	420.00
1.25 m high	nr	470.00
1.50 m high	nr	520.00
1.80 m high	nr	580.00
2.00 m high	nr	620.00
Brick piers on concrete footings 1000 × 1000 × 600 mm thick; inclusive of excavations and disposal off site; coping Haddonstone S215C classically moulded corbelled pyramidal piercap PC £214.00		
Pier 600 × 600 mm; three bricks thick; brick PC £800/1000; height of brickwork above ground		
1.50 m high	nr	1025.00
1.80 m high	nr	1150.00
2.00 m high	nr	1225.00

F BRICK/BLOCK WALLING

Item Excluding site overheads and profit	Unit	Total rate £
BLOCK PIERS		
Block piers; blocks 440 × 215 × 100 mm thick laid on flat piers on concrete footings 600 × 600 × 250 mm thick; inclusive of excavations and disposal off site; coping of engineering brick on edge; 2 courses underground		
Pier 450 × 450 mm; to receive cladding treatment priced separately		
500 mm high	nr	44.00
750 mm high	nr	57.00
1.00 m high	nr	75.00
1.25 m high	nr	88.00
1.50 m high	nr	100.00
1.80 m high	nr	120.00
FACED BLOCK WALLING		
Concrete block wall with brick face 112.5 mm thick to bond pattern using snap headers; including excavation of foundation trench 450 mm deep; remove spoil off site; lay GEN 1 concrete foundations 600 × 300 mm thick; place stainless steel ties at 4 nr/m² of wall face; lay coping of brick on edge in engineering brick		
Solid blocks 7 N/mm² with snap header brick face; bricks PC £800.00/1000		
500 mm high	m²	115.00
750 mm high	m²	150.00
1.00 m high	m²	170.00
1.50 m high	m²	230.00
1.80 m high	m²	290.00

Q PAVINGS

Item Excluding site overheads and profit	Unit	Total rate £
BASES FOR PAVING		
Excavate ground and reduce levels to receive 38 mm thick slab and 25 mm		
mortar bed; dispose of excavated material off site		
Lay granular fill Type 1; limestone aggregate 150 mm thick laid to falls and compacted		
all by machine	m²	**13.10**
all by hand except disposal by grab	m²	**45.00**
Lay granular fill Type 1; recycled material; 150 mm thick laid to falls and compacted		
all by machine	m²	**11.40**
all by hand except disposal by grab	m²	**43.50**
Lay 1:2:4 concrete base 150 mm thick laid to falls		
all by machine	m²	**21.00**
all by hand except disposal by grab	m²	**57.00**
150 mm hardcore base with concrete base 150 mm deep		
concrete 1:2:4 site mixed	m²	**32.00**
concrete PAV 1 35 ready mixed	m²	**33.00**
Hardcore base 150 mm deep; concrete reinforced with A142 mesh		
site mixed concrete 1:2:4; 150 mm deep	m²	**48.50**
site mixed concrete 1:2:4; 250 mm deep	m²	**66.00**
concrete PAV 1 35 N/mm² ready mixed; 150 mm deep	m²	**38.50**
concrete PAV 1 35 N/mm² ready mixed; 250 mm deep	m²	**55.00**
PRECAST CONCRETE KERBS		
Note: excavation is by machine unless otherwise mentioned		
Excavate trench and construct concrete foundation 300 mm wide × 150 mm		
deep; lay precast concrete kerb units bedded in semi-dry concrete; slump 35 mm		
maximum; haunching one side; disposal of arisings off site		
Kerbs laid straight		
125 mm high × 125 mm thick; bullnosed; type BN	m	**26.00**
125 × 255 mm; ref HB2; SP	m	**28.00**
150 × 305 mm; ref HB1	m	**33.50**
Straight kerbs 125 mm × 255 mm laid to radius		
0.9 m radius	m	**39.00**
1.8 m radius	m	**37.00**
3.0 m radius	m	**36.00**
6.1 m radius	m	**34.50**
9.15 m radius	m	**34.00**
12.20 m radius	m	**33.50**
Dropper kerbs		
125 mm × 255–150 mm; left or right handed	m	**31.00**
Quadrants		
305 mm radius	nr	**34.50**
455 mm radius	nr	**36.00**

Q PAVINGS

Item Excluding site overheads and profit	Unit	Total rate £
Excavate trench and construct concrete foundation 300 mm wide × 150 mm deep; lay precast concrete kerb units bedded in semi-dry concrete; slump 35 mm maximum; haunching one side; disposal of arisings off site		
Conservation kerbs laid straight		
225 mm wide × 150 mm high;	m	**50.00**
155 mm wide × 255 mm high	m	**49.00**
Conservation kerbs laid to radius; 145 mm × 255 mm		
radius 3.25 m	m	**55.00**
radius 6.50 m	m	**50.00**
radius 9.80 m	m	**49.50**
solid quadrant; 305 × 305 mm	nr	**56.00**
STONE KERBS		
Excavate and construct concrete foundation 450 mm wide × 150 mm deep; lay precast concrete kerb units bedded in semi-dry concrete; slump 35 mm maximum; haunching one side; lay channel units bedded in 1:3 sand mortar; jointed in cement: sand mortar 1:3		
Granite kerbs		
straight; 125 × 250 mm	m	**57.00**
curved; 125 × 250 mm	m	**69.00**
CHANNELS		
Excavate and construct concrete foundation 450 mm wide × 150 mm deep; lay precast concrete kerb units bedded in semi-dry concrete; slump 35 mm maximum; haunching one side; lay channel units bedded in 1:3 sand mortar; jointed in cement: sand mortar 1:3		
Kerbs; 125 mm high × 125 mm thick; bullnosed; type BN		
dished channel; 125 × 225 mm; Ref CS	m	**48.50**
square channel; 125 × 150 mm; Ref CS2	m	**46.00**
bullnosed channel; 305 × 150 mm; Ref CBN	m	**56.00**
Brick channel; class B engineering bricks		
Excavate and construct concrete foundation 600 mm wide × 200 mm deep; lay channel to depths and falls bricks to be laid as headers along the channel; bedded in 1:3 sand mortar bricks close jointed in cement: sand mortar 1:3		
3 courses wide	m	**63.00**
3 coursed granite setts; dished		
Excavate and construct concrete foundation 400 mm wide × 200 mm deep; lay channel to depths and falls ; bedded in 1:3: sand mortar		
3 courses wide	m	**66.00**

Q PAVINGS

Item Excluding site overheads and profit	Unit	Total rate £
EDGINGS		
Precast concrete edging on concrete foundation 100 × 150 mm deep and haunching one side 1:2:4 including all necessary excavation disposal and formwork		
Rectangular, chamfered or bullnosed; to one side of straight path		
50 × 150 mm	m	24.00
50 × 200 mm	m	29.00
50 × 250 mm	m	30.00
Rectangular, chamfered or bullnosed as above but to both sides of straight paths		
50 × 150 mm	m	51.00
50 × 200 mm	m	59.00
50 × 250 mm	m	60.00
Timber edgings; softwood		
Straight		
150 × 38 mm	m	6.75
150 × 50 mm	m	8.80
Curved; over 5 m radius		
150 × 38 mm	m	8.40
150 × 50 mm	m	11.20
Curved; 4–5 m radius		
150 × 38 mm	m	9.10
150 × 50 mm	m	11.90
Curved; 3–4 m radius		
150 × 38 mm	m	10.10
150 × 50 mm	m	13.20
Curved; 1–3 m radius		
150 × 38 mm	m	11.80
150 × 50 mm	m	14.90
Brick or concrete block edge restraint; excavate for foundation; lay concrete 1:2:4 250 mm wide × 150 mm deep; on 50 mm thick sharp sand bed; lay blocks or bricks		
Blocks 200 × 100 × 60 mm; PC £8.17/m²; butt jointed		
header course	m	28.00
stretcher course	m	24.50
Bricks 215 × 112.5 × 50 mm; PC £300.00/1000; with mortar joints		
header course	m	29.50
header course; brick on edge	m	37.50
stretcher course; flat or brick on edge	m	28.50
stretcher course; 2 rows brick on edge	m	39.00
Bricks 215 × 112.5 × 50 mm; PC £600.00/1000; with mortar joints		
header course	m	32.00
header course; brick on edge	m	40.50
stretcher course; flat or on edge	m	23.00
stretcher course; 2 rows brick on edge	m	58.00

Q PAVINGS

Item Excluding site overheads and profit	Unit	Total rate £
Sawn yorkstone edgings; excavate for foundations; lay concrete 1:2:4 150 mm deep × 33.3% wider than the edging; on 35 mm thick mortar bed		
Yorkstone 50 mm thick		
100 mm wide × random lengths	m	38.50
100 × 100 mm	m	39.50
100 mm wide × 200 mm long	m	47.00
250 mm wide × random lengths	m	46.50
500 mm wide × random lengths	m	72.00
Granite edgings; excavate for footings; lay concrete 1:2:4 150 mm deep × 33.3% wider than the edging; on 35 mm thick mortar bed;		
Granite 50 mm thick		
100 mm wide × 200 mm long	m	43.00
100 mm wide × random lengths	m	45.00
250 mm wide × random lengths	m	51.00
setts; 100 × 100 mm	m	56.00
300 mm wide × random lengths	m	56.00
CAR PARKS		
Excavate 350 mm for pathways or roadbed to receive surface 100 mm thick priced separately; bring to grade; lay 100 mm well-rolled hardcore; lay 150 mm Type 1 granular material; lay kerbs 125 × 255 mm on both sides; including foundations haunched one side in 11.5 N/mm² concrete with all necessary formwork		
Work to falls, crossfalls and cambers not exceeding 15°		
1.50 m wide	m	87.00
2.00 m wide	m	99.00
3.00 m wide	m	120.00
4.00 m wide	m	140.00
5.00 m wide	m	160.00
6.00 m wide	m	180.00
7.00 m wide	m	200.00
As above but excavation 450 mm deep and base of Type 1 at 250 mm thick		
1.50 m wide	m	160.00
2.00 m wide	m	195.00
3.00 m wide	m	265.00
4.00 m wide	m	335.00
5.00 m wide	m	405.00
6.00 m wide	m	475.00
7.00 m wide	m	540.00

Q PAVINGS

Item Excluding site overheads and profit	Unit	Total rate £
CAR PARKS – cont		
To excavated and prepared base above, lay roadbase of 40 mm size dense bitumen macadam 70 mm thick; lay wearing course of 10 mm size dense bitumen macadam 30 mm thick; mark out car parking bays 5.0 × 2.4 m with thermoplastic road paint; surfaces all mechanically laid		
Per bay; 5.0 × 2.4 m	each	265.00
Gangway	m²	18.60
As above but stainless steel road studs 100 × 100 mm; two per bay in lieu of thermoplastic paint		
per bay; 5.0 × 2.4 m	each	220.00
Car park as above but with interlocking concrete blocks 200 × 100 × 80 mm; grey		
per bay; 5.0 × 2.4 m	each	485.00
gangway	m²	40.50
Car park as above but with interlocking concrete blocks 200 × 100 × 80 mm; colours		
per bay; 5.0 × 2.4 m	each	500.00
gangway	m²	41.50
Car park as above but with interlocking concrete blocks 200 × 100 × 60 mm; grey		
per bay; 5.0 × 2.4 m	each	470.00
gangway	m²	39.50
Car park as above but with interlocking concrete blocks 200 × 100 × 60 mm; colours		
per bay; 5.0 × 2.4 m	each	480.00
gangway	m²	40.00
Car park as above but with laying of Grass Concrete Grasscrete in situ continuously reinforced cellular surfacing; including expansion joints at 10 m centres; fill with topsoil and peat (5:1) and fertilizer at 35 g/m²; seed with dwarf rye grass at 35 g/m²		
GC1; 100 mm thick for cars and light traffic		
per bay; 5.0 × 2.4 m	each	450.00
gangway	m²	38.00
GC2; 150 mm thick for HGV traffic including dust carts		
per bay; 5.0 × 2.4 m	each	590.00
gangway	m²	49.00
CAR PARKING FOR DISABLED PEOPLE		
To excavated and prepared base above, lay roadbase of 20 mm size dense bitumen macadam 80 mm thick; lay wearing course of 10 mm size dense bitumen macadam 30 mm thick; mark out car parking bays with thermoplastic road paint		
per bay; 5.80 × 3.25 m ambulant	each	350.00
per bay; 6.55 × 3.80 m wheelchair	each	460.00
gangway	m²	18.60
Car park as above but with interlocking concrete blocks 200 × 100 × 60 mm; grey; but mark out bays		
per bay; 5.80 × 3.25 m ambulant	each	740.00
per bay; 6.55 × 3.80 m wheelchair	each	980.00
gangway	m²	39.50

Q PAVINGS

Item Excluding site overheads and profit	Unit	Total rate £
INTERLOCKING BLOCK PAVING		
Edge restraint to block paving; excavate for groundbeam; lay concrete 1:2:4 200 mm wide × 150 mm deep; on 50 mm thick sharp sand bed; inclusive of haunching one side		
Blocks 200 × 100 × 60 mm; PC £8.17/m²; butt jointed		
header course	m	15.40
stretcher course	m	12.10
Bricks 215 × 112.5 × 50 mm; PC £300.00/1000; with mortar joints		
header course	m	16.80
stretcher course	m	16.00
Excavate ground; supply and lay granular fill Type 1 150 mm thick laid to falls and compacted; supply and lay block pavers; laid on 50 mm compacted sharp sand; vibrated; joints filled with loose sand excluding edgings or kerbs measured separately		
Concrete blocks		
200 × 100 × 60 mm	m²	54.00
200 × 100 × 80 mm	m²	55.00
Reduce levels; lay 150 mm granular material Type 1; lay vehicular block paving to 90° herringbone pattern; on 50 mm compacted sand bed; vibrated; jointed in sand and vibrated; excavate and lay precast concrete edging 50 × 150 mm; on concrete foundation 1:2:4		
Blocks 200 × 100 × 60 mm		
1.0 m wide clear width between edgings	m	100.00
1.5 m wide clear width between edgings	m	130.00
2.0 m wide clear width between edgings	m	155.00
Blocks 200 × 100 × 60 mm but blocks laid 45° herringbone pattern including cutting edging blocks		
1.0 m wide clear width between edgings	m	110.00
1.5 m wide clear width between edgings	m	135.00
2.0 m wide clear width between edgings	m	160.00
Blocks 200 × 100 × 60 mm but all excavation by hand disposal off site by grab		
1.0 m wide clear width between edgings	m	210.00
1.5 m wide clear width between edgings	m	245.00
2.0 m wide clear width between edgings	m	285.00
BRICK PAVING		
WORKS BY MACHINE		
Excavate and lay base Type 1 150 mm thick remove arisings; all by machine; lay clay brick paving		
200 × 100 × 50 mm thick; butt jointed on 50 mm sharp sand bed		
PC £300.00/1000	m²	63.00
PC £600.00/1000	m²	79.00

Q PAVINGS

Item Excluding site overheads and profit	Unit	Total rate £
BRICK PAVING – cont		
Excavate and lay base Type 1 150 mm thick remove arisings – cont		
200 × 100 × 50 mm thick; 10 mm mortar joints on 35 mm mortar bed		
PC £300.00/1000	m²	79.00
PC £600.00/1000	m²	92.00
Excavate and lay base Type 1 250 mm thick all by machine; remove arisings; lay clay brick paving		
200 × 100 × 50 mm thick; 10 mm mortar joints on 35 mm mortar bed		
PC £300.00/1000	m²	86.00
Excavate and lay 150 mm readymix concrete base reinforced with A393 mesh; all by machine; remove arisings; lay clay brick paving		
200 × 100 × 50 mm thick; 10 mm mortar joints on 35 mm mortar bed; running or stretcher bond		
PC £300.00/1000	m²	97.00
PC £600.00/1000	m²	110.00
200 × 100 × 50 mm thick; 10 mm mortar joints on 35 mm mortar bed; butt jointed; herringbone bond		
PC £300.00/1000	m²	81.00
PC £600.00/1000	m²	96.00
Excavate and lay base readymix concrete base 150 mm thick reinforced with A393 mesh; all by machine; remove arisings; lay clay brick paving		
215 × 102.5 × 50 mm thick; 10 mm mortar joints on 35 mm mortar bed		
PC £300.00/1000; herringbone	m²	99.00
PC £600.00/1000; herringbone	m²	110.00
WORKS BY HAND		
Excavate and lay base Type 1 150 mm thick by hand; arisings barrowed to spoil heap maximum distance 25 m and removal off site by grab; lay clay brick paving		
200 × 100 × 50 mm thick; butt jointed on 50 mm sharp sand bed		
PC £300.00/1000	m²	96.00
PC £600.00/1000	m²	125.00
Excavate and lay 150 mm concrete base; 1:3:6: site mixed concrete reinforced with A393 mesh; remove arisings to stockpile and then off site by grab; lay clay brick paving		
215 × 102.5 × 50 mm thick; 10 mm mortar joints on 35 mm mortar bed		
PC £300.00/1000	m²	97.00
PC £600.00/1000	m²	125.00
FLAG PAVING TO PEDESTRIAN AREAS Prices are inclusive of all mechanical excavation and disposal.		

Q PAVINGS

Item Excluding site overheads and profit	Unit	Total rate £
Supply and lay precast concrete flags; excavate ground and reduce levels; treat substrate with total herbicide; supply and lay granular fill Type 1 150 mm thick laid to falls and compacted		
Standard precast concrete flags bedded and jointed in lime: sand mortar (1:3)		
450 × 450 × 70 mm; chamfered	m²	**45.00**
450 × 450 × 50 mm; chamfered	m²	**41.00**
600 × 300 × 50 mm	m²	**39.00**
600 × 450 × 50 mm	m²	**38.50**
600 × 600 × 50 mm	m²	**35.50**
750 × 600 × 50 mm	m²	**35.00**
900 × 600 × 50 mm	m²	**33.50**
Coloured flags bedded and jointed in lime: sand mortar (1:3)		
600 × 600 × 50 mm	m²	**37.50**
450 × 450 × 70 mm; chamfered	m²	**51.00**
400 × 400 × 65 mm	m²	**51.00**
750 × 600 × 50 mm	m²	**37.00**
900 × 600 × 50 mm	m²	**35.50**
Marshalls Saxon; textured concrete flags; reconstituted yorkstone in colours; butt jointed bedded in lime: sand mortar (1:3)		
300 × 300 × 35 mm	m²	**70.00**
450 × 450 × 50 mm	m²	**59.00**
600 × 300 × 35 mm	m²	**55.00**
600 × 600 × 50 mm	m²	**51.00**
Tactile flags; Marshalls blister tactile pavings; red or buff; for blind pedestrian guidance laid to designed pattern		
450 × 450 mm	m²	**51.00**
400 × 400 mm	m²	**57.00**
PEDESTRIAN DETERRENT PAVING		
Excavate ground and bring to levels; treat substrate with total herbicide; supply and lay granular fill Type 1 150 mm thick laid to falls and compacted; supply and lay precast deterrent paving units bedded in lime: sand mortar (1:3) and jointed in lime: sand mortar (1:3)		
Marshalls Mono		
Lambeth pyramidal paving; 600 × 600 × 75 mm	m²	**42.00**
Townscape		
Abbey square cobble pattern pavings; reinforced; 600 × 600 × 60 mm	m²	**47.00**
Geoset chamfered studs; 600 × 600 × 60 mm	m²	**42.00**
IMITATION YORKSTONE PAVING		
Excavate ground and bring to levels; treat substrate with total herbicide; supply and lay granular fill Type 1 150 mm thick laid to falls and compacted; supply and lay imitation yorkstone paving laid to coursed patterns bedded in lime: sand mortar (1:3) and jointed in lime: sand mortar (1:3)		
Marshalls Heritage; square or rectangular		
300 × 300 × 38 mm	m²	**75.00**
600 × 300 × 30 mm	m²	**61.00**
600 × 450 × 38 mm	m²	**58.00**
450 × 450 × 38 mm	m²	**55.00**
600 × 600 × 38 mm	m²	**52.00**

Q PAVINGS

Item Excluding site overheads and profit	Unit	Total rate £
IMITATION YORKSTONE PAVING – cont		
Excavate ground and bring to levels – cont		
As above but laid to random rectangular patterns		
various sizes selected from the above	m²	64.00
Imitation yorkstone laid random rectangular as above but on concrete base 150 mm thick		
by machine	m²	72.00
by hand	m²	100.00
NATURAL STONE SLAB PAVING		
WORKS BY MACHINE (For works by hand please see Minor Works)		
Excavate ground by machine and reduce levels; to receive 65 mm thick slab and 35 mm mortar bed; dispose of excavated material off site; treat substrate with total herbicide; lay granular fill Type 1 150 mm thick laid to falls and compacted; lay to random rectangular pattern on 35 mm mortar bed		
New riven slabs		
laid random rectangular	m²	150.00
New riven slabs; but to 150 mm plain concrete base		
laid random rectangular	m²	160.00
New riven slabs; disposal by grab		
laid random rectangular	m²	150.00
Reclaimed Cathedral grade riven slabs		
laid random rectangular	m²	175.00
Reclaimed Cathedral grade riven slabs; but to 150 mm plain concrete base		
laid random rectangular	m²	180.00
Reclaimed Cathedral grade riven slabs; disposal by grab		
laid random rectangular	m²	175.00
New slabs sawn 6 sides		
laid random rectangular	m²	110.00
three sizes; laid to coursed pattern	m²	110.00
New slabs sawn 6 sides; but to 150 mm plain concrete base		
laid random rectangular	m²	115.00
three sizes; laid to coursed pattern	m²	120.00
New slabs sawn 6 sides; disposal by grab		
laid random rectangular	m²	110.00
3 sizes, laid to coursed pattern	m²	110.00

Q PAVINGS

Item Excluding site overheads and profit	Unit	Total rate £
WORKS BY HAND		
Excavate ground by hand and reduce levels, to receive 65 mm thick slab and 35 mm mortar bed; barrow all materials and arisings 25 m; dispose of excavated material off site by grab; treat substrate with total herbicide; lay granular fill Type 1 150 thick laid to falls and compacted; lay to random rectangular pattern on 35 mm mortar bed		
New riven slabs		
laid random rectangular	m²	185.00
New riven slabs laid random rectangular; but to 150 mm plain concrete base		
laid random rectangular	m²	195.00
New riven slabs; but disposal to skip		
laid random rectangular	m²	200.00
Reclaimed Cathedral grade riven slabs		
laid random rectangular	m²	210.00
Reclaimed Cathedral grade riven slabs; but to 150 mm plain concrete base		
laid random rectangular	m²	220.00
Reclaimed Cathedral grade riven slabs; but disposal to skip		
laid random rectangular	m²	225.00
New slabs sawn 6 sides		
laid random rectangular	m²	140.00
three sizes; sawn 6 sides laid to coursed pattern	m²	145.00
New slabs sawn 6 sides; but to 150 mm plain concrete base		
laid random rectangular	m²	150.00
three sizes; sawn 6 sides laid to coursed pattern	m²	160.00
GRANITE SETT PAVING – PEDESTRIAN		
Excavate ground and bring to levels; lay 100 mm hardcore to falls; compacted with 5 tonne roller; blind with compacted Type 1 50 mm thick; lay 100 mm concrete 1:2:4; supply and lay granite setts 100 × 100 × 100 mm bedded in cement: sand mortar (1:3) 25 mm thick minimum; close butted and jointed in fine sand; all excluding edgings or kerbs measured separately		
Setts laid to bonded pattern and jointed		
new setts 100 × 100 × 100 mm	m²	110.00
second-hand cleaned 100 × 100 × 100 mm	m²	115.00
Setts laid in curved pattern		
new setts 100 × 100 × 100 mm	m²	115.00
second-hand cleaned 100 × 100 × 100 mm	m²	120.00

Q PAVINGS

Item Excluding site overheads and profit	Unit	Total rate £
GRANITE SETT PAVING – TRAFFICKED AREAS		
Excavate ground and bring to levels; lay 100 mm hardcore to falls; compacted with 5 tonne roller; blind with compacted Type 1 50 mm thick; lay 150 mm site mixed concrete 1:2:4 reinforced with steel fabric; supply and lay granite setts 100 × 100 × 100 mm bedded in cement: sand mortar (1:3) 25 mm thick minimum; close butted and jointed in fine sand; all excluding edgings or kerbs measured separately		
Site mixed concrete		
new setts 100 × 100 × 100 mm	m²	**120.00**
second-hand cleaned 100 × 100 × 100 mm	m²	**120.00**
Ready mixed concrete		
new setts 100 × 100 × 100 mm	m²	**115.00**
second-hand cleaned 100 × 100 × 100 mm	m²	**120.00**
CONCRETE PAVING		
Pedestrian areas		
Excavate to reduce levels; lay 100 mm Type 1; lay PAV 1 air entrained concrete; joints at maximum width of 6.0 m cut out and sealed with sealant; inclusive of all formwork and stripping		
100 mm thick	m²	**47.00**
Trafficked areas		
Excavate to reduce levels; lay 150 mm Type 1 lay PAV 2 40 N/mm² air entrained concrete; reinforced with steel mesh 200 × 200 mm square at 2.22 kg/m²; joints at max width of 6.0 m cut out and sealed with sealant		
150 mm thick	m²	**65.00**
BEACH COBBLE PAVING		
Excavate ground and bring to levels and fill with compacted Type 1 fill 100 mm thick; lay GEN 1 concrete base 100 mm thick; supply and lay cobbles individually laid by hand bedded in cement: sand mortar (1:3) 25 mm thick minimum; dry grout with 1:3 cement: sand grout; brush off surplus grout and water in; sponge off cobbles as work proceeds; all excluding formwork edgings or kerbs measured separately		
Scottish beach cobbles 50–75 mm	m²	**115.00**
Scottish beach cobbles 100–200 mm	m²	**140.00**
Kidney flint cobbles 75–100 mm	m²	**110.00**

Q PAVINGS

Item Excluding site overheads and profit	Unit	Total rate £
CONCRETE SETT PAVING		
Excavate ground and bring to levels; supply and lay 150 mm granular fill Type 1 laid to falls and compacted; supply and lay setts bedded in 50 mm sand and vibrated; joints filled with dry sand and vibrated; all excluding edgings or kerbs measured separately		
Marshalls Mono; Tegula precast concrete setts		
random sizes 60 mm thick	m²	49.00
single size 60 mm thick	m²	46.50
random size 80 mm thick	m²	53.00
single size 80 mm thick	m²	67.00
GRASS CONCRETE PAVING		
Reduce levels; lay 150 mm Type 1 granular fill compacted; supply and lay precast grass concrete blocks on 20 sand and level by hand; fill blocks with 3 mm sifted topsoil and pre-seeding fertilizer at 50 g/m²; sow with perennial ryegrass/ chewings fescue seed at 35 g/m²		
Grass concrete		
GB103 406 × 406 × 103 mm	m²	41.50
GB83 406 × 406 × 83 mm	m²	38.50
extra for geotextile fabric underlayer	m²	0.70
Firepaths		
Excavate to reduce levels; lay 300 mm well rammed hardcore; blinded with 100 mm type 1; supply and lay Marshalls Mono Grassguard 180 precast grass concrete blocks on 50 mm sand and level by hand; fill blocks with sifted topsoil and pre-seeding fertilizer at 50 g/m²		
firepath 3.8 m wide	m	265.00
firepath 4.4 m wide	m	310.00
firepath 5.0 m wide	m	350.00
turning areas	m²	70.00
Charcon Hard Landscaping Grassgrid; 366 × 274 × 100 mm	m²	36.00
SLAB/BRICK PATHS		
Stepping stone path inclusive of hand excavation and mechanical disposal, 100 mm Type 1 and sand blinding; slabs laid 100 mm apart to existing turf		
600 mm wide; 600 × 600 × 50 mm slabs		
natural finish	m	49.00
coloured	m	49.50
exposed aggregate	m	60.00
900 mm wide; 600 × 900 × 50 mm slabs		
natural finish	m	54.00
coloured	m	56.00
exposed aggregate	m	72.00

Q PAVINGS

Item Excluding site overheads and profit	Unit	Total rate £
SLAB/BRICK PATHS – cont		
Pathway inclusive of mechanical excavation and disposal; 100 mm Type 1 and sand blinding; slabs close butted		
Straight butted path 900 mm wide; 600 × 900 × 50 mm slabs		
natural finish	m	22.00
coloured	m	23.00
exposed aggregate	m	34.00
Straight butted path 1200 mm wide; double row; 600 × 900 × 50 mm slabs; laid stretcher bond		
natural finish	m	39.00
coloured	m	41.50
exposed aggregate	m	65.00
Straight butted path 1200 mm wide; one row of 600 × 600 × 50 mm; one row 600 × 900 × 50 mm slabs		
natural finish	m	38.00
coloured	m	40.00
exposed aggregate	m	72.00
Straight butted path 1500 mm wide; slabs of 600 × 900 × 50 mm and 600 × 600 × 50 mm; laid to bond		
natural finish	m	47.00
coloured	m	49.00
exposed aggregate	m	72.00
Straight butted path 1800 mm wide; two rows of 600 × 900 × 50 mm slabs; laid bonded		
natural finish	m	52.00
coloured	m	62.00
exposed aggregate	m	98.00
Straight butted path 1800 mm wide; three rows of 600 × 900 × 50 mm slabs; laid stretcher bond		
natural finish	m	56.00
coloured	m	67.00
exposed aggregate	m	100.00
Brick paved paths; bricks 215 × 112.5 × 65 mm with mortar joints; prices are inclusive of excavation, disposal off site, 100 mm hardcore, 100 mm 1:2:4 concrete bed; edgings of brick 215 mm wide, haunched; all jointed in cement: lime: sand mortar (1:1:6)		
Path 1015 mm wide laid stretcher bond; edging course of headers		
rough stocks PC £450.00/1000	m	89.00
engineering brick	m	84.00
Path 1015 mm wide laid stack bond		
rough stocks PC £450.00/1000	m	94.00
engineering brick	m	89.00
Path 1115 mm wide laid header bond		
rough stocks PC £450.00/1000	m	98.00
engineering brick	m	93.00
Path 1330 mm wide laid basketweave bond		
rough stocks PC £450.00/1000	m	110.00
engineering brick	m	105.00
Path 1790 mm wide laid basketweave bond		
rough stocks PC £450.00/1000	m	150.00
engineering brick	m	110.00

Q PAVINGS

Item Excluding site overheads and profit	Unit	Total rate £
Brick paved paths; brick paviors 200 × 100 × 50 mm chamfered edge with butt joints; prices are inclusive of excavation, 100 mm Type 1 and 50 mm sharp sand; jointing in kiln dried sand brushed in; exclusive of edge restraints		
Path 1000 mm wide laid stretcher bond; edging course of headers		
rough stocks PC £450.00/1000	m	80.00
engineering brick	m	85.00
Path 1330 mm wide laid basketweave bond		
rough stocks PC £450.00/1000	m	115.00
engineering brick	m	110.00
Path 1790 mm wide laid basketweave bond		
rough stocks PC £450.00/1000	m	150.00
engineering brick	m	110.00
GRAVEL PATHS		
Reduce levels and remove spoil to dump on site; lay 150 mm hardcore well rolled; lay geofabric; fix timber edge 150 × 38 mm to both sides of straight paths		
Lay Cedec gravel 50 mm thick; watered and rolled		
1.0 m wide	m	32.50
1.5 m wide	m	43.00
2.0 m wide	m	59.00
Lay Breedon gravel 50 mm thick; watered and rolled		
1.0 m wide	m	30.50
1.5 m wide	m	41.00
2.0 m wide	m	51.00
Reduce levels and remove spoil to dump on site; lay 150 mm Type 1 well rolled; lay geofabric; fix timber edge 150 × 38 mm to both sides of straight paths		
Lay Cedec gravel 50 mm thick; watered and rolled		
1.0 m wide	m	35.00
1.5 m wide	m	47.50
2.0 m wide	m	65.00
Lay Breedon gravel 50 mm thick; watered and rolled		
1.0 m wide	m	33.00
1.5 m wide	m	45.00
2.0 m wide	m	57.00
Lay 20 mm shingle; 30 mm thick		
1.0 m wide	m	23.00
1.5 m wide	m	29.00
2.0 m wide	m	39.50

Q PAVINGS

Item Excluding site overheads and profit	Unit	Total rate £
BARK PAVING		
Excavate to reduce levels; remove all topsoil to dump on site; treat area with herbicide; lay 150 mm clean hardcore; blind with sand; lay 0.7 mm geotextile filter fabric water flow 50 l/m²/sec; supply and fix treated softwood edging boards 50 × 150 mm to bark area on hardcore base extended 150 mm beyond the bark area; boards fixed with galvanized nails to treated softwood posts 750 × 50 × 75 mm driven into firm ground at 1.0 m centres; edging boards to finish 25 mm above finished bark surface; tops of posts to be flush with edging boards and once weathered		
Supply and lay 100 mm Melcourt conifer walk chips 10–40 mm size		
1.00 m wide	m	28.00
2.00 m wide	m	43.00
3.00 m wide	m	58.00
4.00 m wide	m	70.00
Extra over for wood fibre	m²	−1.55
Extra over for hardwood chips	m²	−1.36
FOOTPATHS		
Note: For pathway surfaces; select appropriate finish from the Measured Works section		
Pathway edgings; excavate 250 mm; bring to grade; lay 150 mm Type 1 granular material; all disposal off site		
Lay Permaloc Asphalt edge 51 mm metal edgings; path width		
1.50 m wide	m	41.00
3.00 m wide	m	58.00
Lay timber edging 150 × 38 mm; path width		
1.50 m wide	m	27.00
3.00 m wide	m	44.50
Lay precast edgings kerbs 50 × 150 mm on both sides; including foundations haunched one side in 11.5 N/mm² concrete with all necessary formwork; path width		
1.50 m wide	m	67.00
3.00 m wide	m	85.00
Lay brick on flat header edging; brick £300.00/1000; path width		
1.50 m wide	m	69.00
3.00 m wide	m	86.00
Lay brick on flat header edging; brick £600.00/1000; path width		
1.50 m wide	m	74.00
3.00 m wide	m	91.00
Lay brick on edge header edging; brick £300.00/1000; path width		
1.50 m wide	m	81.00
3.00 m wide	m	98.00
Lay brick on edge header edging; brick £600.00/1000; path width		
1.50 m wide	m	89.00
3.00 m wide	m	105.00

Q PAVINGS

Item Excluding site overheads and profit	Unit	Total rate £
Lay brick on flat or brick on edge stretcher edging; brick £300.00/1000; path width		
1.50 m wide	m	59.00
3.00 m wide	m	76.00
Lay brick on edge strecher edging; brick £600.00/1000; path width		
1.50 m wide	m	62.00
3.00 m wide	m	79.00
Lay new granite setts 100 × 100 × 100 mm single row; path width		
1.50 m wide	m	105.00
3.00 m wide	m	120.00
Lay new granite setts 200 × 100 × 100 mm single row; path width		
1.50 m wide	m	83.00
3.00 m wide	m	100.00
Excavate footpath to reduce level; remove arisings to tip on site maximum distance 25 m; lay Type 1 granular fill 100 mm thick; lay base course of 28 mm size dense bitumen macadam 50 mm thick; wearing course of 10 mm size dense bitumen macadam 20 mm thick; timber edge 150 × 38 mm		
Areas over 1000 m^2		
1.0 m wide	m	37.00
1.5 m wide	m	50.00
2.0 m wide	m	64.00
Areas 400–1000 m^2		
1.0 m wide	m	41.00
1.5 m wide	m	56.00
2.0 m wide	m	89.00
Reinforced concrete paths; excavate path and dispose of excavated material off site; bring to grade; lay reinforcing mesh A142 lapped and joined; lay in situ reinforced concrete PAV1 35 N/mm^2; with 15 impregnated fibreboard expansion joints sealant at 50 m centres; on 100 mm hardcore blinded sand; edgings 150 × 50 mm to both sides; including foundations haunched one side in 11.5 N/mm^2 concrete with all necessary formwork		
Concrete 150 mm thick		
1.00 m wide	m	60.00
2.00 m wide	m	95.00
3.00 m wide	m	130.00
WORKS BY HAND; Ready mixed concrete – mixed on site		
Reinforced concrete paths; excavate path by hand barrow to spoil heap 25 m and dispose by grab off site; bring to grade; lay reinforcing mesh A142 lapped and joined; lay in situ reinforced ready mixed concrete mixed on site (1:2:4:); with 15 impregnated fibreboard expansion joints sealant at 50 m centres; on 100 mm hardcore blinded sand; edgings 150 × 50 mm to both sides; including foundations haunched one side in 11.5 N/mm^2 concrete with all necessary formwork		
Concrete 150 mm thick		
1.00 m wide	m	85.00
2.00 m wide	m	155.00
3.00 m wide	m	220.00

Q PAVINGS

Item Excluding site overheads and profit	Unit	Total rate £
FOOTPATHS – cont		
Resin pathways; excavate 230 mm; bring to grade; lay 150 mm Type 1 granular material; lay precast edgings kerbs 50 × 150 mm on both sides; including foundations haunched one side in 11.5 N/mm² concrete with all necessary formwork; hand lay base course of 60 mm macadam and resin bound wearing course; all disposal off site		
Natratex resin bonded macadam wearing course		
1.50 m wide	m	125.00
3.00 m wide	m	200.00
Addagrip; resin bonded aggregate; 1–3 mm golden pea gravel with buff adhesive		
1.50 m wide	m	140.00
3.00 m wide	m	225.00
Addagrip; resin bonded aggregate; 6 mm chocolate gravel 18 mm depth		
1.50 m wide	m	185.00
3.00 m wide	m	320.00
Addagrip; resin bonded aggregate; Cobalt blue recycled glass 15 mm depth		
1.50 m wide	m	215.00
3.00 m wide	m	380.00
Sureset; resin bound permeable surfacing; natural gravel 6 mm aggregate 18 mm thick		
1.50 m wide	m	170.00
3.00 m wide	m	290.00
Sureset; resin bound permeable surfacing; crushed rock 6 mm aggregate 18 mm thick		
1.50 m wide	m	170.00
3.00 m wide	m	455.00
Sureset; resin bound permeable surfacing; marble 6 mm aggregate 18 mm thick		
1.50 m wide	m	175.00
3.00 m wide	m	300.00
Sureset; resin bound permeable surfacing; recycled glass 6 mm aggregate 18 mm thick		
1.50 m wide	m	195.00
3.00 m wide	m	340.00
STEPS		
Excavate for new steps; grade to correct levels; remove excavated material from site and lay concrete base; Construct formwork; supply and lay site mixed concrete; Fix treads and risers to steps		
Stone clad steps; 1.20 m wide; tread depth 225 mm; Granite cut off site		
1 riser	nr	205.00
2 risers	nr	335.00
3 risers	nr	480.00
4 risers	nr	650.00
6 risers	nr	910.00
Stone clad steps; 1.20 m wide; tread depth 225 mm; Riven Yorkstone cut on site		
1 riser	nr	240.00
2 risers	nr	400.00
3 risers	nr	570.00
4 risers	nr	780.00
6 risers	nr	1100.00

Q PAVINGS

Item Excluding site overheads and profit	Unit	Total rate £
Stone clad steps; 2.40 m wide; tread depth 225 mm; Granite cut off site		
1 riser	nr	390.00
2 risers	nr	640.00
3 risers	nr	930.00
4 risers	nr	1275.00
6 risers	nr	1825.00

Q ROADS

Item Excluding site overheads and profit	Unit	Total rate £
MACADAM SURFACES		
Macadam roadway over 1000 m² Excavate 350 mm for pathways or roadbed; bring to grade; lay 100 mm well-rolled hardcore; lay 150 mm Type 1 granular material; lay precast edgings kerbs 50 × 150 mm on both sides; including foundations haunched one side in 11.5 N/mm² concrete with all necessary formwork; machine lay surface of 90 mm macadam 60 mm base course and 30 mm wearing course; all disposal off site		
1.50 m wide	m	110.00
2.00 m wide	m	115.00
3.00 m wide	m	170.00
4.00 m wide	m	210.00
5.00 m wide	m	250.00
6.00 m wide	m	290.00
7.00 m wide	m	330.00
Macadam roadway 400–1000 m² Excavate 350 mm for pathways or roadbed; bring to grade; lay 100 mm well-rolled hardcore; lay 150 mm Type 1 granular material; lay precast edgings kerbs 50 × 150 mm on both sides; including foundations haunched one side in 11.5 N/mm² concrete with all necessary formwork; machine lay surface of 90 mm macadam 60 mm base course and 30 mm wearing course; all disposal off site		
1.50 m wide	m	115.00
2.00 m wide	m	140.00
3.00 m wide	m	180.00
4.00 m wide	m	230.00
5.00 m wide	m	270.00
6.00 m wide	m	315.00
7.00 m wide	m	360.00
CONCRETE ROADS		
Excavate weak points of excavation by hand and fill with well rammed Type 1 granular material		
100 mm thick	m²	10.50
200 mm thick	m²	13.40
Reinforced concrete roadbed; excavate road bed and dispose excavated material off site; bring to grade; lay reinforcing mesh A142 lapped and joined; lay in situ reinforced concrete roadbed PAV1 35 N/mm²; with 15 impregnated fibreboard expansion joints sealant at 50 m centres; on 150 mm hardcore blinded sand; kerbs 155 × 255 mm to both sides; including foundations haunched one side in 11.5 N/mm² concrete with all necessary formwork		
Concrete 150 mm thick		
4.00 m wide	m	190.00
5.00 m wide	m	220.00
6.00 m wide	m	295.00
7.00 m wide	m	310.00

Q ROADS

Item Excluding site overheads and profit	Unit	Total rate £
Concrete 250 mm thick		
4.00 m wide	m	**250.00**
5.00 m wide	m	**290.00**
6.00 m wide	m	**340.00**
7.00 m wide	m	**395.00**

Q TIMBER DECKING

Item Excluding site overheads and profit	Unit	Total rate £
HARDWOOD DECKS		
Timber decking; Exterior Decking Limited; timber decking to roof gardens; cranage or movement of materials to roof top not included; coated one coat with Seasonite		
Joist size 47 × 47 mm		
Ipe; pre-drilled and countersunk; screw fixed	m^2	**150.00**
Ipe; Exterpark; 21 mm invisibly fixed	m^2	**180.00**
Ipe; Exterpark; 28 mm invisibly fixed	m^2	**235.00**
Ipe; Exterpark; 35 mm invisibly fixed	m^2	**300.00**
Rustic Teak; 21 mm invisibly fixed	m^2	**190.00**
Merbau; 21 mm invisibly fixed	m^2	**155.00**
Merbau; 28 mm invisibly fixed	m^2	**130.00**
Timber decking to commercial applications; double beams at 300 mm centres; coated one coat with Seasonite		
Joist size 47 × 150 mm		
Ipe; pre-drilled and countersunk; screw fixed	m^2	**205.00**
Ipe; Exterpark; 21 mm invisibly fixed	m^2	**230.00**
Ipe; Exterpark; 28 mm invisibly fixed	m^2	**290.00**
Ipe; Exterpark; 35 mm invisibly fixed	m^2	**350.00**
Rustic Teak; 21 mm invisibly fixed	m^2	**240.00**
Merbau; 21 mm invisibly fixed	m^2	**210.00**
Merbau; 28 mm invisibly fixed	m^2	**180.00**
SOFTWOOD DECKS		
Timber decking; AVS Ltd support structure of timber joists for decking laid on blinded base (measured separately)		
joists 50 × 150 mm	10 m^2	**96.00**
joists 50 × 200 mm	10 m^2	**95.00**
joists 50 × 250 mm	10 m^2	**90.00**
As above but boards 141 × 28 mm thick		
joists 50 × 200 mm	10 m^2	**95.00**
joists 50 × 250 mm	10 m^2	**90.00**
As above but decking boards in red cedar 131 mm wide × 42 mm thick		
joists 50 × 150 mm	10 m^2	**110.00**
joists 50 × 200 mm	10 m^2	**110.00**
joists 50 × 250 mm	10 m^2	**110.00**
Add to all of the above for handrails fixed to posts 100 × 100 × 1370 mm high		
square balusters at 100 mm centres	m	**73.00**
square balusters at 300 mm centres	m	**47.00**
turned balusters at 100 mm centres	m	**93.00**
turned balusters at 300 mm centres	m	**54.00**

Q SPECIAL SURFACES FOR SPORT/PLAYGROUNDS

Item Excluding site overheads and profit	Unit	Total rate £
BARK PLAY AREA		
Excavate playground area to 450 mm depth; lay 150 mm broken stone or clean hardcore; lay filter membrane; lay bark surface		
Melcourt Industries		
Playbark 10/50; 300 mm thick	100 m²	**3700.00**
Playbark 8/25; 300 mm thick	100 m²	**4400.00**
BOWLING GREEN CONSTRUCTION		
Agripower Ltd; bowling green; complete		
Excavate 300 mm deep and grade to level; excavate and install 100 mm land drain to perimeter and backfill with shingle; install 60 mm land drain to surface at 4.5 m centres backfilled with shingle; install 50 m non-perforated pipe 50 m long; install 100 mm compacted and levelled drainage stone, blind with grit and sand; spread 150 mm imported 70:30 sand: soil accurately compacted and levelled; lay bowling green turf and top dress twice luted into surface; exclusive of perimeter ditches and bowls protection		
6 rink green 38.4 × 38.4 m	each	**54000.00**
install Toro automatic bowling green irrigation system with pump, tank, controller, electrics, pipework and 8 nr Toro 780 sprinklers; excluding pumphouse (optional)	each	**8600.00**
Supply and install to perimeter of green; Sportsmark preformed bowling green ditch channels		
Ultimate Design 99 steel reinforced concrete channel; 600 mm long section	each	**8000.00**
Supply and fit bowls protection material to rear hitting face of Ultimate channels 1 and 2 above		
Curl Grass artificial grass; 0.45 m wide	each	**2200.00**
Astroturf artificial grass; 0.45 m wide	each	**2050.00**
bowls protection ditch liner laid loose; 300 mm wide	each	**1250.00**
JOGGING TRACK		
Excavate track 250 mm deep; lay filter membrane; lay 100 mm depth gravel waste or similar; lay 100 mm compacted gravel	100 m²	**7100.00**
Extra for treated softwood edging 50 × 150 mm on 50 × 50 mm posts		
both sides	m	**9.70**
PLAYGROUNDS		
Excavate playground area and dispose of arisings to tip; lay Type 1 granular fill; lay macadam surface two coat work 80 mm thick; base course of 28 mm size dense bitumen macadam 50 mm thick; wearing course of 10 mm size dense bitumen macadam 30 mm thick		
Areas over 1000 m²		
excavation 180 mm base 100 mm thick	m²	**30.00**
excavation 225 mm base 150 mm thick	m²	**33.00**
Areas 400–1000 m²		
excavation 180 mm base 100 mm thick	m²	**34.00**
excavation 225 mm base 150 mm thick	m²	**37.00**

Q SPECIAL SURFACES FOR SPORT/PLAYGROUNDS

Item Excluding site overheads and profit	Unit	Total rate £
PLAYGROUNDS – cont		
Excavate playground area to given levels and falls; remove soil off site and backfill with compacted Type 1 granular fill; lay ready mixed concrete to fall 2% in all direction		
Base 150 mm thick; surface 100 mm thick	m²	25.50
Base 150 mm thick; surface 150 mm thick	m²	32.00
SAFETY SURFACING		
Excavate ground and reduce levels to receive safety surface; dispose of excavated material off site; treat substrate with total herbicide; lay granular fill Type 1 150 mm thick laid to falls and compacted; lay macadam base 40 mm thick; supply and lay Rubaflex wet pour safety system to thicknesses as specified		
All by machine except macadam by hand		
black		
15 mm thick	100 m²	5400.00
35 mm thick	100 m²	6900.00
60 mm thick	100 m²	8200.00
coloured		
15 mm thick	100 m²	9100.00
35 mm thick	100 m²	9700.00
60 mm thick	100 m²	10500.00
All by hand except disposal by grab		
black		
15 mm thick	100 m²	8500.00
35 mm thick	100 m²	10000.00
60 mm thick	100 m²	11500.00
coloured		
15 mm thick	100 m²	12000.00
35 mm thick	100 m²	13000.00
60 mm thick	100 m²	14000.00
PLAY AREA FOR BALL GAMES		
Excavate for playground; bring to grade; lay 225 mm consolidated hardcore; blind with 100 mm type 1; lay macadam base of 40 mm size dense bitumen macadam to 75 mm thick; lay wearing course 30 mm thick		
Over 1000 m²	100 m²	4650.00
400–1000 m²	100 m²	4800.00
Excavate for playground; bring to grade; lay 100 mm consolidated hardcore; blind with 100 mm type 1; lay macadam base of dense bitumen macadam to 50 mm thick; lay wearing course 20 mm thick		
Over 1000 m²	100 m²	3500.00
400–1000 m²	100 m²	4300.00

Q SPECIAL SURFACES FOR SPORT/PLAYGROUNDS

Item Excluding site overheads and profit	Unit	Total rate £
Excavate for playground; bring to grade; lay 100 mm consolidated hardcore; blind with 100 mm type 1; lay macadam base of dense bitumen macadam to 50 mm thick; lay wearing course of Addagrip resin coated aggregate 3 mm diameter × 6 mm thick		
Over 1000 m²	100 m²	5700.00
400–1000 m²	100 m²	6300.00
Safety surfaced play area		
Excavate for playground; bring to grade; lay with 150 mm type 1; lay macadam base of dense bitumen macadam to 50 mm thick; lay safety surface		
Wetpour coloured; 0.50 m critical fall height	m	96.00
Wetpour black; 0.50 m critical fall height	m	53.00
Wetpour coloured; 1.50 m critical fall height	m	105.00
Wetpour black; 1.50 m critical fall height	m	82.00
Excavate for playground; bring to grade; lay with 150 mm type 1; lay safety surface tiles 'Play Matta'		
natural colours; 0.50 m critical fall height	m	81.00
bright colours; 0.50 m critical fall height	m	93.00
natural colours; 1.70 m critical fall height	m	86.00
bright colours; 1.70 m critical fall height	m	96.00
natural colours; 3.20 m critical fall height	m	100.00
bright colours; 3.20 m critical fall height	m	110.00
natural colours; 3.20 m critical fall height	m	110.00
bright colours; 3.20 m critical fall height	m	120.00
SPORTSGROUND CONSTRUCTION		
Plain sports pitches; site clearance, grading and drainage not included		
Agripower Ltd; cultivate ground and grade to levels; apply pre-seeding fertilizer at 900 kg/ha; apply pre-seeding selective weedkiller; seed in two operations with sports pitch type grass seed at 350 kg/ha; harrow and roll lightly; including initial cut; size		
association football; senior 114 × 72 m	each	3300.00
association football; junior 106 × 58 m	each	2450.00
rugby union pitch; 156 × 81 m	each	5100.00
rugby league pitch; 134 × 60 m	each	3200.00
hockey pitch; 95 × 60 m	each	2275.00
shinty pitch; 186 × 96 m	each	7300.00
men's lacrosse pitch; 100 × 55 m	each	2500.00
women's lacrosse pitch; 110 × 73 m	each	3200.00
target archery ground; 150 × 50 m	each	2950.00
cricket outfield; 160 × 142 m	each	3100.00
cycle track outfield; 160 × 80 m	each	5100.00
polo ground; 330 × 220 m	each	29000.00
Agripower Ltd; cricket square; excavate to depth of 100 mm; lay imported marl or clay loam; bring to accurate levels; apply pre-seeding fertilizer at 50 g/m; apply selective weedkiller; seed with cricket square type g grass seed at 50 g/m²; rake in and roll lightly; erect and remove temporary protective chestnut fencing; allow for initial cut and watering three times		
22.8 × 22.8 m	each	14000.00

Q PREPARATION FOR PLANTING/TURFING

Item Excluding site overheads and profit	Unit	Total rate £
SURFACE PREPARATION BY MACHINE		
Treat area with systemic non-selective herbicide one month before starting cultivation operations; rip up subsoil using subsoiling machine to a depth of 250 mm below topsoil at 1.20 m centres in light to medium soils; rotavate to 200 mm deep in two passes; cultivate with chain harrow; roll lightly; clear stones over 50 mm		
By tractor	100 m²	11.50
As above but carrying out operations in clay or compacted gravel	100 m²	13.00
As above but ripping by tractor rotavation by pedestrian operated rotavator; clearance and raking by hand; herbicide application by knapsack sprayer	100 m²	29.00
As above but carrying out operations in clay or compacted gravel	100 m²	30.00
Spread and lightly consolidate topsoil brought from spoil heap not exceeding 100 m; in layers not exceeding 150 mm; grade to specified levels; remove stones over 25 mm; all by machine		
100 mm thick	100 m²	76.00
150 mm thick	100 m²	115.00
300 mm thick	100 m²	230.00
450 mm thick	100 m²	340.00
Extra to the above for imported topsoil		
topsoil PC £28.00 m³ allowing for 20% settlement		
100 mm thick	100 m²	335.00
150 mm thick	100 m²	500.00
300 mm thick	100 m²	1000.00
450 mm thick	100 m²	1500.00
500 mm thick	100 m²	1675.00
600 mm thick	100 m²	2025.00
750 mm thick	100 m²	2500.00
1.00 m thick	100 m²	3350.00
topsoil to BS3882 PC £33.00 m³ allowing for 20% settlement		
100 mm thick	100 m²	620.00
150 mm thick	100 m²	920.00
300 mm thick	100 m²	1850.00
450 mm thick	100 m²	2800.00
500 mm thick	100 m²	3100.00
600 mm thick	100 m²	3700.00
750 mm thick	100 m²	4600.00
1.00 m thick	100 m²	62.00
Extra for incorporating mushroom compost at 50 mm/m² into the top 150 mm of topsoil (compost delivered in 20 m³) loads		
manually spread; mechanically rotavated	100 m²	175.00
mechanically spread and rotavated	100 m²	130.00
Extra for incorporating manure at 50 mm/m² into the top 150 mm of topsoil loads		
manually spread; mechanically rotavated; 20 m³ loads	100 m²	280.00
mechanically spread and rotavated; 60 m³ loads	100 m²	170.00

Q PREPARATION FOR PLANTING/TURFING

Item Excluding site overheads and profit	Unit	Total rate £
Excavate existing topsoil and remove off site; fill with new good quality topsoil to BS3882; add 50 mm mushroom compost (delivered in 65 m³ loads) and cultivate in; grade to level removing stones		
For planting		
300 mm deep	m²	29.00
400 mm deep	m²	40.00
500 mm deep	m²	48.00
For turfing		
200 mm deep	m²	21.00
300 mm deep	m²	29.00
SURFACE PREPARATION BY HAND		
To area preavious cleared and sprayed; cultivate existing topsoil to receive new ornamental planting; rotavate using pedestrian rotavator. dog over by hand 1 spit deep; rake and grade to level to receive new planting		
For ornamental planting beds	m²	0.55
For areas to be seeded or turfed	m²	0.80
Spread only and lightly consolidate topsoil brought from spoil heap in layers not exceeding 150 mm; grade to specified levels; remove stones over 25 mm; all by hand		
100 mm thick	100 m²	385.00
150 mm thick	100 m²	580.00
300 mm thick	100 m²	1150.00
450 mm thick	100 m²	1725.00
As above but inclusive of loading to location by barrow maximum distance 25 m; finished topsoil depth		
100 mm thick	100 m²	850.00
150 mm thick	100 m²	1275.00
300 mm thick	100 m²	2550.00
450 mm thick	100 m²	3800.00
500 mm thick	100 m²	4200.00
600 mm thick	100 m²	5100.00
As above but loading to location by barrow maximum distance 100 m; finished topsoil depth		
100 mm thick	100 m²	870.00
150 mm thick	100 m²	1275.00
300 mm thick	100 m²	2600.00
450 mm thick	100 m²	3900.00
500 mm thick	100 m²	4300.00
600 mm thick	100 m²	5200.00
Extra to the above for incorporating mushroom compost at 50 mm/m² into the top 150 mm of topsoil; by hand	100 m²	200.00
Extra to above for imported topsoil PC £28.00 m³ allowing for 20% settlement		
100 mm thick	100 m²	335.00
150 mm thick	100 m²	500.00
300 mm thick	100 m²	1000.00
450 mm thick	100 m²	1500.00
500 mm thick	100 m²	1675.00
600 mm thick	100 m²	2025.00
750 mm thick	100 m²	2500.00
1.00 m thick	100 m²	3350.00

Q SEEDING AND TURFING

Item Excluding site overheads and profit	Unit	Total rate £
SEEDING		
Bring top 200 mm of topsoil to a fine tilth using tractor drawn implements; remove stones over 25 mm by mechanical stone rake and bring to final tilth by harrow; apply pre-seeding fertilizer at 50 g/m² and work into top 50 mm during final cultivation; seed with certified grass seed in two operations; roll seedbed lightly after sowing		
General amenity grass at 35 g/m²; BSH A3	100 m²	66.00
General amenity grass at 25 g/m²; DLF Trifolium J Court	100 m²	55.00
Shaded areas; BSH A6 at 50 g/m²	100 m²	78.00
Motorway and road verges; DLF Trifolium Promaster 120 at 25–35 g/m²	100 m²	57.00
Bring top 200 mm of topsoil to a fine tilth using pedestrian operated rotavator; remove stones over 25 mm; apply pre-seeding fertilizer at 50 mm g/m² and work into top 50 mm during final hand cultivation; seed with certified grass seed in two operations; rake and roll seedbed lightly after sowing		
General amenity grass at 35 g/m²; BSH A3	100 m²	110.00
General amenity grass at 25 g/m²; DLF Trifolium J Court	100 m²	100.00
Shaded areas; BSH A6 at 50 g/m²	100 m²	125.00
Motorway and road verges; DLF Trifolium Promaster 120 at 25–35 g/m²	100 m²	105.00
Extra to above for using imported topsoil spread by machine		
100 mm minimum depth	100 m²	370.00
150 mm minimum depth	100 m²	550.00
extra for slopes over 30°	50%	–
Extra to above for using imported topsoil spread by hand; maximum distance for transporting soil 100 m		
100 mm minimum depth	m²	720.00
150 mm minimum depth	m²	1075.00
extra for slopes over 30°	50%	–
Extra for mechanically screening top 25 mm of topsoil through 6 mm screen and spreading on seedbed; debris carted to dump on site not exceeding 100 m	m³	6.55
Bring top 200 mm of topsoil to a fine tilth; remove stones over 50 mm; apply pre-seeding fertilizer at 50 g/m² and work into top 50 mm of topsoil during final cultivation; seed with certified grass seed in two operations; harrow and roll seedbed lightly after sowing		
Areas inclusive of fine levelling by specialist machinery		
outfield grass at 350 kg/ha; DLF Trifolium Promaster 40	ha	6400.00
outfield grass at 350 kg/ha; DLF Trifolium Promaster 70	ha	6100.00
sportsfield grass at 300 kg/ha; DLF Trifolium J Pitch	ha	6200.00
Seeded areas prepared by chain harrow		
low maintenance grass at 350 kg/ha	ha	7100.00
verge mixture grass at 150 kg/ha	ha	6600.00
Extra for wild flora mixture at 30 kg/ha		
BSH WSF 75 kg/ha	ha	3250.00
extra for slopes over 30°	50%	–
Cut existing turf to 1.0 × 1.0 × 0.5 m turves; roll up and move to stack not exceeding 100 m		
by pedestrian operated machine; roll up and stack by hand	100 m²	73.00
all works by hand	100 m²	210.00
Extra for boxing and cutting turves	100 m²	3.80

Q SEEDING AND TURFING

Item Excluding site overheads and profit	Unit	Total rate £
TURFING		
Bring top 200 mm of topsoil to a fine tilth in two passes; remove stones over 25 mm and bring to final tilth; apply pre-seeding fertilizer at 50 g/m² and work into top 50 mm during final cultivation; roll turf bed lightly		
Using tractor drawn implements and mechanical stone rake	m²	0.55
Cultivation by pedestrian rotavator; all other operations by hand	m²	0.60
As above but bring turf from stack not exceeding 100 m; lay turves to stretcher bond using plank barrow runs; firm turves using wooden turf beater		
using tractor drawn implements and mechanical stone rake	m²	2.20
cultivation by pedestrian rotavator; all other operations by hand	m²	2.10
As above but including imported turf; Rolawn Medallion		
using tractor drawn implements and mechanical stone rake	m²	3.95
cultivation by pedestrian rotavator; all other operations by hand	m²	3.90
Extra over to all of the above for watering on two occasions and carrying out initial cut		
by ride on triple mower	100 m²	1.15
by pedestrian mower	100 m²	4.95
by pedestrian mower; box cutting	100 m²	5.55
Extra for using imported topsoil spread and graded by machine		
25 mm minimum depth	m²	1.00
75 mm minimum depth	m²	2.80
100 mm minimum depth	m²	3.70
150 mm minimum depth	m²	5.50
Extra for using imported topsoil spread and graded by hand; distance of barrow run 25 m		
25 mm minimum depth	m²	1.90
75 mm minimum depth	m²	4.90
100 mm minimum depth	m²	6.25
150 mm minimum depth	m²	9.35
Extra for work on slopes over 30° including pegging with 200 galvanized wire pins	m²	1.85
Inturf Big Roll		
Supply, deliver in one consignment, fully prepare the area and install in Big Roll format Inturf 553, a turfgrass comprising dwarf perennial ryegrass, smooth stalked rneadowgrass and fescues; installation by tracked machine		
Preparation by tractor drawn rotavator	m²	3.10
Cultivation by pedestrian rotavator; all other operations by hand	m²	3.22
Erosion control		
On ground previously cultivated, bring area to level, treat with herbicide; lay 20 mm thick open texture erosion control mat with 100 mm laps, fixed with 8 × 400 mm steel pegs at 1.0 m centres; sow with low maintenance grass suitable for erosion control on slopes at 35 g/m²; spread imported topsoil 25 mm thick and fertilizer at 35 g/m²; water lightly using sprinklers on two occasions	m²	7.30
as above but hand-watering by hose pipe maximum distance from mains supply 50 m	m²	7.40

Q SEEDING AND TURFING

Item Excluding site overheads and profit	Unit	Total rate £
ARTIFICIAL TURF		
Excavate ground and reduce levels to receive 20 mm thick artifial grass surface on a compacted 150 thick Type 1 base; dispose of excavated material off site		
Lay artificial turf; by machine		
Trulawn Value; 16 mm	m²	48.00
Trulawn Play; 20 mm; entry level; lawn and play areas	m²	50.00
Trulawn Continental; 20 mm; realistic appearance; lawns and patios	m²	46.00
Trulawn Optimum; 26 mm; softest pile; lush green; lawns and patios	m²	58.00
Trulawn Luxury; 30 mm; deep pile; realistic lawn	m²	60.00
All by hand except disposal by grab		
Trulawn Value; 16 mm	m²	75.00
Trulawn Play; 20 mm; entry level; lawn and play areas	m²	77.00
Trulawn Continental; 20 mm; realistic appearance; lawns and patios	m²	73.00
Trulawn Optimum; 26 mm; softest pile; lush green; lawns and patios	m²	85.00
Trulawn Luxury; 30 mm; deep pile; realistic lawn	m²	88.00

Q PLANTING

Item Excluding site overheads and profit	Unit	Total rate £
HEDGE PLANTING		
Works by machine; excavate trench for hedge 300 mm wide × 300 mm deep; deposit spoil alongside and plant hedging plants in single row at 200 mm centres; backfill with excavated material		
Trench 300 mm × 300 mm; bare root hedging plants PC £0.48 per plant; single row		
200 mm centres	m	6.15
300 mm centres	m	5.20
400 mm centres	m	4.55
600 mm centres	m	3.90
800 mm centres	m	3.55
Trench 600 mm × 300 mm deep; bare root hedging plants PC £0.48 per plant; double staggered row		
300 mm centres	m	9.60
400 mm centres	m	3.85
500 mm centres	m	7.05
600 mm centres	m	6.70
800 mm centres	m	5.85
1.00 mm centres	m	5.45
Works by hand; excavate trench for hedge 300 mm wide × 450 mm deep; deposit spoil alongside and plant hedging plants in single row at 200 mm centres; backfill with excavated material incorporating organic manure at 1 m³ per 5 m³; carry out initial cut; including delivery of plants from nursery		
Trench 300 mm × 300 mm; bare root hedging plants PC £0.48 per plant; single row		
200 mm centres	m	8.40
300 mm centres	m	7.50
400 mm centres	m	6.80
600 mm centres	m	6.15
800 mm centres	m	5.80
Trench 600 mm × 300 mm deep; bare root hedging plants PC £0.48 per plant; double staggered row		
300 mm centres	m	15.40
400 mm centres	m	13.60
500 mm centres	m	12.80
600 mm centres	m	12.50
800 mm centres	m	11.60
1.00 mm centres	m	11.20
Practicality Brown Ltd; individual mature hedge planting; excavate trench by machine 700 mm wide × 500 mm deep; add compost at 100 mm/m² mixed to excavated material; plant mature hedge plants; backfill with excavated material and compost; allow for disposal of 50% of excavated material to spoil heaps 50 m distant		
Beech or hornbeam hedging		
1.50 m × 500 mm wide at 500 mm centres	m	260.00
1.50 m × 500 mm wide at 750 mm centres	m	170.00
2.00 m × 500 mm wide at 500 mm centres	m	350.00
2.00 m × 500 mm wide at 750 mm centres	m	250.00
1.50 m × 500 mm wide at 500 mm centres	m	260.00

Q PLANTING

Item Excluding site overheads and profit	Unit	Total rate £
HEDGE PLANTING – cont		
Practicality Brown Ltd – cont		
Yew (Taxus) hedging		
1.50 m × 500 mm wide at 500 mm centres	m	340.00
1.50 m × 500 mm wide at 750 mm centres	m	220.00
1.50 m × 500 mm wide at 1.00 m centres	m	175.00
1.75 m × 500 mm wide at 500 mm centres	m	385.00
1.75 m × 500 mm wide at 750 mm centres	m	250.00
1.75 m × 500 mm wide at 1.00 m centres	m	200.00
2.00 m × 500 mm wide at 500 mm centres	m	450.00
2.00 m × 500 mm wide at 750 mm centres	m	310.00
2.00 m × 500 mm wide at 1.00 mm centres	m	230.00
Laurel (Prunus) hedging		
2.00 m × 500 mm wide at 1.00 m centres	m	10.70
1.40/1.60 m high × 400 mm wide	m	170.00
1.60/1.80 m high × 500 mm wide	m	185.00
1.80/2.00 m high × 500 mm wide	m	200.00
Box (Buxus) hedging		
800 mm/1.00 m high × 300 mm wide	m	150.00
1.00 mm/1.20 m high × 300 mm wide	m	170.00
Practicality Brown Ltd; feathered hedge planting; excavate trench by machine 500 mm wide × 500 mm deep; add compost at 100 mm/m² mixed to excavated material; plant mature hedge plants; backfill with excavated material and compost; allow for disposal of 50% of excavated material to spoil heaps 50 m distant		
Beech or Hornbeam hedging		
1.75 m high × 300 mm wide at 400 mm centres	m	94.00
1.75 m high × 300 mm wide at 500 mm centres	m	77.00
1.75 m high × 300 mm wide at 750 mm centres	m	57.00
1.75 m high × 300 mm wide at 900 mm centres	m	46.00
2.00 m high × 300 mm wide at 400 mm centres	m	170.00
2.00 m high × 300 mm wide at 500 mm centres	m	140.00
Elveden hedges; strip hedge planting; excavate trench 700 mm wide × 500 mm deep by machine; add compost at 100 mm/m² mixed to excavated material; plant instant strip hedging; backfill with excavated material and compost; allow for disposal of 50% of excavated material to spoil heaps 50 m distant		
Beech hedging		
1.40/1.60 m high × 400 mm wide	m	170.00
1.60/1.80 m high × 500 mm wide	m	190.00
Yew (Taxus) hedging		
1.40/1.60 m high × 400 mm wide	m	210.00
1.60/1.80 m high × 500 mm wide	m	240.00
1.80/2.00 m high × 500 mm wide	m	265.00
Laurel (Prunus) hedging		
1.40/1.60 m high × 400 mm wide	m	170.00
1.60/1.80 m high × 500 mm wide	m	185.00
1.80/2.00 m high × 500 mm wide	m	200.00
Box (Buxus) hedging		
800 mm/1.00 m high × 300 mm wide	m	150.00
1.00/1.20 m high × 300 mm wide	m	170.00

Q PLANTING

Item Excluding site overheads and profit	Unit	Total rate £
TREE PLANTING		
Excavate tree pit by hand; fork over bottom of pit; plant tree with roots well spread out; backfill with excavated material incorporating treeplanting compost at 1 m³ per 3 m³ of soil; one tree stake and two ties; tree pits square in sizes shown		
Light standard bare root tree in pit; PC £9.75		
600 × 600 mm deep	each	34.50
900 × 600 mm deep	each	50.00
Standard bare root tree in pit; PC £16.00		
600 × 600 mm deep	each	46.50
900 × 600 mm deep	each	55.00
Standard root balled tree in pit; PC £28.00		
600 × 600 mm deep	each	60.00
900 × 600 mm deep	each	69.00
1.00 × 1.00 m deep	each	91.00
Selected standard bare root tree in pit; PC £22.50		
900 × 900 mm deep	each	60.00
1.00 × 1.00 m deep	each	88.00
Selected standard root ball tree in pit; PC £40.00		
900 × 900 mm deep	each	94.00
1.00 × 1.00 m deep	each	135.00
Heavy standard bare root tree in pit; PC £39.25		
900 × 900 mm deep	each	99.00
1.00 × 1.00 m deep	each	110.00
1.50 m × 750 mm deep	each	150.00
Heavy standard root ball tree in pit; PC £60.26		
900 × 900 mm deep	each	120.00
1.00 × 1.00 m deep	each	135.00
1.50 m × 750 mm deep	each	170.00
Extra heavy standard bare root tree in pit; PC £63.00		
1.00 × 1.00 m deep	each	140.00
1.20 × 1.00 m deep	each	140.00
1.50 m × 750 mm deep	each	185.00
Extra heavy standard root ball tree in pit; PC £72.00		
1.00 × 1.00 m deep	each	155.00
1.50 m × 750 mm deep	each	190.00
1.50 × 1.00 m deep	each	220.00
Excavate tree pit by machine; fork over bottom of pit; plant tree with roots well spread out; backfill with excavated material, incorporating organic manure at 1 m³ per 3 m³ of soil; one tree stake and two ties; tree pits square in sizes shown		
Light standard bare root tree in pit; PC £9.75		
600 × 600 mm deep	each	29.50
900 × 900 mm deep	each	43.00
Standard bare root tree in pit; PC £16.00		
600 × 600 mm deep	each	42.00
900 × 600 mm deep	each	44.00
900 × 900 mm deep	each	50.00

Q PLANTING

Item Excluding site overheads and profit	Unit	Total rate £
TREE PLANTING – cont		
Excavate tree pit by machine – cont		
Standard root balled tree in pit; PC £28.00		
600 × 600 mm deep	each	55.00
900 × 600 mm deep	each	57.00
900 × 900 mm deep	each	63.00
Selected standard bare root tree in pit; PC £22.50		
900 × 900 mm deep	each	60.00
1.00 × 1.00 m deep	each	71.00
Selected standard root ball tree in pit; PC £40.00		
900 × 900 mm deep	each	94.00
1.00 × 1.00 m deep	each	120.00
Heavy standard bare root tree in pit; PC £39.25		
900 × 900 mm deep	each	99.00
1.00 × 1.00 m deep	each	91.00
1.20 × 1.00 m deep	each	110.00
Heavy standard root ball tree in pit; PC £60.26		
900 × 900 mm deep	each	120.00
1.00 × 1.00 m deep	each	120.00
1.20 × 1.00 m deep	each	130.00
Extra heavy standard bare root tree in pit; PC £63.00		
1.00 × 1.00 m deep	each	125.00
1.20 × 1.00 m deep	each	140.00
1.50 × 1.00 m deep	each	170.00
Extra heavy standard root ball tree in pit; PC £72.00		
1.00 × 1.00 m deep	each	140.00
1.20 × 1.00 m deep	each	150.00
1.50 × 1.00 m deep	each	180.00
SEMI-MATURE TREE PLANTING		
Excavate tree pit deep by machine; fork over bottom of pit; plant rootballed tree Acer platanoides 'Emerald Queen' using telehandler where necessary; backfill with excavated material, incorporating Melcourt Topgrow bark/manure mixture at 1 m³ per 3 m³ of soil; Platipus underground guying system; tree pits 1500 × 1500 × 1500 mm deep inclusive of Platimats; excavated material not backfilled to treepit spread to surrounding area		
16–18 cm girth; PC £85.00	each	260.00
18–20 cm girth; PC £100.00	each	285.00
20–25 cm girth; PC £125.00	each	380.00
25–30 cm girth; PC £165.00	each	480.00
30–35 cm girth; PC £300.00	each	720.00
As above but treepits 2.00 × 2.00 × 1.5 m deep		
40–45 cm girth; PC £500.00	each	1150.00
45–50 cm girth; PC £700.00	each	1425.00
55–60 cm girth; PC £1,200.00	each	1850.00
67–70 cm girth; PC £2,200.00	each	3050.00
75–80 cm girth; PC £4,500.00	each	5700.00

Q PLANTING

Item Excluding site overheads and profit	Unit	Total rate £
Extra to the above for imported topsoil moved 25 m from tipping area and disposal off site of excavated material		
Tree pits 1500 × 1500 × 1500 mm deep		
16–18 cm girth	each	210.00
18–20 cm girth	each	200.00
20–25 cm girth	each	190.00
25–30 cm girth	each	180.00
30–35 cm girth	each	170.00
TREE PLANTING WITH MOBILE CRANES		
Excavate treepit 1.50 × 1.50 × 1.00 m deep; supply and plant semi-mature trees delivered in full loads; trees lifted by crane; inclusive of backfilling tree pit with imported topsoil, compost, fertilizers and underground guying using Platipus anchors		
Self-managed lift; local authority applications, health and safety, traffic management or road closures not included; tree size and distance of lift; 35 tonne crane		
25–30 cm; maximum 25 m distance	each	670.00
30–35 cm; maximum 25 m distance	each	780.00
35–40 cm; maximum 25 m distance	each	1150.00
55–60 cm; maximum 15 m distance	each	1850.00
80–90 cm; maximum 10 m distance	each	6900.00
Managed lift; inclusive of all local authority applications, health and safety, traffic management or road closures all by crane hire company; tree size and distance of lift; 35 tonne crane		
25–30 cm; maximum 25 m distance	each	680.00
30–35 cm; maximum 25 m distance	each	800.00
35–40 cm; maximum 25 m distance	each	1150.00
55–60 cm; maximum 15 m distance	each	1875.00
80–90 cm; maximum 10 m distance	each	7000.00
Self-managed lift; local authority applications, health and safety, traffic management or road closures not included; tree size and distance of lift; 80 tonne crane		
25–30 cm; maximum 40 m distance	each	680.00
30–35 cm; maximum 40 m distance	each	790.00
35–40 cm; maximum 40 m distance	each	1150.00
55–60 cm; maximum 33 m distance	each	1850.00
80–90 cm; maximum 23 m distance	each	7000.00
Managed lift; inclusive of all local authority applications, health and safety, traffic management or road closures all by crane hire company; tree size and distance of lift; 35 tonne crane		
25–30 cm; maximum 40 m distance	each	690.00
30–35 cm; maximum 40 m distance	each	800.00
35–40 cm; maximum 40 m distance	each	1175.00
55–60 cm; maximum 33 m distance	each	1900.00
80–90 cm; maximum 23 m distance	each	7100.00

Q PLANTING

Item Excluding site overheads and profit	Unit	Total rate £
SHRUBS, GROUND COVERS AND BULBS		
Excavate planting holes 250 × 250 × 300 mm deep to area previously ripped and rotavated; excavated material left alongside planting hole		
By mechanical auger		
250 mm centres (16 plants per m²)	m²	11.00
300 mm centres (11.11 plants per m²)	m²	7.65
400 mm centres (6.26 plants per m²)	m²	4.30
450 mm centres (4.93 plants per m²)	m²	3.40
500 mm centres (4 plants per m²)	m²	2.75
600 mm centres (2.77 plants per m²)	m²	1.90
750 mm centres (1.77 plants per m²)	m²	1.20
900 mm centres (1.23 plants per m²)	m²	0.85
1.00 m centres (1 plant per m²)	m²	0.70
1.50 m centres (0.44 plants per m²)	m²	0.30
As above but excavation by hand		
250 mm centres (16 plants per m²)	m²	11.00
300 mm centres (11.11 plants per m²)	m²	7.65
400 mm centres (6.26 plants per m²)	m²	4.30
450 mm centres (4.93 plants per m²)	m²	3.40
500 mm centres (4 plants per m²)	m²	2.75
600 mm centres (2.77 plants per m²)	m²	1.90
750 mm centres (1.77 plants per m²)	m²	1.20
900 mm centres (1.23 plants per m²)	m²	0.85
1.00 m centres (1 plant per m²)	m²	0.70
1.50 m centres (0.44 plants per m²)	m²	0.30
Clear light vegetation from planting area and remove to dump on site; dig planting holes; plant whips with roots well spread out; backfill with excavated topsoil; including one 38 × 38 mm treated softwood stake, two tree ties and mesh guard 1.20 m high; planting matrix 1.5 × 1.5 m; allow for beating up once at 10% of original planting, cleaning and weeding round whips once, applying fertilizer once at 35 gm/m²; using the following mix of whips, bare rooted		
Plant bare root plants average PC £0.41 each to a required matrix		
plant mix as above	100 m²	270.00
Plant bare root plants average PC £0.75 each to a required matrix		
plant mix as above	100 m²	290.00
Cultivate and grade shrub bed; bring top 300 mm of topsoil to a fine tilth, incorporating mushroom compost at 50 mm and Enmag slow release fertilizer; rake and bring to given levels; remove all stones and debris over 50 mm; dig planting holes average 300 × 300 × 300 mm deep; supply and plant specified shrubs in quantities as shown below; backfill with excavated material as above; water to field capacity and mulch 50 mm bark chips 20–40 mm size; water and weed regularly for 12 months and replace failed plants		
Shrubs 3 l PC £3.40; ground covers 9 cm PC £1.50		
100% shrub area		
300 mm centres	100 m²	6600.00
400 mm centres	100 m²	4000.00
500 mm centres	100 m²	2700.00
600 mm centres	100 m²	2075.00

Q PLANTING

Item Excluding site overheads and profit	Unit	Total rate £
100% groundcovers		
200 mm centres	100 m²	7400.00
300 mm centres	100 m²	3600.00
400 mm centres	100 m²	2275.00
500 mm centres	100 m²	1650.00
groundcover 30%/shrubs 70% at the distances shown below		
200/300 mm	100 m²	6800.00
300/400 mm	100 m²	3800.00
300/500 mm	100 m²	3000.00
400/500 mm	100 m²	2600.00
groundcover 50%/shrubs 50% at the distances shown below		
200/300 mm	100 m²	7000.00
300/400 mm	100 m²	3800.00
300/500 mm	100 m²	3200.00
400/500 mm	100 m²	2475.00
Cultivate ground by machine and rake to level; plant bulbs as shown; bulbs PC £25.00/100		
15 bulbs per m²	100 m²	660.00
25 bulbs per m²	100 m²	1075.00
50 bulbs per m²	100 m²	2150.00
Cultivate ground by machine and rake to level; plant bulbs as shown; bulbs PC £13.00/100		
15 bulbs per m²	100 m²	460.00
25 bulbs per m²	100 m²	750.00
50 bulbs per m²	100 m²	1475.00
Form holes in grass areas and plant bulbs using bulb planter, backfill with organic manure and turf plug; bulbs PC £13.00/100		
15 bulbs per m²	100 m²	680.00
25 bulbs per m²	100 m²	1125.00
50 bulbs per m²	100 m²	2250.00
BEDDING		
Spray surface with glyphosate; lift and dispose of turf when herbicide action is complete; cultivate new area for bedding plants to 400 mm deep; spread compost 100 mm deep and chemical fertilizer Enmag and rake to fine tilth to receive new bedding plants; remove all arisings		
Existing turf area		
disposal to skip	m²	5.15
disposal to compost area on site; distance 25 m	m²	5.70

Q PLANTING

Item Excluding site overheads and profit	Unit	Total rate £
BEDDING – cont		
Plant bedding to existing planting area; bedding planting PC £0.25 each		
Clear existing bedding; cultivate soil to 230 mm deep; incorporate compost 75 mm and rake to fine tilth; collect bedding from nursery and plant at 100 mm centres; irrigate on completion; maintain weekly for 12 weeks		
mass planted; 100 mm centres	m²	33.00
to patterns; 100 mm centres	m²	37.00
mass planted; 150 mm centres	m²	19.50
to patterns; 150 mm centres	m²	23.00
mass planted; 200 mm centres	m²	19.50
to patterns; 200 mm centres	m²	23.00
Extra for watering by hand held hose pipe		
Flow rate 25 litres/minute		
10 litres/m²	100 m²	0.10
15 litres/m²	100 m²	0.20
20 litres/m²	100 m²	0.20
25 litres/m²	100 m²	0.30
Flow rate 40 litres/minute		
10 litres/m²	100 m²	0.10
15 litres/m²	100 m²	0.10
20 litres/m²	100 m²	0.15
25 litres/m²	100 m²	0.20
PLANTING PLANTERS		
To brick planter; coat insides with 2 coats RIW liquid asphaltic composition; fill with 50 mm shingle and cover with geofabric; fill with screened topsoil incorporating 25% Topgrow compost and Enmag		
Planters 1.00 m deep		
1.00 × 1.00 m	each	110.00
1.00 × 2.00 m	each	185.00
1.00 × 3.00 m	each	260.00
Planters 1.50 m deep		
1.00 × 1.00 m	each	160.00
1.00 × 2.00 m	each	240.00
1.00 × 3.00 m	each	390.00
Container planting; fill with 50 mm shingle and cover with geofabric; fill with screened topsoil incorporating 25% Topgrow compost and Enmag		
Planters 1.00 m deep		
400 × 400 × 400 mm deep	each	8.10
400 × 400 × 600 mm deep	each	10.20
1.00 m × 400 mm wide × 400 mm deep	each	17.80
1.00 m × 600 mm wide × 600 mm deep	each	26.50
1.00 m × 100 mm wide × 400 mm deep	each	28.00
1.00 m × 100 mm wide × 600 mm deep	each	40.00
1.00 m × 100 mm wide × 1.00 m deep	each	66.00
1.00 m diameter × 400 mm deep	each	22.00
1.00 m diameter × 1.00 m deep	each	52.00
2.00 m diameter × 1.00 m deep	each	200.00

Q LANDSCAPE MAINTENANCE

Item Excluding site overheads and profit	Unit	Total rate £
MAINTENANCE OF GRASSED AREAS		
Maintenance executed as part of a landscape construction contract		
Grass cutting		
Grass cutting; fine turf; using pedestrian guided machinery; arisings boxed and disposed of off site		
per occasion	100 m²	5.70
per annum 26 cuts	100 m²	150.00
per annum 18 cuts	100 m²	110.00
Grass cutting; standard turf; using self-propelled three gang machinery		
per occasion	100 m²	0.40
per annum 26 cuts	100 m²	10.70
per annum 18 cuts	100 m²	7.40
Grass cutting for one year; recreation areas, parks, amenity grass areas; using tractor drawn machinery		
per occasion	ha	30.50
per annum 26 cuts	ha	800.00
per annum 18 cuts	ha	550.00
Maintain grass area inclusive of grass cutting; fertilization twice per annum; leaf and litter clearance		
using pedestrian guided machinery; arisings boxed and disposed of off site		
26 cuts; public areas	100 m²	180.00
26 cuts; private areas	100 m²	170.00
18 cuts; public areas	100 m²	140.00
18 cuts; private areas	100 m²	130.00
Maintain grass area inclusive of grass cutting; fertilization twice per annum; leaf and litter clearance		
using ride-on three cylinder machinery; arisings left on site		
26 cuts; public areas	100 m²	42.00
26 cuts; private areas	100 m²	29.00
18 cuts; public areas	100 m²	39.00
18 cuts; private areas	100 m²	26.00
Aeration of turfed areas		
Aerate ground with spiked aerator; apply spring/summer fertilizer once; apply autumn/winter fertilizer once; cut grass, 16 cuts; sweep up leaves twice		
as part of a landscape contract, defects liability	ha	1975.00
as part of a long-term maintenance contract	ha	1325.00

Q LANDSCAPE MAINTENANCE

Item Excluding site overheads and profit	Unit	Total rate £
ORNAMENTAL SHRUB BEDS		
Ornamental shrub beds; private gardens or business parks and resticted commercial environments		
Hand weed ornamental shrub bed during the growing season; clear leaves and litter; edge beds when necessary; prune shrubs or hedges and remove replace mulch to 50 mm at end of each year; (mulched areas only) planting less than 2 years old		
Mulched beds; weekly visits; 12 month period planting centres shown; public areas; shopping areas		
600 mm centres	100 m²	140.00
400 mm centres	100 m²	120.00
300 mm centres	100 m²	110.00
ground covers	100 m²	93.00
Mulched beds; weekly visits; 12 month period planting centres shown; private areas; business or office parks or similar		
600 mm centres	100 m²	125.00
400 mm centres	100 m²	100.00
300 mm centres	100 m²	85.00
ground covers	100 m²	68.00
Non-mulched beds; weekly visits; planting centres		
600 mm centres	100 m²	110.00
400 mm centres	100 m²	105.00
300 mm centres	100 m²	100.00
ground covers	100 m²	99.00
Non-mulched beds; monthly visits; planting centres		
600 mm centres	100 m²	90.00
400 mm centres	100 m²	83.00
300 mm centres	100 m²	79.00
ground covers	100 m²	76.00
Maintain rose bed		
Weed around base of plants; dead head regularly; fertilize once per annum with Enmag; prune once per annum; edge with half moon edging tool; add compost to 50 mm once per annum; clear of litter and leaves; private or restricted access beds		
private or restricted access beds	100 m²	240.00
public areas	100 m²	265.00
Remulch planting bed at the start of the planting season; top up mulch 25 mm thick; Melcourt Ltd		
Larger areas maximum distance 25 m; 80 m³ loads		
Ornamental bark mulch	100 m²	150.00
Melcourt Bark Nuggets	100 m²	140.00
Amenity Bark	100 m²	105.00
Forest biomulch	100 m²	99.00
Smaller areas; maximum distance 25 m; 25 m³ loads		
Ornamental bark mulch	100 m²	185.00
Melcourt Bark Nuggets	100 m²	175.00
Amenity Bark	100 m²	140.00
Forest biomulch	100 m²	130.00

Q LANDSCAPE MAINTENANCE

Item Excluding site overheads and profit	Unit	Total rate £
Shrub beds in ornamental gardens Maintain mature shrub bed; shrubs average 1.00 m–3.00 m high with occassional mature trees in bed or in proximity; prune one third of all planting by 30% every year in rotation; fertilse; add compost; annually; add fertilizers; cleaqr leaves; edge bed with half moon tool; water with hand held hose on 6 occassions		
massed plants; 1.00 m high	m²	370.00
plants up to 2 m high;	m²	415.00
plants 2 m–3 m high;	m²	540.00
VEGETATION CONTROL		
Native planting; roadside railway or forestry planted areas.		
Post planting maintenance; control of weeds and grass; herbicide spray applications; maintain weed free circles 1.00 m diameter to planting less than 5 years old in roadside, rail or forestry planting environments and the like; strim grass to 50–75 mm; prices per occasion (three applications of each operation normally required) Knapsack spray application; glyphosate; planting at		
1.50 mm centres	ha	990.00
1.75 mm centres	ha	610.00
2.00 mm centres	ha	660.00
Maintain planted areas; control of weeds and grass; maintain weed free circles 1.00 m diameter to planting less than 5 years old in roadside, rail or forestry planting environments and the like; strim surrounding grass to 50–75 mm; prices per occasion (three applications of each operation normally required) Herbicide spray applications; CDA (controlled droplet application) glyphosate and strimming; plants planted at the following centres		
1.50 mm centres	ha	540.00
1.75 mm centres	ha	610.00
2.00 mm centres	ha	660.00
Post planting maintenance; control of weeds and grass; herbicide spray applications; CDA (controlled droplet application); Xanadu glyphosate/diuron; maintain weed free circles 1.00 m diameter to planting less than 5 years old in roadside, rail or forestry planting environments and the like; strim grass to 50–75 mm; prices per occasion (1.5 applications of herbicide and three strim operations normally required) Plants planted at the following centres		
1.50 mm centres	ha	570.00
1.75 mm centres	ha	640.00
2.00 mm centres	ha	690.00

Q FENCING

Item Excluding site overheads and profit	Unit	Total rate £
TEMPORARY FENCING		
Site fencing; supply and erect temporary protective fencing and remove at completion of works		
Cleft chestnut paling		
1.20 m high 75 mm larch posts at 3 m centres	100 m	840.00
1.50 m high 75 mm larch posts at 3 m centres	100 m	1025.00
Heras fencing		
2 week hire period	100 m	930.00
4 week hire period	100 m	1400.00
8 week hire period	100 m	2350.00
12 week hire period	100 m	3300.00
16 week hire period	100 m	4200.00
24 week hire period	100 m	6100.00
CHAIN LINK AND WIRE FENCING		
Chain link fencing; supply and erect chain link fencing; form post holes and erect concrete posts and straining posts with struts at 50 m centres all set in 1:3:6 concrete; fix line wires		
3 mm galvanized wire 50 mm chainlink fencing		
900 mm high	m	24.00
1200 mm high	m	27.00
1800 mm high	m	37.00
Plastic coated 3.15 gauge galvanized wire mesh		
900 mm high	m	23.00
1200 mm high	m	25.50
1800 mm high	m	36.50
Extra for additional concrete straining posts with 1 strut set in concrete		
900 mm high	each	69.00
1200 mm high	each	73.00
1800 mm high	each	88.00
Extra for additional concrete straining posts with 2 struts set in concrete		
900 mm high	each	100.00
1200 mm high	each	100.00
1800 mm high	each	130.00
Extra for additional angle iron straining posts with 2 struts set in concrete		
900 mm high	each	77.00
1200 mm high	each	88.00
1400 mm high	each	105.00
1800 mm high	each	110.00
2400 mm high	each	120.00

Q FENCING

Item Excluding site overheads and profit	Unit	Total rate £
TIMBER FENCING		
Clear fence line of existing evergreen shrubs 3 m high average; grub out roots by machine chip on site and remove off site; erect closeboarded timber fence in treated softwood, pales 100 × 22 mm lapped, 150 × 22 mm gravel boards		
Concrete posts 100 × 100 mm at 3.0 m centres set into ground in 1:3:6 concrete		
900 mm high	m	**65.00**
1500 mm high	m	**71.00**
1800 mm high	m	**75.00**
As above but with softwood posts; three arris rails		
1350 mm high	m	**68.00**
1650 mm high	m	**75.00**
1800 mm high	m	**77.00**
Erect chestnut pale fencing; cleft chestnut pales; two lines galvanized wire, galvanized tying wire, treated softwood posts at 3.0 m centres and straining posts and struts at 50 m centres driven into firm ground		
900 mm high; posts 75 mm diameter × 1200 mm long	m	**5.90**
1200 mm high; posts 75 mm diameter × 1500 mm long	m	**8.15**
Construct timber rail; horizontal hit and miss type; rails 150 × 25 mm; posts 100 × 100 mm at 1.8 m centres; twice stained with coloured wood preservative; including excavation for posts and concreting into ground (C7P)		
In treated softwood		
1800 mm high	m	**58.00**
Construct cleft oak rail fence with rails 300 mm minimum girth tennoned both ends; 125 × 100 mm treated softwood posts double mortised for rails; corner posts 125 × 125 mm, driven into firm ground at 2.5 m centres		
3 rails	m	**17.80**
4 rails	m	**21.50**
DEER STOCK RABBIT FENCING		
Construct rabbit-stop fencing; erect galvanized wire netting; mesh 31 mm; 900 mm above ground, 150 mm below ground turned out and buried; on 75 mm diameter treated timber posts 1.8 m long driven 700 mm into firm ground at 4.0 m centres; netting clipped to top and bottom straining wires 2.63 mm diameter; straining post 150 mm diameter × 2.3 m long driven into firm ground at 50 m intervals		
turned in 150 mm	100 m	**1025.00**
buried 150 mm	100 m	**1100.00**
Deer fence		
Construct deer-stop fencing; erect five 4 mm diameter plain galvanized wires and five 2 ply galvanized barbed wires at 150 mm spacing on 45 × 45 × 5 mm angle iron posts 2.4 m long driven into firm ground at 3.0 m centres, driven into firm ground at 3.0 m centres, with timber droppers 25 × 38 mm × 1.0 m long at 1.5 m centres	100 m	**1750.00**

Q FENCING

Item Excluding site overheads and profit	Unit	Total rate £
DEER STOCK RABBIT FENCING – cont		
Forestry fencing		
Supply and erect forestry fencing of three lines of 3 mm plain galvanized wire tied to 1700 × 65 mm diameter angle iron posts at 2750 m centres with 1850 × 100 mm diameter straining posts and 1600 × 80 mm diameter struts at 50.0 m centres driven into firm ground		
1800 mm high; three wires	100 m	990.00
1800 mm high; three wires including cattle fencing	100 m	1300.00
CONCRETE FENCING		
Supply and erect precast concrete post and panel fence in 2 m bays; panels to be shiplap profile, aggregate faced one side; posts set 600 mm in ground in concrete		
1500 mm high	m	28.50
1800 mm high	m	33.00
2100 mm high	m	38.00
SECURITY FENCING		
Supply and erect chainlink fence; 51 × 3 mm mesh; with line wires and stretcher bars bolted to concrete posts at 3.0 m centres and straining posts at 10 m centres; posts set in concrete 450 × 450 mm × 33% of height of post deep; fit straight extension arms of 45 × 45 × 5 mm steel angle with three lines of barbed wire and droppers; all metalwork to be factory hot-dip galvanized for painting on site		
Galvanized 3 mm mesh		
900 mm high	m	29.00
1200 mm high	m	33.00
1800 mm high	m	44.00
As above but with PVC coated 3.15 mm mesh (diameter of wire 2.5 mm)		
900 mm high	m	27.50
1200 mm high	m	31.00
1800 mm high	m	43.50
Add to fences above for base of fence to be fixed with hairpin staples cast into concrete ground beam 1:3:6 site mixed concrete; mechanical excavation disposal to on site spoil heaps		
125 × 225 mm deep	m	5.30
Add to fences above for straight extension arms of 45 × 45 × 5 mm steel angle with three lines of barbed wire and droppers	m	5.25

Q FENCING

Item Excluding site overheads and profit	Unit	Total rate £
Supply and erect palisade security fence Jacksons Barbican 2500 mm high with rectangular hollow section steel pales at 150 mm centres on three 50 × 50 × 6 mm rails; rails bolted to 80 × 60 mm posts set in concrete 450 × 450 × 750 mm deep at 2750 mm centres; tops of pales to points and set at 45° angle; all metalwork to be hot-dip factory galvanized for painting on site	m	**120.00**
Supply and erect single gate to match above complete with welded hinges and lock		
1.0 m wide	each	**1475.00**
4.0 m wide	each	**3150.00**
8.0 m wide	pair	**3500.00**
Supply and erect Orsogril proprietary welded steel mesh panel fencing on steel posts set 750 mm deep in concrete foundations 600 × 600 mm; supply and erect proprietary single gate 2.0 m wide to match fencing		
930 mm high	100 m	**6700.00**
1326 mm high	100 m	**8600.00**
1722 mm high	100 m	**10500.00**
RAILINGS		
Conservation of historic railings; Eura Conservation Ltd		
Remove railings to workshop off site; shotblast and repair mildly damaged railings; remove rust and paint with three coats; transport back to site and re-erect		
railings with finials 1.80 m high	m	**570.00**
railings ornate cast or wrought iron	m	**890.00**
Supply and erect mild steel bar railing of 19 mm balusters at 115 mm centres welded to mild steel top and bottom rails 40 × 10 mm; bays 2.0 m long bolted to 51 × 51 mm ms hollow section posts set in C15P concrete; all metal work galvanized after fabrication		
900 mm high	m	**82.00**
1200 mm high	m	**110.00**
1500 mm high	m	**120.00**
Supply and erect mild steel pedestrian guard rail Class A; rails to be rectangular hollow sealed section 50 × 30 × 2.5 mm, vertical support 25 × 19 mm central between intermediate and top rail; posts to be set 300 mm into paving base; all components factory welded and factory primed for painting on site		
Panels 1000 mm high × 2000 mm wide with 150 mm toe space and 200 mm visibility gap at top	m	**91.00**
BALLSTOP FENCING		
Supply and erect plastic coated 30 × 30 mm netting fixed to 60.3 mm diameter 12 mm solid bar lattice galvanized dual posts; top, middle and bottom rails with 3 horizontal rails on 60.3 mm diameter nylon coated tubular steel posts at 3.0 m centres and 60.3 mm diameter straining posts with struts at 50 m centres set 750 mm into FND2 concrete footings 300 × 300 × 600 mm deep; include framed chain link gate 900 × 1800 mm high to match complete with hinges and locking latch		
4500 mm high	100 m	**11000.00**
5000 mm high	100 m	**13500.00**
6000 mm high	100 m	**15000.00**

Q FENCING

Item Excluding site overheads and profit	Unit	Total rate £
STOCK GATE		
Erect stock gate, ms tubular field gate, diamond braced 1.8 m high hung on tubular steel posts set in concrete (C7P); complete with ironmongery; all galvanized		
Width 3.00 m	each	330.00
Width 4.20 m	each	360.00
TRIP RAILS		
Steel trip rail		
Erect trip rail of galvanized steel tube 38 mm internal diameter with sleeved joint fixed to 38 mm diameter steel posts as above 700 mm long; set in C7P concrete at 1.20 m centres; metalwork primed and painted two coats metal preservative paint	m	160.00
Birdsmouth fencing; timber		
600 mm high	m	21.00
900 mm high	m	22.50

Q STREET FURNITURE

Item Excluding site overheads and profit	Unit	Total rate £
BENCHES, SEATS AND BOLLARDS		
Bollards and access restriction Supply and install 10 nr cast iron Doric bollards 920 mm high above ground × 170 mm diameter bedded in concrete base 400 mm diameter × 400 mm deep	10 nr	1475.00
Benches and seating In grassed area excavate for base 2500 × 1575 mm and lay 100 mm hardcore, 100 mm concrete, brick pavers in stack bond bedded in 25 mm cement: lime: sand mortar; supply and fix where shown on drawing proprietary seat, hardwood slats on black powder coated steel frame, bolted down with 4 nr 24 × 90 mm recessed hex-head stainless steel anchor bolts set into concrete	set	1075.00
Cycle stand Supply and fix cycle stand 1250 m × 550 mm of 60.3 mm black powder coated hollow steel sections, one-piece with rounded top corners; set 250 mm into paving	each	400.00

R AGRICULTURAL DRAINAGE

Item Excluding site overheads and profit	Unit	Total rate £
DRAINAGE DITCHES		
Excavate and form ditch and bank with 45° sides in light to medium soils; all widths taken at bottom of ditch		
300 mm wide × 600 mm deep	100 m	105.00
600 mm wide × 900 mm deep	100 m	100.00
1.20 m wide × 900 mm deep	100 m	920.00
1.50 m wide × 1.20 m deep	100 m	1550.00
Clear and bottom existing ditch average 1.50 m deep, trim back vegetation and remove debris to licensed tip not exceeding 13 km, lay jointed concrete pipes; including bedding, haunching and topping with 150 mm concrete; 11.50 N/mm² – 40 mm aggregate; backfill with approved spoil from site		
Pipes 300 mm diameter	100 m	6300.00
Pipes 450 mm diameter	100 m	8000.00
Pipes 600 mm diameter	100 m	10500.00
SUBSOIL DRAINAGE – BY MACHINE		
Main drain; remove 150 mm topsoil and deposit alongside trench, excavate drain trench by machine and lay flexible perforated drain; lay bed of gravel rejects 100 mm; backfill with gravel rejects or similar to within 150 mm of finished ground level; complete fill with topsoil; remove surplus spoil to approved dump on site		
Main drain 160 mm supplied in 35 m lengths		
450 mm deep	100 m	700.00
600 mm deep	100 m	810.00
900 mm deep	100 m	1350.00
Extra for couplings	each	2.30
As above but with 100 mm main drain supplied in 100 m lengths		
450 mm deep	100 m	1025.00
600 mm deep	100 m	1375.00
900 mm deep	100 m	2225.00
Extra for couplings	each	1.90
Laterals to mains above; herringbone pattern; excavation and backfilling as above; inclusive of connecting lateral to main drain		
160 mm pipe to 450 mm deep trench		
laterals at 1.0 m centres	100 m²	700.00
laterals at 2.0 m centres	100 m²	350.00
laterals at 3.0 m centres	100 m²	230.00
laterals at 5.0 m centres	100 m²	140.00
laterals at 10.0 m centres	100 m²	70.00
160 mm pipe to 600 mm deep trench		
laterals at 1.0 m centres	100 m²	810.00
laterals at 2.0 m centres	100 m²	410.00
laterals at 3.0 m centres	100 m²	270.00
laterals at 5.0 m centres	100 m²	160.00
laterals at 10.0 m centres	100 m²	82.00

R AGRICULTURAL DRAINAGE

Item Excluding site overheads and profit	Unit	Total rate £
160 mm pipe to 900 mm deep trench		
laterals at 1.0 m centres	100 m²	1350.00
laterals at 2.0 m centres	100 m²	680.00
laterals at 3.0 m centres	100 m²	450.00
laterals at 5.0 m centres	100 m²	270.00
laterals at 10.0 m centres	100 m²	135.00
Extra for 160/160 mm couplings connecting laterals to main drain		
laterals at 1.0 m centres	10 m	95.00
laterals at 2.0 m centres	10 m	47.50
laterals at 3.0 m centres	10 m	31.50
laterals at 5.0 m centres	10 m	19.00
laterals at 10.0 m centres	10 m	9.50
100 mm pipe to 450 mm deep trench		
laterals at 1.0 m centres	100 m²	1025.00
laterals at 2.0 m centres	100 m²	520.00
laterals at 3.0 m centres	100 m²	340.00
laterals at 5.0 m centres	100 m²	205.00
laterals at 10.0 m centres	100 m²	100.00
100 mm pipe to 600 mm deep trench		
laterals at 1.0 m centres	100 m²	1375.00
laterals at 2.0 m centres	100 m²	680.00
laterals at 3.0 m centres	100 m²	450.00
laterals at 5.0 m centres	100 m²	270.00
laterals at 10.0 m centres	100 m²	135.00
100 mm pipe to 900 mm deep trench		
laterals at 1.0 m centres	100 m²	2225.00
laterals at 2.0 m centres	100 m²	1125.00
laterals at 3.0 m centres	100 m²	740.00
laterals at 5.0 m centres	100 m²	445.00
laterals at 10.0 m centres	100 m²	220.00
80 mm pipe to 450 mm deep trench		
laterals at 1.0 m centres	100 m²	990.00
laterals at 2.0 m centres	100 m²	490.00
laterals at 3.0 m centres	100 m²	325.00
laterals at 5.0 m centres	100 m²	200.00
laterals at 10.0 m centres	100 m²	99.00
80 mm pipe to 600 mm deep trench		
laterals at 1.0 m centres	100 m²	1325.00
laterals at 2.0 m centres	100 m²	660.00
laterals at 3.0 m centres	100 m²	435.00
laterals at 5.0 m centres	100 m²	260.00
laterals at 10.0 m centres	100 m²	130.00
80 mm pipe to 900 mm deep trench		
laterals at 1.0 m centres	100 m²	2175.00
laterals at 2.0 m centres	100 m²	1075.00
laterals at 3.0 m centres	100 m²	720.00
laterals at 5.0 m centres	100 m²	435.00
laterals at 10.0 m centres	100 m²	220.00

R AGRICULTURAL DRAINAGE

Item Excluding site overheads and profit	Unit	Total rate £
SUBSOIL DRAINAGE – BY MACHINE – cont		
Laterals to mains above – cont		
Extra for 100/80 mm junctions connecting laterals to main drain		
laterals at 1.0 m centres	10 m	44.00
laterals at 2.0 m centres	10 m	22.00
laterals at 3.0 m centres	10 m	14.60
laterals at 5.0 m centres	10 m	8.70
laterals at 10.0 m centres	10 m	4.35
Clay land drain		
Excavate trench by excavator to 450 mm deep; lay 100 mm vitrified clay drain with butt joints, bedding Class B; backfill with excavated material screened to remove stones over 40 mm; backfill to be laid in layers not exceeding 150 mm; top with 150 mm topsoil remove surplus material to approved dump on site not exceeding 100 m; final level of fill to allow for settlement	100 m	1650.00
SUBSOIL DRAINAGE – BY HAND		
Main drain; remove 150 mm topsoil and deposit alongside trench; excavate drain trench by machine and lay flexible perforated drain; lay bed of gravel rejects 100 mm; backfill with gravel rejects or similar to within 150 mm of finished ground level; complete fill with topsoil; remove surplus spoil to approved dump on site		
Main drain 160 mm in supplied in 35 m lengths		
450 mm deep	100 m	1975.00
600 mm deep	100 m	2450.00
900 mm deep	100 m	3300.00
Extra for couplings	each	2.30
As above but with 100 mm main drain supplied in 100 m lengths		
450 mm deep	100 m	1275.00
600 mm deep	100 m	1625.00
900 mm deep	100 m	2275.00
Extra for couplings	each	1.90
Laterals to mains above; herringbone pattern; excavation and backfilling as above; inclusive of connecting lateral to main drain		
160 mm pipe to 450 mm deep trench		
laterals at 1.0 m centres	100 m²	1975.00
laterals at 2.0 m centres	100 m²	980.00
laterals at 3.0 m centres	100 m²	650.00
laterals at 5.0 m centres	100 m²	390.00
laterals at 10.0 m centres	100 m²	195.00
160 mm pipe to 600 mm deep trench		
laterals at 1.0 m centres	100 m²	2450.00
laterals at 2.0 m centres	100 m²	1225.00
laterals at 3.0 m centres	100 m²	810.00
laterals at 5.0 m centres	100 m²	490.00
laterals at 10.0 m centres	100 m²	245.00

R AGRICULTURAL DRAINAGE

Item Excluding site overheads and profit	Unit	Total rate £
160 mm pipe to 900 mm deep trench		
laterals at 1.0 m centres	100 m²	3300.00
laterals at 2.0 m centres	100 m²	1650.00
laterals at 3.0 m centres	100 m²	1075.00
laterals at 5.0 m centres	100 m²	660.00
laterals at 10.0 m centres	100 m²	330.00
Extra for 160/160 mm couplings connecting laterals to main drain		
laterals at 1.0 m centres	10 m	95.00
laterals at 2.0 m centres	10 m	47.50
laterals at 3.0 m centres	10 m	31.50
laterals at 5.0 m centres	10 m	19.00
laterals at 10.0 m centres	10 m	9.50
100 mm pipe to 450 mm deep trench		
laterals at 1.0 m centres	100 m²	1275.00
laterals at 2.0 m centres	100 m²	640.00
laterals at 3.0 m centres	100 m²	425.00
laterals at 5.0 m centres	100 m²	260.00
laterals at 10 m centres	100 m²	130.00
100 mm pipe to 600 mm deep trench		
laterals at 1.0 m centres	100 m²	1625.00
laterals at 2.0 m centres	100 m²	820.00
laterals at 3.0 m centres	100 m²	540.00
laterals at 5.0 m centres	100 m²	325.00
laterals at 10.0 m centres	100 m²	160.00
100 mm pipe to 900 mm deep trench		
laterals at 1.0 m centres	100 m²	2275.00
laterals at 2.0 m centres	100 m²	1150.00
laterals at 3.0 m centres	100 m²	750.00
laterals at 5.0 m centres	100 m²	455.00
laterals at 10.0 m centres	100 m²	230.00
80 mm pipe to 450 mm deep trench		
laterals at 1.0 m centres	100 m²	1250.00
laterals at 2.0 m centres	100 m²	620.00
laterals at 3.0 m centres	100 m²	410.00
laterals at 5.0 m centres	100 m²	250.00
laterals at 10.0 m centres	100 m²	125.00
80 mm pipe to 600 mm deep trench		
laterals at 1.0 m centres	100 m²	1575.00
laterals at 2.0 m centres	100 m²	790.00
laterals at 3.0 m centres	100 m²	520.00
laterals at 5.0 m centres	100 m²	320.00
laterals at 10.0 m centres	100 m²	160.00
80 mm pipe to 900 mm deep trench		
laterals at 1.0 m centres	100 m²	2250.00
laterals at 2.0 m centres	100 m²	1125.00
laterals at 3.0 m centres	100 m²	740.00
laterals at 5.0 m centres	100 m²	450.00
laterals at 10.0 m centres	100 m²	220.00

R AGRICULTURAL DRAINAGE

Item Excluding site overheads and profit	Unit	Total rate £
SUBSOIL DRAINAGE – BY HAND – cont		
Laterals to mains above – cont		
Extra for 100/80 mm couplings connecting laterals to main drain		
laterals at 1.0 m centres	10 m	**44.00**
laterals at 2.0 m centres	10 m	**22.00**
laterals at 3.0 m centres	10 m	**14.60**
laterals at 5.0 m centres	10 m	**8.70**
laterals at 10.0 m centres	10 m	**4.35**
SOAKAWAYS		
Construct soakaway from perforated concrete rings; excavation, casting in situ concrete ring beam base; filling with gravel 250 mm deep; placing perforated concrete rings; surrounding with geofabric and backfilling to external surround of soakaway with 250 mm wide granular surround and excavated material; step irons and cover slab; inclusive of all earthwork retention and disposal off site of surplus material		
900 mm diameter		
1.00 m deep	nr	**610.00**
2.00 m deep	nr	**1025.00**
1200 mm diameter		
1.00 m deep	nr	**820.00**
2.00 m deep	nr	**1375.00**
2400 mm diameter		
1.00 m deep	nr	**2325.00**
2.00 m deep	nr	**3800.00**
Aquacell Infiltration unit soakaway; 1.00 m × 500 × 400 mm; 200 litre volume each in trench		
Excavate for new soakaway; grade bottom of excavation; Lay bedding layer of 100 mm sharp sand; Install geotextile to line excavation and bedding; Install Aquacell units in bonded formation; wrap entire installation in geotextile; back filling to surround in MOT Type 2 150 mm thick. Place sharp sand over infiltration units; inclusive of connection only of in-flow pipework; Cover over 500 deep with Type 2 granular material.		
8 units in trench; 1520 litres	nr	**760.00**
12 units in trench; 2280 litres	nr	**1225.00**
16 units in trench; 3040 litres	nr	**1550.00**
20 units in trench; 3800 litres	nr	**1900.00**
30 units; in trench 5700 litres	nr	**3100.00**
60 units; in trench 11400 litres	nr	**5700.00**
LAND DRAINS		
Excavate trench by excavator; lay Type 2 bedding; backfill to 150 mm above pipe with gravel rejects; lay non-woven geofabric and fill with topsoil to ground level		
Trench 600 deep		
160 mm PVC-u drainpipe	100 m	**2600.00**
110 mm PVC-u drainpipe	100 m	**1575.00**
150 mm vitrified clay	100 m	**4100.00**
100 mm vitrified clay	100 m	**2375.00**

R SURFACE WATER DRAINAGE

Item Excluding site overheads and profit	Unit	Total rate £
MANHOLES		
Brick manhole; excavate pit including earthwork support and working space disposal of surplus spoil to dump on site not exceeding 100 m; lay concrete (1:2:4) base 1500 mm diameter × 200 mm thick; reinforced with mesh reinforcement; 110 mm vitrified clay channels; benching in concrete (1:3:6) allowing one outlet and two inlets for 110 mm diameter pipe; construct inspection chamber 1 brick thick walls of engineering brick Class B; backfill with excavated material; complete with 2 nr cast iron step irons		
1200 × 1200 × 1200 mm		
cover slab of precast concrete	each	**1025.00**
access cover; Group 2; 600 × 450 mm	each	**1050.00**
1200 × 1200 × 1500 mm		
access cover; Group 2; 600 × 450 mm	each	**1275.00**
recessed cover 5 tonne load; 600 × 450 mm; filled with block paviors	each	**1425.00**
1200 × 1200 × 2000 mm; Walls 327.5 thick		
access cover; Group 2; 600 × 450 mm	each	**2000.00**
recessed cover 5 tonne load; 600 × 450 mm; filled with block paviors	each	**2150.00**
1500 × 1500 × 2500 mm; Walls 327.5 thick		
access cover; Group 2; 600 × 450 mm	each	**3200.00**
recessed cover 5 tonne load; 600 × 450 mm; filled with block paviors	each	**3300.00**
ACCESS CHAMBERS		
Excavate inspection chamber by machine; lay base of concrete 150 thick; allow for half section pipework and benchings		
Precast concrete inspection chamber		
600 × 400 × 600 mm deep	nr	**350.00**
600 × 400 × 900 mm deep	nr	**380.00**
Polypropylene inspection chamber		
Mini access chamber 600 deep	nr	**280.00**
475 mm diameter × 900 mm deep; polymer cover	nr	**445.00**
475 mm diameter × 900 mm deep; ductile iron cover with screw down lid	nr	**475.00**
GULLIES		
Gullies; vitrified clay; excavate by hand; supply and set gully in concrete (C10P); connect to drainage system with flexible joints; backfilling with excavated material to 250 mm below finished level. Lay 150 mm type 1 to receive surface treatments (not included)		
Trapped mud (dirt) gully; complete with galvanized bucket and cast iron hinged locking grate and frame		
loading to 1 tonne; 100 or 150 mm outlet	each	**345.00**
loading to 5 tonne; 100 mm outlet	each	**390.00**
loading to 5 tonne; 150 mm outlet	each	**405.00**
Trapped mud (dirt) gully with rodding eye; complete with galvanized bucket and cast iron hinged locking grate and frame		
100 outlet; 300 mm internal diameter; 300 mm internal depth	each	**340.00**
150 outlet; 400 mm internal diameter; 750 mm internal depth	each	**365.00**

R SURFACE WATER DRAINAGE

Item Excluding site overheads and profit	Unit	Total rate £
GULLIES – cont		
Concrete road gully		
Excavate and lay 100 mm concrete base (1:3:6) 150 x150 mm to suit given invert level of drain; supply and connect trapped precast concrete road gully set in concrete surround; connect to vitrified clay drainage system with flexible joints; supply and fix straight bar dished top cast iron grating and frame; bedded in cement: sand mortar (1:3)		
450 mm diameter × 1.07 m deep with 160 mm outlet	each	210.00
Gullies PVC-u		
Excavate and lay 100 mm concrete (C20P) base 150 × 150 mm to suit given invert level of drain; connect to drainage system; backfill with DoT Type 1 granular fill; install gully; complete with cast iron grate and frame		
yard gully 300 mm diameter × 600 mm deep	each	340.00
trapped PVC-u gully; 110 mm diameter × 215 mm deep; ductile iron frame	each	170.00
bottle gully 228 × 228 × 642 mm deep	each	150.00
bottle gully 228 × 228 × 317 mm deep	each	135.00
PIPE LAYING		
PVC-u pipe laying; excavate trench 300 mm wide by excavator; lay bedding; backfill to 150 mm above pipe with gravel rejects; backfill with excavated material to ground level; lay surface water drain pipe to depth shown		
Trench 900 mm deep; pipe 640–590 mm deep; bedding on reject sand		
110 mm	m	17.80
110 mm; short lengths	m	19.50
160 mm	m	28.00
160 mm; short lengths	m	45.00
Trench 900 mm deep; pipe 640–590 mm deep; bedding on 20 mm shingle		
110 mm	m	22.00
110 mm; short lengths	m	24.00
160 mm	m	32.50
160 mm; short lengths	m	49.00
Trench 1.20 m deep; pipe 940–890 mm deep; bedding on reject sand		
110 mm	m	29.50
110 mm; short lengths	m	42.00
160 mm	m	50.00
160 mm; short lengths	m	67.00
Trench 1.20 m deep; pipe 940–890 mm deep; bedding on 20 mm shingle		
110 mm	m	29.50
110 mm; short lengths	m	42.00
160 mm	m	50.00
160 mm; short lengths	m	67.00
Vitrified clay pipe laying; excavate trench by excavator; lay bedding; backfill to 150 mm above pipe with gravel rejects; backfill with excavated material to ground level; lay surface water drain pipe to depth shown		
Trench 900 mm deep; pipe 640–590 mm deep; bedding on reject sand		
100 mm	m	25.50
100 mm; short lengths	m	27.00
160 mm	m	43.00
160 mm; short lengths	m	49.00

R SURFACE WATER DRAINAGE

Item Excluding site overheads and profit	Unit	Total rate £
Trench 900 mm deep; pipe 640–590 mm deep; bedding on 20 mm shingle		
110 mm	m	30.00
110 mm; short lengths	m	31.00
160 mm	m	47.00
160 mm; short lengths	m	53.00
Trench 1.20 m deep; pipe 940–890 mm deep; bedding on reject sand		
110 mm	m	37.50
110 mm; short lengths	m	49.00
160 mm	m	65.00
160 mm; short lengths	m	71.00
Trench 1.20 m deep; pipe 940–890 mm deep; bedding on 20 mm shingle		
110 mm	m	37.00
110 mm; short lengths	m	49.00
160 mm	m	65.00
160 mm; short lengths	m	71.00
LINEAR DRAINAGE		
Linear drainage to design sensitive areas		
Excavate trench by machine; lay Aco Brickslot channel drain on concrete base and surround to falls; all to manufacturers specifications		
paving surround to both sides of channel	m	200.00
Linear drainage to pedestrian area		
Excavate trench by machine; lay Aco MultiDrain MD polymer concrete channel drain on concrete base and surround to falls; all to manufacturers specifications; paving surround to channel with brick paving PC £300.00/1000		
Brickslot galvanized grating; paving to both sides	m	200.00
Brickslot stainless steel grating; paving to both sides	m	370.00
slotted galvanized steel grating; paving surround to one side of channel	m	140.00
Excavate trench by machine; lay Aco MultiDrain PPD recycled polypropylene channel drain on concrete base and surround to falls; all to manufacturers specifications; paving surround to channel with brick paving PC £300.00/1000		
Heelguard composite black; paving surround to both sides of channel	m	140.00
Linear drainage to light vehicular area		
Excavate trench by machine; lay Aco MultiDrain PPD recycled polypropylene channel drain on concrete base and surround to falls; all to manufacturers specifications; paving surround to channel with brick paving PC £300.00/1000		
Heelguard composite black; paving surround to both sides of channel	m	170.00
ductile iron; paving surround to both sides of channel	m	240.00
Accessories for channel drain		
Sump unit with sediment bucket	nr	180.00
End cap; inlet/outlet	nr	17.20

S IRRIGATION

Item Excluding site overheads and profit	Unit	Total rate £
LANDSCAPE IRRIGATION		
Automatic irrigation; KAR UK Ltd		
Large garden consisting of 24 stations; 7000 m² irrigated area		
turf only	nr	11000.00
70/30% turf/shrub beds	nr	14000.00
Medium sized garden consisting of 12 stations; 3500 m² irrigated area		
turf only	nr	7700.00
70/30% turf/shrub beds	nr	11000.00
Medium sized garden consisting of 6 stations of irrigated area; 1000 m²		
turf only	nr	5200.00
70/30% turf/shrub beds	nr	5800.00
50/50% turf/shrub beds	nr	6300.00
Leaky pipe irrigation; Leaky Pipe Ltd		
Works by machine; main supply and connection to laterals; excavate trench for main or ring main 450 mm deep; supply and lay pipe; backfill and lightly compact trench		
20 mm LDPE	100 m	270.00
16 mm LDPE	100 m	235.00
Works by hand; main supply and connection to laterals; excavate trench for main or ring main 450 mm deep; supply and lay pipe; backfill and lightly compact trench		
20 mm LDPE	100 m	1375.00
16 mm LDPE	100 m	1350.00
Turf area irrigation; laterals to mains; to cultivated soil; excavate trench 150 mm deep using hoe or mattock; lay moisture leaking pipe laid 150 mm below ground at centres of 350 mm		
low leak	100 m²	630.00
high leak	100 m²	560.00
Landscape area irrigation; laterals to mains; moisture leaking pipe laid to the surface of irrigated areas at 600 mm centres		
low leak	100 m²	330.00
high leak	100 m²	290.00
Landscape area irrigation; laterals to mains; moisture leaking pipe laid to the surface of irrigated areas at 900 mm centres		
low leak	100 m²	220.00
high leak	100 m²	195.00
multi-station controller	each	415.00
solenoid valves connected to automatic controller	each	530.00

S IRRIGATION

Item Excluding site overheads and profit	Unit	Total rate £
RAINWATER HARVESTING		
Combined Harvesters Ltd; rainwater tanks; soft landscape areas installed below ground		
Excavate in soft landscape area by machine to install rainwater harvesting tank; grade base and lay compacted granular base to level; connect filters to rainwater down pipe of building 10 m distance and lay 110 mm uPVC pipe in trench. Install rainwater tank complete with filters for irrigation and submersible pump. cover with geofabric and backfill with excavated material compacting lightly. remove surplus material off -site		
Columbus tank 3700 litres	nr	2900.00
Columbus tank 4500 litres	nr	3100.00
Columbus tank 6500 litres	nr	3400.00
Columbus tank 9000 litres	nr	4200.00
Columbus tank 13000 litres; 2 nr 6500 tanks	nr	5200.00
Excavate in soft landscape area by hand to install rainwater harvesting tank; grade base and lay compacted granular base to level; connect filters to rainwater down pipe of building 10 m distance and lay 110 mm uPVC pipe in trench. Install low profile – rainwater tank complete with filters for irrigation and submersible pump. cover with geofabric and backfill with excavated material compacting lightly. remove surplus material off-site		
Flat tank 1500 litres	nr	3250.00
Flat tank 3000 litres	nr	4450.00

S WATER FEATURES

Item Excluding site overheads and profit	Unit	Total rate £
LAKES AND PONDS		
FAIRWATER LTD		
Excavate for small pond or lake maximum depth 1.00 m; remove arisings off site; grade and trim to shape; lay 75 mm sharp sand; line with 0.75 mm butyl liner 75 mm sharp sand and geofabric; cover over with sifted topsoil; anchor liner to anchor trench; install balancing tank and automatic top-up system		
Pond or lake of organic shape		
100 m²; perimeter 50 m	each	5600.00
250 m²; perimeter 90 m	each	8700.00
500 m²; perimeter 130 m	each	20500.00
1000 m²; perimeter 175 m	each	41000.00
Excavate for lake average depth 1.0 m, allow for bringing to specified levels; reserve topsoil; remove spoil to approved dump on site; remove all stones and debris over 75 mm; lay polythene sheet including welding all joints and seams by specialist; screen and replace topsoil 200 mm thick		
Prices are for lakes of regular shape		
500 micron sheet	1000 m	16500.00
1000 micron sheet	1000 m	20500.00
Extra for removing spoil to tip	m³	22.00
Extra for 25 mm sand blinding to lake bed	100 m²	110.00
Extra for screening topsoil	m²	1.15
Extra for spreading imported topsoil	100 m²	760.00
Plant aquatic plants in lake topsoil		
Aponogeton distachyum PC £312.00/100	100	77.00
Acorus calamus PC £198.00/100	100	77.00
Butomus umbellatus PC £198.00/100	100	77.00
Typha latifolia PC £198.00/100	100	77.00
Nymphaea PC £816.00/100	100	77.00
Formal water features		
Excavate and construct water feature of regular shape; lay 100 mm hardcore base and 150 mm concrete 1:2:4 site mixed; line base and vertical face with butyl liner 0.75 micron and construct vertical sides of reinforced blockwork; rendering two coats; anchor the liner behind blockwork; install pumps balancing tanks and all connections to mains supply		
1.00 × 1.00 × 1.00 m deep	nr	2475.00
2.00 × 1.00 × 1.00 m deep	nr	3100.00

PART 4

Prices for Measured Works – Major Works

INTRODUCTION

Typical Project Profile

Contract value	£100,000.00–£900,000.00
Labour rate (see page 3)	£19.25 per hour
Labour rate for maintenance contracts	£16.50 per hour
Number of site staff	35
Project area	6000 m²
Project location	Outer London
Project components	50% hard landscape
	50% soft landscape and planting
Access to works areas	Very good
Contract	Main contract
Delivery of materials	Full loads
Profit and site overheads	Excluded

NEW ITEMS 2013

Item Excluding site overheads and profit	PC £	Labour hours	Labour £	Plant £	Material £	Unit	Total rate £
NEW ITEMS 2013							
B10 GARDEN BUILDINGS							
Garden rooms; 3rd Space; Sawhorse Ltd;							
fully insulated pent roof modular garden							
buildings installed on granular base with							
jacking legs							
Douglas fir framed with western red cedar							
cladding; glazed with fully glazed doors;							
integral floor and internal finishes including							
birch ply cladding							
15.25 m²; 6.5 × 2.5 × 2.5 m high	–	–	–	–	–	nr	18330.00
12.5 m²; 5.0 × 2.5 × 2.5 m high	–	–	–	–	–	nr	16000.00
9.25 m²; 3.7 × 2.5 × 2.5 m high	–	–	–	–	–	nr	14000.00
6.25 m²; 2.5 × 2.5 × 2.5 m high	–	–	–	–	–	nr	11500.00
D11 WATERSIDE REVETMENTS							
Waterside revetments; Willowbank							
Services Ltd; natural engineering							
solutions; ecological sustainable waterside							
stabilization systems							
Live willow spiling; 1 m high uprights at 0.50 m							
centres; woven geotextile backed; included							
filling with site based topsoil behind spiling to							
top of bank							
less than 50 m	–	–	–	–	–	m	161.83
50–100 m	–	–	–	–	–	m	126.42
Brushwood faggot and pre-planted coir							
revetments; 900 mm high revetment; 300 mm							
diameter faggot; 300 mm diameter pre-planted							
coir roll; staked in position with 1.8 m tanalized							
posts; 2 posts per m; faggots and coir fixed to							
posts with wire							
less than 50 m	–	–	–	–	–	m	145.02
50–100 m	–	–	–	–	–	m	124.84
Brushwood faggot and pre-planted coir							
revetments; 900 mm high revetment; 2 nr							
300 mm diameter faggot staked in position							
with 1.8 m tanalized posts; 2 posts per m;							
faggots and coir fixed to posts with wire							
less than 50 m	–	–	–	–	–	m	124.48
50–100 m	–	–	–	–	–	m	94.64

NEW ITEMS 2013

Item Excluding site overheads and profit	PC £	Labour hours	Labour £	Plant £	Material £	Unit	Total rate £
F10 MORTAR MIXES							
Mortar mixes for brickwork and pavings; all mortars mixed on site and priced at delivery to waiting wheelbarrows at the cement mixer; cement and lime in 25 kg bags; aggregates as described							
Mortar grade 1:3; aggregates delivered in 20 tonne load; cement: sand volumes shown below (bags: litres)							
85 litre mixer (1.84:93.5)	–	0.75	14.44	–	119.56	m³	**134.00**
150 litre mixer (3.24:165)	–	0.50	9.63	–	119.56	m³	**129.19**
Mortar grade 1:3; aggregates delivered in 10 tonne load; cement: sand volumes shown below (bags: litres)							
85 litre mixer (1.84:93.5)	–	0.75	14.44	–	128.85	m³	**143.29**
150 litre mixer (3.24:165)	–	0.50	9.63	–	128.85	m³	**138.48**
Mortar grade 1:3; aggregates delivered in 850 kg bulk bags; cement: sand volumes shown below (bags: litres)							
85 litre mixer (1.84:93.5)	–	0.75	14.44	–	157.57	m³	**172.01**
150 litre mixer (3.24:165)	–	0.50	9.63	–	157.57	m³	**167.20**
Mortar grade 1:4; aggregates delivered in 20 tonne load; cement: sand volumes shown below (bags: litres)							
85 litre mixer (1.36:102)	–	0.75	14.44	–	99.95	m³	**114.39**
150 litre mixer (2.4:180)	–	0.50	9.63	–	99.95	m³	**109.58**
Mortar grade 1:4; aggregates delivered in 10 tonne load; cement: sand volumes shown below (bags: litres)							
85 litre mixer (1.36:102)	–	0.75	14.44	–	110.08	m³	**124.52**
150 litre mixer (2.4:180)	–	0.50	9.63	–	110.08	m³	**119.71**
Mortar grade 1:4; aggregates delivered in 850 kg bulk bags; cement: sand volumes shown below (bags: litres)							
85 litre mixer (1.36:102)	–	0.75	14.44	–	141.48	m³	**155.92**
150 litre mixer (2.4:180)	–	0.50	9.63	–	141.48	m³	**151.11**
Mortar grade 1:1:6; aggregates delivered in 20 tonne load; cement: lime: sand volumes shown below (bags: bags: litres)							
85 litre mixer;(0.9:0.44:93.5)	–	0.75	14.44	–	128.56	m³	**143.00**
150 litre mixer; (1.6: 0.8: 165)	–	0.50	9.63	–	128.56	m³	**138.19**
Mortar grade 1:1:6; aggregates delivered in 10 tonne load; cement: lime: sand volumes shown below (bags: bags: litres)							
85 litre mixer (0.9:0.44:93.5)	–	0.75	14.44	–	141.64	m³	**156.08**
150 litre mixer (1.6:0.8:165)	–	0.50	9.63	–	141.64	m³	**151.27**
Mortar grade 1:1:6; aggregates delivered in 850 kg bulk bags; cement: lime: sand volumes shown below (bags: bags: litres)							
85 litre mixer (0.9:0.44:93.5)	–	0.75	14.44	–	166.57	m³	**181.01**
150 litre mixer (1.6:0.8:165)	–	0.50	9.63	–	166.57	m³	**176.20**

NEW ITEMS 2013

Item Excluding site overheads and profit	PC £	Labour hours	Labour £	Plant £	Material £	Unit	Total rate £
F10 MORTAR MIXES – cont							
Mortar grade 1:2:9; aggregates delivered in 20 tonne load; cement: lime: sand volumes shown below (bags: bags: litres)							
85 litre mixer (0.7:0.5:102)	–	0.75	14.44	–	128.31	m³	**142.75**
150 litre mixer (1.2:0.9:180)	–	0.50	9.63	–	125.36	m³	**134.99**
Mortar grade 1:2:9; aggregates delivered in 10 tonne load; cement: lime: sand volumes shown below (bags: bags: litres)							
85 litre mixer (0.7:0.5:102)	–	0.75	14.44	–	138.44	m³	**152.88**
150 litre mixer (1.2:0.9:180)	–	0.50	9.63	–	138.44	m³	**148.07**
Mortar grade 1:2:9; aggregates delivered in 850 kg bulk bags; cement: lime: sand volumes shown below (bags: bags: litres)							
85 litre mixer (0.7:0.5:102)	–	0.75	14.44	–	169.84	m³	**184.28**
150 litre mixer (1.2:0.9:180)	–	0.50	9.63	–	169.50	m³	**179.13**
Q20 LOAD SUPPORT (BASES); CELLULAR CONFINEMENT SYSTEMS							
Nidagravel; Cedar Nursery; honeycomb shaped cellular polypropylene containment grids for retaining filled decorative aggregate surfaces (not included) for driveways, cycle paths and footpaths; to graded compacted substrate (not included) and 50 mm sharp sand base; integrated geofabric base							
Foot traffic, wheelchairs, bikes and pushchairs							
Nidagravel 125; 1200 × 800 × 25 mm deep	1.67	0.08	2.41	–	13.67	m²	**16.08**
Driveways: vehicles, wheelchairs, bikes and pushchairs (requires compacted granular base to support vehicle loadings)							
Nidagravel 140; 2400 × 1200 × 40 mm deep	1.67	0.08	2.61	–	16.62	m²	**19.23**
Aggregate filling to Nidagravel 125							
shingle	0.83	0.03	1.92	1.99	0.83	m²	**4.74**
decorative aggregate; CED Ltd; PC £48.00/tonne	2.16	0.03	1.92	1.99	2.16	m²	**6.07**
Aggregate filling to Nidagravel 140							
shingle	–	0.03	1.62	2.42	1.33	m²	**5.37**
decorative aggregate; CED Ltd; PC £48.00/tonne	3.46	0.03	1.62	2.42	3.46	m²	**7.50**

NEW ITEMS 2013

Item Excluding site overheads and profit	PC £	Labour hours	Labour £	Plant £	Material £	Unit	Total rate £
StableDRIVE; Cedar Nursery; honeycomb shaped polypropylene cellular containment system for retaining filled decorative aggregate surfaces for driveways, cyclepaths and footpaths; on compacted subgrade and subbase (both not included)							
Driveways: vehicles, wheelchairs, bikes and pushchairs (requires compacted granular base to support vehicle loadings)							
StableDRIVE; 600 × 500 × 32 mm deep	1.67	0.08	2.77	–	17.62	m²	**20.39**
Aggregate filling to StableDRIVE							
shingle	1.07	0.03	1.81	2.96	1.07	m²	**5.84**
decorative aggregate; CED Ltd;							
PC £48.00/tonne	2.76	0.03	1.81	2.96	2.76	m²	**7.53**
Cellweb; Geosynthetics Ltd; cellular confinement system for load support or permeable bases below pavings							
To graded and compacted substrate (not included); on 50 mm sharp sand base filled with angular drainage aggregate 20–5 mm							
100 mm deep	9.62	0.05	2.89	2.07	17.58	m²	**22.54**
Q31 BRITISH SUGAR TOPSOIL							
Imported topsoil; market prices; British Sugar Topsoil; sustainably sourced graded topsoil to British standards and with independent certification; rates shown allow for 20% settlement							
Landscape 20; BS3882:2007; sandy loam							
8 wheel delivery; 20 tonne loads							
(approximately 13.33 m³)	30.90	–	–	–	30.90	m³	**30.90**
articulated delivery	24.60	–	–	–	24.60	m³	**24.60**
Sports 10; for topdressing to sports pitches; application rate 40–80 t/6000 m² pitch							
8 wheel delivery; 20 tonne loads							
(approximately 13.33 m³)	33.60	–	–	–	33.60	m³	**33.60**
articulated delivery	27.60	–	–	–	27.60	m³	**27.60**
Clay loam for use as a moisture retentive topsoil							
8 wheel delivery; 20 tonne loads							
(approximately 13.33 m³)	30.90	–	–	–	30.90	m³	**30.90**
articulated delivery	–	–	–	–	24.60	m³	**24.60**

NEW ITEMS 2013

Item Excluding site overheads and profit	PC £	Labour hours	Labour £	Plant £	Material £	Unit	Total rate £
Q40 METAL ESTATE FENCING							
Metal estate and parkland fencing; Stonebank Ironcraft Ltd; continously welded, mild steel, 40 × 10 mm uprights manually driven into normal soil to a minimum depth of 600 mm at 1.00 m centres; 40 × 10 mm flat steel supports welded to uprights beneath topsoil at typically every third upright; horizontal rails threaded through holes within uprights; 20 mm solid steel top rail; 4 nr 25 × 8 mm × 16 mm diameter solid steel lower rails; painted fencing shot-blasted, primed and finished with two coats of paint							
Bare metal; 1200 mm high							
10–50 m	–	–	–	–	–	m	65.33
51–100 m	–	–	–	–	–	m	55.17
101–500 m	–	–	–	–	–	m	51.50
above 500 m	–	–	–	–	–	m	49.50
Primed and painted							
10–50 m	–	–	–	–	–	m	78.04
51–100 m	–	–	–	–	–	m	66.08
101–500 m	–	–	–	–	–	m	61.77
above 500 m	–	–	–	–	–	m	59.41
Galvanized and painted							
10–50 m	–	–	–	–	–	m	98.55
51–100 m	–	–	–	–	–	m	83.60
101–500 m	–	–	–	–	–	m	78.21
above 500 m	–	–	–	–	–	m	75.26
End posts and corner posts; 40 × 40 mm; standard hollow section; end or corner posts with capped tops set into concrete							
bare metal	–	–	–	–	–	nr	65.00
primed and painted	–	–	–	–	–	nr	85.00
galvanized and painted	–	–	–	–	–	nr	100.00
Gates for metal estate fencing; Stonebank Ironcraft Ltd; fully welded gates to match fence; scroll at hinge end; turnover at latch end; posts topped with finials and set into concrete							
Pedestrian gates; 1.0–1.5 m wide							
bare metal	–	–	–	–	–	nr	450.00
galvanized and painted	–	–	–	–	–	nr	600.00
Field gates; 1.50–4.0 m wide							
bare metal	–	–	–	–	–	nr	600.00
galvanized and painted	–	–	–	–	–	nr	800.00
Kissing gates							
bare metal	–	–	–	–	–	nr	550.00
galvanized and painted	–	–	–	–	–	nr	745.00

NEW ITEMS 2013

Item Excluding site overheads and profit	PC £	Labour hours	Labour £	Plant £	Material £	Unit	Total rate £
Quadrant; 4 nr uprights with rails welded to fence							
bare metal	–	–	–	–	–	nr	230.00
primed and painted	–	–	–	–	–	nr	300.00
Step stile; 2 treads							
bare metal	–	–	–	–	–	nr	140.00
primed and painted	–	–	–	–	–	nr	200.00
S10 SUSTAINABILITY/ RAINWATER HARVESTING							
Rainwater harvesting tanks							
Notes: Rainwater harvesting tanks must be installed on granular or concrete bases. If installed below ground the tanks may require construction of drained underground chambers to support them. Please see the appropriate sections in this book for excavation, disposal, bases, retaining walls and drainage.							
Excavation for underground tanks; excavation inclusive of earthwork retention for self-supporting tanks							
7 tonne tracked excavator (bucket volume 0.28 m^3)							
maximum depth not exceeding 1.00 m	–	0.06	1.20	4.28	–	m^3	5.48
maximum depth not exceeding 2.00 m	–	0.07	1.37	4.89	–	m^3	6.26
maximum depth not exceeding 3.00 m	–	0.09	1.75	6.23	–	m^3	7.98
Disposal							
Excavated material; off site; to tip; mechanically loaded (JCB)							
inert	–	–	–	–	–	m^3	17.14
Type 1 granular fill base; PC £16.50/tonne (£36.30/m^3 compacted)							
By machine							
100 mm thick	3.63	0.03	0.54	0.40	3.63	m^2	4.57
150 mm thick	5.45	0.03	0.48	0.61	5.45	m^2	6.54
Backfilling to surround of rainwater tank; carefully compacting as work proceeds							
Arising from the excavations							
average thickness exceeding 0.25 m; depositing in layers 150 mm maximum thickness	–	0.03	0.64	5.25	–	m^3	5.89

NEW ITEMS 2013

Item Excluding site overheads and profit	PC £	Labour hours	Labour £	Plant £	Material £	Unit	Total rate £
S10 SUSTAINABILITY/ RAINWATER **HARVESTING – cont**							
Rainwater harvesting tanks; Combined **Harvesters Ltd; self-supporting** **underground or above ground tanks**							
Columbus rainwater tank with max 1.00 m cover in pedestrian areas; complete with tank dome and pedestrian lid, submersible automatic pump system, supra filtration system, 25 m black and green rainwater pipe, rainwater labelling kit, 125 mm fine filter (excavation, backfilling base, trenching and inlet pipework all not included)							
3700 litres; 2.44 × 1.65 × 1.95	1374.00	3.00	115.50	–	1514.00	nr	**1629.50**
4500 litres; 2.44 × 1.84 × 1.84	1508.00	3.50	134.75	–	1648.00	nr	**1782.75**
6500 litres; 2.68 × 2.02 × 2.29	1768.00	4.00	154.00	–	1908.00	nr	**2062.00**
9000 litres; 2.44 × 1.84 × 1.88	2494.00	4.50	173.25	–	2634.00	nr	**2807.25**
13000 litres – 2 × 6500 l tanks;							
2.68 × 4.20 × 2.29	3014.00	7.00	269.50	–	3154.00	nr	**3423.50**
Cristall rainwater tank; lighter weight tank; complete with tank dome and pedestrian lid, submersible automatic pump system, supra filtration system, 25 m black and green rainwater pipe, rainwater labelling kit, 125 mm fine filter (excavation, backfilling base, trenching and inlet pipework all not included)							
1650 litres; 2.10 × 1.05 × 1.22	1069.00	5.00	192.50	–	1209.00	nr	**1401.50**
2650 litres; 2.10 × 1.30 × 1.50	1182.00	6.00	231.00	–	1322.00	nr	**1553.00**
Hercules rainwater tank; above or below ground (pedestrian areas only) installation; extendable to multiple tanks							
1600 litres; 1.35 m diameter × 1.60 m high	1383.00	4.50	173.25	–	1523.00	nr	**1696.25**
Lilo low profile tanks for reduced excavation; complete with rainwater filters and pressure pump							
1500 litres; 2.10 × 1.25 × 1.02	1124.00	–	–	–	1124.00	nr	**1124.00**
3000 litres; 2.46 × 2.1 × 1.05 m	1529.00	6.00	231.00	–	1669.00	nr	**1900.00**
5000 litres; 2.89 × 2.30 × 1.26	1906.00	6.50	250.25	–	2046.00	nr	**2296.25**
10000 litres 2.89 × 4,60 × 1.26 m	3289.00	7.00	269.50	–	3429.00	nr	**3698.50**
Combined Harvesters Ltd; accessories for **rainwater harvesting tanks**							
Downpipe filter for connection to roof downpipes.							
Quattro; for roof areas up to 50 m²	7.88	0.25	4.81	–	7.88	nr	**12.69**
Regendieb; for roof areas up to 80 m²	27.84	0.25	4.81	–	27.84	nr	**32.65**
Regendieb Deluxe; for roof areas up to 100 m²	31.84	–	–	–	31.84	nr	**31.84**

NEW ITEMS 2013

Item Excluding site overheads and profit	PC £	Labour hours	Labour £	Plant £	Material £	Unit	Total rate £
Pumps; inclusive of electrical installation and connections							
Raincatcher direct pump package	318.90	0.25	4.81	–	458.90	nr	**463.71**
Garden Comfort pump package	292.90	0.75	14.44	–	292.90	nr	**307.34**
Filters for rainwater tanks; complete with overflow siphons and rodent guard	153.16	–	–	–	153.16	nr	**153.16**

A PRELIMINARIES

Item Excluding site overheads and profit	PC £	Labour hours	Labour £	Plant £	Material £	Unit	Total rate £
A PRELIMINARIES							
A DESIGN COSTS FOR LANDSCAPE WORKS							
Landscape design; Craft Pegg (landscape architects); mixed hard and soft landscape; indicative prices for design project stages; brief; planning; detailed design and supervision during construction stage; the complexity ratings shown refer to the Landscape Institute tables published in this book							
Commercial urban development; complexity rating 2; landscape construction value							
£100,000.00	–	–	–	–	–	nr	10000.00
£250,000.00	–	–	–	–	–	nr	20000.00
£500,000.00	–	–	–	–	–	nr	35000.00
£1,000,000.00	–	–	–	–	–	nr	65000.00
Hospitals and education; complexity rating 3; landscape construction value							
£100,000.00	–	–	–	–	–	nr	11000.00
£250,000.00	–	–	–	–	–	nr	23000.00
£500,000.00	–	–	–	–	–	nr	40000.00
£1,000,000.00	–	–	–	–	–	nr	74000.00
Public urban open space; complexity rating 4; landscape construction value							
£100,000.00	–	–	–	–	–	nr	12000.00
£250,000.00	–	–	–	–	–	nr	26000.00
£500,000.00	–	–	–	–	–	nr	47500.00
£1,000,000.00	–	–	–	–	–	nr	85000.00
Countryside or leisure open space; complexity rating 1							
£100,000.00	–	–	–	–	–	nr	9000.00
£250,000.00	–	–	–	–	–	nr	18000.00
£500,000.00	–	–	–	–	–	nr	32000.00
£1,000,000.00	–	–	–	–	–	nr	60000.00
Reclamation project; complexity rating 4							
£100,000.00	–	–	–	–	–	nr	12000.00
£250,000.00	–	–	–	–	–	nr	26000.00
£500,000.00	–	–	–	–	–	nr	45000.00
£1,000,000.00	–	–	–	–	–	nr	80000.00
Housing; complexity rating 3							
£100,000.00	–	–	–	–	–	nr	11000.00
£250,000.00	–	–	–	–	–	nr	23000.00
£500,000.00	–	–	–	–	–	nr	40000.00
£1,000,000.00	–	–	–	–	–	nr	74000.00

A PRELIMINARIES

Item Excluding site overheads and profit	PC £	Labour hours	Labour £	Plant £	Material £	Unit	Total rate £
A11 TENDER AND CONTRACT DOCUMENTS							
Note: The preliminary requirements are different for all projects. Many are time related and many are set costs. Others vary depending on the project requirements. We have attempted to provide a guide to the cost related items in Section A of the National Building Specification. These costs are a guide only. Users of the book should evaluate each project separately. There are other preliminary cost items which may be added in to the preliminary costs of any project which may not be featured here.							
Method statements							
Provide detailed method statements on all aspects of the works; project value							
£30,000	–	2.50	48.13	–	–	nr	48.13
£50,000	–	3.50	67.38	–	–	nr	67.38
£75,000	–	4.00	77.00	–	–	nr	77.00
£100,000	–	5.00	96.25	–	–	nr	96.25
£200,000	–	6.00	115.50	–	–	nr	115.50
Tender and contract documents							
Drawing and plan printing and distributioncosts; plans issued by employer on CD; project value							
£30,000	–	–	–	–	19.00	nr	19.00
£30,000 to £80,000	–	–	–	–	47.50	nr	47.50
£80,000 to £150,000	–	–	–	–	95.00	nr	95.00
£150,000 to £300,000	–	–	–	–	133.00	nr	133.00
£300,000 to £1,000,000	–	–	–	–	228.00	nr	228.00
Contract drawings for project management and distribution to suppliers, subcontractors and site staff							
£30,000	–	–	–	–	38.00	nr	38.00
£30,000 to £80,000	–	–	–	–	95.00	nr	95.00
£80,000 to £150,000	–	–	–	–	190.00	nr	190.00
£150,000 to £300,000	–	–	–	–	266.00	nr	266.00
£300,000 to £1,000,000	–	–	–	–	456.00	nr	456.00

A PRELIMINARIES

Item Excluding site overheads and profit	PC £	Labour hours	Labour £	Plant £	Material £	Unit	Total rate £
A12 THE SITE/EXISTING BUILDINGS							
A12/140 – Existing mains and services;							
mark positions of existing mains and							
services; locate and mark; site area							
less than 500 m²	–	1.50	28.88	–	–	nr	28.88
up to 1000 m²	–	3.00	57.75	–	–	nr	57.75
up to 2000 m²	–	8.00	154.00	–	–	nr	154.00
up to 4000 m²	–	16.00	308.00	–	–	nr	308.00
A12/200 – Access to the site							
For pedestrians							
security kiosk	–	–	–	65.00	–	week	65.00
security guard	–	40.00	400.00	–	–	week	400.00
protected walkways; Heras fencing on							
2 sides	–	–	–	4.70	–	m	4.70
A12/210 – Parking							
Parking expenses where vehicles do not park							
on the site area; per vehicle							
metropolitan area; city centre	–	–	–	–	–	week	200.00
metropolitan area; outer areas	–	–	–	–	–	week	160.00
suburban restricted parking areas	–	–	–	–	–	week	40.00
A12/210 – Congestion charging							
London only	–	–	–	–	–	week	40.00
A12/250 – Site visit; pre-tender for							
purposes of understanding site and tender							
requirements; prices below are based on a							
single senior manager attending site							
City location; average distance of travel							
20 miles inclusive of travel and parking costs							
less than 500 m²	–	4.00	77.00	–	36.00	nr	113.00
up to 1000 m²	–	5.00	96.25	–	38.00	nr	134.25
up to 2000 m²	–	6.00	115.50	–	44.00	nr	159.50
up to 4000 m²	–	8.00	154.00	–	52.00	nr	206.00
Town or rural location; average distance of							
travel 20 miles inclusive of travel and parking							
costs							
less than 500 m²	–	3.00	57.75	–	22.00	nr	79.75
up to 1000 m²	–	4.00	77.00	–	23.00	nr	100.00
up to 2000 m²	–	5.00	96.25	–	24.00	nr	120.25
up to 4000 m²	–	6.00	115.50	–	24.00	nr	139.50

A PRELIMINARIES

Item Excluding site overheads and profit	PC £	Labour hours	Labour £	Plant £	Material £	Unit	Total rate £
A20 THE CONTRACT/SUBCONTRACT							
Contract/subcontract evaluation							
Due diligence on evaluation or examination of							
clauses in contract documents; evaluation and							
report by a suitably qualified quantity surveyor							
or legal advisor where necessary							
minor works contract	–	1.00	19.25	–	–	nr	19.25
JCLI	–	–	–	–	70.00	nr	70.00
JCT intermediate contract or subcontract	–	1.00	19.25	–	140.00	nr	159.25
A30 EMPLOYER'S REQUIREMENTS: **TENDERING/SUBLETTING/SUPPLY**							
Tendering costs for an employed estimator **for the acquisition of a landscape main or** **subcontract; inclusive of measurement,** **sourcing of materials and suppliers, cost** **and area calculations, pre and post tender** **meetings, bid compliance method** **statements and submission documents**							
Remeasureable contract from prepared bills;							
contract value							
£25,000	–	–	–	–	–	nr	216.00
£50,000	–	–	–	–	–	nr	360.00
£100,000	–	–	–	–	–	nr	864.00
£200,000	–	–	–	–	–	nr	1440.00
£500,000	–	–	–	–	–	nr	1872.00
£1,000,000	–	–	–	–	–	nr	2880.00
Lump sum contract from specifications and							
drawings only; contract value							
£25,000	–	–	–	–	–	nr	576.00
£50,000	–	–	–	–	–	nr	720.00
£100,000	–	–	–	–	–	nr	1440.00
£200,000	–	–	–	–	–	nr	2592.00
£500,000	–	–	–	–	–	nr	3600.00
£1,000,000	–	–	–	–	–	nr	5760.00
Material costs for tendering							
Drawing and plan printing and							
distributioncosts; plans issued by employer on							
CD; project value							
£30,000	–	–	–	–	19.00	nr	19.00
£30,000 to £80,000	–	–	–	–	47.50	nr	47.50
£80,000 to £150,000	–	–	–	–	95.00	nr	95.00
£150,000 to £300,000	–	–	–	–	133.00	nr	133.00
£300,000 to £1,000,000	–	–	–	–	228.00	nr	228.00

A PRELIMINARIES

Item Excluding site overheads and profit	PC £	Labour hours	Labour £	Plant £	Material £	Unit	Total rate £
A30 EMPLOYER'S REQUIREMENTS: **TENDERING/SUBLETTING/SUPPLY – cont**							
A30/480 – Programmes; tender stage **programmes**							
Allow for production of an outline works programmes for submission with the tenders; project value							
£30,000	–	1.50	28.88	–	–	nr	28.88
£50,000	–	2.00	38.50	–	–	nr	38.50
£75,000	–	2.50	48.13	–	–	nr	48.13
£100,000	–	3.00	57.75	–	–	nr	57.75
£200,000	–	4.00	77.00	–	–	nr	77.00
£500,000	–	4.50	86.63	–	–	nr	86.63
£1,000,000	–	6.00	115.50	–	–	nr	115.50
Method statements for private or commercial projects where award is not points based; provide detailed method statements on all aspects of the works; project value							
£30,000	–	2.50	48.13	–	–	nr	48.13
£50,000	–	3.50	67.38	–	–	nr	67.38
£75,000	–	4.00	77.00	–	–	nr	77.00
£100,000	–	5.00	96.25	–	–	nr	96.25
£200,000	–	6.00	115.50	–	–	nr	115.50
£500,000	–	7.00	134.75	–	–	nr	134.75
£1,000,000	–	8.00	154.00	–	–	nr	154.00
A30/500 – Method statements for local **authority type projects where award is** **points rated; provide detailed method** **statements on all aspects of the works;** **project value**							
£30,000	–	5.00	96.25	–	–	nr	96.25
£50,000	–	5.00	96.25	–	–	nr	96.25
£75,000	–	6.00	115.50	–	–	nr	115.50
£100,000	–	7.00	134.75	–	–	nr	134.75
£200,000	–	8.00	154.00	–	–	nr	154.00
£500,000	–	9.00	173.25	–	–	nr	173.25
£1,000,000	–	16.00	308.00	–	–	nr	308.00
A30/550–570 – Health and safety							
Produce health and safety file including preliminary meeting and subsequent progress meetings with external planning officer in connection with health and safety; project value							
£35,000	–	8.00	154.00	–	–	nr	154.00
£75,000	–	12.00	231.00	–	–	nr	231.00
£100,000	–	16.00	308.00	–	–	nr	308.00
£200,000 to £500,000	–	40.00	770.00	–	–	nr	770.00

A PRELIMINARIES

Item Excluding site overheads and profit	PC £	Labour hours	Labour £	Plant £	Material £	Unit	Total rate £
Maintain health and safety file for project duration; project value							
£35,000	–	4.00	77.00	–	–	week	**77.00**
£75,000	–	4.00	77.00	–	–	week	**77.00**
£100,000	–	8.00	154.00	–	–	week	**154.00**
£200,000 to £500,000	–	8.00	154.00	–	–	week	**154.00**
Produce written risk assessments on all areas of operations within the scope of works of the contract; project value							
£35,000	–	2.00	38.50	–	–	nr	**38.50**
£75,000	–	3.00	57.75	–	–	nr	**57.75**
£100,000	–	5.00	96.25	–	–	nr	**96.25**
£200,000 to £500,000	–	8.00	154.00	–	–	nr	**154.00**
Produce COSHH assessments on all substances to be used in connection with the contract; project value							
£35,000	–	1.50	28.88	–	–	nr	**28.88**
£75,000	–	2.00	38.50	–	–	nr	**38.50**
£100,000	–	3.00	57.75	–	–	nr	**57.75**
£200,000 to £500,000	–	3.00	57.75	–	–	nr	**57.75**
A31 EMPLOYERS REQUIREMENTS: PROVISION CONTENT AND USE OF DOCUMENTS							
Provision, content and use of documents Supply as built drawings for elements of the project that may have carried from the original design drawings							
£35,000	–	2.00	70.00	–	–	nr	**70.00**
£75,000	–	2.00	70.00	–	–	nr	**70.00**
£100,000	–	4.00	140.00	–	–	nr	**140.00**
£200,000 to £500,000	–	5.00	175.00	–	–	nr	**175.00**
A31/155 – Fees for Considerate Contractors Scheme (CCS) Project value							
up to £100,000	–	–	–	–	100.00	nr	**100.00**
£100,000–£500000	–	–	–	–	200.00	nr	**200.00**
£500,000–£5,000,000	–	–	–	–	400.00	nr	**400.00**
over £5,000,000	–	–	–	–	600.00	nr	**600.00**
Climatic conditions – keep records of temperature and rainfall; delays due to weather including descriptions of weather							
daily cost	–	0.08	1.60	–	–	day	**1.60**
weekly cost	–	0.42	8.02	–	–	day	**8.02**

A PRELIMINARIES

Item Excluding site overheads and profit	PC £	Labour hours	Labour £	Plant £	Material £	Unit	Total rate £
A31 EMPLOYERS REQUIREMENTS: **PROVISION CONTENT AND USE OF** **DOCUMENTS – cont**							
A31/211 – Programmes; master programme **for the contract works**							
Allow for production of works programmes							
prior to the start of the works; project value							
£30,000	–	3.00	57.75	–	–	nr	57.75
£50,000	–	6.00	115.50	–	–	nr	115.50
£75,000	–	8.00	154.00	–	–	nr	154.00
£100,000	–	10.00	192.50	–	–	nr	192.50
£200,000	–	14.00	269.50	–	–	nr	269.50
£500,000	–	15.00	288.75	–	–	nr	288.75
£1,000,000	–	18.00	346.50	–	–	nr	346.50
A31/212 – Indicative staff resource chart							
Project value							
up to £100000	–	1.00	19.25	–	–	nr	19.25
up to £500000	–	1.50	28.88	–	–	nr	28.88
up to £1,000,000	–	2.00	38.50	–	–	nr	38.50
A31/213 – Curriculum vitae of staff							
Prepare and submit curriculum vitae of							
pre-construction and construction phase staff							
per staff member	–	0.75	14.44	–	–	nr	14.44
A32 EMPLOYER'S REQUIREMENTS: **MANAGEMENT OF THE WORKS**							
Allow for updating the works programme							
during the course of the works; project value							
£30,000	–	1.00	19.25	–	–	nr	19.25
£50,000	–	1.50	28.88	–	–	nr	28.88
£75,000	–	2.00	38.50	–	–	nr	38.50
£100,000	–	3.00	57.75	–	–	nr	57.75
£200,000	–	5.00	96.25	–	–	nr	96.25
Setting out							
Setting out for external works operations							
comprising hard and soft works elements;							
placing of pegs and string lines to Landscape							
Architect's drawings; surveying levels and							
placing level pegs; obtaining approval from the							
Landscape Architect to commence works;							
areas of entire site							
1,000 m²	–	5.00	96.25	–	7.80	nr	104.05
2,500 m²	–	8.00	154.00	–	15.60	nr	169.60
5,000 m²	–	8.00	154.00	–	15.60	nr	169.60
10,000 m²	–	32.00	616.00	–	39.00	nr	655.00

A PRELIMINARIES

Item Excluding site overheads and profit	PC £	Labour hours	Labour £	Plant £	Material £	Unit	Total rate £
A33 SUBMISSION OF SAMPLES							
Set up sample panels for the works							
inclusive of arranging delivery of sample							
materials; paving samples on 100 mm base							
Costs of labours only							
brick paving panel (pointed)	–	6.00	115.50	–	–	m²	**115.50**
brick paving panel (butt jointed)	–	4.00	77.00	–	–	m²	**77.00**
block paving panel	–	4.00	77.00	–	–	m²	**77.00**
stone slab paving panel	–	6.00	115.50	–	–	m²	**115.50**
cladding panel	–	7.00	134.75	–	–	m²	**134.75**
brick wall panel	–	5.00	96.25	–	–	m²	**96.25**
render panel	–	3.50	67.38	–	–	m²	**67.38**
paint panel	–	3.00	57.75	–	–	m²	**57.75**
A34 EMPLOYER'S REQUIREMENTS:							
SECURITY/SAFETY/PROTECTION							
Temporary security fence; HSS Hire; mesh							
framed unclimbable fencing; including							
precast concrete supports and couplings							
Weekly hire; 2.85 × 2.00 m high							
weekly hire rate	–	–	–	2.35	–	m	**2.35**
erection of fencing; labour only	–	0.10	1.93	–	–	m	**1.93**
removal of fencing loading to collection							
vehicle	–	0.07	1.28	–	–	m	**1.28**
delivery charge	–	–	–	0.80	–	m	**0.80**
return haulage charge	–	–	–	0.60	–	m	**0.60**
A41 CONTRACTOR'S GENERAL COST							
ITEMS: SITE ACCOMODATION							
General							
The following items are instances of the							
commonly found preliminary costs associated							
with external works contracts. The assumption							
is made that the external works contractor is							
subcontracted to a main contractor.							
Elliot Hire; erect temporary							
accommodation and storage on concrete							
base measured separately							
Prefabricated office hire; jackleg; open plan							
3.6 × 2.4 m	–	–	–	26.00	–	week	**26.00**
4.8 × 2.4 m	–	–	–	30.00	–	week	**30.00**
Armoured store; erect temporary secure							
storage container for tools and equipment							
3.0 × 2.4 m	–	–	–	12.00	–	week	**12.00**
3.6 × 2.4 m	–	–	–	12.00	–	week	**12.00**
Delivery and collection charges on site offices							
delivery charge	–	–	–	140.00	–	load	**140.00**
collection charge	–	–	–	140.00	–	load	**140.00**

A PRELIMINARIES

Item Excluding site overheads and profit	PC £	Labour hours	Labour £	Plant £	Material £	Unit	Total rate £
A41 CONTRACTOR'S GENERAL COST ITEMS: SITE ACCOMODATION – cont							
Toilet facilities							
HSS Hire; serviced self-contained toilet delivered to and collected from site; maintained by toilet supply company							
single chemical toilet including wash hand basin and water	–	–	–	27.27	–	week	**27.27**
delivery and collection; each way	–	–	–	25.25	–	nr	**25.25**
A43 CONTRACTOR'S GENERAL COST ITEMS: MECHANICAL PLANT							
Southern Conveyors; moving only of granular material by belt conveyor; conveyors fitted with troughed belts, receiving hopper, front and rear undercarriages and driven by electrical and air motors; support work for conveyor installation (scaffolding), delivery, collection and installation all excluded							
Conveyor belt width 400 mm; mechanically loaded and removed at offload point; conveyor length							
up to 5 m	–	1.50	28.88	19.98	–	m^3	**48.86**
10 m	–	1.50	28.88	20.44	–	m^3	**49.32**
12.5 m	–	1.50	28.88	20.99	–	m^3	**49.87**
15 m	–	1.50	28.88	21.38	–	m^3	**50.26**
20 m	–	1.50	28.88	22.32	–	m^3	**51.20**
25 m	–	1.50	28.88	23.49	–	m^3	**52.37**
30 m	–	1.50	28.88	24.66	–	m^3	**53.54**
Conveyor belt width 600 mm; mechanically loaded and removed at offload point; conveyor length							
up to 5 m	–	1.00	19.25	13.58	–	m^3	**32.83**
10 m	–	1.00	19.25	14.10	–	m^3	**33.35**
12.5 m	–	1.00	19.25	14.47	–	m^3	**33.72**
15 m	–	1.00	19.25	14.88	–	m^3	**34.13**
20 m	–	1.00	19.25	15.66	–	m^3	**34.91**
25 m	–	1.00	19.25	16.97	–	m^3	**36.22**
30 m	–	1.00	19.25	18.01	–	m^3	**37.26**
Conveyor belt width 400 mm; mechanically loaded and removed by hand at offload point; conveyor length							
up to 5 m	–	3.19	61.42	5.62	–	m^3	**67.04**
10 m	–	3.19	61.42	6.25	–	m^3	**67.67**
12.5 m	–	3.19	61.42	6.98	–	m^3	**68.40**
15 m	–	3.19	61.42	7.50	–	m^3	**68.92**
20 m	–	3.19	61.42	8.75	–	m^3	**70.17**
25 m	–	3.19	61.42	10.31	–	m^3	**71.73**
30 m	–	3.19	61.42	11.87	–	m^3	**73.29**

A PRELIMINARIES

Item Excluding site overheads and profit	PC £	Labour hours	Labour £	Plant £	Material £	Unit	Total rate £
Terranova Cranes Ltd; crane hire; materials handling and lifting; telescopic cranes supply and management; exclusive of roadway management or planning applications; prices below illustrate lifts of 1 tonne at maximum crane reach; lift cycle of 0.25 hours; mechanical filling of material skip if appropriate							
35 tonne mobile crane; lifting capacity of 1 tonne at 26 m; C.P.A hire (all management by hirer)							
granular materials including concrete	–	0.75	14.44	15.81	–	tonne	**30.25**
palletized or packed materials	–	0.50	9.63	13.59	–	tonne	**23.22**
35 tonne mobile crane; lifting capacity of 1 tonne at 26 m; contract lift							
granular materials or concrete	–	0.75	14.44	44.28	–	tonne	**58.72**
palletized or packed materials	–	0.50	9.63	42.06	–	tonne	**51.69**
50 tonne mobile crane; lifting capacity of 1 tonne at 34 m; C.P.A hire (all management by hirer)							
granular materials including concrete	–	0.75	14.44	20.02	–	tonne	**34.46**
palletized or packed materials	–	1.00	19.25	17.80	–	tonne	**37.05**
50 tonne mobile crane; lifting capacity of 1 tonne at 34 m; contract lift							
granular materials or concrete	–	0.75	14.44	50.69	–	tonne	**65.13**
palletized or packed materials	–	0.50	9.63	48.47	–	tonne	**58.10**
80 tonne mobile crane; lifting capacity of 1 tonne at 44 m; C.P.A. hire (all management by hirer)							
granular materials including concrete	–	0.75	14.44	28.11	–	tonne	**42.55**
palletized or packed materials	–	0.50	9.63	25.88	–	tonne	**35.51**
80 tonne mobile crane; lifting capacity of 1 tonne at 44 m; contract lift							
granular materials or concrete	–	0.75	14.44	57.23	–	tonne	**71.67**
palletized or packed materials	–	0.50	9.63	55.00	–	tonne	**64.63**
A44 CONTRACTOR'S GENERAL COST ITEMS: TEMPORARY WORKS							
Eve Trakway; portable roadway systems Temporary roadway system laid directly onto existing surface or onto PVC matting to protect existing surface; most systems are based on a weekly hire charge with transportation, installation and recovery charges included							
Heavy Duty Trakpanel; per panel (3.05 × 2.59 m per week	–	–	–	–	–	m²	**5.32**
Outrigger Mats for use in conjunction with Heavy Duty Trakpanels; per set of 4 mats per week	–	–	–	–	–	set	**85.00**
Medium Duty Trakpanel; per panel (2.44 × 3.00 m per week	–	–	–	–	–	m²	**5.74**
LD20 Eveolution; Light Duty Trakway; roll out system minimum delivery 50 m	–	–	–	–	–	m²	**5.25**
Terraplas Walkways; turf protection system; per section (1 × 1 m); per week	–	–	–	–	–	m²	**5.50**

B COMPLETE BUILDINGS/STRUCTURES/UNITS

Item Excluding site overheads and profit	PC £	Labour hours	Labour £	Plant £	Material £	Unit	Total rate £
B COMPLETE BUILDINGS/STRUCTURES/ UNITS							
B10 PREFABRICATED BUILDINGS/ STRUCTURES							
Cast stone buildings; Haddonstone Ltd; ornamental garden buildings in Portland Bath or Terracotta finished cast stone; prices for stonework and facades only; excavations, foundations, reinforcement, concrete infill, roofing and floors all priced separately							
Pavilion Venetian Folly L9400; Tuscan columns, pedimented arch, quoins and optional balustrading							
4184 mm high × 4728 mm wide × 3147 mm deep	14100.00	175.00	3368.75	241.22	14256.08	nr	**17866.05**
Pavilion L9300; Tuscan columns							
3496 mm high × 3634 mm wide	7298.00	144.00	2772.00	206.76	7454.08	nr	**10432.84**
Small Classical Temple L9250; 6 column with fibreglass lead effect finish dome roof							
overall height 3610 mm; diameter 2540 mm	6300.00	130.00	2502.50	172.30	6456.08	nr	**9130.88**
Large Classical Temple L9100; 8 column with fibreglass lead effect finish dome roof							
overall height 4664 mm; diameter 3190 mm	11823.00	165.00	3176.25	172.30	11979.08	nr	**15327.63**
Stepped floors to temples							
single step; Small Classical Temple	1160.00	24.00	462.00	–	1222.26	nr	**1684.26**
single step; Large Classical Temple	1638.00	26.00	500.50	–	1712.71	nr	**2213.21**
Stone structures; Architectural Heritage Ltd; hand carved from solid natural limestone with wrought iron domed roof, decorated frieze and base and integral seats; supply and erect only; excavations and concrete bases priced separately							
The Park Temple; 5 columns; 3500 mm high × 1650 mm diameter	9800.00	96.00	1848.00	75.00	9800.00	nr	**11723.00**
The Estate Temple; 6 columns; 4000 mm high × 2700 mm diameter	17600.00	120.00	2310.00	150.00	17600.00	nr	**20060.00**
Stone structures; Architectural Heritage Ltd; hand carved from solid natural limestone with solid oak trelliage; prices for stonework and facades only; supply and erect only; excavations and concrete bases priced separately							
The Pergola; 2240 mm high × 2640 mm wide × 6990 mm long	14000.00	96.00	1848.00	300.00	14032.41	nr	**16180.41**

B COMPLETE BUILDINGS/STRUCTURES/UNITS

Item Excluding site overheads and profit	PC £	Labour hours	Labour £	Plant £	Material £	Unit	Total rate £
Ornamental timber buildings; Architectural Heritage Ltd; hand built in English oak with fire retardant wheat straw thatch; hand painted, fired lead glass windows The Thatched Edwardian Summer House; 3710 mm high × 3540 mm wide × 2910 mm deep overall; 2540 × 1910 mm internal	22000.00	32.00	616.00	–	22000.00	nr	**22616.00**
Ornamental stone structures; Architectural Heritage Ltd; setting to bases or plinths (not included) The Obelisk; classic natural stone obelisk; tapering square form on panelled square base; 1860 mm high × 360 mm square	1200.00	2.00	38.50	–	1200.00	nr	**1238.50**
The Narcissus Column; natural limestone column on pedestal surmounted by bronze sculpture; overall height 3440 mm with 440 mm square base	3200.00	2.00	38.50	–	3200.00	nr	**3238.50**
The Armillary Sundial; artificial stone baluster pedestal base surmounted by verdigris copper armillary sphere calibrated sundial	2200.00	1.00	19.25	–	2200.00	nr	**2219.25**
B10 GARDEN STUDIOS							
Garden rooms; 3rd Space; Sawhorse Ltd; fully insulated pent roof modular garden buildings installed on granular base with jacking legs Douglas fir framed with western red cedar cladding; glazed with fully glazed doors; integral floor and internal finishes including birch ply cladding							
12.5 m² ; 5.0 × 2.5 × 2.5 m high	–	–	–	–	–	nr	**16000.00**
9.25 m² ; 3.7 × 2.5 × 2.5 m high	–	–	–	–	–	nr	**14000.00**
6.25 m² ; 2.5 × 2.5 × 2.5 m high	–	–	–	–	–	nr	**11500.00**
15.25 m² ; 6.5 × 2.5 × 2.5 m high	–	–	–	–	–	nr	**18330.00**

C EXISTING SITE/BUILDINGS/SERVICES

Item Excluding site overheads and profit	PC £	Labour hours	Labour £	Plant £	Material £	Unit	Total rate £
C EXISTING SITE/BUILDINGS/SERVICES							
C12 UNDERGROUND SERVICES SURVEY							
Ground penetrating radar surveys for artefacts, services and cable avoidance							
Areas not exceeding 1000 m^2	–	–	–	–	–	nr	**4900.00**
C13 BUILDING FABRIC SURVEY							
Surveys							
Asbestos surveys							
type 3 asbestos survey (destructive testing) to single storey free-standing buildings such as park buildings, storage sheds and the like; total floor area less than 100 m^2	–	–	–	–	–	nr	**950.00**
extra over for the removal only of asbestos off site to tip (tipping charges only)	–	–	–	–	–	tonne	**80.00**
C20 DEMOLITION							
Demolish existing structures; disposal off site; mechanical demolition; with 3 tonne excavator and dumper							
Brick wall							
112.5 mm thick	–	0.07	1.28	2.32	3.42	m^2	**7.02**
225 mm thick	–	0.08	1.60	2.90	6.83	m^2	**11.33**
337.5 mm thick	–	–	–	9.28	11.46	m^2	**20.74**
450 mm thick	–	–	–	12.38	13.66	m^2	**26.04**
Demolish existing structures; disposal off site mechanically loaded; all other works by hand							
Brick wall							
112.5 mm thick	–	0.33	6.42	3.56	3.42	m^2	**13.40**
225 mm thick	–	0.50	9.63	–	6.83	m^2	**16.46**
337.5 mm thick	–	0.67	12.83	–	11.46	m^2	**24.29**
450 mm thick	–	1.00	19.25	–	13.66	m^2	**32.91**
Demolish existing structures; disposal off site mechanically loaded; by diesel or electric breaker; all other works by hand							
Brick wall							
112.5 mm thick	–	0.17	3.21	0.34	3.42	m^2	**6.97**
225 mm thick	–	0.20	3.85	0.40	6.83	m^2	**11.08**
337.5 mm thick	–	0.25	4.81	0.50	10.13	m^2	**15.44**
450 mm thick	–	0.33	6.42	0.67	13.66	m^2	**20.75**

C EXISTING SITE/BUILDINGS/SERVICES

Item Excluding site overheads and profit	PC £	Labour hours	Labour £	Plant £	Material £	Unit	Total rate £
Break out concrete footings associated with free-standing walls; inclusive of all excavation and backfilling with excavated material; disposal off site mechanically loaded							
By mechanical breaker; diesel or electric							
plain concrete	–	1.50	28.87	11.42	17.14	m^3	**57.43**
reinforced concrete	–	2.50	48.13	16.12	30.37	m^3	**94.62**
Remove existing free-standing buildings; demolition by hand							
Timber building with suspended timber floor; hardstanding or concrete base not included; disposal off site							
shed 6.0 m^2	–	2.00	38.50	–	31.63	nr	**70.13**
shed 10.0 m^2	–	3.00	57.76	–	56.94	nr	**114.70**
shed 15.0 m^2	–	3.50	67.38	–	88.57	nr	**155.95**
Timber building; insulated; with timber or concrete posts set in concrete, felt covered timber or tiled roof; internal walls cladding with timber or plasterboard; load arisings to skip							
timber structure 6.0 m^2	–	2.50	48.13	–	136.64	nr	**184.77**
timber structure 12.0 m^2	–	3.50	67.38	–	191.05	nr	**258.43**
timber structure 20.0 m^2	–	8.00	154.00	323.72	202.44	nr	**680.16**
Demolition of free-standing brick buildings with tiled or sheet roof; concrete foundations measured separately; mechanical demolition; maximum distance to stockpile 25 m; inclusive for all access scaffolding and the like; maximum height of roof 4.0 m; inclusive of all doors, windows, guttering and down pipes; including disposal by grab							
Half brick thick							
10 m^2	–	8.00	154.00	163.88	147.39	nr	**465.27**
20 m^2	–	16.00	308.00	293.25	189.56	nr	**790.81**
1 brick thick							
10 m^2	–	10.00	192.50	258.78	294.78	nr	**746.06**
20 m^2	–	16.00	308.00	388.26	379.12	nr	**1075.38**
Cavity wall with blockwork inner skin and brick outer skin; insulated							
10 m^2	–	12.00	231.09	353.79	409.19	nr	**994.07**
20 m^2	–	20.00	385.00	483.33	527.17	nr	**1395.50**

C EXISTING SITE/BUILDINGS/SERVICES

Item Excluding site overheads and profit	PC £	Labour hours	Labour £	Plant £	Material £	Unit	Total rate £
C20 DEMOLITION – cont							
Extra over to the above for disconnection of services							
Electrical							
disconnection	–	2.00	38.50	–	–	nr	**38.50**
grub out cables and dispose; backfilling; by machine	–	–	–	0.62	0.51	m	**1.13**
grub out cables and dispose; backfilling; by hand	–	0.50	9.63	–	0.51	m	**10.14**
Water supply, foul or surface water drainage							
disconnection; capping off	–	1.00	19.25	–	38.50	nr	**57.75**
grub out pipes and dispose; backfilling; by machine	–	–	–	0.62	0.51	m	**1.13**
grub out pipes and dispose; backfilling; by hand	–	0.50	9.63	–	0.51	m	**10.14**
Demolish existing fence; remove stakes or grub out posts as appropriate; remove to stock pile for removal off site; (not included)							
Mechanical demolition; clear fence line							
chain link fence 1.20–1.50 m high	–	0.01	0.39	0.70	–	m	**1.09**
closeboard or timber panel fence 1.80 m high	–	0.03	0.96	1.31	–	m	**2.27**
Mechanical demolition; light shrubs or creepers in chainlinks							
chain link fence 1.20–1.50 m high	–	0.01	0.48	0.89	–	m	**1.37**
closeboard or timber panel fence 1.80 m high	–	0.03	1.28	1.76	–	m	**3.04**
Mechanical demolition; heavy shrubs or creepers bramble and the like requiring clearance to enable removal							
fence 1.20–1.50 m high	–	0.05	1.93	2.17	–	m	**4.10**
closeboard or timber panel fence 1.80 m high	–	0.05	1.93	2.84	–	m	**4.77**
Demolition and site transport by hand; clear fence lines							
fence 1.20–1.50 m high	–	0.03	1.28	–	–	m	**1.28**
closeboard or timber panel fence 1.80 m high	–	0.07	2.57	–	–	m	**2.57**
Demolition and site transport by hand; light vegetation and creepers							
fence 1.20–1.50 m high	–	0.05	1.93	–	–	m	**1.93**
closeboard or timber panel fence 1.80 m high	–	0.13	4.81	–	–	m	**4.81**
Demolition and site transport by hand; heavy shrubs or creepers bramble and the like requiring clearance to enable removal							
fence 1.20–1.50 m high	–	0.13	4.81	–	–	m	**4.81**
closeboard or timber panel fence 1.80 m high	–	0.17	6.42	–	–	m	**6.42**

C EXISTING SITE/BUILDINGS/SERVICES

Item Excluding site overheads and profit	PC £	Labour hours	Labour £	Plant £	Material £	Unit	Total rate £
Demolition of posts and straining posts							
Break out straining post; grub out concrete							
footings; remove to stockpile on site							
single straining post by machine	–	0.25	4.81	16.79	–	nr	21.60
double straining post by machine	–	0.30	5.78	19.56	–	nr	25.34
single straining post by hand	–	1.00	38.50	–	–	nr	38.50
double straining post by hand	–	1.25	48.13	–	–	nr	48.13
Break out gatepost; concrete steel or timber;							
girth not exceeding 150 mm; grub out concrete							
footings							
by machine	–	0.25	4.81	16.79	–	nr	21.60
by hand	–	1.00	19.25	–	–	nr	19.25
C50 REPAIRING/RENOVATING/							
CONSERVING METAL							
Eura Conservation Ltd; conservation of							
metal railings; works to heritage							
conservation standards							
Taking down and transporting existing							
metalwork to an off site workshop for							
conservation							
pair of gates; maximum overall width							
4.00 m	–	–	–	–	–	pair	596.00
side screens to gates	–	–	–	–	–	pair	485.00
railings; 1.20 m high; plain	–	–	–	–	–	m	52.00
railings; 1.80 m high; with finials	–	–	–	–	–	m	98.00
ornate cast or wrought iron railings	–	–	–	–	–	m	108.00
Conserving metalwork off site; inclusive of							
repairs to metalwork; rust removal, rubbing							
down and preparing for re-erection on site							
gates; 2.50 m high; width not exceeding							
4.00 m	–	–	–	–	–	m	1100.00
railings; 1.20 m high; plain	–	–	–	–	–	m	145.00
railings; 1.80 m high; with finials	–	–	–	–	–	m	220.00
ornate cast or wrought iron railings	–	–	–	–	–	m	435.00
Transporting from store and re-erection of							
existing gates into locations recorded with							
position orientation and features of the original							
installation							
gates; 2.50 m high; width not exceeding							
4.00 m	–	–	–	–	–	nr	1018.00
Refixing existing railings previously taken							
down and repaired on site							
railings; 1.20 m high; plain	–	–	–	–	–	m	104.00
railings; 1.80 m high; with finials	–	–	–	–	–	m	200.00
ornate cast or wrought iron railings	–	–	–	–	–	m	248.00

C EXISTING SITE/BUILDINGS/SERVICES

Item Excluding site overheads and profit	PC £	Labour hours	Labour £	Plant £	Material £	Unit	Total rate £
C50 REPAIRING/RENOVATING/ **CONSERVING METAL – cont**							
Eura Conservation Ltd; supply of **conservation grade railings to match** **existing materials on site; inclusive of all** **surveys, measurements and analysis of** **materials, installation methods and the like**							
Wavy bar railings							
1.20 m high; plain	–	–	–	–	–	m	**97.00**
1.70 m high; straight	–	–	–	–	–	m	**310.00**
1.70 m high; curved	–	–	–	–	–	m	**375.00**
1.80 m high; with finials	–	–	–	–	–	m	**300.00**
ornate cast or wrought iron railings	–	–	–	–	–	m	**825.00**
Painting of railings: see section M60							

D11 SOIL STABILIZATION

Item Excluding site overheads and profit	PC £	Labour hours	Labour £	Plant £	Material £	Unit	Total rate £
D11 SOIL STABILIZATION							
CLARIFICATION NOTES ON LABOUR COSTS IN THIS SECTION							
General groundworks team Generally a three man team is used in this section; The column 'Labour hours' reports team hours. The column 'Labour £' reports the total cost of the team for the unit of work shown							
3 man team	–	1.00	57.75	–	–	hr	**57.75**
GENERALLY							
Soil stabilization – General Preamble: Earth-retaining and stabilizing materials are often specified as part of the earth-forming work in landscape contracts, therefore this section lists a number of products specially designed for large-scale earth control. There are two types: rigid units for structural retention of earth on steep slopes; and flexible meshes and sheets for control of soil erosion where structural strength is not required. Prices for these items depend on quantity, difficulty of access to the site and availability of suitable filling material; estimates should be obtained from the manufacturer when the site conditions have been determined.							
RETAINING WALLS							
Preamble Earth retaining and stabilization can take numerous forms depending on the height, soil type and loadings. The main types include gravity walls, reinforced soil and soil nailing. Prices for these items depend on quantity, access for the installation and the strength of the soils. Estimates should be obtained from the manufacturer when site conditions and layout have been determined.							
Retaining walls; Milton Precast Concrete Retaining walls of units with plain concrete finish; prices based on 24 tonne loads but other quantities available (excavation, temporary shoring, foundations and backfilling not included)							
1000 mm wide × 1000 mm high	106.00	0.50	28.88	5.59	144.23	m	**178.70**
1000 mm wide × 1500 mm high	131.00	0.42	24.43	5.59	177.98	m	**208.00**
1000 mm wide × 2000 mm high	237.00	0.42	24.46	5.59	302.30	m	**332.35**
1000 mm wide × 2500 mm high	321.00	0.42	24.45	5.59	411.18	m	**441.22**
1000 mm wide × 3000 mm high	379.00	0.42	24.43	52.50	480.70	m	**557.63**

D11 SOIL STABILIZATION

Item Excluding site overheads and profit	PC £	Labour hours	Labour £	Plant £	Material £	Unit	Total rate £
RETAINING WALLS – cont							
Retaining walls; Tensar International							
Tensartech TW1 retaining wall system;							
modular dry laid concrete blocks;							
220 × 400 mm long × 150 mm high connected							
to Tensar RE geogrid with proprietary							
connectors; geogrid laid horizontally within the							
fill at 300 mm centres; on 150 × 450 mm							
concrete foundation; filling with imported							
granular material							
1.00 m high	84.05	1.00	57.75	10.31	91.12	m²	**159.18**
2.00 m high	84.05	1.33	77.00	10.31	91.12	m²	**178.43**
3.00 m high	84.05	1.50	86.63	10.31	91.12	m²	**188.06**
Retaining walls; Grass Concrete Ltd							
Betoflor precast concrete landscape retaining							
walls including soil filling to pockets							
(excavation, concrete foundations, backfilling							
stones to rear of walls and planting not							
included)							
Betoflor interlocking units; 250 mm long ×							
250 × 200 mm modular deep; in walls							
250 mm wide	–	–	–	–	–	m²	**85.50**
extra over Betoflor interlocking units for							
colours	–	–	–	–	–	m²	**7.99**
Betoatlas earth retaining walls; 250 mm							
long × 500 mm wide × 200 mm modular							
deep; in walls 500 mm wide	–	–	–	–	–	m²	**137.28**
extra over Betoatlas interlocking units for							
colours	–	–	–	–	–	m²	**15.38**
Retaining walls; Forticrete Ltd							
Keystone precast concrete block retaining							
wall; geogrid included for walls over 1.00 m							
high; excavation, concrete foundation, stone							
backfill to rear of wall all measured separately							
1.0 m high	88.15	0.80	46.20	–	88.15	m²	**134.35**
2.0 m high	88.15	0.80	46.20	8.62	101.27	m²	**156.09**
3.0 m high	88.15	0.80	46.20	11.49	110.39	m²	**168.08**
4.0 m high	88.15	0.80	46.20	13.78	117.45	m²	**177.43**
Retaining walls; Forticrete Ltd							
Stepoc Blocks; interlocking blocks; 10 mm							
reinforcing laid loose horizontally to preformed							
notches and vertical reinforcing nominal size							
10 mm fixed to starter bars; infilling with							
concrete; foundations and starter bars							
measured separately							
type 325; 400 × 225 × 325 mm	47.08	0.47	26.95	–	66.68	m²	**93.63**
type 256; 400 × 225 × 256 mm	44.61	0.40	23.10	–	62.56	m²	**85.66**
type 190; 400 × 225 × 200 mm	34.21	0.33	19.25	–	48.84	m²	**68.09**

D11 SOIL STABILIZATION

Item Excluding site overheads and profit	PC £	Labour hours	Labour £	Plant £	Material £	Unit	Total rate £
Timber log retaining walls; AVS Fencing Supplies Ltd; timber posts are kiln dried redwood with 15 year guarantee							
Machine rounded softwood logs to trenches priced separately; disposal of excavated material priced separately; inclusive of 75 mm hardcore blinding to trench and backfilling trench with site mixed concrete 1:3:6; geofabric pinned to rear of logs; heights of logs above ground							
500 mm (constructed from 1.80 m lengths)	30.85	0.50	28.88	–	40.48	m	**69.36**
1.20 m (constructed from 1.80 m lengths)	61.70	0.43	25.02	–	83.67	m	**108.69**
1.60 m (constructed from 2.40 m lengths)	82.40	0.83	48.11	–	109.96	m	**158.07**
2.00 m (constructed from 3.00 m lengths)	103.00	1.17	67.37	–	136.16	m	**203.53**
As above but with 150 mm machine rounded timbers							
500 mm	42.28	0.83	48.11	–	51.91	m	**100.02**
1.20 m	140.79	0.58	33.69	–	162.76	m	**196.45**
1.60 m	140.93	1.00	57.75	–	168.50	m	**226.25**
As above but with 200 mm machine rounded timbers							
1.80 m (constructed from 2.40 m lengths)	187.90	1.33	77.00	–	221.06	m	**298.06**
Railway sleeper walls; AVS Fencing Supplies Ltd; retaining wall from railway sleepers; fixed with steel galvanized pins 12 mm driven into the ground; sleepers laid flat							
Grade 1 softwood; 2590 × 250 × 125 mm							
150 mm; 1 sleeper high	7.49	0.17	9.62	–	7.91	m	**17.53**
300 mm; 2 sleepers high	14.98	0.33	19.25	–	15.83	m	**35.08**
450 mm; 3 sleepers high	22.29	0.50	28.88	–	23.40	m	**52.28**
600 mm; 4 sleepers high	30.03	0.67	38.50	–	31.14	m	**69.64**
Grade 1 softwood as above but with 2 nr galvanized angle iron stakes set into concrete internally and screwed to the inside face of the sleepers							
750 mm; 5 sleepers high	37.15	0.83	48.12	–	65.74	m²	**113.86**
900 mm; 6 sleepers high	44.47	0.92	52.94	–	76.04	m²	**128.98**
Grade 1 hardwood; 2590 × 250 × 150 mm							
150 mm; 1 sleeper high	7.70	0.17	9.62	–	8.13	m	**17.75**
300 mm; 2 sleepers high	15.41	0.33	19.25	–	16.26	m	**35.51**
450 mm; 3 sleepers high	22.93	0.50	28.88	–	24.04	m	**52.92**
600 mm; 4 sleepers high	30.89	0.67	38.50	–	32.00	m	**70.50**
Grade 1 hardwood as above but with 2 nr galvanized angle iron stakes set into concrete internally and screwed to the inside face of the sleepers							
750 mm; 5 sleepers high	38.21	0.83	48.12	–	66.80	m²	**114.92**
900 mm; 6 sleepers high	45.74	0.92	52.94	–	77.31	m²	**130.25**

D11 SOIL STABILIZATION

Item Excluding site overheads and profit	PC £	Labour hours	Labour £	Plant £	Material £	Unit	Total rate £
RETAINING WALLS – cont							
Railway sleeper walls – cont							
New pine softwood; 2400 × 250 × 125 mm							
120 mm; 1 sleeper high	8.14	0.17	9.63	–	8.56	m	**18.19**
240 mm; 2 sleepers high	16.28	0.33	19.25	–	17.13	m	**36.38**
360 mm; 3 sleepers high	24.42	0.50	28.88	–	25.53	m	**54.41**
480 mm; 4 sleepers high	32.56	0.67	38.50	–	33.67	m	**72.17**
New pine softwood as above but with 2 nr							
galvanized angle iron stakes set into concrete							
internally and screwed to the inside face of the							
sleepers							
600 mm; 5 sleepers high	40.70	0.83	48.12	–	69.29	m²	**117.41**
720 mm; 6 sleepers high	48.84	0.92	52.94	–	80.41	m²	**133.35**
New oak hardwood 2600 × 220 × 130 mm							
130 mm; 1 sleeper high	10.78	0.17	9.62	–	11.20	m	**20.82**
260 mm; 2 sleepers high	21.56	0.33	19.25	–	22.41	m	**41.66**
390 mm; 3 sleepers high	32.34	5.00	288.75	–	33.45	m	**322.20**
520 mm; 4 sleepers high	43.12	0.67	38.50	–	44.23	m	**82.73**
New oak hardwood as above but with 2 nr							
galvanized angle iron stakes set into concrete							
internally and screwed to the inside face of the							
sleepers							
640 mm; 5 sleepers high	53.90	0.83	48.12	–	82.49	m²	**130.61**
760 mm; 6 sleepers high	64.68	9.00	519.75	–	96.25	m²	**616.00**
Excavate foundation trench; set railway							
sleepers vertically on end in concrete 1:3:6							
continuous foundation to 33.3% of their							
length to form retaining wall							
Grade 1 hardwood; finished height above							
ground level							
300 mm	20.49	1.00	57.75	1.88	25.48	m	**85.11**
500 mm	29.70	1.00	57.75	1.88	34.69	m	**94.32**
600 mm	41.28	1.00	57.75	1.88	50.50	m	**110.13**
750 mm	44.55	1.17	67.38	1.88	53.76	m	**123.02**
1.00 m	50.74	1.25	72.19	1.88	60.79	m	**134.86**
Excavate and place vertical steel universal							
beams 165 mm wide in concrete base at							
2.59 m centres; fix railway sleepers set							
horizontally between beams to form							
horizontal fence or retaining wall							
Grade 1 hardwood; bay length 2.59 m							
500 mm high; 2 sleepers	15.05	0.83	48.12	–	57.72	bay	**105.84**
750 mm high; 3 sleepers	22.12	0.87	50.05	–	85.78	bay	**135.83**
1.00 m high; 4 sleepers	29.64	1.00	57.75	–	111.14	bay	**168.89**
1.25 m high; 5 sleepers	37.85	1.17	67.38	–	142.82	bay	**210.20**
1.50 m high; 6 sleepers	44.46	1.00	57.75	–	170.08	bay	**227.83**
1.75 m high; 7 sleepers	51.76	1.00	57.75	–	177.38	bay	**235.13**

D11 SOIL STABILIZATION

Item Excluding site overheads and profit	PC £	Labour hours	Labour £	Plant £	Material £	Unit	Total rate £
EMBANKMENTS							
Embankments; Grass Concrete Ltd							
Grasscrete; in situ reinforced concrete surfacing; to 20 mm thick sand blinding layer (not included); including soiling and seeding							
GC1; 100 mm thick	–	–	–	–	41.99	m²	**41.99**
GC2; 150 mm thick	–	–	–	–	55.10	m²	**55.10**
Grassblock 103; solid matrix precast concrete blocks; to 20 mm thick sand blinding layer; excluding edge restraint; including soiling and seeding							
406 × 406 × 103 mm; fully interlocking	–	–	–	–	38.80	m²	**38.80**
Grass reinforcement; Farmura Environmental Ltd							
Matrix grass paver; recycled polyethylene and polypropylene mixed interlocking erosion control and grass reinforcement system laid to rootzone prepared separately and filled with screened topsoil and seeded with grass seed; green							
Golpla; 640 × 330 × 38 mm	13.50	0.04	2.41	–	15.96	m²	**18.37**
Embankments; Cooper Clarke Civils and Lintels							
Ecoblock polyethylene; 925 × 310 × 50 mm; heavy duty for car parking and fire paths							
to firm subsoil (not included)	19.57	0.01	0.77	–	30.07	m²	**30.84**
to 100 mm granular fill and geotextile	19.57	0.03	1.54	0.13	34.47	m²	**36.14**
to 250 mm granular fill and geotextile	19.57	0.07	3.85	0.22	40.01	m²	**44.08**
Extra for filling Ecoblock with topsoil; seeding with rye grass at 50 g/m²	1.70	0.07	3.87	0.18	1.70	m²	**5.75**
Neoweb polyethylene soil stabilizing panels; to soil surfaces brought to grade (not included); filling with excavated material							
panels; 2.5 × 8.0 m × 100 mm deep	8.22	0.03	1.92	2.56	8.22	m²	**12.70**
panels; 2.5 × 8.0 m × 200 mm deep	16.48	0.04	2.56	3.06	16.48	m²	**22.10**
Neoweb polyethylene soil stabilizing panels; to soil surfaces brought to grade (not included); filling with ballast							
panels; 2.5 × 8.0 m × 100 mm deep	8.22	0.04	2.14	2.56	10.12	m²	**14.82**
panels; 2.5 × 8.0 m × 200 mm deep	16.48	0.05	2.96	3.06	20.29	m²	**26.31**
Neoweb polyethylene soil stabilizing panels; to soil surfaces brought to grade (not included); filling with ST2 10 N/mm² concrete							
panels; 2.5 × 8.0 m × 100 mm deep	8.22	0.05	3.08	–	15.81	m²	**18.89**
panels; 2.5 × 8.0 m × 200 mm deep	16.48	0.07	3.81	–	31.66	m²	**35.47**
Extra over for filling Neoweb with imported topsoil; seeding with rye grass at 50 g/m²	3.53	0.02	1.28	0.30	3.53	m²	**5.11**

D11 SOIL STABILIZATION

Item Excluding site overheads and profit	PC £	Labour hours	Labour £	Plant £	Material £	Unit	Total rate £
EMBANKMENTS – cont							
Willow walling to riverbanks; LDC Ltd							
Woven willow walling as retention to							
riverbanks; driving or concreting posts in at							
2 m centres; intermediate posts at 500 mm							
centres							
1.20 m high	–	–	–	–	–	m	**132.67**
1.50 m high	–	–	–	–	–	m	**169.13**
Flexible sheet materials; Tensar							
International							
Tensar Mat 400 erosion mats; 3.0–4.5 m wide;							
securing with Tensar pegs; lap rolls 100 mm;							
anchors at top and bottom of slopes; in trenches	4.40	0.01	0.32	–	4.40	m²	**4.72**
Topsoil filling to Tensar Mat; including brushing							
and raking	0.88	0.01	0.32	0.08	0.88	m²	**1.28**
Tensar Bi-axial Geogrid; to graded compacted							
base; filling with 200 mm granular fill;							
compacting (turf or paving to surfaces not							
included); 400 mm laps							
TriAx 150; 39 × 39 mm mesh	1.40	–	0.26	0.11	6.87	m²	**7.24**
TriAx 160; 39 × 39 mm mesh	1.82	–	0.26	0.11	7.29	m²	**7.66**
TriAx 170; 39 × 39 mm mesh	2.15	–	0.26	0.11	7.62	m²	**7.99**
Flexible sheet materials; Terram Ltd							
Terram synthetic fibre filter fabric; to graded							
base (not included)							
Terram 1000; 0.70 mm thick; mean water							
flow 50 l/m²/s	0.41	0.01	0.32	–	0.41	m²	**0.73**
Terram 2000; 1.00 mm thick; mean water							
flow 33 l/m²/s	0.62	0.01	0.32	–	0.62	m²	**0.94**
Terram Minipack	0.71	–	0.06	–	0.71	m²	**0.77**
Flexible sheet materials; Greenfix Ltd							
Greenfix; erosion control mats; 10–15 mm							
thick; fixing with 4 nr crimped pins in							
accordance with manufacturer's instructions;							
to graded surface (not included)							
unseeded Eromat 1; 2.40 m wide	103.47	0.67	38.50	–	167.47	100 m²	**205.97**
unseeded Eromat 2; 2.40 m wide	130.69	0.67	38.50	–	194.69	100 m²	**233.19**
unseeded Eromat 3; 2.40 m wide	147.03	0.67	38.50	–	211.03	100 m²	**249.53**
seeded Covamat 1; 2.40 m wide	166.25	0.67	38.50	–	230.25	100 m²	**268.75**
seeded Covamat 2; 2.40 m wide	210.00	0.67	38.50	–	274.00	100 m²	**312.50**
seeded Covamat 3; 2.40 m wide	227.50	0.67	38.50	–	291.50	100 m²	**330.00**
Bioroll; 300 mm diameter; to river banks and							
revetments	19.00	0.04	2.41	–	21.18	m	**23.59**

D11 SOIL STABILIZATION

Item Excluding site overheads and profit	PC £	Labour hours	Labour £	Plant £	Material £	Unit	Total rate £
Extra over Greenfix erosion control mats for fertilizer applied at 70 g/m²	9.05	0.07	3.85	–	9.05	100 m²	**12.90**
Extra over Greenfix erosion control mats for Geojute; fixing with steel pins	1.32	0.01	0.43	–	1.32	m²	**1.75**
Extra over Greenfix erosion control mats for laying to slopes exceeding 30°	–	–	–	–	–	25%	**–**
Extra for the following operations							
Spreading 25 mm approved topsoil							
by machine	0.84	–	0.12	0.13	0.84	m²	**1.09**
by hand	0.84	0.01	0.68	–	0.84	m²	**1.52**
Grass seed; PC £4.50/kg; spreading in two operations; by hand							
35 g/m²	15.75	0.06	3.21	–	15.75	100 m²	**18.96**
50 g/m²	22.50	0.06	3.21	–	22.50	100 m²	**25.71**
70 g/m²	31.50	0.06	3.21	–	31.50	100 m²	**34.71**
100 g/m²	45.00	0.07	3.85	–	45.00	100 m²	**48.85**
125 g/m²	56.25	0.67	38.50	–	56.25	100 m²	**94.75**
Extra over seeding by hand for slopes over 30° (allowing for the actual area but measured in plan)							
35 g/m²	2.34	–	0.03	–	2.34	100 m²	**2.37**
50 g/m²	3.38	–	0.03	–	3.38	100 m²	**3.41**
70 g/m²	4.72	–	0.03	–	4.72	100 m²	**4.75**
100 g/m²	6.75	–	0.03	–	6.75	100 m²	**6.78**
125 g/m²	8.41	–	0.03	–	8.41	100 m²	**8.44**
Grass seed; PC £4.50/kg; spreading in two operations; by machine							
35 g/m²	15.75	–	–	0.51	15.75	100 m²	**16.26**
50 g/m²	22.50	–	–	0.51	22.50	100 m²	**23.01**
70 g/m²	31.50	–	–	0.51	31.50	100 m²	**32.01**
100 g/m²	45.00	–	–	0.51	45.00	100 m²	**45.51**
125 kg/ha	562.50	–	–	50.95	562.50	ha	**613.45**
150 kg/ha	675.00	–	–	50.95	675.00	ha	**725.95**
200 kg/ha	900.00	–	–	50.95	900.00	ha	**950.95**
250 kg/ha	1125.00	–	–	50.95	1125.00	ha	**1175.95**
300 kg/ha	1350.00	–	–	50.95	1350.00	ha	**1400.95**
350 kg/ha	1575.00	–	–	50.95	1575.00	ha	**1625.95**
400 kg/ha	1800.00	–	–	50.95	1800.00	ha	**1850.95**
500 kg/ha	2250.00	–	–	50.95	2250.00	ha	**2300.95**
700 kg/ha	3150.00	–	–	50.95	3150.00	ha	**3200.95**
1400 kg/ha	6300.00	–	–	50.95	6300.00	ha	**6350.95**
Extra over seeding by machine for slopes over 30° (allowing for the actual area but measured in plan)							
35 g/m²	2.36	–	–	0.08	2.36	100 m²	**2.44**
50 g/m²	3.38	–	–	0.08	3.38	100 m²	**3.46**
70 g/m²	4.72	–	–	0.08	4.72	100 m²	**4.80**
100 g/m²	6.75	–	–	0.08	6.75	100 m²	**6.83**
125 kg/ha	84.38	–	–	7.64	84.38	ha	**92.02**
150 kg/ha	101.25	–	–	7.64	101.25	ha	**108.89**
200 kg/ha	135.00	–	–	7.64	135.00	ha	**142.64**

D11 SOIL STABILIZATION

Item Excluding site overheads and profit	PC £	Labour hours	Labour £	Plant £	Material £	Unit	Total rate £
EMBANKMENTS – cont							
Extra for the following operations – cont							
Extra over seeding by machine for slopes over							
30° (allowing for the actual area but measured							
in plan) – cont							
250 kg/ha	168.75	–	–	7.64	168.75	ha	**176.39**
300 kg/ha	202.50	–	–	7.64	202.50	ha	**210.14**
350 kg/ha	236.25	–	–	7.64	236.25	ha	**243.89**
400 kg/ha	270.00	–	–	7.64	270.00	ha	**277.64**
500 kg/ha	337.50	–	–	7.64	337.50	ha	**345.14**
700 kg/ha	472.50	–	–	7.64	472.50	ha	**480.14**
1400 kg/ha	945.00	–	–	7.64	945.00	ha	**952.64**
REVETMENTS							
Waterside revetments; Willowbank							
Services Ltd; natural engineering							
solutions; ecological sustainable waterside							
stabilization systems							
Live willow spiling; 1 m high uprights at 0.50 m							
centres; woven geotextile backed; included							
filling with site based topsoil behind spiling to							
top of bank							
less than 50 m	–	–	–	–	–	m	**161.83**
50–100 m	–	–	–	–	–	m	**126.42**
Brushwood faggot and pre-planted coir							
revetments; 900 mm high revetment; 300 mm							
diameter faggot; 300 mm diameter pre-planted							
coir roll; staked in position with 1.8 m tanalized							
posts; 2 posts per m; faggots and coir fixed to							
posts with wire							
less than 50 m	–	–	–	–	–	m	**145.02**
50–100 m	–	–	–	–	–	m	**124.84**
Brushwood faggot and pre-planted coir							
revetments; 900 mm high revetment; 2 nr							
300 mm diameter faggot staked in position							
with 1.8 m tanalized posts; 2 posts per m;							
faggots and coir fixed to posts with wire							
less than 50 m	–	–	–	–	–	m	**124.48**
50–100 m	–	–	–	–	–	m	**94.64**

D20 EXCAVATION AND FILLING

Item Excluding site overheads and profit	PC £	Labour hours	Labour £	Plant £	Material £	Unit	Total rate £
D20 EXCAVATION AND FILLING							
MACHINE SELECTION							
MACHINE SELECTION TABLE							
Road Equipment Ltd; machine volumes for excavating/filling only and placing excavated material alongside or to a dumper; no bulkages are allowed for in the material volumes; these rates should be increased by user-preferred percentages to suit prevailing site conditions; the figures in the next section for 'Excavation mechanical' and filling allow for the use of banksmen within the rates shown below							
1.5 tonne excavators: digging volume							
1 cycle/minute; $0.04\,m^3$	–	0.42	8.02	2.40	–	m^3	**10.42**
2 cycles/minute; $0.08\,m^3$	–	0.21	4.01	1.70	–	m^3	**5.71**
3 cycles/minute; $0.12\,m^3$	–	0.14	2.67	1.48	–	m^3	**4.15**
3 tonne excavators; digging volume							
1 cycle/minute; $0.13\,m^3$	–	0.13	2.47	1.67	–	m^3	**4.14**
2 cycles/minute; $0.26\,m^3$	–	0.06	1.23	3.13	–	m^3	**4.36**
3 cycles/minute; $0.39\,m^3$	–	0.04	0.82	1.31	–	m^3	**2.13**
5 tonne excavators; digging volume							
1 cycle/minute; $0.28\,m^3$	–	0.06	1.15	2.48	–	m^3	**3.63**
2 cycles/minute; $0.56\,m^3$	–	0.03	0.57	2.37	–	m^3	**2.94**
3 cycles/minute; $0.84\,m^3$	–	0.02	0.38	2.51	–	m^3	**2.89**
7 tonne excavators; supplied with operator; digging volume							
1 cycle/minute; $0.28\,m^3$	–	0.06	1.15	4.05	–	m^3	**5.20**
2 cycles/minute; $0.56\,m^3$	–	0.03	0.57	1.98	–	m^3	**2.55**
3 cycles/minute; $0.84\,m^3$	–	0.02	0.38	2.42	–	m^3	**2.80**
21 tonne excavators; supplied with operator; digging volume							
1 cycle/minute; $1.21\,m^3$	–	–	–	0.83	–	m^3	**0.83**
2 cycles/minute; $2.42\,m^3$	–	–	–	0.34	–	m^3	**0.34**
3 cycles/minute; $3.63\,m^3$	–	–	–	0.23	–	m^3	**0.23**
Backhoe loader; excavating; JCB 3 CX rear bucket capacity $0.28\,m^3$							
1 cycle/minute; $0.28\,m^3$	–	–	–	2.23	–	m^3	**2.23**
2 cycles/minute; $0.56\,m^3$	–	–	–	1.12	–	m^3	**1.12**
3 cycles/minute; $0.84\,m^3$	–	–	–	0.74	–	m^3	**0.74**
Backhoe loader; loading from stockpile; JCB 3 CX front bucket capacity $1.00\,m^3$							
1 cycle/minute; $1.00\,m^3$	–	–	–	0.63	–	m^3	**0.63**
2 cycles/minute; $2.00\,m^3$	–	–	–	0.31	–	m^3	**0.31**

D20 EXCAVATION AND FILLING

Item Excluding site overheads and profit	PC £	Labour hours	Labour £	Plant £	Material £	Unit	Total rate £
MACHINE SELECTION – cont							
Note: All volumes below are based on excavated 'earth' moist at 1,997 kg/m³ solid or 1,598 kg/m³ loose; a 25% bulkage factor has been used; the weight capacities below exceed the volume capacities of the machine in most cases; see the memorandum section at the back of this book for further weights of materials.							
Dumpers; Road Equipment Ltd							
1 tonne high tip skip loader; volume 0.485 m³ (775 kg)							
5 loads per hour	–	0.41	7.94	1.61	–	m³	9.55
7 loads per hour	–	0.29	5.67	1.22	–	m³	6.89
10 loads per hour	–	0.21	3.97	0.94	–	m³	4.91
2.5 tonne dumper; excavated material volume 1.3 m³							
4 loads per hour	–	–	–	5.25	–	m³	5.25
5 loads per hour	–	–	–	4.20	–	m³	4.20
7 loads per hour	–	–	–	3.73	–	m³	3.73
10 loads per hour	–	–	–	2.05	–	m³	2.05
6 tonne dumper; maximum volume 3.40 m³ (5.4 t); available volume 3.77 m³							
4 loads per hour	–	0.07	1.28	0.33	–	m³	1.61
5 loads per hour	–	0.05	1.02	0.28	–	m³	1.30
7 loads per hour	–	0.04	0.73	0.22	–	m³	0.95
10 loads per hour	–	0.03	0.51	0.18	–	m³	0.69
CLARIFICATION NOTES ON LABOUR COSTS IN THIS SECTION							
General groundworks team							
Generally a three man team is used in this section; The column 'Labour hours' reports team hours. The column 'Labour £' reports the total cost of the team for the unit of work shown							
3 man team	–	1.00	57.75	–	–	hr	57.75
banksman	–	1.00	19.25	–	–	hr	19.25
SITE PREPARATION AND CLEARANCE							
Site preparation							
Felling and removing trees off site							
girth 600 mm–1.50 m (95–240 mm trunk diameter)	–	2.50	144.38	19.04	–	nr	163.42
girth 1.50–3.00 m (240–475 mm trunk diameter)	–	10.00	577.50	76.15	–	nr	653.65
girth 3.00–4.00 m (475–630 mm trunk diameter)	–	16.00	924.00	106.62	–	nr	1030.62

D20 EXCAVATION AND FILLING

Item Excluding site overheads and profit	PC £	Labour hours	Labour £	Plant £	Material £	Unit	Total rate £
Removing tree stumps							
girth 600 mm–1.50 m	–	2.00	38.50	82.50	–	nr	**121.00**
girth 1.50–3.00 m	–	7.00	134.75	78.20	–	nr	**212.95**
girth over 3.00 m	–	12.00	231.00	134.06	–	nr	**365.06**
Stump grinding; disposing to spoil heaps							
girth 600 mm–1.50 m	–	2.00	38.50	32.20	–	nr	**70.70**
girth 1.50–3.00 m	–	2.50	48.13	64.41	–	nr	**112.54**
girth over 3.00 m	–	4.00	77.00	56.35	–	nr	**133.35**
Clearing site vegetation							
mechanical clearance	–	0.25	4.81	13.59	–	100 m²	**18.40**
hand clearance	–	2.00	38.50	–	–	100 m²	**38.50**
Lifting turf for preservation							
sod cutter machine lift and stack	–	0.75	14.44	9.96	–	100 m²	**24.40**
hand lift and stack	–	8.33	160.41	–	–	100 m²	**160.41**
Lifting turf for disposal (disposal not included)							
by excavator; 5 tonne	–	–	–	41.41	–	100 m²	**41.41**
by excavator; 7 tonne	–	–	–	31.93	–	100 m²	**31.93**
by excavator; 7 tonne; working with two dumpers	–	–	–	29.61	–	100 m²	**29.61**
by excavator; 7 tonne; working with two dumpers	–	–	–	25.06	–	100 m²	**25.06**
hand lift and stack	–	8.33	160.41	–	–	100 m²	**160.41**
Site clearance; by machine; clear site of mature shrubs from existing cultivated beds; dig out roots by machine							
Mixed shrubs in beds; planting centres 500 mm average							
height less than 1 m	–	0.03	0.48	0.78	–	m²	**1.26**
1.00–1.50 m	–	0.04	0.77	1.25	–	m²	**2.02**
1.50–2.00 m; pruning to ground level by hand	–	0.10	1.93	3.13	–	m²	**5.06**
2.00–3.00 m; pruning to ground level by hand	–	0.10	1.93	6.27	–	m²	**8.20**
3.00–4.00 m; pruning to ground level by hand	–	0.20	3.85	10.45	–	m²	**14.30**
Site clearance; by hand; clear site of mature shrubs from existing cultivated beds; dig out roots							
Mixed shrubs in beds; planting centres 500 mm average							
height less than 1 m	–	0.33	6.42	–	–	m²	**6.42**
1.00–1.50 m	–	0.50	9.63	–	–	m²	**9.63**
1.50–2.00 m	–	1.00	19.25	–	–	m²	**19.25**
2.00–3.00 m	–	2.00	38.50	–	–	m²	**38.50**
3.00–4.00 m	–	3.00	57.76	–	–	m²	**57.76**

D20 EXCAVATION AND FILLING

Item Excluding site overheads and profit	PC £	Labour hours	Labour £	Plant £	Material £	Unit	Total rate £
SITE PREPARATION AND CLEARANCE – **cont**							
Mechanical ground clearance to large **areas; Cleartrack (Evl) Limited; vegetation** **clearance of high capacity mobile mulching** **machines; 500 HP; 2.50 m head width;** **tracked; wheeled; skid steer and 360°** **operating**							
Site clearance and in situ mulching and incorporating to soil surface to existing area of scrub and small trees up to 150 mm diameter; cleared mulched material left on surface for clearance to stockpile (not included)							
Areas of existing grass, bramble and weed; average height not exceeding 1.00 m	–	–	–	–	–	100 m²	**10.00**
Areas of existing small shrubs and weed; average height not exceeding 1.80 m	–	–	–	–	–	100 m²	**15.00**

Areas of existing grass, bramble, weed and
  regular shrubs; varying heights to 3.00 m;
  max stem diameter 100 mm
	–	–	–	–	–	100 m²	**21.00**

Areas of existing grass, bramble, weed,
  regular shrubs and small trees; varying
  heights up and over 3.00 m; max stem
  diameter 150 mm
	–	–	–	–	–	100 m²	**54.00**

Spraying of vegetation; Glyphosate at 5l/ha by tractor drawn spray equipment	–	–	–	–	–	100 m²	**2.50**
Excavated material; on site							
In spoil heaps							
average 25 m distance	–	–	–	2.87	–	m³	**2.87**
average 50 m distance	–	–	–	3.31	–	m³	**3.31**
average 100 m distance (1 dumper)	–	–	–	4.14	–	m³	**4.14**
average 100 m distance (2 dumpers)	–	–	–	4.22	–	m³	**4.22**
average 200 m distance	–	–	–	5.00	–	m³	**5.00**
average 200 m distance (2 dumpers)	–	–	–	5.08	–	m³	**5.08**
EXCAVATING							
Market prices of topsoil; prices shown **include for 20% settlement**							
Multiple source screened topsoil	–	–	–	–	33.60	m³	**33.60**
Single source topsoil; British Sugar PLC	–	–	–	–	33.60	m³	**33.60**
High grade topsoil for planting	–	–	–	–	60.00	m³	**60.00**

D20 EXCAVATION AND FILLING

Item Excluding site overheads and profit	PC £	Labour hours	Labour £	Plant £	Material £	Unit	Total rate £
Note: The figures in this section relate to the machine capacities shown earlier in this section. The figures below however allow for dig efficiency based on depth. A banksman is allowed for in all excavation build and disposal costs shown. Bulkages of 25% allowed; adjustments should be made for different soil types.							
Excavating; mechanical; topsoil for preservation							
3 tonne tracked excavator (bucket volume 0.13 m³)							
average depth 100 mm	–	0.02	0.32	1.09	–	m²	**1.41**
average depth 150 mm	–	0.02	0.37	1.23	–	m²	**1.60**
average depth 200 mm	–	–	–	1.64	–	m²	**1.64**
average depth 250 mm	–	0.03	0.54	1.82	–	m²	**2.36**
average depth 300 mm	–	0.03	0.64	2.18	–	m²	**2.82**
JCB Sitemaster 3CX (bucket volume 0.28 m³)							
average depth 100 mm	–	1.00	19.25	41.25	–	100 m²	**60.50**
average depth 150 mm	–	1.25	24.06	51.56	–	100 m²	**75.62**
average depth 200 mm	–	1.78	34.27	73.42	–	100 m²	**107.69**
average depth 250 mm	–	1.90	36.58	78.38	–	100 m²	**114.96**
average depth 300 mm	–	2.00	38.50	82.50	–	100 m²	**121.00**
Excavating; mechanical; to reduce levels							
5 tonne excavator (bucket volume 0.28 m³)							
maximum depth not exceeding 0.25 m	–	0.07	1.35	3.89	–	m³	**5.24**
maximum depth not exceeding 1.00 m	–	0.05	0.92	2.64	–	m³	**3.56**
maximum depth not exceeding 2.00 m	–	0.06	1.15	3.30	–	m³	**4.45**
JCB Sitemaster 3CX (bucket volume 0.28 m³)							
maximum depth not exceeding 0.25 m	–	0.07	1.35	2.89	–	m³	**4.24**
maximum depth not exceeding 1.00 m	–	0.06	1.15	2.45	–	m³	**3.60**
7 tonne tracked excavator (bucket volume 0.28 m³)							
maximum depth not exceeding 1.00 m	–	0.06	1.20	4.28	–	m³	**5.48**
maximum depth not exceeding 2.00 m	–	0.07	1.37	4.89	–	m³	**6.26**
maximum depth not exceeding 3.00 m	–	0.09	1.75	6.23	–	m³	**7.98**
21 tonne 360 tracked excavator (bucket volume 1.21 m³)							
maximum depth not exceeding 1.00 m	–	0.01	0.21	0.55	–	m³	**0.76**
maximum depth not exceeding 2.00 m	–	0.02	0.32	0.82	–	m³	**1.14**
maximum depth not exceeding 3.00 m	–	0.03	0.64	1.67	–	m³	**2.31**
Pits; 3 tonne tracked excavator							
maximum depth not exceeding 0.25 m	–	0.33	6.42	15.43	–	m³	**21.85**
maximum depth not exceeding 1.00 m	–	0.25	4.81	11.57	–	m³	**16.38**
maximum depth not exceeding 2.00 m	–	0.40	7.70	18.52	–	m³	**26.22**
Trenches; width not exceeding 0.30 m; 3 tonne excavator							
maximum depth not exceeding 0.25 m	–	1.00	19.25	4.50	–	m³	**23.75**
maximum depth not exceeding 1.00 m	–	0.69	13.19	3.08	–	m³	**16.27**
maximum depth not exceeding 2.00 m	–	0.60	11.55	2.70	–	m³	**14.25**

D20 EXCAVATION AND FILLING

Item Excluding site overheads and profit	PC £	Labour hours	Labour £	Plant £	Material £	Unit	Total rate £
EXCAVATING – cont							
Excavating – cont							
Trenches; width exceeding 0.30 m; 3 tonne excavator							
maximum depth not exceeding 0.25 m	–	0.60	11.55	2.70	–	m³	14.25
maximum depth not exceeding 1.00 m	–	0.50	9.63	2.25	–	m³	11.88
maximum depth not exceeding 2.00 m	–	0.38	7.40	1.73	–	m³	9.13
Extra over any types of excavating irrespective of depth for breaking out existing materials; JCB with breaker attachment							
hard rock	–	0.50	9.63	78.13	–	m³	87.76
concrete	–	0.50	9.63	29.30	–	m³	38.93
reinforced concrete	–	1.00	19.25	48.47	–	m³	67.72
brickwork, blockwork or stonework	–	0.25	4.81	29.30	–	m³	34.11
Extra over any types of excavating irrespective of depth for breaking out existing hard pavings; JCB with breaker attachment							
concrete; 100 mm thick	–	–	–	2.15	–	m²	2.15
concrete; 150 mm thick	–	–	–	3.58	–	m²	3.58
concrete; 200 mm thick	–	–	–	4.30	–	m²	4.30
concrete; 300 mm thick	–	–	–	6.45	–	m²	6.45
reinforced concrete; 100 mm thick	–	0.08	1.60	2.54	–	m²	4.14
reinforced concrete; 150 mm thick	–	0.08	1.44	3.46	–	m²	4.90
reinforced concrete; 200 mm thick	–	0.10	1.93	4.61	–	m²	6.54
reinforced concrete; 300 mm thick	–	0.15	2.89	6.92	–	m²	9.81
tarmacadam; 75 mm thick	–	–	–	2.15	–	m²	2.15
tarmacadam and hardcore; 150 mm thick	–	–	–	3.44	–	m²	3.44
Extra over any types of excavating irrespective of depth for taking up							
precast concrete paving slabs	–	0.07	1.28	0.71	–	m²	1.99
natural stone paving	–	0.10	1.93	1.06	–	m²	2.99
cobbles	–	0.13	2.41	1.32	–	m²	3.73
brick paviors	–	0.13	2.41	1.32	–	m²	3.73
Excavating; hand							
Topsoil for preservation; loading to barrows							
average depth 100 mm	–	0.08	4.57	–	–	m²	4.57
average depth 150 mm	–	0.12	6.86	–	–	m²	6.86
average depth 200 mm	–	0.19	10.98	–	–	m²	10.98
average depth 250 mm	–	0.24	13.72	–	–	m²	13.72
average depth 300 mm	–	0.29	16.46	–	–	m²	16.46
Excavating; hand							
Topsoil to reduce levels							
maximum depth not exceeding 0.25 m	–	0.79	45.74	–	–	m³	45.74
maximum depth not exceeding 1.00 m	–	1.03	59.46	–	–	m³	59.46
Pits							
maximum depth not exceeding 0.25 m	–	0.88	50.82	–	–	m³	50.82
maximum depth not exceeding 1.00 m	–	1.14	66.07	–	–	m³	66.07
maximum depth not exceeding 2.00 m (includes earthwork support)	–	2.29	132.13	–	–	m³	188.44

D20 EXCAVATION AND FILLING

Item Excluding site overheads and profit	PC £	Labour hours	Labour £	Plant £	Material £	Unit	Total rate £
Trenches; width not exceeding 0.30 m							
maximum depth not exceeding 0.25 m	–	0.94	54.45	–	–	m³	54.45
maximum depth not exceeding 1.00 m	–	1.23	70.91	–	–	m³	70.91
maximum depth not exceeding 2.00 m							
(includes earthwork support)	–	1.23	70.91	–	–	m³	99.06
Trenches; width exceeding 0.30 m wide							
maximum depth not exceeding 0.25 m	–	0.94	54.45	–	–	m³	54.45
maximum depth not exceeding 1.00 m	–	1.32	76.23	–	–	m³	76.23
maximum depth not exceeding 2.00 m							
(includes earthwork support)	–	1.98	114.34	–	–	m³	170.65
Extra over any types of excavating irrespective							
of depth for breaking out existing materials;							
hand held pneumatic breaker							
rock	–	1.65	95.29	35.32	–	m³	130.61
concrete	–	0.82	47.64	17.66	–	m³	65.30
reinforced concrete	–	1.32	76.23	32.96	–	m³	109.19
brickwork, blockwork or stonework	–	0.50	28.59	10.60	–	m³	39.19
FILLING							
Filling to make up levels; mechanical (JCB rear bucket)							
Arising from the excavations							
average thickness not exceeding 0.25 m	–	0.07	1.41	2.76	–	m³	4.17
average thickness less than 500 mm	–	0.06	1.20	2.34	–	m³	3.54
average thickness 1.00 m	–	0.05	0.89	1.74	–	m³	2.63
Obtained from on site spoil heaps; average							
25 m distance; multiple handling							
average thickness less than 250 mm	–	0.07	1.37	3.26	–	m³	4.63
average thickness less than 500 mm	–	0.06	1.13	2.68	–	m³	3.81
average thickness 1.00 m	–	0.05	0.96	2.28	–	m³	3.24
Obtained off site; planting quality topsoil PC							
£28.00/m³							
average thickness less than 250 mm	–	0.07	1.37	3.26	33.60	m³	38.23
average thickness less than 500 mm	–	0.06	1.13	2.68	33.60	m³	37.41
average thickness 1.00 m	–	0.05	0.96	2.28	33.60	m³	36.84
Obtained off site; hardcore; PC £24.00/m³							
average thickness less than 250 mm	–	0.08	1.60	3.80	21.60	m³	27.00
average thickness less than 500 mm	–	0.06	1.20	2.85	21.60	m³	25.65
average thickness 1.00 m	–	0.05	1.01	2.40	21.60	m³	25.01
Filling to make up levels; hand							
Arising from the excavations							
average thickness exceeding 0.25 m;							
depositing in layers 150 mm maximum							
thickness	–	0.60	11.55	–	–	m³	11.55
Obtained from on site spoil heaps; average							
25 m distance; multiple handling							
average thickness exceeding 0.25 m thick;							
depositing in layers 150 mm maximum							
thickness	–	1.00	19.25	–	–	m³	19.25

D20 EXCAVATION AND FILLING

Item Excluding site overheads and profit	PC £	Labour hours	Labour £	Plant £	Material £	Unit	Total rate £
DISPOSAL							
Disposal							
Note: Most commercial site disposal is carried out by 20 tonne, 8 wheeled vehicles. It has been customary to calculate disposal from construction sites in terms of full 15 m³ loads. Spon's research has found that based on weights of common materials such as clean hardcore and topsoil, vehicles could not load more than 12 m³ at a time. Most hauliers do not make it apparent that their loads are calculated by weight and not by volume. The rates below reflect these lesser volumes which are limited by the 20 tonne limit.							
The volumes shown below are based on volumes 'in the solid'. Weight and bulking factors have been applied. For further information please see the weights of typical materials in the Earthworks section of the Memoranda at the back of this book.							
Disposal; mechanical (all rates include banksman)							
Excavated material; off site; to tip; mechanically loaded (JCB)							
inert (clean landfill, rubble concrete); loose	–	0.03	0.54	–	17.14	m³	**17.68**
inert (clean excavated material);							
compacted	–	0.03	0.54	–	21.43	m³	**21.97**
hardcore	–	0.02	0.40	0.78	14.40	m³	**15.58**
macadam	–	0.02	0.40	0.78	18.00	m³	**19.18**
clean concrete	–	0.03	0.54	1.04	10.80	m³	**12.38**
broken out compacted materials such as							
road bases and the like	–	0.03	0.54	1.04	25.72	m³	**27.30**
Light soils and loams (bulking factor of 1.25); 40 tonnes (2 loads per hour)							
soil (sandy and loam); dry	–	0.03	0.54	–	20.57	m³	**21.11**
soil (sandy and loam); wet	–	0.03	0.54	–	22.29	m³	**22.83**
soil (clay); dry	–	0.03	0.54	–	24.00	m³	**24.54**
soil (clay); wet	–	0.03	0.54	1.04	25.72	m³	**27.30**
Other materials							
rubbish (mixed loads)	–	0.03	0.64	1.25	24.00	m³	**25.89**
green waste	–	0.03	0.64	1.25	24.00	m³	**25.89**
As above but allowing for 3 loads (36 m³, 60 tonne) per hour removed							
inert material	–	0.03	0.54	1.04	33.95	m³	**35.53**
Excavated material; off site to tip; mechanically loaded by grab; capacity of load 13 m³ (18 tonne)							
inert material	–	–	–	–	–	m³	**28.60**
hardcore	–	–	–	–	–	m³	**38.57**
macadam	–	–	–	–	–	m³	**23.20**

D20 EXCAVATION AND FILLING

Item Excluding site overheads and profit	PC £	Labour hours	Labour £	Plant £	Material £	Unit	Total rate £
clean concrete	–	–	–	–	–	m³	26.00
soil (sandy and loam); dry	–	–	–	–	–	m³	28.60
soil (sandy and loam); wet	–	–	–	–	–	m³	30.33
broken out compacted materials such as							
road bases and the like	–	–	–	–	–	m³	31.78
soil (clay); dry	–	–	–	–	–	m³	37.12
soil (clay); wet	–	–	–	–	–	m³	38.57
rubbish (mixed loads)	–	–	–	–	–	m³	33.85
green waste	–	–	–	–	–	m³	33.33
Disposal by skip; 9 yd³ (6 m³)							
Excavated material loaded to skip							
by machine	–	–	–	43.23	–	m³	43.23
by hand	–	3.00	57.75	39.58	–	m³	97.33
Excavated material; on site							
In spoil heaps							
average 25 m distance	–	–	–	2.87	–	m³	2.87
average 50 m distance	–	–	–	3.31	–	m³	3.31
average 100 m distance (1 dumper)	–	–	–	4.14	–	m³	4.14
average 100 m distance (2 dumpers)	–	–	–	4.22	–	m³	4.22
average 200 m distance	–	–	–	5.00	–	m³	5.00
average 200 m distance (2 dumpers)	–	–	–	5.08	–	m³	5.08
Spreading on site							
average 25 m distance	–	0.02	0.43	5.03	–	m³	5.46
average 50 m distance	–	0.04	0.75	5.63	–	m³	6.38
average 100 m distance	–	0.06	1.19	6.87	–	m³	8.06
average 200 m distance	–	0.12	2.31	6.41	–	m³	8.72
Disposal; hand							
Excavated material; on site; in spoil heaps							
average 25 m distance	–	2.40	46.20	–	–	m³	46.20
average 50 m distance	–	2.64	50.82	–	–	m³	50.82
average 100 m distance	–	3.00	57.75	–	–	m³	57.75
average 200 m distance	–	3.60	69.30	–	–	m³	69.30
Excavated material; spreading on site							
average 25 m distance	–	2.64	50.82	–	–	m³	50.82
average 50 m distance	–	3.00	57.75	–	–	m³	57.75
average 100 m distance	–	3.60	69.30	–	–	m³	69.30
average 200 m distance	–	4.20	80.85	–	–	m³	80.85
Disposal of material from site clearance **operations**							
Shrubs and groundcovers less than 1.00 m							
height; disposal by 15 m³ self-loaded truck							
deciduous shrubs; not chipped; winter	–	0.05	0.96	0.63	8.00	m²	9.59
deciduous shrubs; chipped; winter	–	0.05	0.96	1.53	1.20	m²	3.69
evergreen or deciduous shrubs; chipped;							
summer	–	0.08	1.44	0.91	4.80	m²	7.15

D20 EXCAVATION AND FILLING

Item Excluding site overheads and profit	PC £	Labour hours	Labour £	Plant £	Material £	Unit	Total rate £
DISPOSAL – cont							
Disposal of material from site clearance operations – cont							
Shrubs 1.00–2.00 m height; disposal by 15 m^3 self-loaded truck							
deciduous shrubs; not chipped; winter	–	0.17	3.21	0.63	24.00	m^2	**27.84**
deciduous shrubs; chipped; winter	–	0.25	4.81	0.91	4.80	m^2	**10.52**
evergreen or deciduous shrubs; chipped; summer	–	0.30	5.78	1.01	9.60	m^2	**16.39**
Shrubs or hedges 2.00–3.00 m height; disposal by 15 m^3 self-loaded truck							
deciduous plants non-woody growth; not chipped; winter	–	0.25	4.81	0.63	24.00	m^2	**29.44**
deciduous shrubs; chipped; winter	–	0.50	9.63	0.91	12.00	m^2	**22.54**
evergreen or deciduous shrubs non-woody growth; chipped; summer	–	0.25	4.81	0.91	4.80	m^2	**10.52**
evergreen or deciduous shrubs woody growth; chipped; summer	–	0.67	12.83	1.20	14.40	m^2	**28.43**
Shrubs and groundcovers less than 1.00 m height; disposal to spoil heap							
deciduous shrubs; not chipped; winter	–	0.05	0.96	–	–	m^2	**0.96**
deciduous shrubs; chipped; winter	–	0.05	0.96	0.29	–	m^2	**1.25**
evergreen or deciduous shrubs; chipped; summer	–	0.08	1.44	0.29	–	m^2	**1.73**
Shrubs 1.00–2.00 m height; disposal to spoil heap							
deciduous shrubs; not chipped; winter	–	0.17	3.21	–	–	m^2	**3.21**
deciduous shrubs; chipped; winter	–	0.25	4.81	0.29	–	m^2	**5.10**
evergreen or deciduous shrubs; chipped; summer	–	0.30	5.78	0.38	–	m^2	**6.16**
Shrubs or hedges 2.00–3.00 m height; disposal to spoil heaps							
deciduous plants non-woody growth; not chipped; winter	–	0.25	4.81	–	–	m^2	**4.81**
deciduous shrubs; chipped; winter	–	0.50	9.63	0.29	–	m^2	**9.92**
evergreen or deciduous shrubs non-woody growth; chipped; summer	–	0.67	12.83	0.57	–	m^2	**13.40**
evergreen or deciduous shrubs woody growth; chipped; summer	–	1.50	28.87	1.14	–	m^2	**30.01**

D20 EXCAVATION AND FILLING

Item Excluding site overheads and profit	PC £	Labour hours	Labour £	Plant £	Material £	Unit	Total rate £
GRADING AND SURFACE PREPARATION							
Grading operations; surface previously excavated to reduce levels to prepare to receive subsequent treatments; grading to accurate levels and falls 20 mm tolerances							
Clay or heavy soils or hardcore							
5 tonne excavator	–	0.04	0.77	0.18	–	m²	**0.95**
JCB Sitemaster 3CX	–	0.02	0.39	0.82	–	m²	**1.21**
21 tonne 360 tracked excavator	–	0.01	0.19	0.50	–	m²	**0.69**
by hand	–	0.10	1.93	–	–	m²	**1.93**
Loamy topsoils							
5 tonne excavator	–	0.03	0.51	0.12	–	m²	**0.63**
JCB Sitemaster 3CX	–	0.01	0.26	0.55	–	m²	**0.81**
21 tonne 360 tracked excavator	–	0.01	0.14	0.37	–	m²	**0.51**
by hand	–	0.05	0.96	–	–	m²	**0.96**
Sand or graded granular materials							
5 tonne excavator	–	0.02	0.39	0.09	–	m²	**0.48**
JCB Sitemaster 3CX	–	0.01	0.19	0.41	–	m²	**0.60**
21 tonne 360 tracked excavator	–	0.01	0.11	0.28	–	m²	**0.39**
by hand	–	0.03	0.64	–	–	m²	**0.64**
Grading operations to bottoms of excavations or trenches to receive foundations or bases; surface previously excavated to reduce levels to prepare to receive subsequent treatments; grading to accurate levels and falls 20 mm tolerances							
Trenches							
5 tonne excavator	–	0.32	10.91	0.59	–	m²	**11.50**
by hand	–	0.33	12.83	–	–	m²	**12.83**
Excavated areas to receive bases or formwork							
5 tonne excavator	–	0.13	4.49	0.30	–	m²	**4.79**
by hand	–	0.17	6.42	–	–	m²	**6.42**
Grading operations; surface recently filled to raise levels to prepare to receive subsequent treatments; grading to accurate levels and falls 20 mm tolerances							
Clay or heavy soils or hardcore							
5 tonne excavator	–	0.03	0.58	0.13	–	m²	**0.71**
JCB Sitemaster 3CX	–	0.02	0.29	0.62	–	m²	**0.91**
21 tonne 360 tracked excavator	–	0.01	0.13	0.34	–	m²	**0.47**
by hand	–	0.08	1.60	–	–	m²	**1.60**
Loamy topsoils							
5 tonne excavator	–	0.02	0.40	0.09	–	m²	**0.49**
JCB Sitemaster 3CX	–	0.01	0.20	0.42	–	m²	**0.62**
21 tonne 360 tracked excavator	–	0.01	0.13	0.34	–	m²	**0.47**
by hand	–	0.04	0.77	–	–	m²	**0.77**

D20 EXCAVATION AND FILLING

Item Excluding site overheads and profit	PC £	Labour hours	Labour £	Plant £	Material £	Unit	Total rate £
GRADING AND SURFACE PREPARATION – **cont**							
Grading operations – cont							
Sand or graded granular materials							
5 tonne excavator	–	0.02	0.29	0.07	–	m²	**0.36**
JCB Sitemaster 3CX	–	0.01	0.14	0.31	–	m²	**0.45**
21 tonne 360 tracked excavator	–	0.01	0.10	0.25	–	m²	**0.35**
by hand	–	0.03	0.55	–	–	m²	**0.55**
Surface grading; Agripower Ltd; using **laser controlled equipment**							
Maximum variation of existing ground level							
75 mm; laser box grader and tractor	–	–	–	–	–	ha	**1200.01**
100 mm; laser controlled low ground							
pressure bulldozer	–	–	–	–	–	ha	**1900.02**
Fraise mowing; scarification of existing turf **layer to rootzone level; Agripower Ltd**							
Cart arisings to tip on site							
average thickness 50–75 mm	–	–	–	–	–	ha	**2963.03**
Cultivating							
Ripping up subsoil; using approved subsoiling							
machine; minimum depth 250 mm below							
topsoil; at 1.20 m centres; in							
gravel or sandy clay	–	–	–	2.16	–	100 m²	**2.16**
soil compacted by machines	–	–	–	2.40	–	100 m²	**2.40**
clay	–	–	–	3.59	–	100 m²	**3.59**
chalk or other soft rock	–	–	–	3.59	–	100 m²	**3.59**
Extra for subsoiling at 1 m centres	–	–	–	1.08	–	100 m²	**1.08**
Breaking up existing ground; using pedestrian							
operated tine cultivator or rotavator; loam or							
sandy soil							
100 mm deep	–	0.22	4.24	2.37	–	100 m²	**6.61**
150 mm deep	–	0.28	5.29	2.96	–	100 m²	**8.25**
200 mm deep	–	0.37	7.06	3.94	–	100 m²	**11.00**
As above but in heavy clay or wet soils							
100 mm deep	–	0.44	8.47	4.73	–	100 m²	**13.20**
150 mm deep	–	0.66	12.71	7.09	–	100 m²	**19.80**
200 mm deep	–	0.82	15.88	8.87	–	100 m²	**24.75**
Breaking up existing ground; using tractor							
drawn tine cultivator or rotavator							
100 mm deep	–	–	–	0.83	–	100 m²	**0.83**
150 mm deep	–	–	–	1.04	–	100 m²	**1.04**
200 mm deep	–	–	–	1.39	–	100 m²	**1.39**
400 mm deep	–	–	–	4.17	–	100 m²	**4.17**
Cultivating ploughed ground; using disc, drag,							
or chain harrow							
4 passes	–	–	–	4.55	–	100 m²	**4.55**
Rolling cultivated ground lightly; using							
self-propelled agricultural roller	–	0.06	1.07	0.60	–	100 m²	**1.67**

D41 CRIB WALLS/GABIONS/REINFORCED EARTH

Item Excluding site overheads and profit	PC £	Labour hours	Labour £	Plant £	Material £	Unit	Total rate £
Importing and storing selected and approved topsoil; 20 tonne load = 11.88 m³ average							
20 tonne loads	33.60	–	–	–	33.60	m³	**33.60**
Surface treatments							
Compacting							
bottoms of excavations	–	0.01	0.10	0.04	–	m²	**0.14**
Surface preparation							
Trimming surfaces of cultivated ground to final levels; removing roots, stones and debris exceeding 50 mm in any direction to tip off site; slopes less than 15°							
clean ground with minimal stone content	–	0.25	4.81	–	–	100 m²	**4.81**
slightly stony; 0.5 kg stones per m²	–	0.33	6.41	–	0.01	100 m²	**6.42**
very stony; 1.0–3.0 kg stones per m²	–	0.50	9.63	–	0.02	100 m²	**9.65**
clearing mixed, slightly contaminated rubble; inclusive of roots and vegetation	–	0.50	9.63	–	0.06	100 m²	**9.69**
clearing brick-bats, stones and clean rubble	–	0.60	11.55	–	0.04	100 m²	**11.59**
D41 CRIB WALLS/GABIONS/REINFORCED EARTH							
CRIB WALLS							
Crib walls; Phi Group							
Permacrib (timber) or Andacrib (concrete) crib walling system; machine filled infill with crushed rock; inclusive of concrete footing and rear wall drain; excluding excavation and backfill material							
retaining walls up to 2.0 m	–	–	–	–	–	m²	**160.00**
retaining walls up to 4.0 m	–	–	–	–	–	m²	**180.00**
retaining walls up to 6.0 m	–	–	–	–	–	m²	**200.00**
GABIONS							
Retaining walls; Maccaferri Ltd							
Wire mesh gabions; galvanized mesh 80 × 100 mm; filling with broken stones 125–200 mm size; wire down securely to manufacturer's instructions; filling front face by hand							
2 × 1 × 0.50 m	19.09	0.67	38.50	3.72	66.97	nr	**109.19**
2 × 1 × 1.0 m	26.82	1.33	77.00	7.45	122.58	nr	**207.03**
PVC coated gabions							
2 × 1 × 0.5 m	24.84	0.67	38.50	3.72	72.72	nr	**114.94**
2 × 1 × 1.0 m	34.16	1.33	77.00	7.45	129.92	nr	**214.37**

D41 CRIB WALLS/GABIONS/REINFORCED EARTH

Item Excluding site overheads and profit	PC £	Labour hours	Labour £	Plant £	Material £	Unit	Total rate £
GABIONS – cont							
Retaining walls – cont							
Reno mattress gabions							
6 × 2 × 0.17 m	80.78	1.00	57.75	5.59	178.46	nr	**241.80**
6 × 2 × 0.23 m	87.62	1.50	86.63	7.45	219.77	nr	**313.85**
6 × 2 × 0.30 m	102.82	2.00	115.50	8.38	275.19	nr	**399.07**
REINFORCED EARTH							
Embankments; Tensar International							
Embankments; reinforced with Tensar Geogrid RE520; geogrid laid horizontally within fill to 100% of vertical height of slope at 1.00 m centres; filling with excavated material							
slopes less than 45°; Tensartech Natural Green System	2.36	0.61	35.38	1.60	2.36	m³	**39.34**
slopes exceeding 45°; Tensartech Greenslope System	5.08	0.06	3.57	2.31	5.08	m³	**10.96**
extra over to slopes exceeding 45° for temporary shuttering; shuttering measured per m² of vertical elevation	–	–	–	–	12.00	m²	**12.00**
Embankments; reinforced with Tensar Geogrid RE540; geogrid laid horizontally within fill to 100% of vertical height of slope at 1.00 m centres; filling with excavated material							
slopes less than 45°; Tensartech Natural Green System	2.92	0.61	35.38	1.60	2.92	m³	**39.90**
slopes exceeding 45°; Tensartech Greenslope System	6.28	0.06	3.57	2.31	6.28	m³	**12.16**
extra over to slopes exceeding 45° for temporary shuttering; shuttering measured per m² of vertical elevation	–	–	–	–	12.00	m²	**12.00**
Extra for Tensar Mat 400; erosion control mats; to faces of slopes of 45° or less; filling with 20 mm fine topsoil; seeding	3.95	0.01	0.38	0.21	5.26	m²	**5.85**
Embankments; reinforced with Tensar Geogrid RE540; geogrid laid horizontally within fill to 50% of horizontal length of slopes at specified centres; filling with imported fill PC £11.20/m³							
300 mm centres; slopes up to 45°; Tensartech Natural Green System	4.85	0.08	4.49	1.80	18.29	m³	**24.58**
600 mm centres; slopes up to 45°; Tensartech Natural Green System	2.42	0.06	3.57	2.31	15.86	m³	**21.74**
extra over to slopes exceeding 45° for temporary shuttering; shuttering measured per m² of vertical elevation	–	–	–	–	12.00	m²	**12.00**
Extra for Tensar Mat 400; erosion control mats; to faces of slopes of 45° or less; filling with 20 mm fine topsoil; seeding	3.95	0.01	0.38	0.21	5.26	m²	**5.85**

D41 CRIB WALLS/GABIONS/REINFORCED EARTH

Item Excluding site overheads and profit	PC £	Labour hours	Labour £	Plant £	Material £	Unit	Total rate £
Embankments; reinforced with Tensar Geogrid RE540; geogrid laid horizontally within fill; filling with excavated material; wrapping around at faces							
300 mm centres; slopes exceeding 45°; Tensartech Natural Green System	9.72	0.06	3.57	1.60	9.72	m³	14.89
600 mm centres; slopes exceeding 45°; Tensartech Greenslope System	4.85	0.06	3.57	1.60	4.85	m³	10.02
extra over to slopes exceeding 45° for temporary shuttering; shuttering measured per m² of vertical elevation	–	–	–	–	12.00	m²	12.00
Extra over embankments for bagwork face supports for slopes exceeding 45°	–	0.17	9.62	–	14.00	m²	23.62
Extra over embankments for seeding of bags	–	0.03	1.92	–	0.32	m²	2.24
Extra over embankments for Bodkin joints	–	–	–	–	1.10	m	1.10
Anchoring systems; surface stabilization; Platipus Anchors Ltd; centres and depths shown should be verified with a design engineer; they may vary either way dependent on circumstances							
Concrete revetments; hard anodized stealth anchors with stainless steel accessories installed to a depth of 1 m							
SO4 anchors to a depth of 1.00–1.50 m at 1.00 m centres	26.63	0.17	6.42	1.77	26.63	m²	34.82
Loadlocking and crimping tool; for stressing and crimping the anchors							
hire rate per week	72.93	–	–	–	72.93	nr	72.93
Surface erosion; geotextile anchoring on slopes; aluminium alloy Stealth anchors							
SO4/SO6 to a depth of 1.0–2.0 m at 2.00 m centres	9.10	1.00	38.50	1.77	9.10	m²	49.37
Brick block or in situ concrete distressed retaining walls; anchors installed in two rows at varying tensions between 18–50 kN along the length of the retaining wall; core drilling of wall not included							
SO8/BO6/BO8 bronze Stealth and Bat anchors installed in combination with stainless steel accessories to a depth of 6 m; average cost base on 1.50 m centres; long-term solution	382.02	0.25	9.63	1.77	382.02	m²	393.42
SO8/BO6/BO8 cast SG iron Stealth and Bat anchors installed in combination with galvanized steel accessories to a depth of 6 m; average cost base on 1.50 m centres; short term solution	163.22	0.25	9.63	1.77	163.22	m²	174.62

D41 CRIB WALLS/GABIONS/REINFORCED EARTH

Item Excluding site overheads and profit	PC £	Labour hours	Labour £	Plant £	Material £	Unit	Total rate £
REINFORCED EARTH – cont							
Anchoring systems – cont							
Timber retaining wall support; anchors installed 19 kN along the length of the retaining wall							
SO6/SO8 bronze Stealth anchors including load plate and wege grip; installed in combination with stainless steel accessories to a depth of 3.0 m; 1.00 m centres; average cost	184.80	0.25	14.44	1.41	184.80	m²	200.65
Gabions; anchors for support and stability to moving, overturning or rotating gabion retaining walls							
SO8/BO6 anchors including base plate at 1.00 m centres to a depth of 4.00 m	289.41	0.17	9.63	1.77	289.41	m²	300.81
Soil nailing of geofabrics; anchors fixed through surface of fabric or erosion control surface treatment (not included)							
SO8/BO6 anchors including base plate at 1.50 m centres to a depth of 3.00 m	115.76	0.08	4.81	1.18	115.76	m²	121.75
Green faced reinforced soil; Phi Group							
Textomur reinforced soil system embankments; to reinforced soil slopes 55–70° to the horizontal; including facing and geogrid; excluding backfill material							
retaining walls up to 3.0 m	–	–	–	–	–	m²	120.00
retaining walls up to 6.0 m	–	–	–	–	–	m²	125.00
Split face concrete blocks reinforced soil; Phi Group							
Modular blocks reinforced soil system embankments; to reinforced soil slopes near vertical; including geogrid; excluding backfill material							
retaining walls up to 3.0 m	–	–	–	–	–	m²	180.00
retaining walls up to 6.0 m	–	–	–	–	–	m²	210.00

E10 IN SITU CONCRETE

Item Excluding site overheads and profit	PC £	Labour hours	Labour £	Plant £	Material £	Unit	Total rate £
E10 IN SITU CONCRETE							
CLARIFICATION NOTES ON LABOUR COSTS IN THIS SECTION							
In situ concrete team Generally a three man team is used in this section; The column 'Labour hours' reports team hours. The column 'Labour £' reports the total cost of the team for the unit of work shown							
3 man team	–	1.00	57.75	–	–	hr	**57.75**
2 man team	–	1.00	38.50	–	–	hr	**38.50**
CONCRETE MIX INFORMATION							
General The concrete mixes used here are referred to as 'Designed', 'Standard' and 'Designated' mixes.							
Designed mix User specifies the performance of the concrete. Producer is responsible for selecting the appropriate mix. Strength testing is essential.							
Prescribed mix User specifies the mix constituents and is responsible for ensuring that the concrete meets performance requirements. Accurate mix proportions are essential.							
Standard mix Made with a restricted range of materials. Specification to include the proposed use of the material as well as the standard mix reference, type of cement, type and size of aggregate, and slump (workability). Quality assurance is required.							
Designated mix Producer to hold current product conformity certification and quality approval. Quality assurance is essential. The mix must not be modified.							

E10 IN SITU CONCRETE

Item Excluding site overheads and profit	PC £	Labour hours	Labour £	Plant £	Material £	Unit	Total rate £
CONCRETE MIXES							
Note: The mixes below are all mixed in 113 litre mixers. Please see the Minor Works section of this book for mixes containing aggregates delivered in bulk bags.							
Concrete mixes; mixed on site; costs for producing concrete; prices for commonly used mixes for various types of work; based on bulk load 20 tonne rates for aggregates							
Roughest type mass concrete such as footings and road haunchings; 300 mm thick							
1:3:6	85.28	–	–	–	85.28	m³	**85.28**
As above but aggregates delivered in 10 tonne loads							
1:3:6	99.11	–	–	–	99.11	m³	**99.11**
As above but aggregates delivered in 850 kg bulk bags							
1:3:6	128.24	–	–	–	128.24	m³	**128.24**
Most ordinary use of concrete such as mass walls above ground, road slabs, etc. and general reinforced concrete work							
1:2:4	–	–	–	–	99.67	m³	**99.67**
As above but aggregates delivered in 10 tonne loads							
1:2:4	109.08	–	–	–	109.08	m³	**109.08**
As above but aggregates delivered in 850 kg bulk bags							
1:2:4	–	–	–	–	138.22	m³	**138.22**
Watertight floors, pavements, walls, tanks, pits, steps, paths, surface of two course roads; reinforced concrete where extra strength is required							
1:1.5:3	96.05	1.20	23.04	–	96.05	m³	**119.09**
As above but aggregates delivered in 10 tonne loads							
1:1.5:3	105.46	1.20	23.04	–	105.46	m³	**128.50**
As above but aggregates delivered in 850 kg bulk bags							
1:1.5:3	134.59	1.20	23.04	–	134.59	m³	**157.63**

E10 IN SITU CONCRETE

Item Excluding site overheads and profit	PC £	Labour hours	Labour £	Plant £	Material £	Unit	Total rate £
IN SITU CONCRETE							
Plain in situ concrete; site mixed; 10 N/mm² –							
40 mm aggregate (1:3:6) (aggregate							
delivery indicated)							
Foundations							
ordinary portland cement; 20 tonne ballast							
loads	87.42	0.33	19.06	–	87.42	m³	**106.48**
ordinary portland cement; 10 tonne ballast							
loads	101.58	0.33	19.06	–	101.58	m³	**120.64**
ordinary portland cement; 850 kg bulk bags	131.45	0.33	19.06	–	131.45	m³	**150.51**
Foundations; poured on or against earth or							
unblinded hardcore							
ordinary portland cement; 20 tonne ballast							
loads	89.55	0.33	19.06	–	89.55	m³	**108.61**
ordinary portland cement; 10 tonne ballast							
loads	104.06	0.33	19.06	–	104.06	m³	**123.12**
ordinary portland cement; 850 kg bulk bags	134.66	0.33	19.06	–	134.66	m³	**153.72**
Isolated foundations							
ordinary portland cement; 20 tonne ballast							
loads	87.42	0.33	19.06	–	87.42	m³	**106.48**
ordinary portland cement; 850 kg bulk bags	131.45	0.33	19.06	–	131.45	m³	**150.51**
Plain in situ concrete; site mixed; 21 N/mm² –							
20 mm aggregate (1:2:4)							
Foundations							
ordinary portland cement; 20 tonne ballast							
loads	102.16	0.33	19.06	–	102.16	m³	**121.22**
ordinary portland cement; 850 kg bulk bags	141.67	0.33	19.06	–	141.67	m³	**160.73**
Foundations; poured on or against earth or							
unblinded hardcore							
ordinary portland cement; 20 tonne ballast							
loads	104.65	0.33	19.06	–	104.65	m³	**123.71**
ordinary portland cement; 850 kg bulk bags	145.13	0.33	19.06	–	145.13	m³	**164.19**
Isolated foundations							
ordinary portland cement; 20 tonne ballast							
loads	102.16	0.33	19.06	–	102.16	m³	**121.22**
ordinary portland cement; 850 kg bulk bags	141.67	0.33	19.06	–	141.67	m³	**160.73**
Ready mix concrete; concrete mixed on							
site; Euromix							
Ready mix concrete mixed on site; placed by							
barrow not more than 25 m distance (16							
barrows/m³)							
concrete 1:3:6	85.00	0.33	19.06	–	85.00	m³	**104.06**
concrete 1:2:4	87.50	0.33	19.06	–	87.50	m³	**106.56**
concrete C30	90.00	0.33	19.06	–	90.00	m³	**109.06**

E10 IN SITU CONCRETE

Item Excluding site overheads and profit	PC £	Labour hours	Labour £	Plant £	Material £	Unit	Total rate £
IN SITU CONCRETE – cont							
Reinforced in situ concrete; site mixed; **21 N/mm² – 20 mm aggregate (1:2:4);** **aggregates delivered in 10 tonne loads**							
Foundations							
ordinary portland cement	102.16	0.67	38.46	–	102.16	m³	**140.62**
Foundations; poured on or against earth or							
unblinded hardcore							
ordinary portland cement	102.16	0.67	38.46	–	102.16	m³	**140.62**
Isolated foundations							
ordinary portland cement	102.16	0.83	48.11	–	102.16	m³	**150.27**
Plain in situ concrete; ready mixed; Tarmac **Southern; 10 N/mm mixes; suitable for** **mass concrete fill and blinding**							
Foundations							
GEN 1; Designated mix	72.70	0.50	28.88	–	72.70	m³	**101.58**
ST 2; Standard mix	79.69	0.50	28.88	–	79.69	m³	**108.57**
Foundations; poured on or against earth or							
unblinded hardcore							
GEN 1; Designated mix	72.70	0.50	28.88	–	72.70	m³	**101.58**
ST 2; Standard mix	81.59	0.50	28.88	–	81.59	m³	**110.47**
Isolated foundations							
GEN 1; Designated mix	72.70	0.67	38.46	–	72.70	m³	**111.16**
ST 2; Standard mix	79.69	0.67	38.46	–	79.69	m³	**118.15**
Plain in situ concrete; ready mixed; Tarmac **Southern; 15 N/mm mixes; suitable for** **oversite below suspended slabs and strip** **footings in non-aggressive soils**							
Foundations							
GEN 2; Designated mix	77.35	0.50	28.88	–	77.35	m³	**106.23**
ST 3; Standard mix	78.64	0.50	28.88	–	78.64	m³	**107.52**
Foundations; poured on or against earth or							
unblinded hardcore							
GEN 2; Designated mix	73.67	0.50	28.88	–	73.67	m³	**102.55**
ST 3; Standard mix	74.90	0.50	28.88	–	74.90	m³	**103.78**
Isolated foundations							
GEN 2; Designated mix	73.67	0.67	38.46	–	73.67	m³	**112.13**
ST 3; Standard mix	74.90	0.67	38.46	–	74.90	m³	**113.36**
Plain in situ concrete; ready mixed; Tarmac **Southern; air entrained mixes suitable for** **paving**							
Beds or slabs; house drives, parking and							
external paving							
PAV 1; 35 N/mm²; Designated mix	82.42	0.50	28.88	–	82.42	m³	**111.30**

E20 FORMWORK FOR IN SITU CONCRETE

Item Excluding site overheads and profit	PC £	Labour hours	Labour £	Plant £	Material £	Unit	Total rate £
Beds or slabs; heavy duty external paving							
PAV 2; 40 N/mm²; Designated mix	81.23	0.50	28.88	–	81.23	m³	**110.11**
Reinforced in situ concrete; ready mixed;							
Tarmac Southern; 35 N/mm² mix; suitable							
for foundations in class 2 sulphate							
conditions							
Foundations							
RC 35; Designated mix	82.90	0.67	38.46	–	82.90	m³	**121.36**
Foundations; poured on or against earth or							
unblinded hardcore							
RC 35; Designated mix	84.87	0.67	38.46	–	84.87	m³	**123.33**
Isolated foundations							
RC 35; Designated mix	78.95	0.67	38.46	–	78.95	m³	**117.41**
Reinforced in situ concrete; site mixed;							
21 N/mm² – 20 mm aggregate (1:2:4);							
aggregates delivered in 20 tonne loads							
Walls; thickness not exceeding 150 mm							
Site mixed concrete 1:2:4: 21 N/mm²;							
20 mm aggregate	102.16	0.67	38.46	2.21	102.16	m³	**142.83**
GEN 1; Designated mix	72.70	0.33	19.25	2.21	72.70	m³	**94.16**
GEN 2; Designated mix	77.35	0.33	19.25	2.21	77.35	m³	**98.81**
ST 2; Standard mix	79.69	0.33	19.25	2.21	79.69	m³	**101.15**
ST 3; Standard mix	78.64	0.33	19.25	2.21	78.64	m³	**100.10**
RC 35; Designated mix	82.90	0.33	19.25	2.21	82.90	m³	**104.36**
Walls; thickness not exceeding 150–450 mm							
Site mixed concrete 1:2:4: 21 N/mm²;							
20 mm aggregate	102.16	0.50	28.88	2.21	102.16	m³	**133.25**
GEN 1; Designated mix	72.70	0.20	11.55	2.21	72.70	m³	**86.46**
GEN 2; Designated mix	77.35	0.20	11.55	2.21	77.35	m³	**91.11**
ST 2; Standard mix	79.69	0.20	11.55	2.21	79.69	m³	**93.45**
ST 3; Standard mix	78.64	0.20	11.55	2.21	78.64	m³	**92.40**
RC 35; Designated mix	82.90	0.20	11.55	2.21	82.90	m³	**96.66**
Adjustments for site mixed concrete; add the							
following amounts for aggregates delivered in							
different load sizes							
850 kg bulk bags	–	–	–	–	38.55	m³	**38.55**
10 tonne loads	–	–	–	–	9.41	m³	**9.41**
E20 FORMWORK FOR IN SITU CONCRETE							
PLAIN VERTICAL FORMWORK							
Plain vertical formwork; basic finish							
Sides of foundations							
height not exceeding 250 mm	–	0.25	9.63	–	1.32	m	**10.95**
height 250–500 mm	–	0.25	9.63	–	2.26	m	**11.89**
height 500 mm–1.00 m	–	0.33	12.83	–	6.01	m	**18.84**
height exceeding 1.00 m	–	0.50	19.25	–	7.51	m²	**26.76**

E30 REINFORCEMENT FOR IN SITU CONCRETE

Item Excluding site overheads and profit	PC £	Labour hours	Labour £	Plant £	Material £	Unit	Total rate £
PLAIN VERTICAL FORMWORK – cont							
Plain vertical formwork – cont							
Sides of foundations; left in							
height not exceeding 250 mm	–	0.20	7.70	–	4.72	m	**12.42**
height 250–500 mm	–	0.20	7.70	–	9.06	m	**16.76**
height 500 mm–1.00 m	–	0.25	9.63	–	18.13	m	**27.76**
height over 1.00 m	–	0.40	15.40	–	13.03	m²	**28.43**
Walls							
height not exceeding 250 mm	–	0.31	12.03	–	3.32	m	**15.35**
height 250–500 mm	–	0.33	12.83	–	6.26	m	**19.09**
height 500 mm–1.00 m	–	0.42	16.04	–	14.01	m	**30.05**
height exceeding 1.00 m	–	0.75	28.88	–	19.51	m²	**48.39**
Walls curved to 6.00 radius							
height not exceeding 250 mm	–	0.38	14.44	–	3.32	m	**17.76**
height 250–500 mm	–	0.40	15.40	–	6.26	m	**21.66**
height 500 mm–1.00 m	–	0.50	19.25	–	14.01	m	**33.26**
height exceeding 1.00 m	–	0.90	34.65	–	19.51	m²	**54.16**
Walls curved to 3.00 radius							
height not exceeding 250 mm	–	0.47	18.05	–	3.32	m	**21.37**
height 250–500 mm	–	0.50	19.25	–	6.26	m	**25.51**
height 500 mm–1.00 m	–	0.62	24.06	–	14.01	m	**38.07**
height exceeding 1.00 m	–	1.13	43.31	–	19.51	m²	**62.82**
E30 REINFORCEMENT FOR IN SITU **CONCRETE**							
BAR REINFORCEMENT							
The rates for reinforcement shown for steel							
bars below are based on prices which would							
be supplied on a typical landscape contract.							
The steel prices shown have been priced on a							
selection of steel delivered to site where the							
total order quantity is in the region of 2 tonnes.							
The assumption is that should larger quantities							
be required, the work would fall outside the							
scope of the typical landscape contract							
defined in the front of this book. Keener rates							
can be obtained for larger orders.							
Reinforcement bars; hot rolled plain round **mild steel; straight or bent**							
Bars							
8 mm nominal size	530.00	13.50	519.75	–	530.00	tonne	**1049.75**
10 mm nominal size	530.00	13.00	500.50	–	530.00	tonne	**1030.50**
12 mm nominal size	530.00	12.50	481.25	–	530.00	tonne	**1011.25**
16 mm nominal size	530.00	12.00	462.00	–	530.00	tonne	**992.00**
20 mm nominal size	530.00	11.50	442.75	–	530.00	tonne	**972.75**

E30 REINFORCEMENT FOR IN SITU CONCRETE

Item Excluding site overheads and profit	PC £	Labour hours	Labour £	Plant £	Material £	Unit	Total rate £
Reinforcement bar to concrete formwork							
8 mm bar							
100 ccs	2.07	0.17	6.35	–	2.07	m²	**8.42**
200 ccs	1.01	0.13	4.81	–	1.01	m²	**5.82**
300 ccs	0.69	0.08	2.89	–	0.69	m²	**3.58**
10 mm bar							
100 ccs	3.23	0.17	6.35	–	3.23	m²	**9.58**
200 ccs	1.59	0.13	4.81	–	1.59	m²	**6.40**
300 ccs	1.06	0.08	2.89	–	1.06	m²	**3.95**
12 mm bar							
100 ccs	4.66	0.17	6.35	–	4.66	m²	**11.01**
200 ccs	2.33	0.13	4.81	–	2.33	m²	**7.14**
300 ccs	1.54	0.08	2.89	–	1.54	m²	**4.43**
16 mm bar							
100 ccs	8.32	0.20	7.70	–	8.32	m²	**16.02**
200 ccs	4.19	0.17	6.35	–	4.19	m²	**10.54**
300 ccs	2.81	0.13	4.81	–	2.81	m²	**7.62**
25 mm bar							
100 ccs	20.41	0.20	7.70	–	20.41	m²	**28.11**
200 ccs	10.18	0.13	4.81	–	10.18	m²	**14.99**
300 ccs	6.78	0.13	4.81	–	6.78	m²	**11.59**
32 mm bar							
100 ccs	33.44	0.20	7.70	–	33.44	m²	**41.14**
200 ccs	16.70	0.17	6.35	–	16.70	m²	**23.05**
300 ccs	11.13	0.13	4.81	–	11.13	m²	**15.94**
Reinforcement fabric; lapped; in beds or							
suspended slabs							
Fabric							
A98 (1.54 kg/m²)	0.98	0.11	4.24	–	0.98	m²	**5.22**
A142 (2.22 kg/m²)	1.42	0.11	4.24	–	1.42	m²	**5.66**
A193 (3.02 kg/m²)	1.94	0.11	4.24	–	1.94	m²	**6.18**
A252 (3.95 kg/m²)	2.51	0.12	4.62	–	2.51	m²	**7.13**
A393 (6.16 kg/m²)	3.90	0.14	5.39	–	3.90	m²	**9.29**

F10 BRICK/BLOCK WALLING

Item Excluding site overheads and profit	PC £	Labour hours	Labour £	Plant £	Material £	Unit	Total rate £
F10 BRICK/BLOCK WALLING							
MARKET PRICES OF MATERIALS							
Cement; Builder Centre							
Portland cement	–	–	–	–	4.03	25 kg	**4.03**
Sulphate-resistant cement	–	–	–	–	5.20	25 kg	**5.20**
White cement	–	–	–	–	14.15	25 kg	**14.15**
Sand; Builder Centre							
Building sand							
loose	33.34	–	–	–	33.34	m³	**33.34**
850 kg bulk bags	–	–	–	–	34.07	nr	**34.07**
850 kg bulk bags	–	–	–	–	66.44	m³	**66.44**
Sharp sand							
loose	–	–	–	–	33.34	m³	**33.34**
850 kg bulk bags	–	–	–	–	34.07	nr	**34.07**
850 kg bulk bags	–	–	–	–	66.44	m³	**66.44**
Sand; Yeoman Aggregates Ltd							
Sharp sand	–	–	–	–	18.52	tonne	**18.52**
Sharp sand	–	–	–	–	29.63	m³	**29.63**
Soft sand	–	–	–	–	18.52	tonne	**18.52**
Soft sand	–	–	–	–	29.56	m³	**29.56**
BRICK/BLOCK WALLING DATA							
Note: Batching quantities for these mortar mixes may be found in the Tables and Memoranda section of this book.							
Mortar mixes for brickwork and pavings; all **mortars mixed on site and priced at** **delivery to waiting wheelbarrows at the** **cement mixer; cement and lime in 25 kg** **bags; aggregates as described**							
Mortar grade 1:3; aggregates delivered in 20 tonne load; cement: sand volumes shown below (bags: litres)							
85 litre mixer (1.84:93.5)	–	0.75	14.44	–	119.56	m³	**134.00**
150 litre mixer (3.24:165)	–	0.50	9.63	–	119.56	m³	**129.19**
Mortar grade 1:3; aggregates delivered in 10 tonne load; cement: sand volumes shown below (bags: litres)							
85 litre mixer (1.84:93.5)	–	0.75	14.44	–	128.85	m³	**143.29**
150 litre mixer (3.24:165)	–	0.50	9.63	–	128.85	m³	**138.48**
Mortar grade 1:3; aggregates delivered in 850 kg bulk bags; cement: sand volumes shown below (bags: litres)							
85 litre mixer (1.84:93.5)	–	0.75	14.44	–	157.57	m³	**172.01**
150 litre mixer (3.24:165)	–	0.50	9.63	–	157.57	m³	**167.20**

F10 BRICK/BLOCK WALLING

Item Excluding site overheads and profit	PC £	Labour hours	Labour £	Plant £	Material £	Unit	Total rate £
Mortar grade 1:4; aggregates delivered in 20 tonne load; cement: sand volumes shown below (bags: litres)							
85 litre mixer (1.36:102)	–	0.75	14.44	–	99.95	m³	**114.39**
150 litre mixer (2.4:180)	–	0.50	9.63	–	99.95	m³	**109.58**
Mortar grade 1:4; aggregates delivered in 10 tonne load; cement: sand volumes shown below (bags: litres)							
85 litre mixer (1.36:102)	–	0.75	14.44	–	110.08	m³	**124.52**
150 litre mixer (2.4:180)	–	0.50	9.63	–	110.08	m³	**119.71**
Mortar grade 1:4; aggregates delivered in 850 kg bulk bags; cement: sand volumes shown below (bags: litres)							
85 litre mixer (1.36:102)	–	0.75	14.44	–	141.48	m³	**155.92**
150 litre mixer (2.4:180)	–	0.50	9.63	–	141.48	m³	**151.11**
Mortar grade 1:1:6; aggregates delivered in 20 tonne load; cement: lime: sand volumes shown below (bags: bags: litres)							
85 litre mixer;(0.9:0.44:93.5)	–	0.75	14.44	–	128.56	m³	**143.00**
150 litre mixer; (1.6: 0.8: 165)	–	0.50	9.63	–	128.56	m³	**138.19**
Mortar grade 1:1:6; aggregates delivered in 10 tonne load; cement: lime: sand volumes shown below (bags: bags: litres)							
85 litre mixer (0.9:0.44:93.5)	–	0.75	14.44	–	141.64	m³	**156.08**
150 litre mixer (1.6:0.8:165)	–	0.50	9.63	–	141.64	m³	**151.27**
Mortar grade 1:1:6; aggregates delivered in 850 kg bulk bags; cement: lime: sand volumes shown below (bags: bags: litres)							
85 litre mixer (0.9:0.44:93.5)	–	0.75	14.44	–	166.57	m³	**181.01**
150 litre mixer (1.6:0.8:165)	–	0.50	9.63	–	166.57	m³	**176.20**
Mortar grade 1:2:9; aggregates delivered in 20 tonne load; cement: lime: sand volumes shown below (bags: bags: litres)							
85 litre mixer (0.7:0.5:102)	–	0.75	14.44	–	128.31	m³	**142.75**
150 litre mixer (1.2:0.9:180)	–	0.50	9.63	–	125.36	m³	**134.99**
Mortar grade 1:2:9; aggregates delivered in 10 tonne load; cement: lime: sand volumes shown below (bags: bags: litres)							
85 litre mixer (0.7:0.5:102)	–	0.75	14.44	–	138.44	m³	**152.88**
150 litre mixer (1.2:0.9:180)	–	0.50	9.63	–	138.44	m³	**148.07**
Mortar grade 1:2:9; aggregates delivered in 850 kg bulk bags; cement: lime: sand volumes shown below (bags: bags: litres)							
85 litre mixer (0.7:0.5:102)	–	0.75	14.44	–	169.84	m³	**184.28**
150 litre mixer (1.2:0.9:180)	–	0.50	9.63	–	169.50	m³	**179.13**

F10 BRICK/BLOCK WALLING

Item Excluding site overheads and profit	PC £	Labour hours	Labour £	Plant £	Material £	Unit	Total rate £
BRICK/BLOCK WALLING DATA – cont							
Variation in brick prices (area × price difference × value shown below)							
Add or subtract the following amounts for every £1.00/1000 difference in the PC price of the measured items below							
half brick thick	–	–	–	–	6.30	m²	**6.30**
one brick thick	–	–	–	–	12.60	m²	**12.60**
one and a half brick thick	–	–	–	–	18.90	m²	**18.90**
two brick thick	–	–	–	–	25.20	m²	**25.20**
Mortar (1:3) required per m² of brickwork; brick size 215 × 102.5 × 65 mm							
Half brick wall (103 mm); 58 bricks/m²							
no frog	–	–	–	–	2.87	m²	**2.87**
single frog	–	–	–	–	3.31	m²	**3.31**
double frog	–	–	–	–	3.91	m²	**3.91**
2 × half brick cavity wall (270 mm); 116 bricks/ m²							
no frog	–	–	–	–	5.72	m²	**5.72**
single frog	–	–	–	–	6.78	m²	**6.78**
double frog	–	–	–	–	8.28	m²	**8.28**
One brick wall (215 mm); 116 bricks/m²							
no frog	–	–	–	–	6.92	m²	**6.92**
single frog	–	–	–	–	8.28	m²	**8.28**
double frog	–	–	–	–	9.63	m²	**9.63**
One and a half brick wall (328 mm) 174 bricks/m²							
no frog	–	–	–	–	9.49	m²	**9.49**
single frog	–	–	–	–	11.13	m²	**11.13**
double frog	–	–	–	–	13.24	m²	**13.24**
Mortar (1:3) required per m² of blockwork; blocks 440 × 215 mm							
Block thickness							
100 mm	–	–	–	–	1.06	m²	**1.06**
140 mm	–	–	–	–	1.36	m²	**1.36**
hollow blocks 440 × 215 mm	–	–	–	–	0.34	m²	**0.34**
Movement of materials							
Loading to wheelbarrows and transporting to location; per 215 mm thick walls; maximum distance 25 m	–	0.42	8.02	–	–	m²	**8.02**
Adjustments for mortars in the brick prices below							
Difference for cement mortar (1:3) in lieu of gauged mortar							
half brick thick	–	–	–	–	-0.29	m²	**-0.29**
one brick thick	–	–	–	–	-0.72	m²	**-0.72**
one and a half brick thick	–	–	–	–	-0.97	m²	**-0.97**
two brick thick	–	–	–	–	-1.44	m²	**-1.44**

F10 BRICK/BLOCK WALLING

Item Excluding site overheads and profit	PC £	Labour hours	Labour £	Plant £	Material £	Unit	Total rate £
CLARIFICATION NOTES ON LABOUR COSTS IN THIS SECTION							
Brick/Block walling team							
Generally a three man team is used in this section; The column 'Labour hours' reports team hours. The column 'Labour £' reports the total cost of the team for the unit of work shown							
3 man team	–	1.00	57.75	–	–	hr	**57.75**
2 man team	–	1.00	38.50	–	–	hr	**38.50**
BRICK WALLING							
Note: frog types and mortar quantities used in the calculations below are based on bricks with single frogs. Please make adjustments for mortar volumes used with varying brick frog formats. See data above for mortar quantities for various frog types. Delivery volumes of aggregates for mortars are based on 10 tonne deliveries. The figures below show the approximate areas of walling which would be constructed mixed with building sand from a 10 tonne (6.25 m³) load of building sand.							
half brick wall = 256 m²							
one brick wall = 124 m²							
Class B engineering bricks; PC £276.00/ 1000; double Flemish bond in cement mortar (1:3)							
Offloading mechanically; loading to wheelbarrows; transporting to location maximum 25 m distance per 215 mm thick walls	–	0.42	8.02	–	–	m²	**8.02**
Construct walls							
half brick thick	–	0.60	34.65	–	20.54	m²	**55.19**
one brick thick	–	1.20	69.30	–	42.66	m²	**111.96**
one and a half brick thick	–	1.80	103.95	–	68.65	m²	**172.60**
two brick thick	–	2.40	138.54	–	85.31	m²	**223.85**
Construct walls; curved; mean radius 6 m							
half brick thick	–	0.90	51.98	–	20.51	m²	**72.49**
one brick thick	–	1.80	103.95	–	42.66	m²	**146.61**
Construct walls; curved; mean radius 1.50 m							
half brick thick	–	1.20	69.29	–	20.54	m²	**89.83**
one brick thick	–	3.60	138.60	–	42.66	m²	**181.26**

F10 BRICK/BLOCK WALLING

Item Excluding site overheads and profit	PC £	Labour hours	Labour £	Plant £	Material £	Unit	Total rate £
BRICK WALLING – cont							
Class B engineering bricks – cont							
Construct walls; tapering; one face battering; average							
one and a half brick thick	–	2.22	127.99	–	62.77	m²	190.76
two brick thick	–	2.96	170.65	–	85.31	m²	255.96
Construct walls; battering (retaining)							
one and a half brick thick	–	2.22	127.99	–	62.77	m²	190.76
two brick thick	–	2.96	170.65	–	85.31	m²	255.96
Projections; vertical							
one brick × half brick	–	2.33	134.73	–	4.36	m	139.09
one brick × one brick	–	0.47	26.91	–	8.73	m	35.64
one and a half brick × one brick	–	0.70	40.42	–	13.10	m	53.52
two brick by one brick	–	0.77	44.24	–	17.46	m	61.70
Walls; half brick thick							
in honeycomb bond	–	0.60	34.65	–	14.18	m²	48.83
in quarter bond	–	0.56	32.05	–	19.12	m²	51.17
Facing bricks; PC £300.00/1000; English garden wall bond; in gauged mortar (1:1:6); facework one side							
Mechanically offloading; maximum 25 m distance; loading to wheelbarrows and transporting to location; walls	–	0.42	8.02	–	–	m²	8.02
Construct walls							
half brick thick	–	0.60	34.59	–	22.33	m²	56.92
half brick thick (using site cut snap headers to form bond)	–	0.80	46.43	–	22.33	m²	68.76
one brick thick	–	1.20	69.30	–	123.64	m²	192.94
one and a half brick thick	–	1.80	103.95	–	68.25	m²	172.20
two brick thick	–	2.40	138.60	–	92.77	m²	231.37
Walls; curved; mean radius 6 m							
half brick thick	–	0.90	51.98	–	23.23	m²	75.21
one brick thick	–	1.80	103.95	–	125.44	m²	229.39
Walls; curved; mean radius 1.50 m							
half brick thick	–	1.20	69.30	–	22.78	m²	92.08
one brick thick	–	2.40	138.60	–	124.54	m²	263.14
Walls; tapering; one face battering; average							
one and a half brick thick	–	2.20	127.05	–	70.95	m²	198.00
two brick thick	–	3.00	173.25	–	96.37	m²	269.62
Walls; battering (retaining)							
one and a half brick thick	–	2.00	115.50	–	70.95	m²	186.45
two brick thick	–	2.70	155.93	–	88.56	m²	244.49
Projections; vertical							
one brick × half brick	–	0.23	13.28	–	4.75	m	18.03
one brick × one brick	–	0.50	28.88	–	9.49	m	38.37
one and a half brick × one brick	–	0.70	40.42	–	14.24	m	54.66
two brick × one brick	–	0.80	46.20	–	19.92	m	66.12

F10 BRICK/BLOCK WALLING

Item Excluding site overheads and profit	PC £	Labour hours	Labour £	Plant £	Material £	Unit	Total rate £
Brickwork fair faced both sides; facing bricks in gauged mortar (1:1:6)							
extra for fair face both sides; flush, struck, weathered, or bucket-handle pointing	–	0.67	12.83	–	–	m²	**12.83**
Bricks; PC £800.00/1000; English garden wall bond; in gauged mortar (1:1:6)							
Walls							
half brick thick (stretcher bond)	–	0.60	34.65	–	53.83	m²	**88.48**
half brick thick (using site cut snap headers to form bond)	–	0.80	46.20	–	53.83	m²	**100.03**
one brick thick	–	1.20	69.30	–	107.46	m²	**176.76**
one and a half brick thick	–	1.80	103.95	–	159.87	m²	**263.82**
two brick thick	–	2.40	138.60	–	207.12	m²	**345.72**
Walls; curved; mean radius 6 m							
half brick thick	–	0.90	51.98	–	56.23	m²	**108.21**
one brick thick	–	2.40	138.60	–	114.18	m²	**252.78**
Walls; curved; mean radius 1.50 m							
half brick thick	–	0.53	30.61	–	58.63	m²	**89.24**
one brick thick	–	1.07	61.62	–	118.98	m²	**180.60**
Walls; stretcher bond; wall ties at 450 mm centres vertically and horizontally							
one brick thick	1.62	1.09	62.95	–	109.08	m²	**172.03**
two brick thick	4.86	1.64	94.71	–	211.98	m²	**306.69**
Brickwork fair faced both sides; facing bricks in gauged mortar (1:1:6)							
extra for fair face both sides; flush, struck, weathered or bucket-handle pointing	–	0.67	12.83	–	–	m²	**12.83**
Brick copings							
Copings; all brick headers-on-edge; two angles rounded 53 mm radius; flush pointing top and both sides as work proceeds; one brick wide; horizontal							
machine-made specials	48.05	0.16	9.47	–	48.99	m	**58.46**
handmade specials	61.80	0.16	9.47	–	62.74	m	**72.21**
Extra over copings for two courses machine-made tile creasings; projecting 25 mm each side; 260 mm wide copings; horizontal	6.11	0.17	9.82	–	6.58	m	**16.40**
Copings; all brick headers-on-edge; flush pointing top and both sides as work proceeds; one brick wide; horizontal							
facing bricks PC £300.00/1000	4.30	0.16	9.47	–	4.30	m	**13.77**
engineering bricks PC £276.00/1000	3.86	0.16	9.47	–	3.86	m	**13.33**

F10 BRICK/BLOCK WALLING

Item Excluding site overheads and profit	PC £	Labour hours	Labour £	Plant £	Material £	Unit	Total rate £
BLOCK WALLING							
Dense aggregate concrete blocks; Tarmac Topblock or other equal and approved; in gauged mortar (1:2:9); one course underground							
Walls							
Solid blocks 7 N/mm²							
440 × 215 × 100 mm thick	8.55	0.40	23.10	–	9.70	m²	**32.80**
440 × 215 × 140 mm thick	13.75	0.44	25.58	–	15.04	m²	**40.62**
Solid blocks 7 N/mm² laid flat							
440 × 100 × 215 mm thick	17.44	1.07	61.79	–	18.73	m²	**80.52**
Hollow concrete blocks							
440 × 215 × 215 mm thick	22.27	0.43	25.01	–	23.56	m²	**48.57**
Filling of hollow concrete blocks with concrete as work proceeds; tamping and compacting							
440 × 215 × 215 mm thick	18.22	0.07	3.81	–	18.22	m²	**22.03**
BRICK PIERS							
Isolated brick piers in English bond; in guaged mortar 1:1:6							
Engineering brick PC £276.00/1000							
one brick thick (225 mm)	–	2.33	134.73	–	44.14	m²	**178.87**
one and a half brick thick (337.5 mm)	–	3.00	173.25	–	66.21	m²	**239.46**
two brick thick (450 mm)	–	3.33	192.48	–	89.84	m²	**282.32**
three brick thick (675 mm)	–	4.13	238.68	–	132.42	m²	**371.10**
Engineering brick PC £276.00/1000 (not SMM)							
one brick (225 mm)	–	0.54	31.45	–	10.12	m	**41.57**
one and a half brick thick (337.5 mm)	–	1.00	57.75	–	30.48	m	**88.23**
two brick thick (450 mm)	–	1.50	86.63	–	48.47	m	**135.10**
three brick thick (675 mm)	–	2.79	161.12	–	98.52	m	**259.64**
Brick PC £300.00/1000							
one brick thick (225 mm)	–	2.33	134.73	–	47.16	m²	**181.89**
one and a half brick thick (337.5 mm)	–	3.00	173.25	–	70.75	m²	**244.00**
two brick thick (450 mm)	–	3.33	192.48	–	95.89	m²	**288.37**
three brick thick (675 mm)	–	4.13	238.68	–	141.49	m²	**380.17**
Brick PC £300.00/1000 (not SMM)							
one brick (225 mm)	–	0.54	31.42	–	10.82	m	**42.24**
one and a half brick thick (337.5 mm)	–	1.00	57.75	–	32.01	m	**89.76**
two brick thick (450 mm)	–	1.50	86.63	–	51.19	m	**137.82**
three brick thick (675 mm)	–	2.79	161.12	–	104.64	m	**265.76**
Brick PC £800.00/1000							
one brick thick (225 mm)	–	2.33	134.73	–	110.16	m²	**244.89**
one and a half brick thick (337.5 mm)	–	3.00	173.25	–	165.25	m²	**338.50**
two brick thick (450 mm)	–	3.33	192.48	–	221.89	m²	**414.37**
three brick thick (675 mm)	–	4.13	238.68	–	330.49	m²	**569.17**
Brick PC £800.00/1000 (not SMM)							
one brick (225 mm)	–	0.54	31.42	–	25.53	m	**56.95**
one and a half brick thick (337.5 mm)	–	1.00	57.75	–	63.91	m	**121.66**
two brick thick (450 mm)	–	1.50	86.63	–	107.89	m	**194.52**
three brick thick (675 mm)	–	2.79	161.12	–	232.21	m	**393.33**

F20 NATURAL STONE WALLING

Item Excluding site overheads and profit	PC £	Labour hours	Labour £	Plant £	Material £	Unit	Total rate £
BLOCK PIERS							
Isolated blockwork piers; to receive facing treatments (not included)							
Solid blocks; 440 × 100 × 215 mm thick; laid on-flat; 7 N/mm^2							
450 × 450 mm (SMM measurement)	34.31	2.00	77.00	–	35.31	m^2	**112.31**
450 × 450 mm (not SMM)	15.45	0.90	34.65	–	15.91	m	**50.56**
F20 NATURAL STONE WALLING							
CLARIFICATION NOTES ON LABOUR COSTS IN THIS SECTION							
Expert stonework team							
Generally a three man team is used in this section; The column 'Labour hours' reports team hours. The column 'Labour £' reports the total cost of the team for the unit of work shown							
2 man team	–	1.00	38.50	–	–	hr	**38.50**
GRANITE WALLS							
Granite walls							
Granite random rubble walls; laid dry							
200 mm thick; single faced	48.09	3.33	128.21	–	48.09	m^2	**176.30**
Granite walls; one face battering to 50°; pointing faces							
450 mm (average) thick	116.32	2.50	96.25	–	117.22	m^2	**213.47**
DRY STONE WALLING							
Dry stone walling – General							
Preamble: In rural areas where natural stone is a traditional material, it may be possible to use dry stone walling or dyking as an alternative to fences or brick walls. Many local authorities are willing to meet the extra cost of stone walling in areas of high landscape value, and they may hold lists of available craftsmen. DSWA Office, Westmorland County Showground, Lane Farm, Crooklands, Milnthorpe, Cumbria, LA7 7NH; Tel: 01539 567953; E-mail: information@dswa.org.uk							
Note: Traditional walls are not built on concrete foundations.							

F22 CAST STONE/ASHLAR WALLING/DRESSINGS

Item Excluding site overheads and profit	PC £	Labour hours	Labour £	Plant £	Material £	Unit	Total rate £
DRY STONE WALLING – cont							
Dry stone walling – General – cont							
Dry stone wall; wall on concrete foundation							
(not included); dry stone coursed wall inclusive							
of locking stones and filling to wall with broken							
stone or rubble; walls up to 1.20 m high;							
battered; 2 sides fair faced							
Yorkstone	–	3.25	125.13	–	66.73	m²	**191.86**
Cotswold stone	–	3.25	125.13	–	81.57	m²	**206.70**
Purbeck	–	3.25	125.13	–	88.98	m²	**214.11**
F22 CAST STONE/ASHLAR WALLING/							
DRESSINGS							
RECONSTITUTED STONE WALLING							
Haddonstone Ltd; cast stone piers;							
ornamental masonry in Portland Bath or							
Terracotta finished cast stone							
Gate Pier S120; to foundations and							
underground work measured separately;							
concrete infill							
S120G base unit to pier;							
699 × 699 × 172 mm	178.00	0.25	14.44	–	182.86	nr	**197.30**
S120F/F shaft base unit;							
533 × 533 × 280 mm	141.60	0.33	19.23	–	146.65	nr	**165.88**
S120E/E main shaft unit;							
533 × 533 × 280 mm; nr of units required							
dependent on height of pier	141.60	0.33	19.23	–	146.65	nr	**165.88**
S120D/D top shaft unit; 33 × 533 × 280 mm	141.60	0.33	19.23	–	146.65	nr	**165.88**
S120C pier cap unit; 737 × 737 × 114 mm	168.00	0.17	9.59	–	173.05	nr	**182.64**
S120B pier block unit; base for finial;							
533 × 533 × 64 mm	72.00	0.11	6.35	–	72.19	nr	**78.54**
Pier blocks; flat to receive gate finial							
S100B; 440 × 440 × 63 mm	46.80	0.11	6.35	–	47.24	nr	**53.59**
S120B; 546 × 546 × 64 mm	72.00	0.11	6.35	–	72.44	nr	**78.79**
S150B; 330 × 330 × 51 mm	25.20	0.11	6.35	–	25.64	nr	**31.99**
Pier caps; part weathered							
S100C; 915 × 915 × 150 mm	399.60	0.25	14.44	–	400.47	nr	**414.91**
S120C; 737 × 737 × 114 mm	168.00	0.17	9.59	–	168.44	nr	**178.03**
S150C; 584 × 584 × 120 mm	122.40	0.17	9.59	–	122.84	nr	**132.43**
Pier caps; weathered							
S230C; 1029 × 1029 × 175 mm	524.40	0.17	9.59	–	524.84	nr	**534.43**
S215C; 687 × 687 × 175 mm	230.40	0.17	9.59	–	230.84	nr	**240.43**
S210C; 584 × 584 × 175 mm	144.00	0.17	9.59	–	144.44	nr	**154.03**
Pier strings							
S100S; 800 × 800 × 55 mm	136.80	1.66	95.86	–	137.07	nr	**232.93**
S120S; 555 × 555 × 44 mm	62.40	0.17	9.59	–	62.67	nr	**72.26**
S150S; 457 × 457 × 48 mm	38.40	0.17	9.59	–	38.67	nr	**48.26**

F22 CAST STONE/ASHLAR WALLING/DRESSINGS

Item Excluding site overheads and profit	PC £	Labour hours	Labour £	Plant £	Material £	Unit	Total rate £
Balls and bases							
E150 A Ball 535 mm and E150C collared base	381.00	0.17	9.59	–	381.27	nr	**390.86**
E120 A Ball 330 mm and E120C collared base	144.00	0.17	9.59	–	144.27	nr	**153.86**
E110 A Ball 230 mm and E110C collared base	101.00	0.17	9.59	–	101.27	nr	**110.86**
E100 A Ball 170 mm and E100B plain base	64.00	0.17	9.59	–	64.27	nr	**73.86**
Haddonstone Ltd; cast stone copings; ornamental masonry in Portland Bath or Terracotta finished cast stone Copings for walls; bedded, jointed and pointed in approved coloured cement-lime mortar 1:2:9							
T100 weathered coping; 102 mm high × 178 mm wide × 914 mm	41.86	0.11	6.41	–	42.23	m	**48.64**
T140 weathered coping; 102 mm high × 337 mm wide × 914 mm	69.32	0.11	6.41	–	69.70	m	**76.11**
T200 weathered coping; 127 mm high × 508 mm wide × 750 mm	104.64	0.11	6.41	–	105.01	m	**111.42**
T170 weathered coping; 108 mm high × 483 mm wide × 914 mm	112.49	0.11	6.41	–	112.86	m	**119.27**
T340 raked coping; 75–100 mm high × 290 mm wide × 900 mm	64.09	0.11	6.41	–	64.47	m	**70.88**
T310 raked coping; 76–89 mm high × 381 mm wide × 914 mm	74.56	0.11	6.41	–	74.93	m	**81.34**
Bordeaux walling; Forticrete Ltd Dry stacked random sized units 150–400 mm long × 150 mm high cast stone wall mechanically interlocked with fibreglass pins; constructed to levelling pad of coarse compacted granular material back filled behind the elevation with 300 mm wide granular drainage material; walls to 5.00 m high (retaining walls over heights shown below require individual design)							
gravity wall; near vertical; 250 mm thick	81.00	0.33	19.23	–	81.00	m²	**100.23**
gravity wall; 9.5°; battered	81.00	0.67	38.46	–	81.00	m²	**119.46**
copings to Bordeaux wall; 70 mm thick; random lengths	13.00	0.08	4.81	–	13.00	m	**17.81**
Retaining wall; as above but reinforced with Tensar Geogrid RE520 laid between every two courses horizontally into the face of the excavation (excavation not included) Near vertical; 250 mm thick; 1.50 m of geogrid length							
up to 1.20 m high; 2 layers of geogrid	81.00	0.50	28.88	–	87.75	m²	**116.63**
1.20–1.50 m high; 3 layers of geogrid	81.00	0.67	38.46	–	91.13	m²	**129.59**
1.50–1.80 m high; 4 layers of geogrid	81.00	0.75	43.31	–	94.50	m²	**137.81**

F31 PRECAST CONCRETE SILLS/LINTELS/COPINGS/FEATURES

Item Excluding site overheads and profit	PC £	Labour hours	Labour £	Plant £	Material £	Unit	Total rate £
RECONSTITUTED STONE WALLING – cont							
Retaining wall – cont							
Battered; maximum 1:3 slope							
up to 1.20 m high; 3 layers of geogrid	81.00	0.83	48.11	–	91.13	m²	**139.24**
1.20–1.50 m high; 4 layers of geogrid	81.00	0.83	48.11	–	94.50	m²	**142.61**
1.50–1.80 m high; 5 layers of geogrid	81.00	1.00	57.75	–	97.31	m²	**155.06**
F30 ACCESSORIES/SUNDRY ITEMS FOR BRICK/BLOCK/STONE WALLING							
DAMP-PROOF COURSES							
Damp-proof courses; pitch polymer; 150 mm laps							
Horizontal							
width not exceeding 225 mm	4.86	1.14	21.95	–	6.44	m²	**28.39**
width exceeding 225 mm	5.42	0.58	11.16	–	7.00	m²	**18.16**
Vertical							
width not exceeding 225 mm	4.86	1.72	33.11	–	6.44	m²	**39.55**
Two courses slates in cement mortar (1:3)							
Horizontal							
width exceeding 225 mm	9.31	3.46	66.61	–	13.61	m²	**80.22**
Vertical							
width exceeding 225 mm	9.31	5.18	99.72	–	13.61	m²	**113.33**
F31 PRECAST CONCRETE SILLS/LINTELS/ COPINGS/FEATURES							
COPINGS AND PIERCAPS							
Precast concrete coping							
Copings; once weathered; twice grooved							
152 × 75 mm	5.62	0.40	7.70	–	5.88	m	**13.58**
178 × 65 mm	7.77	0.40	7.70	–	8.13	m	**15.83**
305 × 75 mm	11.03	0.50	9.63	–	11.60	m	**21.23**
Pier caps; four sides weathered							
305 × 305 mm	7.04	1.00	19.25	–	7.21	nr	**26.46**
381 × 381 mm	7.46	1.00	19.25	–	7.63	nr	**26.88**
533 × 533 mm	7.25	1.20	23.10	–	7.42	nr	**30.52**

H CLADDING/COVERING

Item Excluding site overheads and profit	PC £	Labour hours	Labour £	Plant £	Material £	Unit	Total rate £
H CLADDING/COVERING							
H51 NATURAL STONE SLAB CLADDING/ FEATURES							
Sawn Yorkstone cladding; Johnsons Wellfield Quarries Ltd							
Six sides sawn stone; rubbed face; sawn and jointed edges; fixed to blockwork (not included) with stainless steel fixings; Ancon Ltd; grade 304 stainless steel frame cramp and dowel 7 mm; cladding units drilled 4 × to receive dowels							
440 × 200 × 50 mm thick	84.50	1.87	36.00	–	87.90	m²	**123.90**
H52 CAST STONE SLAB CLADDING/ FEATURES							
Cast stone cladding; Haddonstone Ltd							
Reconstituted stone in Portland Bath or Terracotta; fixed to blockwork or concrete (not included) with stainless steel fixings; stainless steel frame cramp and dowel M6 mm; cladding units drilled 4 × to receive dowels							
1000 × 300 × 50 mm thick	45.00	1.87	36.00	–	48.40	m²	**84.40**

J20 MASTIC ASPHALT TANKING/DAMP-PROOFING

Item Excluding site overheads and profit	PC £	Labour hours	Labour £	Plant £	Material £	Unit	Total rate £
J10 SPECIALIST WATERPROOF RENDERING							
SIKA							
Sika 1 waterproof rendering; steel trowelled							
Walls; 18 mm thick; three coats; to concrete base							
width exceeding 300 mm	–	–	–	–	–	m²	45.42
width not exceeding 300 mm	–	–	–	–	–	m²	47.92
Walls; 18 mm thick; three coats; to concrete base							
width exceeding 300 mm	–	–	–	–	–	m²	42.43
width not exceeding 300 mm	–	–	–	–	–	m²	44.93
J20 MASTIC ASPHALT TANKING/DAMP-PROOFING							
TANKING AND DAMP-PROOFING							
Tanking and damp-proofing; mastic asphalt; type T1097; Bituchem							
13 mm thick; one coat covering; to concrete base; flat; work subsequently covered							
width exceeding 300 mm	–	–	–	–	–	m²	11.24
20 mm thick; two coat coverings; to concrete base; flat; work subsequently covered							
width exceeding 300 mm	–	–	–	–	–	m²	14.09
30 mm thick; three coat coverings; to concrete base; flat; work subsequently covered							
width exceeding 300 mm	–	–	–	–	–	m²	19.13
13 mm thick; two coat coverings; to brickwork base; vertical; work subsequently covered							
width exceeding 300 mm	–	–	–	–	–	m²	37.34
20 mm thick; three coat coverings; to brickwork base; vertical; work subsequently covered							
width exceeding 300 mm	–	–	–	–	–	m²	50.98
Internal angle fillets; work subsequently covered	–	–	–	–	–	m	3.97
Turning asphalt nibs into grooves; 20 mm deep	–	–	–	–	–	m	2.52

J40 FLEXIBLE SHEET TANKING/DAMP-PROOFING

Item Excluding site overheads and profit	PC £	Labour hours	Labour £	Plant £	Material £	Unit	Total rate £
J30 LIQUID APPLIED TANKING/DAMP-PROOFING							
COLD APPLIED BITUMOUS EMULSION							
Tanking and damp-proofing; Ruberoid Building Products; Synthaprufe cold applied bituminous emulsion waterproof coating Synthaprufe; to smooth finished concrete or screeded slabs; flat; blinding with sand							
two coats	3.58	0.22	4.28	–	3.73	m²	**8.01**
three coats	5.37	0.31	5.97	–	5.52	m²	**11.49**
Synthaprufe; to fair faced brickwork with flush joints, rendered brickwork or smooth finished concrete walls; vertical							
two coats	4.03	0.29	5.50	–	4.17	m²	**9.67**
three coats	5.90	0.40	7.70	–	6.05	m²	**13.75**
LIQUID ASPHALT							
Tanking and damp-proofing; RIW Ltd Liquid asphaltic composition; to smooth finished concrete screeded slabs or screeded slabs; flat							
two coats	5.46	0.33	6.42	–	5.46	m²	**11.88**
Liquid asphaltic composition; fair faced brickwork with flush joints, rendered brickwork or smooth finished concrete walls; vertical							
two coats	5.46	0.50	9.63	–	5.46	m²	**15.09**
Heviseal; to smooth finished concrete or screeded slabs; to surfaces of ponds, tanks or planters; flat							
two coats	9.11	0.33	6.42	–	9.11	m²	**15.53**
Heviseal; to fair faced brickwork with flush joints, rendered brickwork or smooth finished concrete walls; to surfaces of retaining walls, ponds, tanks or planters; vertical							
two coats	9.11	0.50	9.63	–	9.11	m²	**18.74**
J40 FLEXIBLE SHEET TANKING/DAMP-PROOFING							
SELF ADHESIVE							
Tanking and damp-proofing; Grace Construction Products Bitu-thene 3000; 1.50 mm thick; overlapping and bonding; including sealing all edges							
to concrete slabs; flat	9.13	0.25	4.81	–	9.82	m²	**14.63**
to brick/concrete walls; vertical	9.13	0.40	7.70	–	9.82	m²	**17.52**

M SURFACE FINISHES

Item Excluding site overheads and profit	PC £	Labour hours	Labour £	Plant £	Material £	Unit	Total rate £
M SURFACE FINISHES							
M20 PLASTERED/RENDERED/ ROUGHCAST COATINGS							
Cement: lime: sand (1:1:6); 19 mm thick; two coats; wood floated finish Walls							
width exceeding 300 mm; to brickwork or blockwork base	–	–	–	–	–	m²	**14.25**
Extra over cement: sand: lime (1:1:6) coatings for decorative texture finish with water repellent cement							
combed or floated finish	–	–	–	–	–	m²	**3.80**
M40 STONE/CONCRETE/QUARRY/ CERAMIC TILING/MOSAIC							
Ceramic tiles; unglazed slip-resistant; various colours and textures; jointing Floors							
level or to falls only not exceeding 15° from horizontal; 150 × 150 × 8 mm thick	19.03	1.00	19.25	–	21.89	m²	**41.14**
level or to falls only not exceeding 15° from horizontal; 150 × 150 × 12 mm thick	23.18	1.00	19.25	–	26.05	m²	**45.30**
Clay tiles – General Preamble: Typical specification – Clay tiles shall be reasonably true to shape, flat, free from flaws, frost resistant and true to sample approved by the Landscape Architect prior to laying. Quarry tiles (or semi-vitrified tiles) shall be of external quality, either heather brown or blue, to size specified, laid on 1:2:4 concrete, with 20 mm maximum aggregate 100 mm thick, on 100 mm hardcore. The hardened concrete should be well wetted and the surplus water taken off. Clay tiles shall be thoroughly wetted immediately before laying and then drained and shall be bedded to 19 mm thick cement: sand (1:3) screed. Joints should be approximately 4 mm (or 3 mm for vitrified tiles) grouted in cement: sand (1:2) and cleaned off immediately.							

M SURFACE FINISHES

Item Excluding site overheads and profit	PC £	Labour hours	Labour £	Plant £	Material £	Unit	Total rate £
Quarry tiles; external quality; including bedding; jointing							
Floors							
level or to falls only not exceeding 15° from horizontal; 150 × 150 × 12.5 mm thick; heather brown	23.78	0.80	15.40	–	26.65	m²	**42.05**
level or to falls only not exceeding 15° from horizontal; 225 × 225 × 29 mm thick; heather brown	23.78	0.67	12.83	–	26.65	m²	**39.48**
level or to falls only not exceeding 15° from horizontal; 150 × 150 × 12.5 mm thick; blue/black	15.10	1.00	19.25	–	17.96	m²	**37.21**
level or to falls only not exceeding 15° from horizontal; 194 × 194 × 12.5 mm thick; heather brown	13.91	0.80	15.40	–	16.78	m²	**32.18**
M60 PAINTING/CLEAR FINISHING							
Eura Conservation Ltd; restoration of railings; works carried out off site; excludes removal of railings from site							
Shotblast railings; remove all traces of paint and corrosion; apply temporary protective primer; measured overall 2 sides							
to plain railings	–	–	–	–	–	m²	**42.00**
to decorative railings	–	–	–	–	–	m²	**84.00**
Paint new galvanized railings in situ with 3 coats system							
vertical bar railings or gates	–	–	–	–	–	m²	**11.00**
vertical bar railings or gates with dog mesh panels	–	–	–	–	–	m²	**12.00**
gate posts not exceeding 300 mm girth	–	–	–	–	–	m²	**6.00**
Paint previously painted railings; shotblasted off site							
vertical bar railings or gates	–	–	–	–	–	m²	**14.00**
gate posts not exceeding 300 mm girth	–	–	–	–	–	m²	**6.00**
Prepare; touch up primer; two undercoats and one finishing coat of gloss oil paint; on metal surfaces							
General surfaces							
girth exceeding 300 mm	1.55	0.67	12.83	–	1.55	m²	**14.38**
isolated surfaces; girth not exceeding 300 mm	4.70	0.40	7.70	–	4.70	m	**12.40**
isolated areas not exceeding 0.50 m² irrespective of girth	3.10	0.67	12.83	–	3.10	nr	**15.93**
Ornamental railings; each side measured separately							
girth exceeding 300 mm	1.55	0.57	11.00	–	1.55	m²	**12.55**

M SURFACE FINISHES

Item Excluding site overheads and profit	PC £	Labour hours	Labour £	Plant £	Material £	Unit	Total rate £
M60 PAINTING/CLEAR FINISHING – cont							
Prepare; one coat primer; two finishing							
coats of gloss paint; on wood surfaces							
New wood surfaces							
girth exceeding 300 mm	2.28	0.80	15.40	–	2.28	m²	**17.68**
isolated surfaces; girth not exceeding							
300 mm	0.76	1.00	19.25	–	0.76	m	**20.01**
isolated areas not exceeding 0.50 m²							
irrespective of girth	3.36	0.50	9.63	–	3.36	nr	**12.99**
Previously painted wood surfaces							
girth exceeding 300 mm	0.80	0.50	9.63	–	0.80	m²	**10.43**
isolated surfaces; girth not exceeding							
300 mm	0.26	0.67	12.83	–	0.26	m	**13.09**
isolated areas not exceeding 0.50 m²							
irrespective of girth	1.59	0.40	7.70	–	1.59	nr	**9.29**
Fences and sheds; prepare; two coats of							
Protek wood preserver on wood surfaces							
Planed surfaces							
girth exceeding 300 mm	0.60	0.07	1.28	–	0.60	m²	**1.88**
isolated surfaces; girth not exceeding							
300 mm	0.20	0.17	3.21	–	0.20	m	**3.41**
isolated areas not exceeding 0.50 m²							
irrespective of girth	0.60	0.25	4.81	–	0.60	nr	**5.41**
Sawn surfaces							
girth exceeding 300 mm	0.22	0.10	1.93	–	0.22	m²	**2.15**
isolated surfaces; girth not exceeding							
300 mm	2.97	0.17	3.21	–	2.97	m	**6.18**
isolated areas not exceeding 0.50 m²							
irrespective of girth	0.89	0.25	4.81	–	0.89	nr	**5.70**
Fences and sheds; prepare; two coats of							
Sadolin wood preserver on wood surfaces							
Planed surfaces							
girth exceeding 300 mm	1.89	0.07	1.28	–	1.89	m²	**3.17**
isolated surfaces; girth not exceeding							
300 mm	0.63	0.17	3.21	–	0.63	m	**3.84**
isolated areas not exceeding 0.50 m²							
irrespective of girth	0.95	0.25	4.81	–	0.95	nr	**5.76**
Sawn surfaces							
girth exceeding 300 mm	1.43	0.10	1.93	–	1.43	m²	**3.36**
isolated surfaces; girth not exceeding							
300 mm	1.01	0.17	3.21	–	1.01	m	**4.22**
isolated areas not exceeding 0.50 m²							
irrespective of girth	1.51	0.25	4.81	–	1.51	nr	**6.32**

P BUILDING FABRIC SUNDRIES

Item Excluding site overheads and profit	PC £	Labour hours	Labour £	Plant £	Material £	Unit	Total rate £
Prepare; proprietary solution primer; two coats of dark stain; on wood surfaces							
General surfaces							
girth exceeding 300 mm	1.80	0.17	3.21	–	1.80	m²	**5.01**
isolated surfaces; girth not exceeding							
300 mm	0.60	0.13	2.41	–	0.60	m	**3.01**
Two coats Weathershield; to clean, dry surfaces; in accordance with manufacturer's instructions							
Brick or block walls							
girth exceeding 300 mm	1.42	0.40	7.70	–	1.42	m²	**9.12**
Cement render or concrete walls							
girth exceeding 300 mm	1.42	0.33	6.42	–	1.42	m²	**7.84**
P BUILDING FABRIC SUNDRIES							
P30 TRENCHES/PIPEWAYS/PITS FOR BURIED ENGINEERING SERVICES							
Excavating trenches; using 3 tonne tracked excavator; to receive pipes; grading bottoms; earthwork support; filling with excavated material to within 150 mm of finished surfaces and compacting; completing fill with topsoil; disposal of surplus soil							
Services not exceeding 200 mm nominal size							
average depth of run not exceeding 0.50 m	1.51	0.12	2.31	1.05	1.51	m	**4.87**
average depth of run not exceeding 0.75 m	1.51	0.16	3.14	1.44	1.51	m	**6.09**
average depth of run not exceeding 1.00 m	1.51	0.28	5.45	2.50	1.51	m	**9.46**
average depth of run not exceeding 1.25 m	1.51	0.38	7.38	3.37	1.51	m	**12.26**
Excavating trenches; using 3 tonne tracked excavator; to receive pipes; grading bottoms; earthwork support; filling with imported granular material and compacting; disposal of surplus soil							
Services not exceeding 200 mm nominal size							
average depth of run not exceeding 0.50 m	1.37	0.09	1.67	0.75	3.94	m	**6.36**
average depth of run not exceeding 0.75 m	2.05	0.11	2.08	0.94	5.91	m	**8.93**
average depth of run not exceeding 1.00 m	2.74	0.14	2.69	1.21	7.88	m	**11.78**
average depth of run not exceeding 1.25 m	3.42	0.23	4.40	2.01	9.85	m	**16.26**

P BUILDING FABRIC SUNDRIES

Item Excluding site overheads and profit	PC £	Labour hours	Labour £	Plant £	Material £	Unit	Total rate £
P30 TRENCHES/PIPEWAYS/PITS FOR **BURIED ENGINEERING SERVICES – cont**							
Excavating trenches; using 3 tonne tracked **excavator; to receive pipes; grading** **bottoms; earthwork support; filling with** **lean mix concrete; disposal of surplus soil**							
Services not exceeding 200 mm nominal size							
average depth of run not exceeding 0.50 m	11.38	0.11	2.05	0.36	14.47	m	**16.88**
average depth of run not exceeding 0.75 m	17.08	0.13	2.50	0.45	21.71	m	**24.66**
average depth of run not exceeding 1.00 m	22.77	0.17	3.21	0.60	28.94	m	**32.75**
average depth of run not exceeding 1.25 m	28.46	0.23	4.33	0.90	36.18	m	**41.41**
Earthwork support; providing support to **opposing faces of excavation; moving** **along as work proceeds**							
Maximum depth not exceeding 2.00 m							
trenchbox; distance between opposing faces not exceeding 2.00 m	–	0.80	15.40	21.00	–	m	**36.40**
timber; distance between opposing faces not exceeding 500 mm	–	0.20	7.70	–	0.03	m	**7.73**

Q10 KERBS/EDGINGS/CHANNELS/PAVING ACCESSORIES

Item	PC	Labour	Labour	Plant	Material	Unit	Total
Excluding site overheads and profit	£	hours	£	£	£		rate £
Q10 KERBS/EDGINGS/CHANNELS/PAVING ACCESSORIES							
CLARIFICATION NOTES ON LABOUR COSTS IN THIS SECTION							
General groundworks team							
Generally a three man team is used in this section; The column 'Labour hours' reports team hours. The column 'Labour £' reports the total cost of the team for the unit of work shown							
3 man team	–	1.00	57.75	–	–	hr	**57.75**
2 man team	–	1.00	38.50	–	–	hr	**38.50**
FOUNDATIONS TO KERBS							
Foundations to kerbs							
Excavating trenches; width 300 mm; 3 tonne excavator; disposal off site							
depth 300 mm	–	–	–	3.96	1.54	m	**5.50**
depth 400 mm	–	–	–	4.33	2.06	m	**6.39**
By hand; barrowing to spoil heap and disposal off site							
depth 300	–	0.06	3.46	–	17.14	m	**20.60**
depth 400	–	0.08	4.62	–	17.14	m	**21.76**
Excavating trenches; width 450 mm; 3 tonne excavator; disposal off site							
depth 300 mm	–	–	–	4.94	2.31	m	**7.25**
depth 400 mm	–	–	–	5.65	3.09	m	**8.74**
Foundations; to kerbs, edgings or channels; in situ concrete; 21 N/mm² – 20 mm aggregate ((1:2:4) site mixed); one side against earth face, other against formwork (not included); site mixed concrete							
Site mixed concrete							
150 mm wide × 100 mm deep	–	0.04	2.57	–	1.50	m	**4.07**
150 mm wide × 150 mm deep	–	0.06	3.21	–	2.24	m	**5.45**
200 mm wide × 150 mm deep	–	0.07	3.85	–	2.99	m	**6.84**
300 mm wide × 150 mm deep	–	0.06	3.37	–	4.49	m	**7.86**
600 mm wide × 200 mm deep	–	0.06	3.37	–	11.96	m	**15.33**
Ready mixed concrete							
150 mm wide × 100 mm deep	–	0.04	2.57	–	1.13	m	**3.70**
150 mm wide × 150 mm deep	–	0.06	3.21	–	1.69	m	**4.90**
200 mm wide × 150 mm deep	–	0.07	3.85	–	2.25	m	**6.10**
300 mm wide × 150 mm deep	–	0.08	4.49	–	3.38	m	**7.87**
600 mm wide × 200 mm deep	–	0.10	5.50	–	9.01	m	**14.51**

Q10 KERBS/EDGINGS/CHANNELS/PAVING ACCESSORIES

Item Excluding site overheads and profit	PC £	Labour hours	Labour £	Plant £	Material £	Unit	Total rate £
FOUNDATIONS TO KERBS – cont							
Foundations – cont							
Formwork; sides of foundations (this will usually be required to one side of each kerb foundation adjacent to road subbases)							
100 mm deep	–	0.01	0.80	–	0.28	m	1.08
150 mm deep	–	0.01	0.80	–	0.42	m	1.22
PRECAST CONCRETE KERBS							
Precast concrete kerbs, channels, edgings etc.; Marshalls Mono; bedding, jointing and pointing in cement mortar (1:3); including haunching with in situ concrete; 11.50 N/mm² – 40 mm aggregate one side							
Kerbs; straight							
150 × 305 mm; HB1	7.99	0.17	9.63	–	10.55	m	20.18
125 × 255 mm; HB2; SP	3.82	0.15	8.55	–	6.38	m	14.93
125 × 150 mm; BN	2.59	0.13	7.70	–	5.15	m	12.85
Dropper kerbs; left and right handed							
125 × 255–150 mm; DL1 or DR1	5.78	0.17	9.63	–	8.34	m	17.97
125 × 255–150 mm; DL2 or DR2	5.78	0.17	9.63	–	8.34	m	17.97
Quadrant kerbs							
305 mm radius	9.72	0.17	9.63	–	11.43	nr	21.06
455 mm radius	10.46	0.17	9.63	–	13.02	nr	22.65
Straight kerbs or channels; to radius; 125 × 255 mm							
0.90 m radius (2 units per quarter circle)	7.31	0.27	15.40	–	9.95	m	25.35
1.80 m radius (4 units per quarter circle)	7.31	0.24	14.00	–	9.95	m	23.95
2.40 m radius (5 units per quarter circle)	7.31	0.23	13.05	–	9.95	m	23.00
3.00 m radius (5 units per quarter circle)	7.31	0.22	12.83	–	9.95	m	22.78
4.50 m radius (2 units per quarter circle)	7.31	0.20	11.56	–	9.95	m	21.51
6.10 m radius (11 units per quarter circle)	7.31	0.19	11.16	–	9.95	m	21.11
7.60 m radius (14 units per quarter circle)	7.31	0.19	11.00	–	9.95	m	20.95
9.15 m radius (17 units per quarter circle)	7.31	0.19	10.70	–	9.95	m	20.65
10.70 m radius (20 units per quarter circle)	7.31	0.19	10.70	–	9.95	m	20.65
12.20 m radius (22 units per quarter circle)	7.31	0.18	10.14	–	9.95	m	20.09
Kerbs; Conservation Kerb units; to simulate natural granite kerbs							
255 × 150 × 914 mm; laid flat	21.14	0.19	11.00	–	25.72	m	36.72
150 × 255 × 914 mm; laid vertical	21.14	0.19	11.00	–	24.97	m	35.97
145 × 255 mm; radius internal 3.25 m	22.21	0.28	16.04	–	26.05	m	42.09
150 × 255 mm; radius external 3.40 m	21.21	0.28	16.04	–	25.05	m	41.09
145 × 255 mm; radius internal 6.50 m	19.85	0.22	12.83	–	23.68	m	36.51
145 × 255 mm; radius external 6.70 m	19.73	0.22	12.83	–	23.57	m	36.40
150 × 255 mm; radius internal 9.80 m	19.62	0.22	12.83	–	23.46	m	36.29
150 × 255 mm; radius external 10.00 m	18.60	0.22	12.83	–	22.44	m	35.27
305 × 305 × 255 mm; solid quadrants	30.61	0.19	11.00	–	34.45	nr	45.45

Q10 KERBS/EDGINGS/CHANNELS/PAVING ACCESSORIES

Item Excluding site overheads and profit	PC £	Labour hours	Labour £	Plant £	Material £	Unit	Total rate £
Marshalls Mono; small element precast concrete kerb system; Keykerb Large (KL) upstand of 100–125 mm; on 150 mm concrete foundation including haunching with in situ concrete (1:3:6) 1 side							
Bullnosed or half battered; 100 × 127 × 200 mm							
laid straight	12.16	0.27	15.40	–	13.06	m	**28.46**
radial blocks laid to curve; 8 blocks/1/4 circle – 500 mm radius	9.36	0.33	19.25	–	10.26	m	**29.51**
radial blocks laid to curve; 8 radial blocks, alternating 8 standard blocks/1/4 circle – 1000 mm radius	15.44	0.42	24.06	–	16.34	m	**40.40**
radial blocks, alternating 16 standard blocks/1/4 circle – 1500 mm radius	14.20	0.50	28.88	–	15.10	m	**43.98**
internal angle 90°	5.14	0.07	3.85	–	5.14	nr	**8.99**
external angle	5.14	0.07	3.85	–	5.14	nr	**8.99**
drop crossing kerbs; KL half battered to KL Splay; LH and RH	21.48	0.33	19.25	–	22.38	pair	**41.63**
drop crossing kerbs; KL half battered to KS bullnosed	20.90	0.33	19.25	–	21.80	pair	**41.05**
Marshalls Mono; small element precast concrete kerb system; Keykerb Small (KS) upstand of 25–50 mm; on 150 mm concrete foundation including haunching with in situ concrete (1:3:6) 1 side							
Half battered							
laid straight	8.54	0.27	15.40	–	9.44	m	**24.84**
radial blocks laid to curve; 8 blocks/1/4 circle; 500 mm radius	7.01	0.33	19.25	–	7.91	m	**27.16**
radial blocks laid to curve; 8 radial blocks, alternating 8 standard blocks/1/4 circle; 1000 mm radius	11.28	0.42	24.06	–	12.18	m	**36.24**
radial blocks, alternating 16 standard blocks/1/4 circle; 1500 mm radius	12.07	0.50	28.88	–	12.97	m	**41.85**
internal angle 90°	5.14	0.07	3.85	–	5.14	nr	**8.99**
external angle	5.14	0.07	3.85	–	5.14	nr	**8.99**
PRECAST CONCRETE EDGINGS							
Precast concrete edging units; including haunching with in situ concrete; 11.50 N/mm² – 40 mm aggregate both sides							
Edgings; rectangular, bullnosed or chamfered							
50 × 150 mm	1.75	0.11	6.42	–	6.01	m	**12.43**
125 × 150 mm bullnosed	1.81	0.11	6.42	–	6.07	m	**12.49**
50 × 200 mm	2.63	0.11	6.42	–	6.89	m	**13.31**
50 × 250 mm	3.05	0.11	6.42	–	7.31	m	**13.73**
50 × 250 mm flat top	3.22	0.11	6.42	–	7.48	m	**13.90**

Q10 KERBS/EDGINGS/CHANNELS/PAVING ACCESSORIES

Item Excluding site overheads and profit	PC £	Labour hours	Labour £	Plant £	Material £	Unit	Total rate £
STONE KERBS							
Dressed natural stone kerbs – General							
Preamble: The kerbs are to be good, sound							
and uniform in texture and free from defects;							
worked straight or to radius, square and out of							
wind, with the top front and back edges							
parallel or concentric to the dimensions							
specified. All drill and pick holes shall be							
removed from dressed faces. Standard							
dressings shall be designated as either fine							
picked, single axed or nidged or rough							
punched.							
Dressed natural stone kerbs; CED Ltd; on							
concrete foundations (not included);							
including haunching with in situ concrete;							
11.50 N/mm² – 40 mm aggregate one side							
Granite kerbs; 125 × 250 mm							
special quality; straight; random lengths	22.88	0.27	15.40	–	26.00	m	**41.40**
Granite kerbs; 125 × 250 mm; curved to mean							
radius 3 m							
special quality; random lengths	32.89	0.30	17.52	–	36.01	m	**53.53**
STONE EDGINGS							
New granite sett edgings; CED Ltd;							
100 × 100 × 100 mm; bedding in cement							
mortar (1:4); including haunching with in							
situ concrete; 11.50 N/mm² – 40 mm							
aggregate one side; concrete foundations							
(not included)							
Edgings 100 × 100 × 100 mm; ref 1R 'better							
quality' with 20 mm pointing gaps							
100 mm wide	3.06	0.44	25.60	–	5.95	m	**31.55**
2 rows; 220 mm wide	6.12	0.74	42.73	–	9.86	m	**52.59**
3 rows; 340 mm wide	9.19	1.00	57.75	–	12.12	m	**69.87**
Edgings 100 × 100 × 200 mm; ref 1R 'better							
quality' with 20 mm pointing gaps							
100 mm wide	3.12	0.25	14.44	–	6.01	m	**20.45**
2 rows; 220 mm wide	6.25	0.50	28.88	–	9.99	m	**38.87**
3 rows; 340 mm wide	9.37	0.75	43.31	–	12.31	m	**55.62**
Edgings 100 × 100 × 100 mm; ref 8R 'standard							
quality' with 20 mm pointing gaps							
100 mm wide	2.51	0.44	25.60	–	5.40	m	**31.00**
2 rows; 220 mm wide	5.03	0.74	42.73	–	8.76	m	**51.49**
3 rows; 340 mm wide	7.54	1.00	57.75	–	10.48	m	**68.23**

Q10 KERBS/EDGINGS/CHANNELS/PAVING ACCESSORIES

Item Excluding site overheads and profit	PC £	Labour hours	Labour £	Plant £	Material £	Unit	Total rate £
Edgings 100 × 100 × 200 mm; ref 8R 'standard quality' with 20 mm pointing gaps							
100 mm wide	2.76	0.25	14.44	–	5.65	m	**20.09**
2 rows; 220 mm wide	5.53	0.50	28.88	–	9.26	m	**38.14**
3 rows; 340 mm wide	8.29	0.75	43.31	–	11.23	m	**54.54**
Reclaimed granite setts edgings; CED Ltd; 100 × 100 mm; bedding in cement mortar (1:4); including haunching with in situ concrete; 11.50 N/mm² – 40 mm aggregate one side; on concrete foundations (not included)							
Edgings; with 20 mm pointing gaps							
100 mm wide	3.58	0.44	25.60	–	6.47	m	**32.07**
2 rows; 220 mm wide	7.17	0.74	42.73	–	10.90	m	**53.63**
3 rows; 340 mm wide	10.75	1.00	57.75	–	13.69	m	**71.44**
Natural yorkstone edgings; Johnsons Wellfield Quarries; sawn 6 sides; 50 mm thick; on prepared base measured separately; bedding on 25 mm cement: sand (1:3); cement: sand (1:3) joints							
Edgings							
100 mm wide × random lengths	6.98	0.17	9.63	–	7.34	m	**16.97**
100 × 100 mm	7.73	0.17	9.63	–	11.31	m	**20.94**
100 × 200 mm	15.46	0.17	9.63	–	19.04	m	**28.67**
250 mm wide × random lengths	15.27	0.13	7.70	–	18.85	m	**26.55**
500 mm wide × random lengths	30.55	0.11	6.42	–	34.14	m	**40.56**
Yorkstone edgings; 600 mm long × 250 mm wide; cut to radius							
1.00 m to 3.00	48.63	0.17	9.63	–	48.99	m	**58.62**
3.00 m to 5.00 m	48.63	0.15	8.55	–	48.99	m	**57.54**
exceeding 5.00 m	48.63	0.13	7.70	–	48.99	m	**56.69**
Natural yorkstone edgings; Johnsons Wellfield Quarries; sawn 6 sides; 75 mm thick; on prepared base measured separately; bedding on 25 mm cement: sand (1:3); cement: sand (1:3) joints							
Edgings							
100 mm wide × random lengths	8.73	0.20	11.55	–	9.08	m	**20.63**
100 × 100 mm	0.90	0.20	11.55	–	4.48	m	**16.03**
100 × 200 mm	1.80	0.17	9.63	–	5.38	m	**15.01**
250 mm wide × random lengths	18.25	0.17	9.63	–	21.83	m	**31.46**
500 mm wide × random lengths	36.49	0.13	7.70	–	40.07	m	**47.77**
Edgings; 600 mm long × 250 mm wide; cut to radius							
1.00 m to 3.00	56.11	0.20	11.55	–	56.47	m	**68.02**
3.00 m to 5.00 m	56.11	0.17	9.63	–	56.47	m	**66.10**
exceeding 5.00 m	56.11	0.15	8.55	–	56.47	m	**65.02**

Q10 KERBS/EDGINGS/CHANNELS/PAVING ACCESSORIES

Item Excluding site overheads and profit	PC £	Labour hours	Labour £	Plant £	Material £	Unit	Total rate £
STONE EDGINGS – cont							
Natural stone, slate or granite flag edgings;							
CED Ltd; on prepared base (not included);							
bedding on 25 mm cement: sand (1:3);							
cement: sand (1:3) joints							
Edgings; silver grey							
100 mm wide × random lengths	6.77	0.33	19.25	–	7.13	m	**26.38**
100 × 100 mm	7.74	0.50	28.88	–	8.10	m	**36.98**
100 mm long × 200 mm wide	12.57	0.20	11.55	–	12.93	m	**24.48**
250 mm wide × random lengths	13.30	0.25	14.44	–	16.88	m	**31.32**
300 mm wide × random lengths	14.51	0.25	14.44	–	18.09	m	**32.53**
BRICK EDGINGS							
Brick or block stretchers; bedding in							
cement mortar (1:4); on 150 mm deep							
concrete foundations (not included);							
including haunching with in situ concrete;							
11.50 N/mm^2 – 40 mm aggregate one side							
Single course							
concrete paving blocks; PC £8.79/m^2;							
200 × 100 × 60 mm	1.76	0.10	5.92	–	2.46	m	**8.38**
engineering bricks; PC £267.00/1000;							
215 × 102.5 × 65 mm	1.29	0.13	7.70	–	1.99	m	**9.69**
paving bricks; PC £450.00/1000;							
215 × 102.5 × 65 mm	2.00	0.13	7.70	–	2.70	m	**10.40**
Two courses							
concrete paving blocks; PC £8.79/m^2;							
200 × 100 × 60 mm	3.52	0.13	7.70	–	4.77	m	**12.47**
engineering bricks; PC £267.00/1000;							
215 × 102.5 × 65 mm	2.58	0.19	11.00	–	3.83	m	**14.83**
paving bricks; PC £450.00/1000;							
215 × 102.5 × 65 mm	4.00	0.19	11.00	–	5.26	m	**16.26**
Three courses							
concrete paving blocks; PC £8.79/m^2;							
200 × 100 × 60 mm	5.27	0.15	8.55	–	6.53	m	**15.08**
engineering bricks; PC £267.00/1000;							
215 × 102.5 × 65 mm	3.86	0.22	12.83	–	5.12	m	**17.95**
paving bricks; PC £450.00/1000;							
215 × 102.5 × 65 mm	6.00	0.22	12.83	–	7.26	m	**20.09**

Q10 KERBS/EDGINGS/CHANNELS/PAVING ACCESSORIES

Item Excluding site overheads and profit	PC £	Labour hours	Labour £	Plant £	Material £	Unit	Total rate £
Bricks on edge; bedding in cement mortar (1:4); on 150 mm deep concrete foundations; including haunching with in situ concrete; 11.50 N/mm² – 40 mm aggregate one side							
One brick wide							
engineering bricks; 215 × 102.5 × 65 mm	3.86	0.19	11.00	–	5.26	m	16.26
paving bricks; 215 × 102.5 × 65 mm	6.00	0.19	11.00	–	7.39	m	18.39
Two courses; stretchers laid on edge; 225 mm wide							
engineering bricks; 215 × 102.5 × 65 mm	7.73	0.40	23.10	–	10.37	m	33.47
paving bricks; 215 × 102.5 × 65 mm	12.00	0.40	23.10	–	14.64	m	37.74
Extra over bricks on edge for standard kerbs to one side; haunching in concrete							
125 × 255 mm; HB2; SP	3.82	0.15	8.55	–	6.38	m	14.93
50 × 150 mm hardwood (iroko) edge boards	9.49	0.13	4.81	–	9.49	m	14.30
Edge restraints; to brick paving; on prepared base (not included); 65 mm thick bricks; PC £300.00/1000; haunching one side							
Header course							
200 × 100 mm; butt joints	3.00	0.09	5.13	–	3.34	m	8.47
200x 100 mm; mortar joints	2.80	0.17	9.63	–	5.01	m	14.64
200 × 100 mm × 50; on edge; mortar joints	4.19	0.23	13.48	–	4.86	m	18.34
215 × 102.5 mm mortar joints	2.67	0.17	9.63	–	3.34	m	12.97
215 × 102.5 mm × 65; on edge; mortar joints	4.00	0.25	14.44	–	4.67	m	19.11
Stretcher course; on flat or on edge							
200 × 100 mm; butt joints	1.50	0.06	3.21	–	1.67	m	4.88
210 × 105 mm; mortar joints	1.36	0.11	6.42	–	1.69	m	8.11
Precast concrete block edgings; PC £8.17/m²; 200 × 100 × 60 mm; on prepared base (not included); haunching one side							
Edgings; butt joints							
stretcher course	0.82	0.06	3.21	–	0.98	m	4.19
header course	1.63	0.09	5.13	–	1.97	m	7.10
Haunching							
Haunching with in situ concrete; 11.50 N/mm² – 40 mm aggregate one side							
one side	–	0.13	4.81	–	2.90	m	7.71
Variation in brick prices; add or subtract the following amounts for every £100/1000 difference in the PC price							
Edgings; mortar jointed							
100 mm wide	0.50	–	–	–	0.50	m	0.50
200 mm wide	0.95	–	–	–	0.95	m	0.95
102.5 mm wide	0.47	–	–	–	0.47	m	0.47
215 mm wide	0.93	–	–	–	0.93	m	0.93

Q10 KERBS/EDGINGS/CHANNELS/PAVING ACCESSORIES

Item Excluding site overheads and profit	PC £	Labour hours	Labour £	Plant £	Material £	Unit	Total rate £
TIMBER EDGINGS							
Timber edging boards; fixed with							
50 × 50 × 750 mm timber pegs at 1000 mm							
centres (excavations and hardcore under							
edgings not included)							
Straight							
38 × 150 mm treated softwood edge boards	2.28	0.07	2.57	–	2.28	m	**4.85**
50 × 150 mm treated softwood edge boards	3.08	0.10	3.85	–	3.08	m	**6.93**
Curved							
38 × 150 mm treated softwood edge boards	2.28	0.13	4.81	–	2.28	m	**7.09**
50 × 150 mm treated softwood edge boards	3.08	0.14	5.50	–	3.08	m	**8.58**
METAL EDGINGS							
Permaloc AshphaltEdge; Kinley Systems;							
extruded aluminium alloy L shaped edging							
with 5.33 mm exposed upper lip; edging							
fixed to roadway base and edge profile							
with 250 mm steel fixing spike; laid to							
straight or curvilinear road edge;							
subsequently filled with macadam (not							
included)							
Depth of macadam							
38 mm	7.52	0.02	0.64	–	7.52	m	**8.16**
51 mm	8.48	0.06	3.31	–	8.48	m	**11.79**
64 mm	9.37	0.01	0.35	–	9.37	m	**9.72**
76 mm	11.05	0.01	0.39	–	11.05	m	**11.44**
102 mm	13.18	0.01	0.54	–	13.18	m	**13.72**
Permaloc Cleanline; Kinley Systems; heavy							
duty straight profile edging; for edgings to							
soft landscape beds or turf areas;							
3.2 mm × 102 mm high; 3.2 mm thick with							
4.75 mm exposed upper lip; fixed to form							
straight or curvilinear edge with 305 mm							
fixing spike							
Milled aluminium							
100 mm deep	8.62	0.01	0.35	–	8.62	m	**8.97**
Black							
100 mm deep	8.62	0.01	0.35	–	8.62	m	**8.97**

Q10 KERBS/EDGINGS/CHANNELS/PAVING ACCESSORIES

Item Excluding site overheads and profit	PC £	Labour hours	Labour £	Plant £	Material £	Unit	Total rate £
Permaloc Permastrip; Kinley Systems; heavy duty L shaped profile maintenance strip; 3.2 mm × 89 mm high with 5.2 mm exposed top lip; for straight or gentle curves on paths or bed turf interfaces; fixed to form straight or curvilinear edge with standard 305 mm stake; other stake lengths available							
Milled aluminium							
89 mm deep	8.62	0.01	0.35	–	8.62	m	**8.97**
Black							
89 mm deep	8.62	0.01	0.35	–	8.62	m	**8.97**
Permaloc Proline; Kinley Systems; medium duty straight profiled maintenance strip; 3.2 mm × 102 mm high with 3.18 mm exposed top lip; for straight or gentle curves on paths or bed turf interfaces; fixed to form straight or curvilinear edge with standard 305 mm stake; other stake lengths available							
Milled aluminium							
89 mm deep	7.50	0.01	0.35	–	7.50	m	**7.85**
Black							
89 mm deep	7.50	0.01	0.35	–	7.50	m	**7.85**
CHANNELS							
Channels; bedding in cement mortar (1:3); joints pointed flush; on concrete foundations (not included)							
Three courses stretchers; 350 mm wide; quarter bond to form dished channels							
engineering bricks; PC £267.00/1000;							
215 × 102.5 × 65 mm	3.86	0.33	19.25	–	5.12	m	**24.37**
paving bricks; PC £450.00/1000;							
215 × 102.5 × 65 mm	6.00	0.33	19.25	–	7.26	m	**26.51**
Three courses granite setts; 340 mm wide; to form dished channels							
340 mm wide	12.03	0.67	38.50	–	14.92	m	**53.42**
Precast concrete channels etc.; Marshalls Mono; bedding, jointing and pointing in cement mortar (1:3); including haunching with in situ concrete; 11.50 N/mm² – 40 mm aggregate one side							
Channels; square							
125 × 225 × 915 mm long; CS1	3.65	0.13	7.70	–	12.86	m	**20.56**
125 × 150 × 915 mm long; CS2	2.95	0.13	7.70	–	10.58	m	**18.28**
Channels; dished							
305 × 150 × 915 mm long; CD	10.32	0.13	7.70	–	20.27	m	**27.97**
150 × 100 × 915 mm	6.05	0.13	7.70	–	11.71	m	**19.41**

Q20 GRANULAR SUB-BASES TO ROADS AND PAVINGS

Item Excluding site overheads and profit	PC £	Labour hours	Labour £	Plant £	Material £	Unit	Total rate £
Q20 GRANULAR SUB-BASES TO ROADS AND PAVINGS							
CLARIFICATION NOTES ON LABOUR COSTS IN THIS SECTION							
General groundworks team							
Generally a three man team is used in this section; The column 'Labour hours' reports team hours. The column 'Labour £' reports the total cost of the team for the unit of work shown							
3 man team	–	1.00	57.75	–	–	hr	**57.75**
HARDCORE BASES							
Hardcore bases; obtained off site; PC £22.80/m³							
By machine							
100 mm thick	2.28	0.05	0.96	1.71	2.28	m²	**4.95**
150 mm thick	3.42	0.07	1.28	2.22	3.42	m²	**6.92**
200 mm thick	4.56	0.08	1.54	2.33	4.56	m²	**8.43**
300 mm thick	6.84	0.08	1.60	2.48	6.84	m²	**10.92**
exceeding 300 mm thick	27.36	0.17	3.21	5.71	27.36	m³	**36.28**
By hand							
100 mm thick	2.28	0.20	3.85	0.19	2.28	m²	**6.32**
150 mm thick	4.10	0.30	5.78	0.19	4.10	m²	**10.07**
200 mm thick	5.47	0.40	7.70	0.30	5.47	m²	**13.47**
300 mm thick	8.21	0.60	11.55	0.19	8.21	m²	**19.95**
exceeding 300 mm thick	27.36	2.00	38.50	0.62	27.36	m³	**66.48**
Hardcore; difference for each £1.00 increase/ decrease in PC price per m³; price will vary with type and source of hardcore							
average 75 mm thick	–	–	–	–	0.07	m²	**0.07**
average 100 mm thick	–	–	–	–	0.10	m²	**0.10**
average 150 mm thick	–	–	–	–	0.14	m²	**0.14**
average 200 mm thick	–	–	–	–	0.19	m²	**0.19**
average 250 mm thick	–	–	–	–	0.24	m²	**0.24**
average 300 mm thick	–	–	–	–	0.28	m²	**0.28**
exceeding 300 mm thick	–	–	–	–	1.04	m³	**1.04**
TYPE 1 BASES							
Type 1 granular fill base; PC £16.50/tonne (£36.30 / m³ compacted)							
By machine							
100 mm thick	3.63	0.02	1.20	0.86	3.63	m²	**5.69**
150 mm thick	5.45	0.03	1.44	0.86	5.45	m²	**7.75**
250 mm thick	9.07	0.03	1.65	1.71	9.07	m²	**12.43**
over 250 mm thick	36.30	0.20	11.55	5.24	36.30	m³	**53.09**

Q20 GRANULAR SUB-BASES TO ROADS AND PAVINGS

Item Excluding site overheads and profit	PC £	Labour hours	Labour £	Plant £	Material £	Unit	Total rate £
By hand (mechanical compaction)							
100 mm thick	2.42	0.07	3.85	0.06	2.42	m²	**6.33**
150 mm thick	5.45	0.08	4.81	0.09	5.45	m²	**10.35**
250 mm thick	9.07	0.13	7.22	0.15	9.07	m²	**16.44**
over 250 mm thick	18.15	0.20	11.55	0.15	18.15	m³	**29.85**
Type 1 (crushed concrete) granular fill base; PC £12.00/tonne (£21.60/m³ compacted)							
By machine							
100 mm thick	2.51	0.02	1.20	0.57	2.51	m²	**4.28**
150 mm thick	3.76	0.03	1.44	0.86	3.76	m²	**6.06**
250 mm thick	6.27	0.03	1.65	1.43	6.27	m²	**9.35**
over 250 mm thick	11.40	0.20	11.55	5.24	11.40	m³	**28.19**
By hand (mechanical compaction)							
100 mm thick	2.51	0.07	3.85	0.06	2.51	m²	**6.42**
150 mm thick	3.76	0.08	4.81	0.09	3.76	m²	**8.66**
250 mm thick	6.27	0.13	7.22	0.15	6.27	m²	**13.64**
over 250 mm thick	11.40	0.20	11.55	0.15	11.40	m³	**23.10**
SURFACE TREATMENTS TO GRANULAR BASES							
Surface treatments							
Sand blinding; to hardcore base (not included); 25 mm thick	0.83	0.03	0.64	–	0.83	m²	**1.47**
Sand blinding; to hardcore base (not included); 50 mm thick	1.67	0.05	0.96	–	1.67	m²	**2.63**
Filter fabrics; to hardcore base (not included)	0.43	0.01	0.19	–	0.43	m²	**0.62**
CELLULAR CONFINED BASES							
Cellweb; Geosynthetics Ltd; cellular confinement system for load support or permeable bases below pavings							
To graded and compacted substrate (not included); on 50 mm sharp sand base filled with angular drainage aggregate 20–5 mm							
100 mm deep	9.62	0.05	2.89	2.07	17.58	m²	**22.54**

Q21 IN SITU CONCRETE ROADS/PAVINGS

Item Excluding site overheads and profit	PC £	Labour hours	Labour £	Plant £	Material £	Unit	Total rate £
Q21 IN SITU CONCRETE ROADS/PAVINGS							
CLARIFICATION NOTES ON LABOUR COSTS IN THIS SECTION							
General groundworks team Generally a three man team is used in this section; The column 'Labour hours' reports team hours. The column 'Labour £' reports the total cost of the team for the unit of work shown							
3 man team	–	1.00	57.75	–	–	hr	**57.75**
IN SITU CONCRETE PAVINGS							
Unreinforced concrete; on prepared subbase (not included) Roads; 21.00 N/mm² – 20 mm aggregate (1:2:4) mechanically mixed on site							
100 mm thick	10.47	0.04	2.41	–	10.47	m²	**12.88**
150 mm thick	15.70	–	0.12	–	15.70	m²	**15.82**
Reinforced in situ concrete; mechanically mixed on site; normal Portland cement; on hardcore base (not included); reinforcement (not included) Roads; 11.50 N/mm² – 40 mm aggregate (1:3:6)							
100 mm thick	8.74	0.03	1.92	–	8.74	m²	**10.66**
150 mm thick	13.12	0.20	11.55	–	13.12	m²	**24.67**
200 mm thick	17.91	0.27	15.40	–	17.91	m²	**33.31**
250 mm thick	21.85	0.33	19.25	0.25	21.85	m²	**41.35**
300 mm thick	26.22	0.40	23.10	0.25	26.22	m²	**49.57**
Roads; 21.00 N/mm² – 20 mm aggregate (1:2:4)							
100 mm thick	10.22	0.13	7.70	–	10.22	m²	**17.92**
150 mm thick	15.33	0.20	11.55	–	15.33	m²	**26.88**
200 mm thick	20.43	0.27	15.40	–	20.43	m²	**35.83**
250 mm thick	25.54	0.33	19.25	0.25	25.54	m²	**45.04**
300 mm thick	30.65	0.40	23.10	0.25	30.65	m²	**54.00**
Roads; 25.00 N/mm² – 20 mm aggregate GEN 4 ready mixed							
100 mm thick	7.89	0.13	7.70	–	7.89	m²	**15.59**
150 mm thick	11.83	0.20	11.55	–	11.83	m²	**23.38**
200 mm thick	15.77	0.27	15.40	–	15.77	m²	**31.17**
250 mm thick	19.71	0.33	19.25	0.25	19.71	m²	**39.21**
300 mm thick	23.66	0.40	23.10	0.25	23.66	m²	**47.01**

Q21 IN SITU CONCRETE ROADS/PAVINGS

Item Excluding site overheads and profit	PC £	Labour hours	Labour £	Plant £	Material £	Unit	Total rate £
Reinforced in situ concrete; ready mixed;							
discharged directly into location from							
supply vehicle; normal Portland cement;							
on hardcore base (not included);							
reinforcement (not included)							
Roads; 11.50 N/mm² – 40 mm aggregate (1:3:6)							
100 mm thick	7.59	0.05	3.08	–	7.59	m²	**10.67**
150 mm thick	11.38	0.08	4.62	–	11.38	m²	**16.00**
200 mm thick	15.18	0.12	6.93	–	15.18	m²	**22.11**
250 mm thick	18.98	0.18	10.39	0.25	18.98	m²	**29.62**
300 mm thick	22.77	0.22	12.71	0.25	22.77	m²	**35.73**
Roads; 21.00 N/mm² – 20 mm aggregate (1:2:4)							
100 mm thick	7.52	2.08	120.31	–	7.52	m²	**127.83**
150 mm thick	11.28	0.08	4.62	–	11.28	m²	**15.90**
200 mm thick	15.04	0.12	6.93	–	15.04	m²	**21.97**
250 mm thick	18.80	0.18	10.39	0.25	18.80	m²	**29.44**
300 mm thick	22.57	0.22	12.71	0.25	22.57	m²	**35.53**
Roads; 26.00 N/mm² – 20 mm aggregate							
(1:1.5:3)							
100 mm thick	7.60	0.05	3.08	–	7.60	m²	**10.68**
150 mm thick	11.40	0.08	4.62	–	11.40	m²	**16.02**
200 mm thick	15.04	0.12	6.93	–	15.04	m²	**21.97**
250 mm thick	19.00	0.18	10.39	0.25	19.00	m²	**29.64**
300 mm thick	22.80	0.22	12.71	0.25	22.80	m²	**35.76**
Roads; PAV1 concrete – 35 N/mm²;							
Designated mix							
100 mm thick	8.05	0.05	3.08	–	8.05	m²	**11.13**
150 mm thick	12.07	0.08	4.62	–	12.07	m²	**16.69**
200 mm thick	16.09	0.12	6.93	–	16.09	m²	**23.02**
250 mm thick	20.11	0.18	10.39	0.25	20.11	m²	**30.75**
300 mm thick	24.14	0.22	12.71	0.25	24.14	m²	**37.10**
Concrete sundries							
Treating surfaces of unset concrete; grading to							
cambers; tamping with 75 mm thick steel shod							
tamper or similar	–	0.13	5.13	–	–	m²	**5.13**
FORMWORK FOR CONCRETE PAVINGS							
Formwork for in situ concrete							
Sides of foundations							
height not exceeding 250 mm	0.69	0.01	0.64	–	0.78	m	**1.42**
height 250–500 mm	0.92	0.01	0.77	–	1.12	m	**1.89**
height 500 mm–1.00 m	0.92	0.02	0.96	–	1.23	m	**2.19**
height exceeding 1.00 m	2.77	0.67	38.50	–	4.51	m²	**43.01**
Extra over formwork for curved work 6 m radius	–	0.25	9.63	–	–	m	**9.63**
Steel road forms; to edges of beds or faces of							
foundations							
150 mm wide	–	0.07	3.85	0.22	–	m	**4.07**

Q22 COATED MACADAM ROADS/PAVINGS

Item Excluding site overheads and profit	PC £	Labour hours	Labour £	Plant £	Material £	Unit	**Total rate £**
EXPANSION JOINTS							
General groundworks team							
Generally a two man team is used in this							
section; The column 'Labour hours' reports							
team hours. The column 'Labour £' reports the							
total cost of the team for the unit of work shown							
2 man team	–	1.00	38.50	–	–	hr	**38.50**
Expansion joints							
13 mm thick joint filler; formwork							
width or depth not exceeding 150 mm	2.05	0.04	1.72	–	3.15	m	**4.87**
width or depth 150–300 mm	4.11	0.06	2.15	–	6.30	m	**8.45**
width or depth 300–450 mm	6.16	0.07	2.58	–	9.45	m	**12.03**
25 mm thick joint filler; formwork							
width or depth not exceeding 150 mm	5.50	0.04	1.72	–	6.60	m	**8.32**
width or depth 150–300 mm	11.00	0.06	2.15	–	13.19	m	**15.34**
width or depth 300–450 mm	16.50	0.07	2.58	–	19.79	m	**22.37**
Sealants; sealing top 25 mm of joint with							
rubberized bituminous compound	1.25	0.06	2.15	–	1.25	m	**3.40**
REINFORCEMENT TO CONCRETE PAVINGS							
Reinforcement; fabric; side laps 150 mm; head laps 300 mm; mesh 200 × 200 mm; in roads, footpaths or pavings							
Fabric							
A142 (2.22 kg/m²)	1.42	0.03	1.60	–	1.42	m²	**3.02**
A193 (3.02 kg/m²)	1.94	0.02	1.02	–	1.94	m²	**2.96**
Q22 COATED MACADAM ROADS/PAVINGS							
MACADAM SURFACING							
Coated macadam/asphalt roads/pavings – General							
Preamble: The prices for all in situ finishings to							
roads and footpaths include for work to falls,							
crossfalls or slopes not exceeding 15° from							
horizontal; for laying on prepared bases (not							
included) and for rolling with an appropriate							
roller.							
Users should note the new terminology for the							
surfaces described below which is to							
European standard descriptions. The now							
redundant descriptions for each course are							
shown in brackets.							

Q22 COATED MACADAM ROADS/PAVINGS

Item Excluding site overheads and profit	PC £	Labour hours	Labour £	Plant £	Material £	Unit	Total rate £
Macadam surfacing; Spadeoak **Construction Co Ltd; surface (wearing)** **course; 20 mm of 6 mm dense bitumen** **macadam**							
Machine lay; areas 1000 m² and over							
limestone aggregate	–	–	–	–	–	m²	6.76
granite aggregate	–	–	–	–	–	m²	6.84
red	–	–	–	–	–	m²	12.16
Hand lay; areas 400 m² and over							
limestone aggregate	–	–	–	–	–	m²	8.61
granite aggregate	–	–	–	–	–	m²	8.69
red	–	–	–	–	–	m²	14.31
Macadam surfacing; Spadeoak **Construction Co Ltd; surface (wearing)** **course; 30 mm of 10 mm dense bitumen** **macadam**							
Machine lay; areas 1000 m² and over							
limestone aggregate	–	–	–	–	–	m²	8.17
granite aggregate	–	–	–	–	–	m²	8.28
red	–	–	–	–	–	m²	17.13
Hand lay; areas 400 m² and over							
limestone aggregate	–	–	–	–	–	m²	10.08
granite aggregate	–	–	–	–	–	m²	10.21
red	–	–	–	–	–	m²	19.61
Macadam surfacing; Spadeoak **Construction Co Ltd; surface (wearing)** **course; 40 mm of 10 mm dense bitumen** **macadam**							
Machine lay; areas 1000 m² and over							
limestone aggregate	–	–	–	–	–	m²	10.39
granite aggregate	–	–	–	–	–	m²	10.54
red	–	–	–	–	–	m²	19.42
Hand lay; areas 400 m² and over							
limestone aggregate	–	–	–	–	–	m²	12.47
granite aggregate	–	–	–	–	–	m²	12.62
red	–	–	–	–	–	m²	22.04
Macadam surfacing; Spadeoak **Construction Co Ltd; binder (base) course;** **50 mm of 20 mm dense bitumen macadam**							
Machine lay; areas 1000 m² and over							
limestone aggregate	–	–	–	–	–	m²	10.42
granite aggregate	–	–	–	–	–	m²	10.59
Hand lay; areas 400 m² and over							
limestone aggregate	–	–	–	–	–	m²	12.49
granite aggregate	–	–	–	–	–	m²	12.68

Q22 COATED MACADAM ROADS/PAVINGS

Item Excluding site overheads and profit	PC £	Labour hours	Labour £	Plant £	Material £	Unit	Total rate £
MACADAM SURFACING – cont							
Macadam surfacing; Spadeoak							
Construction Co Ltd; binder (base) course;							
60 mm of 20 mm dense bitumen macadam							
Machine lay; areas 1000 m² and over							
limestone aggregate	–	–	–	–	–	m²	11.46
granite aggregate	–	–	–	–	–	m²	11.66
Hand lay; areas 400 m² and over							
limestone aggregate	–	–	–	–	–	m²	13.60
granite aggregate	–	–	–	–	–	m²	13.80
Macadam surfacing; Spadeoak							
Construction Co Ltd; base (roadbase)							
course; 75 mm of 28 mm dense bitumen							
macadam							
Machine lay; areas 1000 m² and over							
limestone aggregate	–	–	–	–	–	m²	14.06
granite aggregate	–	–	–	–	–	m²	14.31
Hand lay; areas 400 m² and over							
limestone aggregate	–	–	–	–	–	m²	16.37
granite aggregate	–	–	–	–	–	m²	16.65
Macadam surfacing; Spadeoak							
Construction Co Ltd; base (roadbase)							
course; 100 mm of 28 mm dense bitumen							
macadam							
Machine lay; areas 1000 m² and over							
limestone aggregate	–	–	–	–	–	m²	17.42
granite aggregate	–	–	–	–	–	m²	17.76
Hand lay; areas 400 m² and over							
limestone aggregate	–	–	–	–	–	m²	19.96
granite aggregate	–	–	–	–	–	m²	20.31

Q22 COATED MACADAM ROADS/PAVINGS

Item Excluding site overheads and profit	PC £	Labour hours	Labour £	Plant £	Material £	Unit	Total rate £
Macadam surfacing; Spadeoak **Construction Co Ltd; base (roadbase)** **course; 150 mm of 28 mm dense bitumen** **macadam in two layers**							
Machine lay; areas 1000 m² and over							
limestone aggregate	–	–	–	–	–	m²	**27.47**
granite aggregate	–	–	–	–	–	m²	**27.98**
Hand lay; areas 400 m² and over							
limestone aggregate	–	–	–	–	–	m²	**32.07**
granite aggregate	–	–	–	–	–	m²	**32.60**
Base (roadbase) course; 200 mm of 28 mm **dense bitumen macadam in two layers**							
Machine lay; areas 1000 m² and over							
limestone aggregate	–	–	–	–	–	m²	**34.91**
granite aggregate	–	–	–	–	–	m²	**35.59**
Hand lay; areas 400 m² and over							
limestone aggregate	–	–	–	–	–	m²	**27.12**
granite aggregate	–	–	–	–	–	m²	**40.71**
Resin bound macadam pavings; Bituchem; **to pedestrian or vehicular hard landscape** **areas; laid to base course (not included)**							
Natratex wearing course; clear resin bound macadam							
25 mm thick to pedestrian areas	16.60	0.04	2.41	–	16.60	m²	**19.01**
30 mm thick to vehicular areas	20.00	0.01	0.64	–	20.00	m²	**20.64**
Colourtex coloured resin bound macadam							
25 mm thick	16.60	0.04	2.41	–	16.60	m²	**19.01**

Q22 COATED MACADAM ROADS/PAVINGS

Item Excluding site overheads and profit	PC £	Labour hours	Labour £	Plant £	Material £	Unit	Total rate £
RESIN BONDED SURFACING							
Bonded aggregates; Addagrip Surface Treatments UK Ltd; natural decorative resin bonded surface dressing laid to concrete, macadam or to plywood panels priced separately							
Primer coat to macadam or concrete base	–	0.04	2.41	–	4.00	m²	6.41
Golden pea gravel; 1–3 mm							
buff adhesive	–	0.04	2.41	–	20.00	m²	22.41
red adhesive	–	–	–	–	–	m²	20.00
green adhesive	–	–	–	–	–	m²	20.00
Golden pea gravel; 2–5 mm							
buff adhesive	–	–	–	–	–	m²	24.00
Chinese bauxite; 1–3 mm							
buff adhesive	–	–	–	–	–	m²	20.00
Cobalt Blue Glass; 6 mm							
15 mm depth	–	–	–	–	–	m²	75.00
Midnight Grey; 6 mm							
15 mm depth	–	–	–	–	–	m²	45.00
Chocolate; 6 mm							
15 mm depth	–	–	–	–	–	m²	45.00
18 mm depth	–	–	–	–	–	m²	55.00
RESIN BOUND SURFACING							
Resin bound paving; Sureset Ltd; fully mixed permeable decorative aggregate paving; laid to macadam binder and base course (not included); thickness dependent on loading application and aggregate size							
18 mm thick – 6 mm aggregate							
Natural gravel							
areas 100–300 m²	–	–	–	–	–	m²	52.00
areas 300–500 m²	–	–	–	–	–	m²	49.50
areas over 500 m²	–	–	–	–	–	m²	44.00
Crushed rock							
areas 100–300 m²	–	–	–	–	–	m²	54.50
areas 300–500 m²	–	–	–	–	–	m²	51.50
areas over 500 m²	–	–	–	–	–	m²	46.00
Marble							
areas 100–300 m²	–	–	–	–	–	m²	56.50
areas 300–500 m²	–	–	–	–	–	m²	54.50
areas over 500 m²	–	–	–	–	–	m²	47.50
Recycled glass							
areas 100–300 m²	–	–	–	–	–	m²	67.05
areas 300–500 m²	–	–	–	–	–	m²	67.00
areas over 500 m²	–	–	–	–	–	m²	63.00
Spectrum							
areas 100–300 m²	–	–	–	–	–	m²	80.50
areas 300–500 m²	–	–	–	–	–	m²	76.00
areas over 500 m²	–	–	–	–	–	m²	73.50

Q23 GRAVEL/HOGGIN/WOODCHIP ROADS/PAVINGS

Item Excluding site overheads and profit	PC £	Labour hours	Labour £	Plant £	Material £	Unit	Total rate £
Resin bound macadam pavings; Bituchem; **to pedestrian or vehicular hard landscape** **areas; laid to base course (not included)** Natratex wearing course; clear resin bound macadam							
25 mm thick to pedestrian areas	16.60	0.04	2.41	–	16.60	m²	**19.01**
30 mm thick to vehicular areas	20.00	0.01	0.64	–	20.00	m²	**20.64**
Colourtex coloured resin bound macadam							
25 mm thick	16.60	0.04	2.41	–	16.60	m²	**19.01**
Resin bound paving to individual treepits; **SureSet Ltd; fully mixed permeable** **decorative aggregate paving; laid to well** **compacted stone Type 3 (not included)** 30–40 mm thick – average tree pit size 1.50 × 1.50 tree pit aggregate							
Natural gravel							
areas 100–300 m²	–	–	–	–	–	nr	**98.00**
MARKING CAR PARKS							
Marking car parks Car parking space division strips; laid hot at 115° C; on bitumen macadam surfacing							
minimum daily rate	–	–	–	–	–	item	**500.00**
Stainless metal road studs							
100 × 100 mm	5.78	0.08	4.81	–	5.78	nr	**10.59**
Q23 GRAVEL/HOGGIN/WOODCHIP ROADS/ **PAVINGS**							
CLARIFICATION NOTES ON LABOUR **COSTS IN THIS SECTION**							
General groundworks team Generally a three man team is used in this section; The column 'Labour hours' reports team hours. The column 'Labour £' reports the total cost of the team for the unit of work shown							
3 man team	–	1.00	57.75	–	–	hr	**57.75**
EXCAVATION OF PATHWAYS							
Excavation and path preparation Excavating; 300 mm deep; to width of path; depositing excavated material at sides of excavation							
width 1.00 m	–	–	–	2.75	–	m²	**2.75**
width 1.50 m	–	–	–	2.29	–	m²	**2.29**
width 2.00 m	–	–	–	1.97	–	m²	**1.97**
width 3.00 m	–	–	–	1.65	–	m²	**1.65**

Q23 GRAVEL/HOGGIN/WOODCHIP ROADS/PAVINGS

Item Excluding site overheads and profit	PC £	Labour hours	Labour £	Plant £	Material £	Unit	Total rate £
EXCAVATION OF PATHWAYS – cont							
Excavation and path preparation – cont							
Excavating trenches; in centre of pathways; 100 mm flexible drain pipes; filling with clean broken stone or gravel rejects							
300 × 450 mm deep	5.05	0.03	1.92	1.37	5.05	m	**8.34**
Hand trimming and compacting reduced surface of pathway; by machine							
width 1.00 m	–	0.02	0.96	0.19	–	m	**1.15**
width 1.50 m	–	0.01	0.85	0.16	–	m	**1.01**
width 2.00 m	–	0.01	0.77	0.15	–	m	**0.92**
width 3.00 m	–	0.01	0.77	0.15	–	m	**0.92**
Permeable membranes; to trimmed and compacted surface of pathway							
Terram 1000	0.43	0.01	0.39	–	0.43	m²	**0.82**
FILLING TO MAKE UP LEVELS							
Filling to make up levels							
Obtained off site; hardcore; PC £22.40/m³							
150 mm thick	3.65	0.01	0.77	0.56	3.65	m²	**4.98**
Type 1 granular fill base; PC £16.50/tonne (£36.30/m³ compacted)							
100 mm thick	3.63	0.01	0.54	0.40	3.63	m²	**4.57**
150 mm thick	5.45	0.01	0.48	0.61	5.45	m²	**6.54**
Surface treatments							
Sand blinding; to hardcore (not included)							
50 mm thick	1.83	0.01	0.77	–	1.83	m²	**2.60**
Filter fabric; to hardcore (not included)	0.43	–	0.19	–	0.43	m²	**0.62**
CELLULAR CONFINED GRANULAR PAVINGS							
Nidagravel; Cedar Nursery; honeycomb shaped cellular polypropylene containment grids for retaining filled decorative aggregate surfaces (not included) for driveways, cycle paths and footpaths; to graded compacted substrate (not included) and 50 mm sharp sand base; integrated geofabric base							
Foot traffic, wheelchairs, bikes and pushchairs							
Nidagravel 125; 1200 × 800 × 25 mm deep	1.67	0.08	2.41	–	13.67	m²	**16.08**
Driveways: vehicles, wheelchairs, bikes and pushchairs (requires compacted granular base to support vehicle loadings)							
Nidagravel 140; 2400 × 1200 × 40 mm deep	1.67	0.08	2.61	–	16.62	m²	**19.23**

Q23 GRAVEL/HOGGIN/WOODCHIP ROADS/PAVINGS

Item Excluding site overheads and profit	PC £	Labour hours	Labour £	Plant £	Material £	Unit	Total rate £
Aggregate filling to Nidagravel 125							
shingle	–	0.03	1.92	1.99	0.83	m²	4.74
decorative aggregate; CED Ltd;							
PC £48.00/tonne	–	0.03	1.92	1.99	2.16	m²	6.07
Aggregate filling to Nidagravel 140							
shingle	–	0.03	1.62	2.42	1.33	m²	5.37
decorative aggregate; CED Ltd;							
PC £48.00/tonne	3.46	0.03	1.62	2.42	3.46	m²	7.50
StableDRIVE; Cedar Nursery; honeycomb shaped polypropylene cellular containment system for retaining filled decorative aggregate surfaces for driveways, cyclepaths and footpaths; on compacted subgrade and subbase (both not included)							
Driveways: vehicles, wheelchairs, bikes and pushchairs (requires compacted granular base to support vehicle loadings)							
StableDRIVE; 600 × 500 × 32 mm deep	1.67	0.08	2.77	–	17.62	m²	20.39
Aggregate filling to StableDRIVE							
shingle	–	0.03	1.81	2.96	1.07	m²	5.84
decorative aggregate; CED Ltd;							
PC £48.00/tonne	–	0.03	1.81	2.96	2.76	m²	7.53
GRANULAR PAVINGS							
Footpath gravels; porous self-binding gravel							
CED Ltd; Cedec gravel; self-binding; laid to inert (non-limestone) base measured separately; compacting							
red, silver or gold; 50 mm thick	13.01	0.01	0.48	0.41	13.01	m²	13.90
Grundon Ltd; Coxwell self-binding path gravels laid and compacted to excavation or base measured separately							
50 mm thick	3.97	0.01	0.48	0.41	3.97	m²	4.86
Breedon Special Aggregates; Golden Gravel or equivalent; rolling wet; on hardcore base (not included); for pavements; to falls and crossfalls and to slopes not exceeding 15° from horizontal; over 300 mm wide							
50 mm thick	11.10	0.01	0.48	0.34	11.10	m²	11.92
75 mm thick	16.65	0.03	1.92	0.42	16.65	m²	18.99
Breedon Special Aggregates; Wayfarer specially formulated fine gravel for use on golf course pathways							
50 mm thick	10.00	0.01	0.48	0.34	10.00	m²	10.82
75 mm thick	14.99	0.02	0.96	0.61	14.99	m²	16.56

Q23 GRAVEL/HOGGIN/WOODCHIP ROADS/PAVINGS

Item Excluding site overheads and profit	PC £	Labour hours	Labour £	Plant £	Material £	Unit	Total rate £
GRANULAR PAVINGS – cont							
Footpath gravels – cont							
Hoggin (stabilized); PC £26.00/m³ on hardcore							
base (not included); to falls and crossfalls and							
to slopes not exceeding 15° from horizontal;							
over 300 mm wide							
100 mm thick	4.27	0.01	0.64	0.55	4.27	m²	**5.46**
150 mm thick	6.40	0.02	0.96	0.82	6.40	m²	**8.18**
Ballast; as dug; watering; rolling; on hardcore							
base (not included)							
100 mm thick	4.17	0.01	0.64	0.55	4.17	m²	**5.36**
150 mm thick	6.25	0.02	0.96	0.82	6.25	m²	**8.03**
Footpath gravels; porous loose gravels							
Breedon Special Aggregates; Breedon Buff							
decorative limestone chippings							
50 mm thick	5.14	–	0.19	0.09	5.14	m²	**5.42**
75 mm thick	14.58	–	0.24	0.11	14.58	m²	**14.93**
Breedon Special Aggregates; Breedon Buff							
decorative limestone chippings							
50 mm thick	5.14	–	0.19	0.09	5.14	m²	**5.42**
75 mm thick	7.72	–	0.24	0.11	7.72	m²	**8.07**
Breedon Special Aggregates; Brindle or							
Moorland Black chippings							
50 mm thick	5.78	–	0.19	0.09	5.78	m²	**6.06**
75 mm thick	8.67	–	0.24	0.11	8.67	m²	**9.02**
Breedon Special Aggregates; slate chippings;							
plum/blue							
50 mm thick	9.32	–	0.19	0.09	9.32	m²	**9.60**
75 mm thick	13.99	–	0.24	0.11	13.99	m²	**14.34**
Washed shingle; on prepared base (not							
included)							
25–50 mm size; 25 mm thick	0.88	0.01	0.48	0.08	0.88	m²	**1.44**
25–50 mm size; 75 mm thick	2.63	0.03	1.44	0.24	2.63	m²	**4.31**
50–75 mm size; 25 mm thick	0.88	0.01	0.39	0.09	0.88	m²	**1.36**
50–75 mm size; 75 mm thick	2.63	0.02	1.28	0.30	2.63	m²	**4.21**
Pea shingle; on prepared base (not included)							
10–15 mm size; 25 mm thick	0.88	0.01	0.34	0.08	0.88	m²	**1.30**
5–10 mm size; 75 mm thick	2.63	0.02	1.01	0.24	2.63	m²	**3.88**

Q23 GRAVEL/HOGGIN/WOODCHIP ROADS/PAVINGS

Item Excluding site overheads and profit	PC £	Labour hours	Labour £	Plant £	Material £	Unit	Total rate £
WOODCHIP PATHWAYS							
Wood chip surfaces; Melcourt Industries Ltd (items labelled FSC are Forest Stewardship Council certified) Wood chips; to surface of pathways by machine; material delivered in 80 m³ loads; levelling and spreading by hand (excavation and preparation not included)							
Walk Chips; 100 mm thick; FSC (25 m³ loads)	4.06	0.01	0.64	0.34	4.06	m²	**5.04**
Walk Chips; 100 mm thick; FSC (80 m³ loads)	2.70	0.01	0.64	0.34	2.70	m²	**3.68**
Woodfibre; 100 mm thick; FSC (80 m³ loads)	2.51	0.01	0.64	0.34	2.51	m²	**3.49**
PATH EDGINGS							
Permaloc AshphaltEdge; Kinley Systems; extruded aluminium alloy L shaped edging with 5.33 mm exposed upper lip; edging fixed to roadway base and edge profile with 250 mm steel fixing spike; laid to straight or curvilinear road edge; subsequently filled with macadam (not included) Depth of macadam							
38 mm	7.52	0.01	0.33	–	7.52	m	**7.85**
51 mm	8.48	0.01	0.33	–	8.48	m	**8.81**
64 mm	9.37	0.01	0.35	–	9.37	m	**9.72**
76 mm	11.05	0.01	0.39	–	11.05	m	**11.44**
102 mm	13.18	0.01	0.40	–	13.18	m	**13.58**
Permaloc Cleanline; Kinley Systems; heavy duty straight profile edging; for edgings to soft landscape beds or turf areas; 3.2 mm × 102 mm high; 3.2 mm thick with 4.75 mm exposed upper lip; fixed to form straight or curvilinear edge with 305 mm fixing spike Milled aluminium							
100 mm deep	8.62	0.01	0.35	–	8.62	m	**8.97**
Black							
100 mm deep	8.62	0.01	0.35	–	8.62	m	**8.97**

Q24 INTERLOCKING BRICK/BLOCK ROADS/PAVINGS

Item Excluding site overheads and profit	PC £	Labour hours	Labour £	Plant £	Material £	Unit	Total rate £
PATH EDGINGS – cont							
Permaloc Permastrip; Kinley Systems;							
heavy duty L shaped profile maintenance							
strip; 3.2 mm × 89 mm high with 5.2 mm							
exposed top lip; for straight or gentle							
curves on paths or bed turf interfaces;							
fixed to form straight or curvilinear edge							
with standard 305 mm stake; other stake							
lengths available							
Milled aluminium							
89 mm deep	8.62	0.01	0.35	–	8.62	m	**8.97**
Black							
89 mm deep	8.62	0.01	0.35	–	8.62	m	**8.97**
Permaloc Proline; Kinley Systems; medium							
duty straight profiled maintenance strip;							
3.2 mm × 102 mm high with 3.18 mm							
exposed top lip; for straight or gentle							
curves on paths or bed turf interfaces;							
fixed to form straight or curvilinear edge							
with standard 305 mm stake; other stake							
lengths available							
Milled aluminium							
89 mm deep	7.50	0.01	0.35	–	7.50	m	**7.85**
Black							
89 mm deep	7.50	0.01	0.35	–	7.50	m	**7.85**
Q24 INTERLOCKING BRICK/BLOCK							
ROADS/PAVINGS							
CLARIFICATION NOTES ON LABOUR							
COSTS IN THIS SECTION							
General groundworks team							
Generally a three man team is used in this							
section; The column 'Labour hours' reports							
team hours. The column 'Labour £' reports the							
total cost of the team for the unit of work							
shown							
3 man team	–	1.00	57.75	–	–	hr	**57.75**

Q24 INTERLOCKING BRICK/BLOCK ROADS/PAVINGS

Item Excluding site overheads and profit	PC £	Labour hours	Labour £	Plant £	Material £	Unit	Total rate £
PRECAST CONCRETE BLOCK PAVINGS							
Precast concrete block edgings; PC £8.17/m²;							
200 × 100 × 60 mm; on prepared base (not							
included); haunching one side							
Edgings; butt joints							
stretcher course	0.82	0.06	3.21	–	3.20	m	**6.41**
header course	1.63	0.09	5.13	–	4.19	m	**9.32**
Precast concrete vehicular paving blocks;							
Marshalls Plc; on prepared base (not							
included); on 50 mm compacted sharp							
sand bed; blocks laid in 7 mm loose sand							
and vibrated; joints filled with sharp sand							
and vibrated; level and to falls only							
Trafica paving blocks; 450 × 450 × 70 mm							
Perfecta finish; colour natural	29.76	0.17	9.63	0.12	31.60	m²	**41.35**
Perfecta finish; colour buff	34.46	0.17	9.63	0.12	36.30	m²	**46.05**
Saxon finish; colour natural	26.18	0.17	9.63	0.12	28.01	m²	**37.76**
Saxon finish; colour buff	29.11	0.17	9.63	0.12	30.94	m²	**40.69**
Precast concrete vehicular paving blocks;							
Keyblok Marshalls Plc; on prepared base							
(not included); on 50 mm compacted sharp							
sand bed; blocks laid in 7 mm loose sand							
and vibrated; joints filled with sharp sand							
and vibrated; level and to falls only							
Herringbone bond							
200 × 100 × 60 mm; natural grey	8.58	0.50	28.88	0.12	10.46	m²	**39.46**
200 × 100 × 60 mm; colours	9.23	0.50	28.88	0.12	11.11	m²	**40.11**
200 × 100 × 80 mm; natural grey	9.55	0.50	28.88	0.12	11.44	m²	**40.44**
200 × 100 × 80 mm; colours	10.78	0.50	28.88	0.12	12.67	m²	**41.67**
Basketweave bond							
200 × 100 × 60 mm; natural grey	8.58	0.40	23.10	0.12	10.46	m²	**33.68**
200 × 100 × 60 mm; colours	9.23	0.40	23.10	0.12	11.11	m²	**34.33**
200 × 100 × 80 mm; natural grey	9.55	0.40	23.10	0.12	11.44	m²	**34.66**
200 × 100 × 80 mm; colours	10.78	0.40	23.10	0.12	12.67	m²	**35.89**
Precast concrete vehicular paving blocks;							
Charcon Hard Landscaping; on prepared							
base (not included); on 50 mm compacted							
sharp sand bed; blocks laid in 7 mm loose							
sand and vibrated; joints filled with sharp							
sand and vibrated; level and to falls only							
Europa concrete blocks							
200 × 100 × 60 mm; natural grey	11.34	0.50	28.88	0.12	13.22	m²	**42.22**
200 × 100 × 60 mm; colours	12.37	0.50	28.88	0.12	14.25	m²	**43.25**
Parliament concrete blocks							
200 × 100 × 65 mm; natural grey	30.55	0.50	28.88	0.12	32.43	m²	**61.43**
200 × 100 × 65 mm; colours	30.55	0.50	28.88	0.12	32.43	m²	**61.43**

Q24 INTERLOCKING BRICK/BLOCK ROADS/PAVINGS

Item Excluding site overheads and profit	PC £	Labour hours	Labour £	Plant £	Material £	Unit	Total rate £
SUDS PERMEABLE BLOCK PAVINGS							
SUDS paving; Concrete Block Permeable Paving (CBPP) Note: Sustainable Drainage Systems (SUDS) aim to reduce flood risk by controlling the rate and volume of surface water run off from developments. Permeable Paving is a SUDS technique. In addition to managing the quantity of surface water run-off more effectively, one of the primary benefits of CBPP is the potential enhancement of water quality. Permeable pavements improve water quality by mirroring nature in providing filtration and allowing for natural biodegradation of hydrocarbons and the dilution of other contaminants as water passes through the system.							
Bases for Sustainable Urban Drainage System (SUDS) subbase for use below impervious paving; Aggregate Industries Ltd; Bardon DrainAgg 20 4/20 mm open graded material laid loose on prepared subgrade; within edge restraints to paving area (not included); graded to levels and falls as per paving manufacturer's instructions							
150 mm thick	5.87	0.01	0.48	0.61	5.87	m²	6.96
200 mm thick	7.83	–	0.03	0.81	7.83	m²	8.67
250 mm thick	9.78	0.01	0.40	1.01	9.78	m²	11.19
350 mm thick	13.68	0.01	0.55	1.39	13.68	m²	15.62
450 mm thick	17.60	0.01	0.77	1.95	17.60	m²	20.32
Subbase replacement system for SUDS paving Charcon Permavoid; Aggregate Industries Ltd; geocellular, interlocking, high strength; high void capacity (95%); typically a 4:1 ratio when compared with alternative granular systems; allows for water storage to be confined at shallow depth within the subbase layer; inclusive of connection, ties, pins; does not include for encapsulation in geotextile/geomembranes 708 × 354 × 150 mm							
150 mm thick	59.39	0.06	3.22	–	59.39	m²	62.61
300 mm thick	123.22	0.11	6.42	–	123.22	m²	129.64

Q24 INTERLOCKING BRICK/BLOCK ROADS/PAVINGS

Item Excluding site overheads and profit	PC £	Labour hours	Labour £	Plant £	Material £	Unit	Total rate £
Encapsulation of Permavoid; 150 mm thick systems; geomembrane geofabric or combination of both for filtration or attenuation of water; please see the manufacturers design guide							
Charcon geomembrane, heat welded	14.60	0.02	0.96	–	14.60	m²	**15.56**
Charcon non-reinforced geotextile	9.05	0.11	6.42	–	9.05	m²	**15.47**
Charcon reinforced geotextile	4.86	0.01	0.64	–	4.86	m²	**5.50**
Encapsulation of Permavoid; 300 mm thick systems; geomembrane geofabric or combination of both for filtration or attenuation of water; please see the manufacturers design guide							
Charcon geomembrane, heat welded	16.50	0.02	1.07	–	16.50	m²	**17.57**
Charcon non-reinforced geotextile	10.23	0.01	0.64	–	10.23	m²	**10.87**
Charcon reinforced geotextile	4.58	0.01	0.69	–	4.58	m²	**5.27**
Pervious surfacing materials; Aggregate Industries Ltd; Charcon Infilta CBPP Rectangular permeable block paving system; incorporating a 5 mm spacer design that provides a 5 mm void allowing ingress of water through to the subbase storage system (priced separately); various colours; bedded on 50 mm thick 2–6.3 mm clean, angular, free draining uncompacted aggregate; joints infilled with 3 mm clean grit							
Infilta; 200 × 100 × 80 mm thick; natural grey	16.39	0.15	8.55	0.12	19.32	m²	**27.99**
Infilta; 200 × 100 × 80 mm thick; coloured	16.96	0.15	8.55	0.12	19.89	m²	**28.56**
CELLULAR PAVINGS							
Recycled polyethylene grassblocks; Fiberweb Reinforcement Solutions; interlocking units laid to prepared base or rootzone (not included) BodPave85; load bearing <400 tonnes per m²; 500 × 500 × 50 mm deep; 35 mm ground spike							
1–50 m²	16.50	0.07	3.85	0.18	19.67	m²	**23.70**
51–500 m²	13.75	0.07	3.85	0.18	16.92	m²	**20.95**
501–1000 m²	13.20	0.07	3.85	0.18	16.37	m²	**20.40**
1001–1559 m²	12.50	0.07	3.85	0.18	15.67	m²	**19.70**
1560 m² or over	12.00	0.07	3.85	0.18	15.17	m²	**19.20**
GrassProtecta; extruded expanded polyethylene flexible mesh laid to existing grass surface or newly seeded areas to provide heavy surface protection from traffic and pedestrians							
Standard; 1.2 kg/m²; 20 × 2 m; up to 320 m²	7.50	–	0.16	–	7.50	m²	**7.66**
Standard; 1.2 kg/m²; 20 × 2 m; 321–3000 m²	6.50	–	0.16	–	6.50	m²	**6.66**
Standard; 1.2 kg/m²; 20 × 2 m; over 3000 m²	6.00	–	0.16	–	6.00	m²	**6.16**

Q24 INTERLOCKING BRICK/BLOCK ROADS/PAVINGS

Item Excluding site overheads and profit	PC £	Labour hours	Labour £	Plant £	Material £	Unit	Total rate £
CELLULAR PAVINGS – cont							
Recycled polyethylene grassblocks – cont							
GrassProtecta – cont							
Heavy; 2 kg/m²; 20 × 2 m; up to 320 m²	8.50	–	0.16	–	8.50	m²	**8.66**
Heavy; 2 kg/m²; 20 × 2 m; 321–3000 m²	8.00	–	0.16	–	8.00	m²	**8.16**
Heavy; 2 kg/m²; 20 × 2 m; over 3000 m²	7.50	–	0.16	–	7.50	m²	**7.66**
TurfProtecta; extruded polyethylene flexible mesh laid to existing grass surface or newly seeded areas to provide surface protection from traffic including vehicle or animal wear and tear							
Standard; 30 × 2 m; up to 300 m²	2.90	–	0.16	–	2.90	m²	**3.06**
Standard; 30 × 2 m; 301–600 m²	2.49	–	0.16	–	2.49	m²	**2.65**
Standard; 30 × 2 m; 601–1440 m²	2.41	–	0.16	–	2.41	m²	**2.57**
Standard; 30 × 2 m; over 1440 m²	2.24	–	0.16	–	2.24	m²	**2.40**
Premium; 30 × 2 m; up to 300 m²	2.99	–	0.16	–	2.99	m²	**3.15**
Premium; 30 × 2 m; 301–600 m²	2.66	–	0.16	–	2.66	m²	**2.82**
Premium; 30 × 2 m; 601–1440 m²	2.57	–	0.16	–	2.57	m²	**2.73**
Premium; 30 × 2 m; over 1440 m²	2.41	–	0.16	–	2.41	m²	**2.57**
Grassroad; Cooper Clarke Civils and Lintels; heavy duty for car parking and fire paths verge hardening and shallow embankments							
Honeycomb cellular polyproylene interconnecting paviors with integral downstead anti-shear cleats including topsoil but excluding edge restraints; to granular subbase (not included)							
635 × 330 × 42 mm overall laid to a module of 622 × 311 × 32 mm	–	–	–	–	–	m²	**28.72**
extra over for green colour	–	–	–	–	–	m²	**1.20**

Q25 SLAB/BRICK/STONE/TIMBER PAVINGS

Item Excluding site overheads and profit	PC £	Labour hours	Labour £	Plant £	Material £	Unit	Total rate £
Q25 SLAB/BRICK/STONE/TIMBER PAVINGS							
CLARIFICATION NOTES ON LABOUR COSTS IN THIS SECTION							
General and Expert paving team							
Generally a three man team is used in this section; The column 'Labour hours' reports team hours. The column 'Labour £' reports the total cost of the team for the unit of work shown							
3 man team	–	1.00	57.75	–	–	hr	**57.75**
BRICK PAVINGS							
Bricks – General							
Preamble: Bricks shall be hard, well burnt, non-dusting, resistant to frost and sulphate attack and true to shape, size and sample.							
Movement of materials							
Mechanically offloading bricks; loading wheelbarrows; transporting maximum 25 m distance	–	0.07	3.85	–	–	m²	**3.85**
Edge restraints; to brick paving; on prepared base (not included); 65 mm thick bricks; PC £300.00/1000; haunching one side							
Header course							
200 × 100 mm; butt joints	3.00	0.09	5.13	–	5.55	m	**10.68**
210 × 105 mm; mortar joints	2.67	0.17	9.63	–	5.56	m	**15.19**
Stretcher course							
200 × 100 mm; butt joints	1.50	0.06	3.21	–	3.88	m	**7.09**
210 × 105 mm; mortar joints	1.36	0.11	6.42	–	3.90	m	**10.32**
Variation in brick prices; add or subtract the following amounts for every £100/1000 difference in the PC price							
Edgings; mortar jointed							
100 mm wide	0.50	–	–	–	0.50	m	**0.50**
200 mm wide	0.95	–	–	–	0.95	m	**0.95**
102.5 mm wide	0.47	–	–	–	0.47	m	**0.47**
215 mm wide	0.93	–	–	–	0.93	m	**0.93**
Clay brick pavings; on prepared base (not included); bedding on 50 mm sharp sand; kiln dried sand joints							
Pavings; 200 × 100 × 65 mm wirecut chamfered paviors							
brick; PC £450.00/1000	23.06	0.46	26.66	0.23	25.13	m²	**52.02**

Q25 SLAB/BRICK/STONE/TIMBER PAVINGS

Item Excluding site overheads and profit	PC £	Labour hours	Labour £	Plant £	Material £	Unit	Total rate £
BRICK PAVINGS – cont							
Clay brick pavings; 200 × 100 × 50 mm; laid **to running stretcher or stack bond only; on** **prepared base (not included); bedding on** **cement: sand (1:4) pointing mortar as work** **proceeds**							
PC £600.00/1000							
laid on edge	48.81	1.59	91.67	–	56.50	m²	**148.17**
laid on edge but pavior 65 mm thick	41.00	1.27	73.33	–	48.69	m²	**122.02**
laid flat	26.62	0.73	42.35	–	31.21	m²	**73.56**
PC £500.00/1000							
laid on edge	40.67	1.59	91.67	–	48.37	m²	**140.04**
laid on edge but pavior 65 mm thick	34.16	1.27	73.33	–	41.86	m²	**115.19**
laid flat	22.19	0.73	42.35	–	26.77	m²	**69.12**
PC £400.00/1000							
laid flat	17.75	0.73	42.35	–	22.33	m²	**64.68**
laid on edge	32.54	1.59	91.67	–	40.23	m²	**131.90**
laid on edge but pavior 65 mm thick	27.33	1.27	73.33	–	35.02	m²	**108.35**
PC £300.00/1000							
laid on edge	24.40	1.59	91.67	–	32.10	m²	**123.77**
laid on edge but pavior 65 mm thick	20.50	1.27	73.33	–	28.19	m²	**101.52**
laid flat	13.31	0.73	42.35	–	17.90	m²	**60.25**
Clay brick pavings; 200 × 100 × 50 mm; butt **jointed laid herringbone or basketweave** **pattern only; on prepared base (not** **included); bedding on 50 mm sharp sand**							
PC £600.00/1000							
laid flat	30.75	0.46	26.66	0.30	32.82	m²	**59.78**
PC £500.00/1000							
laid flat	25.63	0.46	26.66	0.30	27.69	m²	**54.65**
PC £400.00/1000							
laid flat	20.50	0.46	26.66	0.30	22.57	m²	**49.53**
PC £300.00/1000							
laid flat	15.38	0.46	26.66	0.30	17.44	m²	**44.40**
Clay brick pavings; 215 × 102.5 × 65 mm; on **prepared base (not included); bedding on** **cement: sand (1:4) pointing mortar as work** **proceeds**							
Paving bricks; PC £600.00/1000; herringbone bond							
laid on edge	36.44	1.19	68.43	–	43.08	m²	**111.51**
laid flat	23.70	0.79	45.66	–	30.34	m²	**76.00**
Paving bricks; PC £600.00/1000; basketweave bond							
laid on edge	36.44	0.79	45.62	–	43.08	m²	**88.70**
laid flat	23.70	0.53	30.42	–	30.34	m²	**60.76**

Q25 SLAB/BRICK/STONE/TIMBER PAVINGS

Item Excluding site overheads and profit	PC £	Labour hours	Labour £	Plant £	Material £	Unit	Total rate £
Paving bricks; PC £600.00/1000; running or stack bond							
laid on edge	36.44	0.63	36.51	–	43.08	m²	79.59
laid flat	23.70	0.42	24.34	–	30.34	m²	54.68
Paving bricks; PC £500.00/1000; herringbone bond							
laid on edge	30.37	1.19	68.43	–	34.18	m²	102.61
laid flat	19.75	0.79	45.62	–	26.39	m²	72.01
Paving bricks; PC £500.00/1000; basketweave bond							
laid on edge	30.37	0.79	45.63	–	34.18	m²	79.81
laid flat	19.75	0.53	30.42	–	26.39	m²	56.81
Paving bricks; PC £500.00/1000; running or stack bond							
laid on edge	30.37	0.63	36.51	–	34.18	m²	70.69
laid flat	19.75	0.42	24.34	–	26.39	m²	50.73
Paving bricks; PC £400.00/1000; herringbone bond							
laid on edge	24.29	1.19	68.43	–	28.10	m²	96.53
laid flat	15.80	0.79	45.63	–	22.31	m²	67.94
Paving bricks; PC £400.00/1000; basketweave bond							
laid on edge	24.29	0.79	45.63	–	28.10	m²	73.73
laid flat	15.80	0.53	30.42	–	22.31	m²	52.73
Paving bricks; PC £400.00/1000; running or stack bond							
laid on edge	24.29	0.63	36.51	–	28.10	m²	64.61
laid flat	15.80	0.42	24.34	–	22.31	m²	46.65
Paving bricks; PC £300.00/1000; herringbone bond							
laid on edge	17.77	1.19	68.43	–	21.59	m²	90.02
laid flat	11.85	0.79	45.63	–	18.49	m²	64.12
Paving bricks; PC £300.00/1000; basketweave bond							
laid on edge	17.77	0.79	45.63	–	21.59	m²	67.22
laid flat	11.85	0.53	30.42	–	18.49	m²	48.91
Paving bricks; PC £300.00/1000; running or stack bond							
laid on edge	17.77	0.63	36.50	–	21.59	m²	58.09
laid flat	11.85	0.42	24.34	–	18.49	m²	42.83
Cutting							
curved cutting	–	0.15	8.55	6.25	–	m	14.80
raking cutting	–	0.11	6.42	4.98	–	m	11.40
Add or subtract the following amounts for every £10.00/1000 difference in the prime cost of bricks							
Butt joints							
200 × 100 mm	–	–	–	–	0.50	m²	0.50
215 × 102.5 mm	–	–	–	–	0.45	m²	0.45

Q25 SLAB/BRICK/STONE/TIMBER PAVINGS

Item Excluding site overheads and profit	PC £	Labour hours	Labour £	Plant £	Material £	Unit	Total rate £
BRICK PAVINGS – cont							
Add or subtract the following amounts for every £10.00/1000 difference in the prime cost of bricks – cont							
10 mm mortar joints							
200 × 100 mm	–	–	–	–	0.43	m²	0.43
215 × 102.5 mm	–	–	–	–	0.40	m²	0.40
PRECAST CONCRETE SLAB PAVINGS							
Slab paving; precast concrete pavings; Charcon Hard Landscaping; on prepared subbase (not included); bedding on 25 mm thick cement: sand mortar (1:4); butt joints; straight both ways; on 50 mm thick sharp sand base							
Pavings; natural grey							
450 × 450 × 70 mm chamfered	18.47	0.15	8.55	–	23.25	m²	**31.80**
450 × 450 × 50 mm chamfered	14.42	0.15	8.55	–	19.20	m²	**27.75**
600 × 300 × 50 mm	12.56	0.15	8.55	–	17.34	m²	**25.89**
400 × 400 × 65 mm chamfered	25.06	0.13	7.70	–	29.84	m²	**37.54**
450 × 600 × 50 mm	12.04	0.15	8.55	–	16.82	m²	**25.37**
600 × 600 × 50 mm	9.86	0.13	7.70	–	14.64	m²	**22.34**
750 × 600 × 50 mm	9.42	0.13	7.70	–	14.20	m²	**21.90**
900 × 600 × 50 mm	8.04	0.13	7.70	–	12.82	m²	**20.52**
Pavings; coloured							
450 × 450 × 70 mm chamfered	24.59	0.15	8.55	–	29.37	m²	**37.92**
450 × 600 × 50 mm	14.85	0.15	8.55	–	19.63	m²	**28.18**
400 × 400 × 65 mm chamfered	25.06	0.13	7.70	–	29.84	m²	**37.54**
600 × 600 × 50 mm	11.94	0.13	7.70	–	16.72	m²	**24.42**
750 × 600 × 50 mm	11.07	0.13	7.70	–	15.85	m²	**23.55**
900 × 600 × 50 mm	9.51	0.13	7.70	–	14.29	m²	**21.99**
Precast concrete pavings; Charcon Hard Landscaping; on prepared subbase (not included); bedding on 25 mm thick cement: sand mortar (1:4); butt joints; straight both ways; jointing in cement: sand (1:3) brushed in; on 50 mm thick sharp sand base							
Appalacian rough textured exposed aggregate pebble paving							
600 × 600 × 65 mm	31.48	0.17	9.63	–	34.59	m²	**44.22**

Q25 SLAB/BRICK/STONE/TIMBER PAVINGS

Item Excluding site overheads and profit	PC £	Labour hours	Labour £	Plant £	Material £	Unit	Total rate £
Pavings; Marshalls Plc; spot bedding on 5 nr pads of cement: sand mortar (1:4); on sharp sand							
Blister Tactile pavings; specially textured slabs for guidance of blind pedestrians; red or buff							
400 × 400 × 50 mm	31.71	0.17	9.63	–	33.44	m²	**43.07**
450 × 450 × 50 mm	25.82	0.17	9.63	–	27.54	m²	**37.17**
Metric Four Square pavings							
496 × 496 × 50 mm; exposed river gravel aggregate	99.71	0.17	9.63	–	101.44	m²	**111.07**
Metric Four Square cycle blocks							
496 × 496 × 50 mm; exposed aggregate	99.71	0.08	4.81	–	102.17	m²	**106.98**
Precast concrete pavings; Marshalls Plc; Heritage imitation riven yorkstone paving; on prepared subbase measured separately; bedding on 25 mm thick cement: sand mortar (1:4); pointed straight both ways cement: sand (1:3)							
Square and rectangular paving							
450 × 300 × 38 mm	38.27	0.33	19.25	–	42.34	m²	**61.59**
450 × 450 × 38 mm	23.57	0.25	14.44	–	27.64	m²	**42.08**
600 × 300 × 38 mm	28.25	0.27	15.40	–	32.32	m²	**47.72**
600 × 450 × 38 mm	26.01	0.25	14.44	–	30.08	m²	**44.52**
600 × 600 × 38 mm	24.75	0.17	9.63	–	28.89	m²	**38.52**
Extra labours for laying the a selection of the above sizes to random rectangular pattern	–	0.33	6.42	–	–	m²	**6.42**
Radial paving for circles							
circle with centre stone and first ring (8 slabs); 450 × 230/560 × 38 mm; diameter 1.54 m (total area 1.86 m²)	66.85	0.50	28.88	–	72.60	nr	**101.48**
circle with second ring (16 slabs); 450 × 300/460 × 38 mm; diameter 2.48 m (total area 4.83 m²)	66.85	1.33	77.00	–	187.07	nr	**264.07**
circle with third ring (16 slabs); 450 × 470/ 625 × 38 mm; diameter 3.42 m (total area 9.18 m²)	66.85	2.67	154.00	–	341.41	nr	**495.41**
Stepping stones							
380 mm diameter × 38 mm	4.81	0.07	3.85	–	11.04	nr	**14.89**
asymmetrical; 560 × 420 × 38 mm	6.43	0.07	3.85	–	12.66	nr	**16.51**
Marshalls Plc; La Linia Paving; fine textured exposed aggregate 80 mm thick pavings in various sizes to designed laying patterns; laid to 50 mm sharp sand bed on Type 1 base all priced separately							
Bonded laying patterns; 300 × 300 mm							
light granite/anthracite basalt	32.08	0.25	14.44	–	32.08	m²	**46.52**
Indian granite/yellow	32.38	0.25	14.44	–	32.38	m²	**46.82**

Q25 SLAB/BRICK/STONE/TIMBER PAVINGS

Item Excluding site overheads and profit	PC £	Labour hours	Labour £	Plant £	Material £	Unit	Total rate £
PRECAST CONCRETE SLAB PAVINGS – **cont**							
Marshalls Plc – cont							
Random scatter pattern incorporating							
100 × 200 mm, 200 × 200 mm and							
300 × 200 mm units							
light granite/anthracite basalt	31.30	0.33	19.25	–	31.30	m²	**50.55**
Indian granite/yellow	31.59	0.33	19.25	–	31.59	m²	**50.84**
Marshalls Plc; La Linia Grande Paving; as **above but 140 mm thick**							
Bonded laying patterns; 300 × 300 mm							
all colours	61.48	0.25	14.44	–	61.48	m²	**75.92**
Random scatter pattern incorporating							
100 × 200 mm, 200 × 200 mm and							
300 × 200 mm units							
light granite/anthracite basalt	61.48	0.33	19.25	–	61.48	m²	**80.73**
Extra over to the above for incorporating inlay							
stones to patterns; blue, light granite or							
anthracite basalt							
to prescribed patterns	387.63	0.50	28.88	–	387.63	m²	**416.51**
as individual units; triangular							
200 × 200 × 282 mm	2.19	0.02	0.96	–	2.19	nr	**3.15**
DETERRRENT SLAB PAVINGS							
Pedestrian deterrent pavings; Marshalls **Plc; on prepared base (not included);** **bedding on 25 mm cement: sand (1:3);** **cement: sand (1:3) joints**							
Lambeth pyramidal pavings							
600 × 600 × 75 mm	18.98	0.08	4.81	–	23.05	m²	**27.86**
Thaxted pavings; granite sett appearance							
600 × 600 × 75 mm	22.38	0.11	6.42	–	26.45	m²	**32.87**
Pedestrian deterrent pavings; Townscape **Products Ltd; on prepared base (not** **included); bedding on 25 mm cement: sand** **(1:3); cement: sand (1:3) joints**							
Strata striated textured slab pavings; giving							
bonded appearance; grey							
600 × 600 × 60 mm	14.11	0.17	9.63	–	18.18	m²	**27.81**
Geoset raised chamfered studs pavings; grey							
600 × 600 × 60 mm	14.11	0.17	9.63	–	18.18	m²	**27.81**
Abbey square cobble pattern pavings;							
reinforced							
600 × 600 × 65 mm	13.35	0.27	15.40	–	17.42	m²	**32.82**

Q25 SLAB/BRICK/STONE/TIMBER PAVINGS

Item Excluding site overheads and profit	PC £	Labour hours	Labour £	Plant £	Material £	Unit	Total rate £
Edge restraints; to block paving; on **prepared base (not included);** **200 × 100 × 80 mm; PC £8.66/m²; haunching** **one side**							
Header course							
200 × 100 mm; butt joints	1.82	0.09	5.13	–	4.37	m	**9.50**
Stretcher course							
200 × 100 mm; butt joints	0.91	0.06	3.21	–	3.29	m	**6.50**
PRECAST CONCRETE BLOCK PAVINGS							
Concrete paviors; Marshalls Plc; on **prepared base (not included); bedding on** **50 mm sand; kiln dried sand joints swept in**							
Keyblok paviors							
200 × 100 × 60 mm; grey	8.17	0.13	7.70	0.12	10.05	m²	**17.87**
200 × 100 × 60 mm; colours	8.79	0.13	7.70	0.12	10.67	m²	**18.49**
200 × 100 × 80 mm; grey	9.10	0.15	8.55	0.12	10.98	m²	**19.65**
200 × 100 × 80 mm; colours	10.27	0.15	8.55	0.12	12.15	m²	**20.82**
Concrete cobble paviors; Charcon Hard **Landscaping; Concrete Products; on** **prepared base (not included); bedding on** **50 mm sand; kiln dried sand joints swept in**							
Paviors							
Woburn blocks; 100–201 × 134 × 80 mm; random sizes	32.11	0.22	12.83	0.12	34.00	m²	**46.95**
Woburn blocks; 100–201 × 134 × 80 mm; single size	32.11	0.17	9.63	0.12	34.00	m²	**43.75**
Woburn blocks; 100–201 × 134 × 60 mm; random sizes	26.46	0.22	12.83	0.12	28.34	m²	**41.29**
Woburn blocks; 100–201 × 134 × 60 mm; single size	26.43	0.17	9.63	0.12	28.31	m²	**38.06**
Concrete setts; on 25 mm sand; **compacted; vibrated; joints filled with** **sand; natural or coloured; well rammed** **hardcore base (not included)**							
Marshalls Plc; Tegula cobble paving							
60 mm thick; random sizes	21.99	0.19	11.00	0.12	24.04	m²	**35.16**
60 mm thick; single size	21.99	0.15	8.75	0.12	24.04	m²	**32.91**
80 mm thick; random sizes	25.33	0.19	11.00	0.12	27.38	m²	**38.50**
80 mm thick; single size	25.33	0.15	8.75	0.12	27.38	m²	**36.25**
cobbles; 80 × 80 × 60 mm thick; traditional	39.35	0.19	10.70	0.12	41.40	m²	**52.22**
Cobbles							
Charcon Hard Landscaping; Country setts							
100 mm thick; random sizes	42.77	0.33	19.25	0.12	44.82	m²	**64.19**
100 mm thick; single size	43.84	0.22	12.83	0.12	45.89	m²	**58.84**

Q25 SLAB/BRICK/STONE/TIMBER PAVINGS

Item Excluding site overheads and profit	PC £	Labour hours	Labour £	Plant £	Material £	Unit	Total rate £
PRECAST CONCRETE CYCLE BLOCKS							
Concrete cycle blocks; Marshalls Plc;							
bedding in cement: sand (1:4)							
Metric 4 Square cycle stand blocks; smooth							
grey concrete							
496 × 496 × 100 mm	43.00	0.25	4.81	–	43.59	nr	**48.40**
Concrete cycle blocks; Townscape							
Products Ltd; on 100 mm concrete (1:2:4);							
on 150 mm hardcore; bedding in cement:							
sand (1:4)							
Cycle blocks							
Cycle Bloc; in white concrete	32.66	0.25	4.81	–	33.44	nr	**38.25**
Mountain Cycle Bloc; in white concrete	45.18	0.25	4.81	–	45.96	nr	**50.77**
GRASS CONCRETE							
Grass concrete – General							
Preamble: Grass seed should be a perennial							
ryegrass mixture, with the proportion							
depending on expected traffic. Hardwearing							
winter sportsground mixtures are suitable for							
public areas. Loose gravel, shingle or sand is							
liable to be kicked out of the blocks; rammed							
hoggin or other stabilized material should be							
specified.							
Grass concrete; Grass Concrete Ltd; on							
blinded granular Type 1 subbase (not							
included)							
Grasscrete in situ concrete continuously							
reinforced surfacing; including expansion							
joints at 10 m centres; soiling; seeding							
GC2; 150 mm thick; traffic up to 40.00							
tonnes	–	–	–	–	–	m²	**48.87**
GC1; 100 mm thick; traffic up to 13.30							
tonnes	–	–	–	–	–	m²	**37.79**
GC3; 76 mm thick; traffic up to 4.30 tonnes	–	–	–	–	–	m²	**32.87**
Grass concrete; Grass Concrete Ltd;							
406 × 406 mm blocks; on 20 mm sharp							
sand; on blinded MOT type 1 subbase (not							
included); level and to falls only; including							
filling with topsoil; seeding with dwarf rye							
grass at PC £4.50/kg							
Pavings							
GB103; 103 mm thick	16.72	0.40	7.70	0.18	19.34	m²	**27.22**
GB83; 83 mm thick	14.42	0.36	7.00	0.18	17.04	m²	**24.22**

Q25 SLAB/BRICK/STONE/TIMBER PAVINGS

Item Excluding site overheads and profit	PC £	Labour hours	Labour £	Plant £	Material £	Unit	Total rate £
Grass concrete; Charcon Hard Landscaping; on 25 mm sharp sand; including filling with topsoil; seeding with dwarf rye grass at PC £4.50/kg							
Grassgrid grass/concrete paving blocks							
366 × 274 × 100 mm thick	19.24	0.38	7.22	0.18	21.86	m²	29.26
Grass concrete; Marshalls Plc; on 25 mm sharp sand; including filling with topsoil; seeding with dwarf rye grass at PC £4.50/kg							
Concrete grass pavings							
Grassguard 130; for light duty applications (80 mm prepared base not included)	18.54	0.38	7.22	0.18	21.35	m²	28.75
Grassguard 160; for medium duty applications (80–150 mm prepared base not included)	21.86	0.46	8.88	0.18	24.67	m²	33.73
Grassguard 180; for heavy duty applications (150 mm prepared base not included)	24.48	0.60	11.55	0.18	27.28	m²	39.01
Full mortar bedding							
Extra over pavings for bedding on 25 mm cement: sand (1:4); in lieu of spot bedding on sharp sand	–	0.03	0.48	–	1.57	m²	2.05
NATURAL STONE SLAB PAVINGS							
Natural stone, slab or granite paving – General							
Preamble: Provide paving slabs of the specified thickness in random sizes but not less than 25 slabs per 10 m² of surface area, to be laid in parallel courses with joints alternately broken and laid to falls.							
Natural stone, slate or granite flag pavings; CED Ltd; on prepared base (not included); bedding on 25 mm cement: sand (1:3); cement: sand (1:3) joints							
Yorkstone; riven laid random rectangular							
new slabs; 40–60 mm thick	99.33	0.57	33.00	–	103.10	m²	136.10
reclaimed slabs; Cathedral grade; 50–75 mm thick	102.96	0.93	53.90	–	106.73	m²	160.63
Donegal quartzite slabs; standard tiles							
200 mm × random lengths × 15–25 mm	89.18	1.17	67.38	–	92.95	m²	160.33
250 mm × random lengths × 15–25 mm	89.18	1.10	63.52	–	92.95	m²	156.47
300 mm × random lengths × 15–25 mm	89.18	1.03	59.67	–	92.95	m²	152.62
350 mm × random lengths × 15–25 mm	89.18	0.93	53.90	–	92.95	m²	146.85
400 mm × random lengths × 15–25 mm	89.18	0.83	48.12	–	92.95	m²	141.07
450 mm × random lengths × 15–25 mm	89.18	0.78	44.92	–	92.95	m²	137.87

Q25 SLAB/BRICK/STONE/TIMBER PAVINGS

Item Excluding site overheads and profit	PC £	Labour hours	Labour £	Plant £	Material £	Unit	Total rate £
NATURAL STONE SLAB PAVINGS – cont							
Natural yorkstone pavings; Johnsons							
Wellfield Quarries; sawn 6 sides; 50 mm							
thick; on prepared base measured							
separately; bedding on 25 mm cement:							
sand (1:3); cement: sand (1:3) joints							
Paving							
laid to random rectangular pattern	56.65	0.57	33.00	–	60.23	m²	**93.23**
laid to coursed laying pattern; 3 sizes	61.10	0.57	33.19	–	64.68	m²	**97.87**
Paving; single size							
600 × 600 mm	66.09	0.28	16.36	–	69.67	m²	**86.03**
600 × 400 mm	66.09	0.33	19.25	–	69.67	m²	**88.92**
300 × 200 mm	74.81	0.67	38.50	–	78.39	m²	**116.89**
215 × 102.5 mm	77.31	0.83	48.12	–	80.89	m²	**129.01**
Paving; cut to template off site; 600 × 600 mm;							
radius							
1.00 m	161.50	1.11	64.17	–	165.08	m²	**229.25**
2.50 m	161.50	0.67	38.50	–	165.08	m²	**203.58**
5.00 m	161.50	0.67	38.50	–	165.08	m²	**203.58**
Natural yorkstone pavings; Johnsons							
Wellfield Quarries; sawn 6 sides; 75 mm							
thick; on prepared base measured							
separately; bedding on 25 mm cement:							
sand (1:3); cement: sand (1:3) joints							
Paving							
laid to random rectangular pattern	67.98	0.32	18.29	–	71.56	m²	**89.85**
laid to coursed laying pattern; 3 sizes	72.96	0.32	18.29	–	76.55	m²	**94.84**
Paving; single size							
600 × 600 mm	77.95	0.32	18.29	–	81.53	m²	**99.82**
600 × 400 mm	77.95	0.32	18.29	–	81.53	m²	**99.82**
300 × 200 mm	87.78	0.25	14.44	–	91.36	m²	**105.80**
215 × 102.5 mm	89.78	0.83	48.12	–	93.36	m²	**141.48**
Paving; cut to template off site; 600 × 600 mm;							
radius							
1.00 m	190.00	1.33	77.00	–	193.58	m²	**270.58**
2.50 m	190.00	0.83	48.12	–	193.58	m²	**241.70**
5.00 m	190.00	0.83	48.12	–	193.58	m²	**241.70**
CED Ltd; Indian sandstone, riven pavings							
or edgings; calibrated +/-26 mm thick; on							
prepared base measured separately;							
bedding on 25 mm cement: sand (1:3);							
cement: sand (1:3) joints							
Paving							
laid to random rectangular pattern	28.39	0.80	46.21	–	31.97	m²	**78.18**
laid to coursed laying pattern; 3 sizes	28.39	0.67	38.50	–	31.97	m²	**70.47**
Paving; single size							
600 × 600 mm	28.39	0.33	19.25	–	31.97	m²	**51.22**
600 × 400 mm	28.39	0.42	24.06	–	31.97	m²	**56.03**
400 × 400 mm	28.39	0.56	32.09	–	31.97	m²	**64.06**

Q25 SLAB/BRICK/STONE/TIMBER PAVINGS

Item Excluding site overheads and profit	PC £	Labour hours	Labour £	Plant £	Material £	Unit	Total rate £
Natural stone, slate or granite flag pavings;							
CED Ltd; on prepared base (not included);							
bedding on 25 mm cement: sand (1:3);							
cement: sand (1:3) joints							
Granite paving; sawn 6 sides; textured top							
new slabs; silver grey; 50 mm thick	38.35	0.57	33.00	–	42.11	m²	**75.11**
new slabs; blue grey; 50 mm thick	44.96	0.57	33.00	–	48.73	m²	**81.73**
new slabs; yellow grey; 50 mm thick	47.61	0.57	33.00	–	51.38	m²	**84.38**
new slabs; black; 50 mm thick	70.09	0.57	33.00	–	73.86	m²	**106.86**
Edgings; silver grey							
100 mm wide × random lengths	6.77	0.33	19.25	–	7.13	m	**26.38**
100 × 100 mm	7.74	0.50	28.88	–	8.10	m	**36.98**
100 mm long × 200 mm wide	12.57	0.20	11.55	–	12.93	m	**24.48**
250 mm wide × random lengths	13.30	0.25	14.44	–	16.88	m	**31.32**
300 mm wide × random lengths	14.51	0.25	14.44	–	18.09	m	**32.53**
RECONSTITUTED STONE SLAB PAVINGS							
Reconstituted yorkstone aggregate							
pavings; Marshalls Plc; Saxon on prepared							
subbase measured separately; bedding on							
25 mm thick cement: sand mortar (1:4) ;on							
50 mm thick sharp sand base							
Square and rectangular paving in buff; butt							
joints straight both ways							
300 × 300 × 35 mm	35.41	0.27	15.40	–	40.83	m²	**56.23**
600 × 300 × 35 mm	23.18	0.22	12.51	–	28.61	m²	**41.12**
450 × 450 × 50 mm	25.23	0.25	14.44	–	30.66	m²	**45.10**
600 × 600 × 35 mm	17.36	0.17	9.63	–	22.78	m²	**32.41**
600 × 600 × 50 mm	21.70	0.18	10.59	–	27.12	m²	**37.71**
Square and rectangular paving in natural; butt							
joints straight both ways							
300 × 300 × 35 mm	28.97	0.27	15.40	–	34.40	m²	**49.80**
450 × 450 × 50 mm	21.46	0.23	13.47	–	26.89	m²	**40.36**
600 × 300 × 35 mm	20.38	0.25	14.44	–	25.81	m²	**40.25**
600 × 600 × 35 mm	14.98	0.17	9.63	–	20.40	m²	**30.03**
600 × 600 × 50 mm	18.09	0.20	11.55	–	23.51	m²	**35.06**
Radial paving for circles; 20 mm joints							
circle with centre stone and first ring							
(8 slabs); 450 × 230/560 × 35 mm; diameter							
1.54 m (total area 1.86 m²)	69.04	0.50	28.88	–	75.12	nr	**104.00**
circle with second ring (16 slabs);							
450 × 300/460 × 35 mm; diameter 2.48 m							
(total area 4.83 m²)	168.24	1.33	77.00	–	183.93	nr	**260.93**
circle with third ring (24 slabs); 450 × 310/							
430 × 35 mm; diameter 3.42 m (total area							
9.18 m²)	317.04	2.67	154.00	–	347.05	nr	**501.05**

Q25 SLAB/BRICK/STONE/TIMBER PAVINGS

Item Excluding site overheads and profit	PC £	Labour hours	Labour £	Plant £	Material £	Unit	Total rate £
NATURAL STONE SETTS							
Granite setts; bedding on 25 mm cement: **sand (1:3)** Natural granite setts; 100 × 100 mm to 125 × 150 mm × 150–250 mm length; riven surface; silver grey							
new; standard grade	27.42	0.67	38.50	–	36.02	m²	**74.52**
new; high grade	31.00	0.67	38.50	–	39.59	m²	**78.09**
reclaimed; cleaned	36.26	0.67	38.50	–	44.85	m²	**83.35**
COBBLE PAVINGS							
Cobble pavings – General Cobbles should be embedded by hand, tight-butted, endwise to a depth of 60% of their length. A dry grout of rapid-hardening cement: sand (1:2) shall be brushed over the cobbles until the interstices are filled to the level of the adjoining paving. Surplus grout shall then be brushed off and a light, fine spray of water applied over the area.							
Cobble pavings Cobbles; to present a uniform colour in panels; or varied in colour as required							
Scottish Beach Cobbles; 50–75 mm	23.57	1.11	64.17	–	29.99	m²	**94.16**
Scottish Beach Cobbles; 75–100 mm	33.58	0.83	48.12	–	39.99	m²	**88.11**
Scottish Beach Cobbles; 100–200 mm	34.55	2.67	77.00	–	40.97	m²	**117.97**
TIMBER DECKING							
Market prices of timber decking materials; **Exterior Decking Ltd** Hardwood decking; stainless steel screw fixings included in rates shown							
Ipe; pre-drilled and countersunk	75.09	–	–	–	–	m²	–
Hardwood decking; 'Exterpark' invisible fixings included in rates shown							
Ipe; 21 mm thick	82.19	–	–	–	–	m²	–
Ipe; 28 mm thick	124.87	–	–	–	–	m²	–
Ipe; 35 mm thick	170.00	–	–	–	–	m²	–
Teak Rustic FSC 21 mm thick	93.31	–	–	–	–	m²	–
Merbau 21 mm	66.64	–	–	–	–	m²	–
Merbau 28 mm	87.35	–	–	–	–	m²	–
Merbau 35 mm	102.31	–	–	–	–	m²	–
Composite decking; stainless steel screw fixings included in rates shown							
Millboard 176 mm wide	66.00	–	–	–	–	m²	–
Millboard 196 mm wide	59.39	–	–	–	59.39	m²	**59.39**
Softwood decking							
Pine smooth; 145 × 38 mm	15.16	–	–	–	–	m²	–

Q25 SLAB/BRICK/STONE/TIMBER PAVINGS

Item Excluding site overheads and profit	PC £	Labour hours	Labour £	Plant £	Material £	Unit	Total rate £
Timber decking; Exterior Decking Ltd;							
timber decking substructure/carcassing							
costs							
Roof decks exclusive of beams; timber craned							
(not included); joists at 400 mm centres							
47 × 47 mm	–	–	–	–	–	m²	33.79
47 × 100 mm	–	–	–	–	–	m²	36.57
47 × 150 mm	–	–	–	–	–	m²	39.40
pedestals up to 150 mm	–	–	–	–	–	m²	58.90
Commercial applications; decks with double							
beams at 300 mm centres							
47 × 100 mm	–	–	–	–	–	m²	79.65
47 × 150 mm	–	–	–	–	–	m²	85.59
Domestic garden applications; decks with							
double beams at 300 mm centres							
47 × 100 mm	–	–	–	–	–	m²	69.50
47 × 150 mm	–	–	–	–	–	m²	73.68
Structural posts to decks; tanalized treated							
softwood; rebated including excavation and							
setting in concrete							
100 × 100 mm; single beam; single rebate	–	–	–	–	–	m	10.17
150 × 150 mm; double beam; double rebate	–	–	–	–	–	m	13.90
Timber decking; Exterior Decking Ltd;							
timber decking surfaces; fixed to							
substructure (not included)							
Softwood deck timbers; screw fixed							
pine; smooth; 145 × 38 mm fixed with							
65 × 5 mm coated deck screws	–	–	–	–	–	m²	39.34
Hardwood deck timbers; screw fixed							
Ipe; S4S E4E; 145 × 21 mm; pre-drilled	–	–	–	–	–	m²	112.49
Hardwood deck timbers; invisibly fixed							
Ipe; 21 mm thick	–	–	–	–	–	m²	137.16
Ipe; 28 mm thick	–	–	–	–	–	m²	195.69
Ipe; 35 mm thick	–	–	–	–	–	m²	257.37
Teak Rustic FSC 21 mm	–	–	–	–	–	m²	151.62
Merbau 21 mm	–	–	–	–	–	m²	116.95
Merbau 28 mm	–	–	–	–	–	m²	146.90
Merbau 35 mm	–	–	–	–	–	m²	169.38
Composite deck surface; Millboard							
146 mm wide; fixed with stainless screws							
4.5 × 70 mm	–	–	–	–	–	m²	85.80
196 mm wide; fixed with stainless screws							
4.5 × 70 mm	–	–	–	–	–	m²	85.80
Surface treatments to timber decks; Owatrol							
decking oil							
Seasonite	–	–	–	–	–	m²	6.24
Textrol	–	–	–	–	–	m²	4.55

Q25 SLAB/BRICK/STONE/TIMBER PAVINGS

Item Excluding site overheads and profit	PC £	Labour hours	Labour £	Plant £	Material £	Unit	Total rate £
TIMBER DECKING – cont							
Timber decking							
Supports for timber decking; softwood joists to receive decking boards; joists at 400 mm centres; Southern Yellow pine							
38 × 88 mm	15.62	1.00	19.25	–	17.33	m²	**36.58**
47 × 150 mm	14.69	1.00	19.25	–	16.39	m²	**35.64**
47 × 100 mm	9.79	1.00	19.25	–	11.50	m²	**30.75**
Hardwood decking; Yellow Balau; grooved or smooth; 6 mm joints							
deck boards; 90 mm wide × 19 mm thick	56.21	1.00	19.25	–	58.16	m²	**77.41**
deck boards; 145 mm wide × 21 mm thick	37.87	1.00	19.25	–	39.81	m²	**59.06**
deck boards; 145 mm wide × 28 mm thick	38.52	1.00	19.25	–	40.47	m²	**59.72**
Hardwood decking; Ipe; smooth; 6 mm joints							
deck boards; 90 mm wide × 19 mm thick	48.50	1.00	19.25	–	50.45	m²	**69.70**
deck boards; 145 mm wide × 19 mm thick	49.50	1.00	19.25	–	51.45	m²	**70.70**
AVS; Profile; Western Red Cedar; 6 mm joints smooth or reeded							
ex 125 mm wide × 38 mm thick	51.68	1.00	19.25	–	53.62	m²	**72.87**
ex 150 mm wide × 32 mm thick	16.39	1.00	19.25	–	18.34	m²	**37.59**
Handrails and base rail; fixed to posts at 2.00 m centres							
posts; 100 × 100 × 1370 mm high	9.29	1.00	19.25	–	9.74	m	**28.99**
posts; turned; 1220 mm high	15.40	1.00	19.25	–	15.85	m	**35.10**
Handrails; balusters							
square balusters at 100 mm centres	34.20	0.50	9.63	–	34.65	m	**44.28**
square balusters at 300 mm centres	11.39	0.33	6.41	–	11.75	m	**18.16**
turned balusters at 100 mm centres	54.00	0.50	9.63	–	54.45	m	**64.08**
turned balusters at 300 mm centres	17.98	0.33	6.35	–	18.34	m	**24.69**
STEPS							
In situ concrete steps; ready mixed concrete; Gen 1; on 100 mm hardcore base; base concrete thickness 100 mm treads and risers cast simultaneously; inclusive of A142 mesh within the concrete; step riser 170 mm high; inclusive of all shuttering and concrete labours; floated finish; to graded ground (not included)							
Treads 250 mm wide							
one riser	4.91	1.32	76.23	–	8.56	m	**84.79**
two risers	7.86	1.98	114.34	–	13.89	m	**128.23**
three risers	11.38	2.64	152.46	–	25.95	m	**178.41**
four risers	15.28	3.96	228.69	–	30.91	m	**259.60**
six risers	19.35	5.28	304.92	–	35.55	m	**340.47**

Q25 SLAB/BRICK/STONE/TIMBER PAVINGS

Item Excluding site overheads and profit	PC £	Labour hours	Labour £	Plant £	Material £	Unit	Total rate £
Lay surface to concrete in situ steps above granite treads; pre-cut off site; 305 mm wide × 50 mm thick bullnose finish inclusive of 20 mm overhang; granite riser 20 mm thick all laid on 15 mm mortar bed	25.15	0.66	38.12	–	26.22	m	64.34
riven yorkstone tread and riser 50 mm thick; cut on site	41.54	0.82	47.64	–	42.61	m	90.25
riven yorkstone treads 225 mm wide × 50 mm thick cut on site; two courses brick risers laid on flat stretcher bond; bricks PC £500.00/1000	20.32	0.99	57.17	–	26.16	m	83.33
Marshalls Saxon paving slab 450 × 450 mm cut to 225 mm wide × 38 mm thick; cut on site; two courses brick risers laid on flat stretcher bond; bricks PC £500.00/1000	4.86	0.99	57.17	–	10.70	m	67.87
Treads 450 mm wide							
one riser	8.83	1.65	95.29	–	13.37	m	108.66
two risers	18.04	2.31	133.40	–	27.30	m	160.70
three risers	25.50	2.97	171.52	–	41.59	m	213.11
four risers	44.11	4.29	247.75	–	62.60	m	310.35
six risers	56.07	5.94	343.04	–	73.94	m	416.98
Lay surface to concrete in situ steps granite treads; pre-cut off site; 470 mm wide × 50 mm thick bullnose finish inclusive of 20 mm overhang; granite riser 20 mm thick all laid on 15 mm mortar bed	36.17	1.32	76.23	–	37.52	m	113.75
riven yorkstone tread and riser 50 mm thick; cut on site	56.44	1.49	85.76	–	57.78	m	143.54
riven yorkstone treads 460 mm wide 50 mm thick cut on site; two courses brick risers laid on flat stretcher bond; bricks PC £500.00/1000	41.54	1.65	95.29	–	47.59	m	142.88
Marshalls Saxon paving slab 450 × 450 × 38 mm thick cut on site; two courses brick risers laid on flat stretcher bond; bricks PC £500.00/1000	9.72	0.91	52.41	–	15.56	m	67.97
SPECIALIZED MORTARS FOR PAVINGS							
SteinTec Ltd							
Primer for specialized mortar; SteinTec Tuffbond priming mortar immediately prior to paving material being placed on bedding mortar							
maximum thickness 1.5 mm (1.5 kg/m²)	0.71	0.02	0.39	–	0.71	m²	1.10
Tuffbed; 2 pack hydraulic mortar for paving applications							
30 mm thick	7.18	0.17	3.21	–	7.18	m²	10.39
40 mm thick	9.57	0.20	3.85	–	9.57	m²	13.42
50 mm thick	11.96	0.25	4.81	–	11.96	m²	16.77

Q26 SPECIAL SURFACINGS/PAVINGS FOR SPORT/GENERAL AMENITY

Item Excluding site overheads and profit	PC £	Labour hours	Labour £	Plant £	Material £	Unit	Total rate £
SPECIALIZED MORTARS FOR PAVINGS – cont							
SteinTec Ltd – cont							
SteinTec jointing mortar; joint size							
100 × 100 × 100 mm; 10 mm joints;							
31.24 kg/m²	10.73	0.50	9.63	–	10.73	m²	**20.36**
100 × 100 × 100 mm; 20 mm joints;							
31.24 kg/m²	18.89	1.00	19.25	–	18.89	m²	**38.14**
200 × 100 × 65 mm; 10 mm joints;							
15.7 kg/m²	5.39	0.10	1.93	–	5.39	m²	**7.32**
215 × 112.5 × 65 mm; 10 mm joints;							
14.3 kg/m²	4.91	0.08	1.60	–	4.91	m²	**6.51**
300 × 450 × 65 mm; 10 mm joints;							
6.24 kg/m²	2.14	0.04	0.77	–	2.14	m²	**2.91**
300 × 450 × 65 mm; 20 mm joints;							
11.98 kg/m²	4.11	0.10	1.93	–	4.11	m²	**6.04**
450 × 450 × 65 mm; 20 mm joints;							
9.75 kg/m²	3.35	0.11	2.14	–	3.35	m²	**5.49**
450 × 600 × 65 mm; 20 mm joints;							
8.6 kg/m²	2.95	0.09	1.75	–	2.95	m²	**4.70**
600 × 600 × 65 mm; 20 mm joints;							
7.43 kg/m²	2.55	0.07	1.37	–	2.55	m²	**3.92**
Q26 SPECIAL SURFACINGS/PAVINGS FOR SPORT/GENERAL AMENITY							
MARKET PRICES OF SPORTS AND PLAY SURFACING MATERIALS							
Market prices of surfacing materials							
Surfacings; Melcourt Industries Ltd (items labelled FSC are Forest Stewardship Council certified)							
Playbark® 10/50; per 25 m³ load	–	–	–	–	63.60	m³	**63.60**
Playbark® 10/50; per 80 m³ load	–	–	–	–	50.60	m³	**50.60**
Playbark® 8/25; per 25 m³ load	–	–	–	–	62.10	m³	**62.10**
Playbark® 8/25; per 80 m³ load	–	–	–	–	49.10	m³	**49.10**
Playchips®; per 25 m³ load; FSC	–	–	–	–	41.75	m³	**41.75**
Playchips®; per 80 m³ load; FSC	–	–	–	–	28.75	m³	**28.75**
Kushyfall; per 25 m³ load; FSC	–	–	–	–	39.70	m³	**39.70**
Kushyfall; per 80 m³ load; FSC	–	–	–	–	26.70	m³	**26.70**
Softfall; per 25 m³ load	–	–	–	–	31.40	m³	**31.40**
Softfall; per 80 m³ load	–	–	–	–	18.40	m³	**18.40**
Playsand; per 10 t load	–	–	–	–	101.16	m³	**101.16**
Playsand; per 20 t load	–	–	–	–	88.92	m³	**88.92**
Walk Chips; per 25 m³ load; FSC	–	–	–	–	38.70	m³	**38.70**
Walk Chips; per 80 m³ load; FSC	–	–	–	–	25.70	m³	**25.70**
Woodfibre; per 25 m³ load; FSC	–	–	–	–	36.95	m³	**36.95**
Woodfibre; per 80 m³ load; FSC	–	–	–	–	23.95	m³	**23.95**

Q26 SPECIAL SURFACINGS/PAVINGS FOR SPORT/GENERAL AMENITY

Item Excluding site overheads and profit	PC £	Labour hours	Labour £	Plant £	Material £	Unit	Total rate £
NATURAL SPORTS PITCHES							
Natural sports pitches; Agripower Ltd;							
sports pitches; gravel raft construction							
Professional standard							
football pitch; 6500 m²	–	–	–	–	–	nr	150001.50
cricket wicket; county standard	–	–	–	–	–	nr	30000.30
Playing fields; Sport England Compliant							
football pitch; 6500 m²	–	–	–	–	–	nr	65000.65
cricket wicket; playing field standard	–	–	–	–	–	nr	15000.15
track and infield	–	–	–	–	–	nr	92000.92
ARTIFICIAL SPORTS PITCHES							
Artificial surfaces and finishes – General							
Preamble: Advice should also be sought from							
the Technical Unit for Sport, the appropriate							
regional office of the Sports Council or the							
National Playing Fields Association. Some of							
the following prices include base work							
whereas others are for a specialist surface							
only on to a base prepared and costed							
separately.							
Artificial sports pitches; Agripower Ltd;							
third generation rubber crumb							
Sports pitches to respective National							
governing body standards; inclusive of all							
excavation, drainage, lighting, fencing, etc.							
football pitch; 6500 m²	–	–	–	–	–	nr	465004.65
rugby union multi-use pitch; 120 × 75 m	–	–	–	–	–	nr	590005.90
hockey pitch; water-based; 101.4 × 63 m	–	–	–	–	–	nr	550005.50
athletic track; 8 lane; International amateur							
athletic federation	–	–	–	–	–	nr	550005.50
Multi-sport pitches; inclusive of all excavation,							
drainage, lighting, fencing, etc.							
sand dressed; 101.4 × 63 m	–	–	–	–	–	nr	390003.90
sand filled; 101.4 × 63 m	–	–	–	–	–	nr	340003.40
polymeric (synthetic bound rubber)	–	–	–	–	–	m²	130.00
macadam	–	–	–	–	–	m²	80.00
Sports areas; Agripower Ltd							
Sports tracks; polyurethane rubber surfacing;							
on bitumen-macadam (not included); prices for							
5500 m² minimum							
International	–	–	–	–	–	m²	40.00
Club Grade	–	–	–	–	–	m²	31.00

Q26 SPECIAL SURFACINGS/PAVINGS FOR SPORT/GENERAL AMENITY

Item Excluding site overheads and profit	PC £	Labour hours	Labour £	Plant £	Material £	Unit	Total rate £
ARTIFICIAL SPORTS PITCHES – cont							
Sports areas – cont							
Sports areas; multi-component polyurethane rubber surfacing; on bitumen-macadam; finished with polyurethane or acrylic coat; green or red							
Permaprene	–	–	–	–	–	m²	41.00
Cricketweave artificial wicket system; including proprietary subbase (all by specialist subcontractor)							
28 × 2.74 m; hard porous base	–	–	–	–	–	nr	6702.00
28 × 2.74 m; tarmac base	–	–	–	–	–	nr	8470.00
Cricketweave practice batting end							
11 × 2.74 m; hard porous base	–	–	–	–	–	each	3515.00
11 × 2.74 m; tarmac base	–	–	–	–	–	each	4378.00
11 × 3.65 m; hard porous base	–	–	–	–	–	each	4584.00
11 × 3.65 m; tarmac base	–	–	–	–	–	each	5732.00
Cricketweave practice bowling end							
8 × 2.74 m; hard porous base	–	–	–	–	–	each	2639.00
8 × 2.74 m; tarmac base	–	–	–	–	–	each	3265.00
Supply and erection of single bay cricket cage							
18.3 × 3.65 × 3.6 m high	–	–	–	–	–	each	2599.00
TENNIS COURTS							
Tennis courts; Spadeoak Ltd							
Hard playing surfaces to SAPCA Code of Practice minimum requirements; laid on 65 mm thick macadam base on 150 mm thick stone foundation; to include lines, nets, posts, brick edging, 2.75 m high fence and perimeter drainage (excluding excavation, levelling or additional foundation); based on court size 36.6 × 18.3 m (670 m²)							
Durapore; all weather porous macadam and acrylic colour coating	–	–	–	–	–	m²	58.34
Tiger Turf Advantage Pro; sand filled artificial grass	–	–	–	–	–	m²	80.00
Tiger Turf Baseline; short pile sand filled artificial grass	–	–	–	–	–	m²	79.55
DecoColour; impervious acrylic hardcourt (200 mm foundation)	–	–	–	–	–	m²	74.25
DecoTurf; cushioned impervious acrylic tournament surface (200 mm foundation)	–	–	–	–	–	m²	84.85
Porous Kushion Kourt; porous cushioned acrylic surface	–	–	–	–	–	m²	83.08
EasiClay; synthetic clay system	–	–	–	–	–	m²	86.61
Canada Tenn; American green clay; fast dry surface (no macadam but including irrigation)	–	–	–	–	–	m²	88.38

Q26 SPECIAL SURFACINGS/PAVINGS FOR SPORT/GENERAL AMENITY

Item Excluding site overheads and profit	PC £	Labour hours	Labour £	Plant £	Material £	Unit	Total rate £
PLAY SURFACES							
Playgrounds; Rubaflex in situ playground surfacing, porous; on prepared stone/ granular base Type 1 (not included) and macadam base course (not included)							
Black							
15 mm thick (0.50 m critical fall height)	–	–	–	–	–	m²	27.15
35 mm thick (1.00 m critical fall height)	–	–	–	–	–	m²	41.85
60 mm thick (1.50 m critical fall height)	–	–	–	–	–	m²	55.45
Playgrounds; Wetpour safety surfacing in situ playground surfacing, porous; on prepared stone/granular base Type 1 (not included) and macadam base course (not included)							
Coloured							
15 mm thick (0.50 m critical fall height)	–	–	–	–	–	m²	64.48
35 mm thick (1.00 m critical fall height)	–	–	–	–	–	m²	70.15
60 mm thick (1.50 m critical fall height)	–	–	–	–	–	m²	79.19
Playgrounds; Matta Ltd; Play Matta safety tiles; porous; on prepared Type 1 base (not included)							
Critical fall height 500 mm							
natural colours	–	–	–	–	–	m²	67.00
bright colours	–	–	–	–	–	m²	79.00
Critical fall height 1.70 m							
natural colours	–	–	–	–	–	m²	72.00
bright colours	–	–	–	–	–	m²	82.00
Critical fall height 3.20 m							
natural colours	–	–	–	–	–	m²	88.00
bright colours	–	–	–	–	–	m²	98.00
Critical fall height 4.20 m							
natural colours	–	–	–	–	–	m²	99.00
bright colours	–	–	–	–	–	m²	106.00
Playgrounds; Matta Ltd; Safety Matta tiles; supplied and fitted on to grass (turf) or seeded soil base (not included)							
Green or black							
critical fall height 2.70 m	–	–	–	–	–	m²	66.95
critical fall height 2.90 m	–	–	–	–	–	m²	75.00

Q30 SEEDING/TURFING

Item Excluding site overheads and profit	PC £	Labour hours	Labour £	Plant £	Material £	Unit	Total rate £
PLAY SURFACES – cont							
Playgrounds; Melcourt Industries Ltd; **specifiers and users should contact the** **supplier for performance specifications of** **the materials below**							
Play surfaces; on drainage layer (not included); minimum 300 mm settled depth; prices for 80 m³ loads unless specified							
Playbark® 8/25; 8–25 mm particles; red/ brown; 25 m³ loads	18.63	0.35	6.75	–	18.63	m²	**25.38**
Playbark® 8/25; 8–25 mm particles; red/ brown	16.20	0.35	6.75	–	16.20	m²	**22.95**
Playbark® 10/50; 10–50 mm particles; red/ brown	18.55	0.35	6.75	–	18.55	m²	**25.30**
Playchips®; FSC graded woodchips	9.58	0.35	6.75	–	9.58	m²	**16.33**
Kushyfall; fiberised woodchips	8.90	0.35	6.75	–	8.90	m²	**15.65**
Softfall; conifer shavings	6.13	0.35	6.75	–	6.13	m²	**12.88**
Q30 SEEDING/TURFING							
GENERALLY							
Seeding/turfing – General							
Preamble: The following market prices generally reflect the manufacturer's recommended retail prices. Trade and bulk discounts are often available on the prices shown. The manufacturers of these products generally recommend application rates. Note: the following rates reflect the average rate for each product.							
MARKET PRICES OF SEEDING **MATERIALS**							
Market prices of pre-seeding materials							
Rigby Taylor Ltd							
turf fertilizer; Mascot Outfield 16+6+6	–	–	–	–	3.04	100 m²	**3.04**
Boughton Loam							
screened topsoil; 100 mm	–	–	–	–	50.40	m³	**50.40**
screened Kettering loam; 3 mm	–	–	–	–	117.00	m³	**117.00**
screened Kettering loam; sterilized; 3 mm	–	–	–	–	120.60	m³	**120.60**
top dressing; sand soil mixtures; 90/10 to 50/50	–	–	–	–	93.60	m³	**93.60**

Q30 SEEDING/TURFING

Item Excluding site overheads and profit	PC £	Labour hours	Labour £	Plant £	Material £	Unit	Total rate £
Market prices of turf fertilizers;							
recommended average application rates							
Everris; turf fertilizers							
grass fertilizer; slow release; Sierraform GT							
Preseeder 18+22+05	–	–	–	–	5.19	100 m²	**5.19**
grass fertilizer; slow release; Sierraform GT							
All Season 18+06+18	–	–	–	–	4.33	100 m²	**4.33**
grass fertilizer; slow release; Sierraform GT							
Anti-Stress 15+00+26	–	–	–	–	5.19	100 m²	**5.19**
grass fertilizer; slow release; Sierraform GT							
Momentum 22+05+11	–	–	–	–	3.46	100 m²	**3.46**
grass fertilizer; slow release; Sierraform GT							
Spring Start 16+00+16	–	–	–	–	4.33	100 m²	**4.33**
grass fertilizer; controlled release;							
Sierrablen 27+05+05 (8–9 months)	–	–	–	–	6.02	100 m²	**6.02**
grass fertilizer; controlled release;							
Sierrablen 28+05+05 (5–6 months)	–	–	–	–	5.75	100 m²	**5.75**
grass fertilizer; controlled release;							
Sierrablen 15+00+22 (5–6 months)	–	–	–	–	7.19	100 m²	**7.19**
grass fertilizer; Sierrablen Fine 38+00+00							
(2–3 months)	–	–	–	–	2.06	100 m²	**2.06**
grass fertilizer; Sierrablen Mini 00+00+37							
(5–6 months)	–	–	–	–	3.92	100 m²	**3.92**
grass fertilizer; Greenmaster Pro-Lite Turf							
Tonic 08+00+00	–	–	–	–	2.56	100 m²	**2.56**
grass fertilizer; Greenmaster Pro-Lite							
Spring & Summer 14+05+10	–	–	–	–	3.66	100 m²	**3.66**
grass fertilizer; Greenmaster Pro-Lite Zero							
Phosphate 14+00+10	–	–	–	–	3.66	100 m²	**3.66**
grass fertilizer; Greenmaster Pro-Lite							
Mosskiller 14+00+00	–	–	–	–	4.08	100 m²	**4.08**
grass fertilizer; Greenmaster Pro-Lite							
Invigorator 04+00+08	–	–	–	–	2.77	100 m²	**2.77**
grass fertilizer; Greenmaster Pro-Lite							
Autumn 06+05+10	–	–	–	–	3.94	100 m²	**3.94**
grass fertilizer; Greenmaster Pro-Lite							
Double K 07+00+14	–	–	–	–	3.94	100 m²	**3.94**
grass fertilizer; Greenmaster Pro-Lite NK							
12+00+12	–	–	–	–	3.94	100 m²	**3.94**
outfield turf fertilizer; Sportsmaster							
Standard Spring & Summer 09+07+07	–	–	–	–	2.99	100 m²	**2.99**
outfield turf fertilizer; Sportsmaster							
Standard Autumn 04+12+12	–	–	–	–	3.26	100 m²	**3.26**
outfield turf fertilizer; Sportsmaster							
Standard Zero Phosphate 12+00+09	–	–	–	–	3.41	100 m²	**3.41**
TPMC; tree planting and mulching compost	–	–	–	–	3.50	75 l	**3.50**

Q30 SEEDING/TURFING

Item Excluding site overheads and profit	PC £	Labour hours	Labour £	Plant £	Material £	Unit	Total rate £
MARKET PRICES OF SEEDING **MATERIALS – cont**							
Market prices of turf fertilizers – cont							
Rigby Taylor Ltd; 35 g/m²							
grass fertilizer; Mascot Microfine 12+0+10							
+ 2% Mg + 2% Fe	–	–	–	–	6.21	100 m²	6.21
grass fertilizer; Mascot Microfine 8+0+6 +							
2% Mg + 4% Fe	–	–	–	–	5.64	100 m²	5.64
grass fertilizer; Mascot Microfine Organic							
OC1 8+0+0 + 2% Fe	–	–	–	–	5.86	100 m²	5.86
grass fertilizer; Mascot Microfine Organic							
OC2 5+2+10	–	–	–	–	7.08	100 m²	7.08
grass fertilizer; Mascot Guardian 6+1+12 +							
2% Mg + 2% Fe + seaweed	–	–	–	–	7.88	100 m²	7.88
grass fertilizer; Mascot Delta Sport 12+4+8							
+ 0.5% Fe	–	–	–	–	4.88	100 m²	4.88
grass fertilizer; Mascot Fine Turf 12+0+9 +							
1% Mg + 1% Fe + seaweed	–	–	–	–	5.93	100 m²	5.93
outfield fertilizer; Mascot Outfield 16+6+6	–	–	–	–	4.25	100 m²	4.25
outfield fertilizer; Mascot Outfield 4+10+10	–	–	–	–	3.75	100 m²	3.75
outfield fertilizer; Mascot Outfield 9+5+5	–	–	–	–	3.67	100 m²	3.67
outfield fertilizer; Mascot Outfield 12+4+4	–	–	–	–	3.69	100 m²	3.69
liquid fertilizer; Mascot Microflow-C 25+0+0							
+ trace elements; 200–1400 ml/100 m²	–	–	–	–	6.98	100 m²	6.98
liquid fertilizer; Mascot Microflow-CX 4+2							
+18 + trace elements; 200–1400 ml/100 m²	–	–	–	–	6.95	100 m²	6.95
liquid fertilizer; Mascot Microflow-CX 12+0							
+8 + trace elements; 200–1400 ml/100 m²	–	–	–	–	6.70	100 m²	6.70
liquid fertilizer; Mascot Microflow-CX 17+2							
+5 + trace elements; 200–1400 ml/100 m²	–	–	–	–	6.64	100 m²	6.64
Market prices of grass seed							
Preamble: The prices shown are for supply							
only at one number 20 kg or 25 kg bag							
purchase price unless otherwise stated. Rates							
shown are based on the manufacturer's							
maximum recommendation for each seed							
type. Trade and bulk discounts are often							
available on the prices shown for quantities of							
more than one bag.							
Bowling greens; fine lawns; ornamental turf;							
croquet lawns							
British Seed Houses; A1 Green; 35 g/m²	–	–	–	–	25.93	100 m²	25.93
DLF Trifolium; J Green; 34–50 g/m²	–	–	–	–	41.45	100 m²	41.45
DLF Trifolium; J Premier Green; 34–50 g/m²	–	–	–	–	69.50	100 m²	69.50
DLF Trifolium; Promaster 20; 35–50 g/m²	–	–	–	–	20.20	100 m²	20.20
DLF Trifolium; Promaster 10; 35–50 g/m²	–	–	–	–	34.05	100 m²	34.05

Q30 SEEDING/TURFING

Item Excluding site overheads and profit	PC £	Labour hours	Labour £	Plant £	Material £	Unit	Total rate £
Tennis courts; cricket squares							
British Seed Houses; A2 Lawns; 35 g/m²	–	–	–	–	19.35	100 m²	**19.35**
British Seed Houses; A5 Cricket; 35 g/m²	–	–	–	–	29.43	100 m²	**29.43**
DLF Trifolium; J Premier Green; 34–50 g /m²	–	–	–	–	69.50	100 m²	**69.50**
DLF Trifolium; J Court; 18–25 g/m²	–	–	–	–	12.18	100 m²	**12.18**
DLF Trifolium; Promaster 35; 35 g/m²	–	–	–	–	14.11	100 m²	**14.11**
Amenity grassed areas; general purpose lawns							
British Seed Houses; A3 Banks; 25—50 g/m²	–	–	–	–	23.77	100 m²	**23.77**
DLF Trifolium; J Rye Fairway; 18–25 g/m²	–	–	–	–	14.40	100 m²	**14.40**
DLF Trifolium; J Court; 18–25 g/m²	–	–	–	–	12.18	100 m²	**12.18**
DLF Trifolium; Promaster 50; 25–35 g/m²	–	–	–	–	14.95	100 m²	**14.95**
DLF Trifolium; Promaster 120; 25–35 g/m²	–	–	–	–	14.91	100 m²	**14.91**
Conservation; country parks; slopes and banks							
British Seed Houses; A4 Parkland; 17–35 g/m²	–	–	–	–	17.79	100 m²	**17.79**
British Seed Houses; A16 Country Park; 8–19 g/m²	–	–	–	–	12.68	100 m²	**12.68**
British Seed Houses; A17 Legume; 2 g/m²	–	–	–	–	24.62	100 m²	**24.62**
Shaded areas							
British Seed Houses; A6 Shade; 50 g/m²	–	–	–	–	35.65	100 m²	**35.65**
DLF Trifolium; J Green; 34–50 g/m²	–	–	–	–	41.45	100 m²	**41.45**
DLF Trifolium; Promaster 60; 35–50 g/m²	–	–	–	–	28.70	100 m²	**28.70**
Sports pitches; rugby; soccer pitches							
British Seed Houses; A7 Sportsground; 20 g/m²	–	–	–	–	13.19	100 m²	**13.19**
DLF Trifolium; J Pitch; 18–30 g/m²	–	–	–	–	14.07	100 m²	**14.07**
DLF Trifolium; Promaster 70; 15–35 g/m²	–	–	–	–	13.05	100 m²	**13.05**
DLF Trifolium; Promaster 75; 15–35 g/m²	–	–	–	–	17.61	100 m²	**17.61**
DLF Trifolium; Promaster 80; 17–35 g/m²	–	–	–	–	13.86	100 m²	**13.86**
Rigby Taylor; Mascot R11 Football & Rugby; 35 g/m²	–	–	–	–	23.19	100 m²	**23.19**
Rigby Taylor; Mascot R12 General Playing Fields; 35 g/m²	–	–	–	–	25.20	100 m²	**25.20**
Outfields							
British Seed Houses; A7 Sportsground; 20 g/m²	–	–	–	–	13.19	100 m²	**13.19**
British Seed Houses; A9 Outfield; 17–35 g/m²	–	–	–	–	19.37	100 m²	**19.37**
DLF Trifolium; J Fairway; 12–25 g/m²	–	–	–	–	13.90	100 m²	**13.90**
DLF Trifolium; Promaster 40; 35 g/m²	–	–	–	–	15.44	100 m²	**15.44**
DLF Trifolium; Promaster 70; 15–35 g/m²	–	–	–	–	13.05	100 m²	**13.05**
Rigby Taylor; Mascot R4 Cricket Outfields; 35 g/m²	–	–	–	–	33.60	100 m²	**33.60**
Hockey pitches							
DLF Trifolium; J Fairway; 12–25 g/m²	–	–	–	–	13.90	100 m²	**13.90**
DLF Trifolium; Promaster 70; 15–35 g/m²	–	–	–	–	13.05	100 m²	**13.05**
Rigby Taylor; Mascot R10 Cricket & Hockey Outfield; 35 g/m²	–	–	–	–	27.40	100 m²	**27.40**

Q30 SEEDING/TURFING

Item Excluding site overheads and profit	PC £	Labour hours	Labour £	Plant £	Material £	Unit	Total rate £
MARKET PRICES OF SEEDING **MATERIALS – cont**							
Market prices of grass seed – cont							
Parks							
British Seed Houses; A7 Sportsground							
20 g/m²	–	–	–	–	13.19	100 m²	**13.19**
British Seed Houses; A9 Outfield; 17–35 g/m²	–	–	–	–	19.37	100 m²	**19.37**
DLF Trifolium; Promaster 120; 25–35 g/m²	–	–	–	–	14.91	100 m²	**14.91**
Informal playing fields							
DLF Trifolium; J Rye Fairway; 18–25 g/m²	–	–	–	–	14.40	100 m²	**14.40**
DLF Trifolium; Promaster 45; 35 g/m²	–	–	–	–	14.21	100 m²	**14.21**
Caravan sites							
British Seed Houses; A9 Outfield; 17–35 g/m²	–	–	–	–	19.37	100 m²	**19.37**
Sports pitch re-seeding and repair							
British Seed Houses; A8 Pitch Renovator;							
20–35 g/m²	–	–	–	–	18.26	100 m²	**18.26**
British Seed Houses; A20 Ryesport;							
20–35 g/m²	–	–	–	–	20.90	100 m²	**20.90**
DLF Trifolium; J Pitch; 18–30 g/m²	–	–	–	–	14.07	100 m²	**14.07**
DLF Trifolium; Promaster 80; 17–35 g/m²	–	–	–	–	13.86	100 m²	**13.86**
DLF Trifolium; Promaster 81; 17–35 g/m²	–	–	–	–	15.86	100 m²	**15.86**
Rigby Taylor; Mascot R14 Premier Winter							
Games Renovation + ESP coating; 35 g/m²	–	–	–	–	25.03	100 m²	**25.03**
Racecourses; gallops; polo grounds; horse							
rides							
British Seed Houses; Racecourse;							
25–30 g/m²	–	–	–	–	17.84	100 m²	**17.84**
DLF Trifolium; J Court; 18–25 g/m²	–	–	–	–	12.18	100 m²	**12.18**
DLF Trifolium; Promaster 65; 17–35 g/m²	–	–	–	–	17.39	100 m²	**17.39**
Motorway and road verges							
British Seed Houses; A18 Road Verge;							
6–15 g/m²	–	–	–	–	9.14	100 m²	**9.14**
DLF Trifolium; Promaster 85; 10 g/m²	–	–	–	–	4.51	100 m²	**4.51**
DLF Trifolium; Promaster 120; 25–35 g/m²	–	–	–	–	14.91	100 m²	**14.91**
Golf courses; tees							
British Seed Houses; A10 Tees, 35–50 g/m²	–	–	–	–	33.52	100 m²	**33.52**
DLF Trifolium; J Premier Fairway; 18–30 g/m²	–	–	–	–	19.59	100 m²	**19.59**
DLF Trifolium; J Court; 18–25 g/m²	–	–	–	–	12.18	100 m²	**12.18**
DLF Trifolium; Promaster 40; 35 g/m²	–	–	–	–	15.44	100 m²	**15.44**
DLF Trifolium; Promaster 45; 35 g/m²	–	–	–	–	14.21	100 m²	**14.21**
Golf courses; greens							
British Seed Houses; A11 Golf Greens;							
35 g/m²	–	–	–	–	26.45	100 m²	**26.45**
British Seed Houses; A13 Roughs; 8 g/m²	–	–	–	–	4.09	100 m²	**4.09**
DLF Trifolium; J All Bent; 8 g/m²	–	–	–	–	14.32	100 m²	**14.32**
DLF Trifolium; J Green; 34–50 g/m²	–	–	–	–	41.45	100 m²	**41.45**
DLF Trifolium; Promaster 5; 35–50 g/m²	–	–	–	–	24.55	100 m²	**24.55**
Rigby Taylor; Mascot R1 Greenkeeper;							
35 g/m²	–	–	–	–	41.16	100 m²	**41.16**

Q30 SEEDING/TURFING

Item Excluding site overheads and profit	PC £	Labour hours	Labour £	Plant £	Material £	Unit	Total rate £
Golf courses; fairways							
British Seed Houses; A12 Fairways;							
15–25 g/m²	–	–	–	–	13.71	100 m²	13.71
DLF Trifolium; J Rye Fairway; 18–30 g/m²	–	–	–	–	17.28	100 m²	17.28
DLF Trifolium; J Premier Fairway; 18–30 g/m²	–	–	–	–	19.59	100 m²	19.59
DLF Trifolium; Promaster 40; 35 g/m²	–	–	–	–	15.44	100 m²	15.44
DLF Trifolium; Promaster 45; 35 g/m²	–	–	–	–	14.21	100 m²	14.21
Rigby Taylor; Mascot R6 Fescue Rye							
Fairway; 35 g/m²	–	–	–	–	65.63	100 m²	65.63
Golf courses; roughs							
DLF Trifolium; Promaster 25; 17–35 g/m²	–	–	–	–	16.80	100 m²	16.80
Rigby Taylor; Mascot R127 Golf Links &							
Rough; 35 g/m²	–	–	–	–	6.16	100 m²	6.16
Waste land; spoil heaps; quarries							
British Seed Houses; A15 Reclamation;							
15–20 g/m²	–	–	–	–	12.68	100 m²	12.68
DLF Trifolium; Promaster 95; 12–35 g/m²	–	–	–	–	21.45	100 m²	21.45
DLF Trifolium; Promaster 105; 5 g/m²	–	–	–	–	5.30	100 m²	5.30
Low maintenance; housing estates; amenity							
grassed areas							
British Seed Houses; A19 Housing;							
25–35 g/m²	–	–	–	–	17.92	100 m²	17.92
British Seed Houses; A22 Low							
Maintenance; 25–35 g/m²	–	–	–	–	22.18	100 m²	22.18
DLF Trifolium; Promaster 120; 25–35 g/m²	–	–	–	–	14.91	100 m²	14.91
Saline coastal; roadside areas							
British Seed Houses; A21 Saline; 15–20 g/m²	–	–	–	–	10.97	100 m²	10.97
DLF Trifolium; Promaster 90; 15–35 g/m²	–	–	–	–	15.54	100 m²	15.54
Turf production							
British Seed Houses; A25 Meadow Ley;							
160 kg/ha	–	–	–	–	1189.50	ha	1189.50
British Seed Houses; A24 Wear & Tear;							
185 kg/ha	–	–	–	–	994.32	ha	994.32
Market prices of wild flora seed mixtures							
Acid soils							
British Seed Houses; WF1 (Annual							
Flowering); 1–2 g/m²	–	–	–	–	20.63	100 m²	20.63
Neutral soils							
British Seed Houses; WF3 (Neutral Soils);							
0.5–1 g/m²	–	–	–	–	10.31	100 m²	10.31
Market prices of wild flora and grass seed							
mixtures							
General purpose							
DLF Trifolium; Pro Flora 8 Old English							
Country Meadow Mix; 5 g/m²	–	–	–	–	12.50	100 m²	12.50
DLF Trifolium; Pro Flora 9 General							
purpose; 5 g/m²	–	–	–	–	10.00	100 m²	10.00

Q30 SEEDING/TURFING

Item Excluding site overheads and profit	PC £	Labour hours	Labour £	Plant £	Material £	Unit	Total rate £
MARKET PRICES OF SEEDING **MATERIALS – cont**							
Market prices of wild flora and grass seed **mixtures – cont**							
Acid soils							
British Seed Houses; WFG2 (Annual Meadow); 5 g/m²	–	–	–	–	22.50	100 m²	**22.50**
DLF Trifolium; Pro Flora 2 Acidic soils; 5 g/m²	–	–	–	–	11.25	100 m²	**11.25**
Neutral soils							
British Seed Houses; WFG4 (Neutral Meadow); 5 g/m²	–	–	–	–	24.70	100 m²	**24.70**
British Seed Houses; WFG13 (Scotland); 5 g/m²	–	–	–	–	21.80	100 m²	**21.80**
DLF Trifolium; Pro Flora 3 Damp loamy soils; 5 g/m²	–	–	–	–	15.75	100 m²	**15.75**
Calcareous soils							
British Seed Houses; WFG5 (Calcareous Soils); 5 g/m²	–	–	–	–	23.10	100 m²	**23.10**
DLF Trifolium; Pro Flora 4 Calcareous soils; 5 g/m²	–	–	–	–	13.50	100 m²	**13.50**
Heavy clay soils							
British Seed Houses; WFG6 (Clay Soils); 5 g/m²	–	–	–	–	27.75	100 m²	**27.75**
British Seed Houses; WFG12 (Ireland); 5 g/m²	–	–	–	–	21.88	100 m²	**21.88**
DLF Trifolium; Pro Flora 5 Wet loamy soils; 5 g/m²	–	–	–	–	24.75	100 m²	**24.75**
Sandy soils							
British Seed Houses; WFG7 (Free Draining Soils); 5 g/m²	–	–	–	–	32.50	100 m²	**32.50**
British Seed Houses; WFG11 (Ireland); 5 g/m²	–	–	–	–	23.00	100 m²	**23.00**
British Seed Houses; WFG14 (Scotland); 5 g/m²	–	–	–	–	25.25	100 m²	**25.25**
DLF Trifolium; Pro Flora 6 Dry free draining loamy soils; 5 g/m²	–	–	–	–	29.25	100 m²	**29.25**
Shaded areas							
British Seed Houses; WFG8 (Woodland and Hedgerow); 5 g/m²	–	–	–	–	24.25	100 m²	**24.25**
DLF Trifolium; Pro Flora 7 Hedgerow and light shade; 5 g/m²	–	–	–	–	33.75	100 m²	**33.75**
Educational							
British Seed Houses; WFG15 (Schools and Colleges); 5 g/m²	–	–	–	–	34.00	100 m²	**34.00**
Wetlands							
British Seed Houses; WFG9 (Wetlands and Ponds); 5 g/m²	–	–	–	–	30.00	100 m²	**30.00**
DLF Trifolium; Pro Flora 5 Wet loamy soils; 5 g/m²	–	–	–	–	24.75	100 m²	**24.75**

Q30 SEEDING/TURFING

Item Excluding site overheads and profit	PC £	Labour hours	Labour £	Plant £	Material £	Unit	Total rate £
Scrub and moorland							
British Seed Houses; WFG10 (Cornfield Annuals); 5 g/m²	–	–	–	–	33.05	100 m²	**33.05**
Hedgerow							
DLF Trifolium; Pro Flora 7 Hedgerow and light shade; 5 g/m²	–	–	–	–	33.75	100 m²	**33.75**
Vacant sites							
DLF Trifolium; Pro Flora 1 Cornfield annuals; 5 g/m²	–	–	–	–	30.00	100 m²	**30.00**
Regional Environmental mixes							
British Seed Houses; RE1 (Traditional Hay); 5 g/m²	–	–	–	–	24.95	100 m²	**24.95**
British Seed Houses; RE2 (Lowland Meadow); 5 g/m²	–	–	–	–	32.75	100 m²	**32.75**
British Seed Houses; RE3 (Riverflood Plain/Water Meadow); 5 g/m²	–	–	–	–	32.25	100 m²	**32.25**
British Seed Houses; RE4 (Lowland Limestone); 5 g/m²	–	–	–	–	31.75	100 m²	**31.75**
British Seed Houses; RE5 (Calcareous Sub-mountain Restoration); 5 g/m²	–	–	–	–	36.10	100 m²	**36.10**
British Seed Houses; RE6 (Upland Limestone); 5 g/m²	–	–	–	–	46.45	100 m²	**46.45**
British Seed Houses; RE7 (Acid Sub-mountain Restoration); 5 g/m²	–	–	–	–	31.25	100 m²	**31.25**
British Seed Houses; RE8 (Coastal Reclamation); 5 g/m²	–	–	–	–	40.45	100 m²	**40.45**
British Seed Houses; RE9 (Farmland Mixture); 5 g/m²	–	–	–	–	27.85	100 m²	**27.85**
British Seed Houses; RE10 (Marginal Land); 5 g/m²	–	–	–	–	41.75	100 m²	**41.75**
British Seed Houses; RE11 (Heath Scrubland); 5 g/m²	–	–	–	–	12.80	100 m²	**12.80**
British Seed Houses; RE12 (Drought Land); 5 g/m²	–	–	–	–	39.50	100 m²	**39.50**
CLARIFICATION NOTES ON LABOUR COSTS IN THIS SECTION							
General landscape team							
Generally a three man team is used in this section; The column 'Labour hours' reports team hours. The column 'Labour £' reports the total cost of the team for the unit of work shown							
3 man team	–	1.00	57.75	–	–	hr	**57.75**

Q30 SEEDING/TURFING

Item Excluding site overheads and profit	PC £	Labour hours	Labour £	Plant £	Material £	Unit	Total rate £
CULTIVATION AND SOIL PREPARATION							
Cultivation by tractor; Agripower Ltd							
Ripping up subsoil; using approved subsoiling machine; minimum depth 250 mm below topsoil; at 1.20 m centres; in							
gravel or sandy clay	–	–	–	2.16	–	100 m²	**2.16**
soil compacted by machines	–	–	–	2.40	–	100 m²	**2.40**
clay	–	–	–	3.08	–	100 m²	**3.08**
chalk or other soft rock	–	–	–	3.59	–	100 m²	**3.59**
Extra for subsoiling at 1 m centres	–	–	–	0.62	–	100 m²	**0.62**
Breaking up existing ground; using tractor drawn tine cultivator or rotavator							
Single pass							
100 mm deep	–	–	–	8.35	–	100 m²	**8.35**
150 mm deep	–	–	–	10.43	–	100 m²	**10.43**
200 mm deep	–	–	–	11.93	–	100 m²	**11.93**
Cultivating ploughed ground; using disc, drag or chain harrow							
4 passes	–	–	–	9.49	–	100 m²	**9.49**
Rolling cultivated ground lightly; using self-propelled agricultural roller	–	0.07	1.28	0.72	–	100 m²	**2.00**
Cultivation by pedestrian operated rotavator							
Breaking up existing ground; tine cultivator or rotavator							
100 mm deep	–	0.22	4.24	2.37	–	100 m²	**6.61**
150 mm deep	–	0.28	5.29	2.96	–	100 m²	**8.25**
200 mm deep	–	0.37	7.06	3.94	–	100 m²	**11.00**
As above but in heavy clay or wet soils							
100 mm deep	–	0.44	8.47	4.73	–	100 m²	**13.20**
150 mm deep	–	0.66	12.71	7.09	–	100 m²	**19.80**
200 mm deep	–	0.82	15.88	8.87	–	100 m²	**24.75**
Importing and storing selected and approved topsoil; from source not exceeding 13 km from site; inclusive of settlement							
small quantities (less than 15 m³)	60.00	–	–	–	60.00	m³	**60.00**
over 15 m³	33.60	–	–	–	33.60	m³	**33.60**
Spreading and lightly consolidating approved topsoil (imported or from spoil heaps); in layers not exceeding 150 mm; travel distance from spoil heaps not exceeding 100 m; by machine (imported topsoil not included)							
minimum depth 100 mm	–	1.55	29.84	45.34	–	100 m²	**75.18**
minimum depth 150 mm	–	2.33	44.92	68.23	–	100 m²	**113.15**
minimum depth 300 mm	–	4.67	89.83	136.45	–	100 m²	**226.28**
minimum depth 450 mm	–	6.99	134.56	204.46	–	100 m²	**339.02**

Q30 SEEDING/TURFING

Item Excluding site overheads and profit	PC £	Labour hours	Labour £	Plant £	Material £	Unit	Total rate £
Spreading and lightly consolidating approved topsoil (imported or from spoil heaps); in layers not exceeding 150 mm; travel distance from spoil heaps not exceeding 100 m; by hand (imported topsoil not included)							
minimum depth 100 mm	–	20.00	385.08	–	–	100 m²	**385.08**
minimum depth 150 mm	–	30.01	577.62	–	–	100 m²	**577.62**
minimum depth 300 mm	–	60.01	1155.23	–	–	100 m²	**1155.23**
minimum depth 450 mm	–	90.02	1732.85	–	–	100 m²	**1732.85**
Extra over for spreading topsoil to slopes 15–30°; by machine or hand	–	–	–	–	–	10%	**–**
Extra over for spreading topsoil to slopes over 30°; by machine or hand	–	–	–	–	–	25%	**–**
Extra over for spreading topsoil from spoil heaps; travel exceeding 100 m; by machine							
100–150 m	–	0.01	0.24	0.06	–	m³	**0.30**
150–200 m	–	0.02	0.36	0.10	–	m³	**0.46**
200–300 m	–	0.03	0.54	0.15	–	m³	**0.69**
Extra over spreading topsoil for travel exceeding 100 m; by hand							
100 m	–	0.83	16.04	–	–	m³	**16.04**
200 m	–	1.67	32.09	–	–	m³	**32.09**
300 m	–	2.50	48.13	–	–	m³	**48.13**
Evenly grading; to general surfaces to bring to finished levels							
by machine (tractor mounted rotavator)	–	–	–	0.04	–	m²	**0.04**
by pedestrian operated rotavator	–	–	0.08	0.05	–	m²	**0.13**
by hand	–	0.01	0.19	–	–	m²	**0.19**
Extra over grading for slopes 15–30°; by machine or hand	–	–	–	–	–	10%	**–**
Extra over grading for slopes over 30°; by machine or hand	–	–	–	–	–	25%	**–**
Apply screened topdressing to grass surfaces; spread using Tru-Lute							
sand soil mixes 90/10 to 50/50	–	–	0.04	0.03	0.14	m²	**0.21**
Spread only existing cultivated soil to final levels using Tru-Lute							
cultivated soil	–	–	0.04	0.03	–	m²	**0.07**
Clearing stones; disposing off site							
by hand; stones not exceeding 50 mm in any direction; loading to skip 4.6 m³	–	0.01	0.19	0.03	–	m²	**0.22**
by mechanical stone rake; stones not exceeding 50 mm in any direction; loading to 15 m³ truck by mechanical loader	–	–	0.04	0.08	–	m²	**0.12**
Lightly cultivating; weeding; to fallow areas; disposing debris off site							
by hand	–	0.01	0.27	–	0.08	m²	**0.35**

Q30 SEEDING/TURFING

Item Excluding site overheads and profit	PC £	Labour hours	Labour £	Plant £	Material £	Unit	Total rate £
SOIL ADDITIVES AND IMPROVEMENTS							
Surface applications; soil additives; pre-seeding; material delivered to a maximum of 25 m from area of application; applied; by machine							
Soil conditioners; to cultivated ground; mushroom compost; delivered in 25 m³ loads; including turning in							
1 m³ per 40 m² = 25 mm thick	0.59	0.02	0.32	0.13	0.59	m²	**1.04**
1 m³ per 20 m² = 50 mm thick	1.19	0.03	0.59	0.19	1.19	m²	**1.97**
1 m³ per 13.33 m² = 75 mm thick	1.78	0.04	0.77	0.26	1.78	m²	**2.81**
1 m³ per 10 m² = 100 mm thick	2.37	0.05	0.96	0.33	2.37	m²	**3.66**
Soil conditioners; to cultivated ground; mushroom compost; delivered in 35 m³ loads; including turning in							
1 m³ per 40 m² = 25 mm thick	0.47	0.02	0.32	0.13	0.47	m²	**0.92**
1 m³ per 20 m² = 50 mm thick	0.95	0.03	0.59	0.19	0.95	m²	**1.73**
1 m³ per 13.33 m² = 75 mm thick	1.43	0.04	0.77	0.26	1.43	m²	**2.46**
1 m³ per 10 m² = 100 mm thick	1.90	0.05	0.96	0.33	1.90	m²	**3.19**
Surface applications and soil additives; pre-seeding; material delivered to a maximum of 25 m from area of application; applied; by hand							
Soil conditioners; to cultivated ground; mushroom compost; delivered in 25 m³ loads; including turning in							
1 m³ per 40 m² = 25 mm thick	0.59	0.02	0.43	–	0.59	m²	**1.02**
1 m³ per 20 m² = 50 mm thick	1.19	0.04	0.85	–	1.19	m²	**2.04**
1 m³ per 13.33 m² = 75 mm thick	1.78	0.07	1.28	–	1.78	m²	**3.06**
1 m³ per 10 m² = 100 mm thick	2.37	0.08	1.54	–	2.37	m²	**3.91**
Soil conditioners; to cultivated ground; mushroom compost; delivered in 60 m³ loads; including turning in							
1 m³ per 40 m² = 25 mm thick	0.30	0.02	0.43	–	0.30	m²	**0.73**
1 m³ per 20 m² = 50 mm thick	0.53	0.04	0.85	–	0.53	m²	**1.38**
1 m³ per 13.33 m² = 75 mm thick	0.90	0.07	1.28	–	0.90	m²	**2.18**
1 m³ per 10 m² = 100 mm thick	1.20	0.08	1.54	–	1.20	m²	**2.74**
PREPARATION OF SEEDBEDS							
Preparation of seedbeds – General							
Preamble: For preliminary operations see 'Cultivation' section.							
Preparation of seedbeds; soil preparation							
Lifting selected and approved topsoil from spoil heaps; passing through 6 mm screen; removing debris	–	0.08	1.60	4.92	0.02	m³	**6.54**

Q30 SEEDING/TURFING

Item Excluding site overheads and profit	PC £	Labour hours	Labour £	Plant £	Material £	Unit	Total rate £
Topsoil; supply only; PC £28.00/m³; allowing for 20% settlement							
25 mm	–	–	–	–	0.84	m²	0.84
50 mm	–	–	–	–	1.68	m²	1.68
100 mm	–	–	–	–	3.36	m²	3.36
150 mm	–	–	–	–	5.04	m²	5.04
200 mm	–	–	–	–	6.72	m²	6.72
250 mm	–	–	–	–	8.40	m²	8.40
300 mm	–	–	–	–	10.08	m²	10.08
400 mm	–	–	–	–	13.44	m²	13.44
450 mm	–	–	–	–	15.12	m²	15.12
Spreading topsoil to form seedbeds (topsoil not included); by machine							
25 mm deep	–	–	0.05	0.12	–	m²	0.17
50 mm deep	–	–	0.06	0.15	–	m²	0.21
75 mm deep	–	–	0.07	0.18	–	m²	0.25
100 mm deep	–	0.01	0.10	0.23	–	m²	0.33
150 mm deep	–	0.01	0.14	0.35	–	m²	0.49
Spreading only topsoil to form seedbeds (topsoil not included); by hand							
25 mm deep	–	0.03	0.48	–	–	m²	0.48
50 mm deep	–	0.03	0.64	–	–	m²	0.64
75 mm deep	–	0.04	0.82	–	–	m²	0.82
100 mm deep	–	0.05	0.96	–	–	m²	0.96
150 mm deep	–	0.08	1.44	–	–	m²	1.44
Bringing existing topsoil to a fine tilth for seeding; by raking or harrowing; stones not to exceed 6 mm; by machine	–	–	0.08	0.04	–	m²	0.12
Bringing existing topsoil to a fine tilth for seeding; by raking or harrowing; stones not to exceed 6 mm; by hand	–	0.01	0.17	–	–	m²	0.17
Preparation of seedbeds; soil treatments							
For the following operations add or subtract the following amounts for every £0.10 difference in the material cost price							
35 g/m²	–	–	–	–	0.35	100 m²	0.35
50 g/m²	–	–	–	–	0.50	100 m²	0.50
70 g/m²	–	–	–	–	0.70	100 m²	0.70
100 g/m²	–	–	–	–	1.00	100 m²	1.00
125 kg/ha	–	–	–	–	12.50	ha	12.50
150 kg/ha	–	–	–	–	15.00	ha	15.00
175 kg/ha	–	–	–	–	17.50	ha	17.50
200 kg/ha	–	–	–	–	20.00	ha	20.00
225 kg/ha	–	–	–	–	22.50	ha	22.50
250 kg/ha	–	–	–	–	25.00	ha	25.00
300 kg/ha	–	–	–	–	30.00	ha	30.00
350 kg/ha	–	–	–	–	35.00	ha	35.00
400 kg/ha	–	–	–	–	40.00	ha	40.00

Q30 SEEDING/TURFING

Item	PC	Labour	Labour	Plant	Material	Unit	Total
Excluding site overheads and profit	£	hours	£	£	£		rate £
PREPARATION OF SEEDBEDS – cont							
Preparation of seedbeds – cont							
For the following operations add or subtract							
the following amounts for every £0.10							
difference in the material cost price – cont							
500 kg/ha	–	–	–	–	50.00	ha	**50.00**
700 kg/ha	–	–	–	–	70.00	ha	**70.00**
1000 kg/ha	–	–	–	–	100.00	ha	**100.00**
1250 kg/ha	–	–	–	–	125.00	ha	**125.00**
Pre-seeding fertilizers (12+00+09);							
PC £0.65/kg; to seedbeds; by machine							
35 g/m²	2.28	–	–	0.19	2.28	100 m²	**2.47**
50 g/m²	3.25	–	–	0.19	3.25	100 m²	**3.44**
70 g/m²	4.55	–	–	0.19	4.55	100 m²	**4.74**
100 g/m²	6.50	–	–	0.19	6.50	100 m²	**6.69**
Pre-seeding fertilizers (12+00+09);							
PC £0.65/kg; to seedbeds; by hand							
35 g/m²	2.28	0.17	3.21	–	2.28	100 m²	**5.49**
50 g/m²	3.25	0.17	3.21	–	3.25	100 m²	**6.46**
70 g/m²	4.55	0.17	3.21	–	4.55	100 m²	**7.76**
100 g/m²	6.50	0.20	3.85	–	6.50	100 m²	**10.35**
SEEDING							
Seeding							
Seeding labours only in two operations; by							
machine (for seed prices see above)							
35 g/m²	–	–	–	0.51	–	100 m²	**0.51**
Grass seed; spreading in two operations;							
PC £4.50/kg (for changes in material prices							
please refer to table above); by machine							
35 g/m²	–	–	–	0.51	15.75	100 m²	**16.26**
50 g/m²	–	–	–	0.51	22.50	100 m²	**23.01**
70 g/m²	–	–	–	0.51	31.50	100 m²	**32.01**
100 g/m²	–	–	–	0.51	45.00	100 m²	**45.51**
125 kg/ha	–	–	–	50.95	562.50	ha	**613.45**
150 kg/ha	–	–	–	50.95	675.00	ha	**725.95**
200 kg/ha	–	–	–	50.95	900.00	ha	**950.95**
250 kg/ha	–	–	–	50.95	1125.00	ha	**1175.95**
300 kg/ha	–	–	–	50.95	1350.00	ha	**1400.95**
350 kg/ha	–	–	–	50.95	1575.00	ha	**1625.95**
400 kg/ha	–	–	–	50.95	1800.00	ha	**1850.95**
500 kg/ha	–	–	–	50.95	2250.00	ha	**2300.95**
700 kg/ha	–	–	–	50.95	3150.00	ha	**3200.95**
Extra over seeding by machine for slopes over							
30° (allowing for the actual area but measured							
in plan)							
35 g/m²	–	–	–	0.08	2.36	100 m²	**2.44**
50 g/m²	–	–	–	0.08	3.38	100 m²	**3.46**
70 g/m²	–	–	–	0.08	4.72	100 m²	**4.80**

Q30 SEEDING/TURFING

Item Excluding site overheads and profit	PC £	Labour hours	Labour £	Plant £	Material £	Unit	Total rate £
100 g/m^2	–	–	–	0.08	6.75	100 m^2	6.83
125 kg/ha	–	–	–	7.64	84.38	ha	92.02
150 kg/ha	–	–	–	7.64	101.25	ha	108.89
200 kg/ha	–	–	–	7.64	135.00	ha	142.64
250 kg/ha	–	–	–	7.64	168.75	ha	176.39
300 kg/ha	–	–	–	7.64	202.50	ha	210.14
350 kg/ha	–	–	–	7.64	236.25	ha	243.89
400 kg/ha	–	–	–	7.64	270.00	ha	277.64
500 kg/ha	–	–	–	7.64	337.50	ha	345.14
700 kg/ha	–	–	–	7.64	472.50	ha	480.14
Seeding labours only in two operations; by machine (for seed prices see above)							
35 g/m^2	–	0.17	3.21	–	–	100 m^2	3.21
Grass seed; spreading in two operations; PC £4.50/kg (for changes in material prices please refer to table above); by hand							
35 g/m^2	–	0.17	3.21	–	15.75	100 m^2	18.96
50 g/m^2	–	0.17	3.21	–	22.50	100 m^2	25.71
70 g/m^2	–	0.17	3.21	–	31.50	100 m^2	34.71
100 g/m^2	–	0.20	3.85	–	45.00	100 m^2	48.85
125 g/m^2	–	0.20	3.85	–	56.25	100 m^2	60.10
Extra over seeding by hand for slopes over 30° (allowing for the actual area but measured in plan)							
35 g/m^2	2.34	–	0.07	–	2.34	100 m^2	2.41
50 g/m^2	3.38	–	0.07	–	3.38	100 m^2	3.45
70 g/m^2	4.72	–	0.07	–	4.72	100 m^2	4.79
100 g/m^2	6.75	–	0.08	–	6.75	100 m^2	6.83
125 g/m^2	8.41	–	0.08	–	8.41	100 m^2	8.49
Harrowing seeded areas; light chain harrow	–	–	–	0.09	–	100 m^2	0.09
Raking over seeded areas							
by mechanical stone rake	–	–	–	2.18	–	100 m^2	2.18
by hand	–	0.80	15.40	–	–	100 m^2	15.40
Rolling seeded areas; light roller							
by tractor drawn roller	–	–	–	0.54	–	100 m^2	0.54
by pedestrian operated mechanical roller	–	0.08	1.60	0.62	–	100 m^2	2.22
by hand drawn roller	–	0.17	3.21	–	–	100 m^2	3.21
Extra over harrowing, raking or rolling seeded areas for slopes over 30°; by machine or hand	–	–	–	–	–	25%	–
Turf edging; to seeded areas; 300 mm wide	–	0.05	0.92	–	1.85	m^2	2.77
Liquid sod; Turf Management Systems							
Spray on grass system of grass plantlets fertilizer, bio-degradable mulch carrier, root enhancer and water							
to prepared ground	–	–	–	–	–	m^2	1.80

Q30 SEEDING/TURFING

Item Excluding site overheads and profit	PC £	Labour hours	Labour £	Plant £	Material £	Unit	Total rate £
PREPARATION OF TURF BEDS							
Preparation of turf beds							
Rolling turf to be lifted; lifting by hand or mechanical turf stripper; stacks to be not more than 1 m high							
cutting only preparing to lift; pedestrian turf cutter	–	0.75	14.44	9.96	–	100 m²	**24.40**
lifting and stacking; by hand	–	8.33	160.42	–	–	100 m²	**160.42**
Rolling up; moving to stacks							
distance not exceeding 100 m	–	2.50	48.13	–	–	100 m²	**48.13**
extra over rolling and moving turf to stacks to transport per additional 100 m	–	0.83	16.04	–	–	100 m²	**16.04**
Lifting selected and approved topsoil from spoil heaps							
passing through 6 mm screen; removing debris	–	0.17	3.21	9.84	–	m³	**13.05**
Extra over lifting topsoil and passing through screen for imported topsoil; plus 20% allowance for settlement	–	–	–	–	33.60	m³	**33.60**
Topsoil; PC £28.00/m³; plus 20% allowance for settlement							
25 mm deep	–	–	–	–	0.84	m²	**0.84**
50 mm deep	–	–	–	–	1.68	m²	**1.68**
100 mm deep	–	–	–	–	3.36	m²	**3.36**
150 mm deep	–	–	–	–	5.04	m²	**5.04**
200 mm deep	–	–	–	–	6.72	m²	**6.72**
250 mm deep	–	–	–	–	8.40	m²	**8.40**
300 mm deep	–	–	–	–	10.08	m²	**10.08**
400 mm deep	–	–	–	–	13.44	m²	**13.44**
450 mm deep	–	–	–	–	15.12	m²	**15.12**
Spreading topsoil to form turf beds (topsoil not included); by machine							
25 mm deep	–	–	0.05	0.12	–	m²	**0.17**
50 mm deep	–	–	0.06	0.15	–	m²	**0.21**
75 mm deep	–	–	0.07	0.18	–	m²	**0.25**
100 mm deep	–	0.01	0.10	0.23	–	m²	**0.33**
150 mm deep	–	0.01	0.14	0.35	–	m²	**0.49**
Spreading topsoil to form turf beds (topsoil not included); by hand							
25 mm deep	–	0.03	0.48	–	–	m²	**0.48**
50 mm deep	–	0.03	0.64	–	–	m²	**0.64**
75 mm deep	–	0.04	0.82	–	–	m²	**0.82**
100 mm deep	–	0.05	0.96	–	–	m²	**0.96**
150 mm deep	–	0.08	1.44	–	–	m²	**1.44**
Bringing existing topsoil to a fine tilth for turfing by raking or harrowing; stones not to exceed 6 mm; by machine	–	–	0.08	0.04	–	m²	**0.12**
Bringing existing topsoil to a fine tilth for turfing by raking or harrowing; stones not to exceed 6 mm; by hand	–	0.01	0.18	–	–	m²	**0.18**

Q30 SEEDING/TURFING

Item Excluding site overheads and profit	PC £	Labour hours	Labour £	Plant £	Material £	Unit	Total rate £
TURFING							
Turfing							
Turfing; laying only; to stretcher bond; butt joints; including providing and working from barrow plank runs where necessary to surfaces not exceeding 30° from horizontal							
specially selected lawn turves from previously lifted stockpile	–	0.08	1.44	–	–	m²	**1.44**
cultivated lawn turves; to large open areas	–	0.06	1.12	–	–	m²	**1.12**
cultivated lawn turves; to domestic or garden areas	–	0.08	1.50	–	–	m²	**1.50**
road verge quality turf	–	0.04	0.77	–	–	m²	**0.77**
Industrially grown turf; PC prices listed represent the general range of industrial turf prices for sportsfields and amenity purposes; prices will vary with quantity and site location							
Rolawn							
RB Medallion; sports fields, domestic lawns, general landscape; full loads 1720 m²	1.69	0.07	1.35	–	1.69	m²	**3.04**
RB Medallion; sports fields, domestic lawns, general landscape; part loads	1.85	0.07	1.35	–	1.85	m²	**3.20**
Tensar Ltd							
Tensar Mat 400 reinforced turf for embankments; laid to embankments 3.3 m² turves	3.85	0.11	2.12	–	3.85	m²	**5.97**
Inturf							
Inturf Masters; formal lawns, golf greens, bowling greens and low maintenance areas	2.68	0.10	1.84	–	2.68	m²	**4.52**
Inturf Classic; golf tees, surrounds, lawns, parks, general purpose landscaping areas and winter sports	2.08	0.05	1.01	–	2.08	m²	**3.09**
Inturf Ornamental; medium fine turf for ornamental lawns, fairways and green surrounds	2.18	0.05	1.05	–	2.18	m²	**3.23**
Inturf Custom Grown Turf; tailor made for any turfgrass specification	7.08	0.08	1.54	–	7.08	m²	**8.62**
RTF Rhizomatous Tall Fescue; deep rooted self repairing for our changing climate	2.68	0.08	1.54	–	2.68	m²	**4.22**

Q30 SEEDING/TURFING

Item Excluding site overheads and profit	PC £	Labour hours	Labour £	Plant £	Material £	Unit	Total rate £
TURFING – cont							
Reinforced turf; Fiberweb Netlon Advanced Turf; Boddingtons Ltd; blended mesh elements incorporated into root zone; root zone spread and levelled over cultivated, prepared and reduced and levelled ground (not included); compacted with vibratory roller							
Boddingtons ATS 400/B with selected turf and fertilizer 100 mm thick							
100–500 m²	25.30	0.03	0.64	2.44	25.30	m²	28.38
over 500 m²	23.65	0.03	0.64	2.44	23.65	m²	26.73
Boddingtons ATS 400/B with selected turf and fertilizer 150 mm thick							
100–500 m²	31.90	0.04	0.77	2.87	31.90	m²	35.54
over 500 m²	30.80	0.04	0.77	2.87	30.80	m²	34.44
Boddingtons ATS 400/B with selected turf and fertilizer 200 mm thick							
100–500 m²	35.20	0.05	0.96	3.51	35.20	m²	39.67
over 500 m²	33.00	0.05	0.96	3.51	33.00	m²	37.47
Firming turves with wooden beater	–	0.01	0.19	–	–	m²	0.19
Rolling turfed areas; light roller							
by tractor with turf tyres and roller	–	–	–	0.54	–	100 m²	0.54
by pedestrian operated mechanical roller	–	0.08	1.60	0.62	–	100 m²	2.22
by hand drawn roller	–	0.17	3.21	–	–	100 m²	3.21
Dressing with finely sifted topsoil; brushing into joints	0.03	0.05	0.96	–	0.03	m²	0.99
Turfing; laying only							
to slopes over 30°; to diagonal bond (measured as plan area – add 15% to these rates for the incline area of 30° slopes)	–	0.12	2.31	–	–	m²	2.31
Extra over laying turfing for pegging down turves							
wooden or galvanized wire pegs; 200 mm long; 2 pegs per 0.50 m²	1.60	0.01	0.26	–	1.60	m²	1.86
ARTIFICIAL TURF							
Artificial grass; Trulawn Limited							
Artificial lawns laid to cleared and levelled area (not included); laid to 50 mm Type 1 drainage layer on 25 mm sharp sand bed							
Trulawn Value; 16 mm roof; terraces, balconies, general use	–	–	–	–	–	m²	35.95
Trulawn Play; 20 mm; entry level; lawn and play areas	–	–	–	–	–	m²	37.95
Trulawn Continental; 20 mm realistic appearance; lawns and patios	–	–	–	–	–	m²	33.85

Q30 SEEDING/TURFING

Item Excluding site overheads and profit	PC £	Labour hours	Labour £	Plant £	Material £	Unit	Total rate £
Trulawn Optimum; 26 mm; softest pile; lush green; lawns and patios	–	–	–	–	–	m²	**45.50**
Trulawn Luxury; 30 mm; deep pile; realistic lawn	–	–	–	–	–	m²	**47.50**
MAINTENANCE OF SEEDED/TURFED AREAS							
Maintenance operations (Note: the following rates apply to aftercare maintenance executed as part of a landscaping contract only)							
Initial cutting; to turfed areas							
20 mm high; using pedestrian guided power driven cylinder mower; including boxing off cuttings (stone picking and rolling not included)	–	0.18	3.46	0.31	–	100 m²	**3.77**
Repairing damaged grass areas							
scraping out; removing slurry; from ruts and holes; average 100 mm deep	–	0.13	2.57	–	–	m²	**2.57**
100 mm topsoil	–	0.13	2.57	–	3.36	m²	**5.93**
Repairing damaged grass areas; sowing grass seed to match existing or as specified; to individually prepared worn patches							
35 g/m²	0.18	0.01	0.19	–	0.18	m²	**0.37**
50 g/m²	0.26	0.01	0.19	–	0.26	m²	**0.45**
Sweeping leaves; disposing off site; motorized vacuum sweeper or rotary brush sweeper							
areas of maximum 2500 m² with occasional large tree and established boundary planting; 4.6 m³ (1 skip of material to be removed)	–	0.40	7.70	3.08	–	100 m²	**10.78**
Leaf clearance; clearing grassed area of leaves and other extraneous debris							
Using equipment towed by tractor							
large grassed areas with perimeters of mature trees such as sports fields and amenity areas	–	0.01	0.24	0.04	–	100 m²	**0.28**
large grassed areas containing ornamental trees and shrub beds	–	0.03	0.48	0.06	–	100 m²	**0.54**
Using pedestrian operated mechanical equipment and blowers							
grassed areas with perimeters of mature trees such as sports fields and amenity areas	–	0.04	0.77	0.05	–	100 m²	**0.82**
grassed areas containing ornamental trees and shrub beds	–	0.10	1.93	0.13	–	100 m²	**2.06**
verges	–	0.07	1.28	0.09	–	100 m²	**1.37**

Q30 SEEDING/TURFING

Item Excluding site overheads and profit	PC £	Labour hours	Labour £	Plant £	Material £	Unit	Total rate £
MAINTENANCE OF SEEDED/TURFED **AREAS – cont**							
Maintenance operations (Note: the **following rates apply to aftercare** **maintenance executed as part of a** **landscaping contract only) – cont**							
By hand							
grassed areas with perimeters of mature trees such as sports fields and amenity areas	–	0.05	0.96	0.10	–	100 m²	**1.06**
grassed areas containing ornamental trees and shrub beds	–	0.08	1.60	0.17	–	100 m²	**1.77**
verges	–	1.00	19.25	1.99	–	100 m²	**21.24**
Removal of arisings							
areas with perimeters of mature trees	–	0.01	0.11	0.09	1.20	100 m²	**1.40**
areas containing ornamental trees and shrub beds	–	0.02	0.33	0.34	3.00	100 m²	**3.67**
Cutting grass to specified height; per cut							
multi-unit gang mower	–	0.59	11.32	19.35	–	ha	**30.67**
ride-on triple cylinder mower	–	0.01	0.27	0.14	–	100 m²	**0.41**
ride-on triple rotary mower	–	0.01	0.27	–	–	100 m²	**0.27**
pedestrian mower	–	0.18	3.46	0.73	–	100 m²	**4.19**
Cutting grass to banks; per cut							
side arm cutter bar mower	–	0.02	0.45	0.36	–	100 m²	**0.81**
Cutting rough grass; per cut							
power flail or scythe cutter	–	0.04	0.67	–	–	100 m²	**0.67**
Extra over cutting grass for slopes not exceeding 30°	–	–	–	–	–	10%	**–**
Extra over cutting grass for slopes exceeding 30°	–	–	–	–	–	40%	**–**
Cutting fine sward							
pedestrian operated seven-blade cylinder lawn mower	–	0.14	2.70	0.23	–	100 m²	**2.93**
Extra over cutting fine sward for boxing off cuttings							
pedestrian mower	–	0.03	0.54	0.05	–	100 m²	**0.59**
Cutting areas of rough grass							
scythe	–	1.00	19.25	–	–	100 m²	**19.25**
sickle	–	2.00	38.50	–	–	100 m²	**38.50**
petrol operated strimmer	–	0.30	5.78	0.42	–	100 m²	**6.20**
Cutting areas of rough grass which contain trees or whips							
petrol operated strimmer	–	0.40	7.70	0.56	–	100 m²	**8.26**
Extra over cutting rough grass for on site raking up and dumping	–	0.33	6.42	–	–	100 m²	**6.42**
Trimming edge of grass areas; edging tool							
with petrol powered strimmer	–	0.13	2.57	0.19	–	100 m	**2.76**
by hand	–	0.67	12.83	–	–	100 m	**12.83**

Q30 SEEDING/TURFING

Item Excluding site overheads and profit	PC £	Labour hours	Labour £	Plant £	Material £	Unit	Total rate £
Marking out pitches using approved line marking compound; including initial setting out and marking							
discus, hammer, javelin or shot putt area	4.68	2.00	38.50	–	4.68	nr	43.18
cricket square	3.12	2.00	38.50	–	3.12	nr	41.62
cricket boundary	10.92	8.00	154.00	–	10.92	nr	164.92
grass tennis court	4.68	4.00	77.00	–	4.68	nr	81.68
hockey pitch	15.60	8.00	154.00	–	15.60	nr	169.60
football pitch	15.60	8.00	154.00	–	15.60	nr	169.60
rugby pitch	15.60	8.00	154.00	–	15.60	nr	169.60
eight lane running track; 400 m	31.20	16.00	308.00	–	31.20	nr	339.20
Re-marking out pitches using approved line marking compound							
discus, hammer, javelin or shot putt area	3.12	0.50	9.63	–	3.12	nr	12.75
cricket square	3.12	0.50	9.63	–	3.12	nr	12.75
grass tennis court	10.92	1.00	19.25	–	10.92	nr	30.17
hockey pitch	10.92	1.00	19.25	–	10.92	nr	30.17
football pitch	10.92	1.00	19.25	–	10.92	nr	30.17
rugby pitch	10.92	1.00	19.25	–	10.92	nr	30.17
eight lane running track; 400 m	31.20	2.50	48.13	–	31.20	nr	79.33
Rolling grass areas; light roller							
by tractor drawn roller	–	–	–	0.54	–	100 m²	0.54
by pedestrian operated mechanical roller	–	0.08	1.60	0.62	–	100 m²	2.22
by hand drawn roller	–	0.17	3.21	–	–	100 m²	3.21
Aerating grass areas; to a depth of 100 mm							
using tractor-drawn aerator	–	0.06	1.12	1.01	–	100 m²	2.13
using pedestrian-guided motor powered solid or slitting tine turf aerator	–	0.18	3.37	2.96	–	100 m²	6.33
using hollow tine aerator; including sweeping up and dumping corings	–	0.50	9.63	5.92	–	100 m²	15.55
using hand aerator or fork	–	1.67	32.08	–	–	100 m²	32.08
Extra over aerating grass areas for on site sweeping up and dumping corings	–	0.17	3.21	–	–	100 m²	3.21
Switching off dew; from fine turf areas	–	0.20	3.85	–	–	100 m²	3.85
Scarifying grass areas to break up thatch; removing dead grass							
using tractor-drawn scarifier	–	0.07	1.35	0.31	–	100 m²	1.66
using self-propelled scarifier; including removing and disposing of grass on site	–	0.33	6.42	0.15	–	100 m²	6.57
Harrowing grass areas							
using drag mat	–	0.03	0.54	0.30	–	100 m²	0.84
using chain harrow	–	0.04	0.67	0.38	–	100 m²	1.05
using drag mat	–	2.80	53.91	30.25	–	ha	84.16
using chain harrow	–	3.50	67.38	37.81	–	ha	105.19
Extra for scarifying and harrowing grass areas for disposing excavated material off site; to tip not exceeding 13 km; loading by machine							
slightly contaminated	–	–	–	2.06	24.00	m³	26.06
rubbish	–	–	–	2.06	24.00	m³	26.06
inert material	–	–	–	1.37	17.14	m³	18.51

Q30 SEEDING/TURFING

Item Excluding site overheads and profit	PC £	Labour hours	Labour £	Plant £	Material £	Unit	Total rate £
MAINTENANCE OF SEEDED/TURFED AREAS – cont							
Maintenance operations (Note: the following rates apply to aftercare maintenance executed as part of a landscaping contract only) – cont							
For the following topsoil improvement and seeding operations add or subtract the following amounts for every £0.10 difference in the material cost price							
35 g/m²	–	–	–	–	0.35	100 m²	0.35
50 g/m²	–	–	–	–	0.50	100 m²	0.50
70 g/m²	–	–	–	–	0.70	100 m²	0.70
100 g/m²	–	–	–	–	1.00	100 m²	1.00
125 kg/ha	–	–	–	–	12.50	ha	12.50
150 kg/ha	–	–	–	–	15.00	ha	15.00
175 kg/ha	–	–	–	–	17.50	ha	17.50
200 kg/ha	–	–	–	–	20.00	ha	20.00
225 kg/ha	–	–	–	–	22.50	ha	22.50
250 kg/ha	–	–	–	–	25.00	ha	25.00
300 kg/ha	–	–	–	–	30.00	ha	30.00
350 kg/ ha	–	–	–	–	35.00	ha	35.00
400 kg/ha	–	–	–	–	40.00	ha	40.00
500 kg/ha	–	–	–	–	50.00	ha	50.00
700 kg/ha	–	–	–	–	70.00	ha	70.00
1000 kg/ha	–	–	–	–	100.00	ha	100.00
1250 kg/ha	–	–	–	–	125.00	ha	125.00
Top dressing fertilizers (7+7+7); PC £1.29/kg; to seedbeds; by machine							
35 g/m²	4.52	–	–	0.19	4.52	100 m²	4.71
50 g/m²	6.46	–	–	0.19	6.46	100 m²	6.65
300 kg/ha	387.75	–	–	18.76	387.75	ha	406.51
350 kg/ha	452.38	–	–	18.76	452.38	ha	471.14
400 kg/ha	517.00	–	–	18.76	517.00	ha	535.76
500 kg/ha	646.25	–	–	30.01	646.25	ha	676.26
Top dressing fertilizers (7+7+7); PC £1.29/kg; to seedbeds; by hand							
35 g/m²	4.52	0.17	3.21	–	4.52	100 m²	7.73
50 g/m²	6.46	0.17	3.21	–	6.46	100 m²	9.67
70 g/m²	9.05	0.17	3.21	–	9.05	100 m²	12.26
Watering turf; evenly; at a rate of 5 l/m²							
using movable spray lines powering 3 nr sprinkler heads with a radius of 15 m and allowing for 60% overlap (irrigation machinery costs not included)	–	0.02	0.30	–	–	100 m²	0.30
using sprinkler equipment and with sufficient water pressure to run 1 nr 15 m radius sprinkler	–	0.02	0.38	–	–	100 m²	0.38
using hand-held watering equipment	–	0.25	4.81	–	–	100 m²	4.81

Q31 PLANTING

Item Excluding site overheads and profit	PC £	Labour hours	Labour £	Plant £	Material £	Unit	Total rate £
Q31 PLANTING							
PLANTING GENERAL							
Planting – General Preamble: Prices for all planting work are deemed to include carrying out planting in accordance with all good horticultural practice.							
MARKET PRICES OF MATERIALS							
Note: For market prices of landscape **chemicals please see sections Q30 or Q35.**							
Market prices of planting materials (Note: **the rates shown generally reflect the** **manufacturer's recommended retail prices;** **trade and bulk discounts are often** **available on the prices shown)**							
Imported topsoil; market prices; British **Sugar Topsoil; sustainably sourced graded** **topsoil to British standards and with** **independent certification; rates shown** **allow for 20% settlement**							
Landscape 20; BS3882:2007; sandy loam							
8 wheel delivery; 20 tonne loads							
(approximately 13.33 m³)	30.90	–	–	–	30.90	m³	**30.90**
articulated delivery	24.60	–	–	–	24.60	m³	**24.60**
Sports 10; for topdressing to sports pitches; application rate 40–80 t/6000 m² pitch							
8 wheel delivery; 20 tonne loads							
(approximately 13.33 m³)	33.60	–	–	–	33.60	m³	**33.60**
articulated delivery	27.60	–	–	–	27.60	m³	**27.60**
Clay loam for use as a moisture retentive topsoil							
8 wheel delivery; 20 tonne loads							
(approximately 13.33 m³)	30.90	–	–	–	30.90	m³	**30.90**
articulated delivery	24.60	–	–	–	24.60	m³	**24.60**
Topsoil prices; Average locally sourced **prices**							
Topsoil							
general purpose	28.00	–	–	–	28.00	m³	**28.00**
Topsoil enhanced with organic matter	42.00	–	–	–	42.00	m³	**42.00**

Q31 PLANTING

Item Excluding site overheads and profit	PC £	Labour hours	Labour £	Plant £	Material £	Unit	Total rate £
MARKET PRICES OF MATERIALS – cont							
Market prices of mulching materials							
Melcourt Industries Ltd; 25 m³ loads (items labelled FSC are Forest Stewardship Council certified, items marked FT are certified fire tested)							
Ornamental Bark Mulch FT	–	–	–	–	58.40	m³	**58.40**
Bark Nuggets® FT	–	–	–	–	54.70	m³	**54.70**
Amenity Bark Mulch FSC FT	–	–	–	–	39.65	m³	**39.65**
Contract Bark Mulch FSC	–	–	–	–	36.90	m³	**36.90**
Spruce Ornamental FSC FT	–	–	–	–	40.40	m³	**40.40**
Decorative Biomulch®	–	–	–	–	39.55	m³	**39.55**
Rustic Biomulch®	–	–	–	–	43.30	m³	**43.30**
Mulch 2000	–	–	–	–	29.70	m³	**29.70**
Forest BioMulch®	–	–	–	–	37.05	m³	**37.05**
Melcourt Industries Ltd; 50 m³ loads							
Mulch 2000	–	–	–	–	18.95	m³	**18.95**
Melcourt Industries Ltd; 70 m³ loads							
Contract Bark Mulch FSC	–	–	–	–	24.50	m³	**24.50**
Melcourt Industries Ltd; 80 m³ loads							
Ornamental Bark Mulch FT	–	–	–	–	45.40	m³	**45.40**
Bark Nuggets® FT	–	–	–	–	41.70	m³	**41.70**
Amenity Bark Mulch FSC	–	–	–	–	26.65	m³	**26.65**
Spruce Ornamental FSC FT	–	–	–	–	27.40	m³	**27.40**
Decorative Biomulch®	–	–	–	–	26.55	m³	**26.55**
Rustic Biomulch®	–	–	–	–	30.30	m³	**30.30**
Forest BioMulch®	–	–	–	–	24.05	m³	**24.05**
Mulch; Melcourt Industries Ltd; 25 m³ loads							
Composted Fine Bark FSC	–	–	–	–	36.65	m³	**36.65**
Humus 2000	–	–	–	–	30.35	m³	**30.35**
Spent Mushroom Compost	–	–	–	–	23.70	m³	**23.70**
Topgrow	–	–	–	–	36.05	m³	**36.05**
Mulch; Melcourt Industries Ltd; 50 m³ loads							
Humus 2000	–	–	–	–	18.85	m³	**18.85**
Mulch; Melcourt Industries Ltd; 60 m³ loads							
Spent Mushroom Compost	–	–	–	–	12.00	m³	**12.00**
Mulch; Melcourt Industries Ltd; 65 m³ loads							
Composted Fine Bark	–	–	–	–	24.15	m³	**24.15**
Topgrow	–	–	–	–	28.05	m³	**28.05**
Market prices of fertilizers							
Fertilizers; British Seed Houses							
BSH1; 6+9+6	27.79	–	–	–	27.79	kg	**27.79**
BSH4; 11+5+5	27.30	–	–	–	27.30	kg	**27.30**
BSH6; 20+10+10	30.23	–	–	–	30.23	kg	**30.23**
BSH9; 12+2+9+Fe	28.62	–	–	–	28.62	kg	**28.62**
Fertilizers; Everris							
Enmag; controlled release fertilizer (8–9 months); 11+22+09; 70 g/m²	–	–	–	–	16.54	100 m²	**16.54**

Q31 PLANTING

Item Excluding site overheads and profit	PC £	Labour hours	Labour £	Plant £	Material £	Unit	Total rate £
Fertilizers; Everris; granular Osmocote Flora; 15+09+11; controlled release fertilizer; costs for recommended application rates							
transplant	–	–	–	–	0.10	nr	**0.10**
whip	–	–	–	–	0.15	nr	**0.15**
feathered	–	–	–	–	0.20	nr	**0.20**
light standard	–	–	–	–	0.20	nr	**0.20**
standard	–	–	–	–	0.24	nr	**0.24**
selected standard	–	–	–	–	0.34	nr	**0.34**
heavy standard	–	–	–	–	0.39	nr	**0.39**
extra heavy standard	–	–	–	–	0.49	nr	**0.49**
16–18 cm girth	–	–	–	–	0.54	nr	**0.54**
18–20 cm girth	–	–	–	–	0.59	nr	**0.59**
20–22 cm girth	–	–	–	–	0.68	nr	**0.68**
22–24 cm girth	–	–	–	–	0.73	nr	**0.73**
24–26 cm girth	–	–	–	–	0.78	nr	**0.78**
Fertilizers; Farmura Environmental Ltd; Seanure Soilbuilder							
soil amelioration; 70 g/m²	–	–	–	–	10.16	100 m²	**10.16**
to plant pits; 300 × 300 × 300 mm	–	–	–	–	0.04	nr	**0.04**
to plant pits; 600 × 600 × 600 mm	–	–	–	–	0.47	nr	**0.47**
to tree pits; 1.00 × 1.00 × 1.00 m	–	–	–	–	2.18	nr	**2.18**
Fertilizers; Everris							
TPMC tree planting and mulching compost	–	–	–	–	3.50	bag	**3.50**
Fertilizers; Rigby Taylor Ltd; fertilizer application rates 35 g/m² unless otherwise shown							
liquid fertilizer; Mascot Microflow-C; 25+0 +0 + trace elements; 200–1400 ml/100 m²	–	–	–	–	6.98	100 m²	**6.98**
liquid fertilizer; Mascot Microflow-CX; 4+2 +18 + trace elements; 200–1400 ml/100 m²	–	–	–	–	6.95	100 m²	**6.95**
liquid fertilizer; Mascot Microflow-CX; 12+0 +8 + trace elements; 200–1400 ml/100 m²	–	–	–	–	6.70	100 m²	**6.70**
liquid fertilizer; Mascot Microflow-CX; 17+2 +5 + trace elements; 200–1400 ml/100 m²	–	–	–	–	6.64	100 m²	**6.64**
Wetting agents; Rigby Taylor Ltd							
wetting agent; Breaker Advance Granules; 20 kg	–	–	–	–	15.94	100 m²	**15.94**

**CLARIFICATION NOTES ON LABOUR
COSTS IN THIS SECTION**

General landscape team

Generally a three man team is used in this
section; The column 'Labour hours' reports
team hours. The column 'Labour £' reports the
total cost of the team for the unit of work
shown

3 man team	–	1.00	57.75	–	–	hr	**57.75**

Q31 PLANTING

Item Excluding site overheads and profit	PC £	Labour hours	Labour £	Plant £	Material £	Unit	Total rate £
PLANTING PREPARATION AND CULTIVATION							
Site protection; temporary protective fencing							
Cleft chestnut rolled fencing; to 100 mm diameter chestnut posts; driving into firm ground at 3 m centres; pales at 50 mm centres							
900 mm high	3.31	0.11	2.05	–	3.86	m	**5.91**
1100 mm high	4.33	0.11	2.05	–	4.88	m	**6.93**
1500 mm high	6.08	0.11	2.05	–	6.63	m	**8.68**
Extra over temporary protective fencing for removing and making good (no allowance for reuse of material)	–	0.07	1.28	0.20	–	m	**1.48**
Cultivation by tractor; Agripower Ltd							
Ripping up subsoil; using approved subsoiling machine; minimum depth 250 mm below topsoil; at 1.20 m centres; in							
gravel or sandy clay	–	–	–	2.16	–	100 m²	**2.16**
soil compacted by machines	–	–	–	2.40	–	100 m²	**2.40**
clay	–	–	–	3.08	–	100 m²	**3.08**
chalk or other soft rock	–	–	–	3.59	–	100 m²	**3.59**
Extra for subsoiling at 1 m centres	–	–	–	0.62	–	100 m²	**0.62**
Breaking up existing ground; using tractor drawn tine cultivator or rotavator							
Single pass							
100 mm deep	–	–	–	2.85	–	100 m²	**2.85**
150 mm deep	–	–	–	3.56	–	100 m²	**3.56**
200 mm deep	–	–	–	3.56	–	100 m²	**3.56**
Cultivating ploughed ground; using disc, drag or chain harrow							
4 passes	–	–	–	3.56	–	100 m²	**3.56**
Rolling cultivated ground lightly; using self-propelled agricultural roller	–	0.07	1.28	0.72	–	100 m²	**2.00**
Cultivation; pedestrian operated rotavator							
Breaking up existing ground; tine cultivator or rotavator							
100 mm deep	–	0.22	4.24	2.37	–	100 m²	**6.61**
150 mm deep	–	0.28	5.29	2.96	–	100 m²	**8.25**
200 mm deep	–	0.37	7.06	3.94	–	100 m²	**11.00**
As above but in heavy clay or wet soils							
100 mm deep	–	0.44	8.47	4.73	–	100 m²	**13.20**
150 mm deep	–	0.66	12.71	7.09	–	100 m²	**19.80**
200 mm deep	–	0.82	15.88	8.87	–	100 m²	**24.75**
Clearing stones; disposing off site							
by hand; stones not exceeding 50 mm in any direction; loading to skip 4.6 m³	–	0.01	0.19	0.03	–	m²	**0.22**
by mechanical stone rake; stones not exceeding 50 mm in any direction; loading to 15 m³ truck by mechanical loader	–	–	0.04	0.08	–	m²	**0.12**

Q31 PLANTING

Item Excluding site overheads and profit	PC £	Labour hours	Labour £	Plant £	Material £	Unit	Total rate £
Lightly cultivating; weeding; to fallow areas; disposing debris off site							
by hand	–	0.01	0.27	–	0.08	m²	**0.35**
TOPSOIL FILLING							
Topsoil filling							
Market prices of topsoil; price shown includes a 20% settlement factor							
topsoil to BS3882	28.00	–	–	–	61.60	m³	**61.60**
blended topsoil	–	–	–	–	50.40	m³	**50.40**
small quantities (less than 15 m³)	60.00	–	–	–	60.00	m³	**60.00**
Spreading and lightly consolidating approved topsoil (imported or from spoil heaps); in layers not exceeding 150 mm; travel distance from spoil heaps not exceeding 100 m; by machine (imported topsoil not included)							
minimum depth 100 mm	–	1.55	29.84	45.34	–	100 m²	**75.18**
minimum depth 150 mm	–	2.33	44.92	68.23	–	100 m²	**113.15**
minimum depth 300 mm	–	4.67	89.83	136.45	–	100 m²	**226.28**
minimum depth 450 mm	–	6.99	134.56	204.46	–	100 m²	**339.02**
Spreading and lightly consolidating approved topsoil (imported or from spoil heaps); in layers not exceeding 150 mm; travel distance from spoil heaps not exceeding 100 m; by hand (imported topsoil not included)							
minimum depth 100 mm	–	20.00	385.08	–	–	100 m²	**385.08**
minimum depth 150 mm	–	30.01	577.62	–	–	100 m²	**577.62**
minimum depth 300 mm	–	60.01	1155.23	–	–	100 m²	**1155.23**
minimum depth 450 mm	–	90.02	1732.85	–	–	100 m²	**1732.85**
Spreading topsoil; from dump not exceeding 100 m distance; by machine (topsoil not included)							
at 1 m³ per 13 m²; 75 mm thick	–	0.56	10.70	33.86	–	100 m²	**44.56**
at 1 m³ per 10 m²; 100 mm thick	–	0.89	17.11	34.44	–	100 m²	**51.55**
at 1 m³ per 6.50 m²; 150 mm thick	–	1.11	21.37	60.77	–	100 m²	**82.14**
at 1 m³ per 5 m²; 200 mm thick	–	1.48	28.49	80.78	–	100 m²	**109.27**
Spreading topsoil; from dump not exceeding 100 m distance; by hand (topsoil not included)							
at 1 m³ per 13 m²; 75 mm thick	–	15.00	288.81	–	–	100 m²	**288.81**
at 1 m³ per 10 m²; 100 mm thick	–	20.00	385.08	–	–	100 m²	**385.08**
at 1 m³ per 6.50 m²; 150 mm thick	–	30.01	577.62	–	–	100 m²	**577.62**
at 1 m³ per 5 m²; 200 mm thick	–	36.67	705.97	–	–	100 m²	**705.97**
Imported topsoil; tipped 100 m from area of application; by machine							
at 1 m³ per 13 m²; 75 mm thick	252.00	0.56	10.70	33.86	252.00	100 m²	**296.56**
at 1 m³ per 10 m²; 100 mm thick	336.00	0.89	17.11	34.44	336.00	100 m²	**387.55**
at 1 m³ per 6.50 m²; 150 mm thick	504.00	1.11	21.37	60.77	504.00	100 m²	**586.14**
at 1 m³ per 5 m²; 200 mm thick	672.00	1.48	28.49	80.78	672.00	100 m²	**781.27**

Q31 PLANTING

Item Excluding site overheads and profit	PC £	Labour hours	Labour £	Plant £	Material £	Unit	Total rate £
TOPSOIL FILLING – cont							
Imported topsoil; tipped 100 m from area of application; by hand							
at 1 m³ per 13 m²; 75 mm thick	252.00	15.00	288.81	–	252.00	100 m²	**540.81**
at 1 m³ per 10 m²; 100 mm thick	336.00	20.00	385.08	–	336.00	100 m²	**721.08**
at 1 m³ per 6.50 m²; 150 mm thick	504.00	30.01	577.62	–	504.00	100 m²	**1081.62**
at 1 m³ per 5 m²; 200 mm thick	672.00	36.67	705.97	–	672.00	100 m²	**1377.97**
Extra over for spreading topsoil to slopes 15–30°; by machine or hand	–	–	–	–	–	10%	**–**
Extra over for spreading topsoil to slopes over 30°; by machine or hand	–	–	–	–	–	25%	**–**
Extra over for spreading topsoil from spoil heaps; travel exceeding 100 m; by machine							
100–150 m	–	0.01	0.24	0.06	–	m³	**0.30**
150–200 m	–	0.02	0.36	0.10	–	m³	**0.46**
200–300 m	–	0.03	0.54	0.15	–	m³	**0.69**
Extra over spreading topsoil for travel exceeding 100 m; by hand							
100 m	–	0.83	16.04	–	–	m³	**16.04**
200 m	–	1.67	32.09	–	–	m³	**32.09**
300 m	–	2.50	48.13	–	–	m³	**48.13**
Evenly grading; to general surfaces to bring to finished levels							
by machine (tractor mounted rotavator)	–	–	–	0.01	–	m²	**0.01**
by pedestrian operated rotavator	–	–	0.08	0.05	–	m²	**0.13**
by hand	–	0.01	0.19	–	–	m²	**0.19**
Extra over grading for slopes 15–30°; by machine or hand	–	–	–	–	–	10%	**–**
Extra over grading for slopes over 30°; by machine or hand	–	–	–	–	–	25%	**–**
Apply screened topdressing to grass surfaces; spread using Tru-Lute							
sand soil mixes 90/10 to 50/50	–	–	0.04	0.03	0.14	m²	**0.21**
Spread only existing cultivated soil to final levels using Tru-Lute							
cultivated soil	–	–	0.04	0.03	–	m²	**0.07**
SOIL IMPROVEMENTS							
Preparation of planting operations							
For the following topsoil improvement and planting operations add or subtract the following amounts for every £0.10 difference in the material cost price							
35 g/m²	–	–	–	–	0.35	100 m²	**0.35**
50 g/m²	–	–	–	–	0.50	100 m²	**0.50**
70 g/m²	–	–	–	–	0.70	100 m²	**0.70**
100 g/m²	–	–	–	–	1.00	100 m²	**1.00**
150 kg/ha	–	–	–	–	15.00	ha	**15.00**
200 kg/ha	–	–	–	–	20.00	ha	**20.00**
250 kg/ha	–	–	–	–	25.00	ha	**25.00**

Q31 PLANTING

Item Excluding site overheads and profit	PC £	Labour hours	Labour £	Plant £	Material £	Unit	Total rate £
300 kg/ha	–	–	–	–	30.00	ha	**30.00**
400 kg/ha	–	–	–	–	40.00	ha	**40.00**
500 kg/ha	–	–	–	–	50.00	ha	**50.00**
700 kg/ha	–	–	–	–	70.00	ha	**70.00**
1000 kg/ha	–	–	–	–	100.00	ha	**100.00**
1250 kg/ha	–	–	–	–	125.00	ha	**125.00**
Fertilizers; in top 150 mm of topsoil; at 35 g/m²							
Mascot Microfine; controlled release turf fertilizer; 8+0+6 + 2% Mg + 4% Fe	0.06	–	0.03	–	0.06	m²	**0.09**
Enmag; controlled release fertilizer (8–9 months); 11+22+09	0.09	–	0.03	–	0.09	m²	**0.12**
Mascot Outfield; turf fertilizer; 4+10+10	0.04	–	0.03	–	0.04	m²	**0.07**
Mascot Outfield; turf fertilizer; 9+5+5	0.04	–	0.03	–	0.04	m²	**0.07**
Fertilizers; in top 150 mm of topsoil at 70 g/m²							
Mascot Microfine; controlled release turf fertilizer; 8+0+6 + 2% Mg + 4% Fe	0.11	–	0.03	–	0.11	m²	**0.14**
Enmag; controlled release fertilizer (8–9 months); 11+22+09	0.17	–	0.03	–	0.17	m²	**0.20**
Mascot Outfield; turf fertilizer; 4+10+10	0.08	–	0.03	–	0.08	m²	**0.11**
Mascot Outfield; turf fertilizer; 9+5+5	0.07	–	0.03	–	0.07	m²	**0.10**
Topgrow; Melcourt; peat free tree and shrub compost (25 m³ loads); placing on beds by mechanical loader; spreading and rotavating into topsoil by tractor drawn machine							
50 mm thick	180.25	–	–	10.89	180.25	100 m²	**191.14**
100 mm thick	378.52	–	–	16.20	378.52	100 m²	**394.72**
150 mm thick	567.79	–	–	21.73	567.79	100 m²	**589.52**
200 mm thick	757.05	–	–	27.14	757.05	100 m²	**784.19**
Topgrow; Melcourt; peat free tree and shrub compost (65 m³ loads); placing on beds by mechanical loader; spreading and rotavating into topsoil by tractor drawn machine							
50 mm thick	140.25	–	–	10.89	140.25	100 m²	**151.14**
100 mm thick	294.52	–	–	16.20	294.52	100 m²	**310.72**
150 mm thick	441.79	–	–	21.73	441.79	100 m²	**463.52**
200 mm thick	589.05	–	–	27.14	589.05	100 m²	**616.19**
Mushroom compost; Melcourt; (25 m³ loads); from not further than 25 m from location; cultivating into topsoil by pedestrian drawn machine							
50 mm thick	118.50	2.86	55.00	1.97	118.50	100 m²	**175.47**
100 mm thick	237.00	6.05	116.44	1.97	237.00	100 m²	**355.41**
150 mm thick	355.50	8.90	171.42	1.97	355.50	100 m²	**528.89**
200 mm thick	474.00	12.90	248.42	1.97	474.00	100 m²	**724.39**
Mushroom compost; Melcourt (25 m³ loads); placing on beds by mechanical loader; spreading and rotavating into topsoil by tractor drawn cultivator							
50 mm thick	118.50	–	–	10.89	118.50	100 m²	**129.39**
100 mm thick	237.00	–	–	16.20	237.00	100 m²	**253.20**
150 mm thick	355.50	–	–	21.73	355.50	100 m²	**377.23**
200 mm thick	474.00	–	–	27.14	474.00	100 m²	**501.14**

Q31 PLANTING

Item	PC	Labour	Labour	Plant	Material	Unit	Total
Excluding site overheads and profit	£	hours	£	£	£		rate £
SOIL IMPROVEMENTS – cont							
Preparation of planting operations – cont							
Manure (60 m³ loads); from not further than							
25 m from location; cultivating into topsoil by							
pedestrian drawn machine							
50 mm thick	114.90	2.86	55.00	1.97	114.90	100 m²	**171.87**
100 mm thick	241.29	6.05	116.44	1.97	241.29	100 m²	**359.70**
150 mm thick	361.94	8.90	171.42	1.97	361.94	100 m²	**535.33**
200 mm thick	482.58	12.90	248.42	1.97	482.58	100 m²	**732.97**
Surface applications and soil additives;							
pre-planting; from not further than 25 m							
from location; by machine							
Medium bark soil conditioner; Melcourt Ltd;							
including turning in to cultivated ground;							
delivered in 25 m³ loads							
1 m³ per 40 m² = 25 mm thick	0.92	–	–	0.13	0.92	m²	**1.05**
1 m³ per 20 m² = 50 mm thick	1.83	–	–	0.19	1.83	m²	**2.02**
1 m³ per 13.33 m² = 75 mm thick	2.75	–	–	0.26	2.75	m²	**3.01**
1 m³ per 10 m² = 100 mm thick	3.67	–	–	0.33	3.67	m²	**4.00**
Works by hand; surface applications and							
soil additives; pre-planting; from not							
further than 25 m from location; by hand							
Mushroom compost soil conditioner; Melcourt							
Industries Ltd; including turning in to cultivated							
ground; delivered in 25 m³ loads							
1 m³ per 40 m² = 25 mm thick	0.59	0.02	0.43	–	0.59	m²	**1.02**
1 m³ per 20 m² = 50 mm thick	1.19	0.04	0.85	–	1.19	m²	**2.04**
1 m³ per 13.33 m² = 75 mm thick	1.78	0.07	1.28	–	1.78	m²	**3.06**
1 m³ per 10 m² = 100 mm thick	2.37	0.08	1.54	–	2.37	m²	**3.91**
Medium bark soil conditioner; Melcourt Ltd							
Ltd; including turning in to cultivated ground;							
delivered in 25 m³ loads							
1 m³ per 40 m² = 25 mm thick	0.92	0.02	0.43	–	0.92	m²	**1.35**
1 m³ per 20 m² = 50 mm thick	1.83	0.04	0.85	–	1.83	m²	**2.68**
1 m³ per 13.33 m² = 75 mm thick	2.75	0.07	1.28	–	2.75	m²	**4.03**
1 m³ per 10 m² = 100 mm thick	3.67	0.08	1.54	–	3.67	m²	**5.21**
Manure; 20 m³ loads; delivered not further							
than 25 m from location; cultivating into topsoil							
by pedestrian operated machine							
50 mm thick	225.00	2.86	55.00	1.97	225.00	100 m²	**281.97**
100 mm thick	241.29	6.05	116.44	1.97	241.29	100 m²	**359.70**
150 mm thick	361.94	8.90	171.42	1.97	361.94	100 m²	**535.33**
200 mm thick	482.58	12.90	248.42	1.97	482.58	100 m²	**732.97**
Fertilizer (7+7+7); PC £1.29/kg; to beds; by							
hand							
35 g/m²	4.52	0.17	3.21	–	4.52	100 m²	**7.73**
50 g/m²	6.46	0.17	3.21	–	6.46	100 m²	**9.67**
70 g/m²	9.05	0.17	3.21	–	9.05	100 m²	**12.26**

Q31 PLANTING

Item Excluding site overheads and profit	PC £	Labour hours	Labour £	Plant £	Material £	Unit	Total rate £
Fertilizers; Enmag; PC £2.36/kg; controlled release fertilizer; to beds; by hand							
35 g/m²	8.27	0.17	3.21	–	8.27	100 m²	**11.48**
50 g/m²	11.81	0.17	3.21	–	11.81	100 m²	**15.02**
70 g/m²	17.72	0.17	3.21	–	17.72	100 m²	**20.93**
Note: For machine incorporation of fertilizers and soil conditioners see 'Cultivation'.							
TREE PLANTING							
Tree planting; pre-planting operations							
Excavating tree pits; depositing soil alongside pits; by machine							
600 × 600 × 600 mm deep	–	0.15	2.83	0.66	–	nr	**3.49**
900 × 900 × 600 mm deep	–	0.33	6.35	1.49	–	nr	**7.84**
1.00 m × 1.00 m × 600 mm deep	–	0.61	11.81	1.84	–	nr	**13.65**
1.25 m × 1.25 m × 600 mm deep	–	0.96	18.49	2.88	–	nr	**21.37**
1.00 × 1.00 × 1.00 m deep	–	1.02	19.69	3.07	–	nr	**22.76**
1.50 m × 1.50 m × 750 mm deep	–	1.73	33.22	5.18	–	nr	**38.40**
1.50 × 1.50 × 1.00 m deep	–	2.30	44.18	6.89	–	nr	**51.07**
1.75 × 1.75 × 1.00 m deep	–	3.13	60.29	9.40	–	nr	**69.69**
2.00 × 2.00 × 1.00 m deep	–	4.09	78.75	12.27	–	nr	**91.02**
Excavating tree pits; depositing soil alongside pits; by hand							
600 × 600 × 600 mm deep	–	0.44	8.47	–	–	nr	**8.47**
900 × 900 × 600 mm deep	–	1.00	19.25	–	–	nr	**19.25**
1.00 m × 1.00 m × 600 mm deep	–	1.13	21.66	–	–	nr	**21.66**
1.25 m × 1.25 m × 600 mm deep	–	1.93	37.15	–	–	nr	**37.15**
1.00 × 1.00 × 1.00 m deep	–	2.06	39.66	–	–	nr	**39.66**
1.50 m × 1.50 m × 750 mm deep	–	3.47	66.80	–	–	nr	**66.80**
1.75 m × 1.50 m × 750 mm deep	–	4.05	77.96	–	–	nr	**77.96**
1.50 × 1.50 × 1.00 m deep	–	4.63	89.13	–	–	nr	**89.13**
2.00 m × 2.00 m × 750 mm deep	–	6.17	118.77	–	–	nr	**118.77**
2.00 × 2.00 × 1.00 m deep	–	8.23	158.44	–	–	nr	**158.44**
Breaking up subsoil in tree pits; to a depth of 200 mm	–	0.03	0.64	–	–	m²	**0.64**
Spreading and lightly consolidating approved topsoil (imported or from spoil heaps); in layers not exceeding 150 mm; distance from spoil heaps not exceeding 100 m (imported topsoil not included); by machine							
minimum depth 100 mm	–	1.55	29.84	45.34	–	100 m²	**75.18**
minimum depth 150 mm	–	2.33	44.92	68.23	–	100 m²	**113.15**
minimum depth 300 mm	–	4.67	89.83	136.45	–	100 m²	**226.28**
minimum depth 450 mm	–	6.99	134.56	204.46	–	100 m²	**339.02**

Q31 PLANTING

Item Excluding site overheads and profit	PC £	Labour hours	Labour £	Plant £	Material £	Unit	Total rate £
TREE PLANTING – cont							
Tree planting – cont							
Spreading and lightly consolidating approved topsoil (imported or from spoil heaps); in layers not exceeding 150 mm; distance from spoil heaps not exceeding 100 m (imported topsoil not included); by hand							
minimum depth 100 mm	–	20.00	385.08	–	–	100 m²	**385.08**
minimum depth 150 mm	–	30.01	577.62	–	–	100 m²	**577.62**
minimum depth 300 mm	–	60.01	1155.23	–	–	100 m²	**1155.23**
minimum depth 450 mm	–	90.02	1732.85	–	–	100 m²	**1732.85**
Extra for filling tree pits with imported topsoil; PC £28.00/m³; plus allowance for 20% settlement							
depth 100 mm	–	–	–	–	4.03	m²	**4.03**
depth 150 mm	–	–	–	–	6.05	m²	**6.05**
depth 200 mm	–	–	–	–	8.06	m²	**8.06**
depth 300 mm	–	–	–	–	12.10	m²	**12.10**
depth 400 mm	–	–	–	–	16.13	m²	**16.13**
depth 450 mm	–	–	–	–	18.14	m²	**18.14**
depth 500 mm	–	–	–	–	20.16	m²	**20.16**
depth 600 mm	–	–	–	–	24.19	m²	**24.19**
Add or deduct the following amounts for every £1.00 change in the material price of topsoil							
depth 100 mm	–	–	–	–	0.12	m²	**0.12**
depth 150 mm	–	–	–	–	0.18	m²	**0.18**
depth 200 mm	–	–	–	–	0.24	m²	**0.24**
depth 300 mm	–	–	–	–	0.36	m²	**0.36**
depth 400 mm	–	–	–	–	0.48	m²	**0.48**
depth 450 mm	–	–	–	–	0.54	m²	**0.54**
depth 500 mm	–	–	–	–	0.60	m²	**0.60**
depth 600 mm	–	–	–	–	0.72	m²	**0.72**
Structural soils (Amsterdam tree sand); Heicom; non-compressive soil mixture for tree planting in areas to receive compressive surface treatments							
Backfilling and lightly compacting in layers; excavation, disposal, moving of material from delivery position and surface treatments not included; by machine							
individual treepits	61.03	0.50	9.63	2.25	61.03	m³	**72.91**
in trenches	61.03	0.42	8.02	1.88	61.03	m³	**70.93**
Backfilling and lightly compacting in layers; excavation, disposal, moving of material from delivery position and surface treatments not included; by hand							
individual treepits	61.03	1.60	30.80	–	61.03	m³	**91.83**
in trenches	61.03	1.33	25.60	–	61.03	m³	**86.63**

Q31 PLANTING

Item Excluding site overheads and profit	PC £	Labour hours	Labour £	Plant £	Material £	Unit	Total rate £
Tree staking							
J Toms Ltd; extra over trees for tree stake(s); driving 500 mm into firm ground; trimming to approved height; including two tree ties to approved pattern							
one stake; 1.52 m long × 32 × 32 mm	1.50	0.20	3.85	–	1.50	nr	**5.35**
two stakes; 1.21 m long × 25 × 25 mm	1.62	0.30	5.78	–	1.62	nr	**7.40**
two stakes; 1.52 m long × 32 × 32 mm	2.22	0.30	5.78	–	2.22	nr	**8.00**
three stakes; 1.52 m long × 32 × 32 mm	3.33	0.36	6.93	–	3.33	nr	**10.26**
Tree anchors							
Platipus Anchors Ltd; extra over trees for tree anchors							
RF1P rootball kit; for 75–220 mm girth × 2–4.5 m high; inclusive of Plati-Mat PM1	29.67	1.00	19.25	–	29.67	nr	**48.92**
RF2P; rootball kit; for 220–450 mm girth × 4.5–7.5 m high; inclusive of Plati-Mat PM2	50.50	1.33	25.60	–	50.50	nr	**76.10**
RF3P; rootball kit; for 450–750 mm girth × 7.5 to 12 m high; inclusive of Plati-Mat PM3	106.92	1.50	28.88	–	106.92	nr	**135.80**
CG1; guy fixing kit; 75–220 mm girth × 2–4.5 m high	19.08	1.67	32.08	–	19.08	nr	**51.16**
CG2; guy fixing kit; 220–450 mm girth × 4.5–7.5 m high	33.56	2.00	38.50	–	33.56	nr	**72.06**
installation tools; drive rod for RF1/CG1 kits	–	–	–	–	61.85	nr	**61.85**
installation tools; drive rod for RF2/CG2 kits	–	–	–	–	93.20	nr	**93.20**
Extra over trees for land drain to tree pits; 100 mm diameter perforated flexible agricultural drain; including excavating drain trench; laying pipe; backfilling	1.26	1.00	19.25	–	1.26	m	**20.51**
Platipus anchors; deadman anchoring							
Deadman kits; anchoring to concrete kerbs placed in base of tree pit							
trees 12–25 cm girth; 2.5–4.0 m high	26.62	0.75	14.44	–	41.24	nr	**55.68**
trees 25–45 cm girth; 4.0–7.5 m high	45.82	1.50	28.88	–	60.44	nr	**89.32**
trees 45–75 cm girth; 7.5–12.0 m high	100.23	2.00	38.50	–	114.85	nr	**153.35**
Tree planting; tree pit additives							
Melcourt Industries Ltd; Topgrow; incorporating into topsoil at 1 part Topgrow to 3 parts excavated topsoil; supplied in 75 l bags; pit size							
600 × 600 × 600 mm	1.76	0.02	0.39	–	1.76	nr	**2.15**
900 × 900 × 900 mm	5.95	0.06	1.16	–	5.95	nr	**7.11**
1.00 × 1.00 × 1.00	8.16	0.24	4.62	–	8.16	nr	**12.78**
1.25 × 1.25 × 1.25	15.95	0.40	7.70	–	15.95	nr	**23.65**
1.50 × 1.50 × 1.50	27.56	0.90	17.32	–	27.56	nr	**44.88**

Q31 PLANTING

Item Excluding site overheads and profit	PC £	Labour hours	Labour £	Plant £	Material £	Unit	Total rate £
TREE PLANTING – cont							
Tree planting – cont Melcourt Industries Ltd; Topgrow; incorporating into topsoil at 1 part Topgrow to 3 parts excavated topsoil; supplied in 60 m³ loose loads; pit size							
600 × 600 × 600 mm	1.51	0.02	0.32	–	1.51	nr	**1.83**
900 × 900 × 900 mm	5.11	0.05	0.96	–	5.11	nr	**6.07**
1.00 × 1.00 × 1.00	7.01	0.20	3.85	–	7.01	nr	**10.86**
1.25 × 1.25 × 1.25	13.70	0.33	6.42	–	13.70	nr	**20.12**
1.50 × 1.50 × 1.50	23.67	0.75	14.44	–	23.67	nr	**38.11**
Tree planting; root barriers English Woodlands; Root Director; one-piece root control planters for installation at time of planting to divert root growth down away from pavements and out for anchorage; excavation measured separately							
RD 1050; 1050 × 1050 mm to 1300 × 1300 mm at base	99.65	0.25	4.81	–	99.65	nr	**104.46**
RD 640; 640 × 640 mm to 870 × 870 mm at base	64.72	0.25	4.81	–	64.72	nr	**69.53**
Greenleaf Horticulture; linear root deflection barriers; installed to trench measured separately							
Re-Root 2000; 2.0 mm thick × 2 m wide	14.28	0.05	0.96	–	14.28	m	**15.24**
Re-Root 2000; 1.0 mm thick × 1 m wide	4.98	0.05	0.96	–	4.98	m	**5.94**
Re-Root 600; 1.0 mm thick	7.32	0.05	0.96	–	7.32	m	**8.28**
Greenleaf Horticulture; irrigation systems; **Root Rain tree pit irrigation systems** Metro; small; 35 mm pipe diameter × 1.25 m long; for specimen shrubs and standard trees							
plastic	6.48	0.25	4.81	–	6.48	nr	**11.29**
plastic; with chain	8.49	0.25	4.81	–	8.49	nr	**13.30**
metal; with chain	10.52	0.25	4.81	–	10.52	nr	**15.33**
Metro; medium; 35 mm pipe diameter × 1.75 m long; for standard and selected standard trees							
plastic	6.73	0.29	5.50	–	6.73	nr	**12.23**
plastic; with chain	8.91	0.29	5.50	–	8.91	nr	**14.41**
metal; with chain	10.90	0.29	5.50	–	10.90	nr	**16.40**
Metro; large; 35 mm pipe diameter × 2.50 m long; for selected standards and extra heavy standards							
plastic	7.96	0.33	6.42	–	7.96	nr	**14.38**
plastic; with chain	9.97	0.33	6.42	–	9.97	nr	**16.39**
metal; with chain	13.33	0.33	6.42	–	13.33	nr	**19.75**

Q31 PLANTING

Item Excluding site overheads and profit	PC £	Labour hours	Labour £	Plant £	Material £	Unit	Total rate £
Urban; for large capacity general purpose irrigation to parkland and street verge planting							
RRUrb1; 3.0 m pipe	–	0.29	5.50	–	15.98	nr	**21.48**
RRUrb2; 5.0 m pipe	18.35	0.33	6.42	–	18.35	nr	**24.77**
RRUrb3; 8.0 m pipe	21.92	0.40	7.70	–	21.92	nr	**29.62**
Civic; heavy cast aluminium inlet; for heavily trafficked locations							
5.0 m pipe	32.50	0.33	6.42	–	32.50	nr	**38.92**
8.0 m pipe	37.00	0.40	7.70	–	37.00	nr	**44.70**
Tree irrigation kits; Platipus; direct water **delivery to the rootball area of trees**							
Piddler tree irrigation system; permeable membrane delivering targeted irrigation to rootball surround							
root balls up to 550 mm diameter; Irrigation Kit PID0	7.00	0.25	4.81	–	7.00	nr	**11.81**
root balls up to 900 mm diameter; Irrigation Kit PID1	12.50	0.30	5.78	–	12.50	nr	**18.28**
root balls up to 1.55 m diameter; Irrigation Kit PID2	16.50	0.40	7.70	–	16.50	nr	**24.20**
root balls up to 2.40 m diameter; Irrigation Kit PID3	20.50	0.45	8.66	–	20.50	nr	**29.16**
root balls up to 3.10 m diameter; Irrigation Kit PID4	25.50	0.50	9.63	–	25.50	nr	**35.13**
Mulching of tree pits; Melcourt Industries **Ltd; FSC (Forest Stewardship Council) and** **FT (fire tested) certified**							
Spreading mulch; to individual trees; maximum distance 25 m (mulch not included)							
50 mm thick	–	0.05	0.94	–	–	m²	**0.94**
75 mm thick	–	0.07	1.40	–	–	m²	**1.40**
100 mm thick	–	0.10	1.87	–	–	m²	**1.87**
Mulch; Bark Nuggets®; to individual trees; delivered in 80 m³ loads; maximum distance 25 m							
50 mm thick	2.19	0.05	0.94	–	2.19	m²	**3.13**
75 mm thick	3.29	0.07	1.40	–	3.29	m²	**4.69**
100 mm thick	4.38	0.10	1.87	–	4.38	m²	**6.25**
Mulch; Bark Nuggets®; to individual trees; delivered in 25 m³ loads; maximum distance 25 m							
50 mm thick	2.87	0.05	0.94	–	2.87	m²	**3.81**
75 mm thick	4.31	0.05	0.96	–	4.31	m²	**5.27**
100 mm thick	5.74	0.07	1.28	–	5.74	m²	**7.02**
Mulch; Amenity Bark Mulch; to individual trees; delivered in 80 m³ loads; maximum distance 25 m							
50 mm thick	1.40	0.05	0.94	–	1.40	m²	**2.34**
75 mm thick	2.10	0.07	1.40	–	2.10	m²	**3.50**
100 mm thick	2.80	0.10	1.87	–	2.80	m²	**4.67**

Q31 PLANTING

Item Excluding site overheads and profit	PC £	Labour hours	Labour £	Plant £	Material £	Unit	Total rate £
TREE PLANTING – cont							
Mulching of tree pits – cont							
Mulch; Amenity Bark Mulch; to individual							
trees; delivered in 25 m³ loads; maximum							
distance 25 m							
50 mm thick	2.08	0.05	0.94	–	2.08	m²	**3.02**
75 mm thick	3.12	0.07	1.40	–	3.12	m²	**4.52**
100 mm thick	4.16	0.07	1.28	–	4.16	m²	**5.44**
Trees; planting labours only							
Bare root trees; including backfilling with							
previously excavated material (all other							
operations and materials not included)							
light standard; 6–8 cm girth	–	0.35	6.74	–	–	nr	**6.74**
standard; 8–10 cm girth	–	0.40	7.70	–	–	nr	**7.70**
selected standard; 10–12 cm girth	–	0.58	11.16	–	–	nr	**11.16**
heavy standard; 12–14 cm girth	–	0.83	16.04	–	–	nr	**16.04**
extra heavy standard; 14–16 cm girth	–	1.00	19.25	–	–	nr	**19.25**
Root balled trees; including backfilling with							
previously excavated material (all other							
operations and materials not included)							
standard; 8–10 cm girth	–	0.50	9.63	–	–	nr	**9.63**
selected standard; 10–12 cm girth	–	0.60	11.55	–	–	nr	**11.55**
heavy standard; 12–14 cm girth	–	0.80	15.40	–	–	nr	**15.40**
extra heavy standard;14–16 cm girth	–	1.50	28.88	–	–	nr	**28.88**
16–18 cm girth	–	1.30	24.95	26.73	–	nr	**51.68**
18–20 cm girth	–	1.60	30.80	33.00	–	nr	**63.80**
20–25 cm girth	–	4.50	86.63	103.39	–	nr	**190.02**
25–30 cm girth	–	6.00	115.50	136.45	–	nr	**251.95**
30–35 cm girth	–	11.00	211.75	242.75	–	nr	**454.50**
Tree planting; containerized trees; nursery							
stock; James Coles & Sons (Nurseries) Ltd							
Acer platanoides 'Emerald Queen'; including							
backfilling with excavated material (other							
operations not included)							
standard; 8–10 cm girth	44.75	0.48	9.24	–	44.75	nr	**53.99**
selected standard; 10–12 cm girth	56.00	0.56	10.79	–	56.00	nr	**66.79**
heavy standard; 12–14 cm girth	63.00	0.76	14.71	–	63.00	nr	**77.71**
extra heavy standard; 14–16 cm girth	105.00	1.20	23.10	–	105.00	nr	**128.10**
Carpinus betulus; including backfilling with							
excavated material (other operations not							
included)							
standard; 8–10 cm girth	44.75	0.48	9.24	–	44.75	nr	**53.99**
selected standard; 10–12 cm girth	56.00	0.56	10.79	–	56.00	nr	**66.79**
heavy standard; 12–14 cm girth	91.00	0.76	14.71	–	91.00	nr	**105.71**
extra heavy standard; 14–16 cm girth	112.00	1.20	23.10	–	112.00	nr	**135.10**

Q31 PLANTING

Item Excluding site overheads and profit	PC £	Labour hours	Labour £	Plant £	Material £	Unit	Total rate £
Fraxinus excelsior 'Altena'; including backfillling with excavated material (other operations not included)							
standard; 8–10 cm girth	42.00	0.48	9.24	–	42.00	nr	**51.24**
selected standard; 10–12 cm girth	67.25	0.56	10.79	–	67.25	nr	**78.04**
heavy standard; 12–14 cm girth	84.00	0.76	14.71	–	84.00	nr	**98.71**
extra heavy standard; 14–16 cm girth	98.00	1.20	23.10	–	98.00	nr	**121.10**
Prunus avium 'Plena'; including backfillling with excavated material (other operations not included)							
standard; 8–10 cm girth	42.00	0.40	7.70	–	42.00	nr	**49.70**
selected standard; 10–12 cm girth	70.00	0.56	10.79	–	70.00	nr	**80.79**
heavy standard; 12–14 cm girth	84.00	0.76	14.71	–	84.00	nr	**98.71**
extra heavy standard; 14–16 cm girth	98.00	1.20	23.10	–	98.00	nr	**121.10**
Quercus robur; including backfillling with excavated material (other operations not included)							
standard; 8–10 cm girth	49.00	0.48	9.24	–	49.00	nr	**58.24**
selected standard; 10–12 cm girth	77.00	0.56	10.79	–	77.00	nr	**87.79**
heavy standard; 12–14 cm girth	105.00	0.76	14.71	–	105.00	nr	**119.71**
extra heavy standard; 14–16 cm girth	119.00	1.20	23.10	–	119.00	nr	**142.10**
Betula utilis jaquemontii; multistemmed; including backfillling with excavated material (other operations not included)							
175/200 mm high	84.00	0.48	9.24	–	84.00	nr	**93.24**
200/250 mm high	98.00	0.56	10.79	–	98.00	nr	**108.79**
250/300 mm high	84.00	0.76	14.71	–	84.00	nr	**98.71**
300/350 mm high	196.00	1.20	23.10	–	196.00	nr	**219.10**
Tree planting; root balled trees; advanced nursery stock and semi-mature – General Preamble: The cost of planting semi-mature trees will depend on the size and species and on the access to the site for tree handling machines. Prices should be obtained for individual trees and planting.							
Tree planting; bare root trees; nursery stock; James Coles & Sons (Nurseries) Ltd Acer platanoides; including backfillling with excavated material (other operations not included)							
light standard; 6–8 cm girth	9.00	0.35	6.74	–	9.00	nr	**15.74**
standard; 8–10 cm girth	12.00	0.40	7.70	–	12.00	nr	**19.70**
selected standard; 10–12 cm girth	19.50	0.58	11.16	–	19.50	nr	**30.66**
heavy standard; 12–14 cm girth	42.00	0.83	16.04	–	42.00	nr	**58.04**
extra heavy standard; 14–16 cm girth	56.00	1.00	19.25	–	56.00	nr	**75.25**

Q31 PLANTING

Item	PC	Labour	Labour	Plant	Material	Unit	Total
Excluding site overheads and profit	£	hours	£	£	£		rate £
TREE PLANTING – cont							
Tree planting – cont							
Carpinus betulus; including backfillling with excavated material (other operations not included)							
light standard; 6–8 cm girth	13.25	0.35	6.74	–	13.25	nr	**19.99**
standard; 8–10 cm girth	26.50	0.40	7.70	–	26.50	nr	**34.20**
selected standard; 10–12 cm girth	37.75	0.58	11.16	–	37.75	nr	**48.91**
heavy standard; 12–14 cm girth	39.25	0.83	16.05	–	39.25	nr	**55.30**
extra heavy standard; 14–16 cm girth	46.26	1.00	19.25	–	46.26	nr	**65.51**
Fraxinus excelsior; including backfillling with excavated material (other operations not included)							
light standard; 6–8 cm girth	9.75	0.35	6.74	–	9.75	nr	**16.49**
standard; 8–10 cm girth	16.00	0.40	7.70	–	16.00	nr	**23.70**
selected standard; 10–12 cm girth	22.50	0.58	11.16	–	22.50	nr	**33.66**
heavy standard; 12–14 cm girth	39.25	0.83	15.98	–	39.25	nr	**55.23**
extra heavy standard; 14–16 cm girth	47.50	1.00	19.25	–	47.50	nr	**66.75**
Prunus avium 'Plena'; including backfillling with excavated material (other operations not included)							
light standard; 6–8 cm girth	9.75	0.36	7.01	–	9.75	nr	**16.76**
standard; 8–10 cm girth	16.00	0.40	7.70	–	16.00	nr	**23.70**
selected standard; 10–12 cm girth	29.50	0.58	11.16	–	29.50	nr	**40.66**
heavy standard; 12–14 cm girth	47.50	0.83	16.04	–	47.50	nr	**63.54**
extra heavy standard; 14–16 cm girth	63.00	1.00	19.25	–	63.00	nr	**82.25**
Quercus robur; including backfillling with excavated material (other operations not included)							
light standard; 6–8 cm girth	21.00	0.35	6.74	–	21.00	nr	**27.74**
standard; 8–10 cm girth	30.75	0.40	7.70	–	30.75	nr	**38.45**
selected standard; 10–12 cm girth	42.00	0.58	11.16	–	42.00	nr	**53.16**
heavy standard; 12–14 cm girth	58.76	0.83	16.04	–	58.76	nr	**74.80**
Robinia pseudoacacia Frisia; including backfillling with excavated material (other operations not included)							
light standard; 6–8 cm girth	21.00	0.35	6.74	–	21.00	nr	**27.74**
standard; 8–10 cm girth	28.00	0.40	7.70	–	28.00	nr	**35.70**
selected standard; 10–12 cm girth	44.75	0.58	11.16	–	44.75	nr	**55.91**
Tree planting; root balled trees; nursery stock; James Coles & Sons (Nurseries) Ltd							
Acer platanoides; including backfillling with excavated material (other operations not included)							
standard; 8–10 cm girth	24.00	0.48	9.24	–	24.00	nr	**33.24**
selected standard; 10–12 cm girth	37.00	0.56	10.79	–	37.00	nr	**47.79**
heavy standard; 12–14 cm girth	57.00	0.76	14.71	–	57.00	nr	**71.71**
extra heavy standard; 14–16 cm girth	80.50	1.20	23.10	–	80.50	nr	**103.60**

Q31 PLANTING

Item Excluding site overheads and profit	PC £	Labour hours	Labour £	Plant £	Material £	Unit	Total rate £
Carpinus betulus; including backfillling with excavated material (other operations not included)							
standard; 8–10 cm girth	38.50	0.48	9.24	–	38.50	nr	47.74
selected standard; 10–12 cm girth	55.25	0.56	10.79	–	55.25	nr	66.04
heavy standard; 12–14 cm girth	85.50	0.76	14.71	–	85.50	nr	100.21
extra heavy standard; 14–16 cm girth	150.50	1.20	23.10	–	150.50	nr	173.60
Fraxinus excelsior; including backfillling with excavated material (other operations not included)							
standard; 8–10 cm girth	28.00	0.48	9.24	–	28.00	nr	37.24
selected standard; 10–12 cm girth	40.00	0.56	10.79	–	40.00	nr	50.79
heavy standard; 12–14 cm girth	60.26	0.76	14.71	–	60.26	nr	74.97
extra heavy standard; 14–16 cm girth	72.00	1.20	23.10	–	72.00	nr	95.10
Prunus avium 'Plena'; including backfillling with excavated material (other operations not included)							
standard; 8–10 cm girth	35.00	0.40	7.70	–	35.00	nr	42.70
selected standard; 10–12 cm girth	35.00	0.56	10.79	–	35.00	nr	45.79
heavy standard; 12–14 cm girth	68.50	0.76	14.71	–	68.50	nr	83.21
extra heavy standard; 14–16 cm girth	87.50	1.20	23.10	–	87.50	nr	110.60
Quercus robur; including backfillling with excavated material (other operations not included)							
standard; 8–10 cm girth	42.75	0.48	9.24	–	42.75	nr	51.99
selected standard; 10–12 cm girth	59.50	0.56	10.79	–	59.50	nr	70.29
heavy standard; 12–14 cm girth	79.75	0.76	14.71	–	79.75	nr	94.46
extra heavy standard; 14–16 cm girth	112.00	1.20	23.10	–	112.00	nr	135.10
Robinia pseudoacacia 'Frisia'; including backfillling with excavated material (other operations not included)							
standard; 8–10 cm girth	40.00	0.48	9.24	–	40.00	nr	49.24
selected standard; 10–12 cm girth	62.25	0.56	10.79	–	62.25	nr	73.04
heavy standard; 12–14 cm girth	98.00	0.76	14.71	–	98.00	nr	112.71
Tree planting; Airpot container grown trees; advanced nursery stock and semi-mature; Deepdale Trees Ltd							
Acer platanoides 'Emerald Queen'; including backfilling with excavated material (other operations not included)							
16–18 cm girth	95.00	1.98	38.12	41.62	95.00	nr	174.74
18–20 cm girth	130.00	2.18	41.93	44.72	130.00	nr	216.65
20–25 cm girth	190.00	2.38	45.74	49.94	190.00	nr	285.68
25–30 cm girth	250.00	2.97	57.17	76.50	250.00	nr	383.67
30–35 cm girth	450.00	3.96	76.23	83.24	450.00	nr	609.47

Q31 PLANTING

Item Excluding site overheads and profit	PC £	Labour hours	Labour £	Plant £	Material £	Unit	Total rate £
TREE PLANTING – cont							
Tree planting – cont							
Prunus avium 'Flora Plena'; including							
backfilling with excavated material (other							
operations not included)							
16–18 cm girth	95.00	1.98	38.12	41.62	95.00	nr	**174.74**
18–20 cm girth	130.00	1.60	30.80	44.72	130.00	nr	**205.52**
20–25 cm girth	190.00	2.38	45.74	49.94	190.00	nr	**285.68**
25–30 cm girth	250.00	2.97	57.17	76.50	250.00	nr	**383.67**
30–35 cm girth	350.00	3.96	76.23	83.24	350.00	nr	**509.47**
Quercus palustris 'Pin Oak'; including							
backfilling with excavated material (other							
operations not included)							
16–18 cm girth	100.00	1.98	38.12	41.62	100.00	nr	**179.74**
18–20 cm girth	140.00	1.60	30.80	44.72	140.00	nr	**215.52**
20–25 cm girth	210.00	2.38	45.74	49.94	210.00	nr	**305.68**
25–30 cm girth	275.00	2.97	57.17	76.50	275.00	nr	**408.67**
30–35 cm girth	400.00	3.96	76.23	83.24	400.00	nr	**559.47**
Tree planting; Airpot container grown							
trees; semi-mature and mature trees;							
Deepdale Trees Ltd; planting and back							
filling; planted by telehandler or by crane;							
delivery included; all other operations							
priced separately							
Semi-mature trees; indicative prices							
40–45 cm girth	550.00	4.00	77.00	52.05	550.00	nr	**679.05**
45–50 cm girth	750.00	4.00	77.00	52.05	750.00	nr	**879.05**
55–60 cm girth	1350.00	6.00	115.50	52.05	1350.00	nr	**1517.55**
60–70 cm girth	2500.00	7.00	134.75	70.25	2500.00	nr	**2705.00**
70–80 cm girth	3500.00	7.50	144.38	88.44	3500.00	nr	**3732.82**
80–90 cm girth	4500.00	8.00	154.00	104.11	4500.00	nr	**4758.11**
Tree planting; rootballed trees; advanced							
nursery stock and semi-mature; Lorenz							
von Ehren							
Acer platanoides 'Emerald Queen'; including							
backfilling with excavated material (other							
operations not included)							
16–18 cm girth	85.00	1.30	24.95	3.75	85.00	nr	**113.70**
18–20 cm girth	105.00	1.60	30.80	3.75	105.00	nr	**139.55**
20–25 cm girth	130.00	4.50	86.63	16.21	130.00	nr	**232.84**
25–30 cm girth	170.00	6.00	115.50	20.20	170.00	nr	**305.70**
30–35 cm girth	305.00	11.00	211.75	27.13	305.00	nr	**543.88**

Q31 PLANTING

Item Excluding site overheads and profit	PC £	Labour hours	Labour £	Plant £	Material £	Unit	Total rate £
Aesculus carnea 'Briotti'; including backfilling with excavated material (other operations not included)							
16–18 cm girth	150.00	1.30	24.95	3.75	150.00	nr	178.70
18–20 cm girth	180.00	1.60	30.80	3.75	180.00	nr	214.55
20–25 cm girth	215.00	4.50	86.63	16.21	215.00	nr	317.84
25–30 cm girth	290.00	6.00	115.50	20.20	290.00	nr	425.70
30–35 cm girth	390.00	11.00	211.75	27.13	390.00	nr	628.88
Prunus avium 'Plena'; including backfilling with excavated material (other operations not included)							
16–18 cm girth	110.00	1.30	24.95	3.75	110.00	nr	138.70
18–20 cm girth	130.00	1.60	30.80	3.75	130.00	nr	164.55
20–25 cm girth	150.00	4.50	86.63	16.21	150.00	nr	252.84
25–30 cm girth	155.00	6.00	115.50	20.20	155.00	nr	290.70
30–35 cm girth	250.00	11.00	211.75	23.95	250.00	nr	485.70
Quercus palustris 'Pin Oak'; including backfilling with excavated material (other operations not included)							
16–18 cm girth	125.00	1.30	24.95	3.75	125.00	nr	153.70
18–20 cm girth	150.00	1.60	30.80	3.75	150.00	nr	184.55
20–25 cm girth	190.00	4.50	86.63	16.21	190.00	nr	292.84
25–30 cm girth	225.00	6.00	115.50	20.20	225.00	nr	360.70
30–35 cm girth	340.00	11.00	211.75	27.13	340.00	nr	578.88
Tilia cordata 'Greenspire'; including backfilling with excavated material (other operations not included)							
16–18 cm girth	95.00	1.30	24.95	3.75	95.00	nr	123.70
18–20 cm girth	115.00	1.60	30.80	3.75	115.00	nr	149.55
20–25 cm girth	140.00	4.50	86.63	16.21	140.00	nr	242.84
25–30 cm girth; 5 × transplanted; 4.0–5.0 m tall	150.00	6.00	115.50	20.20	150.00	nr	285.70
30–35 cm girth; 5 × transplanted; 5.0–7.0 m tall	220.00	11.00	211.75	27.13	220.00	nr	458.88
Betula pendula (3 stems); including backfilling with excavated material (other operations not included)							
3.0–3.5 m high	60.00	1.98	38.12	41.62	60.00	nr	139.74
3.5–4.0 m high	90.00	1.60	30.80	44.72	90.00	nr	165.52
4.0–4.5 m high	130.00	2.38	45.74	49.94	130.00	nr	225.68
4.5–5.0 m high	150.00	2.97	57.17	76.50	150.00	nr	283.67
5.0–6.0 m high	210.00	3.96	76.23	83.24	210.00	nr	369.47
6.0–7.0 m high	350.00	4.50	86.63	97.58	350.00	nr	534.21
Pinus sylvestris; including backfilling with excavated material (other operations not included)							
3.0–3.5 m high	400.00	1.98	38.12	41.62	400.00	nr	479.74
3.5–4.0 m high	500.00	1.60	30.80	44.72	500.00	nr	575.52
4.0–4.5 m high	690.00	2.38	45.74	49.94	690.00	nr	785.68
4.5–5.0 m high	950.00	2.97	57.17	76.50	950.00	nr	1083.67
5.0–6.0 m high	1500.00	3.96	76.23	83.24	1500.00	nr	1659.47
6.0–7.0 m high	2500.00	4.50	86.63	100.72	2500.00	nr	2687.35

Q31 PLANTING

Item Excluding site overheads and profit	PC £	Labour hours	Labour £	Plant £	Material £	Unit	Total rate £
TREE PLANTING – cont							
Pleached trees; Carpinus betulus;							
specimen trees; Lorenz Von Ehren; supply							
and planting only; excavation, support and							
treepit additives not included							
Box shaped trees to provide 'floating hedge' or							
screen effect; planting distance 1 tree per m							
run; trunk and box alignment; wire root balls							
4 × transplanted; 20–25 cm; 1.00 m centres	300.00	5.00	96.25	16.21	300.00	m	**412.46**
5 × transplanted; 25–30 cm; 1.00 m centres	430.00	6.50	125.13	20.20	430.00	m	**575.33**
5 × transplanted; 30–35 cm; 1.20 m centres	360.00	11.00	211.75	27.13	360.00	m	**598.88**
Pleached trees; Tilia europaea 'Pallida';							
specimen trees; Lorenz Von Ehren; supply							
and planting only; excavation, support and							
treepit additives not included							
Box pleached trees to provide 'floating hedge'							
or screen effect; wire root balls							
4 × transplanted; 20–25 cm; 1.00 m centres	550.00	5.00	96.25	16.21	550.00	m	**662.46**
4 × transplanted; 25–30 cm; 1.00 m centres	900.00	5.00	96.25	16.21	900.00	m	**1012.46**
5 × transplanted; 30–35 cm; 1.20 m centres	249.99	5.00	96.25	16.21	249.99	m	**362.45**
5 × transplanted; 35–40 cm; 1.50 m centres	380.02	5.00	96.25	16.21	380.02	m	**492.48**
6 × transplanted; 40–45 cm; 1.80 m centres	444.48	5.00	96.25	16.21	444.48	m	**556.94**
6 × transplanted; 45–50 cm; 1.80 m centres	550.04	5.00	96.25	16.21	550.04	m	**662.50**
6 × transplanted; 50–60 cm; 2.00 m centres	650.00	5.00	96.25	16.21	650.00	m	**762.46**
Espalier pleached trees; specimen trees;							
Lorenz Von Ehren; supply and planting							
only; excavation, support and treepit							
additives not included							
Frame pleached trees; to provide floating							
screen effects; wire root balls							
Carpinus betulus; 20–25 cm; 1.00 m centres	430.00	5.00	96.25	16.21	430.00	m	**542.46**
Carpinus betulus; 25–30 cm; 1.00 m centres	600.00	5.00	96.25	16.21	600.00	m	**712.46**
Tilia europaea 'Pallida'; 20–25 cm; 1.00 m centres	420.00	5.00	96.25	16.21	420.00	m	**532.46**
Tilia europaea 'Pallida'; 25–30 cm; 1.00 m centres	550.00	5.00	96.25	16.21	550.00	m	**662.46**
Tilia europaea 'Pallida'; 30–35 cm; 1.20 m centres	658.31	5.00	96.25	16.21	658.31	m	**770.77**
Umbrella shaped pleaches							
Umbrella or roof pleached trees to provide							
umbrella effect; wire root balls							
Tilia euchlora; specimen; 4 × transplanted; 20–25 cm	393.90	5.00	96.25	16.21	393.90	each	**506.36**
Tilia euchlora; specimen; 5 × transplanted; 25–30 cm	350.00	5.00	96.25	16.21	350.00	each	**462.46**
Platanus acerifolia; clear stem; 4 × transplanted; 20–25 cm	390.00	5.00	96.25	16.21	390.00	each	**502.46**

Q31 PLANTING

Item Excluding site overheads and profit	PC £	Labour hours	Labour £	Plant £	Material £	Unit	Total rate £
Platanus acerifolia; specimen; 5 × transplanted; 30–35 cm	700.00	5.00	96.25	16.21	700.00	each	812.46
Tilia europaea; specimen; 5 × transplanted; 30–35 cm	690.00	5.00	96.25	16.21	690.00	each	802.46
Platanus acerifolia; specimen; 5 × transplanted; 25–30 cm	480.00	5.00	96.25	16.21	480.00	each	592.46
Tree planting; containerized trees; Lorenz **von Ehren; to the tree prices above add for** **Airpot containerization only**							
Tree size							
20–25 cm	–	–	–	–	45.45	nr	45.45
25–30 cm	–	–	–	–	75.75	nr	75.75
30–35 cm	–	–	–	–	80.80	nr	80.80
35–40 cm	–	–	–	–	131.30	nr	131.30
40–45 cm	–	–	–	–	151.50	nr	151.50
45–50 cm	–	–	–	–	191.90	nr	191.90
50–60 cm	–	–	–	–	252.50	nr	252.50
60–70 cm	–	–	–	–	323.20	nr	323.20
70–80 cm	–	–	–	–	393.90	nr	393.90
80–90 cm	–	–	–	–	454.50	nr	454.50
Tree planting; rootballed trees; **semi-mature and mature trees; Lorenz von** **Ehren; planting and back filling; planted by** **telehandler or by crane; delivery included;** **all other operations priced separately**							
Semi-mature trees							
40–45 cm girth	600.00	8.00	154.00	66.25	600.00	nr	820.25
45–50 cm girth	850.00	8.00	154.00	87.00	850.00	nr	1091.00
50–60 cm girth	1200.00	10.00	192.50	105.63	1200.00	nr	1498.13
60–70 cm girth	2200.00	15.00	288.75	222.11	2200.00	nr	2710.86
70–80 cm girth	3200.00	18.00	346.50	206.73	3200.00	nr	3753.23
80–90 cm girth	4800.00	18.00	346.50	206.73	4800.00	nr	5353.23
90–100 cm girth	5900.00	18.00	346.50	206.73	5900.00	nr	6453.23
SHRUB PLANTING							
Shrub planting – General Preamble: For preparation of planting areas see 'Cultivation' at the beginning of the section on planting.							
Shrub planting Setting out; selecting planting from holding area; loading to wheelbarrows; planting as plan or as directed; distance from holding area maximum 50 m; plants 2–3 litre containers							
single plants not grouped	–	0.04	0.77	–	–	nr	0.77
plants in groups of 3–5 nr	–	0.03	0.48	–	–	nr	0.48
plants in groups of 10–100 nr	–	0.02	0.32	–	–	nr	0.32
plants in groups of 100 nr minimum	–	0.01	0.22	–	–	nr	0.22

Q31 PLANTING

Item Excluding site overheads and profit	PC £	Labour hours	Labour £	Plant £	Material £	Unit	Total rate £
SHRUB PLANTING – cont							
Shrub planting – cont							
Forming planting holes; in cultivated ground (cultivating not included); by mechanical auger; trimming holes by hand; depositing excavated material alongside holes							
250 mm diameter	–	0.03	0.64	0.05	–	nr	0.69
250 × 250 mm	–	0.04	0.77	0.09	–	nr	0.86
300 × 300 mm	–	0.08	1.44	0.11	–	nr	1.55
Hand excavation; forming planting holes; in cultivated ground (cultivating not included); depositing excavated material alongside holes							
100 × 100 × 100 mm deep; with mattock or hoe	–	0.01	0.13	–	–	nr	0.13
250 × 250 × 300 mm deep	–	0.04	0.77	–	–	nr	0.77
300 × 300 × 300 mm deep	–	0.06	1.07	–	–	nr	1.07
400 × 400 × 400 mm deep	–	0.13	2.41	–	–	nr	2.41
500 × 500 × 500 mm deep	–	0.25	4.81	–	–	nr	4.81
600 × 600 × 600 mm deep	–	0.43	8.33	–	–	nr	8.33
900 × 900 × 600 mm deep	–	1.00	19.25	–	–	nr	19.25
1.00 m × 1.00 m × 600 mm deep	–	1.23	23.68	–	–	nr	23.68
1.25 m × 1.25 m × 600 mm deep	–	1.93	37.15	–	–	nr	37.15
Hand excavation; forming planting holes; in uncultivated ground; depositing excavated material alongside holes							
100 × 100 × 100 mm deep; with mattock or hoe	–	0.03	0.48	–	–	nr	0.48
250 × 250 × 300 mm deep	–	0.06	1.07	–	–	nr	1.07
300 × 300 × 300 mm deep	–	0.06	1.20	–	–	nr	1.20
400 × 400 × 400 mm deep	–	0.25	4.81	–	–	nr	4.81
500 × 500 × 500 mm deep	–	0.33	6.27	–	–	nr	6.27
600 × 600 × 600 mm deep	–	0.55	10.59	–	–	nr	10.59
900 × 900 × 600 mm deep	–	1.25	24.06	–	–	nr	24.06
1.00 m × 1.00 m × 600 mm deep	–	1.54	29.60	–	–	nr	29.60
1.25 m × 1.25 m × 600 mm deep	–	2.41	46.44	–	–	nr	46.44
Bare root planting; to planting holes (forming holes not included); including backfilling with excavated material (bare root plants not included)							
bare root 1+1; 30–90 mm high	–	0.02	0.32	–	–	nr	0.32
bare root 1+2; 90–120 mm high	–	0.02	0.32	–	–	nr	0.32
Containerized planting; to planting holes (forming holes not included); including backfilling with excavated material (shrub or ground cover not included)							
9 cm pot	–	0.01	0.19	–	–	nr	0.19
2 litre container	–	0.02	0.39	–	–	nr	0.39
3 litre container	–	0.02	0.43	–	–	nr	0.43

Q31 PLANTING

Item Excluding site overheads and profit	PC £	Labour hours	Labour £	Plant £	Material £	Unit	Total rate £
5 litre container	–	0.03	0.64	–	–	nr	**0.64**
10 litre container	–	0.05	0.96	–	–	nr	**0.96**
15 litre container	–	0.07	1.28	–	–	nr	**1.28**
20 litre container	–	0.08	1.60	–	–	nr	**1.60**
Shrub planting; 2 litre containerized plants; in cultivated ground (cultivating not included); PC £2.90/nr							
average 2 plants per m²	–	0.06	1.08	–	5.80	m²	**6.88**
average 3 plants per m²	–	0.08	1.62	–	8.70	m²	**10.32**
average 4 plants per m²	–	0.11	2.16	–	11.60	m²	**13.76**
average 6 plants per m²	–	0.17	3.24	–	17.40	m²	**20.64**
Shrub planting; 3 litre containerized plants; in cultivated ground (cultivating not included); PC £3.40/nr							
average 2 plants per m²	–	0.07	1.35	–	6.80	m²	**8.15**
average 3 plants per m²	–	0.11	2.02	–	10.20	m²	**12.22**
average 4 plants per m²	–	0.14	2.70	–	13.60	m²	**16.30**
average 6 plants per m²	–	0.21	4.04	–	20.40	m²	**24.44**
Extra over shrubs for stakes	0.70	0.02	0.40	–	0.70	nr	**1.10**
HEDGE PLANTING							
Hedges							
Excavating trench for hedges; depositing soil alongside trench; by machine							
300 mm deep × 300 mm wide	–	0.03	1.17	1.39	–	m	**2.56**
300 mm deep × 600 mm wide	–	0.05	1.75	2.11	–	m	**3.86**
300 mm deep × 450 mm wide	–	0.05	1.73	2.08	–	m	**3.81**
500 mm deep × 500 mm wide	–	0.06	2.41	2.90	–	m	**5.31**
500 mm deep × 700 mm wide	–	0.09	3.38	4.06	–	m	**7.44**
600 mm deep × 750 mm wide	–	0.11	4.34	5.22	–	m	**9.56**
750 mm deep × 900 mm wide	–	0.17	6.51	7.83	–	m	**14.34**
Excavating trench for hedges; depositing soil alongside trench; by hand							
300 mm deep × 300 mm wide	–	0.08	4.81	–	–	m	**4.81**
300 mm deep × 450 mm wide	–	0.13	7.22	–	–	m	**7.22**
300 mm deep × 600 mm wide	–	0.17	9.63	–	–	m	**9.63**
500 mm deep × 500 mm wide	–	0.23	13.06	–	–	m	**13.06**
500 mm deep × 700 mm wide	–	0.32	18.31	–	–	m	**18.31**
600 mm deep × 750 mm wide	–	0.41	23.54	–	–	m	**23.54**
750 mm deep × 900 mm wide	–	0.61	35.24	–	–	m	**35.24**
Setting out; notching out; excavating trench; breaking up subsoil to minimum depth 300 mm							
minimum 400 mm deep	–	0.25	4.81	–	–	m	**4.81**
Disposal of excavated soil to stockpile							
Up to 25 m distant							
By hand	–	2.40	46.20	–	–	m³	**46.20**
By machine	–	–	–	2.87	–	m³	**2.87**

Q31 PLANTING

Item Excluding site overheads and profit	PC £	Labour hours	Labour £	Plant £	Material £	Unit	Total rate £
HEDGE PLANTING – cont							
Instant Hedge planting; Practicality Brown Ltd – Elveden Hedges; continuous strips of instant hedges; delivered in 2.50 m lengths; planting and backfilling with excavated material of hedging plants to trenches (not included); including minor trimming and shaping after planting							
Beech (Fagus) hedges							
height 1.40/1.60 m × 400 mm wide	149.00	0.20	7.70	6.35	149.00	m	163.05
height 1.60/1.80 m × 500 mm wide	162.00	0.22	8.55	7.05	162.00	m	177.60
Yew (Taxus) hedges							
height 1.40/1.60 m × 400 mm wide	189.00	0.20	7.70	6.35	189.00	m	203.05
height 1.60/1.80 m × 500 mm wide	215.00	0.22	8.55	7.05	215.00	m	230.60
height 1.80/2.0 m × 500 mm wide	238.00	0.25	9.63	7.94	238.00	m	255.57
Prunus Lauruscerasus hedges							
height 1.40/1.60 m × 400 mm wide	147.00	0.20	7.70	6.35	147.00	m	161.05
height 1.60/1.80 m × 500 mm wide	160.00	0.22	8.55	7.05	160.00	m	175.60
height 1.80/2.00 m × 500 mm wide	173.00	0.25	9.63	7.94	173.00	m	190.57
Box (Buxus) hedges							
height 0.80/1.00 m × 300 mm wide	135.00	0.10	3.85	3.17	135.00	m	142.02
height 1.00/1.20 m × 300 mm wide	148.00	0.13	4.81	3.97	148.00	m	156.78
Instant Hedge planting; Practicality Brown Ltd; individual pre-clipped hedge plants to form mature hedge; planting and backfilling of hedging plants to trenches excavated separately; including minor trimming and shaping after planting							
Beech and Hornbeam hedges							
1.50 m high × 500 mm wide; 500 mm centres	220.00	0.50	28.88	2.73	220.00	m	251.61
1.50 m high × 500 mm wide; 750 mm centres	146.30	0.19	10.86	2.73	146.30	m	159.89
2.00 m high × 500 mm wide; 500 mm centres	304.00	0.57	33.00	2.73	304.00	m	339.73
2.00 m high × 500 mm wide; 750 mm centres	202.16	0.57	33.00	2.73	202.16	m	237.89
Yew (Taxus) hedges							
1.50 m high × 500 mm wide; 500 mm centres	300.00	0.50	28.88	2.73	300.00	m	331.61
1.50 m high × 500 mm wide; 750 mm centres	199.50	0.19	10.86	2.73	199.50	m	213.09
1.75 m high × 500 mm wide; 500 mm centres	340.00	0.57	33.00	2.73	340.00	m	375.73

Q31 PLANTING

Item Excluding site overheads and profit	PC £	Labour hours	Labour £	Plant £	Material £	Unit	Total rate £
1.75 m high × 500 mm wide; 750 mm centres	226.10	0.21	12.16	2.73	226.10	m	**240.99**
2.00 m high × 500 mm wide; 500 mm centres	400.00	0.67	38.50	2.73	400.00	m	**441.23**
2.00 m high × 500 mm wide; 750 mm centres	266.00	0.57	33.00	2.73	266.00	m	**301.73**
2.00 m high × 500 mm wide; 900 mm centres	222.00	0.27	15.61	2.73	222.00	m	**240.34**
Feathered hedges; Beech or Hornbeam							
1.75 m high × 300 mm wide; 400 mm centres	75.00	0.20	11.55	1.36	75.00	m	**87.91**
1.75 m high × 300 mm wide; 500 mm centres	60.00	0.16	9.24	1.36	60.00	m	**70.60**
1.75 m high × 300 mm wide; 750 mm centres	42.60	0.11	6.60	1.36	42.60	m	**50.56**
1.75 m high × 300 mm wide; 900 mm centres	33.30	0.09	5.13	1.36	33.30	m	**39.79**
1.75–2.00 m high × 300 mm wide; 600 mm centres	50.00	0.33	19.25	20.33	50.00	m	**89.58**
1.75–2.00 m high × 300 mm wide; 900 mm centres	33.30	0.29	16.50	17.68	33.30	m	**67.48**
2.00–2.25 m high 300 wide; 600 mm centres	61.13	0.40	23.10	24.48	61.13	m	**108.71**
2.00–2.25 m high × 300 mm wide; 900 mm centres	40.34	0.33	19.25	20.33	40.34	m	**79.92**
2.00–2.25 m high × 300 mm wide; 1.20 m centres	30.56	0.25	14.44	15.70	30.56	m	**60.70**
Hedge planting; Griffin Nurseries Ltd; feathered hedge plants to form mature hedge; planting and backfilling of hedging plants to trenches excavated separately; including minor trimming and shaping after planting							
(Yew) Taxus hedging; root balled							
600–800 mm high × 300 mm wide; 250 mm centres	50.60	0.20	11.55	1.36	50.60	m	**63.51**
600–800 mm high × 300 mm wide; 500 mm centres	25.30	0.11	6.42	1.36	25.30	m	**33.08**
800 mm–1.00 m high × 300 mm wide; 250 mm centres	59.40	0.20	11.55	1.36	59.40	m	**72.31**
800 mm–1.00 m high × 300 mm wide; 500 mm centres	29.70	0.11	6.42	1.36	29.70	m	**37.48**
1.00–1.20 m high × 300/350 mm wide; 3 nr/m	74.92	0.17	9.63	1.36	74.92	m	**85.91**
1.00–1.20 m high × 300/350 mm wide; 2 nr/m	45.00	0.13	7.22	1.36	45.00	m	**53.58**
1.25–1.50 m high × 300/350 mm wide; 3 nr/m	111.39	0.17	9.63	1.36	111.39	m	**122.38**

Q31 PLANTING

Item Excluding site overheads and profit	PC £	Labour hours	Labour £	Plant £	Material £	Unit	Total rate £
HEDGE PLANTING – cont							
Hedge planting – cont							
(Yew) Taxus hedging – cont							
1.25–1.50 m high × 300/350 mm wide;							
2 nr/m	66.90	0.13	7.70	1.36	66.90	m	**75.96**
1.50–1.75 m high × 400 mm wide; 450 mm							
centres	124.32	0.33	19.25	15.25	124.32	m	**158.82**
1.50–1.75 m high × 400 mm wide; 600 mm							
centres	93.34	0.22	12.83	13.70	93.34	m	**119.87**
1.50–1.75 m high × 400 mm wide; 900 mm							
centres	62.16	0.20	11.55	12.47	62.16	m	**86.18**
1.75–2.00 m high × 500 mm wide; 600 mm							
centres	123.75	0.33	19.25	20.33	123.75	m	**163.33**
1.75–2.00 m high × 500 mm wide; 900 mm							
centres	101.01	0.29	16.50	17.68	183.43	m	**217.61**
1.75–2.00 m high × 500 mm wide; 1.20 m							
centres	61.87	0.25	14.44	15.70	61.87	m	**92.01**
2.00–2.25 m high × 600 mm wide; 600 mm							
centres	151.67	0.40	23.10	24.48	151.67	m	**199.25**
2.00–2.25 m high × 600 mm wide; 900 mm							
centres	101.01	0.33	19.25	20.33	101.01	m	**140.59**
2.00–2.25 m high × 600 mm wide; 1.20 m							
centres	75.83	0.31	17.77	18.91	75.83	m	**112.51**
(Yew) Taxus hedging; wire root balled							
1.50–1.75 m high × 400 mm wide; 450 mm							
centres	193.14	0.33	19.25	15.25	193.14	m	**227.64**
1.50–1.75 m high × 400 mm wide; 600 mm							
centres	145.00	0.22	12.83	13.70	145.00	m	**171.53**
1.50–1.75 m high × 400 mm wide; 900 mm							
centres	95.70	0.20	11.55	12.47	95.70	m	**119.72**
1.75–2.00 m high × 600 mm wide; 600 mm							
centres	169.62	0.33	19.25	20.33	169.62	m	**209.20**
1.75–2.00 m high × 600 mm wide; 900 mm							
centres	112.94	0.29	16.50	17.68	112.94	m	**147.12**
1.75–2.00 m high × 600 mm wide; 1.20 m							
centres	84.79	0.25	14.44	15.70	84.79	m	**114.93**
2.00–2.25 m high × 650 mm wide; 700 mm							
centres	163.30	0.44	25.66	26.95	163.30	m	**215.91**
2.00–2.25 m high × 600 mm wide; 900 mm							
centres	127.65	0.36	21.00	22.01	127.65	m	**170.66**
2.00–2.25 m high × 600 mm wide; 1.20 m							
centres	75.83	0.31	17.77	18.91	75.83	m	**112.51**
Buxus (Box) hedges; rootballed							
400–500 mm high; 200 mm rootball;							
200 mm centres	33.25	0.09	1.71	–	33.25	m	**34.96**
400–500 mm high; 200 mm rootball;							
300 mm centres	22.14	0.07	1.29	–	22.14	m	**23.43**
400–500 mm high; 200 mm rootball;							
500 mm centres	13.30	0.05	0.96	–	13.30	m	**14.26**

Q31 PLANTING

Item Excluding site overheads and profit	PC £	Labour hours	Labour £	Plant £	Material £	Unit	Total rate £
500–600 mm high; 200 mm rootball; 200 mm centres	52.75	0.09	1.71	–	52.75	m	**54.46**
500–600 mm high; 200 mm rootball; 300 mm centres	35.13	0.07	1.29	–	35.13	m	**36.42**
500–600 mm high; 200 mm rootball; 500 mm centres	21.10	0.08	1.48	–	21.10	m	**22.58**
600–800 mm high; 250 mm rootball; 250 mm centres	–	0.09	1.75	–	53.20	m	**54.95**
600–800 mm high; 250 mm rootball; 400 mm centres	33.25	0.08	1.60	–	33.25	m	**34.85**
600–800 mm high; 250 mm rootball; 600 mm centres	17.58	0.08	1.48	–	17.58	m	**19.06**
800 mm–1.00 m high; 300 mm rootball; 300 mm centres	71.43	0.10	1.93	–	71.43	m	**73.36**
800 mm–1.00 m high; 300 mm rootball; 500 mm centres	42.90	0.11	2.14	–	42.90	m	**45.04**
800 mm–1.00 m high; 300 mm rootball; 750 mm centres	28.53	0.13	2.41	–	28.53	m	**30.94**
1.00 mm–1.25 m high; 350 mm rootball; 400 mm centres	74.13	0.13	2.41	–	74.13	m	**76.54**
1.00 mm–1.25 m high; 350 mm rootball; 600 mm centres	49.43	0.11	2.14	–	49.43	m	**51.57**
1.00 mm–1.25 m high; 350 mm rootball; 900 mm centres	32.91	0.10	1.93	–	32.91	m	**34.84**
Native species and bare root hedge planting Hedge planting operations only (excavation of trenches priced separately) including backfill with excavated topsoil; PC £0.48/nr							
single row; 200 mm centres	2.40	0.06	1.20	–	2.40	m	**3.60**
single row; 300 mm centres	1.60	0.06	1.07	–	1.60	m	**2.67**
single row; 400 mm centres	1.20	0.04	0.80	–	1.20	m	**2.00**
single row; 500 mm centres	0.96	0.03	0.64	–	0.96	m	**1.60**
double row; 200 mm centres	4.80	0.17	3.21	–	4.80	m	**8.01**
double row; 300 mm centres	3.20	0.13	2.57	–	3.20	m	**5.77**
double row; 400 mm centres	2.40	0.08	1.60	–	2.40	m	**4.00**
double row; 500 mm centres	1.92	0.07	1.28	–	1.92	m	**3.20**
Extra over hedges for incorporating manure; at 1 m³ per 30 m	0.77	0.03	0.48	–	0.77	m	**1.25**

Q31 PLANTING

Item Excluding site overheads and profit	PC £	Labour hours	Labour £	Plant £	Material £	Unit	Total rate £
TOPIARY							
Clipped topiary; Lorenz von Ehren; German field grown clipped and transplanted as detailed; planted to plant pit; including backfilling with excavated material and TPMC							
Buxus sempervirens (box); balls							
300 mm diameter; 3 × transplanted; container grown or rootballed	15.15	0.25	4.81	–	24.10	nr	**28.91**
500 mm diameter; 4 × transplanted; wire rootballed	45.45	1.50	28.88	4.45	113.58	nr	**146.91**
900 mm diameter; 5 × transplanted; wire rootballed	207.05	2.75	52.94	4.45	320.07	nr	**377.46**
1300 mm diameter; 6 × transplanted; wire rootballed	656.50	3.10	59.67	5.34	767.77	nr	**832.78**
Buxus sempervirens (box); pyramids							
500 mm high; 3 × transplanted; container grown or rootballed	30.30	1.50	28.88	4.45	98.43	nr	**131.76**
900 mm high; 4 × transplanted; wire rootballed	95.95	2.75	52.94	4.45	208.97	nr	**266.36**
1300 mm high; 5 × transplanted; wire rootballed	393.90	3.10	59.67	5.34	505.17	nr	**570.18**
Buxus sempervirens (box); truncated pyramids							
500 mm high; 3 × transplanted; container grown or rootballed	80.79	1.50	28.88	4.45	148.92	nr	**182.25**
900 mm high; 4 × transplanted; wire rootballed	302.97	2.75	52.94	4.45	415.99	nr	**473.38**
Buxus sempervirens (box); truncated cone							
1300 mm high; 5 × transplanted; wire rootballed	959.50	3.10	59.67	5.34	1070.77	nr	**1135.78**
Buxus sempervirens (box); cubes							
500 mm square; 4 × transplanted; rootballed	80.80	1.50	28.88	4.45	148.93	nr	**182.26**
900 mm square; 5 × transplanted; wire rootballed	363.60	2.75	52.94	4.45	476.62	nr	**534.01**
Taxus baccata (yew); balls							
500 mm diameter; 4 × transplanted; rootballed	55.54	1.50	28.88	4.45	123.67	nr	**157.00**
900 mm diameter; 5 × transplanted; wire rootballed	181.78	2.75	52.94	4.45	294.80	nr	**352.19**
1300 mm diameter; 6 × transplanted; wire rootballed	565.54	3.10	59.67	5.34	676.81	nr	**741.82**
Taxus baccata (yew); cones							
800 mm high; 4 × transplanted; wire rootballed	55.54	1.50	28.88	4.45	123.67	nr	**157.00**
1500 mm high; 5 × transplanted; wire rootballed	141.39	2.00	38.50	4.45	225.68	nr	**268.63**
2500 mm high; 6 × transplanted; wire rootballed	474.65	3.10	59.67	5.34	585.92	nr	**650.93**

Q31 PLANTING

Item Excluding site overheads and profit	PC £	Labour hours	Labour £	Plant £	Material £	Unit	Total rate £
Taxus baccata (yew); cubes							
500 mm square; 4 × transplanted; rootballed	65.64	1.50	28.88	4.45	133.77	nr	167.10
900 mm square; 5 × transplanted; wire rootballed	242.38	2.75	52.94	4.45	355.40	nr	412.79
Taxus baccata (yew); pyramids							
900 mm high; 4 × transplanted; wire rootballed	121.19	1.50	28.88	4.45	189.32	nr	222.65
1500 mm high; 5 × transplanted; wire rootballed	242.38	3.10	59.67	5.34	353.65	nr	418.66
2500 mm high; 6 × transplanted; wire rootballed	696.83	4.00	77.00	8.91	838.42	nr	924.33
Carpinus betulus (common hornbeam); columns; round base							
800 mm wide × 2000 mm high; 4 × transplanted; wire rootballed	151.49	3.10	59.67	5.34	262.76	nr	327.77
800 mm wide × 2750 mm high; 4 × transplanted; wire rootballed	302.97	4.00	77.00	8.91	444.56	nr	530.47
Carpinus betulus (common hornbeam); cones							
3 m high; 5 × transplanted; wire rootballed	272.67	4.00	77.00	8.91	414.26	nr	500.17
4 m high; 5 × transplanted; wire rootballed	565.54	5.00	96.25	8.91	734.12	nr	839.28
Carpinus betulus 'Fastigiata'; pyramids							
4 m high; 6 × transplanted; wire rootballed	969.50	5.00	96.25	8.91	1138.08	nr	1243.24
7 m high; 7 × transplanted; wire rootballed	2625.74	7.50	144.38	13.36	2834.71	nr	2992.45
LIVING WALLS							
ANS Living Wall; soil-based system; Scotscape Ltd; self-contained, irrigated vertical planting system; HDPE Modules, fully-established upon installation; saturated weight 76 kg per m²; access costs excluded							
ANS System units 250 × 500 × 100 mm overall; fixed to horizontal hanging rails on softwood batten and DPM; planting depth 150 mm at 30° angle to vertical surface; inclusive of automatic dripline irrigation system and guttering with running outlets to drain (not included)							
areas above 35 m²	–	–	–	–	–	m²	562.00
areas below 35 m²	–	–	–	–	–	m²	669.50
Fully automatic irrigation systems to the above; comprising breaktank, station controller, solenoid valves, plant feeders and dripline emitters							
areas above 35–60 m²	–	–	–	–	–	nr	4550.00
areas below 35 m²	–	–	–	–	–	nr	1560.00

Q31 PLANTING

Item Excluding site overheads and profit	PC £	Labour hours	Labour £	Plant £	Material £	Unit	Total rate £
LIVING WALLS – cont							
Biotecture Living Wall; Scotscape Ltd;							
hydroponic based system; self-contained,							
vertical planting system; for internal or							
load restricted installations							
Biotecture; 73 mm thick backless cladding							
panels on 18 mm Eco sheet backing board fixt							
to timber battens; 4.5 mm void former of							
geocomposite root barrier and waterproof							
backing sheet; all with stainless steel fixings;							
drainage and guttering							
areas above 35 m²	–	–	–	–	–	m²	457.62
areas below 35 m²	–	–	–	–	–	m²	530.80
Fully automatic irrigation systems to the							
above; comprising breaktank, submersible							
pump; filter, regulator station controller,							
solenoid valves, plant feeders and dripline							
emitters							
areas above 35–60 m²	–	–	–	–	–	nr	7462.50
areas below 35 m²	–	–	–	–	–	nr	2611.00
Add to the above for all living walls							
preliminary costs	–	–	–	–	–	nr	350.00
design costs; site visits, consultations and							
presentations	–	–	–	–	–	nr	345.00
Annual maintenance to living walls; regular							
visits to maintain planting and systems;							
inclusive of feeding; pest control and							
calibration of irrigation systems							
areas 35–60 m²	–	–	–	–	–	nr	3663.00
areas below 35 m²	–	–	–	–	–	nr	1898.00
Accessories to living walls							
planters to base of living wall in GRP;							
black; inclusive of drainage and planting	–	–	–	–	–	m	308.75
Green Screen; Mobilane Ltd							
Fully installed vertical green screen of Hedera							
hibernica; 65 plants per linear m; all on 5 mm							
galvanized steel weldmesh; attached to							
existing wall fence or hoarding							
screen 1.80 high	–	–	–	–	170.00	m	170.00
screen 2.20 m high	–	–	–	–	230.00	m	230.00
screen 4.00 m high	–	–	–	–	500.00	m	500.00

Q31 PLANTING

Item Excluding site overheads and profit	PC £	Labour hours	Labour £	Plant £	Material £	Unit	Total rate £
HERBACEOUS/GROUNDCOVER/BULB/ **BEDDING PLANTING**							
Herbaceous and groundcover planting							
Herbaceous plants; PC £1.20/nr; including							
forming planting holes in cultivated ground							
(cultivating not included); backfilling with							
excavated material; 1 litre containers							
average 4 plants per m²; 500 mm centres	5.60	0.09	1.80	–	5.60	m²	**7.40**
average 6 plants per m²; 408 mm centres	8.40	0.14	2.70	–	8.40	m²	**11.10**
average 8 plants per m²; 354 mm centres	11.20	0.19	3.60	–	11.20	m²	**14.80**
Note: For machine incorporation of fertilizers							
and soil conditioners see 'Cultivation'.							
Plant support netting; Bridport Gundry; on							
50 mm diameter stakes; 750 mm long; driving							
into ground at 1.50 m centres							
green extruded plastic mesh; 125 mm							
square	0.53	0.04	0.77	–	0.53	m²	**1.30**
Bulb planting							
Bulbs; including forming planting holes in							
cultivated area (cultivating not included);							
backfilling with excavated material							
small	13.00	0.83	16.04	–	13.00	100 nr	**29.04**
medium	22.00	0.83	16.04	–	22.00	100 nr	**38.04**
large	25.00	0.91	17.50	–	25.00	100 nr	**42.50**
Bulbs; in grassed area; using bulb planter;							
including backfilling with screened topsoil or							
peat and cut turf plug							
small	13.00	1.67	32.08	–	13.00	100 nr	**45.08**
medium	22.00	1.67	32.08	–	22.00	100 nr	**54.08**
large	25.00	2.00	38.50	–	25.00	100 nr	**63.50**
Aquatic planting							
Aquatic plants; in prepared growing medium in							
pool; plant size 2–3 litre containerized (plants							
not included)	–	0.04	0.77	–	–	nr	**0.77**
MARKET PRICES OF NATIVE SPECIES							
Oakover Nurseries Ltd; the following prices							
are typically some of the more popular							
species used in native plantations; prices							
vary between species; readers should							
check the catalogue of the supplier							
1+1; 1 year seedling transplanted and grown							
for a year							
30–40 cm	–	–	–	–	0.48	nr	**0.48**
40–60 cm	–	–	–	–	0.55	nr	**0.55**

Q31 PLANTING

Item Excluding site overheads and profit	PC £	Labour hours	Labour £	Plant £	Material £	Unit	Total rate £
MARKET PRICES OF NATIVE SPECIES – **cont**							
Oakover Nurseries Ltd – cont							
1+2; 1 year seedling transplanted and grown for 2 years							
60–80 cm	–	–	–	–	0.65	nr	**0.65**
Whips							
80–100 cm	–	–	–	–	0.90	nr	**0.90**
100–125 cm	–	–	–	–	1.40	nr	**1.40**
Feathered trees							
125–150 cm	–	–	–	–	2.40	nr	**2.40**
150–180 cm	–	–	–	–	2.70	nr	**2.70**
180–200 cm	–	–	–	–	5.50	nr	**5.50**
200–250 cm	–	–	–	–	8.00	nr	**8.00**
250–300 cm	–	–	–	–	8.93	nr	**8.93**
Native species planting							
Forming planting holes; in cultivated ground (cultivating not included); by mechanical auger; trimming holes by hand; depositing excavated material alongside holes							
250 mm diameter	–	0.03	0.64	0.05	–	nr	**0.69**
250 × 250 mm	–	0.04	0.77	0.09	–	nr	**0.86**
300 × 300 mm	–	0.08	1.44	0.11	–	nr	**1.55**
Hand excavation; forming planting holes; in cultivated ground (cultivating not included); depositing excavated material alongside holes							
100 × 100 × 100 mm deep; with mattock or hoe	–	0.01	0.13	–	–	nr	**0.13**
250 × 250 × 300 mm deep	–	0.04	0.77	–	–	nr	**0.77**
300 × 300 × 300 mm deep	–	0.06	1.07	–	–	nr	**1.07**
TREE/SHRUB PROTECTION							
Tree planting; tree protection – General							
Preamble: Care must be taken to ensure that tree grids and guards are removed when trees grow beyond the specified diameter of guard.							
Tree planting; tree protection							
Tree tube; olive green							
1200 mm high × 80 × 80 mm stakes; 1500 mm high for Crowders Tree Tube; driving into ground	1.15	0.07	1.28	–	1.15	nr	**2.43**
	0.70	0.05	0.96	–	0.70	nr	**1.66**
Expandable plastic tree guards; including 25 mm softwood stakes							
500 mm high	0.91	0.17	3.21	–	0.91	nr	**4.12**
1.00 m high	0.93	0.17	3.21	–	0.93	nr	**4.14**

Q31 PLANTING

Item Excluding site overheads and profit	PC £	Labour hours	Labour £	Plant £	Material £	Unit	Total rate £
English Woodlands; Weldmesh tree guards; nailing to tree stakes (tree stakes not included)							
1800 mm high × 200 mm diameter	15.50	0.33	6.41	–	15.50	nr	21.91
1800 mm high × 250 mm diameter	16.50	0.25	4.81	–	16.50	nr	21.31
1800 mm high × 300 mm diameter	17.50	0.33	6.41	–	17.50	nr	23.91
J. Toms Ltd; spiral rabbit guards; clear or brown							
450 × 38 mm	0.20	0.03	0.64	–	0.20	nr	0.84
610 × 38 mm	0.22	0.03	0.64	–	0.22	nr	0.86
English Woodlands; Plastic Mesh Tree Guards; black; supplied in 50 m rolls							
13 × 13 mm small mesh; roll width 60 cm	0.99	0.04	0.69	–	1.69	nr	2.38
13 × 13 mm small mesh; roll width 120 cm	1.96	0.06	1.13	–	2.66	nr	3.79
English Woodlands; Plastic Mesh Tree Guard in pre-cut pieces; 600 × 150 mm; supplied flat packed	0.86	0.06	1.13	–	1.56	nr	2.69
Tree guards of 3 nr 2.40 m × 100 mm stakes; driving 600 mm into firm ground; bracing with timber braces at top and bottom; including 3 strands barbed wire	0.48	1.00	19.25	–	0.48	nr	19.73
English Woodlands; strimmer guard in heavy duty black plastic; 225 mm high	2.86	0.07	1.28	–	2.86	nr	4.14
Tubex Ltd; Standard Treeshelter inclusive of 25 mm stake; prices shown for quantities of 500 nr							
0.6 m high	0.65	0.05	0.96	–	0.87	each	1.83
0.75 m high	0.77	0.05	0.96	–	1.07	each	2.03
1.2 m high	1.09	0.05	0.96	–	1.45	each	2.41
1.5 m high	1.49	0.05	0.96	–	1.92	each	2.88
Tubex Ltd; Shrubshelter inclusive of 25 mm stake; prices shown for quantities of 500 nr							
Ecostart shelter for forestry transplants and seedlings	0.75	0.07	1.28	–	0.75	nr	2.03
0.6 m high	1.47	0.07	1.28	–	1.69	nr	2.97
0.75 m high	1.61	0.07	1.28	–	1.87	nr	3.15
Extra over trees for spraying with antidesiccant spray; Wiltpruf							
selected standards; standards; light standards	2.62	0.20	3.85	–	2.62	nr	6.47
standards; heavy standards	4.36	0.25	4.81	–	4.36	nr	9.17
FORESTRY PLANTING							
Forestry planting							
Deep ploughing rough ground to form planting ridges at							
2.00 m centres	–	0.63	12.03	13.47	–	100 m²	25.50
3.00 m centres	–	0.59	11.32	12.68	–	100 m²	24.00
4.00 m centres	–	0.40	7.70	8.62	–	100 m²	16.32

Q31 PLANTING

Item Excluding site overheads and profit	PC £	Labour hours	Labour £	Plant £	Material £	Unit	Total rate £
FORESTRY PLANTING – cont							
Forestry planting – cont							
Notching plant forestry seedlings; T or L notch	41.00	0.75	14.44	–	41.00	100 nr	**55.44**
Turf planting forestry seedlings	41.00	2.00	38.50	–	41.00	100 nr	**79.50**
Tree tubes; to young trees	185.00	0.30	5.78	–	185.00	100 nr	**190.78**
Cleaning and weeding around seedlings; once	–	0.50	9.63	–	–	100 nr	**9.63**
Treading in and firming ground around seedlings planted; at 2500 per ha after frost or other ground disturbance; once	–	0.33	6.42	–	–	100 nr	**6.42**
Beating up initial planting; once (including supply of replacement seedlings at 10% of original planting)	4.10	0.25	4.81	–	4.10	100 nr	**8.91**
OPERATIONS AFTER PLANTING							
Operations after planting							
Initial cutting back to shrubs and hedge plants; including disposal of all cuttings	–	1.00	19.25	–	–	100 m²	**19.25**
Mulch; Melcourt Industries Ltd; Bark Nuggets®; to plant beds; delivered in 80 m³ loads; maximum distance 25 m							
50 mm thick	2.19	0.04	0.85	–	2.19	m²	**3.04**
75 mm thick	3.29	0.07	1.28	–	3.29	m²	**4.57**
100 mm thick	4.38	0.09	1.71	–	4.38	m²	**6.09**
Mulch; Melcourt Industries Ltd; Bark Nuggets®; to plant beds; delivered in 25 m³ loads; maximum distance 25 m							
50 mm thick	2.55	0.03	0.56	–	2.55	m²	**3.11**
75 mm thick	4.31	0.07	1.28	–	4.31	m²	**5.59**
100 mm thick	5.74	0.09	1.71	–	5.74	m²	**7.45**
Mulch; Melcourt Industries Ltd; Amenity Bark Mulch FSC; to plant beds; delivered in 80 m³ loads; maximum distance 25 m							
50 mm thick	1.40	0.04	0.85	–	1.40	m²	**2.25**
75 mm thick	2.10	0.07	1.28	–	2.10	m²	**3.38**
100 mm thick	2.80	0.09	1.71	–	2.80	m²	**4.51**
Mulch; Melcourt Industries Ltd; Amenity Bark Mulch FSC; to plant beds; delivered in 25 m³ loads; maximum distance 25 m							
50 mm thick	2.08	0.04	0.85	–	2.08	m²	**2.93**
75 mm thick	3.12	0.07	1.28	–	3.12	m²	**4.40**
100 mm thick	4.16	0.09	1.71	–	4.16	m²	**5.87**

Q31 PLANTING

Item Excluding site overheads and profit	PC £	Labour hours	Labour £	Plant £	Material £	Unit	Total rate £
Mulch mats; English Woodlands; lay mulch mat to planted area or plant station; mat secured with metal J pins 240 × 3 mm							
Mats to individual plants and plant stations							
Hemcore Biodegradable; 50 × 50 cm; square mat	0.90	0.05	0.96	–	1.30	each	**2.26**
Woven Polypropylene; 50 × 50 cm; square mat	0.21	0.05	0.96	–	0.61	each	**1.57**
Woven Polypropylene; 1 × 1 m; square mat	0.78	0.07	1.28	–	1.18	each	**2.46**
Hedging mats							
Woven Polypropylene; 1 × 100 m roll; hedge planting	0.47	0.01	0.24	–	0.67	m²	**0.91**
General planting areas							
Plantex membrane; 1 × 14 m roll; weed control	1.29	0.01	0.19	–	1.49	m²	**1.68**
Fertilizers; application during aftercare period							
Fertilizers; in top 150 mm of topsoil at 35 g/m²							
Mascot Microfine; turf fertilizer; 8+0+6 + 2% Mg + 4% Fe	5.64	0.12	2.36	–	5.64	100 m²	**8.00**
Enmag; controlled release fertilizer (8–9 months); 11+22+09	8.68	0.12	2.36	–	8.68	100 m²	**11.04**
Mascot Outfield; turf fertilizer; 8+12+8	4.52	0.12	2.36	–	4.52	100 m²	**6.88**
Mascot Outfield; turf fertilizer; 9+5+5	3.67	0.12	2.36	–	3.67	100 m²	**6.03**
Fertilizers; in top 150 mm of topsoil at 70 g/m²							
Mascot Microfine; turf fertilizer; 8+0+6 + 2% Mg + 4% Fe	11.29	0.12	2.36	–	11.29	100 m²	**13.65**
Enmag; controlled release fertilizer (8–9 months); 11+22+09	17.37	0.12	2.36	–	17.37	100 m²	**19.73**
Mascot Outfield; turf fertilizer; 8+12+8	9.05	0.12	2.36	–	9.05	100 m²	**11.41**
Mascot Outfield; turf fertilizer; 9+5+5	7.33	0.12	2.36	–	7.33	100 m²	**9.69**
AFTERCARE AS PART OF A LANDSCAPE CONTRACT							
Maintenance operations (Note: the following rates apply to aftercare maintenance executed as part of a landscaping contract only – Please see section Q35 of this publication for further maintenance rates)							
Weeding and hand forking planted areas; including disposing weeds and debris on site; areas maintained weekly	–	–	0.08	–	–	m²	**0.08**
Weeding and hand forking planted areas; including disposing weeds and debris on site; areas maintained monthly	–	0.01	0.19	–	–	m²	**0.19**

Q31 PLANTING

Item Excluding site overheads and profit	PC £	Labour hours	Labour £	Plant £	Material £	Unit	Total rate £
AFTERCARE AS PART OF A LANDSCAPE CONTRACT – cont							
Maintenance operations (Note: the following rates apply to aftercare maintenance executed as part of a landscaping contract only – Please see section Q35 of this publication for further maintenance rates) – cont							
Extra over weeding and hand forking planted areas for disposing excavated material off site; to tip not exceeding 13 km; mechanically loaded							
slightly contaminated	–	–	–	2.06	24.00	m³	**26.06**
rubbish	–	–	–	2.06	24.00	m³	**26.06**
inert material	–	–	–	1.37	17.14	m³	**18.51**
Mulch; Melcourt Industries Ltd; Bark Nuggets®; to plant beds; delivered in 80 m³ loads; maximum distance 25 m							
50 mm thick	2.19	0.04	0.85	–	2.19	m²	**3.04**
75 mm thick	3.29	0.07	1.28	–	3.29	m²	**4.57**
100 mm thick	4.38	0.09	1.71	–	4.38	m²	**6.09**
Mulch; Melcourt Industries Ltd; Bark Nuggets®; to plant beds; delivered in 25 m³ loads; maximum distance 25 m							
50 mm thick	2.87	0.04	0.85	–	2.87	m²	**3.72**
75 mm thick	4.31	0.07	1.28	–	4.31	m²	**5.59**
100 mm thick	5.74	0.09	1.71	–	5.74	m²	**7.45**
Mulch; Melcourt Industries Ltd; Amenity Bark Mulch FSC; to plant beds; delivered in 80 m³ loads; maximum distance 25 m							
50 mm thick	1.40	0.04	0.85	–	1.40	m²	**2.25**
75 mm thick	2.10	0.07	1.28	–	2.10	m²	**3.38**
100 mm thick	2.80	0.09	1.71	–	2.80	m²	**4.51**
Mulch; Melcourt Industries Ltd; Amenity Bark Mulch FSC; to plant beds; delivered in 25 m³ loads; maximum distance 25 m							
50 mm thick	2.08	0.04	0.85	–	2.08	m²	**2.93**
75 mm thick	3.12	0.07	1.28	–	3.12	m²	**4.40**
100 mm thick	4.16	0.09	1.71	–	4.16	m²	**5.87**
Fertilizers; at 35 g/m²							
Mascot Microfine; turf fertilizer; 8+0+6 + 2% Mg + 4% Fe	5.64	0.12	2.36	–	5.64	100 m²	**8.00**
Enmag; controlled release fertilizer (8–9 months); 11+22+09	8.68	0.12	2.36	–	8.68	100 m²	**11.04**
Mascot Outfield; turf fertilizer; 8+12+8	4.52	0.12	2.36	–	4.52	100 m²	**6.88**
Mascot Outfield; turf fertilizer; 9+5+5	3.67	0.12	2.36	–	3.67	100 m²	**6.03**

Q32 PLANTING IN SPECIAL ENVIRONMENTS – GREEN ROOF SYSTEMS

Item Excluding site overheads and profit	PC £	Labour hours	Labour £	Plant £	Material £	Unit	Total rate £
Fertilizers; at 70 g/m²							
Mascot Microfine; turf fertilizer; 8+0+6 +							
2% Mg + 4% Fe	11.29	0.12	2.36	–	11.29	100 m²	**13.65**
Enmag; controlled release fertilizer							
(8–9 months); 11+22+09	17.37	0.12	2.36	–	17.37	100 m²	**19.73**
Mascot Outfield; turf fertilizer; 8+12+8	9.05	0.12	2.36	–	9.05	100 m²	**11.41**
Mascot Outfield; turf fertilizer; 9+5+5	7.33	0.12	2.36	–	7.33	100 m²	**9.69**
Watering planting; evenly; at a rate of 5 l/m²							
using hand-held watering equipment	–	0.25	4.81	–	–	100 m²	**4.81**
using sprinkler equipment and with							
sufficient water pressure to run one 15 m							
radius sprinkler	–	0.14	2.68	–	–	100 m²	**2.68**
using movable spray lines powering three							
sprinkler heads with a radius of 15 m and							
allowing for 60% overlap (irrigation							
machinery costs not included)	–	0.02	0.30	–	–	100 m²	**0.30**

Q32 PLANTING IN SPECIAL ENVIRONMENTS – GREEN ROOF SYSTEMS

PREAMBLE

Preamble: A variety of systems are available which address all the varied requirements for a successful Green Roof. For installation by approved contractors only. The prices shown are for budgeting purposes only, as each installation is site specific and may incorporate some or all of the resources shown. Specifiers should verify that the systems specified include for design liability and inspections by the suppliers. The systems below assume commercial insulation levels are required to the space below the proposed Green Roof. Extensive Green Roofs are those of generally lightweight construction with low maintenance planting and shallow soil designed for aesthetics only; Intensive Green Roofs are designed to allow use for recreation and trafficking. They require more maintenance and allow a greater variety of surfaces and plant types.

Q32 PLANTING IN SPECIAL ENVIRONMENTS – GREEN ROOF SYSTEMS

Item Excluding site overheads and profit	PC £	Labour hours	Labour £	Plant £	Material £	Unit	Total rate £
INTENSIVE GREEN ROOFS							
Intensive Green Roof; Bauder Ltd; soil-based systems able to provide a variety of hard and soft landscaping; laid to the surface of an unprepared roof deck							
Vapour barrier laid to prevent intersticial condensation from spaces below the roof applied by torching to the roof deck							
VB4-Expal aluminium lined	–	–	–	–	–	m²	13.27
Insulation laid and hot bitumen bonded to vapour barrier							
PIR Insulation 100 mm	–	–	–	–	–	m²	33.42
Underlayer to receive root barrier partially bonded to insulation by torching							
G4E	–	–	–	–	–	m²	13.85
Root barrier							
Plant E; chemically treated root resistant capping sheet fully bonded to G4E underlayer by torching	–	–	–	–	–	m²	20.64
Slip layers to absorb differential movement							
PE Foil; 2 layers laid to root barriers	–	–	–	–	–	m²	3.34
Optional protection layer to prevent mechanical damage							
Protection Mat; 6 mm thick rubber matting; loose laid	–	–	–	–	–	m²	12.46
Drainage medium laid to root barrier							
Drainage Board; free draining EPS 50 mm thick	–	–	–	–	–	m²	13.38
Reservoir Board; up to 21.5 litre water storage capacity EPS; 50 mm thick	–	–	–	–	–	m²	15.45
Filtration to prevent soil migration to drainage system							
Filter Fleece; 3 mm thick polyester geotextile; loose laid over drainage/ reservoir layer	–	–	–	–	–	m²	2.89
For hard landscaped areas incorporate rigid drainage board laid to the protection mat							
PLT 60 drainage board	–	–	–	–	–	m²	31.08
Intensive Green Roof Systems; BBS Green roofing Ltd; components laid to the inverted build-up; insulation over waterproof rootproof membrane							
Drainage layer; BBS TDC; recycled polypropylene; providing water reservoir, multi-directional drainage and mechanical damage protection							
BBS TDC40; 40 mm deep	–	–	–	–	–	m²	14.70
BBS TDC60; 60 mm deep	–	–	–	–	–	m²	18.20

Q32 PLANTING IN SPECIAL ENVIRONMENTS – GREEN ROOF SYSTEMS

Item Excluding site overheads and profit	PC £	Labour hours	Labour £	Plant £	Material £	Unit	Total rate £
Optional drainage unfill to BBSTDC layers							
BBS TDC 40; 16 litres/m²	–	–	–	–	–	m²	5.80
BBS TDC60; 26 litres/m²	–	–	–	–	–	m²	7.60
Filter sheet rolled out to drainage layer							
BBSF filter sheet	–	–	–	–	–	m²	2.80
Growing medium and substrates							
BBS INT – Intensive substrate; lightweight growing medium laid to filter sheet 300 mm deep	–	–	–	–	–	m²	38.20
BBS SMi – Semi-intensive substrate; Lightweight growing meduim laid to filter sheet 200 mm deep	–	–	–	–	–	m²	26.76
EXTENSIVE GREEN ROOFS							
Extensive Green Roof System; BBS Green Roofing Ltd; low maintenance Sedum system for flat roof 1:60 min falls up to 15°							
Moisture/protection layer; underlay moisture layer membrane protection							
BBSU	–	–	–	–	–	m²	2.60
Drainage layer; BBS TDC recycled polypropylene; providing water reservoir, multi-directional drainage							
BBS TCD20 20 mm deep	–	–	–	–	–	m²	9.56
Filter sheet rolled to drainage layer							
BBSF filter sheet	–	–	–	–	–	m²	2.80
Growing media							
BBS EXT; extensive substrate; lightweight growing medium laid to filter sheet; 80 mm thick	–	–	–	–	–	m²	14.44
BBS SED; pre grown Sedum turf	–	–	–	–	–	m²	23.09
Extensive Lightweight Green Roof System; BBS Green Roofing Ltd; low maintenance Sedum nutrimat system for flat roof 1:60 min falls up to 15°; 60 kg/m² saturated							
Moisture/Protection layer – Underlay moisture layer which protects the membrane							
BBSU	–	–	–	–	–	m²	2.60
Drainage/reservoir/root-zone/filter – recycled PUR foam hydroponic mat							
BBS Nutrimat	–	–	–	–	–	m²	19.88
BBS EXT – Extensive substrate; lightweight growing medium laid to filter sheet 30 mm	–	–	–	–	–	m²	8.20
BBS EXT – Extensive substrate; lightweight growing medium laid to filter sheet 80 mm	–	–	–	–	–	m²	14.44
BBS SED – Pre grown Sedum turf	–	–	–	–	–	m²	23.09

Q32 PLANTING IN SPECIAL ENVIRONMENTS – GREEN ROOF SYSTEMS

Item Excluding site overheads and profit	PC £	Labour hours	Labour £	Plant £	Material £	Unit	Total rate £
EXTENSIVE PITCH ROOF SYSTEMS							
Extensive Lightweight Green Roof System;							
BBS Green Roofing Ltd; low maintenance							
Sedum system for pitch roof 3°–20°.							
60 kg/m² saturated. NB: 20°–50° requires							
extra restraints							
BBSFG Nutrimat – Fully Grown Sedum Tile –							
modular assembly to form monolithic green							
roof layer low maintenance							
Sedum system; 60 kg/m² saturated	–	–	–	–	–	m²	**68.50**
EXTENSIVE GREEN ROOFS – SOIL FREE							
Extensive Green Roof System; Bauder Ltd;							
low maintenance soil free system							
incorporating single layer growing and							
planting medium							
Vapour barrier laid to prevent intersticial							
condensation applied by torching to the roof							
deck							
VB4-Expal aluminium lined	–	–	–	–	–	m²	**13.27**
Insulation laid and hot bitumen bonded to							
vapour barrier							
PIR Insulation 100 mm	–	–	–	–	–	m²	**33.42**
Underlayer to receive rootbarrier partially							
bonded to insulation by torching							
G4E	–	–	–	–	–	m²	**13.85**
Root barrier							
Plant E; chemically treated root resistant							
capping sheet fully bonded to G4E							
underlayer by torching	–	–	–	–	–	m²	**20.64**
LANDSCAPE OPTIONS TO GREEN ROOF							
SYSTEMS							
Landscape options to the above systems;							
Bauder Ltd							
Hydroplanting system; to Extensive Green							
Roof as detailed above							
SDF mat 20 mm thick drainage layer; loose							
laid	–	–	–	–	–	m²	**10.27**
Xeroflor vegetation blanket; Bauder Ltd; to							
Extensive Green Roof as detailed above							
Xeroflor Xf 301 pre-cultivated sedum							
blanket incorporating 800 gram recycled							
fibre water retention layer; laid loose	–	–	–	–	–	m²	**44.98**

Q32 PLANTING IN SPECIAL ENVIRONMENTS – GREEN ROOF SYSTEMS

Item Excluding site overheads and profit	PC £	Labour hours	Labour £	Plant £	Material £	Unit	Total rate £
Reservoir and drainage boards; Bauder; non-specialist installed products for laying under pavings or planting for collection or drainage							
DSE20 Drainage/Reservoir Board; 20 mm thick drainage board	–	–	–	–	–	m²	10.44
DSE40 Drainage/Reservoir Board; for intensive green roof applications	–	–	–	–	–	m²	19.86
Versicell 20 Drainage Board; specialist installed drainage layer for intensive green roof applications; predominantly hard landscape	–	–	–	–	–	m²	21.88
Bauder Extensive Substrate; 80 mm deep; for use with sedum and hardy perennials and biodiverse vegetation	–	–	–	–	–	m²	18.15
Bauder Intensive Substrate; 200 mm deep; for use in traditional green roof applications e.g. grass and bedding plants	–	–	–	–	–	m²	43.12
Green Roof components; BBS Green roof systems							
Outlet inspection chambers							
BBS CHM; 150 mm deep	–	–	–	–	–	nr	70.00
BBS CHM; 150 mm deep	–	–	–	–	–	nr	85.00
BBS CHM; 150 mm deep	–	–	–	–	–	nr	95.00
WATERPROOFING AND DRAINAGE OPTIONS							
Warm Roof waterproofing systems; to roof surfaces to receive Green Roof systems							
Vapour barriers							
Aluminium lined vapour barrier; 2 mm thick; bonded in hot bitumen to the roof deck	–	–	–	–	–	m²	11.16
Korklite insulation bonded to the vapour barrier in hot bitumen; U value dependent; 80 mm thick	–	–	–	–	–	m²	21.13
Universal 2 mm thick underlayer; fully bonded to the insulation	–	–	–	–	–	m²	9.09
Anti-Root cap sheet impregnated with root resisting chemical bonded to the underlayer	–	–	–	–	–	m²	18.70
Waterproofing to upstands; Bauder Ltd							
Bauder Vapour Barrier; Bauder G4E & Bauder Plant E							
up to 200 mm high	–	–	–	–	–	m	19.11
up to 400 mm high	–	–	–	–	–	m	26.34
up to 600 mm high	–	–	–	–	–	m	31.23

Q35 LANDSCAPE MAINTENANCE

Item Excluding site overheads and profit	PC £	Labour hours	Labour £	Plant £	Material £	Unit	Total rate £
WATERPROOFING AND DRAINAGE OPTIONS – cont							
Inverted waterproofing systems; to roof surfaces to receive Green Roof systems Monolithic hot melt rubberized bitumen; applied in two 3 mm layers incorporating a polyester reinforcing sheet with 4 mm thick protection sheet and chemically impregnated root barrier; fully bonded; applied to plywood or suitably prepared wood float finish and primed concrete deck or screeds							
10 mm thick	–	–	–	–	–	m²	35.62
Roofmate; extruded polystyrene insulation; optional system; thickness to suit required U value; calculated at design stage; indicative thicknesses; laid to Hydrotech 6125							
0.25 U value; average requirement 120 mm	–	–	–	–	–	m²	24.69

Q35 LANDSCAPE MAINTENANCE

LONG-TERM MAINTENANCE

Preamble: Long-term landscape maintenance

Maintenance on long-term contracts differs in cost from that of maintenance as part of a landscape contract. In this section the contract period is generally 3–5 years. Staff are generally allocated to a single project only and therefore productivity is higher whilst overhead costs are lower. Labour costs in this section are lower than the costs used in other parts of the book. Machinery is assumed to be leased over a five year period and written off over the same period. The costs of maintenance and consumables for the various machinery types have been included in the information that follows. Finance costs for the machinery have not been allowed for.

The rates shown below are for machines working in unconfined contiguous areas. Users should adjust the times and rates if working in smaller spaces or spaces with obstructions.

Q35 LANDSCAPE MAINTENANCE

Item Excluding site overheads and profit	PC £	Labour hours	Labour £	Plant £	Material £	Unit	Total rate £
MARKET PRICES OF LANDSCAPE CHEMICALS							
The following table provides the areas of spraying in each planting scenario for plants which require 1.00 m diameter weed free circles							
Area of weed free circles required in each planting density							
plants at 500 mm centres; 400 nr/100 m^2	–	–	–	–	400.00	m^2	**400.00**
plants at 600 m centres; 278 nr/100 m^2	–	–	–	–	278.00	m^2	**278.00**
plants at 750 mm centres; 178 nr/100 m^2	–	–	–	–	178.00	m^2	**178.00**
plants at 1.00 m centres; 100 nr/100 m^2	–	–	–	–	100.00	m^2	**100.00**
plants at 1.50 m centres; 4444 nr/ha	–	–	–	–	3490.00	m^2	**3490.00**
plants at 1.75 m centres; 3265 nr/ha	–	–	–	–	2564.00	m^2	**2564.00**
plants at 2.00 m centres; 2500 nr/ha	–	–	–	–	1963.00	m^2	**1963.00**
MARKET PRICES OF LANDSCAPE CHEMICALS AT SUGGESTED APPLICATION RATES							
Note: All chemicals are standard knapsack or backpack applied unless specifically stated as CDA or TDA.							
Total herbicides							
Herbicide applications; CDA; chemical application via low pressure specialized wands to landscape planting; application to maintain 1.00 m diameter clear circles (0.79 m^2) around new planting							
Vanquish Biactive; ALS Ltd; Bayer; enhanced movement glyphosate; application rate 15 litres/ha							
plants at 1.50 m centres; 4444 nr/ha	–	–	–	–	51.42	ha	**51.42**
plants at 1.75 m centres; 3265 nr/ha	–	–	–	–	37.95	ha	**37.95**
plants at 2.00 m centres; 2500 nr/ha	–	–	–	–	29.10	ha	**29.10**
mass spraying	–	–	–	–	147.49	ha	**147.49**
spot spraying	–	–	–	–	1.47	ha	**1.47**

Q35 LANDSCAPE MAINTENANCE

Item Excluding site overheads and profit	PC £	Labour hours	Labour £	Plant £	Material £	Unit	Total rate £
MARKET PRICES OF LANDSCAPE CHEMICALS – cont							
Total herbicides; CDA; low pressure sustainable applications							
Hilite; Nomix Enviro Ltd; glyphosate oil-emulsion; Total Droplet Control (TDC formerly CDA); low volume sustainable chemical applications; formulation controls annual and perennial grasses and non-woody broad-leaved weeds in amenity and industrial areas, parks, highways, paved areas and factory sites; also effective for clearing vegetation prior to planting of ornamental species, around the bases of trees, shrubs and roses, as well as in forestry areas							
10 l/ha	–	–	–	–	1.63	100 m²	**1.63**
7.5 l/ ha	–	–	–	–	1.22	100 m²	**1.22**
Tribute Plus; Nomix Enviro Ltd; TDA; MCPA/ dicamba/mecaprop-p; TDC; for selective control of broad-leaved weeds in turf							
6 l/ha; white clover, plantains, creeping buttercup, thistle	–	–	–	–	0.66	100 m²	**0.66**
8 l/ha; yarrow, ragwort, parsley, piert, cinquefoil, mouse eared chickweed, selfheal, pearlwort, sorrel, black meddick, dandelion and related weeds, bulbous buttercup	–	–	–	–	0.87	100 m²	**0.87**
Stirrup; Rigby Taylor Ltd; glyphosate; CDA; 192 g							
10 l/ha	–	–	–	–	1.63	100 m²	**1.63**
7.5 l/ha	–	–	–	–	1.63	100 m²	**1.63**
Gallup Biograde Amenity; Rigby Taylor Ltd; glyphosate 360 g/l formulation							
general use; woody weeds; ash, beech, bracken, bramble; 3 l/ha	–	–	–	–	0.40	100 m²	**0.40**
annual and perennial grasses; heather (peat soils); 4 l/ha	–	–	–	–	0.53	100 m²	**0.53**
pre-planting; general clearance; 5 l/ha	–	–	–	–	0.66	100 m²	**0.66**
heather; mineral soils; 6 l/ha	–	–	–	–	0.80	100 m²	**0.80**
rhododendron; 10 l/ha	–	–	–	–	1.33	100 m²	**1.33**
Gallup Hi-aktiv Amenity; Rigby Taylor Ltd; 490 g/l formulation							
general use; woody weeds; ash, beech, bracken, bramble; 2.2 l/ha	–	–	–	–	0.37	100 m²	**0.37**
annual and perennial grasses; heather (peat soils); 2.9 l/ha	–	–	–	–	0.49	100 m²	**0.49**
pre-planting; general clearance; 3.7 l/ha	–	–	–	–	0.62	100 m²	**0.62**
heather; mineral soils; 4.4 l/ha	–	–	–	–	0.74	100 m²	**0.74**
rhododendron; 7.3 l/ha	–	–	–	–	1.22	100 m²	**1.22**

Q35 LANDSCAPE MAINTENANCE

Item Excluding site overheads and profit	PC £	Labour hours	Labour £	Plant £	Material £	Unit	Total rate £
Discman CDA Biograde; Rigby Taylor Ltd							
application rate; 7 l/ha; light vegetation and							
annual weeds	–	–	–	–	1.49	100 m²	**1.49**
application rate; 8.5 l/ha; forestry/							
woodlands; before planting, to control							
broad-leaved and grass weeds	–	–	–	–	1.81	100 m²	**1.81**
application rate; 10 l/ha; established							
deep-rooted annuals, perennial grasses							
and broad-leaved weeds	–	–	–	–	2.13	100 m²	**2.13**
Roundup Pro Biactive 360; Everris							
application rate; 5 l/ha; land not intended for							
cropping	–	–	–	–	0.41	100 m²	**0.41**
application rate; 10 l/ha; forestry; weed control	–	–	–	–	0.83	100 m²	**0.83**
Roundup Pro Biactive 450; Everris; 450 g/							
litre glyphosate							
application rate 4 l/ha; natural surfaces not							
intended to bear vegetation; amenity							
vegetation control; permeable surfaces							
overlaying soil; hard surfaces	–	–	–	–	0.37	100 m²	**0.37**
application rate 8 l/ha; forest; forest nursery							
weed control	–	–	–	–	0.74	100 m²	**0.74**
application rate 16 mm/l; stump application	–	–	–	–	0.15	litre	**0.15**
Roundup Pro Biactive 450; Everris;							
enhanced movement glyphosate;							
application rate 4 litres/ha							
plants at 1.50 m centres; 4444 nr/ha	53.81	–	–	–	53.81	ha	**53.81**
plants at 1.75 m centres; 3265 nr/ha	39.58	–	–	–	39.58	ha	**39.58**
plants at 2.00 m centres; 2500 nr/ha	30.28	–	–	–	30.28	ha	**30.28**
mass spraying	40.92	–	–	–	40.92	ha	**40.92**
spot spraying	1.53	–	–	–	1.53	100 m²	**1.53**
Dual; Nomix Enviro Ltd; glyphosate/							
sulfosulfuron; TDC; low volume							
sustainable chemical applications; residual							
effect lasting up to 6 months; shrub beds,							
gravelled areas, tree bases, parks,							
industrial areas, hard areas and around							
obstacles							
application rate 9 l/ha	–	–	–	–	2.03	100 m²	**2.03**
Glyfos Dakar Pro; Headland Amenity Ltd;							
68% glyphosate as water dispersible							
granules; broad-leaved weeds and grasses							
application rate 2.5 kg/ha	–	–	–	–	0.32	100 m²	**0.32**

Q35 LANDSCAPE MAINTENANCE

Item Excluding site overheads and profit	PC £	Labour hours	Labour £	Plant £	Material £	Unit	Total rate £
MARKET PRICES OF LANDSCAPE **CHEMICALS – cont**							
Total herbicides with residual pre-emergent **action**							
Pistol; ALS; Bayer; barrier prevention for **emergence control and glyphosate for** **control of emerged weeds at a single** **application per season; application rate** **4.5 litres/ha; 40 g/l diflufenican and 250 g/l** **glyphosate**							
plants at 1.50 m centres; 4444 nr/ha	–	–	–	–	35.61	ha	**35.61**
plants at 1.75 m centres; 3265 nr/ha	–	–	–	–	26.09	ha	**26.09**
plants at 2.00 m centres; 2500 nr/ha	–	–	–	–	25.52	ha	**25.52**
Residual herbicides							
Chikara; Rigby Taylor Ltd; Belchim Crop **Protection; flazasulfuron; systemic** **pre-emergent and early post-emergent** **herbicide for the control of annual and** **perennial weeds on natural surfaces not** **intended to bear vegetation and permeable** **surfaces over-lying soil**							
150 g/ha	–	–	–	–	2.64	100 m²	**2.64**
Contact herbicides							
Jewel; Everris; carfentrazone and mecoprop-p; selective weed and mosskiller							
application rate; 1.5 kg/ha	–	–	–	–	1.33	100 m²	**1.33**
Natural Weed and Moss Spray; Headland Amenity Ltd; acetic acid							
application rate; 100 l/ha	–	–	–	–	4.50	100 m²	**4.50**
Selective herbicides CDA							
Hilite; Nomix Enviro; glyphosate oil-emulsion; TD; low volume sustainable chemical applications; formulation controls annual and perennial grasses and non-woody broad- leaved weeds in amenity and industrial areas, parks, highways, paved areas and factory sites; also effective for clearing vegetation prior to planting of ornamental species, around the bases of trees, shrubs and roses, as well as in forestry areas							
10 l/ha	–	–	–	–	1.63	100 m²	**1.63**
7.5 l/ha	–	–	–	–	1.22	100 m²	**1.22**

Q35 LANDSCAPE MAINTENANCE

Item Excluding site overheads and profit	PC £	Labour hours	Labour £	Plant £	Material £	Unit	Total rate £
Tribute Plus; Nomix Enviro; MCPA/dicamba/ mecaprop-p; TDC; for selective control of broad-leaved weeds in turf							
6 l/ha; white clover, plantains, creeping buttercup, thistle	–	–	–	–	0.66	100 m²	**0.66**

8 l/ha; yarrow, ragwort, parsley, piert,
  cinquefoil, mouse eared chickweed,
  selfheal, pearlwort, sorrel, black meddick,
  dandelion and related weeds, bulbous
  buttercup
	–	–	–	–	0.87	100 m²	**0.87**

Contact and systemic combined herbicides Jewel; Everris; carfentrazone and mecoprop-p; selective weed and mosskiller							
application rate; 1.5 kg/ha	–	–	–	–	1.33	100 m²	**1.33**
Finale; ALS; Bayer; glufosinate-ammonium; controls grasses and broad-leaved weeds; it can also be used for weed control around non-edible ornamental crops including trees and shrubs and in tree nurseries; may be used to 'burn off' weeds in shrub beds prior to treatment with a residual herbicide; is also ideal for preparing sports turf for line marking							

application rate; 5 l/ha; seedlings of all
  species; established annual weeds and
  grasses as specified
	–	–	–	–	0.80	100 m²	**0.80**

application rate; 8 l/ha; established annual
  and perennial weeds and grasses as
  specified in the product guide
	–	–	–	–	1.29	100 m²	**1.29**

Selective herbicides; power or knapsack **sprayer** Junction; Rigby Taylor Ltd; selective post-emergence turf herbicide containing 6.25 g/litre florasulam plus 452 g/litre 2,4-D; specially designed for use by contractors for maintenance treatments of a wide range of broad-leaved weeds whilst giving excellent safety to turfgrass							
application rate; 1.2 l/ha	–	–	–	–	0.61	100 m²	**0.61**
Greenor; Rigby Taylor Ltd; fluroxypyr, clopyralid and MCPA; systemic selective treatment of weeds in turfgrass							
application rate; 4 l/ha	–	–	–	–	1.08	100 m²	**1.08**
Super Selective Plus; Rigby Taylor Ltd; combination of three active ingredients (mecoprop-p, MCPA and dicamba) enables this formulation to give good control of most commonly occurring broad-leaved weeds found in sports and amenity turf							
application rate; 3.5 l/ha	–	–	–	–	0.42	100 m²	**0.42**

Q35 LANDSCAPE MAINTENANCE

Item Excluding site overheads and profit	PC £	Labour hours	Labour £	Plant £	Material £	Unit	Total rate £
MARKET PRICES OF LANDSCAPE CHEMICALS – cont							
Selective herbicides – cont							
Crossbar Herbicide; Rigby Taylor Ltd; fluroxypyr, 2,4-D and dicamba for control of weeds in amenity turf including fine turf							
application rate; 2 l/ha	–	–	–	–	0.97	100 m²	**0.97**
Intrepid 2; Everris; for broad-leaved weed control in grass							
application rate; 7700 m²/5 l	–	–	–	–	0.57	100 m²	**0.57**
Re-Act; Everris; MCPA, mecoprop-p and dicamba; selective herbicide; established managed amenity turf and newly seeded grass; controls many annual and perennial weeds; will not vaporize in hot conditions							
application rate; 9090–14,285 m²/5 l	–	–	–	–	0.51	100 m²	**0.51**
Praxys; Everris; fluroxypyr-meptyl, clopyralid and florasulam; selective, systemic post-emergence herbicide for managed amenity turf, lawns and amenity grassland with high selectivity to established and young turf							
application rate; 9090–14,285 m²/5 l	–	–	–	–	1.18	100 m²	**1.18**
Asulox; ALS; Bayer; post-emergence translocated herbicide for the control of bracken and docks							
application rate; docks; 17857 m²/5 l	–	–	–	–	0.36	100 m²	**0.36**
application rate; bracken; 4545 m²/5 l	–	–	–	–	1.41	100 m²	**1.41**
Spearhead; ALS; Bayer; selective herbicide for the control of broad-leaved weeds in established turf							
application rate; 4.5 l/ha	–	–	–	–	0.83	100 m²	**0.83**
Cabadex; Headland Amenity Ltd; fluroxypyr and florasulam; newly sown and established turf (particularly speedwell, yarrow and yellow suckling clover)							
application rate 2 l/ha	–	–	–	–	0.67	100 m²	**0.67**
Relay Turf; Headland Amenity Ltd; mecoprop-p, MCPA and dicamba; many broad-leaved weeds in managed amenity turf							
application rate; 5 l/ha	–	–	–	–	0.45	100 m²	**0.45**
Blaster; Headland Amenity Ltd; triclopyr clopyralid; nettles, docks, thistles, brambles, woody weeds in amenity grassland							
application rate; 4 l/ha	–	–	–	–	1.08	100 m²	**1.08**

Q35 LANDSCAPE MAINTENANCE

Item Excluding site overheads and profit	PC £	Labour hours	Labour £	Plant £	Material £	Unit	Total rate £
Woody weed herbicides; backpack or power sprayers Garlon 4; Nomix Enviro Ltd; 667 g/litre triclopyr; control of brushwood, scrub, woody and perennial herbaceous weeds in amenity, forestry, rail and industrial areas; also suitable for forestry production							
2 l/ha; bramble, briar, broom, gorse, nettle	–	–	–	–	0.20	100 m²	**0.20**
4 l/ha; alder, birch, blackthorn, dogwood, elder, poplar, rosebay, willow-herb, sycamore	–	–	–	–	0.41	100 m²	**0.41**
6 l/ha; beech, box, elm, buckthorn, hazel, hornbeam, horse chestnut, lime, maple, privet, rowan, Spanish chestnut, willow, wild pear	–	–	–	–	0.61	100 m²	**0.61**
8 l/ha; rhododendron, ash, oak	–	–	–	–	0.82	100 m²	**0.82**
Tordon 22 K; Nomix Enviro; picloram; selective action; control of tough woody weeds, invasive species, scrub and deep-rooted perennials; residual action grass cover is maintained							
2.8 l/ha; white clover, creeping thistle, hogweed, plantains	–	–	–	–	0.24	100 m²	**0.24**
4.3–5.6 l/ha; common bird's-foot trefoil, field bindweed, creeping buttercup, burdock, cat's ear, common chickweed, colt's foot, creeping cinquefoil, dandelion, docks, hawkweed, Japanese Knotweed, mugwort, common nettle, silverweed, tansy, cotton thistle, rosebay willowherb, yarrow	–	–	–	–	0.43	100 m²	**0.43**
8.4 l/ha; wild mignonette, common ragwort	–	–	–	–	0.73	100 m²	**0.73**
11.2 l/ha; bracken, brambles	–	–	–	–	1.58	100 m²	**1.58**
Japanese Knotweed and ragwort control Loram 24; ALS Ltd; Bayer; picloram							
4.2 l/ha; Japanese Knotweed and other weeds as listed in the suppliers literature	–	–	–	–	1.60	100 m²	**1.60**
5.6 l/ha; common ragwort, wild mignonette	–	–	–	–	2.13	100 m²	**2.13**
Aquatic herbicides; backpack or power sprayers Roundup Pro Biactive 360; Everris							
application rate; 6 l/ha; aquatic weed control	–	–	–	–	0.50	100 m²	**0.50**
Roundup Pro Biactive 450; Everris; 450 g/litre glyphosate							
application rate 4.8 l/ha; enclosed waters, open waters, land immediately adjacent to aquatic areas	–	–	–	–	0.45	100 m²	**0.45**

Q35 LANDSCAPE MAINTENANCE

Item Excluding site overheads and profit	PC £	Labour hours	Labour £	Plant £	Material £	Unit	Total rate £
MARKET PRICES OF LANDSCAPE CHEMICALS – cont							
Aquatic herbicides; TDC							
Conqueror; Nomix Enviro; low volume sustainable chemical applications glyphosate with aquatic approval							
10 l/ha; established annuals, deep-rooted perennials and broad-leaved weeds	–	–	–	–	0.18	100 m²	**0.18**
7.5 l/ha; light vegetation and small annual weeds, aquatic situations	–	–	–	–	1.32	100 m²	**1.32**
12.5 l/ha; aquatic situations, emergent weeds e.g. reed, grasses and watercress	–	–	–	–	0.22	100 m²	**0.22**
15 l/ha; aquatic situations, floating weeds such as water lilies	–	–	–	–	0.26	100 m²	**0.26**
Fungicides							
Masalon; Rigby Taylor Ltd; systemic							
application rate; 8 l/ha	–	–	–	–	9.10	100 m²	**9.10**
Defender; Rigby Taylor Ltd; broad spectrum fungicide displaying protective and curative activity against fusarium patch and red thread							
application rate; 700 g/ha	–	–	–	–	7.25	100 m²	**7.25**
Fusion; Rigby Taylor Ltd; contact and systemic fungicide with curative and preventative activity against major turf diseases such as fusarium, red thread, dollar spot, anthracnose and rust							
application rate; 700 g/ha	–	–	–	–	7.35	100 m²	**7.35**
Rayzor; Rigby Taylor Ltd; broad spectrum contact fungicide with both protectant and eradicant activity for the control of fusarium patch and red thread in managed amenity turf							
application rate; 5 l/2000 m²	–	–	–	–	9.56	100 m²	**9.56**
Moss control							
Jewel; Everris; carfentrazone and mecoprop-p; selective weed and moss killer							
application rate; 1.5 kg/ha	–	–	–	–	1.33	100 m²	**1.33**
Insect control							
Merit Turf; Bayer; insecticide for the control of chafer grubs and leatherjackets							
application rate; 10 kg/3333 m²	–	–	–	–	5.05	100 m²	**5.05**

Q35 LANDSCAPE MAINTENANCE

Item Excluding site overheads and profit	PC £	Labour hours	Labour £	Plant £	Material £	Unit	Total rate £
CLARIFICATION NOTES ON LABOUR COSTS IN THIS SECTION							
General landscape maintenance team							
Generally a three man team is used in this section; The column 'Labour hours' reports team hours. The column 'Labour £' reports the total cost of the team for the unit of work shown							
3 man team	–	1.00	52.25	–	–	hr	**52.25**
2 man team	–	–	–	–	–	hr	–
LANDSCAPE CHEMICAL APPLICATION							
Chemical applications labours only							
Controlled droplet application (CDA)							
plants at 1.50 m centres; 4444 nr/ha	–	9.26	331.05	–	–	ha	**331.05**
plants at 1.50 m centres; 44.44 nr/100 m²	–	0.93	33.10	–	–	100 m²	**33.10**
plants at 1.75 m centres; 3265 nr/ha	–	6.80	243.10	–	–	ha	**243.10**
plants at 1.75 m centres; 3265 nr/ha	–	0.68	24.31	–	–	100 m²	**24.31**
plants at 2.00 m centres; 2500 nr/ha	–	5.21	186.26	–	–	ha	**186.26**
plants at 2.00 m centres; 2500 nr/ha	–	0.52	18.63	–	–	100 m²	**18.63**
spot spraying to shrub beds	–	0.20	3.85	–	–	100 m²	**3.85**
spot spraying to hard surfaces	–	0.13	2.41	–	–	100 m²	**2.41**
mass spraying	–	2.50	89.38	–	–	ha	**89.38**
mass spraying	–	0.25	8.94	–	–	100 m²	**8.94**
Power spraying							
mass spraying to areas containing vegetation	–	2.00	38.50	60.02	–	ha	**98.52**
mass spraying to hard surfaces	–	–	0.08	0.12	–	100 m²	**0.20**
spraying to kerbs and channels along roadways 300 mm wide	–	1.25	24.06	37.51	–	1 km	**61.57**
General herbicides; in accordance with manufacturer's instructions; Knapsack spray application; see 'Market rates of chemicals' for costs of specific applications							
knapsack sprayer; selective spraying around bases of plants	–	0.18	3.38	–	–	100 m²	**3.38**
knapsack sprayer; spot spraying	–	0.10	1.93	–	–	100 m²	**1.93**
knapsack sprayer; mass spraying	–	0.10	1.93	–	–	100 m²	**1.93**
granular distribution by hand or hand applicator	–	0.20	3.85	–	–	100 m²	**3.85**
Backpack spraying; keeping weed free at bases of plants							
plants at 1.50 m centres; 4444 nr/ha	–	13.89	496.57	–	–	ha	**496.57**
plants at 1.75 m centres; 3265 nr/ha	–	10.20	364.65	–	–	ha	**364.65**
plants at 2.00 m centres; 2500 nr/ha	–	7.82	279.39	–	–	ha	**279.39**
mass spraying	–	7.50	268.13	–	–	ha	**268.13**

Q35 LANDSCAPE MAINTENANCE

Item Excluding site overheads and profit	PC £	Labour hours	Labour £	Plant £	Material £	Unit	Total rate £
LANDSCAPE CHEMICAL APPLICATION – **cont**							
Herbicide applications; standard backpack **spray applicators; application to maintain** **1.00 m diameter clear circles (0.79 m²)** **around new planting**							
Everris; Roundup Pro Biactive 360; glyphosate; enhanced movement glyphosate; application rate 5 litres/ha							
plants at 1.50 m centres; 4444 nr/ha	15.91	12.35	441.33	–	15.91	ha	**457.24**
plants at 1.75 m centres; 3265 nr/ha	11.64	9.07	324.25	–	11.64	ha	**335.89**
plants at 2.00 m centres; 2500 nr/ha	8.98	6.95	248.28	–	8.98	ha	**257.26**
mass spraying	0.01	–	0.03	–	0.01	m²	**0.04**
GRASS CUTTING							
Grass cutting only; ride-on or tractor drawn **equipment; Norris & Gardiner Ltd; works** **carried out on one site only such as large** **fields or amenity areas where labour and** **machinery is present for a full day; grass** **cutting only; strimming not included**							
Using multiple-gang mower with cylindrical cutters; contiguous areas such as playing fields and the like larger than 3000 m²							
3 gang; 2.13 m cutting width	–	0.01	0.19	0.12	–	100 m²	**0.31**
5 gang; 3.40 m cutting width	–	–	0.17	0.14	–	100 m²	**0.31**
7 gang; 4.65 m cutting width	–	–	0.10	0.11	–	100 m²	**0.21**
Using multiple-gang mower with cylindrical cutters; non-contiguous areas such as verges and general turf areas							
3 gang	–	0.01	0.37	0.12	–	100 m²	**0.49**
5 gang	–	0.01	0.24	0.14	–	100 m²	**0.38**
Using multiple-rotary mower with vertical drive shaft and horizontally rotating bar or disc cutters; contiguous areas larger than 3000 m²							
cutting grass, overgrowth or the like using flail mower or reaper	–	0.01	0.36	0.27	–	100 m²	**0.63**
Grass cutting with strimming; ride-on or **tractor drawn equipment; Norris & Gardiner** **Ltd; works carried out on one site only such** **as large fields or amenity areas where labour** **and machinery is present for a full day;** **includes accompanying strimmer operative**							
Using multiple-gang mower with cylindrical cutters; contiguous areas such as playing fields and the like larger than 3000 m²							
3 gang; 2.13 m cutting width	–	0.01	0.37	0.13	–	100 m²	**0.50**
5 gang; 3.40 m cutting width	–	0.01	0.34	0.15	–	100 m²	**0.49**
7 gang; 4.65 m cutting width	–	0.01	0.21	0.14	–	100 m²	**0.35**

Q35 LANDSCAPE MAINTENANCE

Item Excluding site overheads and profit	PC £	Labour hours	Labour £	Plant £	Material £	Unit	Total rate £
Using multiple-gang mower with cylindrical cutters; non-contiguous areas such as verges and general turf areas							
3 gang	–	0.02	0.75	0.18	–	100 m²	**0.93**
5 gang	–	0.01	0.48	0.16	–	100 m²	**0.64**
Using multiple-rotary mower with vertical drive shaft and horizontally rotating bar or disc cutters; contiguous areas larger than 3000 m²							
cutting grass, overgrowth or the like using							
flail mower or reaper	–	0.02	0.71	0.33	–	100 m²	**1.04**
Grass cutting only; ride-on or tractor drawn equipment; Norris & Gardiner Ltd; works carried out on multiple sites; machinery moved by trailer between sites; strimming not included							
Using multiple-gang mower with cylindrical cutters; contiguous areas such as playing fields and the like larger than 3000 m²							
3 gang; 2.13 m cutting width	–	0.01	0.30	0.12	–	100 m²	**0.42**
5 gang; 3.40 m cutting width	–	0.01	0.28	0.14	–	100 m²	**0.42**
7 gang; 4.65 m cutting width	–	–	0.17	0.11	–	100 m²	**0.28**
Using multiple-gang mower with cylindrical cutters; non-contiguous areas such as verges and general turf areas							
3 gang	–	0.02	0.60	0.12	–	100 m²	**0.72**
5 gang	–	0.01	0.39	0.14	–	100 m²	**0.53**
Using multiple-rotary mower with vertical drive shaft and horizontally rotating bar or disc cutters; contiguous areas larger than 3000 m²							
cutting grass, overgrowth or the like using							
flail mower or reaper	–	0.02	0.58	0.27	–	100 m²	**0.85**
Cutting grass, overgrowth or the like; using tractor-mounted side-arm flail mower; in areas inaccessible to alternative machine; on surface							
not exceeding 30° from horizontal	–	0.01	0.42	0.34	–	100 m²	**0.76**
30–50° from horizontal	–	0.02	0.84	0.34	–	100 m²	**1.18**
Grass cutting and strimming; ride-on or tractor drawn equipment; Norris & Gardiner Ltd; works carried out on multiple sites; machinery moved by trailer between sites; includes for accompanying strimmer operative							
Using multiple-gang mower with cylindrical cutters; contiguous areas such as playing fields and the like larger than 3000 m²							
3 gang; 2.13 m cutting width	–	0.02	0.61	0.14	–	100 m²	**0.75**
5 gang; 3.40 m cutting width	–	0.02	0.55	0.16	–	100 m²	**0.71**
7 gang; 4.65 m cutting width	–	0.01	0.34	0.12	–	100 m²	**0.46**

Q35 LANDSCAPE MAINTENANCE

Item Excluding site overheads and profit	PC £	Labour hours	Labour £	Plant £	Material £	Unit	Total rate £
GRASS CUTTING – cont							
Grass cutting and strimming – cont							
Using multiple-gang mower with cylindrical cutters; non-contiguous areas such as verges and general turf areas							
3 gang	–	0.03	1.21	0.17	–	100 m²	**1.38**
5 gang	–	0.02	0.78	0.20	–	100 m²	**0.98**
Using multiple-rotary mower with vertical drive shaft and horizontally rotating bar or disc cutters; contiguous areas larger than 3000 m²							
cutting grass, overgrowth or the like using flail mower or reaper	–	0.03	1.17	0.32	–	100 m²	**1.49**
Cutting grass, overgrowth or the like; using tractor-mounted side-arm flail mower; in areas inaccessible to alternative machine; on surface							
not exceeding 30° from horizontal	–	0.02	0.85	0.37	–	100 m²	**1.22**
30° to 50° from horizontal	–	0.05	1.67	0.40	–	100 m²	**2.07**
Grass cutting; pedestrian operated equipment; Norris & Gardiner Ltd; the following rates are for grass cutting only; there is no allowance for accompanying strimming operations							
Using cylinder lawn mower fitted with not less than five cutting blades, front and rear rollers; on surface not exceeding 30° from horizontal; arisings let fly; width of cut							
51 cm	–	0.03	1.09	0.19	–	100 m	**1.28**
61 cm	–	0.03	0.91	0.17	–	100 m	**1.08**
71 cm	–	0.02	0.78	0.17	–	100 m	**0.95**
91 cm	–	0.02	0.61	0.20	–	100 m	**0.81**
Using rotary self-propelled mower; width of cut							
45 cm	–	0.02	0.71	0.08	–	100 m²	**0.79**
81 cm	–	0.01	0.39	0.08	–	100 m²	**0.47**
91 cm	–	0.01	0.35	0.11	–	100 m²	**0.46**
120 cm	–	0.01	0.26	0.36	–	100 m²	**0.62**
Add for using grass box for collecting and depositing arisings							
removing and depositing arisings	–	0.03	0.89	–	–	100 m²	**0.89**

Q35 LANDSCAPE MAINTENANCE

Item Excluding site overheads and profit	PC £	Labour hours	Labour £	Plant £	Material £	Unit	Total rate £
Grass cutting; pedestrian operated equipment; Norris & Gardiner Ltd; the following rates are for grass cutting with accompanying strimming operations							
Using cylinder lawn mower fitted with not less than five cutting blades, front and rear rollers; on surface not exceeding 30° from horizontal; arisings let fly; width of cut							
51 cm	–	0.06	2.18	0.28	–	100 m	**2.46**
61 cm	–	0.05	1.82	0.24	–	100 m	**2.06**
71 cm	–	0.04	1.56	0.23	–	100 m	**1.79**
91 cm	–	0.03	1.22	0.25	–	100 m	**1.47**
Using rotary self-propelled mower; width of cut							
45 cm	–	0.04	1.42	0.14	–	100 m²	**1.56**
81 cm	–	0.02	0.78	0.11	–	100 m²	**0.89**
91 cm	–	0.02	0.70	0.14	–	100 m²	**0.84**
120 cm	–	0.01	0.53	0.11	–	100 m²	**0.64**
Add for using grass box for collecting and depositing arisings							
removing and depositing arisings	–	0.03	0.89	–	–	100 m²	**0.89**
Add for 30–50° from horizontal	–	–	–	–	–	33%	**–**
Add for slopes exceeding 50°	–	–	–	–	–	100%	**–**
Cutting grass or light woody undergrowth; using trimmer with nylon cord or metal disc cutter; on surface							
not exceeding 30° from horizontal	–	0.10	3.58	0.28	–	100 m²	**3.86**
30–50° from horizontal	–	0.20	7.15	0.56	–	100 m²	**7.71**
exceeding 50° from horizontal	–	0.25	8.94	0.70	–	100 m²	**9.64**
Grass cutting; collecting arisings; Norris & Gardener Ltd							
Extra over for tractor drawn and self-propelled machinery using attached grass boxes; depositing arisings							
22 cuts per year	–	0.03	0.89	–	–	100 m²	**0.89**
18 cuts per year	–	0.04	1.34	–	–	100 m²	**1.34**
12 cuts per year	–	0.05	1.79	–	–	100 m²	**1.79**
4 cuts per year	–	0.13	4.47	–	–	100 m²	**4.47**
bedding plant	–	–	–	–	0.29	each	**0.29**
Arisings collected by trailed sweepers							
22 cuts per year	–	0.01	0.30	0.26	–	100 m²	**0.56**
18 cuts per year	–	0.01	0.40	0.26	–	100 m²	**0.66**
12 cuts per year	–	0.02	0.60	0.26	–	100 m²	**0.86**
4 cuts per year	–	0.03	1.19	0.26	–	100 m²	**1.45**
Disposing arisings on site; 100 m distance maximum							
22 cuts per year	–	–	0.14	0.02	–	100 m²	**0.16**
18 cuts per year	–	–	0.18	0.03	–	100 m²	**0.21**
12 cuts per year	–	0.01	0.26	0.04	–	100 m²	**0.30**
4 cuts per year	–	0.02	0.78	0.13	–	100 m²	**0.91**

Q35 LANDSCAPE MAINTENANCE

Item Excluding site overheads and profit	PC £	Labour hours	Labour £	Plant £	Material £	Unit	Total rate £
GRASS CUTTING – cont							
Grass cutting – cont							
Disposal of arisings off site							
22 cuts per year	–	–	0.15	0.04	0.75	100 m²	**0.94**
18 cuts per year	–	0.01	0.18	0.05	1.00	100 m²	**1.23**
12 cuts per year	–	–	0.12	0.14	1.80	100 m²	**2.06**
4 cuts per year	–	0.01	0.36	0.82	3.00	100 m²	**4.18**
GROUNDCARE OPERATIONS							
Harrowing							
Harrowing grassed area with							
drag harrow	–	0.01	0.23	0.15	–	100 m²	**0.38**
chain or light flexible spiked harrow	–	0.01	0.30	0.15	–	100 m²	**0.45**
Scarifying							
Mechanical							
Sisis ARP4; including grass collection box; towed by tractor; area scarified annually	–	0.01	0.42	0.10	–	100 m²	**0.52**
Sisis ARP4; including grass collection box; towed by tractor; area scarified two years previously	–	0.01	0.50	0.12	–	100 m²	**0.62**
pedestrian operated self-powered equipment	–	0.04	1.25	0.22	–	100 m²	**1.47**
add for disposal of arisings	–	0.01	0.45	1.02	14.98	100 m²	**16.45**
By hand							
hand implement	–	0.25	8.94	–	–	100 m²	**8.94**
add for disposal of arisings	–	0.01	0.45	1.02	14.98	100 m²	**16.45**
Rolling							
Rolling grassed area; equipment towed by tractor; once over; using							
smooth roller	–	0.01	0.26	0.15	–	100 m²	**0.41**
Turf aeration							
By machine							
Vertidrain turf aeration equipment towed by tractor to effect a minimum penetration of 100–250 mm at 100 mm centres	–	0.02	0.70	1.29	–	100 m²	**1.99**
Ryan GA 30; self-propelled turf aerating equipment; to effect a minimum penetration of 100 mm at varying centres	–	0.02	0.70	1.51	–	100 m²	**2.21**
Groundsman; pedestrian operated; self-powered solid or slitting tine turf aerating equipment to effect a minimum penetration of 100 mm	–	0.07	2.51	1.43	–	100 m²	**3.94**
Cushman core harvester; self-propelled; for collection of arisings	–	0.01	0.35	0.66	–	100 m²	**1.01**

Q35 LANDSCAPE MAINTENANCE

Item Excluding site overheads and profit	PC £	Labour hours	Labour £	Plant £	Material £	Unit	Total rate £
By hand							
hand fork; to effect a minimum penetration of 100 mm and spaced 150 mm apart	–	0.67	23.83	–	–	100 m²	**23.83**
hollow tine hand implement; to effect a minimum penetration of 100 mm and spaced 150 mm apart	–	1.00	35.75	–	–	100 m²	**35.75**
collection of arisings by hand	–	1.50	53.63	–	–	100 m²	**53.63**
Turf areas; surface treatments and top dressing;British sugar topsoil							
Apply screened topdressing to grass surfaces; spread using Tru-Lute							
Sand soil mixes 90/10 to 50/50	0.13	–	0.04	0.03	0.13	m²	**0.20**
Apply screened soil 3 mm Kettering loam to goal mouths and worn areas							
20 mm thick	0.81	0.01	0.18	–	0.81	m²	**0.99**
10 mm thick	0.40	0.01	0.18	–	0.40	m²	**0.58**
Leaf clearance; clearing grassed area of leaves and other extraneous debris							
Using equipment towed by tractor							
large grassed areas with perimeters of mature trees such as sports fields and amenity areas	–	0.01	0.23	0.07	–	100 m²	**0.30**
large grassed areas containing ornamental trees and shrub beds	–	–	0.08	0.98	–	100 m²	**1.06**
Using pedestrian operated mechanical equipment and blowers							
grassed areas with perimeters of mature trees such as sports fields and amenity areas	–	0.01	0.30	1.57	–	100 m²	**1.87**
grassed areas containing ornamental trees and shrub beds	–	0.05	1.79	0.67	–	100 m²	**2.46**
verges	–	0.03	1.19	0.09	–	100 m²	**1.28**
By hand							
grassed areas with perimeters of mature trees such as sports fields and amenity areas	–	0.05	1.79	0.20	–	100 m²	**1.99**
grassed areas containing ornamental trees and shrub beds	–	0.08	2.98	0.33	–	100 m²	**3.31**
verges	–	0.17	5.96	0.66	–	100 m²	**6.62**
Removal of arisings							
areas with perimeters of mature trees	–	–	0.12	0.09	1.20	100 m²	**1.41**
areas containing ornamental trees and shrub beds	–	0.01	0.36	0.34	3.00	100 m²	**3.70**
Litter clearance							
Collection and disposal of litter from grassed area	–	0.01	0.18	–	0.07	100 m²	**0.25**
Collection and disposal of litter from isolated grassed area not exceeding 1000 m²	–	0.02	0.71	–	0.07	100 m²	**0.78**

Q35 LANDSCAPE MAINTENANCE

Item Excluding site overheads and profit	PC £	Labour hours	Labour £	Plant £	Material £	Unit	Total rate £
GROUNDCARE OPERATIONS – cont							
Tree guards, stakes and ties							
Adjusting existing tree tie	–	0.02	0.60	–	–	nr	0.60
Taking up single or double tree stake and ties; removing and disposing	–	0.03	0.48	–	–	nr	0.48
SHRUB BED MAINTENANCE							
Edge maintenance							
Maintain edges where lawn abuts pathway or hard surface using							
strimmer	–	–	0.09	0.01	–	m	0.10
shears	–	0.01	0.29	–	–	m	0.29
Maintain edges where lawn abuts plant bed using							
mechanical edging tool	–	–	0.12	0.03	–	m	0.15
shears	–	–	0.19	–	–	m	0.19
half moon edging tool	–	0.01	0.34	–	–	m	0.34
Pruning shrubs							
Trimming ground cover planting							
soft groundcover; vinca ivy and the like	–	0.33	17.24	–	–	100 m^2	17.24
woody groundcover; cotoneaster and the like	–	0.50	25.86	–	–	100 m^2	25.86
Pruning massed shrub border (measure ground area)							
shrub beds pruned annually	–	–	0.17	–	–	m^2	0.17
shrub beds pruned hard every 3 years	–	0.01	0.48	–	–	m^2	0.48
Cutting off dead heads							
bush or standard rose	–	0.02	0.86	–	–	nr	0.86
climbing rose	–	0.03	1.44	–	–	nr	1.44
Pruning roses							
bush or standard rose	–	0.02	0.86	–	–	nr	0.86
climbing rose or rambling rose; tying in as required	–	0.02	1.15	–	–	nr	1.15
Pruning ornamental shrub; height before pruning (increase these rates by 50% if pruning work has not been executed during the previous two years)							
not exceeding 1 m	–	0.01	0.69	–	–	nr	0.69
1 to 2 m	–	0.02	0.96	–	–	nr	0.96
exceeding 2 m	–	0.04	2.16	–	–	nr	2.16
Removing excess growth etc. from face of building etc.; height before pruning							
not exceeding 2 m	–	0.01	0.49	–	–	nr	0.49
2 to 4 m	–	0.02	0.86	–	–	nr	0.86
4 to 6 m	–	0.03	1.44	–	–	nr	1.44
6 to 8 m	–	0.04	2.16	–	–	nr	2.16
8 to 10 m	–	0.05	2.47	–	–	nr	2.47

Q35 LANDSCAPE MAINTENANCE

Item Excluding site overheads and profit	PC £	Labour hours	Labour £	Plant £	Material £	Unit	Total rate £
Removing epicormic growth from base of shrub or trunk and base of tree (any height, any diameter); number of growths							
not exceeding 10	–	0.02	0.86	–	–	nr	0.86
10 to 20	–	0.02	1.15	–	–	nr	1.15
Shrub beds, borders and planters							
Lifting							
bulbs	–	0.17	8.62	–	–	100 nr	8.62
tubers or corms	–	0.13	6.90	–	–	100 nr	6.90
established herbaceous plants; hoeing and depositing for replanting	–	0.66	34.48	–	–	100 nr	34.48
Temporary staking and tying in herbaceous plant	–	0.01	0.57	–	0.12	nr	0.69
Cutting down spent growth of herbaceous plant; clearing arisings							
unstaked	–	0.01	0.34	–	–	nr	0.34
staked; not exceeding 4 stakes per plant; removing stakes and putting into store	–	0.01	0.43	–	–	nr	0.43
Hand weeding; regular visits; costs per occassion							
newly planted areas; mulched beds;	–	0.03	1.44	–	–	100 m²	1.44
newly planted areas; un-mulched	–	0.04	2.30	–	–	100 m²	2.30
established areas	–	0.01	0.57	–	–	100 m²	0.57
Removing grasses from groundcover areas	–	0.02	1.15	–	–	100 m²	1.15
Hand digging with fork between shrubs; not exceeding 150 mm deep; breaking down lumps; leaving surface with a medium tilth	–	0.44	22.99	–	–	100 m²	22.99
Hand digging with fork or spade to an average depth of 230 mm; breaking down lumps; leaving surface with a medium tilth	–	0.66	34.48	–	–	100 m²	34.48
Hand hoeing; not exceeding 50 mm deep; leaving surface with a medium tilth	–	0.13	6.90	–	–	100 m²	6.90
Hand raking to remove stones etc.; breaking down lumps: leaving surface with a fine tilth prior to planting	–	0.22	11.49	–	–	100 m²	11.49
Hand weeding; planter, window box; not exceeding 1.00 m²							
ground level box	–	0.02	0.86	–	–	nr	0.86
box accessed by stepladder	–	0.03	1.44	–	–	nr	1.44
Spreading only compost, mulch or processed bark to a depth of 75 mm							
on shrub bed with existing mature planting	–	0.03	1.57	–	–	m²	1.57
recently planted areas	–	0.02	1.15	–	–	m²	1.15
groundcover and herbaceous areas	–	0.02	1.30	–	–	m²	1.30
Clearing cultivated area of litter and other extraneous debris; using hand implement (exludes winter leaf clearance							
weekly maintenance; private areas	–	0.01	0.29	–	–	100 m²	0.29
weekly maintenance; public areas	–	0.01	0.57	–	–	100 m²	0.57
daily maintenance public areas	–	0.01	0.29	–	–	100 m²	0.29

Q35 LANDSCAPE MAINTENANCE

Item Excluding site overheads and profit	PC £	Labour hours	Labour £	Plant £	Material £	Unit	Total rate £
SHRUB BED MAINTENANCE – cont							
Shrub beds, borders and planters – cont							
Clearing cultivated area of winter leaf fall,							
extraneous debris; using hand implement;							
winter leaf; removal of cleared material to							
stockpile on site							
per occassion; occasional trees and							
deciduous shrubs within 100 m²;	–	0.01	0.72	–	–	100 m²	**0.72**
per occassion; dense tall trees and							
deciduous shrubs in close proximity	–	0.03	1.44	–	–	100 m²	**1.44**
BEDDING MAINTENANCE							
Works to established bedding areas							
Lifting							
bedding plants; hoeing and depositing for							
disposal	–	0.90	46.89	–	–	100 m²	**46.89**
Hand digging with fork; not exceeding 150 mm							
deep: breaking down lumps; leaving surface							
with a medium tilth	–	0.24	12.47	–	–	100 m²	**12.47**
Hand weeding							
newly planted areas	–	0.64	33.26	–	–	100 m²	**33.26**
established areas	–	0.16	8.31	–	–	100 m²	**8.31**
Hand digging with fork or spade to an average							
depth of 230 mm; breaking down lumps;							
leaving surface with a medium tilth	–	0.16	8.31	–	–	100 m²	**8.31**
Hand hoeing: not exceeding 50 mm deep;							
leaving surface with a medium tilth	–	0.13	6.65	–	–	100 m²	**6.65**
Hand raking to remove stones etc.; breaking							
down lumps; leaving surface with a fine tilth							
prior to planting	–	0.21	11.08	–	–	100 m²	**11.08**
Hand weeding; planter, window box; not							
exceeding 1.00 m²							
ground level box	–	0.02	0.84	–	–	nr	**0.84**
box accessed by stepladder	–	0.03	1.38	–	–	nr	**1.38**
Spreading only; compost, mulch or processed							
bark to a depth of 75 mm							
on shrub bed with existing mature planting	–	0.03	1.51	–	–	m²	**1.51**
recently planted areas	–	0.02	1.11	–	–	m²	**1.11**
groundcover and herbaceous areas	–	0.02	1.24	–	–	m²	**1.24**
Collecting bedding from nursery	–	0.95	49.88	13.64	–	100 m²	**63.52**
Setting out							
mass planting single variety	–	0.04	2.08	–	–	m²	**2.08**
pattern	–	0.11	5.54	–	–	m²	**5.54**
Planting only							
massed bedding plants	–	0.06	3.33	–	–	m²	**3.33**

Q35 LANDSCAPE MAINTENANCE

Item Excluding site overheads and profit	PC £	Labour hours	Labour £	Plant £	Material £	Unit	Total rate £
Clearing cultivated area of leaves, litter and other extraneous debris; using hand implement							
weekly maintenance	–	0.04	2.08	–	–	100 m²	**2.08**
daily maintenance	–	–	–	–	–	100 m²	**–**
IRRIGATION AND WATERING							
Irrigation and watering							
Hand held hosepipe; flow rate 25 litres per minute; irrigation requirement							
10 litres/m²	–	0.74	12.16	–	–	100 m²	**12.16**
15 litres/m²	–	1.10	18.15	–	–	100 m²	**18.15**
20 litres/m²	–	1.46	24.14	–	–	100 m²	**24.14**
25 litres/m²	–	1.84	30.31	–	–	100 m²	**30.31**
Hand held hosepipe; flow rate 40 litres per minute; irrigation requirement							
10 litres/m²	–	0.46	7.62	–	–	100 m²	**7.62**
15 litres/m²	–	0.69	11.43	–	–	100 m²	**11.43**
20 litres/m²	–	0.91	15.06	–	–	100 m²	**15.06**
25 litres/m²	–	1.15	18.91	–	–	100 m²	**18.91**
HEDGE MAINTENANCE							
Hedge cutting; field hedges cut once or twice annually							
Trimming sides and top using hand tool or hand held mechanical tools							
not exceeding 2 m high	–	0.05	1.79	0.14	–	10 m²	**1.93**
2 to 4 m high	–	0.17	5.96	0.47	–	10 m²	**6.43**
Hedge cutting; ornamental							
Trimming sides and top using hand tool or hand held mechanical tools							
not exceeding 2 m high	–	0.06	2.23	0.18	–	10 m²	**2.41**
2 to 4 m high	–	0.25	8.94	0.70	–	10 m²	**9.64**
Hedge cutting; reducing width; hand tool or hand held mechanical tools							
Not exceeding 2 m high							
average depth of cut not exceeding							
300 mm	–	0.01	0.45	0.04	–	m²	**0.49**
average depth of cut 300–600 mm	–	0.03	0.99	0.08	–	m²	**1.07**
average depth of cut 600–900 mm	–	0.03	1.12	0.09	–	m²	**1.21**
2–4 m high							
average depth of cut not exceeding							
300 mm	–	0.02	0.71	0.06	–	m²	**0.77**
average depth of cut 300–600 mm	–	0.03	1.12	0.09	–	m²	**1.21**
average depth of cut 600–900 mm	–	0.05	1.79	0.14	–	m²	**1.93**

Q35 LANDSCAPE MAINTENANCE

Item Excluding site overheads and profit	PC £	Labour hours	Labour £	Plant £	Material £	Unit	Total rate £
HEDGE MAINTENANCE – cont							
Hedge cutting – cont							
4–6 m high							
average depth of cut not exceeding							
300 mm	–	0.05	1.79	0.14	–	m²	**1.93**
average depth of cut 300–600 mm	–	0.08	2.98	0.23	–	m²	**3.21**
average depth of cut 600–900 mm	–	0.25	8.94	0.70	–	m²	**9.64**
Hedge cutting; reducing width; tractor							
mounted hedge cutting equipment							
Not exceeding 2 m high							
average depth of cut not exceeding							
300 mm	–	0.02	0.71	0.67	–	10 m²	**1.38**
average depth of cut 300–600 mm	–	0.03	0.89	0.83	–	10 m²	**1.72**
average depth of cut 600–900 mm	–	0.10	3.58	3.33	–	10 m²	**6.91**
2–4 m high							
average depth of cut not exceeding							
300 mm	–	0.01	0.23	0.21	–	10 m²	**0.44**
average depth of cut 300–600 mm	–	0.01	0.45	0.42	–	10 m²	**0.87**
average depth of cut 600–900 mm	–	0.01	0.36	0.33	–	10 m²	**0.69**
Hedge cutting; reducing height; hand tool							
or hand held mechanical tools							
Not exceeding 2 m high							
average depth of cut not exceeding							
300 mm	–	0.03	1.19	0.09	–	10 m²	**1.28**
average depth of cut 300–600 mm	–	0.07	2.38	0.19	–	10 m²	**2.57**
average depth of cut 600–900 mm	–	0.20	7.15	0.56	–	10 m²	**7.71**
2–4 m high							
average depth of cut not exceeding							
300 mm	–	0.02	0.60	0.05	–	m²	**0.65**
average depth of cut 300–600 mm	–	0.03	1.19	0.09	–	m²	**1.28**
average depth of cut 600–900 mm	–	0.10	3.58	0.28	–	m²	**3.86**
4–6 m high							
average depth of cut not exceeding							
300 mm	–	0.03	1.19	0.09	–	m²	**1.28**
average depth of cut 300–600 mm	–	0.06	2.23	0.18	–	m²	**2.41**
average depth of cut 600–900 mm	–	0.13	4.47	0.35	–	m²	**4.82**
Hedge cutting; removal and disposal of							
arisings							
Sweeping up and depositing arisings							
300 mm cut	–	0.03	0.89	–	–	10 m²	**0.89**
600 mm cut	–	0.10	3.58	–	–	10 m²	**3.58**
900 mm cut	–	0.20	7.15	–	–	10 m²	**7.15**
Chipping arisings							
300 mm cut	–	0.01	0.36	0.11	–	10 m²	**0.47**
600 mm cut	–	0.04	1.49	0.48	–	10 m²	**1.97**
900 mm cut	–	0.10	3.58	1.14	–	10 m²	**4.72**

Q35 LANDSCAPE MAINTENANCE

Item Excluding site overheads and profit	PC £	Labour hours	Labour £	Plant £	Material £	Unit	Total rate £
Disposal of unchipped arisings							
300 mm cut	–	0.01	0.30	0.51	2.00	10 m^2	2.81
600 mm cut	–	0.02	0.60	1.02	3.00	10 m^2	4.62
900 mm cut	–	0.04	1.49	2.56	7.49	10 m^2	11.54
Disposal of chipped arisings							
300 mm cut	–	–	0.06	0.17	3.00	10 m^2	3.23
600 mm cut	–	0.01	0.30	0.17	5.99	10 m^2	6.46
900 mm cut	–	0.02	0.60	0.34	14.98	10 m^2	15.92
JAPANESE KNOTWEED							
Japanese Knotweed Management Plan;							
PBA Solutions Ltd							
Formulate Japanese Knotweed Management							
Plan (KMP) as recommended by the							
Environment Agency Code of Practice 2006							
(EA CoP)	–	–	–	–	–	nr	1200.00
Treatment of Japanese Knotweed; PBA							
Solutions Ltd; removal and/or treatment in							
accordance with 'The Knotweed Code of							
Practice' published by the Environment							
Agency 2006 (EA CoP); Japanese							
Knotweed must be treated by qualified							
practitioners							
Commercial chemical treatments based on							
areas less than 50 m^2 knotweed area; allows							
for the provision of contractual arrangements							
including method statements and spraying							
records; also allows for the provision of risk,							
environment and Coshh assessments							
rate per visit	–	–	–	–	–	nr	225.00
rate per year	–	–	–	–	–	nr	900.00
rate for complete treatment programme							
lasting 3 years	–	–	–	–	–	nr	2700.00
Commercial chemical treatments based on							
areas between 50–300 m^2 knotweed area;							
allows for the provision of contractual							
arrangements including method statements							
and spraying records also allows for the							
provision of risk, environment and Coshh							
assessments							
rate per visit	–	–	–	–	–	nr	300.00
rate per year	–	–	–	–	–	nr	1200.00
rate for complete treatment programme							
lasting 3 years	–	–	–	–	–	nr	3600.00
Extra over minimum rate above; for areas							
300–900 m^2							
rate per visit	–	–	–	–	–	m^2	0.48
rate per year	–	–	–	–	–	m^2	1.92
rate for complete treatment programme							
lasting 3 years	–	–	–	–	–	m^2	5.76

Q35 LANDSCAPE MAINTENANCE

Item Excluding site overheads and profit	PC £	Labour hours	Labour £	Plant £	Material £	Unit	Total rate £
JAPANESE KNOTWEED – cont							
Treatment of Japanese Knotweed – cont							
Extra over minimum rate above; for areas over 900 m²							
rate per visit	–	–	–	–	–	m²	0.40
rate per year	–	–	–	–	–	m²	1.60
rate for complete treatment programme lasting 3 years	–	–	–	–	–	m²	4.80
Note: Taking waste off site will require pre-treatment. Landfill operators are required by the Environment Agency to obtain evidence of pre-treatment to show that the waste has been treated to either reduce its volume or hazardousness nature, or facilitate its handling, or enhance its recovery by the waste produce (site of origin) prior being taken to landfill. Generally pre-treatment of Japanese Knotweed can include chemical treatment, cutting or burning. The cost to implement this operation would be similar to item 4/5 with a minimum charge of £300.00. To enable waste to go to landfill soil testing would need to be completed and consultation with a landfill operator or invasive weed specialist is recommended in the first instance.							
Japanese Knotweed contaminated waste removal; PBA Solutions Ltd (see note above)							
Removal off site to a licensed landfill site using 20 tonne gross vehicle weight lorries; price is for haulage and tipping only; the mileage stated below is the distance from point of collection to licensed landfill site; price excludes landfill tax and loading of material into lorries; rates will fluctuate between landfill operators and the figures given are for initial guidance only							
less than 20 miles	–	–	–	–	–	load	575.00
20–40 miles	–	–	–	–	–	load	637.00
40–60 miles	–	–	–	–	–	load	835.00

Q40 FENCING

Item Excluding site overheads and profit	PC £	Labour hours	Labour £	Plant £	Material £	Unit	Total rate £
Rootbarriers; PBA Solutions Ltd; used in conjunction with on site control methods such as bund treatment, cell burial, vertical boundary protection and capping. The following barriers have been reliably used in connection with Japanese Knotweed control and as such are effective when correctly installed as part of a control programme which would usually include for chemical control.							
Linear root deflection barriers; installed to trench measured separately							
Flexiroot non-permeable; reinforced LDPE coated polypropylene geomembrane	5.89	0.05	0.96	–	5.89	m	**6.85**
Bioroot-barrier X permeable root barrier; mechanically bonded geocomposite consisting of a cooper-foil bonded between 2 layers of geotextile	10.76	0.05	0.96	–	10.76	m²	**11.72**
Landfill tax rates; based on waste being within landfill operator thresholds for controlled waste and with Japanese Knotweed being the only form of contamination; the rates below are based on visual inspections of the load at the gate of the tip							
Rates to April 2013							
higher rate (where knotweed is visible at more than 5% on inspection at landfill)	–	–	–	–	64.00	tonne	**64.00**
lower rate (where knotweed is visible at less than 5% on inspection at landfill)	–	–	–	–	2.50	tonne	**2.50**
Q40 FENCING							
CLARIFICATION NOTES ON LABOUR COSTS IN THIS SECTION							
Fencing team							
Generally a two man fencing team is used in this section; The column 'Labour hours' reports team hours. The column 'Labour £' reports the total cost of the team for the unit of fencing shown							
2 man team	–	1.00	38.50	–	–	hr	**38.50**

Q40 FENCING

Item Excluding site overheads and profit	PC £	Labour hours	Labour £	Plant £	Material £	Unit	Total rate £
TEMPORARY FENCING							
Temporary security fence; HSS Hire; mesh framed unclimbable fencing; including precast concrete supports and couplings							
Weekly hire; 2.85 × 2.00high							
weekly hire rate	–	–	–	2.35	–	m	**2.35**
erection of fencing; labour only	–	0.07	2.57	–	–	m	**2.57**
removal of fencing loading to collection							
vehicle	–	0.03	1.29	–	–	m	**1.29**
delivery charge	–	–	–	0.80	–	m	**0.80**
return haulage charge	–	–	–	0.60	–	m	**0.60**
Protective fencing; AVS Fencing Supplies Ltd							
Cleft chestnut rolled fencing; fixing to 100 mm diameter chestnut posts; driving into firm ground at 3 m centres							
900 mm high	3.31	0.05	2.05	–	3.86	m	**5.91**
1200 mm high	4.54	0.08	3.08	–	5.09	m	**8.17**
1500 mm high; 3 strand	6.08	0.11	4.11	–	6.63	m	**10.74**
Enclosures; Earth Anchors							
Rootfast anchored galvanized steel enclosures post ADP 20–1000; 1000 mm high × 20 mm diameter with AA25–750							
socket and padlocking ring	35.00	0.05	1.93	–	37.30	nr	**39.23**
steel cable, orange plastic coated	1.25	–	0.04	–	1.25	m	**1.29**
TIMBER FENCING							
Timber fencing; AVS Fencing Supplies Ltd; tanalized softwood fencing; all timber posts are kiln dried redwood with 15 year guarantee							
Timber lap panels; pressure treated; fixed to timber posts 75 × 75 mm in 1:3:6 concrete; at 1.90 m centres							
900 mm high	7.47	0.33	12.83	–	14.76	m	**27.59**
1200 mm high	7.76	0.38	14.44	–	14.52	m	**28.96**
1500 mm high	7.76	0.40	15.40	–	14.84	m	**30.24**
1800 mm high	8.46	0.45	17.32	–	16.52	m	**33.84**
Timber lap panels; fixed to slotted concrete posts 100 × 100 mm 1:3:6 concrete at 1.88 m centres							
900 mm high	7.47	0.38	14.44	–	17.64	m	**32.08**
1200 mm high	7.71	0.40	15.40	–	17.88	m	**33.28**
1500 mm high	8.83	0.42	16.36	–	20.70	m	**37.06**
1800 mm high	9.53	0.50	19.25	–	21.60	m	**40.85**
extra for corner posts	19.68	0.45	17.32	–	26.50	nr	**43.82**
extra over panel fencing for 300 mm high trellis tops; slats at 100 mm centres; including additional length of posts	4.48	0.50	19.25	–	4.48	m	**23.73**

Q40 FENCING

Item Excluding site overheads and profit	PC £	Labour hours	Labour £	Plant £	Material £	Unit	Total rate £
Closeboarded fencing; to concrete posts 100 × 100 mm; 2 nr softwood arris rails; 100 × 22 mm softwood pales lapped 13 mm; including excavating and backfilling into firm ground at 3.00 m centres; setting in concrete 1:3:6; concrete gravel board 150 × 150 mm							
900 mm high	10.18	0.50	19.25	–	12.73	m	**31.98**
1050 mm high	10.41	0.55	21.18	–	12.97	m	**34.15**
1500 mm high	12.57	0.57	22.14	–	15.13	m	**37.27**
Closeboarded fencing; to concrete posts 100 × 100 mm; 3 nr softwood arris rails; 100 × 22 mm softwood pales lapped 13 mm; including excavating and backfilling into firm ground at 3.00 m centres; setting in concrete 1:3:6							
1650 mm high	13.91	0.63	24.06	–	16.47	m	**40.53**
1800 mm high	14.03	0.65	25.02	–	16.59	m	**41.61**
Closeboarded fencing; to timber redwood posts 100 × 100 mm; 2 nr softwood arris rails; 100 × 22 mm softwood pales lapped 13 mm; including excavating and backfilling into firm ground at 3.00 m centres; setting in concrete 1:3:6							
1350 mm high	12.28	0.50	19.25	–	14.84	m	**34.09**
1650 mm high	13.24	0.63	24.06	–	15.80	m	**39.86**
1800 mm high	13.75	0.65	25.02	–	16.31	m	**41.33**
Closeboarded fencing; to timber posts 100 × 100 mm; 3 nr softwood arris rails; 100 × 22 mm softwood pales lapped 13 mm; including excavating and backfilling into firm ground at 3.00 m centres; setting in concrete 1:3:6							
1350 mm high	13.26	0.48	18.33	–	15.82	m	**34.15**
1650 mm high	14.22	0.65	25.02	–	16.78	m	**41.80**
1800 mm high	14.73	0.68	25.99	–	17.29	m	**43.28**
extra over for post 125 × 100 mm; for 1800 mm high fencing	8.82	–	–	–	8.82	m	**8.82**
extra over to the above for counter rail	0.62	–	–	–	0.62	m	**0.62**
extra over to the above for capping rail	0.62	0.05	1.93	–	0.62	m	**2.55**
extra over to the above for concrete gravel board 150 × 50 mm	5.20	–	–	–	5.20	m	**5.20**
Feather edge board fencing; to timber posts 100 × 100 mm; 2 nr softwood cant rails; 125 × 22 mm feather edge boards lapped 13 mm; including excavating and backfilling into firm ground at 3.00 m centres; setting in concrete 1:3:6							
1350 mm high	11.08	0.28	10.78	–	12.23	m	**23.01**
1650 mm high	12.54	0.28	10.78	–	13.69	m	**24.47**
1800 mm high	12.90	0.28	10.78	–	14.05	m	**24.83**

Q40 FENCING

Item Excluding site overheads and profit	PC £	Labour hours	Labour £	Plant £	Material £	Unit	Total rate £
TIMBER FENCING – cont							
Timber fencing – cont							
Feather edge board fencing; to timber posts							
100 × 100 mm; 3 nr softwood cant rails;							
125 × 22 mm feather edge boards lapped							
13 mm; including excavating and backfilling							
into firm ground at 3.00 m centres; setting in							
concrete 1:3:6							
1350 mm high	12.14	0.29	11.00	–	13.29	m	**24.29**
1650 mm high	13.60	0.29	11.00	–	14.75	m	**25.75**
1800 mm high	12.84	0.29	11.00	–	15.11	m	**26.11**
extra over for post 125 × 100 mm; for							
1800 mm high fencing	8.82	–	–	–	8.82	nr	**8.82**
extra over to the above for counter rail and							
capping	1.89	0.05	1.93	–	1.89	m	**3.82**
Palisade fencing; 22 × 75 mm softwood vertical							
palings with pointed tops; nailing to two							
50 × 100 mm horizontal softwood arris rails;							
morticed to 100 × 100 mm softwood posts with							
weathered tops at 3.00 m centres; setting in							
concrete							
900 mm high	12.71	0.45	17.32	–	15.27	m	**32.59**
1050 mm high	10.21	0.45	17.32	–	12.77	m	**30.09**
1200 mm high	10.61	0.47	18.29	–	13.17	m	**31.46**
extra over for rounded tops	0.66	–	–	–	0.66	m	**0.66**
Post-and-rail fencing; 90 × 38 mm softwood							
horizontal rails; fixing with galvanized nails to							
150 × 75 mm softwood posts; including							
excavating and backfilling into firm ground at							
1.80 m centres; all treated timber							
1200 mm high; 3 horizontal rails	8.60	0.18	6.74	–	8.69	m	**15.43**
1200 mm high; 4 horizontal rails	9.95	0.18	6.74	–	9.95	m	**16.69**
Cleft rail fencing; oak or chestnut adze tapered							
rails 2.80 m long; morticed into joints; to							
125 × 100 mm softwood posts 1.95 m long;							
including excavating and backfilling into firm							
ground at 2.80 m centres							
two rails	9.91	0.13	4.81	–	9.91	m	**14.72**
three rails	12.37	0.14	5.39	–	12.37	m	**17.76**
four rails	14.83	0.18	6.74	–	14.83	m	**21.57**
Hit and Miss horizontal rail fencing;							
87 × 38 mm top and bottom rails; 100 × 22 mm							
vertical boards arranged alternately on							
opposite side of rails; to 100 × 100 mm posts;							
including excavating and backfilling into firm							
ground; setting in concrete at 1.8 m centres							
treated softwood; 1600 mm high	26.84	0.60	23.10	–	29.14	m	**52.24**
treated softwood; 1800 mm high	29.47	0.67	25.66	–	32.01	m	**57.67**
treated softwood; 2000 mm high	32.12	0.70	26.95	–	34.65	m	**61.60**

Q40 FENCING

Item Excluding site overheads and profit	PC £	Labour hours	Labour £	Plant £	Material £	Unit	Total rate £
Trellis tops to fencing (see more detailed items in the Minor Works section of this book)							
Extra over screen fencing for 300 mm high trellis tops; slats at 100 mm centres; including additional length of posts	4.48	0.05	1.93	–	4.48	m	**6.41**
TRELLIS							
Traditional trellis panels; The Garden Trellis Company; bespoke trellis panels for decorative, screening or security applications; timber planed all round; height of trellis 1800 mm							
Free-standing panels; posts 70 × 70 mm set in concrete; timber frames mitred and grooved 45 × 34 mm; heights and widths to suit; slats 32 × 10 mm joinery quality tanalized timber at 100 mm ccs; elements fixed by galvanized staples; capping rail 70 × 34 mm							
HV68 horizontal and vertical slats; softwood	97.20	1.00	19.25	–	103.54	m	**122.79**
D68 diagonal slats; softwood	111.60	1.00	19.25	–	117.94	m	**137.19**
HV68 horizontal and vertical slats; hardwood iroko	234.00	1.00	19.25	–	240.34	m	**259.59**
D68 diagonal slats; hardwood iroko or Western Red Cedar	255.60	1.00	19.25	–	261.94	m	**281.19**
Trellis panels fixed to face of existing wall or railings; timber frames mitred and grooved 45 × 34 mm; heights and widths to suit; slats 32 × 10 mm joinery quality tanalized timber at 100 mm ccs; elements fixed by galvanized staples; capping rail 70 × 34 mm							
HV68 horizontal and vertical slats; softwood	95.40	0.50	9.63	–	97.17	m	**106.80**
D68 diagonal slats; softwood	104.40	0.50	9.63	–	106.17	m	**115.80**
HV68 horizontal and vertical slats; iroko or Western Red Cedar	234.00	0.50	9.63	–	235.77	m	**245.40**
D68 diagonal slats; iroko or Western Red Cedar	237.60	0.50	9.63	–	239.37	m	**249.00**
Contemporary style trellis panels; The Garden Trellis Company; bespoke trellis panels for decorative, screening or security applications; timber planed all round							
Free-standing panels 30/15; posts 70 × 70 mm set in concrete with 90 × 30 mm top capping; slats 30 × 14 mm with 15 mm gaps; vertical support at 450 mm centres							
joinery treated softwood	147.60	1.00	19.25	–	153.94	m	**173.19**
hardwood iroko or Western Red Cedar	252.00	1.00	19.25	–	258.34	m	**277.59**

Q40 FENCING

Item Excluding site overheads and profit	PC £	Labour hours	Labour £	Plant £	Material £	Unit	Total rate £
TRELLIS – cont							
Contemporary style trellis panels – cont							
Panels 30/15 face fixed to existing wall or							
fence; posts 70 × 70 mm set in concrete with							
90 × 30 mm top capping; slats 30 × 14 mm with							
15 mm gaps; vertical support at 450 mm							
centres							
joinery treated softwood	144.00	0.50	9.63	–	145.77	m	**155.40**
hardwood iroko or Western Red Cedar	230.40	0.50	9.63	–	232.17	m	**241.80**
Integral arches to ornamental trellis panels;							
The Garden Trellis Company							
Arches to trellis panels in 45 × 34 mm grooved							
timbers to match framing; fixed to posts							
R450 ¼ circle; joinery treated softwood	45.00	1.50	28.88	–	51.34	nr	**80.22**
R450 ¼ circle; hardwood iroko or Western							
Red Cedar	60.00	1.50	28.88	–	66.34	nr	**95.22**
Arches to span 1800 mm wide							
joinery treated softwood	60.00	1.50	28.88	–	66.34	nr	**95.22**
hardwood iroko or Western Red Cedar	80.00	1.50	28.88	–	86.34	nr	**115.22**
Painting or staining of trellis panels; high							
quality coatings							
microporous opaque paint or spirit based							
stain	–	–	–	–	28.00	m²	**28.00**
STRAINED WIRE FENCING/FIELD							
FENCING							
Posts for strained wire and wire mesh; AVS							
Fencing Supplies Ltd							
Strained wire fencing; concrete inter posts							
only at 2750 mm centres; 610 mm below							
ground; excavating holes; filling with concrete;							
replacing topsoil; disposing surplus soil off site							
900 mm high	3.09	0.28	10.70	0.68	4.44	m	**15.82**
1200 mm high	3.87	0.28	10.70	0.68	5.22	m	**16.60**
1800 mm high	5.29	0.39	14.97	1.71	6.64	m	**23.32**
Extra over strained wire fencing for concrete							
straining posts with one strut; posts and struts							
610 mm below ground; struts, cleats,							
stretchers, winders, bolts and eye bolts;							
excavating holes; filling to within 150 mm of							
ground level with concrete (1:12) – 40 mm							
aggregate; replacing topsoil; disposing surplus							
soil off site							
900 mm high	25.91	0.34	12.90	1.54	54.96	nr	**69.40**
1200 mm high	28.83	0.33	12.84	1.54	58.97	nr	**73.35**
1800 mm high	41.05	0.33	12.84	1.54	73.90	nr	**88.28**

Q40 FENCING

Item Excluding site overheads and profit	PC £	Labour hours	Labour £	Plant £	Material £	Unit	Total rate £
Extra over strained wire fencing for concrete straining posts with two struts; posts and struts 610 mm below ground; excavating holes; filling to within 150 mm of ground level with concrete (1:12) – 40 mm aggregate; replacing topsoil; disposing surplus soil off site							
900 mm high	37.90	0.46	17.52	1.98	81.27	nr	**100.77**
1200 mm high	42.26	0.42	16.36	1.54	85.61	nr	**103.51**
1800 mm high	60.43	0.42	16.36	1.54	114.67	nr	**132.57**
Strained wire fencing; galvanized steel angle posts only at 2750 mm centres; 610 mm below ground; driving in							
900 mm high; 40 × 40 × 5 mm	3.07	0.02	0.59	–	3.07	m	**3.66**
1200 mm high; 40 × 40 × 5 mm	3.50	0.02	0.64	–	3.50	m	**4.14**
1500 mm high; 40 × 40 × 5 mm	4.65	0.03	1.17	–	4.65	m	**5.82**
1800 mm high; 40 × 40 × 5 mm	5.01	0.04	1.40	–	5.01	m	**6.41**
1800 mm high; 45 × 45 × 5 mm with extension for three rows barbed wire	6.42	–	–	–	–	m	**–**
Galvanized steel straining posts with two struts for strained wire fencing; setting in concrete							
900 mm high; 50 × 50 × 6 mm	54.77	0.50	19.25	–	57.45	nr	**76.70**
1200 mm high; 50 × 50 × 6 mm	66.15	0.50	19.25	–	68.83	nr	**88.08**
1500 mm high; 50 × 50 × 6 mm	82.65	0.50	19.25	–	85.33	nr	**104.58**
1800 mm high; 50 × 50 × 6 mm	85.66	0.50	19.25	–	88.34	nr	**107.59**
2400 mm high; 60 × 60 × 6 mm	97.00	0.50	19.25	–	99.68	nr	**118.93**
Fixing of fencing to posts							
Strained wire; to posts (posts not included); 3 mm galvanized wire; fixing with galvanized stirrups							
900 mm high; 2 wire	0.19	0.02	0.64	–	0.36	m	**1.00**
1200 mm high; 3 wire	0.28	0.02	0.90	–	0.45	m	**1.35**
1400 mm high; 3 wire	0.28	0.02	0.90	–	0.45	m	**1.35**
1800 mm high; 3 wire	0.28	0.02	0.90	–	0.45	m	**1.35**
Barbed wire; to posts (posts not included); 3 mm galvanized wire; fixing with galvanized stirrups							
900 mm high; 2 wire	0.29	0.03	1.28	–	0.46	m	**1.74**
1200 mm high; 3 wire	0.43	0.05	1.79	–	0.60	m	**2.39**
1400 mm high; 3 wire	0.43	0.05	1.79	–	0.60	m	**2.39**
1800 mm high; 3 wire	0.43	0.05	1.79	–	0.60	m	**2.39**

Q40 FENCING

Item Excluding site overheads and profit	PC £	Labour hours	Labour £	Plant £	Material £	Unit	Total rate £
STRAINED WIRE FENCING/FIELD FENCING – cont							
Fencing only to posts and strained wire							
Chain link fencing; AVS Fencing Supplies Ltd; to strained wire and posts priced separately; 3 mm galvanized wire; 51 mm mesh; galvanized steel components; fixing to line wires threaded through posts and strained with eye-bolts; posts (not included)							
900 mm high	4.51	0.03	1.29	–	4.68	m	5.97
1200 mm high	6.31	0.03	1.29	–	6.48	m	7.77
1800 mm high	8.75	0.05	1.93	–	8.92	m	10.85
Chain link fencing; to strained wire and posts priced separately; 3.15 mm plastic coated galvanized wire (wire only 2.5 mm); 51 mm mesh; galvanized steel components; fencing with line wires threaded through posts and strained with eye-bolts; posts (not included) (Note: plastic coated fencing can be cheaper than galvanized finish as wire of a smaller cross-sectional area can be used)							
900 mm high	3.19	0.03	1.29	–	3.53	m	4.82
1200 mm high	4.40	0.03	1.29	–	4.74	m	6.03
1800 mm high	6.26	0.06	2.41	–	7.62	m	10.03
Extra over strained wire fencing for cranked arms and galvanized barbed wire							
1 row	4.00	0.01	0.32	–	4.00	m	4.32
2 rows	4.14	0.03	0.96	–	4.14	m	5.10
3 rows	4.29	0.03	0.96	–	4.29	m	5.25
Field fencing; AVS Fencing Supplies Ltd; welded wire mesh; fixed to posts and straining wires measured separately							
cattle fence; 1200 m high; 114 × 300 mm at bottom to 230 × 300 mm at top	0.98	0.05	1.93	–	1.18	m	3.11
sheep fence; 800 mm high; 140 × 300 mm at bottom to 230 × 300 mm at top	0.84	0.05	1.93	–	1.04	m	2.97
deer fence; 2.0 m high; 89 × 150 mm at bottom to 267 × 300 mm at top	1.97	0.06	2.41	–	2.17	m	4.58
Extra for concreting in posts	–	0.25	9.63	–	4.09	nr	13.72
Extra for straining post	10.19	0.38	14.44	–	10.19	nr	24.63
Boundary fencing; strained wire and wire mesh; Jacksons Fencing							
Tubular chain link fencing; galvanized; plastic coated; 60.3 mm diameter posts at 3.0 m centres; setting 700 mm into ground; choice of 10 mesh colours; including excavating holes, backfilling and removing surplus soil; with top rail only							
900 mm high	–	–	–	–	–	m	37.04
1200 mm high	–	–	–	–	–	m	39.35
1800 mm high	–	–	–	–	–	m	43.85
2000 mm high	–	–	–	–	–	m	46.31

Q40 FENCING

Item Excluding site overheads and profit	PC £	Labour hours	Labour £	Plant £	Material £	Unit	Total rate £
Tubular chain link fencing; galvanized; plastic coated; 60.3 mm diameter posts at 3.0 m centres; cranked arms and three lines barbed wire; setting 700 mm into ground; including excavating holes; backfilling and removing surplus soil; with top rail only							
2000 mm high	–	–	–	–	–	m	**46.95**
1800 mm high	–	–	–	–	–	m	**45.62**
RABBIT NETTING							
Rabbit netting; AVS Fencing Supplies Ltd Timber stakes; peeled kiln dried pressure treated; pointed; 1.8 m posts driven 900 mm into ground at 3 m centres (line wires and netting priced separately)							
75–100 mm stakes	2.36	0.13	4.81	–	2.36	m	**7.17**
Corner posts or straining posts 150 mm diameter × 2.3 m high set in concrete; centres to suit local conditions or changes of direction							
1 strut	12.58	0.50	19.25	4.39	18.96	each	**42.60**
2 strut	15.45	0.50	19.25	4.39	21.83	each	**45.47**
Strained wire; to posts (posts not included); 3 mm galvanized wire; fixing with galvanized stirrups							
900 mm high; 2 wire	0.19	0.02	0.64	–	0.36	m	**1.00**
1200 mm high; 3 wire	0.28	0.02	0.90	–	0.45	m	**1.35**
Rabbit netting; 31 mm 19 gauge 1050 mm high netting fixed to posts line wires and straining posts or corner posts all priced separately							
900 mm high; turned in	1.22	0.02	0.77	–	1.22	m	**1.99**
900 mm high; buried 150 mm in trench	1.22	0.04	1.61	–	1.22	m	**2.83**
TUBULAR FENCING							
Boundary fencing; Steelway-Fensecure Ltd Classic two rail tubular fencing; top and bottom with stretcher bars and straining wires in between; comprising 60.3 mm tubular posts at 3.00 m centres; setting in concrete; 35 mm top rail tied with aluminium and steel fittings; 50 × 50 × 355/2.5 mm PVC coated chain link; all components galvanized and coated in green nylon							
964 mm high	–	–	–	–	–	m	**38.75**
1269 mm high	–	–	–	–	–	m	**42.66**
1574 mm high	–	–	–	–	–	m	**48.50**
1878 mm high	–	–	–	–	–	m	**50.93**
2188 mm high	–	–	–	–	–	m	**54.29**
2458 mm high	–	–	–	–	–	m	**60.72**
2948 mm high	–	–	–	–	–	m	**76.94**
3562 mm high	–	–	–	–	–	m	**76.74**

Q40 FENCING

Item Excluding site overheads and profit	PC £	Labour hours	Labour £	Plant £	Material £	Unit	Total rate £
TUBULAR FENCING – cont							
Boundary fencing – cont							
End posts; Classic range; 60.3 mm diameter;							
setting in concrete							
964 mm high	–	–	–	–	–	nr	57.20
1269 mm high	–	–	–	–	–	nr	57.20
1574 mm high	–	–	–	–	–	nr	63.70
1878 mm high	–	–	–	–	–	nr	69.58
2188 mm high	–	–	–	–	–	nr	75.68
2458 mm high	–	–	–	–	–	nr	86.96
2948 mm high	–	–	–	–	–	nr	99.44
3562 mm high	–	–	–	–	–	nr	113.69
Corner posts; 60.3 mm diameter; setting in							
concrete							
964 mm high	–	–	–	–	–	nr	43.50
1269 mm high	–	–	–	–	–	nr	26.07
1574 mm high	–	–	–	–	–	nr	50.88
1878 mm high	–	–	–	–	–	nr	53.31
2188 mm high	–	–	–	–	–	nr	62.00
2458 mm high	–	–	–	–	–	nr	66.06
2948 mm high	–	–	–	–	–	nr	79.32
3562 mm high	–	–	–	–	–	nr	84.45
WELDED MESH FENCING							
Boundary fencing; McArthur Group							
Paladin welded mesh colour coated green							
fencing; fixing to metal posts at 2.975 m							
centres with manufacturer's fixings; setting							
600 mm deep in firm ground; including							
excavating holes; backfilling and removing							
surplus excavated material; includes 10 year							
product guarantee							
1800 mm high	39.65	0.33	12.71	–	39.65	m	52.36
2000 mm high	37.58	0.33	12.71	–	45.63	m	58.34
2400 mm high	47.57	0.33	12.71	–	56.94	m	69.65
Extra over welded galvanized plastic coated							
mesh fencing for concreting in posts	8.19	0.06	2.14	–	8.19	m	10.33
Boundary fencing; Lang and Fulton							
Orsogril rectangular steel bar mesh fence							
panels; pleione pattern; bolting to 60 × 8 mm							
uprights at 2 m centres; mesh 62 × 66 mm;							
setting in concrete							
930 mm high panels	49.00	0.08	4.81	–	61.91	m	66.72
1326 mm high panels	67.00	0.10	5.78	–	79.91	m	85.69
1722 mm high panels	83.00	0.17	9.62	–	95.91	m	105.53

Q40 FENCING

Item Excluding site overheads and profit	PC £	Labour hours	Labour £	Plant £	Material £	Unit	Total rate £
SECURITY FENCING							
Security fencing; Jacksons Fencing							
Barbican galvanized steel paling fencing; on							
60 × 60 mm posts at 3 m centres; setting in							
concrete							
1250 mm high	–	–	–	–	–	m	70.63
1500 mm high	–	–	–	–	–	m	80.34
2000 mm high	–	–	–	–	–	m	99.33
2500 mm high	–	–	–	–	–	m	120.27
Gates; to match Barbican galvanized steel							
paling fencing							
width 1 m	–	–	–	–	–	nr	1480.00
width 2 m	–	–	–	–	–	nr	2749.00
width 3 m	–	–	–	–	–	nr	2802.00
width 4 m	–	–	–	–	–	nr	3141.00
width 8 m	–	–	–	–	–	pair	3513.00
width 9 m	–	–	–	–	–	pair	3765.00
width 10 m	–	–	–	–	–	pair	4218.00
Security fencing; Jacksons Fencing;							
intruder guards							
Viper Spike Intruder Guards; to existing							
structures; including fixing bolts							
Viper 1; 40 × 5 mm × 1.1 m long; with base							
plate	34.50	1.00	38.50	–	34.50	nr	73.00
Viper 3; 160 × 190 mm wide; U shape; to							
prevent intruders climbing pipes	38.95	0.50	19.25	–	38.95	nr	58.20
Security fencing; Jacksons Fencing							
Razor Barb Concertina; spiral wire security							
barriers; fixing to 600 mm steel ground stakes							
Ref 3275; 450 mm diameter roll; medium							
barb; galvanized	2.85	0.05	1.93	–	3.05	m	4.98
Ref 3276; 730 mm diameter roll; medium							
barb; galvanized	4.28	0.05	1.93	–	4.48	m	6.41
Ref 3277; 950 mm diameter roll; medium							
barb; galvanized	5.22	0.05	1.93	–	5.42	m	7.35
Three lines Barbed Tape; medium barb; on							
50 × 50 mm mild steel angle posts; setting in							
concrete							
Ref 3283; barbed tape; medium barb;							
galvanized	1.14	0.21	8.11	–	11.82	m	19.93
Five lines Barbed Tape; medium barb; on							
50 × 50 mm mild steel angle posts; setting in							
concrete							
Ref 3283; barbed tape; medium barb;							
galvanized	1.90	0.17	6.42	–	12.58	m	19.00

Q40 FENCING

Item Excluding site overheads and profit	PC £	Labour hours	Labour £	Plant £	Material £	Unit	Total rate £
TRIP RAILS							
Trip rails; Birdsmouth; Jacksons Fencing							
Diamond rail fencing; Birdsmouth; posts							
100 × 100 mm softwood planed at 1.35 m							
centres set in 1:3:6 concrete; rail 75 × 75 mm							
nominal secured with galvanized straps nailed							
to posts							
posts 900 mm (600 mm above ground); at							
1.35 m centres	9.57	0.25	9.63	0.29	11.28	m	**21.20**
posts 1.20 m (900 mm above ground); at							
1.35 m centres	11.02	0.25	9.63	0.29	12.72	m	**22.64**
posts 900 mm (600 mm above ground); at							
1.80 m centres	8.08	0.19	7.24	0.22	9.38	m	**16.84**
posts 1.20 m (900 mm above ground); at							
1.80 m centres	9.17	0.19	7.24	0.22	10.45	m	**17.91**
Trip rails; Townscape Products Ltd							
Hollow steel section knee rails; galvanized;							
500 mm high; setting in concrete							
1000 mm bays	118.75	1.00	38.50	–	126.17	m	**164.67**
1200 mm bays	25.02	1.00	38.50	–	109.62	m	**148.12**
CONCRETE FENCING							
Concrete fencing							
Panel fencing; to precast concrete posts; in							
2 m bays; setting posts 600 mm into ground;							
sandfaced finish							
900 mm high	11.74	0.13	4.81	–	13.35	m	**18.16**
1200 mm high	15.68	0.13	4.81	–	17.29	m	**22.10**
Panel fencing; to precast concrete posts; in							
2 m bays; setting posts 750 mm into ground;							
sandfaced finish							
1500 mm high	20.46	0.17	6.41	–	22.07	m	**28.48**
1800 mm high	24.52	0.18	7.00	–	26.13	m	**33.13**
2100 mm high	28.39	0.20	7.70	–	30.00	m	**37.70**
2400 mm high	32.63	0.20	7.70	–	34.24	m	**41.94**
extra over capping panel to all of the above	6.72	–	–	–	6.72	m	**6.72**
WINDBREAK FENCING							
Windbreak fencing							
Fencing; English Woodlands; Shade and							
Shelter Netting windbreak fencing; green; to							
100 mm diameter treated softwood posts;							
setting 450 mm into ground; fixing with							
50 × 25 mm treated softwood battens nailed to							
posts; including excavating and backfilling into							
firm ground; setting in concrete at 3 m centres							
1200 mm high	1.25	0.08	3.00	–	5.32	m	**8.32**
1800 mm high	1.87	0.08	3.00	–	5.94	m	**8.94**

Q40 FENCING

Item Excluding site overheads and profit	PC £	Labour hours	Labour £	Plant £	Material £	Unit	Total rate £
BALLSTOP FENCING							
Ball stop fencing; Steelway-Fensecure Ltd							
Ball stop net; 30 × 30 mm netting fixed to 60.3 mm diameter 12 mm solid bar lattice; galvanized; dual posts; top, middle and bottom rails							
4.5 m high	–	–	–	–	–	m	107.69
5.0 m high	–	–	–	–	–	m	135.96
6.0 m high	–	–	–	–	–	m	151.48
7.0 m high	–	–	–	–	–	m	167.02
8.0 m high	–	–	–	–	–	m	182.52
9.0 m high	–	–	–	–	–	m	198.06
10.0 m high	–	–	–	–	–	m	180.97
Ball stop net; corner posts							
4.5 m high	–	–	–	–	–	m	73.89
5.0 m high	–	–	–	–	–	m	80.87
6.0 m high	–	–	–	–	–	m	125.63
7.0 m high	–	–	–	–	–	m	111.92
8.0 m high	–	–	–	–	–	m	127.46
9.0 m high	–	–	–	–	–	m	132.48
10 m high	–	–	–	–	–	m	158.50
Ball stop net; end posts							
4.5 m high	–	–	–	–	–	m	61.47
5.0 m high	–	–	–	–	–	m	68.46
6.0 m high	–	–	–	–	–	m	83.98
7.0 m high	–	–	–	–	–	m	99.52
8.0 m high	–	–	–	–	–	m	115.02
9.0 m high	–	–	–	–	–	m	130.57
10 m high	–	–	–	–	–	m	132.48
RAILINGS							
Railings; Steelway-Fensecure Ltd							
Mild steel bar railings of balusters at 115 mm centres welded to flat rail top and bottom; bays 2.00 m long; bolting to 51 × 51 mm hollow square section posts; setting in concrete							
galvanized; 900 mm high	–	–	–	–	–	m	82.28
galvanized; 1200 mm high	–	–	–	–	–	m	99.66
galvanized; 1500 mm high	–	–	–	–	–	m	110.76
galvanized; 1800 mm high	–	–	–	–	–	m	132.87
primed; 900 mm high	–	–	–	–	–	m	77.55
primed; 1200 mm high	–	–	–	–	–	m	94.16
primed; 1500 mm high	–	–	–	–	–	m	104.09
primed; 1800 mm high	–	–	–	–	–	m	124.06

Q40 FENCING

Item Excluding site overheads and profit	PC £	Labour hours	Labour £	Plant £	Material £	Unit	Total rate £
RAILINGS – cont							
Railings – cont							
Mild steel blunt top railings of 19 mm balusters							
at 130 mm centres welded to bottom rail;							
passing through holes in top rail and welded;							
top and bottom rails 40 × 10 mm; bolting to							
51 × 51 mm hollow square section posts;							
setting in concrete							
galvanized; 800 mm high	–	–	–	–	–	m	**88.59**
galvanized; 1000 mm high	–	–	–	–	–	m	**93.04**
galvanized; 1300 mm high	–	–	–	–	–	m	**110.76**
galvanized; 1500 mm high	–	–	–	–	–	m	**119.61**
primed; 800 mm high	–	–	–	–	–	m	**85.28**
primed; 1000 mm high	–	–	–	–	–	m	**88.59**
primed; 1300 mm high	–	–	–	–	–	m	**105.22**
primed; 1500 mm high	–	–	–	–	–	m	**112.99**
Railings; traditional pattern; 16 mm diameter							
verticals at 127 mm intervals with horizontal							
bars near top and bottom; balusters with							
spiked tops; 51 × 20 mm standards; including							
setting 520 mm into concrete at 2.75 m centres							
primed; 1200 mm high	–	–	–	–	–	m	**99.66**
primed; 1500 mm high	–	–	–	–	–	m	**112.99**
primed; 1800 mm high	–	–	–	–	–	m	**121.83**
Interlaced bow-top mild steel railings;							
traditional park type; 16 mm diameter verticals							
at 80 mm intervals; welded at bottom to							
50 × 10 mm flat and slotted through 38 × 8 mm							
top rail to form hooped top profile; 50 × 10 mm							
standards; setting 560 mm into concrete at							
2.75 m centres							
galvanized; 900 mm high	–	–	–	–	–	m	**150.63**
galvanized; 1200 mm high	–	–	–	–	–	m	**183.87**
galvanized; 1500 mm high	–	–	–	–	–	m	**219.30**
galvanized; 1800 mm high	–	–	–	–	–	m	**254.74**
primed; 900 mm high	–	–	–	–	–	m	**146.19**
primed; 1200 mm high	–	–	–	–	–	m	**177.21**
primed; 1500 mm high	–	–	–	–	–	m	**210.45**
primed; 1800 mm high	–	–	–	–	–	m	**243.65**

Q40 FENCING

Item Excluding site overheads and profit	PC £	Labour hours	Labour £	Plant £	Material £	Unit	Total rate £
ESTATE FENCING/HORIZONTAL BAR FENCING							
Metal estate and parkland fencing; Stonebank Ironcraft Ltd; continously welded, mild steel, 40 × 10 mm uprights manually driven into normal soil to a minimum depth of 600 mm at 1.00 m centres; 40 × 10 mm flat steel supports welded to uprights beneath topsoil at typically every third upright; horizontal rails threaded through holes within uprights; 20 mm solid steel top rail; 4 nr 25 × 8 mm × 16 mm diameter solid steel lower rails; painted fencing shot-blasted, primed and finished with two coats of paint							
Bare metal; 1200 mm high							
10–50 m	–	–	–	–	–	m	65.33
51–100 m	–	–	–	–	–	m	55.17
101–500 m	–	–	–	–	–	m	51.50
above 500 m	–	–	–	–	–	m	49.50
Primed and painted							
10–50 m	–	–	–	–	–	m	78.04
51–100 m	–	–	–	–	–	m	66.08
101–500 m	–	–	–	–	–	m	61.77
above 500 m	–	–	–	–	–	m	59.41
Galvanized and painted							
10–50 m	–	–	–	–	–	m	98.55
51–100 m	–	–	–	–	–	m	83.60
101–500 m	–	–	–	–	–	m	78.21
above 500 m	–	–	–	–	–	m	75.26
End posts and corner posts; 40 × 40 mm; standard hollow section; end or corner posts with capped tops set into concrete							
bare metal	–	–	–	–	–	nr	65.00
primed and painted	–	–	–	–	–	nr	85.00
galvanized and painted	–	–	–	–	–	nr	100.00
Gates for metal estate fencing; Stonebank Ironcraft Ltd; fully welded gates to match fence; scroll at hinge end; turnover at latch end; posts topped with finials and set into concrete							
Pedestrian gates; 1.0–1.5 m wide							
bare metal	–	–	–	–	–	nr	450.00
galvanized and painted	–	–	–	–	–	nr	600.00
Field gates; 1.50–4.0 m wide							
bare metal	–	–	–	–	–	nr	600.00
galvanized and painted	–	–	–	–	–	nr	800.00

Q40 FENCING

Item Excluding site overheads and profit	PC £	Labour hours	Labour £	Plant £	Material £	Unit	Total rate £
ESTATE FENCING/HORIZONTAL BAR FENCING – cont							
Gates for metal estate fencing – cont							
Kissing gates							
bare metal	–	–	–	–	–	nr	550.00
galvanized and painted	–	–	–	–	–	nr	745.00
Quadrant; 4 nr uprights with rails welded to fence							
bare metal	–	–	–	–	–	nr	230.00
primed and painted	–	–	–	–	–	nr	300.00
Step stile; 2 treads							
bare metal	–	–	–	–	–	nr	140.00
primed and painted	–	–	–	–	–	nr	200.00
GATES							
Gates – General							
Preamble: Gates in fences; see specification for fencing as gates in traditional or proprietary fencing systems are usually constructed of the same materials and finished as the fencing itself.							
Gates; hardwood; AVS Fencing Supplies Ltd							
Hardwood entrance gate; five bar diamond braced; curved hanging stile; planed iroko; fixed to 150 × 150 mm softwood posts; inclusive of hinges and furniture							
Ref 1100 040; 0.9 m wide	193.56	2.50	96.25	–	246.11	nr	342.36
Ref 1100 041; 1.2 m wide	206.38	2.50	96.25	–	258.93	nr	355.18
Ref 1100 042; 1.5 m wide	269.00	2.50	96.25	–	321.55	nr	417.80
Ref 1100 043; 1.8 m wide	284.85	2.50	96.25	–	337.40	nr	433.65
Ref 1100 044; 2.1 m wide	311.88	2.50	96.25	–	364.43	nr	460.68
Ref 1100 045; 2.4 m wide	293.52	2.50	96.25	–	346.07	nr	442.32
Ref 1100 047; 3.0 m wide	358.51	2.50	96.25	–	411.06	nr	507.31
Ref 1100 048; 3.3 m wide	374.45	2.50	96.25	–	427.00	nr	523.25
Ref 1100 049; 3.6 m wide	387.23	2.50	96.25	–	439.78	nr	536.03
Hardwood field gate; five bar diamond braced; planed iroko; fixed to 150 × 150 mm softwood posts; inclusive of hinges and furniture							
Ref 1100 100; 0.9 m wide	131.54	2.00	77.00	–	184.09	nr	261.09
Ref 1100 101; 1.2 m wide	142.92	2.00	77.00	–	195.47	nr	272.47
Ref 1100 102; 1.5 m wide	170.37	2.00	77.00	–	222.92	nr	299.92
Ref 1100 103; 1.8 m wide	183.12	2.00	77.00	–	235.67	nr	312.67
Ref 1100 104; 2.1 m wide	225.50	2.00	77.00	–	278.05	nr	355.05
Ref 1100 105; 2.4 m wide	238.91	2.00	77.00	–	291.46	nr	368.46
Ref 1100 106; 2.7 m wide	252.37	2.00	77.00	–	304.92	nr	381.92
Ref 1100 108; 3.3 m wide	279.24	2.00	77.00	–	331.79	nr	408.79

Q40 FENCING

Item Excluding site overheads and profit	PC £	Labour hours	Labour £	Plant £	Material £	Unit	Total rate £
Gates; treated softwood; Jacksons Fencing							
Timber field gates; including wrought iron							
ironmongery; five bar type; diamond braced;							
1.80 m high; to 200 × 200 mm posts; setting							
750 mm into firm ground							
width 2400 mm	109.00	3.00	115.50	–	242.44	nr	**357.94**
width 2700 mm	111.65	3.00	115.50	–	245.09	nr	**360.59**
width 3000 mm	119.70	3.50	134.75	–	253.14	nr	**387.89**
width 3300 mm	128.25	3.75	144.38	–	261.69	nr	**406.07**
Featherboard garden gates; including							
ironmongery; to 100 × 120 mm posts; one							
diagonal brace							
1.0 × 1.2 m high	29.60	1.50	57.75	–	122.82	nr	**180.57**
1.0 × 1.5 m high	39.00	1.50	57.75	–	140.27	nr	**198.02**
1.0 × 1.8 m high	101.25	1.50	57.75	–	152.92	nr	**210.67**
Picket garden gates; including ironmongery; to							
match picket fence; width 1000 mm; to							
100 × 120 mm posts; one diagonal brace							
950 mm high	75.80	1.50	57.75	–	113.47	nr	**171.22**
1200 mm high	79.90	1.50	57.75	–	117.57	nr	**175.32**
1800 mm high	93.75	1.50	57.75	–	145.42	nr	**203.17**
Gates; tubular mild steel; Jacksons							
Fencing							
Field gates; galvanized; including							
ironmongery; diamond braced; 1.80 m high; to							
tubular steel posts; setting in concrete							
width 3000 mm	126.27	2.50	96.25	–	234.13	nr	**330.38**
width 3300 mm	133.88	2.50	96.25	–	241.73	nr	**337.98**
width 3600 mm	141.66	2.50	96.25	–	249.52	nr	**345.77**
width 4200 mm	157.23	2.50	96.25	–	265.09	nr	**361.34**
Gates; sliding; Jacksons Fencing							
Sliding Gate; including all galvanized rails and							
vertical rail infill panels; special guide and							
shutting frame posts (Note: foundations							
installed by suppliers)							
access width 4.00 m; 1.5 m high gates	–	–	–	–	–	nr	**4820.00**
access width 4.00 m; 2.0 m high gates	–	–	–	–	–	nr	**4953.00**
access width 4.00 m; 2.5 m high gates	–	–	–	–	–	nr	**5083.00**
access width 6.00 m; 1.5 m high gates	–	–	–	–	–	nr	**5669.00**
access width 6.00 m; 2.0 m high gates	–	–	–	–	–	nr	**5840.00**
access width 6.00 m; 2.5 m high gates	–	–	–	–	–	nr	**6011.80**
access width 8.00 m; 1.5 m high gates	–	–	–	–	–	nr	**6462.00**
access width 8.00 m; 2.0 m high gates	–	–	–	–	–	nr	**6675.00**
access width 8.00 m; 2.5 m high gates	–	–	–	–	–	nr	**6886.00**
access width 10.00 m; 2.0 m high gates	–	–	–	–	–	nr	**7450.00**
access width 10.00 m; 2.5 m high gates	–	–	–	–	–	nr	**7649.00**

Q50 SITE/STREET FURNITURE/EQUIPMENT

Item Excluding site overheads and profit	PC £	Labour hours	Labour £	Plant £	Material £	Unit	Total rate £
KISSING GATES/STILES							
Kissing gates							
Kissing gates; Jacksons Fencing; in galvanized metal bar; fixing to fencing posts (posts not included); 1.65 × 1.30 × 1.00high	270.00	2.50	96.25	–	270.00	nr	**366.25**
Stiles							
Stiles; two posts; setting into firm ground; three rails; two treads	80.00	1.50	57.75	–	90.92	nr	**148.67**
GUARD RAILS							
Pedestrian guard rails and barriers							
Mild steel pedestrian guard rails; Broxap Street Furniture; 1.00 m high with 150 mm toe space; to posts at 2.00 m centres; galvanized finish							
vertical in-line bar infill panel	44.00	1.00	38.50	–	50.38	m	**88.88**
vertical staggered bar infill with 230 mm visibility gap at top	46.00	1.00	38.50	–	52.38	m	**90.88**
Q50 SITE/STREET FURNITURE/ EQUIPMENT							
GENERALLY							
Note: The costs of delivery from the supplier are generally not included in these items. Readers should check with the supplier to verify the delivery costs.							
Furniture/equipment – General							
Preamble: The following items include fixing to manufacturer's instructions; holding down bolts or other fittings and making good (excavating, backfilling and tarmac, concrete or paving bases not included).							
BASES FOR STREET FURNITURE							
Bases for street furniture							
Excavating; filling with concrete 1:3:6; bases for street furniture							
300 × 450 × 500 mm deep	–	0.75	14.44	–	7.15	nr	**21.59**
300 × 600 × 500 mm deep	–	1.11	21.37	–	9.53	nr	**30.90**
300 × 900 × 500 mm deep	–	1.67	32.15	–	14.29	nr	**46.44**
1750 × 900 × 300 mm deep	–	2.33	44.85	–	50.02	nr	**94.87**
2000 × 900 × 300 mm deep	–	2.67	51.40	–	57.16	nr	**108.56**
2400 × 900 × 300 mm deep	–	3.20	61.60	–	68.59	nr	**130.19**
2400 × 1000 × 300 mm deep	–	3.56	68.53	–	76.21	nr	**144.74**

Q50 SITE/STREET FURNITURE/EQUIPMENT

Item Excluding site overheads and profit	PC £	Labour hours	Labour £	Plant £	Material £	Unit	Total rate £
Precast concrete flags; to concrete bases (not included); bedding and jointing in cement: mortar (1:4)							
450 × 600 × 50 mm	12.04	1.17	22.46	–	18.26	m²	**40.72**
Precast concrete paving blocks; to concrete bases (not included); bedding in sharp sand; butt joints							
200 × 100 × 65 mm	8.37	0.50	9.63	–	10.28	m²	**19.91**
200 × 100 × 80 mm	9.10	0.50	9.63	–	11.07	m²	**20.70**
Engineering paving bricks; to concrete bases (not included); bedding and jointing in sulphate-resisting cement: lime: sand mortar (1:1:6)							
over 300 mm wide	11.43	0.56	10.81	–	24.77	m²	**35.58**
Edge restraints to pavings; haunching in concrete (1:3:6)							
200 × 300 mm	5.12	0.10	1.93	–	5.12	m	**7.05**
Bases; Earth Anchors Ltd							
Rootfast ancillary anchors; A1; 500 mm long × 25 mm diameter and top strap F2; including bolting to site furniture (site furniture not included)	9.50	0.50	9.63	–	9.50	set	**19.13**
installation tool for above	23.50	–	–	–	23.50	nr	**23.50**
Rootfast ancillary anchors; A4; heavy duty 40 mm square fixed head anchors; including bolting to site furniture (site furniture not included)	29.00	0.33	6.42	–	29.00	set	**35.42**
Rootfast ancillary anchors; F4; vertical socket; including bolting to site furniture (site furniture not included)	12.75	0.05	0.96	–	12.75	set	**13.71**
Rootfast ancillary anchors; F3; horizontal socket; including bolting to site furniture (site furniture not included)	14.00	0.05	0.96	–	14.00	set	**14.96**
installation tools for the above	68.00	–	–	–	68.00	nr	**68.00**
Rootfast ancillary anchors; A3; anchored bases; including bolting to site furniture (site furniture not included)	46.00	0.33	6.42	–	46.00	set	**52.42**
installation tools for the above	68.00	–	–	–	68.00	nr	**68.00**

BARRIERS

Barriers – General

Preamble: The provision of car and truck control barriers may form part of the landscape contract. Barriers range from simple manual counterweighted poles to fully automated remote-control security gates, and the exact degree of control required must be specified. Complex barriers may need special maintenance and repair.

Q50 SITE/STREET FURNITURE/EQUIPMENT

Item Excluding site overheads and profit	PC £	Labour hours	Labour £	Plant £	Material £	Unit	Total rate £
BARRIERS – cont							
Barriers; Autopa Ltd							
Manually operated pole barriers;							
counterbalance; to tubular steel supports;							
bolting to concrete foundation (foundation not							
included); aluminium boom; various finishes							
clear opening up to 3.00 m	817.00	6.00	115.50	–	838.53	nr	**954.03**
clear opening up to 4.00 m	879.00	6.00	115.50	–	900.53	nr	**1016.03**
clear opening 5.00 m	941.00	6.00	115.50	–	962.53	nr	**1078.03**
clear opening 6.00 m	1003.00	6.00	115.50	–	1024.53	nr	**1140.03**
clear opening 7.00 m	1063.00	6.00	115.50	–	1084.53	nr	**1200.03**
catch pole; arm rest for all manual barriers	116.00	–	–	–	118.69	nr	**118.69**
Electrically operated pole barriers; enclosed							
fan-cooled motor; double worm reduction gear;							
overload clutch coupling with remote controls;							
aluminium boom; various finishes (exclusive of							
electrical connections by electrician)							
Autopa AU 3.0; 3 m boom	2365.00	6.00	115.50	–	2386.53	nr	**2502.03**
Autopa AU 4.5; 4.5 m boom with catchpost	2465.00	6.00	115.50	–	2486.53	nr	**2602.03**
Autopa AU 6.0; 6.0 m boom with catchpost	2565.00	6.00	115.50	–	2586.53	nr	**2702.03**
BOLLARDS							
Excavating; for bollards and barriers; by hand							
Holes for bollards							
400 × 400 × 400 mm; disposing off site	–	0.42	8.02	–	1.32	nr	**9.34**
600 × 600 × 600 mm; disposing off site	–	0.97	18.67	–	3.70	nr	**22.37**
Concrete bollards – General							
Preamble: Precast concrete bollards are							
available in a very wide range of shapes and							
sizes. The bollards listed here are the most							
commonly used sizes and shapes;							
manufacturer's catalogues should be consulted							
for the full range. Most manufacturers produce							
bollards to match their suites of street furniture,							
which may include planters, benches, litter							
bins and cycle stands. Most parallel sided							
bollards can be supplied in removable form,							
with a reduced shank, precast concrete socket							
and lifting hole to permit removal with a bar.							
Concrete bollards							
Marshalls Plc; cylinder; straight or tapered;							
200–400 mm diameter; plain grey concrete;							
setting into firm ground (excavating and							
backfilling not included)							
Bridgford; 915 mm high above ground	68.00	2.00	38.50	–	78.66	nr	**117.16**

Q50 SITE/STREET FURNITURE/EQUIPMENT

Item Excluding site overheads and profit	PC £	Labour hours	Labour £	Plant £	Material £	Unit	Total rate £
Marshalls Plc; cylinder; straight or tapered; 200–400 mm diameter; Beadalite reflective finish; setting into firm ground (excavating and backfilling not included)							
Wexham concrete bollard; exposed silver grey	159.00	2.00	38.50	–	164.73	nr	**203.23**
Wexham Major concrete bollard; exposed silver grey	255.00	2.00	38.50	–	265.66	nr	**304.16**
Precast concrete verge markers; various shapes; 450 mm high							
plain grey concrete	54.42	1.00	19.25	–	58.51	nr	**77.76**
white concrete	57.06	1.00	19.25	–	61.15	nr	**80.40**
exposed aggregate	60.41	1.00	19.25	–	64.50	nr	**83.75**
Bollards							
Recycled plastic bollards; Furnitubes International Ltd							
Aberdeen Circular ABR150;							
1000 × 150 mm diameter	65.00	2.00	38.50	–	70.73	each	**109.23**
Aberdeen Square ABR140;							
1000 × 140 × 140 mm	47.00	2.00	38.50	–	52.73	each	**91.23**
Service bollards; Furnitubes International Ltd; stainless steel; for housing services connection points to electricity, water etc. (services connections not included)							
Kenton KEN717; 900 × 250 mm diameter	699.00	2.00	38.50	–	704.73	each	**743.23**
Zenith ZEN707; 900 × 250 mm diameter	630.00	2.00	38.50	–	635.73	each	**674.23**
Other bollards							
Removable parking posts; Marshalls Plc							
RT/RD4; domestic telescopic bollard	155.00	2.00	38.50	–	160.73	nr	**199.23**
RT/R8; heavy duty telescopic bollard	206.00	2.00	38.50	–	211.73	nr	**250.23**
Plastic bollards; Marshalls Plc							
Lismore; three ring recycled plastic bollard	75.00	2.00	38.50	–	80.73	nr	**119.23**
Cast iron bollards – General							
Preamble: The following bollards are particularly suitable for conservation areas. Logos for civic crests can be incorporated to order.							
Cast iron bollards							
Bollards; Furnitubes International Ltd (excavating and backfilling not included)							
Doric Round; 920 mm high × 170 mm diameter	96.00	2.00	38.50	–	101.46	nr	**139.96**
Gunner Round; 750 mm high × 165 mm diameter	63.00	2.00	38.50	–	68.46	nr	**106.96**
Manchester Round; 975 mm high; 225 mm square base	107.00	2.00	38.50	–	112.46	nr	**150.96**
Cannon; 1140 mm × 210 mm diameter	155.00	2.00	38.50	–	160.46	nr	**198.96**
Kenton; heavy duty galvanized steel; 900 mm high; 350 mm diameter	197.00	2.00	38.50	–	202.46	nr	**240.96**

Q50 SITE/STREET FURNITURE/EQUIPMENT

Item Excluding site overheads and profit	PC £	Labour hours	Labour £	Plant £	Material £	Unit	Total rate £
BOLLARDS – cont							
Cast iron bollards – cont							
Bollards; Marshalls Plc (excavating and backfilling not included)							
MSF103; cast iron bollard; Small Manchester	115.00	2.00	38.50	–	120.73	nr	**159.23**
MSF102; cast iron bollard; Manchester	128.00	2.00	38.50	–	133.73	nr	**172.23**
Cast iron bollards with rails – General							
Preamble: The following cast iron bollards are suitable for conservation areas.							
Cast iron bollards with rails							
Cast iron posts with steel tubular rails; Broxap Street Furniture; setting into firm ground (excavating not included)							
Sheffield Short; 420 mm high; one rail, type A	70.00	1.50	28.88	–	72.69	nr	**101.57**
Type C mild steel tubular rail; including connector	8.50	0.03	0.64	–	10.05	m	**10.69**
Steel bollards							
Steel bollards; Marshalls Plc (excavating and backfilling not included)							
SSB01; stainless steel bollard; 101 × 1250 mm	98.00	2.00	38.50	–	103.73	nr	**142.23**
RS001; stainless steel bollard; 114 × 1500 mm	112.00	2.00	38.50	–	117.73	nr	**156.23**
RB119 Brunel; steel bollard; 168 × 1500 mm	112.00	2.00	38.50	–	122.66	nr	**161.16**
Timber bollards							
Woodscape Ltd; durable hardwood							
RP 250/1500; 250 mm diameter × 1500 mm long	234.95	1.00	19.25	–	235.19	nr	**254.44**
SP 250/1500; 250 mm square × 1500 mm long	234.95	1.00	19.25	–	235.19	nr	**254.44**
SP 150/1200; 150 mm square × 1200 mm long	84.70	1.00	19.25	–	84.94	nr	**104.19**
SP 125/750; 125 mm square × 750 mm long	46.75	1.00	19.25	–	46.99	nr	**66.24**
RP 125/750; 125 mm diameter × 750 mm long	46.75	1.00	19.25	–	46.99	nr	**66.24**
Deterrent bollards							
Semi-mountable vehicle deterrent and kerb protection bollards; Furnitubes International Ltd (excavating and backfilling not included)							
bell decorative	500.00	2.00	38.50	–	507.77	nr	**546.27**
half bell	555.00	2.00	38.50	–	562.77	nr	**601.27**
full bell	708.00	2.00	38.50	–	715.77	nr	**754.27**
three quarter bell	611.00	2.00	38.50	–	618.77	nr	**657.27**

Q50 SITE/STREET FURNITURE/EQUIPMENT

Item Excluding site overheads and profit	PC £	Labour hours	Labour £	Plant £	Material £	Unit	Total rate £
Security bollards							
Security bollards; Furnitubes International Ltd							
(excavating and backfilling not included)							
Gunner; reinforced with steel insert and tie							
bars; 750 mm high above ground; 600 mm							
below ground	93.00	2.00	38.50	–	101.19	nr	**139.69**
Burr Bloc Type 6; removable steel security							
bollard; 750 mm high above ground;							
410 × 285 mm	487.00	2.00	38.50	–	505.42	nr	**543.92**
SEATING							
Outdoor seats							
CED Ltd; stone bench; Sinuous bench;							
stone type bench to organic S pattern; laid							
to concrete base; 500 mm high × 400 mm							
wide (not included)							
2.00 m long	1335.00	4.00	77.00	–	1335.00	nr	**1412.00**
5.00 m long	3277.00	8.00	154.00	–	3277.00	nr	**3431.00**
10.00 m long	6552.00	11.00	211.75	–	6552.00	nr	**6763.75**
Outdoor seats; concrete framed – General							
Preamble: Prices for the following concrete							
framed seats and benches with hardwood							
slats include for fixing (where necessary) by							
bolting into existing paving or concrete bases							
(bases not included) or building into walls or							
concrete foundations (walls and foundations							
not included). Delivery generally not included.							
Outdoor seats; concrete framed							
Outdoor seats; Townscape Products Ltd							
Oxford benches; 1800 × 430 × 440 mm	388.92	2.00	38.50	–	401.68	nr	**440.18**
Maidstone seats; 1800 × 610 × 785 mm	581.46	2.00	38.50	–	594.22	nr	**632.72**
Outdoor seats; concrete							
Outdoor seats; Marshalls Ltd							
Boulevard 2000 concrete seat	695.00	2.00	38.50	–	710.96	nr	**749.46**
Outdoor seats; Furnitubes International Ltd							
Amesbury concrete seat, 2.2 m long	540.00	2.00	38.50	–	555.96	each	**594.46**
Marlborough concrete seat, 1.8 m long	503.00	2.00	38.50	–	518.96	each	**557.46**
Outdoor seats; metal framed – General							
Preamble: Metal framed seats with hardwood							
backs to various designs can be bolted to							
ground anchors.							

Q50 SITE/STREET FURNITURE/EQUIPMENT

Item Excluding site overheads and profit	PC £	Labour hours	Labour £	Plant £	Material £	Unit	Total rate £
SEATING – cont							
Outdoor seats; metal framed							
Outdoor seats; Furnitubes International Ltd							
NS6 Newstead; steel standards with iroko slats; 1.80 m long	302.00	2.00	38.50	–	317.96	nr	356.46
NEB6 New Forest Single Bench; cast iron standards with iroko slats; 1.83 m long	249.00	2.00	38.50	–	264.96	nr	303.46
EA6 Eastgate; cast iron standards with iroko slats; 1.86 m long	313.00	2.00	38.50	–	328.96	nr	367.46
NE6 New Forest Seat; cast iron standards with iroko slats; 1.83 m long	329.00	2.00	38.50	–	344.96	nr	383.46
Zenith long bench, satin stainless steel, iroko slats, 2.43 m long	1450.00	2.00	38.50	–	1465.96	each	1504.46
Ashburton; steel with recycled plastic slats; 1.8 m long	322.00	2.00	38.50	–	337.96	each	376.46
Outdoor seats; Orchard Street Furniture Ltd; Bramley; broad iroko slats to steel frame							
1.20 m long	182.90	2.00	38.50	–	191.48	nr	229.98
1.80 m long	223.05	2.00	38.50	–	231.63	nr	270.13
2.40 m long	256.09	2.00	38.50	–	264.67	nr	303.17
Outdoor seats; Orchard Street Furniture Ltd; Laxton; narrow iroko slats to steel frame							
1.20 m long	195.40	2.00	38.50	–	203.98	nr	242.48
1.80 m long	240.11	2.00	38.50	–	248.69	nr	287.19
2.40 m long	272.87	2.00	38.50	–	281.45	nr	319.95
Outdoor seats; Orchard Street Furniture Ltd; Lambourne; iroko slats to cast iron frame							
1.80 m long	507.58	2.00	38.50	–	516.16	nr	554.66
2.40 m long	607.83	2.00	38.50	–	616.41	nr	654.91
Outdoor seats; Broxap Street Furniture; metal and timber							
Eastgate	384.00	2.00	38.50	–	392.58	nr	431.08
Outdoor seats; Earth Anchors Ltd; Forest-Saver; steel frame and recycled slats							
bench; 1.8 m	199.00	1.00	19.25	–	199.00	nr	218.25
seat; 1.8 m	322.00	1.00	19.25	–	322.00	nr	341.25
Outdoor seats; Earth Anchors Ltd; Evergreen; cast iron frame and recycled slats							
bench; 1.8 m	426.00	1.00	19.25	–	426.00	nr	445.25
seat; 1.8 m	539.00	1.00	19.25	–	539.00	nr	558.25
Outdoor seats; all steel							
Outdoor seats; Marshalls Plc							
MSF Central; stainless steel seat	1150.00	0.50	9.63	–	1150.00	nr	1159.63
MSF Central; steel seat	570.00	2.00	38.50	–	570.00	nr	608.50
MSF 502; cast iron Heritage seat	485.00	2.00	38.50	–	485.00	nr	523.50
Outdoor seats; Earth Anchors Ltd; Ranger							
bench; 1.8 m	275.00	1.00	19.25	–	275.00	nr	294.25
seat; 1.8 m	431.00	1.00	19.25	–	431.00	nr	450.25

Q50 SITE/STREET FURNITURE/EQUIPMENT

Item Excluding site overheads and profit	PC £	Labour hours	Labour £	Plant £	Material £	Unit	Total rate £
Outdoor seats; all timber							
Outdoor seats; Lister Lutyens Co Ltd; Mendip; teak							
1.524 m long	488.00	1.00	19.25	–	507.94	nr	**527.19**
1.829 m long	589.00	1.00	19.25	–	608.94	nr	**628.19**
2.438 m long (including centre leg)	738.00	1.00	19.25	–	757.94	nr	**777.19**
Outdoor seats; Lister Lutyens Co Ltd; Sussex; hardwood							
1.5 m	194.00	2.00	38.50	–	213.94	nr	**252.44**
Outdoor seats; Woodscape Ltd; solid hardwood							
seat type 3 with back; 2.00 m long; free-standing	829.50	2.00	38.50	–	849.44	nr	**887.94**
seat type 3 with back; 2.00 m long; building in	871.50	4.00	77.00	–	891.44	nr	**968.44**
seat type 4; 2.00 m long; fixing to wall	383.25	2.00	38.50	–	403.19	nr	**441.69**
seat type 4 with back; 2.00 m long; fixing to wall	509.25	3.00	57.75	–	514.05	nr	**571.80**
seat type 4 with back; 2.50 m long; fixing to wall	593.25	3.00	57.75	–	598.05	nr	**655.80**
seat type 5 with back; 2.00 m long; free-standing	845.25	2.00	38.50	–	865.19	nr	**903.69**
seat type 5 with back; 2.50 m long; free-standing	929.25	2.00	38.50	–	949.19	nr	**987.69**
seat type 5 with back; 2.00 m long; building in	887.25	2.00	38.50	–	907.19	nr	**945.69**
seat type 5 with back; 2.50 m long; building in	971.25	3.00	57.75	–	991.19	nr	**1048.94**
bench type 1; 2.00 m long; free-standing	656.25	2.00	38.50	–	676.19	nr	**714.69**
bench type 1; 2.00 m long; building in	698.25	4.00	77.00	–	718.19	nr	**795.19**
bench type 2; 2.00 m long; free-standing	630.00	2.00	38.50	–	649.94	nr	**688.44**
bench type 2; 2.50 m long; free-standing	714.00	2.00	38.50	–	733.94	nr	**772.44**
bench type 2; 2.00 m long; building in	672.00	4.00	77.00	–	691.94	nr	**768.94**
bench type 2; 2.50 m long; building in	756.00	4.00	77.00	–	775.94	nr	**852.94**
bench type 2; 2.00 m long overall; curved to 5 m radius; building in	861.00	4.00	77.00	–	880.94	nr	**957.94**
Outdoor seats; Orchard Street Furniture Ltd; Allington; all iroko							
1.20 m long	272.30	2.00	38.50	–	292.24	nr	**330.74**
1.80 m long	313.03	2.00	38.50	–	332.97	nr	**371.47**
2.40 m long	380.82	2.00	38.50	–	400.76	nr	**439.26**
Outdoor seats; tree benches/seats							
Tree bench; Neptune Outdoor Furniture Ltd; Beaufort; hexagonal; timber							
SF34–15 A; 1500 mm diameter	1103.00	0.50	9.63	–	1103.00	nr	**1112.63**
Tree seat; Neptune Outdoor Furniture Ltd; Beaufort; hexagonal; timber; with back							
SF32–10 A; 720 mm diameter	1268.00	0.50	9.63	–	1268.00	nr	**1277.63**
SF32–20 A; 1720 mm diameter	1579.00	0.50	9.63	–	1579.00	nr	**1588.63**

Q50 SITE/STREET FURNITURE/EQUIPMENT

Item Excluding site overheads and profit	PC £	Labour hours	Labour £	Plant £	Material £	Unit	Total rate £
SEATING – cont							
Outdoor seats; recycled plastic							
Furnitubes International Ltd							
Aberdeen seat, brown recycled plastic;							
1 slat; 2 m long	326.00	2.00	38.50	–	341.96	each	**380.46**
Aberdeen seat, brown recycled plastic;							
2 slats; 2 m long	432.00	2.00	38.50	–	447.96	each	**486.46**
Anti skate board devices							
Stainless steel; Furnitubes International Ltd							
for fitting to seats	20.00	0.20	3.85	–	20.00	each	**23.85**
for fitting to benches	40.00	0.20	3.85	–	40.00	each	**43.85**
SIGNAGE							
Directional signage; cast aluminium							
Signage; Furnitubes International Ltd							
FFL1 Lancer; cast aluminium finials	64.50	0.07	1.28	–	64.50	nr	**65.78**
FAAIS; arrow end type cast aluminium							
directional arms; single line; 90 mm wide	120.00	2.00	38.50	–	120.00	nr	**158.50**
FAAID; arrow end type cast aluminium							
directional arms; double line; 145 mm wide	145.00	0.13	2.57	–	145.00	nr	**147.57**
FAAIS; arrow end type cast aluminium							
directional arms; treble line; 200 mm wide	180.00	0.20	3.85	–	180.00	nr	**183.85**
FCK1211G Kingston; composite standard							
root columns	385.00	2.00	38.50	–	396.57	nr	**435.07**
Park signage							
Entrance signs and map boards; vitreous							
enamel							
entrance map board; 1250 × 1000 mm high							
with two support posts; all associated							
works to post bases	–	–	–	–	–	nr	**2159.00**
information board with two locking							
cabinets; 1250 × 1000 mm high with two							
support posts; all associated works to post							
bases	–	–	–	–	–	nr	**2061.20**
Miscellaneous park signage; vitreous							
enamel							
'No dogs' sign; 200 × 150 mm; fixing to fencing							
or gates	–	–	–	–	–	nr	**319.58**
'Dog exercise area' sign; 300 × 400 mm; fixing							
to fencing or gates	–	–	–	–	–	nr	**397.92**
'Nature conservation area' sign; 900 × 400 mm							
high with two support posts; all associated							
works to post bases	–	–	–	–	–	nr	**1140.00**

Q50 SITE/STREET FURNITURE/EQUIPMENT

Item Excluding site overheads and profit	PC £	Labour hours	Labour £	Plant £	Material £	Unit	Total rate £
Monolith signage boards							
Furnitubes International Ltd; Fulham monolith signage board; stainless steel frame; vinyl graphics; 2600 mm above ground; 600 mm below ground; 350 mm width × 120 mm depth							
without baseplate	5080.00	2.00	38.50	–	5085.38	each	**5123.88**
with baseplate	5230.00	2.00	38.50	–	5235.38	each	**5273.88**
TREE GRILLES/TREE PROTECTION							
Tree grilles; cast iron							
Cast iron tree grilles; Furnitubes International Ltd							
GS 1070 Greenwich; two part; 1000 mm square × 700 mm diameter tree hole	93.00	2.00	38.50	–	93.00	nr	**131.50**
GC 1270 Greenwich; two part, 1200 mm diameter × 700 mm diameter tree hole	75.00	2.00	38.50	–	75.00	nr	**113.50**
Cast iron tree grilles; Marshalls Plc							
Heritage; cast iron grille plus frame; 1 × 1 m	395.00	3.00	57.75	–	406.95	nr	**464.70**
Cast iron tree grilles; Townscape Products Ltd							
Baltimore; 1200 mm square × 460 mm diameter tree hole	531.17	2.00	38.50	–	531.17	nr	**569.67**
Baltimore; hexagonal; maximum width 1440 mm nominal × 600 mm diameter tree hole	854.70	2.00	38.50	–	854.70	nr	**893.20**
Tree grilles; steel							
Steel tree grilles; Furnitubes International Ltd							
GSF 102G Greenwich; steel tree grille frame for GS 1045; one part	123.00	2.00	38.50	–	123.00	nr	**161.50**
GSF 122G Greenwich; steel tree grille frame for GS 1270 and GC 1245; one part	130.00	2.00	38.50	–	130.00	nr	**168.50**
Note: Care must be taken to ensure that tree grids and guards are removed when trees grow beyond the specified diameter of guard.							
BINS–LITTER/WASTE/DOGWASTE/SALT AND GRIT BINS							
Dog waste bins							
Earth Anchors Ltd							
HG45 A; steel; 45 l; earth anchored; post mounted	178.00	0.33	6.42	–	178.00	nr	**184.42**
HG45 A; steel; 45 l; as above with pedal operation	208.00	0.33	6.42	–	208.00	nr	**214.42**

Q50 SITE/STREET FURNITURE/EQUIPMENT

Item Excluding site overheads and profit	PC £	Labour hours	Labour £	Plant £	Material £	Unit	Total rate £
BINS–LITTER/WASTE/DOGWASTE/SALT **AND GRIT BINS – cont**							
Dog waste bins – cont							
Furnitubes International Ltd							
PED 701; Pedigree; post mounted cast iron dog waste bins; 1250 mm total height above ground; 400 mm square bin	457.00	0.75	14.44	–	463.38	nr	**477.82**
LUK745 P; Lucky; steel dog bin; post mounted; 47 l	308.00	1.50	28.88	–	314.38	each	**343.26**
LUK745 W; Lucky; steel dog bin; wall mounted; 47 l	258.00	0.75	14.44	–	264.38	each	**278.82**
TER801 P; Terrier; polythene dog bin; post mounted; 40 l	139.00	1.50	28.88	–	145.38	each	**174.26**
TER801 W; Terrier; polythene dog bin; wall mounted; 40 l	93.00	0.75	14.44	–	99.38	each	**113.82**
Litter bins; precast concrete in textured **white or exposed aggregate finish; with** **wire baskets and drainage holes**							
Bins; Marshalls Plc							
Boulevard 700; concrete circular litter bin	420.00	0.50	9.63	–	420.00	nr	**429.63**
Bins; Neptune Outdoor Furniture Ltd							
SF16; 42 l	211.00	0.50	9.63	–	211.00	nr	**220.63**
SF14; 100 l	295.00	0.50	9.63	–	295.00	nr	**304.63**
Bins; Townscape Products Ltd							
Sutton; 750 mm high × 500 mm diameter; 70 l capacity; including GRP canopy	266.56	0.33	6.42	–	457.04	nr	**463.46**
Braunton; 750 mm high × 500 mm diameter; 70 l capacity; including GRP canopy	282.61	1.50	28.88	–	473.09	nr	**501.97**
Litter bins; metal; stove-enamelled **perforated metal for holder and container**							
Bins; Townscape Products Ltd							
Metro; 440 × 420 × 800 mm high; 62 l capacity	785.37	0.33	6.42	–	785.37	nr	**791.79**
Voltan; large round; 460 mm diameter × 780 mm high; 56 l capacity	703.16	0.33	6.42	–	703.16	nr	**709.58**
Voltan; small round with pedestal; 410 mm diameter × 760 mm high; 31 l capacity	593.54	0.33	6.42	–	593.54	nr	**599.96**
Litter bins; all-steel							
Bins; Marshall Plc							
MSF Central; steel litter bin	295.00	1.00	19.25	–	295.00	nr	**314.25**
MSF Central; stainless steel litter bin	574.00	0.67	12.83	–	574.00	nr	**586.83**

Q50 SITE/STREET FURNITURE/EQUIPMENT

Item Excluding site overheads and profit	PC £	Labour hours	Labour £	Plant £	Material £	Unit	Total rate £
Bins; Furnitubes International Ltd							
Wave Bin; WVB 440; free-standing 55 l							
liners; 440 mm diameter × 850 mm high	358.00	0.50	9.63	–	359.60	nr	369.23
Wave Bin WVB 520; free-standing 85 l							
liners; 520 mm diameter × 850 mm high;							
cast iron plinth	446.00	0.50	9.63	–	447.60	nr	457.23
Wave Bin; WVB 440 S304; free-standing;							
steel with satin finish; cast bronze plinth;							
open top; 55 l	698.00	0.50	9.63	–	698.00	each	707.63
Wave Bin; WVB 520 S304; free-standing;							
steel with satin finish; cast bronze plinth;							
open top; 80 l	783.00	0.50	9.63	–	783.00	each	792.63
Liverpool Bin; LVR520 S304; free-standing;							
stainless steel; side opening; 965 mm high;							
125 l	1350.00	0.50	9.63	–	1350.00	each	1359.63
Bins; Earth Anchors Ltd							
Ranger; 100 l; pedestal mounted	427.00	0.50	9.63	–	427.00	nr	436.63
Big Ben; 82 l; steel frame and liner; colour							
coated finish; earth anchored	285.00	1.00	19.25	–	285.00	nr	304.25
Beau; 42 l; steel frame and liner; colour							
coated finish; earth anchored	232.00	1.00	19.25	–	232.00	nr	251.25
Bins; Townscape Products Ltd							
Baltimore Major with GRP canopy; 560 mm							
diameter × 960 mm high; 140 l capacity	1134.81	1.00	19.25	–	1134.81	nr	1154.06
Litter bins; cast iron							
Bins; Marshalls Plc							
MSF5501 Heritage; cast iron litter bin	595.00	1.00	19.25	3.13	595.00	nr	617.38
Bins; Furnitubes International Ltd							
Covent Garden COV 702; side opening,							
500 mm square × 1050 mm high; 105 l							
capacity	236.50	0.50	9.63	–	240.79	nr	250.42
Covent Garden COV 803; side opening,							
500 mm diameter × 1025 mm high; 85 l							
capacity	248.50	0.50	9.63	–	252.79	nr	262.42
Covent Garden COV 912; open top;							
500 mm A/F octagonal × 820 mm high; 85 l							
capacity	284.00	0.50	9.63	–	288.29	nr	297.92
Albert ALB 800; open top; 400 mm							
diameter × 845 mm high; 55 l capacity	238.00	0.50	9.63	–	242.29	nr	251.92
Bins; Broxap Street Furniture							
Derby Hercules; post mounted; steel; 40 l	161.60	1.00	19.25	–	165.89	nr	185.14
Bins; Townscape Products Ltd							
York Major; 650 mm diameter × 1060 mm							
high; 140 l capacity	1358.70	1.00	19.25	–	1359.10	nr	1378.35

Q50 SITE/STREET FURNITURE/EQUIPMENT

Item Excluding site overheads and profit	PC £	Labour hours	Labour £	Plant £	Material £	Unit	Total rate £
BINS–LITTER/WASTE/DOGWASTE/SALT AND GRIT BINS – cont							
Litter bins; timber faced; hardwood slatted casings with removable metal litter containers; ground or wall fixing							
Bins; Lister Lutyens Co Ltd							
Monmouth; 675 mm high × 450 mm wide; free-standing	130.00	0.33	6.42	–	130.00	nr	**136.42**
Monmouth; 675 mm high × 450 mm wide; bolting to ground (without legs)	120.00	1.00	19.25	–	125.09	nr	**144.34**
Bins; Woodscape Ltd							
square; 580 × 580 × 950 mm high; with lockable lid	693.00	0.50	9.63	–	693.00	nr	**702.63**
round; 580 mm diameter × 950 mm high; with lockable lid	693.00	0.50	9.63	–	693.00	nr	**702.63**
Plastic litter and grit bins; glassfibre reinforced polyester grit bins; yellow body; hinged lids							
Bins; Wybone Ltd; Victoriana glass fibre; cast iron effect litter bins; including lockable liner							
LBV/2; 521 × 521 × 673 mm high; open top; square shape; 0.078 m³ capacity	204.90	1.00	19.25	–	209.19	nr	**228.44**
LVC/3; 457 mm diameter × 648 mm high; open top; drum shape; 0.084 m³ capacity; with lockable liner	223.52	1.00	19.25	–	227.81	nr	**247.06**
Grit bins; Furnitubes International Ltd							
Grit and salt bin; yellow glass fibre; hinged lid; 170 l	249.00	–	–	–	249.00	each	**249.00**
Cigarette bins							
Furnitubes International Ltd							
ZEN 275 Zenith cigarette bin; stainless steel; wall or post mounted; 1.7 L	29.99	0.75	14.44	–	36.37	each	**50.81**
LVR250 PC Liverpool cigarette bin; steel; wall mounted; 1.9 L	16.75	0.50	9.63	–	16.75	each	**26.38**
SMK500F Smoke King cigarette bin; cast aluminium; bolt down; 3.5 L	199.99	0.75	14.44	–	206.37	each	**220.81**

Q50 SITE/STREET FURNITURE/EQUIPMENT

Item Excluding site overheads and profit	PC £	Labour hours	Labour £	Plant £	Material £	Unit	Total rate £
CYCLE HOLDERS/CYCLE SHELTERS							
Cycle holders							
Cycle stands; Marshalls Plc							
Sheffield; steel cycle stand; RCS1	40.00	0.50	9.63	–	40.00	nr	**49.63**
Sheffield; stainless steel cycle stand;							
RSCS1	110.00	0.50	9.63	–	110.00	nr	**119.63**
Cycle holders; Autopa; VELOPA; galvanized steel							
R; fixing to wall or post; making good	34.00	1.00	19.25	–	42.58	nr	**61.83**
SR(V); fixing in ground; making good	43.00	1.00	19.25	–	51.58	nr	**70.83**
Sheffield cycle stands; ragged steel	54.00	1.00	19.25	–	62.58	nr	**81.83**
Cycle holders; Townscape Products Ltd							
Guardian cycle holders; tubular steel frame; setting in concrete;							
1250 × 550 × 775 mm high; making good	372.58	1.00	19.25	–	381.16	nr	**400.41**
Penny cycle stands; 600 mm diameter; exposed aggregate bollards with 8 nr cycle holders; in galvanized steel; setting in concrete; making good	867.31	1.00	19.25	–	875.89	nr	**895.14**
Cycle holders; Broxap Street Furniture							
Neath cycle rack; BX/MW/AG; for 6 nr cycles; semi-vertical; galvanized and polyester powder coated; 1.32 m wide × 2.542 m long × 1.80 m high	432.00	10.00	192.50	–	449.16	nr	**641.66**
Premier Senior economy combined shelter and rack; BX/MW/AW; for 10 nr cycles; horizontal; galvanized only; 2.13 m wide × 3.05 m long × 2.15 m high	870.00	10.00	192.50	–	887.16	nr	**1079.66**
Toast Rack double sided free-standing cycle rack; BX/MW/GH; for 10 nr cycles; galvanized and polyester coated; 3.25 m long	293.00	4.00	77.00	–	303.76	nr	**380.76**
Cycle shelters							
Furnitubes International Ltd; Academy free-standing bolt down shelter, galvanized steel frame and roof; 3450 mm overall width; 2150 mm maximum height; 2670 mm depth							
corrugated roof	1873.00	6.00	115.50	–	1885.76	each	**2001.26**
clear polycarbonate roof	2024.00	6.00	115.50	–	2036.76	each	**2152.26**

Q50 SITE/STREET FURNITURE/EQUIPMENT

Item Excluding site overheads and profit	PC £	Labour hours	Labour £	Plant £	Material £	Unit	Total rate £
PICNIC TABLES/SUNDRY FURNITURE							
Street furniture ranges							
Townscape Products Ltd; Belgrave; natural grey concrete							
bollards; 250 mm diameter × 500 mm high	123.07	1.00	19.25	–	126.42	nr	**145.67**
seats; 1800 × 600 × 736 mm high	479.83	2.00	38.50	–	479.83	nr	**518.33**
Picnic benches – General							
Preamble: The following items include for fixing to ground to manufacturer's instructions or concreting in.							
Picnic tables and benches							
Picnic tables and benches; Broxap Street Furniture							
Eastgate picnic unit	629.00	2.00	38.50	–	637.58	nr	**676.08**
Picnic tables and benches; Woodscape Ltd							
table and benches built in; 2 m long	2042.25	2.00	38.50	–	2050.83	nr	**2089.33**
Picnic table; recycled plastic							
Furnitubes International Ltd							
Dundee picnic table; brown recycled plastic; 1.8 m long	644.00	2.00	38.50	–	659.96	each	**698.46**
Lifebuoy stations							
Lifebuoy stations; Earth Anchors Ltd							
Rootfast lifebuoy station complete with post AP44; SOLAS approved lifebouys 590 mm and lifeline	165.00	2.00	38.50	–	174.97	nr	**213.47**
installation tool for above	–	–	–	–	83.00	nr	**83.00**
SMOKING SHELTERS							
Smoking shelters							
Furnitubes International Ltd; Ashby free-standing bolt down; aluminium frame; PET glazing; 2050 mm length × 2330 mm high							
3290 mm depth; maximum 3 people	2046.33	4.00	77.00	–	2059.09	each	**2136.09**
6050 mm depth; maximum 6 people	2571.03	5.00	96.25	–	2583.79	each	**2680.04**

Q50 SITE/STREET FURNITURE/EQUIPMENT

Item Excluding site overheads and profit	PC £	Labour hours	Labour £	Plant £	Material £	Unit	Total rate £
FLAGPOLES							
Flagpoles							
Flagpoles; ground mounted; Harrison External Display Systems; in glass fibre; smooth white finish; including hinged baseplate, nylon halyard system and revolving gold onion finial; setting in concrete; to manufacturer's recommendations (excavating not included)							
6 m high; external halyard system	177.00	4.00	77.00	–	198.53	nr	**275.53**
6 m high; internal halyard system	279.99	4.00	77.00	–	301.52	nr	**378.52**
10 m high; external halyard system	322.00	5.00	96.25	–	343.53	nr	**439.78**
10 m high; internal halyard system	440.99	5.00	96.25	–	462.52	nr	**558.77**
12 m high; external halyard system	390.00	6.00	115.50	–	411.53	nr	**527.03**
12 m high; internal halyard system	502.99	6.00	115.50	–	524.52	nr	**640.02**
Flagpoles; wall mounted; Harrison External Display Systems; vertical or angled poles; in glass fibre; smooth white finish; including base, top bracket, external nylon halyard rope, cleat and revolving gold onion finial							
3 m pole	187.00	2.00	38.50	–	187.00	nr	**225.50**
Banner flagpoles; Harrison External Display Systems; 90 mm diameter aluminium pole; suitable for 1 × 2 m banner							
6 m high	458.62	4.00	77.00	–	480.15	nr	**557.15**
PLANT CONTAINERS							
Market prices of containers							
Plant containers; terracotta							
Capital Garden Products Ltd							
Large Pot LP63; weathered terracotta; 1170 × 1600 mm diameter	–	–	–	–	755.70	nr	**755.70**
Large Pot LP38; weathered terracotta; 610 × 970 mm diameter	–	–	–	–	342.50	nr	**342.50**
Large Pot LP23; weathered terracotta; 480 × 580 mm diameter	–	–	–	–	174.00	nr	**174.00**
Indian style Shimmer Pot 2322; 585 × 560 mm diameter	–	–	–	–	159.00	nr	**159.00**
Indian style Shimmer Pot 1717; 430 × 430 mm diameter	–	–	–	–	103.00	nr	**103.00**
Indian style Shimmer Pot 1314; 330 × 355 mm diameter	–	–	–	–	94.00	nr	**94.00**

Q50 SITE/STREET FURNITURE/EQUIPMENT

Item Excluding site overheads and profit	PC £	Labour hours	Labour £	Plant £	Material £	Unit	Total rate £
PLANT CONTAINERS – cont							
Plant containers; faux lead							
Capital Garden Products Ltd							
Trough 2508 Tudor Rose;							
620 × 220 × 230 mm high	–	–	–	–	60.00	nr	**60.00**
Tub 2004 Elizabethan; 510 mm square	–	–	–	–	94.00	nr	**94.00**
Tub 1513 Elizabethan; 380 mm square	–	–	–	–	64.00	nr	**64.00**
Tub 1601 Tudor Rose; 420 × 400 mm							
diameter	–	–	–	–	66.00	nr	**66.00**
Plant containers; window boxes							
Capital Garden Products Ltd							
Adam 5401; faux lead;							
1370 × 270 × 210 mm high	–	–	–	–	103.00	nr	**103.00**
Oakleaf OAK24; terracotta;							
610 × 230 × 240 mm high	–	–	–	–	83.00	nr	**83.00**
Swag 2402; faux lead; 610 × 200 × 210 mm							
high	–	–	–	–	53.75	nr	**53.75**
Plant containers; timber							
Plant containers; hardwood; Neptune Outdoor							
Furniture Ltd							
Beaufort T38–4D; 1500 × 1500 × 900 mm							
high	–	–	–	–	1270.00	nr	**1270.00**
Beaufort T38–3C; 1000 × 1500 × 700 mm							
high	–	–	–	–	930.00	nr	**930.00**
Beaufort T38–2 A; 1000 × 500 × 500 mm							
high	–	–	–	–	490.00	nr	**490.00**
Kara T42–4D; 1500 × 1500 × 900 mm high	–	–	–	–	1360.00	nr	**1360.00**
Kara T42–3C; 1000 × 1500 × 700 mm high	–	–	–	–	1000.00	nr	**1000.00**
Kara T42–2 A; 1000 × 500 × 500 mm high	–	–	–	–	530.00	nr	**530.00**
Plant containers; hardwood; Woodscape Ltd							
square; 900 × 900 × 420 mm high	–	–	–	–	359.75	nr	**359.75**
Measured works							
Plant containers; precast concrete							
Plant containers; Marshalls Plc							
Boulevard 700; circular base and ring	574.00	2.00	38.50	20.63	574.00	nr	**633.13**
Boulevard 1200; circular base and ring	728.00	1.00	19.25	20.63	728.00	nr	**767.88**

Q50 SITE/STREET FURNITURE/EQUIPMENT

Item Excluding site overheads and profit	PC £	Labour hours	Labour £	Plant £	Material £	Unit	Total rate £
PERGOLAS							
Pergolas; AVS Fencing Supplies Ltd; construct timber pergola; posts 150 × 150 mm × 2.40 m finished height in 600 mm deep minimum concrete pits; beams of 200 × 50 mm × 2.00 m wide; fixed to posts with dowels; rafters 200 × 38 mm notched to beams; inclusive of all mechanical excavation and disposal off site							
Pergola 2.00 m wide in green oak; posts at 2.00 m centres; dowel fixed							
rafters at 600 mm centres	191.62	8.40	161.70	3.75	255.49	m	**420.94**
rafters at 400 mm centres	237.34	10.80	207.90	3.75	308.18	m	**519.83**
Pergola 3.00 m wide in green oak; posts at 1.50 m centres							
rafters at 600 mm centres	262.72	7.20	138.60	4.50	338.21	m	**481.31**
rafters at 400 mm centres	331.17	8.00	154.00	4.50	417.09	m	**575.59**
Pergola 2.00 wide in prepared softwood; beams fixed with bolts; posts at 2.00 m centres							
rafters at 600 mm centres	86.22	5.00	96.25	3.75	150.59	m	**250.59**
rafters at 400 mm centres	101.67	6.00	115.50	3.75	173.00	m	**292.25**
Pergola 3.00 m wide in prepared softwood; posts at 1.50 m centres							
rafters at 600 mm centres	76.86	7.20	138.60	4.50	152.35	m	**295.45**
rafters at 400 mm centres	87.76	8.00	154.00	4.50	173.67	m	**332.17**
STONES AND BOULDERS							
Standing stones; CED Ltd; erect standing stones; vertical height above ground; in concrete base; including excavation setting in concrete to 1/3 depth and crane offload into position							
Purple schist							
1.00 m high	97.37	1.50	28.88	15.68	114.27	nr	**158.83**
1.25 m high	123.35	1.50	28.88	15.68	135.48	nr	**180.04**
1.50 m high	197.34	2.00	38.50	18.81	211.43	nr	**268.74**
2.00 m high	324.58	3.00	57.75	31.35	365.40	nr	**454.50**
2.50 m high	493.35	1.50	28.88	62.70	572.25	nr	**663.83**
Rockery stone – General Preamble: Rockery stone prices vary considerably with source, carriage, distance and load. Typical PC prices are in the range of £60–80 per tonne collected.							

Q50 SITE/STREET FURNITURE/EQUIPMENT

Item Excluding site overheads and profit	PC £	Labour hours	Labour £	Plant £	Material £	Unit	Total rate £
STONES AND BOULDERS – cont							
Rockery stone; CED Ltd							
Boulders; maximum distance 25 m; by machine							
750 mm diameter	100.29	0.90	17.32	5.59	100.29	nr	**123.20**
1 m diameter	237.74	2.00	38.50	11.17	237.74	nr	**287.41**
1.5 m diameter	802.37	2.00	38.50	49.18	802.37	nr	**890.05**
2 m diameter	1901.94	2.00	38.50	49.18	1901.94	nr	**1989.62**
Boulders; maximum distance 25 m; by hand							
750 mm diameter	100.29	0.75	14.44	–	100.29	nr	**114.73**
1 m diameter	237.74	1.89	36.30	–	237.74	nr	**274.04**
PLAYGROUND EQUIPMENT							
Playground equipment – General							
Preamble: The range of equipment manufactured or available in the UK is so great that comprehensive coverage would be impossible, especially as designs, specifications and prices change fairly frequently. The following information should be sufficient to give guidance to anyone designing or equipping a playground. In comparing prices note that only outline specification details are given here and that other refinements which are not mentioned may be the reason for some difference in price between two apparently identical elements. The fact that a particular manufacturer does not appear under one item heading does not necessarily imply that they do not make it. Landscape designers are advised to check that equipment complies with ever more stringent safety standards before specifying.							
Playground equipment – Installation							
The rates below include for installation of the specified equipment by the manufacturers. Most manufacturers will offer an option to install the equipment they have supplied.							
Play systems; Kompan Ltd; Galaxy; multiple play activity systems for non-prescribed play; for children 6–14 years; galvanized steel and high density polyethylene							
GXY906 Adara; 14 different play activities	–	–	–	–	–	nr	**15971.70**
GXY8011 Sirius; 10 different play activities	–	–	–	–	–	nr	**9977.25**

Q50 SITE/STREET FURNITURE/EQUIPMENT

Item Excluding site overheads and profit	PC £	Labour hours	Labour £	Plant £	Material £	Unit	Total rate £
Sports and social areas; Freegame multi-use games areas; Kompan Ltd; enclosed sports areas; complete with surfacing boundary and goals and targets; galvanized steel framework with high density polyethylene panels, galvanized steel goals and equipment; surfacing priced separately							
FRE1211 Classic Multigoal; 7 m	–	–	–	–	–	nr	5343.70
Pitch; complete; suitable for multiple ball sports; suitable for use with natural artificial or hard landscape surfaces; fully enclosed including two end sports walls							
FRE2110 Cosmos; 12 × 20 m	–	–	–	–	–	nr	31702.40
FRE2116 Cosmos; 19 × 36 m	–	–	–	–	–	nr	46897.00
FRE3000; Meeting Point; shelter or social area	–	–	–	–	–	nr	4365.60
Swings – General							
Preamble: Prices for the following vary considerably. Those given represent the middle of the range and include multiple swings with tubular steel frames and timber or tyre seats; ground fixing and priming only.							
Swings							
Swings; Wicksteed Ltd							
traditional swings; 1850 mm high; 1 bay; 2 seat	–	–	–	–	–	nr	2367.22
traditional swings; 1850 mm high; 2 bay; 4 seat	–	–	–	–	–	nr	3803.96
traditional swings; 2450 mm high; 1 bay; 2 seat	–	–	–	–	–	nr	2287.24
traditional swings; 2450 mm high; 2 bay; 4 seat	–	–	–	–	–	nr	3610.57
traditional swings; 3050 mm high; 1 bay; 2 seat	–	–	–	–	–	nr	2427.97
traditional swings; 3050 mm high; 2 bay; 4 seat	–	–	–	–	–	nr	3784.72
double arch swing; cradle safety seats; 1850 mm high	–	–	–	–	–	nr	2401.65
twin double arch swing; cradle safety seat; 1850 mm high	–	–	–	–	–	nr	4022.66
single arch swing; flat rubber safety seat; 2450 mm high	–	–	–	–	–	nr	1848.83
double arch swing; flat rubber safety seats; 2450 mm high	–	–	–	–	–	nr	2268.00
Swings; Lappset UK Ltd							
020414M; swing frame with two flat seats	–	–	–	–	–	nr	1520.00

Q50 SITE/STREET FURNITURE/EQUIPMENT

Item Excluding site overheads and profit	PC £	Labour hours	Labour £	Plant £	Material £	Unit	Total rate £
PLAYGROUND EQUIPMENT – cont							
Swings – cont							
Swings; Kompan Ltd							
M951P Sunflower swing; 1–3 years	–	–	–	–	–	nr	1330.25
M947P double swings; 1–6 years	–	–	–	–	–	nr	2116.20
M961P double swings; 6–12 years	–	–	–	–	–	nr	1782.00
Slides							
Slides; Wicksteed Ltd							
Pedestal slides; 3.40 m	–	–	–	–	–	nr	4275.79
Pedestal slides; 4.40 m	–	–	–	–	–	nr	3284.55
Pedestal slides; 5.80 m	–	–	–	–	–	nr	4975.43
Embankment slides; 3.40 m	–	–	–	–	–	nr	2516.06
Embankment slides; 4.40 m	–	–	–	–	–	nr	3271.39
Embankment slides; 5.80 m	–	–	–	–	–	nr	4163.40
Embankment slides; 7.30 m	–	–	–	–	–	nr	5290.31
Embankment slides; 9.10 m	–	–	–	–	–	nr	6626.81
Embankment slides; 11.00 m	–	–	–	–	–	nr	7924.84
Slides; Lappset UK Ltd							
142015M slide	–	–	–	–	–	nr	3682.00
141115M Jumbo slide	–	–	–	–	–	nr	5518.00
Slides; Kompan Ltd							
M351P slide	–	–	–	–	–	nr	2972.65
M326P Aladdin's Cave slide	–	–	–	–	–	nr	2754.75
Moving equipment – General							
Preamble: The following standard items of playground equipment vary considerably in quality and price; the following prices are middle of the range.							
Moving equipment							
Roundabouts; Wicksteed Ltd							
Turnstile	–	–	–	–	–	nr	915.30
Speedway (without restrictor)	–	–	–	–	–	nr	3606.53
Spiro Whirl (without restrictor)	–	–	–	–	–	nr	4109.74
Roundabouts; Kompan Ltd							
Supernova GXY 916; multifunctional spinning and balancing disc; capacity approximately 15 children	–	–	–	–	–	nr	4535.75
Seesaws							
Seesaws; Lappset UK Ltd							
010300; seesaws	–	–	–	–	–	nr	962.00
010237; seesaws	–	–	–	–	–	nr	2439.00
Seesaws; Wicksteed Ltd							
Seesaw; non-bump	–	–	–	–	–	nr	2663.89
Jolly Gerald; non-bump	–	–	–	–	–	nr	2872.46
Rocking Rockette; with motion restrictor	–	–	–	–	–	nr	3853.57
Rocking Horse; with motion restrictor	–	–	–	–	–	nr	4395.26

Q50 SITE/STREET FURNITURE/EQUIPMENT

Item Excluding site overheads and profit	PC £	Labour hours	Labour £	Plant £	Material £	Unit	Total rate £
Play sculptures – General Preamble: Many variants on the shapes of playground equipment are available, simulating spacecraft, trains, cars, houses etc., and these designs are frequently changed. The basic principles remain constant but manufacturer's catalogues should be checked for the latest styles.							
Climbing equipment and play structures – General Preamble: Climbing equipment generally consists of individually designed modules. Play structures generally consist of interlinked and modular pieces of equipment and sculptures. These may consist of climbing, play and skill based modules, nets and various other activities. Both are set into either safety surfacing or defined sand pit areas. The equipment below outlines a range from various manufacturers. Individual catalogues should be consulted in each instance. Safety areas should be allowed round all equipment.							
Climbing equipment and play structures Climbing equipment; Wicksteed Ltd							
Funrun Fitness Trail; Under Starter's Orders; set of 12 units	–	–	–	–	–	nr	16887.49
Climbing equipment; Lappset UK Ltd							
138401M Storks Nest	–	–	–	–	–	nr	4368.00
122457M Playhouse	–	–	–	–	–	nr	3765.00
120100M Activity Tower	–	–	–	–	–	nr	13853.00
120124M Tower and Climbing Frame	–	–	–	–	–	nr	13479.00
SMP Playgrounds Ltd							
Nexus – The Core; multi-play structure	–	–	–	–	–	nr	11961.00
Spring equipment Spring based equipment for 1–8 year olds; Kompan Ltd							
M101P Crazy Hen	–	–	–	–	–	nr	594.05
M128P Crazy Daisy	–	–	–	–	–	nr	868.10
M141P Spring Seesaw	–	–	–	–	–	nr	2072.15
M155P Quartet Seesaw	–	–	–	–	–	nr	1422.25
Spring based equipment for under 12's; Lappset UK Ltd							
010501 Horse Springer	–	–	–	–	–	nr	1113.00
Sandpits Market prices							
play pit sand; Boughton Loam Ltd	122.40	–	–	–	–	m³	–
Kompan Ltd							
Basic550; 276 × 154 × 31 cm deep	–	–	–	–	–	nr	1115.40

Q50 SITE/STREET FURNITURE/EQUIPMENT

Item Excluding site overheads and profit	PC £	Labour hours	Labour £	Plant £	Material £	Unit	Total rate £
SPORTS EQUIPMENT							
Sports equipment							
Tennis posts; steel; suitable for hard or grass tennis courts; including winder, sockets and dust cap							
round	257.99	1.00	19.25	–	257.99	set	**277.24**
square	273.99	1.00	19.25	–	273.99	set	**293.24**
Tennis nets; not including posts or fixings							
Tournament	114.99	3.00	57.75	–	122.17	set	**179.92**
Match	89.99	3.00	57.75	–	97.17	set	**154.92**
Club	69.99	3.00	57.75	–	77.17	set	**134.92**
Football goals; full size; socketed; including international net supports and nets							
aluminium	1380.00	4.00	77.00	–	1387.18	set	**1464.18**
steel; heavyweight	1040.00	0.14	2.77	–	1239.34	set	**1242.11**
Football goals; full size; free-standing; including nets							
aluminium	1420.00	4.00	77.00	–	1420.00	set	**1497.00**
steel	1110.00	4.00	77.00	–	1110.00	set	**1187.00**
Mini-soccer goals; free-standing; including nets							
aluminium	645.00	4.00	77.00	–	645.00	set	**722.00**
steel	525.00	4.00	77.00	–	525.00	set	**602.00**
Rugby posts; socketed							
aluminium; 10 m high	1350.50	6.00	115.50	–	1369.64	set	**1485.14**
aluminium; 12 m high	1450.50	6.00	115.50	–	1469.64	set	**1585.14**
steel; 12 m high	1750.00	6.00	115.50	–	1769.14	set	**1884.64**
Hockey goals; steel; including backboards and nets							
socketed	950.00	1.00	19.25	–	950.00	set	**969.25**
free-standing	1130.00	1.00	19.25	–	1130.00	set	**1149.25**
Cricket cages; steel; including netting							
free-standing	895.50	12.00	231.00	–	895.50	set	**1126.50**
wheelaway	1195.50	12.00	231.00	–	1195.50	set	**1426.50**

R12 DRAINAGE

Item Excluding site overheads and profit	PC £	Labour hours	Labour £	Plant £	Material £	Unit	Total rate £
R12 DRAINAGE							
CLARIFICATION NOTES ON LABOUR COSTS IN THIS SECTION							
General groundworks team							
Generally a three man team is used in this section; The column 'Labour hours' reports team hours. The column 'Labour £' reports the total cost of the team for the unit of work shown							
3 man team	–	1.00	57.75	–	–	hr	**57.75**
EXCAVATION FOR DRAINAGE							
Machine excavation							
Excavating pits; starting from ground level; works exclude for earthwork retention							
maximum depth not exceeding 1.00 m	–	0.50	9.63	23.15	–	m³	**32.78**
maximum depth not exceeding 2.00 m	–	0.50	9.63	41.64	–	m³	**51.27**
maximum depth not exceeding 4.00 m	–	0.50	9.63	61.87	–	m³	**71.50**
Disposal of excavated material; depositing on site in permanent spoil heaps; average 50 m	–	0.04	0.80	1.92	–	m³	**2.72**
Filling to excavations; obtained from on site spoil heaps; average thickness not exceeding 0.25 m	–	0.13	2.57	5.50	–	m³	**8.07**
Hand excavation							
Excavating pits; starting from ground level; works exclude for earthwork retention							
maximum depth not exceeding 1.00 m	–	1.20	46.20	–	–	m³	**46.20**
maximum depth not exceeding 2.00 m	–	1.50	57.75	–	–	m³	**57.75**
Works to pits or drainage excavations							
Surface treatments; compacting; bottoms of excavations	–	0.05	0.96	–	–	m²	**0.96**
Earthwork support; distance between opposing faces not exceeding 2.00 m							
maximum depth not exceeding 1.00 m	–	0.20	3.85	–	6.29	m³	**10.14**
maximum depth not exceeding 2.00 m	–	0.30	5.78	–	6.29	m³	**12.07**
maximum depth not exceeding 4.00 m	–	0.67	12.83	–	2.89	m³	**15.72**

R12 DRAINAGE

Item Excluding site overheads and profit	PC £	Labour hours	Labour £	Plant £	Material £	Unit	Total rate £
INSPECTION CHAMBERS AND MANHOLES							
Inspection chambers; in situ concrete							
Beds; plain in situ concrete; 11.50 N/mm^2 –							
40 mm aggregate							
thickness not exceeding 150 mm	–	2.00	38.50	–	99.67	m^3	**138.17**
thickness 150–450 mm	–	1.75	33.69	–	99.67	m^3	**133.36**
Benchings in bottoms; plain in situ concrete;							
25.50 N/mm^2 – 20 mm aggregate							
thickness 150–450 mm	–	2.00	38.50	–	99.67	m^3	**138.17**
Isolated cover slabs; reinforced In situ							
concrete; 21.00 N/mm^2 – 20 mm aggregate							
thickness not exceeding 150 mm	99.67	4.00	77.00	–	99.67	m^3	**176.67**
Fabric reinforcement; A193 (3.02 kg/m^2) in							
cover slabs	1.94	0.06	1.21	–	1.94	m^2	**3.15**
Formwork to reinforced in situ concrete;							
isolated cover slabs							
soffits; horizontal	–	3.28	63.14	–	5.53	m^2	**68.67**
height not exceeding 250 mm	–	0.97	18.67	–	3.15	m	**21.82**
Inspection chambers; precast concrete units							
Precast concrete inspection chamber units; FP							
McCann Ltd; bedding, jointing and pointing in							
cement mortar (1:3); 600 × 450 mm internally							
600 mm deep	51.82	6.00	115.50	–	56.12	nr	**171.62**
900 mm deep	64.78	7.00	134.75	–	69.08	nr	**203.83**
Drainage chambers; FP McCann Ltd;							
1200 × 750 mm reducing to 600 × 600 mm; no							
base unit; depth of invert							
1050 mm deep	392.57	9.00	173.25	–	399.74	nr	**572.99**
1650 mm deep	581.17	11.00	211.75	–	592.64	nr	**804.39**
2250 mm deep	769.77	12.50	240.63	–	784.10	nr	**1024.73**
Cover slabs for chambers or shaft sections;							
FP McCann Ltd; heavy duty							
900 mm diameter internally	62.00	0.67	12.82	–	62.00	nr	**74.82**
1050 mm diameter internally	68.00	2.00	38.50	13.75	68.00	nr	**120.25**
1200 mm diameter internally	82.00	1.00	19.25	13.75	82.00	nr	**115.00**
1500 mm diameter internally	139.04	1.00	19.25	13.75	139.04	nr	**172.04**
1800 mm diameter internally	208.47	2.00	38.50	30.94	208.47	nr	**277.91**
Brickwork							
Walls to manholes; bricks; PC £300.00/1000;							
in cement mortar (1:3)							
one brick thick	37.80	3.00	57.75	–	46.40	m^2	**104.15**
one and a half brick thick	56.70	4.00	77.00	–	69.60	m^2	**146.60**
two brick thick projection of footing or the							
like	75.60	4.80	92.40	–	92.79	m^2	**185.19**

R12 DRAINAGE

Item Excluding site overheads and profit	PC £	Labour hours	Labour £	Plant £	Material £	Unit	Total rate £
Walls to manholes; engineering bricks; in cement mortar (1:3)							
one brick thick	34.78	3.00	57.75	–	43.37	m²	**101.12**
one and a half brick thick	52.16	4.00	77.00	–	65.06	m²	**142.06**
two brick thick projection of footing or the like	69.55	4.80	92.40	–	86.75	m²	**179.15**
Extra over common or engineering bricks in any mortar for fair face; flush pointing as work proceeds; English bond walls or the like	–	0.13	2.57	–	–	m²	**2.57**
In situ finishings; cement: sand mortar (1:3); steel trowelled; 13 mm one coat work to manhole walls; to brickwork or blockwork base; over 300 mm wide	–	0.80	15.40	–	2.87	m²	**18.27**
Building into brickwork; ends of pipes; making good facings or renderings							
small	–	0.20	3.85	–	–	nr	**3.85**
large	–	0.30	5.78	–	–	nr	**5.78**
extra large	–	0.40	7.70	–	–	nr	**7.70**
extra large; including forming ring arch cover	–	0.50	9.63	–	–	nr	**9.63**
Inspection chambers; polypropylene; Hepworth Plc							
Mini access chamber; up to 600 mm deep; including cover and frame							
300 mm diameter × 600 mm deep; three 100/110 mm inlets	151.32	3.00	57.75	–	153.88	nr	**211.63**
Up to 1200 mm deep; including polymer cover and frame with screw down lid							
475 mm diameter × 940 mm deep; five 100/110 mm inlets; supplied with four stoppers in inlets	223.71	4.00	77.00	–	298.60	nr	**375.60**
Extra for square ductile iron cover and frame to 1 tonne load; screw down lid	103.78	–	–	–	31.45	nr	**31.45**
Step irons; drainage systems; malleable cast iron; galvanized; building into joints							
General purpose pattern; for one brick walls	4.14	0.17	3.27	–	4.14	nr	**7.41**
Accessories in PVC-u							
110 mm screwed access cover	15.78	–	–	–	15.78	nr	**15.78**
110 mm rodding eye	30.52	0.50	9.63	–	34.79	nr	**44.42**
gully with P traps; 110 mm; 154 × 154 mm grating	70.26	1.00	19.25	–	71.97	nr	**91.22**
Kerbs; to gullies; in one course Class B engineering bricks; to four sides; rendering in cement: mortar (1:3); dished to gully gratings	2.32	1.00	19.25	–	3.18	nr	**22.43**

R12 DRAINAGE

Item Excluding site overheads and profit	PC £	Labour hours	Labour £	Plant £	Material £	Unit	Total rate £
CHANNELS							
Best quality vitrified clay half section channels; Hepworth Plc; bedding and jointing in cement: mortar (1:2)							
Channels; straight							
100 mm	6.37	0.80	15.40	–	9.27	m	**24.67**
150 mm	10.60	1.00	19.25	–	13.50	m	**32.75**
225 mm	23.81	1.35	25.99	–	26.71	m	**52.70**
300 mm	48.87	1.80	34.65	–	51.77	m	**86.42**
Bends; 15, 30, 45 or 90°							
100 mm bends	5.74	0.75	14.44	–	7.19	nr	**21.63**
150 mm bends	9.91	0.90	17.32	–	12.09	nr	**29.41**
225 mm bends	38.42	1.20	23.10	–	41.32	nr	**64.42**
300 mm bends	78.32	1.10	21.18	–	81.96	nr	**103.14**
Best quality vitrified clay channels; Hepworth Plc; bedding and jointing in cement: mortar (1:2)							
Branch bends; 15, 30, 45 or 90°; left or right hand							
100 mm	5.74	0.75	14.44	–	7.19	nr	**21.63**
150 mm	9.91	0.90	17.32	–	12.09	nr	**29.41**
PIPE LAYING							
Excavating trenches; to receive pipes; grading bottoms to falls; backfilling with excavated material and compacting; disposal of surplus material off site; volumes allow for bedding materials which are priced separately below							
Trenches 300 mm wide							
depth of pipe 750 mm	–	–	–	10.93	2.92	m³	**13.85**
depth of pipe 900 mm	–	–	–	12.49	3.69	m³	**16.18**
depth of pipe 1.20 m	–	–	–	13.11	5.24	m³	**18.35**
depth of pipe 1.50 m	–	–	–	14.57	6.78	m³	**21.35**
depth of pipe 2.00 m	–	–	–	12.34	9.35	m³	**21.69**
Earthwork support; providing support to opposing faces of excavation; moving along as work proceeds							
Maximum depth not exceeding 2.00 m							
distance between opposing faces not exceeding 2.00 m	–	0.80	15.40	21.00	–	m	**36.40**

R12 DRAINAGE

Item Excluding site overheads and profit	PC £	Labour hours	Labour £	Plant £	Material £	Unit	Total rate £
Excavating trenches; using 3 tonne tracked excavator; to receive pipes; grading bottoms; earthwork support; filling with excavated material to within 150 mm of finished surfaces and compacting; completing fill with topsoil; disposal of surplus soil							
Services not exceeding 200 mm nominal size							
average depth of run not exceeding 0.50 m	1.51	0.12	2.31	1.05	1.51	m	**4.87**
average depth of run not exceeding 0.75 m	1.51	0.16	3.14	1.44	1.51	m	**6.09**
average depth of run not exceeding 1.00 m	1.51	0.28	5.45	2.50	1.51	m	**9.46**
average depth of run not exceeding 1.25 m	1.26	0.38	7.38	3.37	1.26	m	**12.01**
Granular beds to trenches; lay granular material to trenches excavated separately; to receive pipes (not included)							
300 mm wide × 100 mm thick							
reject sand	–	0.05	0.96	0.23	3.67	m	**4.86**
reject gravel	–	0.05	0.96	0.23	3.14	m	**4.33**
shingle 40 mm aggregate	–	0.05	0.96	0.23	3.50	m	**4.69**
sharp sand	3.67	0.05	0.96	0.23	3.67	m	**4.86**
300 mm wide × 150 mm thick							
reject sand	–	0.08	1.44	0.34	5.50	m	**7.28**
reject gravel	–	0.08	1.44	0.34	4.71	m	**6.49**
shingle 40 mm aggregate	–	0.08	1.44	0.34	5.25	m	**7.03**
sharp sand	5.50	0.08	1.44	0.34	5.50	m	**7.28**
Excavating trenches; using 3 tonne tracked excavator; to receive pipes; grading bottoms; earthwork support; filling with imported granular material type 2 and compacting; disposal of surplus soil							
Services not exceeding 200 mm nominal size							
average depth of run not exceeding 0.50 m	1.37	0.09	1.67	0.75	4.46	m	**6.88**
average depth of run not exceeding 0.75 m	2.05	0.11	2.08	0.94	6.68	m	**9.70**
average depth of run not exceeding 1.00 m	2.74	0.14	2.69	1.21	8.91	m	**12.81**
average depth of run not exceeding 1.25 m	3.42	0.23	4.40	2.01	11.14	m	**17.55**
Excavating trenches; using 3 tonne tracked excavator; to receive pipes; grading bottoms; earthwork support; filling with concrete, ready mixed ST2; disposal of surplus soil							
Services not exceeding 200 mm nominal size							
average depth of run not exceeding 0.50 m	11.67	0.11	2.05	0.36	14.76	m	**17.17**
average depth of run not exceeding 0.75 m	17.50	0.13	2.50	0.45	22.13	m	**25.08**
average depth of run not exceeding 1.00 m	23.34	0.17	3.21	0.60	29.51	m	**33.32**
average depth of run not exceeding 1.25 m	29.18	0.23	4.33	0.90	36.90	m	**42.13**

R12 DRAINAGE

Item Excluding site overheads and profit	PC £	Labour hours	Labour £	Plant £	Material £	Unit	Total rate £
PIPE LAYING – cont							
Earthwork support; providing support to							
opposing faces of excavation; moving							
along as work proceeds							
Maximum depth not exceeding 2.00 m							
trenchbox; distance between opposing							
faces not exceeding 2.00 m	–	0.80	15.40	21.00	–	m	36.40
timber; distance between opposing faces							
not exceeding 500 mm	–	0.20	7.70	–	0.03	m	7.73
Clay pipes and fittings; Hepworth Plc;							
Supersleve							
100 mm clay pipes; polypropylene slip							
coupling; in trenches (trenches not included)							
laid straight	9.40	0.25	4.81	–	12.11	m	16.92
short runs under 3.00 m	9.40	0.31	6.02	–	12.11	m	18.13
Extra over 100 mm clay pipes for							
bends; 15–90°; single socket	12.23	0.25	4.81	–	12.23	nr	17.04
junction; 45 or 90°; double socket	25.77	0.25	4.81	–	25.77	nr	30.58
slip couplings; polypropylene	4.33	0.08	1.60	–	4.33	nr	5.93
gully with P trap; 100 mm; 154 × 154 mm							
plastic grating	39.93	1.00	19.25	–	58.76	nr	78.01
150 mm clay pipes; polypropylene slip							
coupling; in trenches (trenches not included)							
laid straight	28.66	0.30	5.78	–	28.66	m	34.44
short runs under 3.00 m	28.66	0.60	11.55	–	28.66	m	40.21
Extra over 150 mm clay pipes for							
bends; 15–90°	16.32	0.28	5.39	–	24.20	nr	29.59
junction; 45 or 90°; 100 × 150 mm	23.98	0.40	7.70	–	39.73	nr	47.43
junction; 45 or 90°; 150 × 150 mm	21.85	0.40	7.70	–	37.60	nr	45.30
slip couplings; polypropylene	7.88	0.05	0.96	–	7.88	nr	8.84
tapered pipe; 100–150 mm	24.55	0.50	9.63	–	24.55	nr	34.18
tapered pipe; 150–225 mm	63.05	0.50	9.63	–	63.05	nr	72.68
socket adaptor; connection to traditional							
pipes and fittings	16.56	0.33	6.35	–	24.44	nr	30.79
Accessories in clay							
access pipe; 150 mm	61.56	–	–	–	77.31	nr	77.31
rodding eye; 150 mm	58.88	0.50	9.63	–	63.15	nr	72.78
gully with P traps; 150 mm; 154 × 154 mm							
plastic grating	79.97	0.80	15.40	–	89.56	nr	104.96
PVC-u pipes and fittings; Wavin Plastics							
Ltd; OsmaDrain							
110 mm PVC-u pipes; in trenches (trenches							
not included)							
laid straight	7.67	0.08	1.54	–	7.67	m	9.21
short runs under 3.00 m	8.63	0.12	2.31	–	8.63	m	10.94

R12 DRAINAGE

Item Excluding site overheads and profit	PC £	Labour hours	Labour £	Plant £	Material £	Unit	Total rate £
Extra over 110 mm PVC-u pipes for							
bends; short radius	14.94	0.25	4.81	–	14.94	nr	**19.75**
bends; long radius	27.99	0.25	4.81	–	27.99	nr	**32.80**
junctions; equal; double socket	17.82	0.25	4.81	–	17.82	nr	**22.63**
slip couplings	8.64	0.25	4.81	–	8.64	nr	**13.45**
adaptors to clay	16.84	0.50	9.63	–	16.84	nr	**26.47**
160 mm PVC-u pipes; in trenches (trenches not included)							
laid straight	17.99	0.08	1.54	–	17.99	m	**19.53**
short runs under 3.00 m	34.18	0.12	2.31	–	34.18	m	**36.49**
Extra over 160 mm PVC-u pipes for							
socket bend; double; 90 or 45°	56.31	0.20	3.85	–	56.31	nr	**60.16**
socket bend; double; 15 or 30°	52.65	0.20	3.85	–	52.65	nr	**56.50**
socket bend; single; 87.5 or 45°	32.02	0.20	3.85	–	32.02	nr	**35.87**
socket bend; single; 15 or 30°	28.39	0.20	3.85	–	28.39	nr	**32.24**
bends; short radius	41.97	0.25	4.81	–	56.90	nr	**61.71**
bends; long radius	109.07	0.25	4.81	–	109.07	nr	**113.88**
junctions; single	106.23	0.33	6.42	–	106.23	nr	**112.65**
pipe coupler	20.98	0.05	0.96	–	20.98	nr	**21.94**
slip couplings PVC-u	10.46	0.05	0.96	–	10.46	nr	**11.42**
adaptors to clay	45.76	0.50	9.63	–	45.76	nr	**55.39**
level invert reducer	17.22	0.50	9.63	–	17.22	nr	**26.85**
spiggot	38.21	0.20	3.85	–	38.21	nr	**42.06**
GULLIES AND INTERCEPTION TRAPS							
Intercepting traps; Hepworth Plc							
Vitrified clay; inspection arms; brass stoppers; iron levers; chains and staples; galvanized; staples cut and pinned to brickwork; cement: mortar (1:2) joints to vitrified clay pipes and channels; bedding and surrounding in concrete; 11.50 N/mm^2 – 40 mm aggregate; cutting and fitting brickwork; making good facings							
100 mm inlet; 100 mm outlet	83.33	3.00	57.75	–	93.71	nr	**151.46**
150 mm inlet; 150 mm outlet	120.16	2.00	38.50	–	137.61	nr	**176.11**
Gullies; concrete; FP McCann							
Concrete road gullies; trapped; cement: mortar (1:2) joints to concrete pipes; bedding and surrounding in concrete; 11.50 N/mm^2 – 40 mm aggregate; 450 mm diameter × 1.07 m deep; rodding eye; stoppers	46.00	6.00	115.50	–	67.01	nr	**182.51**

R12 DRAINAGE

Item Excluding site overheads and profit	PC £	Labour hours	Labour £	Plant £	Material £	Unit	Total rate £
GULLIES AND INTERCEPTION TRAPS – **cont**							
Gullies; vitrified clay; Hepworth Plc; **bedding in concrete; 11.50 N/mm² – 40 mm** **aggregate**							
Yard gullies (mud); trapped; domestic duty (up to 1 tonne)							
100 mm outlet; 100 mm diameter; 225 mm internal width; 585 mm internal depth	108.37	3.50	67.38	–	108.97	nr	**176.35**
150 mm outlet; 100 mm diameter; 225 mm internal width; 585 mm internal depth	108.37	3.50	67.38	–	108.97	nr	**176.35**
Yard gullies (mud); trapped; medium duty (up to 5 tonnes)							
100 mm outlet; 100 mm diameter; 225 mm internal width; 585 mm internal depth	153.19	3.50	67.38	–	153.80	nr	**221.18**
150 mm outlet; 100 mm diameter; 225 mm internal width; 585 mm internal depth	167.85	3.50	67.38	–	168.45	nr	**235.83**
Combined filter and silt bucket for yard gullies							
225 mm diameter	39.19	–	–	–	39.19	nr	**39.19**
Road gullies; trapped with rodding eye							
100 mm outlet; 300 mm internal diameter; 600 mm internal depth	–	3.50	67.38	–	104.83	nr	**172.21**
150 mm outlet; 300 mm internal diameter; 600 mm internal depth	106.73	3.50	67.38	–	107.34	nr	**174.72**
150 mm outlet; 400 mm internal diameter; 750 mm internal depth	123.78	3.50	67.38	–	124.38	nr	**191.76**
150 mm outlet; 450 mm internal diameter; 900 mm internal depth	167.47	3.50	67.38	–	168.08	nr	**235.46**
Hinged gratings and frames for gullies; alloy							
193 mm for 150 mm diameter gully	–	–	–	–	25.97	nr	**25.97**
120 × 120 mm	–	–	–	–	9.23	nr	**9.23**
150 × 150 mm	–	–	–	–	16.71	nr	**16.71**
230 × 230 mm	–	–	–	–	30.55	nr	**30.55**
316 × 316 mm	–	–	–	–	80.98	nr	**80.98**
Hinged gratings and frames for gullies; cast iron							
265 mm for 225 mm diameter gully	–	–	–	–	51.89	nr	**51.89**
150 × 150 mm	–	–	–	–	16.71	nr	**16.71**
230 × 230 mm	–	–	–	–	30.55	nr	**30.55**
316 × 316 mm	–	–	–	–	80.98	nr	**80.98**
Universal gully trap PVC-u; Wavin Plastics **Ltd; OsmaDrain system; bedding in** **concrete; 11.50 N/mm² – 40 mm aggregate**							
Universal gully fitting; comprising gully trap only							
110 mm outlet; 110 mm diameter; 205 mm internal depth	12.67	3.50	67.38	–	13.27	nr	**80.65**
Vertical inlet hopper; c\w plastic grate							
272 × 183 mm	17.02	0.25	4.81	–	17.02	nr	**21.83**
Sealed access hopper							
110 × 110 mm	40.02	0.25	4.81	–	40.02	nr	**44.83**

R12 DRAINAGE

Item Excluding site overheads and profit	PC £	Labour hours	Labour £	Plant £	Material £	Unit	Total rate £
Universal gully PVC-u; Wavin Plastics Ltd; **OsmaDrain system; accessories to** **universal gully trap**							
Hoppers; backfilling with clean granular material; tamping; surrounding in lean mix concrete							
plain hopper; with 110 mm spigot; 150 mm long	13.92	0.40	7.70	–	14.27	nr	**21.97**
vertical inlet hopper; with 110 mm spigot; 150 mm long	17.02	0.40	7.70	–	17.02	nr	**24.72**
sealed access hopper; with 110 mm spigot; 150 mm long	40.02	0.40	7.70	–	40.02	nr	**47.72**
plain hopper; solvent weld to trap	9.50	0.40	7.70	–	9.50	nr	**17.20**
vertical inlet hopper; solvent wed to trap	16.33	0.40	7.70	–	16.33	nr	**24.03**
sealed access cover; PVC-u	20.38	0.10	1.93	–	20.38	nr	**22.31**
Gullies PVC-u; Wavin Plastics Ltd; **OsmaDrain system; bedding in concrete;** **11.50 N/mm² – 40 mm aggregate**							
Bottle gully; providing access to the drainage system for cleaning							
bottle gully; 228 × 228 × 317 mm deep	33.79	0.50	9.63	–	34.30	nr	**43.93**
sealed access cover; PVC-u; 217 × 217 mm	26.27	0.10	1.93	–	26.27	nr	**28.20**
grating; ductile iron; 215 × 215 mm	20.25	0.10	1.93	–	20.25	nr	**22.18**
bottle gully riser; 325 mm	4.36	0.50	9.63	–	5.46	nr	**15.09**
Yard gully; trapped; 300 mm diameter × 600 mm deep; including catchment bucket and ductile iron cover and frame; medium duty loading							
305 mm diameter × 600 mm deep	204.90	2.50	48.13	–	208.07	nr	**256.20**
Kerbs to gullies							
One course Class B engineering bricks to four sides; rendering in cement: mortar (1:3); dished to gully gratings							
150 × 150 mm	1.38	0.33	6.42	–	1.74	nr	**8.16**

R12 DRAINAGE

Item Excluding site overheads and profit	PC £	Labour hours	Labour £	Plant £	Material £	Unit	Total rate £
LINEAR DRAINAGE							
Marshalls Plc; Mini Beany combined kerb and channel drainage system; to trenches (not included)							
Precast concrete drainage channel base; 185–385 mm deep; bedding, jointing and pointing in cement mortar (1:3); on 150 mm deep concrete (ready mixed) foundation; including haunching with in situ concrete; 11.50 N/mm^2 – 40 mm aggregate one side; channels 250 mm wide × 1.00 m long							
straight; 1.00 m long	22.23	1.00	19.25	–	25.67	m	**44.92**
straight; 500 mm long	22.61	1.05	20.21	–	26.05	m	**46.26**
radial; 30/10 or 9/6 internal or external	20.89	1.33	25.67	–	24.33	m	**50.00**
angles 45 or 90°	62.33	1.00	19.25	–	65.77	nr	**85.02**
Mini Beany Top Block; perforated kerb unit to drainage channel above; natural grey							
straight	11.44	0.33	6.42	–	12.72	m	**19.14**
radial; 30/10 or 9/6 internal or external	14.94	0.50	9.63	–	16.22	m	**25.85**
angles 45 or 90°	34.80	0.50	9.63	–	36.08	nr	**45.71**
Mini Beany; outfalls; two section concrete trapped outfall with Mini Beany cast iron access cover and frame; to concrete foundation							
high capacity outfalls; silt box 150/225 mm outlet; two section trapped outfall silt box and cast iron access cover	42.23	1.00	19.25	–	261.02	nr	**280.27**
inline side or end outlet; outfall 150 mm; 2 section concrete trapped outfall; cast iron Mini Beany access cover and frame	226.08	1.00	19.25	–	226.80	nr	**246.05**
Ancillaries to Mini Beany							
end cap	12.66	0.25	4.81	–	12.66	nr	**17.47**
end cap outlets	32.90	0.25	4.81	–	32.90	nr	**37.71**
Precast concrete channels; Charcon Hard Landscaping; on 150 mm deep concrete foundations; including haunching with in situ concrete; 21.00 N/mm^2 – 20 mm aggregate; both sides							
Charcon Safeticurb; slotted safety channels; for pedestrians and light vehicles							
DBJ; 305 × 305 mm	80.03	0.67	12.83	–	96.97	m	**109.80**
DBA; 250 × 250 mm	41.20	0.67	12.83	–	58.14	m	**70.97**
Charcon Safeticurb; slotted safety channels; for heavy vehicles							
DBM; 248 × 248 mm	73.36	0.80	15.40	–	90.30	m	**105.70**
Clearway; 324 × 257 mm	84.14	0.80	15.40	–	101.08	m	**116.48**

R12 DRAINAGE

Item Excluding site overheads and profit	PC £	Labour hours	Labour £	Plant £	Material £	Unit	Total rate £
Charcon Safeticurb; inspection units; ductile iron lids; including jointing to drainage channels							
248 × 248 × 914 mm	99.84	1.50	28.88	–	100.84	nr	**129.72**
Silt box tops; concrete frame; cast iron grid lids; type 1; set over gully							
457 × 610 mm	447.48	2.00	38.50	–	448.48	nr	**486.98**
Manhole covers; type K; cast iron; providing inspection to blocks and back gullies	568.54	2.00	38.50	–	570.97	nr	**609.47**
Slot and Channel drains							
Loadings for slot drains							
A15; 1.5 tonne; pedestrian							
B125; 12.5 tonne; domestic use							
C250; 25 tonne; car parks, supermarkets, industrial units							
D400; 40 tonne; highways							
E600; 60 tonne; forklifts							
Slot drains; ACO Technologies; laid to concrete bed C25 on compacted granular base on 200 mm deep concrete bed; haunched with 200 mm concrete surround; all in 750 × 430 mm wide trench with compacted 200 mm granular base surround (excavation and subbase not included)							
ACO MultiDrain M100PPD; recycled polypropylene drainage channel; range of gratings to complement installations which require discreet slot drainage							
142 mm wide × 150 mm deep	71.50	1.20	23.10	–	82.34	m	**105.44**
Accessories for M100PPD							
connectors; vertical outlet 110 mm	14.65	–	–	–	14.65	nr	**14.65**
connectors; vertical outlet 160 mm	16.45	–	–	–	16.45	nr	**16.45**
sump units; 110 mm	130.45	1.50	28.88	–	150.38	nr	**179.26**
universal gully and bucket;							
440 × 440 × 1315 mm deep	762.55	3.00	57.75	–	799.93	nr	**857.68**
Channel drains; ACO Technologies; ACO MultiDrain MD polymer concrete channel drainage system; traditional channel and grate drainage solution							
ACO MultiDrain MD Brickslot; offset galvanized slot drain grating for M100PPD; load class C250							
brickslot; galvanized steel; 1.00 mm	81.75	–	–	–	81.75	m	**81.75**
ACO MultiDrain MD Brickslot; offset slot drain grating; load class C250–400							
brickslot; galvanized steel; 1000 mm	152.70	1.20	23.10	–	152.70	m	**175.80**
brickslot; stainless steel; 1000 mm	167.10	1.20	23.10	–	334.20	m	**357.30**

R12 DRAINAGE

Item Excluding site overheads and profit	PC £	Labour hours	Labour £	Plant £	Material £	Unit	Total rate £
LINEAR DRAINAGE – cont							
Channel drains – cont							
Accessories for MultiDrain MD							
sump unit; complete with sediment bucket							
and access unit	130.45	1.00	19.25	–	262.50	nr	**281.75**
end cap; closing piece	8.50	–	–	–	8.50	nr	**8.50**
end cap; inlet/outlet	17.20	–	–	–	17.20	nr	**17.20**
ACO Drainlock Gratings for M100PPD and							
M100D system							
A15 loading							
slotted galvanized steel	22.15	0.15	2.89	–	22.15	m	**25.04**
perforated galvanized steel	30.45	0.15	2.89	–	30.45	m	**33.34**
C250 loading							
Heelguard composite black; 500 mm long							
with security locking	52.50	–	–	–	52.50	m	**52.50**
Intercept; ductile iron; 500 mm long	121.20	–	–	–	121.20	m	**121.20**
slotted galvanized steel; 1.00 mm long	57.30	–	–	–	57.30	m	**57.30**
perforated galvanized steel; 1.00 mm long	63.10	–	–	–	63.10	m	**63.10**
mesh galvanized steel; 1.00 mm long	42.90	–	–	–	42.90	m	**42.90**
Wade Ltd; stainless steel channel drains;							
specialized applications; bespoke							
manufacture							
Drain in stainless steel; c/w tie in lugs and							
inbuilt falls to 100 mm spigot outlet; stainless							
steel gratings							
NE channel; Ref 12430; secured gratings	420.00	1.20	23.10	–	430.84	m	**453.94**
FIN DRAINS							
Fin drains; Cooper Clarke Civils and							
Lintels; Geofin Shallow Drain; drainage of							
sports fields and grassed landscaped							
areas							
Geofin Shallow Drain; to trenches; excavation							
by trenching machine; backfilled with single							
size aggregate 20 mm and covered with sharp							
sand rootzone 200 mm thick							
Geofin; 25 mm thick × 150 mm deep	5.50	0.05	0.96	0.78	10.11	m	**11.85**
Geofin; 25 mm thick × 450 mm deep	4.20	0.07	1.28	1.04	13.53	m	**15.85**
Geofin; 25 mm thick × 900 mm deep	8.70	0.10	1.93	1.04	26.03	m	**29.00**

R12 DRAINAGE

Item Excluding site overheads and profit	PC £	Labour hours	Labour £	Plant £	Material £	Unit	Total rate £
Fin drains; Cooper Clarke Civils and Lintels; Geofin Geocomposite Finn Drain; laid to slabs							
Geofin fin drain laid horizontally to slab or blinded ground; covered with 20 mm shingle 200 mm thick							
Geofin; 25 mm thick × 900 mm wide; laid flat to falls	9.66	0.02	0.39	0.16	16.32	m²	**16.87**
DRAINAGE TO ROOF DECKS AND PLANTERS							
Maxit Ltd; Leca (light expanded clay aggregate); drainage aggregate to roofdecks and planters							
Placed mechanically to planters; average 100 mm thick; by mechanical plant tipped into planters							
aggregate size 10–20 mm; delivered in 30 m³ loads	47.00	0.40	7.70	6.99	47.00	m³	**61.69**
aggregate size 10–20 mm; delivered in 70 m³ loads	44.00	0.20	3.85	6.04	44.00	m³	**53.89**
Placed by light aggregate blower (maximum 40 m)							
aggregate size 10–20 mm; delivered in 30 m³ loads	47.00	0.14	2.75	–	51.59	m³	**54.34**
aggregate size 10–20 mm; delivered in 55 m³ loads on blower vehicle	67.00	0.14	2.75	–	67.00	m³	**69.75**
aggregate size 10–20 mm; delivered in 70 m³ loads	44.00	0.14	2.75	–	48.59	m³	**51.34**
By hand							
aggregate size 10–20 mm; delivered in 30 m³ loads	47.00	1.33	25.67	–	47.00	m³	**72.67**
aggregate size 10–20 mm; delivered in 70 m³ loads	44.00	1.33	25.67	–	44.00	m³	**69.67**
Drainage boards laid to insulated slabs on roof decks; boards laid below growing medium and granulated drainage layer and geofabric (all not included) to collect and channel water to drainage outlets (not included)							
Alumasc Floradrain; polyethylene irrigation/drainage layer; inclusive of geofabric laid over the surface of the drainage board							
Floradrain FD40; 0.96 × 2.08 panels	12.05	0.05	0.96	–	13.20	m²	**14.16**
Floradrain FD60; 1.00 × 2.00 panel	21.42	0.07	1.28	–	22.57	m²	**23.85**

R12 DRAINAGE

Item Excluding site overheads and profit	PC £	Labour hours	Labour £	Plant £	Material £	Unit	Total rate £
ACCESS COVERS AND FRAMES							
Load Classes for Access Covers							
FACTA (Fabricated Access Cover Trade							
Association) class:							
A – 0.5 tonne maximum slow moving wheel							
load							
AA – 1.5 tonne maximum slow moving							
wheel load							
AAA – 2.5 tonne maximum slow moving							
wheel load							
B – 5 tonne maximum slow moving wheel							
load							
C – 6.5 tonne maximum slow moving							
wheel load							
D – 11 tonne maximum slow moving wheel							
load							
Access covers and frames; solid top;							
galvanized; Steelway Brickhouse; Bristeel;							
bedding frame in cement mortar (1:3);							
cover in grease and sand; clear opening							
sizes; base size shown in brackets (50 mm							
depth)							
FACTA AA; single seal							
450 × 450 mm (520 × 520 mm)	64.26	1.50	28.88	–	74.55	nr	**103.43**
600 × 450 mm (670 × 520 mm)	71.32	1.80	34.65	–	82.75	nr	**117.40**
600 × 600 mm (670 × 670 mm)	76.26	2.00	38.50	–	89.98	nr	**128.48**
FACTA AA; double seal							
450 × 450 mm (560 × 560 mm)	104.39	1.50	28.88	–	114.68	nr	**143.56**
600 × 450 mm (710 × 560 mm)	114.70	1.80	34.65	–	126.14	nr	**160.79**
600 × 600 mm (710 × 710 mm)	139.05	2.00	38.50	–	152.77	nr	**191.27**
FACTA AAA; single seal							
450 × 450 mm (520 × 520 mm)	98.97	1.50	28.88	–	109.27	nr	**138.15**
600 × 450 mm (670 × 520 mm)	106.81	1.80	34.65	–	118.25	nr	**152.90**
600 × 600 mm (670 × 670 mm)	113.81	2.00	38.50	–	127.53	nr	**166.03**
FACTA AAA; double seal							
450 mm × 450 mm (560 × 560)	140.00	1.50	28.88	–	150.29	nr	**179.17**
600 mm × 450 mm (710 × 560)	152.95	1.80	34.65	–	164.39	nr	**199.04**
600 mm × 600 mm (710 × 710)	170.42	2.00	38.50	–	184.14	nr	**222.64**
FACTA B; single seal							
450 × 450 mm (520 × 520 mm)	83.07	1.50	28.88	–	93.36	nr	**122.24**
600 × 450 mm (670 × 520 mm)	103.71	1.80	34.65	–	115.14	nr	**149.79**
600 × 600 mm (670 × 670 mm)	120.56	2.00	38.50	–	134.28	nr	**172.78**
FACTA B; double seal							
450 × 450 mm (560 × 560 mm)	133.19	1.50	28.88	–	143.48	nr	**172.36**
600 × 450 mm (710 × 560 mm)	148.28	1.80	34.65	–	159.72	nr	**194.37**
600 × 600 mm (710 × 710 mm)	175.40	2.00	38.50	–	189.12	nr	**227.62**

R12 DRAINAGE

Item Excluding site overheads and profit	PC £	Labour hours	Labour £	Plant £	Material £	Unit	Total rate £
Access covers and frames; recessed;							
galvanized; Steelway Brickhouse; Bripave;							
bedding frame in cement mortar (1:3);							
cover in grease and sand; clear opening							
sizes; base size shown in brackets (for							
block depths 50 mm, 65 mm, 80 mm or							
100 mm)							
FACTA AA							
450 × 450 mm (562 × 562 mm)	206.95	1.50	28.88	–	217.24	nr	**246.12**
600 × 450 mm (712 × 562 mm)	227.00	1.80	34.65	–	238.44	nr	**273.09**
600 × 600 mm (712 × 712 mm)	235.30	2.00	38.50	–	248.47	nr	**286.97**
750 × 600 mm (862 × 712 mm)	262.56	2.40	46.20	–	277.43	nr	**323.63**
FACTA B							
450 × 450 mm (562 × 562 mm)	207.31	1.50	28.88	–	217.60	nr	**246.48**
600 × 450 mm (712 × 562 mm)	241.00	1.80	34.65	–	252.44	nr	**287.09**
600 × 600 mm (712 × 712 mm)	244.00	2.00	38.50	–	257.72	nr	**296.22**
750 × 600 mm (862 × 712 mm)	244.79	2.40	46.20	–	259.66	nr	**305.86**
FACTA D							
450 × 450 mm (562 × 562 mm)	232.35	1.50	28.88	–	242.64	nr	**271.52**
600 × 450 mm (712 × 562 mm)	246.86	1.80	34.65	–	258.30	nr	**292.95**
600 × 600 mm (712 × 712 mm)	255.05	2.00	38.50	–	268.77	nr	**307.27**
750 × 600 mm (862 × 712 mm)	310.40	2.40	46.20	–	325.27	nr	**371.47**
Access covers and frames; Jones of							
Oswestry; bedding frame in cement mortar							
(1:3); cover in grease and sand; clear							
opening sizes							
Access covers and frames; Suprabloc; to							
paved areas; filling with blocks cut and fitted to							
match surrounding paving							
pedestrian weight; 300 × 300 mm	121.47	2.50	48.13	–	128.35	nr	**176.48**
pedestrian weight; 450 × 450 mm	207.27	2.50	48.13	–	207.27	nr	**255.40**
pedestrian weight; 450 × 600 mm	221.01	2.50	48.13	–	228.17	nr	**276.30**
light vehicular weight; 300 × 300 mm	155.63	2.40	46.20	–	162.51	nr	**208.71**
light vehicular weight; 450 × 450 mm	119.98	3.00	57.75	–	127.86	nr	**185.61**
light vehicular weight; 450 × 600 mm	233.57	4.00	77.00	–	240.45	nr	**317.45**
heavy vehicular weight; 300 × 300 mm	155.97	3.00	57.75	–	163.13	nr	**220.88**
heavy vehicular weight; 450 × 600 mm	271.60	3.00	57.75	–	279.48	nr	**337.23**
heavy vehicular weight; 600 × 600 mm	290.00	4.00	77.00	–	301.46	nr	**378.46**
Extra over manhole frames and covers for							
filling recessed manhole covers with brick							
paviors; PC £305.00/1000	13.42	1.00	19.25	–	13.98	m²	**33.23**
filling recessed manhole covers with							
vehicular paving blocks; PC £8.17/m²	8.17	0.75	14.44	–	8.73	m²	**23.17**
filling recessed manhole covers with							
concrete paving flags; PC £13.43/m²	14.42	0.35	6.74	–	14.76	m²	**21.50**

R13 LAND DRAINAGE

Item Excluding site overheads and profit	PC £	Labour hours	Labour £	Plant £	Material £	Unit	Total rate £
R13 LAND DRAINAGE							
MARKET PRICES OF BACKFILLING MATERIALS							
Market prices of commonly used materials used for backfilling The prices below show the market price with standard settlement factors allowed for							
Sharp sand	–	–	–	–	35.01	m³	**35.01**
Gravel (washed river or pit)	–	–	–	–	34.17	m³	**34.17**
Shingle	–	–	–	–	34.17	m³	**34.17**
Topsoil	–	–	–	–	33.60	m³	**33.60**
Gravel rejects	–	–	–	–	30.68	m³	**30.68**
Selected granular material 6F2	–	–	–	–	13.44	m³	**13.44**
CLARIFICATION NOTES ON LABOUR COSTS IN THIS SECTION							
General groundworks team Generally a three man team is used in this section; The column 'Labour hours' reports team hours. The column 'Labour £' reports the total cost of the team for the unit of work shown							
3 man team	–	1.00	57.75	–	–	hr	**57.75**
SOAKAWAYS							
Preamble: Flat rate hourly rainfall = 50 mm/hr and assumes 100 impermeability of the run-off area. A storage capacity of the soakaway should be ⅓ of the hourly rainfall. Formulae for calculating soakaway depths are provided in the publications mentioned below and in the Tables and Memoranda section of this publication. The design of soakaways is dependent on, amongst other factors, soil conditions, permeability, groundwater level and runoff. The definitive documents for design of soakaways are CIRIA 156 and BRE Digest 365 dated September 1991. The suppliers of the systems below will assist through their technical divisions. Excavation earthwork support of pits and disposal not included.							

R13 LAND DRAINAGE

Item Excluding site overheads and profit	PC £	Labour hours	Labour £	Plant £	Material £	Unit	Total rate £
Soakaway excavation							
Excavating mechanical; to reduce levels							
maximum depth not exceeding 1.00 m;							
JCB sitemaster	–	0.05	0.96	2.06	–	m³	**3.02**
maximum depth not exceeding 1.00 m; 360							
tracked excavator	–	0.04	0.77	2.00	–	m³	**2.77**
maximum depth not exceeding 2.00 m; 360							
tracked excavator	–	0.06	1.16	3.00	–	m³	**4.16**
Disposal							
Excavated material; off site; to tip;							
mechanically loaded (JCB)							
inert	–	–	–	–	–	m³	**17.14**
In situ concrete ring beam foundations to							
base of soakaway; 300 mm wide × 250 mm							
deep; poured on or against earth or							
unblinded hardcore							
Internal diameters of rings							
900 mm	19.90	4.00	77.00	–	19.90	nr	**96.90**
1200 mm	26.52	4.50	86.63	–	26.52	nr	**113.15**
1500 mm	33.15	5.00	96.25	–	33.15	nr	**129.40**
2400 mm	53.05	5.50	105.88	–	53.05	nr	**158.93**
Concrete soakaway rings; Milton Pipes Ltd;							
perforations and step irons to concrete rings							
at manufacturers recommended centres;							
placing of concrete ring soakaways to in							
situ concrete ring beams (1:3:6) (not							
included); filling and surrounding base							
with gravel 225 mm deep (not included)							
Ring diameter 900 mm							
1.00 m deep; volume 636 litres	72.65	1.25	24.06	21.88	191.75	nr	**237.69**
1.50 m deep; volume 954 litres	108.97	1.65	31.76	57.75	284.98	nr	**374.49**
2.00 m deep; volume 1272 litres	145.30	1.65	31.76	57.75	378.20	nr	**467.71**
Ring diameter 1200 mm							
1.00 m deep; volume 1131 litres	94.50	3.00	57.75	52.50	231.42	nr	**341.67**
1.50 m deep; volume 1696 litres	141.75	4.50	86.63	78.75	342.42	nr	**507.80**
2.00 m deep; volume 2261 litres	189.00	4.50	86.63	78.75	453.42	nr	**618.80**
2.50 m deep; volume 2827 litres	236.25	6.00	115.50	105.00	516.47	nr	**736.97**
Ring diameter 1500 mm							
1.00 m deep; volume 1767 litres	161.00	4.50	86.63	52.50	330.63	nr	**469.76**
1.50 m deep; volume 2651 litres	241.50	6.00	115.50	73.50	488.58	nr	**677.58**
2.00 m deep; volume 3534 litres	322.00	6.00	115.50	73.50	646.53	nr	**835.53**
2.50 m deep; volume 4418 litres	236.25	7.50	144.38	131.25	638.23	nr	**913.86**
Ring diameter 2400 mm							
1.00 m deep; volume 4524 litres	616.60	6.00	115.50	52.50	847.88	nr	**1015.88**
1.50 m deep; volume 6786 litres	924.90	7.50	144.38	73.50	1261.03	nr	**1478.91**
2.00 m deep; volume 9048 litres	1233.20	7.50	144.38	73.50	1681.03	nr	**1898.91**
2.50 m deep; volume 11310 litres	1541.50	9.00	173.25	131.25	2094.18	nr	**2398.68**
Extra over for							
250 mm depth chamber ring	–	–	–	–	–	100%	**–**
500 mm depth chamber ring	–	–	–	–	–	50%	**–**

R13 LAND DRAINAGE

Item Excluding site overheads and profit	PC £	Labour hours	Labour £	Plant £	Material £	Unit	Total rate £
SOAKAWAYS – cont							
Cover slabs to soakaways							
Heavy duty precast concrete							
900 mm diameter	114.20	1.00	19.25	26.25	114.20	nr	**159.70**
1200 mm diameter	137.90	1.00	19.25	26.25	137.90	nr	**183.40**
1500 mm diameter	219.45	1.00	19.25	26.25	219.45	nr	**264.95**
2400 mm diameter	823.70	1.00	19.25	26.25	823.70	nr	**869.20**
Step irons to concrete chamber rings	31.60	–	–	–	31.60	m	**31.60**
Extra over soakaways for filter wrapping with a proprietary filter membrane							
900 mm diameter × 1.00 m deep	1.44	1.00	19.25	–	1.44	nr	**20.69**
900 mm diameter × 2.00 m deep	2.88	1.50	28.88	–	2.88	nr	**31.76**
1050 mm diameter × 1.00 m deep	1.68	1.50	28.88	–	1.68	nr	**30.56**
1050 mm diameter × 2.00 m deep	3.36	2.00	38.50	–	3.36	nr	**41.86**
1200 mm diameter × 1.00 m deep	1.92	2.00	38.50	–	1.92	nr	**40.42**
1200 mm diameter × 2.00 m deep	3.83	2.50	48.13	–	3.83	nr	**51.96**
1500 mm diameter × 1.00 m deep	2.39	2.50	48.13	–	2.39	nr	**50.52**
1500 mm diameter × 2.00 m deep	4.78	2.50	48.13	–	4.78	nr	**52.91**
1800 mm diameter × 1.00 m deep	2.87	3.00	57.75	–	2.87	nr	**60.62**
1800 mm diameter × 2.00 m deep	5.74	3.25	62.56	–	5.74	nr	**68.30**
Gravel surrounding to concrete ring soakaway							
40 mm aggregate backfilled to vertical face of soakaway wrapped with geofabric (not included)							
250 mm thick	33.34	0.20	3.85	10.50	33.34	m³	**47.69**
Backfilling to face of soakaway; carefully compacting as work proceeds							
Arising from the excavations							
average thickness exceeding 0.25 m; depositing in layers 150 mm maximum thickness	–	0.03	0.64	5.25	–	m³	**5.89**
Aquacell soakaway; Wavin Plastics Ltd; preformed polypropylene soakaway infiltration crate units; to trenches; surrounded by geotextile and 40 mm aggregate laid 100 mm thick in trenches (excavation, disposal and backfilling not included)							
1.00 m × 500 × 400 mm; internal volume 190 litres							
4 crates; 2.00 m × 1.00 m × 400 mm; 760 litres	181.22	1.60	30.80	17.32	227.79	nr	**275.91**
8 crates; 2.00 m × 1.00 m × 800 mm; 1520 litres	362.43	3.20	61.60	22.27	419.78	nr	**503.65**
12 crates; 6.00 m × 500 mm × 800 mm; 2280 litres	543.65	4.80	92.40	46.20	653.06	nr	**791.66**

R13 LAND DRAINAGE

Item Excluding site overheads and profit	PC £	Labour hours	Labour £	Plant £	Material £	Unit	Total rate £
16 crates; 4.00 m × 1.00 m × 800 mm; 3040 litres	724.86	6.40	123.20	41.25	823.50	nr	987.95
20 crates; 5.00 m × 1.00 m × 800 mm; 3800 litres	906.08	8.00	154.00	40.84	1012.69	nr	1207.53
30 crates; 15.00 m × 1.00 m × 400 mm; 5700 litres	1359.12	12.00	231.00	111.38	1619.25	nr	1961.63
60 crates; 15.00 m × 1.00 m × 800 mm; 11400 litres	2718.24	20.00	385.00	145.61	3070.61	nr	3601.22
Geofabric surround to Aquacell units; Terram Ltd Terram synthetic fibre filter fabric; to face of concrete rings (not included); anchoring whilst backfilling (not included)							
Terram 1000; 0.70 mm thick; mean water flow 50 l/m²/s	0.49	0.05	0.96	–	0.49	m²	1.45
SAND SLITTING/GROOVING							
Sand slitting; Agripower Ltd Drainage slits; at 1.00 m centres; using spinning disc trenching machine; backfilling to 100 mm of surface with pea gravel; finished surface with medium grade sand 100; arisings to be loaded, hauled and tipped on site; minimum pitch size 6000 m²							
250 mm depth	–	–	–	–	–	m	1.80
300 mm depth	–	–	–	–	–	m	1.97
400 mm depth	–	–	–	–	–	m	3.01
Sand grooving or banding; Agripower Ltd Drainage bands at 260 mm m centres; using Blec Sandmaster; to existing grass at 260 mm centres; minimum pitch size 4000 m²							
sand banding or sand grooving	–	–	–	–	–	m²	1.02
LAND DRAINS							
Market prices of backfilling materials							
Sand	40.01	–	–	–	40.01	m³	40.01
Gravel rejects	32.92	–	–	–	32.92	m³	32.92
Imported topsoil (allowing for 20% settlement)	–	–	–	–	33.60	m³	33.60
Land drainage; calculation table For calculation of drainage per hectare the following table can be used; rates show the lengths of drains per unit and not the value							

R13 LAND DRAINAGE

Item Excluding site overheads and profit	PC £	Labour hours	Labour £	Plant £	Material £	Unit	Total rate £
LAND DRAINS – cont							
Land drainage – cont							
Lateral drains							
10.00 m centres	–	–	–	–	–	m/ha	1000.00
15.00 m centres	–	–	–	–	–	m/ha	650.00
25.00 m centres	–	–	–	–	–	m/ha	400.00
30.00 m centres	–	–	–	–	–	m/ha	330.00
Main drains							
1 nr (at 100 m centres)	–	–	–	–	–	m/ha	100.00
2 nr (at 50 m centres)	–	–	–	–	–	m/ha	200.00
3 nr (at 33.3 m centres)	–	–	–	–	–	m/ha	300.00
4 nr (at 25 m centres)	–	–	–	–	–	m/ha	400.00
Land drainage; excavating							
Removing 150 mm depth of topsoil; 300 mm							
wide; depositing beside trench; by machine	–	1.50	28.88	13.50	–	100 m	42.38
Removing 150 mm depth of topsoil; 300 mm							
wide; depositing beside trench; by hand	–	8.00	154.00	–	–	100 m	154.00
Excavated material; disposal on site							
In spoil heaps							
average 25 m distance	–	–	–	2.87	–	m³	2.87
average 50 m distance	–	–	–	3.31	–	m³	3.31
average 100 m distance (1 dumper)	–	–	–	4.14	–	m³	4.14
average 100 m distance (2 dumpers)	–	–	–	4.22	–	m³	4.22
average 200 m distance	–	–	–	5.00	–	m³	5.00
average 200 m distance (2 dumpers)	–	–	–	5.08	–	m³	5.08
Removing excavated material from site to tip;							
mechanically loaded							
excavated material and clean hardcore							
rubble	–	–	–	1.56	15.84	m³	17.40
Land drainage; Agripower Ltd; excavating							
trenches (with minimum project size of							
1000 m) by trenching machine; spreading							
arisings on site; laying perforated pipe; as							
shown below; backfilling with shingle to							
350 mm from surface							
Width 150 mm; 80 mm perforated pipe							
depth 750 mm	–	–	–	–	–	m	5.50
Width 175 mm; 100 mm perforated pipe							
depth 800 mm	–	–	–	–	–	m	6.65
Width 250 mm; 150 mm perforated pipe							
depth 900 mm	–	–	–	–	–	m	9.70

R13 LAND DRAINAGE

Item Excluding site overheads and profit	PC £	Labour hours	Labour £	Plant £	Material £	Unit	Total rate £
Land drainage; excavating for drains; by backacter excavator JCB C3X Sitemaster; including disposing spoil to spoil heaps not exceeding 100 m; boning to levels by laser							
Width 150–225 mm							
depth 450 mm	–	9.67	186.08	310.00	–	100 m	**496.08**
depth 600 mm	–	13.00	250.25	465.00	–	100 m	**715.25**
depth 700 mm	–	15.50	298.38	581.25	–	100 m	**879.63**
depth 900 mm	–	23.00	442.75	930.00	–	100 m	**1372.75**
Width 300 mm							
depth 450 mm	–	9.67	186.08	310.00	–	100 m	**496.08**
depth 700 mm	–	14.00	269.50	511.50	–	100 m	**781.00**
depth 900 mm	–	23.00	442.75	930.00	–	100 m	**1372.75**
depth 1000 mm	–	25.00	481.25	1023.00	–	100 m	**1504.25**
depth 1200 mm	–	25.00	481.25	1162.50	–	100 m	**1643.75**
depth 1500 mm	–	31.00	596.75	1302.00	–	100 m	**1898.75**
Land drainage; excavating for drains; by 7 tonne tracked excavator; including disposing spoil to spoil heaps not exceeding 100 m							
Width 225 mm							
depth 450 mm	–	7.00	134.75	65.69	–	100 m	**200.44**
depth 600 mm	–	7.17	137.96	68.43	–	100 m	**206.39**
depth 700 mm	–	7.35	141.45	71.40	–	100 m	**212.85**
depth 900 mm	–	7.76	149.42	78.20	–	100 m	**227.62**
depth 1000 mm	–	8.00	154.00	82.11	–	100 m	**236.11**
depth 1200 mm	–	8.26	159.07	86.43	–	100 m	**245.50**
Width 300 mm							
depth 450 mm	–	7.35	141.45	71.40	–	100 m	**212.85**
depth 600 mm	–	7.76	149.42	78.20	–	100 m	**227.62**
depth 700 mm	–	8.26	159.07	86.43	–	100 m	**245.50**
depth 900 mm	–	9.25	178.06	102.64	–	100 m	**280.70**
depth 1000 mm	–	10.14	195.25	117.30	–	100 m	**312.55**
depth 1200 mm	–	11.00	211.75	131.38	–	100 m	**343.13**
Width 600 mm							
depth 450 mm	–	13.00	250.25	164.22	–	100 m	**414.47**
depth 600 mm	–	14.11	271.64	182.47	–	100 m	**454.11**
depth 700 mm	–	16.33	314.42	218.96	–	100 m	**533.38**
depth 900 mm	–	17.29	332.75	234.60	–	100 m	**567.35**
depth 1000 mm	–	18.38	353.90	252.65	–	100 m	**606.55**
depth 1200 mm	–	21.18	407.75	298.58	–	100 m	**706.33**
Land drainage; excavating for drains; by hand, including disposing spoil to spoil heaps not exceeding 100 m							
Width 150 mm							
depth 450 mm	–	22.03	424.08	–	–	100 m	**424.08**
depth 600 mm	–	29.38	565.57	–	–	100 m	**565.57**
depth 700 mm	–	34.27	659.70	–	–	100 m	**659.70**
depth 900 mm	–	44.06	848.15	–	–	100 m	**848.15**

R13 LAND DRAINAGE

Item Excluding site overheads and profit	PC £	Labour hours	Labour £	Plant £	Material £	Unit	Total rate £
LAND DRAINS – cont							
Land drainage – cont							
Width 225 mm							
depth 450 mm	–	33.05	636.21	–	–	100 m	**636.21**
depth 600 mm	–	44.06	848.15	–	–	100 m	**848.15**
depth 700 mm	–	51.41	989.64	–	–	100 m	**989.64**
depth 900 mm	–	66.10	1272.42	–	–	100 m	**1272.42**
depth 1000 mm	–	73.44	1413.72	–	–	100 m	**1413.72**
Width 300 mm							
depth 450 mm	–	44.06	848.15	–	–	100 m	**848.15**
depth 600 mm	–	58.75	1130.94	–	–	100 m	**1130.94**
depth 700 mm	–	68.54	1319.39	–	–	100 m	**1319.39**
depth 900 mm	–	88.13	1696.50	–	–	100 m	**1696.50**
depth 1000 mm	–	97.92	1884.96	–	–	100 m	**1884.96**
Width 375 mm							
depth 450 mm	–	55.08	1060.29	–	–	100 m	**1060.29**
depth 600 mm	–	73.44	1413.72	–	–	100 m	**1413.72**
depth 700 mm	–	85.68	1649.34	–	–	100 m	**1649.34**
depth 900 mm	–	110.16	2120.58	–	–	100 m	**2120.58**
depth 1000 mm	–	122.40	2356.20	–	–	100 m	**2356.20**
Width 450 mm							
depth 450 mm	–	66.10	1272.42	–	–	100 m	**1272.42**
depth 600 mm	–	88.13	1696.50	–	–	100 m	**1696.50**
depth 700 mm	–	102.82	1979.29	–	–	100 m	**1979.29**
depth 900 mm	–	132.19	2544.66	–	–	100 m	**2544.66**
depth 1000 mm	–	146.88	2827.44	–	–	100 m	**2827.44**
Width 600 mm							
depth 450 mm	–	88.13	1696.50	–	–	100 m	**1696.50**
depth 600 mm	–	117.50	2261.88	–	–	100 m	**2261.88**
depth 700 mm	–	137.09	2638.98	–	–	100 m	**2638.98**
depth 900 mm	–	176.26	3393.01	–	–	100 m	**3393.01**
depth 1000 mm	–	195.84	3769.92	–	–	100 m	**3769.92**
Width 900 mm							
depth 450 mm	–	132.19	2544.66	–	–	100 m	**2544.66**
depth 600 mm	–	176.26	3393.01	–	–	100 m	**3393.01**
depth 700 mm	–	205.63	3958.38	–	–	100 m	**3958.38**
depth 900 mm	–	264.38	5089.31	–	–	100 m	**5089.31**
depth 1000 mm	–	293.76	5654.88	–	–	100 m	**5654.88**
Earthwork support; moving along as work proceeds							
Maximum depth not exceeding 2.00 m							
distance between opposing faces not exceeding 2.00 m	–	0.80	15.40	21.00	–	m	**36.40**
Land drainage; pipe laying							
Hepworth; agricultural clay drain pipes;							
300 mm length; butt joints; in straight runs							
75 mm diameter	489.80	8.00	154.00	–	489.80	100 m	**643.80**
100 mm diameter	836.25	9.00	173.25	–	836.25	100 m	**1009.50**
150 mm diameter	1713.11	10.00	192.50	–	1713.11	100 m	**1905.61**

R13 LAND DRAINAGE

Item Excluding site overheads and profit	PC £	Labour hours	Labour £	Plant £	Material £	Unit	Total rate £
Extra over clay drain pipes for filter-wrapping pipes with Terram or similar filter fabric							
Terram 700	0.23	0.04	0.77	–	0.23	m²	1.00
Terram 1000	0.20	0.04	0.77	–	0.20	m²	0.97
Junctions between drains in clay pipes							
75 × 75 mm	16.20	0.25	4.81	–	16.56	nr	21.37
100 × 100 mm	21.50	0.25	4.81	–	22.08	nr	26.89
150 × 150 mm	26.47	0.25	4.81	–	27.34	nr	32.15
Wavin Plastics Ltd; flexible plastic perforated pipes in trenches (not included); to a minimum depth of 450 mm (couplings not included)							
OsmaDrain; flexible plastic perforated pipes in trenches (not included); to a minimum depth of 450 mm (couplings not included)							
80 mm diameter; available in 100 m coil	72.75	2.00	38.50	–	72.75	100 m	111.25
100 mm diameter; available in 100 m coil	117.81	2.00	38.50	–	117.81	100 m	156.31
160 mm diameter; available in 35 m coil	284.91	2.00	38.50	–	284.91	100 m	323.41
WavinCoil; plastic pipe junctions							
80 × 80 mm	3.20	0.05	0.96	–	3.20	nr	4.16
100 × 100 mm	3.58	0.05	0.96	–	3.58	nr	4.54
100 × 60 mm	3.41	0.05	0.96	–	3.41	nr	4.37
100 × 80 mm	3.41	0.05	0.96	–	3.41	nr	4.37
160 × 160 mm	8.52	0.05	0.96	–	8.52	nr	9.48
WavinCoil; couplings for flexible pipes							
80 mm diameter	1.13	0.03	0.64	–	1.13	nr	1.77
100 mm diameter	1.25	0.03	0.64	–	1.25	nr	1.89
160 mm diameter	1.69	0.03	0.64	–	1.69	nr	2.33
Land drainage; backfilling trench after laying pipes with gravel rejects or similar; blind filling with ash or sand; topping with 150 mm imported topsoil from dumps not exceeding 100 m; by machine							
Width 150 mm							
depth 450 mm	–	3.30	63.52	66.53	205.82	100 m	335.87
depth 600 mm	–	4.30	82.77	87.15	279.16	100 m	449.08
depth 750 mm	–	4.96	95.48	100.76	318.06	100 m	514.30
depth 900 mm	–	6.30	121.27	128.40	407.85	100 m	657.52
Width 225 mm							
depth 450 mm	–	4.95	95.29	99.79	318.05	100 m	513.13
depth 600 mm	–	6.45	124.16	130.72	418.91	100 m	673.79
depth 750 mm	–	7.95	153.04	161.66	520.07	100 m	834.77
depth 900 mm	–	9.45	181.91	192.60	620.94	100 m	995.45
Width 375 mm							
depth 450 mm	–	8.25	158.81	166.31	529.87	100 m	854.99
depth 600 mm	–	10.75	206.94	217.88	698.07	100 m	1122.89
depth 750 mm	–	13.25	255.06	269.44	866.58	100 m	1391.08
depth 900 mm	–	15.75	303.19	321.00	1034.78	100 m	1658.97

R13 LAND DRAINAGE

Item Excluding site overheads and profit	PC £	Labour hours	Labour £	Plant £	Material £	Unit	Total rate £
LAND DRAINS – cont							
Land drainage; backfilling trench after							
laying pipes with gravel rejects or similar,							
blind filling with ash or sand; topping with							
150 mm topsoil from dumps not exceeding							
100 m; by hand							
Width 150 mm							
depth 450 mm	–	18.63	358.63	–	193.22	100 m	551.85
depth 600 mm	–	24.84	478.17	–	279.16	100 m	757.33
depth 750 mm	–	31.05	597.71	–	255.06	100 m	852.77
depth 900 mm	–	37.26	717.25	–	407.85	100 m	1125.10
Width 225 mm							
depth 450 mm	–	27.94	537.85	–	318.05	100 m	855.90
depth 600 mm	–	37.26	717.25	–	418.91	100 m	1136.16
depth 750 mm	–	46.57	896.47	–	520.07	100 m	1416.54
depth 900 mm	–	55.89	1075.88	–	620.94	100 m	1696.82
Width 375 mm							
depth 450 mm	–	46.57	896.47	–	529.87	100 m	1426.34
depth 600 mm	–	61.10	1176.17	–	698.07	100 m	1874.24
depth 750 mm	–	77.63	1494.38	–	866.58	100 m	2360.96
depth 900 mm	–	93.15	1793.14	–	1034.78	100 m	2827.92
SPORTS AND AMENITY DRAINAGE							
Sports or amenity drainage; Agripower Ltd							
Ltd; excavating trenches (with minimum							
project size of 1000 m) by trenching							
machine; arisings to spoil heap on site							
maximum 100 m; backfill with shingle to							
within 125 mm from surface and with							
rootzone to ground level							
Width 145 mm; 80 mm perforated pipe							
depth 550 mm	–	–	–	–	–	m	7.94
Width 145 mm; 100 mm perforated pipe							
depth 650 mm	–	–	–	–	–	m	9.12
Width 145 mm; 150 mm perforated pipe							
depth 700 mm	–	–	–	–	–	100 m	11.75
CATCHWATER OR FRENCH DRAINS							
Catchwater or french drains; laying pipes							
to drain excavated separately; 100 mm							
diameter non-coilable perforated plastic							
pipes; including straight jointing; pipes laid							
with perforations uppermost; lining trench;							
wrapping pipes with filter fabric; backfilling							
with shingle							

R13 LAND DRAINAGE

Item Excluding site overheads and profit	PC £	Labour hours	Labour £	Plant £	Material £	Unit	Total rate £
Width 300 mm							
depth 450 mm	7918.86	9.10	175.18	37.50	8337.28	100 m	**8549.96**
depth 600 mm	7922.98	9.84	189.42	58.58	8491.42	100 m	**8739.42**
depth 750 mm	7927.28	10.60	204.05	74.25	8645.76	100 m	**8924.06**
depth 900 mm	7931.49	11.34	218.29	89.51	8800.00	100 m	**9107.80**
depth 1000 mm	7934.29	11.84	227.92	99.83	8902.82	100 m	**9230.57**
depth 1200 mm	7939.91	12.84	247.17	120.45	9108.47	100 m	**9476.09**
Width 450 mm							
depth 450 mm	7925.17	10.44	200.97	108.08	8568.64	100 m	**8877.69**
depth 600 mm	7931.49	11.34	218.29	89.51	8800.00	100 m	**9107.80**
depth 750 mm	7937.80	12.46	239.85	112.61	9031.35	100 m	**9383.81**
depth 900 mm	7944.12	13.60	261.80	136.13	9262.71	100 m	**9660.64**
depth 1000 mm	7948.32	14.34	276.05	151.39	9416.95	100 m	**9844.39**
depth 1200 mm	7956.74	15.84	304.92	182.32	9725.43	100 m	**10212.67**
depth 1500 mm	7969.37	18.10	348.43	228.94	10188.15	100 m	**10765.52**
depth 2000 mm	7990.42	21.84	420.42	306.07	10959.35	100 m	**11685.84**
Width 600 mm							
depth 450 mm	7931.49	11.24	216.37	89.51	8800.00	100 m	**9105.88**
depth 600 mm	7939.91	12.84	247.17	120.45	9108.47	100 m	**9476.09**
depth 750 mm	7948.32	14.34	276.05	151.39	9416.95	100 m	**9844.39**
depth 900 mm	7956.74	15.84	304.92	182.32	9725.43	100 m	**10212.67**
depth 1000 mm	7953.96	16.84	324.17	202.95	9922.69	100 m	**10449.81**
depth 1200 mm	7973.58	18.84	362.67	244.20	10342.39	100 m	**10949.26**
depth 1500 mm	7990.42	21.84	420.42	576.08	10959.35	100 m	**11955.85**
depth 2000 mm	8018.48	19.84	381.92	553.58	11987.61	100 m	**12923.11**
Width 900 mm							
depth 450 mm	7944.12	13.60	261.80	136.13	9262.71	100 m	**9660.64**
depth 600 mm	7956.74	7.92	152.46	182.32	9725.43	100 m	**10060.21**
depth 750 mm	7969.37	9.05	174.21	228.94	10188.15	100 m	**10591.30**
depth 900 mm	7982.00	10.17	195.77	275.14	10650.87	100 m	**11121.78**
depth 1000 mm	7990.42	10.92	210.21	306.07	10959.35	100 m	**11475.63**
depth 1200 mm	8007.26	12.42	239.09	367.95	11576.30	100 m	**12183.34**
depth 1500 mm	8032.51	14.67	282.40	460.76	12501.74	100 m	**13244.90**
depth 2000 mm	8074.61	18.92	364.21	705.45	14044.13	100 m	**15113.79**
depth 2500 mm	8116.70	22.17	426.77	860.14	15586.53	100 m	**16873.44**
depth 3000 mm	8158.79	25.92	498.96	924.83	17128.92	100 m	**18552.71**
Catchwater or french drains; laying pipes to drains excavated separately; 160 mm diameter non-coilable perforated plastic pipes; including straight jointing; pipes laid with perforations uppermost; lining trench; wrapping pipes with filter fabric; backfilling with shingle							
Width 600 mm							
depth 450 mm	15121.21	11.20	215.60	86.63	15962.38	100 m	**16264.61**
depth 600 mm	15129.63	12.70	244.47	117.56	16270.85	100 m	**16632.88**
depth 750 mm	15138.05	14.20	273.35	148.50	16579.33	100 m	**17001.18**
depth 900 mm	15146.46	15.70	302.23	179.44	16887.81	100 m	**17369.48**
depth 1000 mm	15152.08	16.70	321.48	200.06	17093.46	100 m	**17615.00**
depth 1200 mm	15163.30	18.70	359.98	241.43	17504.77	100 m	**18106.18**
depth 1500 mm	15180.14	21.70	417.73	303.19	18121.73	100 m	**18842.65**

R13 LAND DRAINAGE

Item Excluding site overheads and profit	PC £	Labour hours	Labour £	Plant £	Material £	Unit	Total rate £
CATCHWATER OR FRENCH DRAINS – cont							
Catchwater or french drains – cont							
Width 600 mm – cont							
depth 2000 mm	15208.20	26.70	513.98	406.31	19149.99	100 m	**20070.28**
depth 2500 mm	15236.26	31.70	610.23	509.44	20178.25	100 m	**21297.92**
depth 3000 mm	15264.33	36.70	706.48	612.56	21206.51	100 m	**22525.55**
Width 900 mm							
depth 450 mm	7944.12	13.60	261.80	136.13	9262.71	100 m	**9660.64**
depth 600 mm	7956.74	15.84	304.92	182.32	9725.43	100 m	**10212.67**
depth 750 mm	7969.37	18.10	348.43	228.94	10188.15	100 m	**10765.52**
depth 900 mm	7982.00	20.34	391.55	275.14	10650.87	100 m	**11317.56**
depth 1000 mm	7990.42	21.84	420.42	306.07	10959.35	100 m	**11685.84**
depth 1200 mm	8007.26	24.84	478.17	367.95	11576.30	100 m	**12422.42**
depth 1500 mm	8032.51	29.34	564.79	460.76	12501.74	100 m	**13527.29**
depth 2000 mm	8074.61	36.84	709.17	615.45	14044.13	100 m	**15368.75**
depth 2500 mm	8116.70	44.34	853.54	770.14	15586.53	100 m	**17210.21**
depth 3000 mm	8158.79	51.84	997.92	924.83	17128.92	100 m	**19051.67**
Catchwater or french drains; Exxon Chemical Geopolymers Ltd; Filtram filter drain; in trenches (trenches not included); comprising filter fabric, liquid conducting core and 110 mm uPVC slitpipes; all in accordance with manufacturer's instructions; backfilling with shingle							
Width 600 mm							
depth 1000 mm	7962.36	16.84	324.17	202.95	9931.08	100 m	**10458.20**
depth 1200 mm	7973.58	18.84	362.67	244.20	10342.39	100 m	**10949.26**
depth 1500 mm	7990.42	21.84	420.42	576.08	10959.35	100 m	**11955.85**
depth 2000 mm	8018.48	19.84	381.92	553.58	11987.61	100 m	**12923.11**
Width 900 mm							
depth 450 mm	7944.12	13.60	261.80	136.13	9262.71	100 m	**9660.64**
depth 600 mm	7956.74	15.84	304.92	182.32	9725.43	100 m	**10212.67**
depth 750 mm	7969.37	18.10	348.43	228.94	10188.15	100 m	**10765.52**
depth 900 mm	7982.00	20.34	391.55	275.14	10650.87	100 m	**11317.56**
depth 1000 mm	7990.42	21.84	420.42	306.07	10959.35	100 m	**11685.84**
depth 1200 mm	8007.26	24.84	478.17	367.95	11576.30	100 m	**12422.42**
depth 1500 mm	8032.51	29.34	564.79	460.76	12501.74	100 m	**13527.29**
depth 2000 mm	8074.61	37.84	728.42	705.45	14044.13	100 m	**15478.00**
depth 2500 mm	8116.70	44.34	853.54	860.14	15586.53	100 m	**17300.21**
depth 3000 mm	8158.79	51.84	997.92	1014.83	17128.92	100 m	**19141.67**
DITCHING							
Ditching; clear silt and bottom ditch not exceeding 1.50 m deep; strim back vegetation; disposing to spoil heaps; by machine							
Up to 1.50 m wide at top	–	6.00	115.50	27.00	–	100 m	**142.50**
1.50–2.50 m wide at top	–	8.00	154.00	36.00	–	100 m	**190.00**
2.50–4.00 m wide at top	–	9.00	173.25	40.50	–	100 m	**213.75**

R13 LAND DRAINAGE

Item Excluding site overheads and profit	PC £	Labour hours	Labour £	Plant £	Material £	Unit	Total rate £
Ditching; clear only vegetation from ditch							
not exceeding 1.50 m deep; disposing to							
spoil heaps; by strimmer							
Up to 1.50 m wide at top	–	5.00	96.25	7.01	–	100 m	103.26
1.50–2.50 m wide at top	–	6.00	115.50	12.61	–	100 m	128.11
2.50–4.00 m wide at top	–	7.00	134.75	19.62	–	100 m	154.37
Ditching; clear silt from ditch not exceeding							
1.50 m deep; trimming back vegetation;							
disposing to spoil heaps; by hand							
Up to 1.50 m wide at top	–	15.00	288.75	7.01	–	100 m	295.76
1.50–2.50 m wide at top	–	27.00	519.75	12.61	–	100 m	532.36
2.50–4.00 m wide at top	–	42.00	808.50	19.62	–	100 m	828.12
Ditching; excavating and forming ditch and							
bank to given profile (normally 45°); in							
loam or sandy loam; by machine							
Width 300 mm							
depth 600 mm	–	3.70	71.22	33.30	–	100 m	104.52
depth 900 mm	–	5.20	100.10	46.80	–	100 m	146.90
depth 1200 mm	–	7.20	138.60	64.80	–	100 m	203.40
depth 1500 mm	–	9.40	180.95	74.03	–	100 m	254.98
Width 600 mm							
depth 600 mm	–	–	–	69.27	–	100 m	69.27
depth 900 mm	–	–	–	102.78	–	100 m	102.78
depth 1200 mm	–	–	–	139.65	–	100 m	139.65
depth 1500 mm	–	–	–	201.09	–	100 m	201.09
Width 900 mm							
depth 600 mm	–	–	–	393.75	–	100 m	393.75
depth 900 mm	–	–	–	693.75	–	100 m	693.75
depth 1200 mm	–	–	–	932.25	–	100 m	932.25
depth 1500 mm	–	–	–	1155.00	–	100 m	1155.00
Width 1200 mm							
depth 600 mm	–	–	–	618.75	–	100 m	618.75
depth 900 mm	–	–	–	918.75	–	100 m	918.75
depth 1200 mm	–	–	–	1216.88	–	100 m	1216.88
depth 1500 mm	–	–	–	1559.25	–	100 m	1559.25
Width 1500 mm							
depth 600 mm	–	–	–	763.13	–	100 m	763.13
depth 900 mm	–	–	–	1163.25	–	100 m	1163.25
depth 1200 mm	–	–	–	1546.88	–	100 m	1546.88
depth 1500 mm	–	–	–	1938.75	–	100 m	1938.75
Extra for ditching in clay	–	–	–	–	–	20%	–
Ditching; excavating and forming ditch and							
bank to given profile (normal 45°); in loam							
or sandy loam; by hand							
Width 300 mm							
depth 600 mm	–	36.00	693.00	–	–	100 m	693.00
depth 900 mm	–	42.00	808.50	–	–	100 m	808.50
depth 1200 mm	–	56.00	1078.00	–	–	100 m	1078.00
depth 1500 mm	–	70.00	1347.50	–	–	100 m	1347.50

R13 LAND DRAINAGE

Item Excluding site overheads and profit	PC £	Labour hours	Labour £	Plant £	Material £	Unit	Total rate £
DITCHING – cont							
Ditching – cont							
Width 600 mm							
depth 600 mm	–	56.00	1078.00	–	–	100 m	**1078.00**
depth 900 mm	–	84.00	1617.00	–	–	100 m	**1617.00**
depth 1200 mm	–	112.00	2156.00	–	–	100 m	**2156.00**
depth 1500 mm	–	140.00	2695.00	–	–	100 m	**2695.00**
Width 900 mm							
depth 600 mm	–	84.00	1617.00	–	–	100 m	**1617.00**
depth 900 mm	–	126.00	2425.50	–	–	100 m	**2425.50**
depth 1200 mm	–	168.00	3234.00	–	–	100 m	**3234.00**
depth 1500 mm	–	210.00	4042.50	–	–	100 m	**4042.50**
Width 1200 mm							
depth 600 mm	–	112.00	2156.00	–	–	100 m	**2156.00**
depth 900 mm	–	168.00	3234.00	–	–	100 m	**3234.00**
depth 1200 mm	–	224.00	4312.00	–	–	100 m	**4312.00**
depth 1500 mm	–	280.00	5390.00	–	–	100 m	**5390.00**
Extra for ditching in clay	–	–	–	–	–	50%	–
Extra for ditching in compacted soil	–	–	–	–	–	90%	–
Piped ditching							
Jointed concrete pipes; FP McCann Ltd; including bedding, haunching and topping with 150 mm concrete; 11.50 N/mm^2 – 40 mm aggregate; to existing ditch							
300 mm diameter	16.13	0.67	12.83	8.25	34.00	m	**55.08**
450 mm diameter	23.97	0.67	12.83	8.25	50.77	m	**71.85**
600 mm diameter	38.66	0.67	12.83	8.25	76.71	m	**97.79**
900 mm diameter	102.05	1.00	19.25	12.38	123.45	m	**155.08**
Jointed concrete pipes; FP McCann Ltd; including bedding, haunching and topping with 150 mm concrete; 11.50 N/mm^2 – 40 mm aggregate; to existing ditch							
1200 mm diameter	175.39	1.00	19.25	12.38	203.40	m	**235.03**
extra over jointed concrete pipes for bends to 45°	161.30	0.67	12.83	10.31	169.84	nr	**192.98**
extra over jointed concrete pipes for single junctions 300 mm diameter	110.22	0.67	12.83	10.31	118.76	nr	**141.90**
extra over jointed concrete pipes for single junctions 450 mm diameter	163.39	0.67	12.83	9.38	171.93	nr	**194.14**
extra over jointed concrete pipes for single junctions 600 mm diameter	264.67	0.67	12.83	9.38	273.21	nr	**295.42**
extra over jointed concrete pipes for single junctions 900 mm diameter	447.99	0.67	12.83	9.38	456.53	nr	**478.74**
extra over jointed concrete pipes for single junctions 1200 mm diameter	684.31	0.67	12.83	9.38	692.85	nr	**715.06**
Outfalls							

R13 LAND DRAINAGE

Item Excluding site overheads and profit	PC £	Labour hours	Labour £	Plant £	Material £	Unit	Total rate £
Reinforced concrete outfalls to water course; flank walls; for 150 mm drain outlets; overall dimensions							
900 × 1050 × 900 mm high	–	–	–	–	–	m	**487.60**
GRC outfall headwalls	–	–	–	–	–	nr	**105.00**
JKH Unit; standard small	–	–	–	–	–	nr	**85.00**

S10 COLD WATER

Item Excluding site overheads and profit	PC £	Labour hours	Labour £	Plant £	Material £	Unit	Total rate £
S10 COLD WATER							
BOREHOLES							
Borehole drilling; Agripower Ltd							
Drilling operations to average 100 m depth;							
lining with 150 mm diameter sieve							
drilling to 100 m	–	–	–	–	–	nr	12000.12
rate per m over 100 m	–	–	–	–	–	m	120.00
pump, rising main, cable and kiosk	–	–	–	–	–	nr	5000.05
COLD WATER							
Blue MDPE polythene pipes; type 50; for cold water services; with compression fittings; bedding on 100 mm DOT type 1 granular fill material							
Pipes							
20 mm diameter	0.57	0.08	1.54	–	3.84	m	5.38
25 mm diameter	0.61	0.08	1.54	–	3.88	m	5.42
32 mm diameter	1.32	0.08	1.54	–	4.59	m	6.13
50 mm diameter	2.17	0.10	1.93	–	5.44	m	7.37
60 mm diameter	3.48	0.10	1.93	–	6.75	m	8.68
Hose union bib taps; including fixing to wall; making good surfaces							
15 mm	13.30	0.75	14.44	–	13.70	nr	28.14
22 mm	18.08	0.75	14.44	–	18.68	nr	33.12
Stopcocks; including fixing to wall; making good surfaces							
15 mm	7.70	0.75	14.44	–	8.10	nr	22.54
22 mm	11.85	0.75	14.44	–	12.25	nr	26.69
Standpipes; to existing 25 mm water mains							
1.00 m high	–	–	–	–	–	nr	225.00
Hose junction bib taps; to standpipes							
19 mm	–	–	–	–	–	nr	80.00
RAINWATER HARVESTING							
Rainwater harvesting tanks							
Notes: Rainwater harvesting tanks must be installed on granular or concrete bases. If installed below ground the tanks may require construction of drained underground chambers to support them. Please see the appropriate sections in this book for excavation, disposal, bases, retaining walls and drainage.							

S10 COLD WATER

Item Excluding site overheads and profit	PC £	Labour hours	Labour £	Plant £	Material £	Unit	Total rate £
Excavation for underground tanks; **excavation inclusive of earthwork retention** **for self-supporting tanks** 7 tonne tracked excavator (bucket volume 0.28 m³)							
maximum depth not exceeding 1.00 m	–	0.06	1.20	4.28	–	m³	5.48
maximum depth not exceeding 2.00 m	–	0.07	1.37	4.89	–	m³	6.26
maximum depth not exceeding 3.00 m	–	0.09	1.75	6.23	–	m³	7.98
Disposal Excavated material; off site; to tip; mechanically loaded (JCB)							
inert	–	–	–	–	–	m³	17.14
Type 1 granular fill base; PC £16.50/tonne **(£36.30/m³ compacted)** By machine							
100 mm thick	3.63	0.03	0.54	0.40	3.63	m²	4.57
150 mm thick	5.45	0.03	0.48	0.61	5.45	m²	6.54
Backfilling to surround of rainwater tank; **carefully compacting as work proceeds** Arising from the excavations average thickness exceeding 0.25 m; depositing in layers 150 mm maximum thickness	–	0.03	0.64	5.25	–	m³	5.89
Rainwater harvesting tanks; Combined **Harvesters Ltd; self-supporting** **underground or above ground tanks** Columbus rainwater tank with max 1.00 m cover in pedestrian areas; complete with tank dome and pedestrian lid, submersible automatic pump system, supra filtration system, 25 m black and green rainwater pipe, rainwater labelling kit, 125 mm fine filter (excavation, backfilling base, trenching and inlet pipework all not included)							
3700 litres; 2.44 × 1.65 × 1.95	1374.00	3.00	115.50	–	1514.00	nr	1629.50
4500 litres; 2.44 × 1.84 × 1.84	1508.00	3.50	134.75	–	1648.00	nr	1782.75
6500 litres; 2.68 × 2.02 × 2.29	1768.00	4.00	154.00	–	1908.00	nr	2062.00
9000 litres; 2.44 × 1.84 × 1.88	2494.00	4.50	173.25	–	2634.00	nr	2807.25
13000 litres – 2 × 6500 l tanks; 2.68 × 4.20 × 2.29	3014.00	7.00	269.50	–	3154.00	nr	3423.50

S10 COLD WATER

Item Excluding site overheads and profit	PC £	Labour hours	Labour £	Plant £	Material £	Unit	Total rate £
RAINWATER HARVESTING – cont							
Rainwater harvesting tanks – cont							
Cristall rainwater tank; lighter weight tank; complete with tank dome and pedestrian lid, submersible automatic pump system, supra filtration system, 25 m black and green rainwater pipe, rainwater labelling kit, 125 mm fine filter (excavation, backfilling base, trenching and inlet pipework all not included)							
1650 litres; 2.10 × 1.05 × 1.22	1069.00	5.00	192.50	–	1209.00	nr	**1401.50**
2650 litres; 2.10 × 1.30 × 1.50	1182.00	6.00	231.00	–	1322.00	nr	**1553.00**
Hercules rainwater tank; above or below ground (pedestrian areas only) installation; extendable to multiple tanks							
1600 litres; 1.35 m diameter × 1.60 m high	1383.00	4.50	173.25	–	1523.00	nr	**1696.25**
Lilo low profile tanks for reduced excavation; complete with rainwater filters and pressure pump							
1500 litres; 2.10 × 1.25 × 1.02	1124.00	–	–	–	1124.00	nr	**1124.00**
3000 litres; 2.46 × 2.1 × 1.05 m	1529.00	6.00	231.00	–	1669.00	nr	**1900.00**
5000 litres; 2.89 × 2.30 × 1.26	1906.00	6.50	250.25	–	2046.00	nr	**2296.25**
10000 litres 2.89 × 4,60 × 1.26 m	3289.00	7.00	269.50	–	3429.00	nr	**3698.50**
Combined Harvesters Ltd; accessories for rainwater harvesting tanks							
Downpipe filter for connection to roof downpipes.							
Quattro; for roof areas up to 50 m²	7.88	0.25	4.81	–	7.88	nr	**12.69**
Regendieb; for roof areas up to 80 m²	27.84	0.25	4.81	–	27.84	nr	**32.65**
Regendieb Deluxe; for roof areas up to 100 m²	31.84	–	–	–	31.84	nr	**31.84**
Pumps; inclusive of electrical installation and connections							
Raincatcher direct pump package	318.90	0.25	4.81	–	458.90	nr	**463.71**
Garden Comfort pump package	292.90	0.75	14.44	–	292.90	nr	**307.34**
Filters for rainwater tanks; complete with overflow siphons and rodent guard	153.16	–	–	–	153.16	nr	**153.16**
Extra over to cover rainwater tanks in geofabric prior to backfilling; 10 m² of terram per tank; average cost	–	0.25	9.63	–	4.07	nr	**13.70**

S14 IRRIGATION

Item Excluding site overheads and profit	PC £	Labour hours	Labour £	Plant £	Material £	Unit	Total rate £
S14 IRRIGATION							
IRRIGATION							
Irrigation Infrastructure and pipework;							
main or ring main supply							
Excavate and lay mains supply pipe to supply							
irrigated area							
PE 80 20 mm	–	–	–	–	–	m	4.87
PE 80 25 mm	–	–	–	–	–	m	5.06
Pro-Flow PEHD 63 mm PE	–	–	–	–	–	m	8.68
Pro-Flow PEHD 50 mm PE	–	–	–	–	–	m	7.09
Pro-Flow PEHD 40 mm PE	–	–	–	–	–	m	6.11
Pro-Flow PEHD 32 mm PE	–	–	–	–	–	m	5.49
Irrigation systems; head control to							
sprinkler stations							
Valves installed at supply point on irrigation							
supply manifold; includes for pressure control							
and filtration							
2 valve unit	–	–	–	–	–	nr	132.03
4 valve unit	–	–	–	–	–	nr	206.11
6 valve unit	–	–	–	–	–	nr	291.49
12 valve unit	–	–	–	–	–	nr	558.92
Solenoid valve; 25 mm with chamber; extra							
over for each active station	–	–	–	–	–	nr	106.16
Cable to solenoid valves (alternative to valves							
on manifold as above)							
4 core	–	–	–	–	–	m	1.57
6 core	–	–	–	–	–	m	7.63
12 core	–	–	–	–	–	m	3.19
Irrigation; water supply and control							
Header tank and submersible pump and							
pressure stat							
1000 litres (500 gallon 25 mm/10 days to							
1500 m^2) 2 m^3/hr pump	–	–	–	–	–	nr	489.67
4540 litres (1000 gallon 25 mm/10 days to							
3500 m^2)	–	–	–	–	–	nr	991.89
9080 litres (2000 gallon 25 mm/10 days to							
7000 m^2)	–	–	–	–	–	nr	1642.27
Electric multi-station controllers 240 V							
WeatherMatic Smartline; 4 station; standard	–	–	–	–	–	nr	149.21
WeatherMatic Smartline; 6 station; standard	–	–	–	–	–	nr	167.62
WeatherMatic Smartline; 12 station; standard	–	–	–	–	–	nr	284.18
WeatherMatic Smartline; 12 station; radio							
controlled	–	–	–	–	–	nr	583.80
WeatherMatic Smartline; 12 station; with							
moisture sensor	–	–	–	–	–	nr	981.43
WeatherMatic Smartline; 12 station; radio							
controlled; with moisture sensor	–	–	–	–	–	nr	1178.61

S14 IRRIGATION

Item Excluding site overheads and profit	PC £	Labour hours	Labour £	Plant £	Material £	Unit	Total rate £
IRRIGATION – cont							
Irrigation sprinklers; station consisting of							
multiple sprinklers; inclusive of all							
trenching, wiring and connections to ring							
main or main supply							
Sprayheads; WeatherMatic LX; including							
nozzle; 21 m² coverage							
100 mm (4')	–	–	–	–	–	nr	15.52
150 mm (6')	–	–	–	–	–	nr	21.92
300 mm (12')	–	–	–	–	–	nr	25.28
Matched precipitation sprinklers;							
WeatherMatic LX							
MP 1000; 16 m² coverage	–	–	–	–	–	nr	35.57
MP 2000; 30 m² coverage	–	–	–	–	–	nr	35.57
MP 3000; 81 m² coverage	–	–	–	–	–	nr	35.57
Gear drive sprinklers; WeatherMatic; placed to							
provide head to head (100%) overlap; average							
T3; 121 m² coverage	–	–	–	–	–	nr	46.72
CT70; 255 m² coverage	–	–	–	–	–	nr	82.36
CT70SS; 361 m² coverage	–	–	–	–	–	nr	91.46
Drip irrigation; driplines; inclusive of							
connections and draindown systems							
Metzerplas TechLand 30 cm drip emitter							
fixed on soil surface	–	–	–	–	–	m	1.07
subsurface	–	–	–	–	–	m	1.57
Metzerplas TechLand 50 cm drip emitter							
fixed on soil surface	–	–	–	–	–	m	0.90
subsurface	–	–	–	–	–	m	1.41
Commissioning and testing of irrigation							
system							
per station	–	–	–	–	–	nr	39.56
Annual maintenance costs of irrigation							
system							
Call out charge per visit	–	–	–	–	–	nr	329.59
Extra over per station	–	–	–	–	–	nr	13.19
Moisture Control moisture sensing control							
system							
Large garden comprising 7000 m² and							
24 stations with four moisture sensors							
turf only	–	–	–	–	–	nr	13333.33
turf/shrub beds; 70/30	–	–	–	–	–	nr	7777.78
Medium garden comprising 3500 m² and							
12 stations with two moisture sensors							
turf only	–	–	–	–	–	nr	8888.89
turf/shrub beds; 70/30	–	–	–	–	–	nr	11555.56

S14 IRRIGATION

Item Excluding site overheads and profit	PC £	Labour hours	Labour £	Plant £	Material £	Unit	Total rate £
Smaller gardens garden comprising 1000 m²							
and 6 stations with one moisture sensor							
turf only	–	–	–	–	–	nr	**5444.44**
turf/shrub beds; 70/30	–	–	–	–	–	nr	**6000.00**
turf/shrub beds; 50/50	–	–	–	–	–	nr	**6444.44**
Leaky Pipe Systems Ltd; Leaky Pipe;							
moisture leaking pipe irrigation system							
Main supply pipe inclusive of machine							
excavation; exclusive of connectors							
20 mm LDPE polytubing	1.12	0.05	0.96	0.61	1.12	m	**2.69**
16 mm LDPE polytubing	0.75	0.05	0.96	0.61	0.75	m	**2.32**
Water filters and cartridges							
No 10; 20 mm	–	–	–	–	54.47	nr	**54.47**
Big Blue and RR30 cartridge; 25 mm	–	–	–	–	139.36	nr	**139.36**
Water filters and pressure regulator sets;							
complete assemblies							
No 10; flow rate 3.1–82 litres per minute	–	–	–	–	92.95	nr	**92.95**
Leaky pipe hose; placed 150 mm sub surface							
for turf irrigation; distance between laterals							
350 mm; excavation and backfilling priced							
separately							
LP12L low leak	4.73	0.04	0.77	–	4.73	m²	**5.50**
LP12H high leak	4.10	0.04	0.77	–	4.10	m²	**4.87**
LP12UH ultra high leak	5.02	0.04	0.77	–	5.02	m²	**5.79**
Leaky pipe hose; laid to surface for landscape							
irrigation; distance between laterals 600 mm							
LP12L low leak	2.82	0.03	0.48	–	2.82	m²	**3.30**
LP12H high leak	2.45	0.03	0.48	–	2.45	m²	**2.93**
LP12UH ultra high leak	2.99	0.03	0.48	–	2.99	m²	**3.47**
Leaky pipe hose; laid to surface for landscape							
irrigation; distance between laterals 900 mm							
LP12L low leak	1.89	0.02	0.32	–	1.89	m²	**2.21**
LP12H high leak	1.64	0.02	0.32	–	1.64	m²	**1.96**
LP12UH ultra high leak	2.00	0.02	0.32	–	2.00	m²	**2.32**
Leaky pipe hose; laid to surface for tree							
irrigation							
laid around circumference of tree pit							
LP12L low leak	2.79	0.13	2.41	–	2.79	nr	**5.20**
LP12H high leak	2.42	0.13	2.41	–	2.42	nr	**4.83**
LP12UH ultra high leak	2.96	0.13	2.41	–	2.96	nr	**5.37**
Accessories							
automatic multi-station controller stations;							
inclusive of connections	375.70	2.00	38.50	–	375.70	nr	**414.20**
solenoid valves; inclusive of wiring and							
connections to a multi-station controller;							
nominal distance from controller 25 m	62.40	0.50	9.63	–	517.40	nr	**527.03**

S15 FOUNTAINS/WATER FEATURES/SWIMMING POOLS

Item Excluding site overheads and profit	PC £	Labour hours	Labour £	Plant £	Material £	Unit	Total rate £
S15 FOUNTAINS/WATER FEATURES/ SWIMMING POOLS							
LAKES AND PONDS							
Lakes and ponds – General							
Preamble: The pressure of water against a retaining wall or dam is considerable, and where water retaining structures form part of the design of water features, the landscape architect is advised to consult a civil engineer. Artificially contained areas of water in raised reservoirs over 25,000 m³ have to be registered with the local authority and their dams will have to be covered by a civil engineer's certificate of safety.							
Typical linings – General							
Preamble: In addition to the traditional methods of forming the linings of lakes and ponds in puddled clay or concrete, there are a number of lining materials available. They are mainly used for reservoirs but can also help to form comparatively economic water features especially in soil which is not naturally water retentive. Information on the construction of traditional clay puddle ponds can be obtained from the British Trust for Conservation Volunteers, 36 St. Mary's Street, Wallingford, Oxfordshire, OX10 0EU. Tel: (01491) 39766. The cost of puddled clay ponds depends on the availability of suitable clay, the type of hand or machine labour that can be used and the use to which the pond is to be put.							
Lake liners; Fairwater Water Garden Design Consultants Ltd; to evenly graded surface of excavations (excavating not included); all stones over 75 mm; removing debris; including all welding and jointing of liner sheets							
Geotextile underlay; inclusive of spot welding to prevent dragging							
to water features	–	–	–	–	–	m²	2.27
to lakes or large features	–	–	–	–	–	1000 m²	2265.70
Butyl rubber liners; Varnamo; inclusive of site vulcanising laid to geotextile above							
0.75 mm thick	–	–	–	–	–	m²	7.57
0.75 mm thick	–	–	–	–	–	1000 m²	5570.00
1.00 mm thick	–	–	–	–	–	m²	8.62
1.00 mm thick	–	–	–	–	–	1000 m²	8620.00

S15 FOUNTAINS/WATER FEATURES/SWIMMING POOLS

Item Excluding site overheads and profit	PC £	Labour hours	Labour £	Plant £	Material £	Unit	Total rate £
Lake liners; Landline Ltd; Landflex or Alkorplan geomembranes; to prepared surfaces (surfaces not included); all joints fully welded; installation by Landline employees							
Landflex HC polyethylene geomembranes							
0.50 mm thick	–	–	–	–	–	1000 m²	3410.00
0.75 mm thick	–	–	–	–	–	1000 m²	3940.00
1.00 mm thick	–	–	–	–	–	1000 m²	4460.00
1.50 mm thick	–	–	–	–	–	1000 m²	5250.00
2.00 mm thick	–	–	–	–	–	1000 m²	6040.00
Alkorplan PVC geomembranes							
0.80 mm thick	–	–	–	–	–	1000 m	6250.00
1.20 mm thick	–	–	–	–	–	1000 m	9080.00
Lake liners; Monarflex							
Low density polyethylene (LDPE) lake and reservoir lining system; welding on site by Monarflex technicians (surface preparation and backfilling not included)							
Blackline 500	–	–	–	–	–	100 m²	931.70
Blackline 750	–	–	–	–	–	100 m²	1023.80
Blackline 1000	–	–	–	–	–	100 m²	1264.95
Operations over surfaces of lake liners							
Dug ballast; evenly spread over excavation already brought to grade							
150 mm thick	550.11	2.00	38.50	22.34	550.11	100 m²	610.95
200 mm thick	733.48	3.00	57.75	33.52	733.48	100 m²	824.75
300 mm thick	1100.22	3.50	67.38	39.10	1100.22	100 m²	1206.70
Imported topsoil; evenly spread over excavation							
100 mm thick	336.00	1.50	28.88	16.76	336.00	100 m²	381.64
150 mm thick	504.00	2.00	38.50	22.34	504.00	100 m²	564.84
200 mm thick	672.00	3.00	57.75	33.52	672.00	100 m²	763.27
Blinding existing subsoil with 50 mm sand	183.37	1.00	19.25	22.34	183.37	100 m²	224.96
Topsoil from excavation; evenly spread over excavation							
100 mm thick	–	–	–	16.76	–	100 m²	16.76
200 mm thick	–	–	–	22.34	–	100 m²	22.34
300 mm thick	–	–	–	33.52	–	100 m²	33.52
Extra over for screening topsoil using a Powergrid screener; removing debris	–	–	–	4.98	0.86	m³	5.84

S15 FOUNTAINS/WATER FEATURES/SWIMMING POOLS

Item Excluding site overheads and profit	PC £	Labour hours	Labour £	Plant £	Material £	Unit	Total rate £
LAKES AND PONDS – cont							
Lake construction; Fairwater Water Garden Design Consultants Ltd; lake construction at ground level; excavation; forming of lake; commissioning; excavated material spread on site							
Lined with existing site clay; natural edging with vegetation meeting the water							
500 m²	–	–	–	–	–	nr	9261.00
1000 m²	–	–	–	–	–	nr	16096.50
1500 m²	–	–	–	–	–	nr	22050.00
2000 m²	–	–	–	–	–	nr	27562.50
Lined with 0.75 mm butyl on geotextile underlay; natural edging with vegetation meeting the water							
500 m²	–	–	–	–	–	nr	14883.75
1000 m²	–	–	–	–	–	nr	25798.50
1500 m²	–	–	–	–	–	nr	35280.00
2000 m²	–	–	–	–	–	nr	44100.00
Extra over to the above for hard block edging to secure and protect liner; measured at lake perimeter	–	–	–	–	–	m	46.31
Lined with imported puddling clay; natural edging with vegetation meeting the water							
500 m²	–	–	–	–	–	nr	24255.00
1000 m²	–	–	–	–	–	nr	41895.00
1500 m²	–	–	–	–	–	nr	57330.00
2000 m²	–	–	–	–	–	nr	71662.50

Ornamental pools – General

Preamble: Small pools may be lined with one of the materials mentioned under lakes and ponds, or may be in rendered brickwork, puddled clay or, for the smaller sizes, fibreglass. Most of these tend to be cheaper than waterproof concrete. Basic prices for various sizes of concrete pools are given in the Approximate Estimates section. Prices for excavation, grading, mass concrete, and precast concrete retaining walls are given in the relevant sections. The manufacturers should be consulted before specifying the type and thickness of pool liner, as this depends on the size, shape and proposed use of the pool. The manufacturer's recommendation on foundations and construction should be followed.

S15 FOUNTAINS/WATER FEATURES/SWIMMING POOLS

Item Excluding site overheads and profit	PC £	Labour hours	Labour £	Plant £	Material £	Unit	Total rate £
Ornamental pools							
Pool liners; to 50 mm sand blinding to							
excavation (excavating not included); all							
stones over 50 mm; removing debris from							
surfaces of excavation; including underlay, all							
welding and jointing of liner sheets							
black polythene; 1000 gauge	–	–	–	–	–	m²	**5.88**
blue polythene; 1000 gauge	–	–	–	–	–	m²	**7.77**
coloured PVC; 1500 gauge	–	–	–	–	–	m²	**8.90**
black PVC; 1500 gauge	–	–	–	–	–	m²	**6.72**
black butyl; 0.75 mm thick	–	–	–	–	–	m²	**8.85**
black butyl; 1.00 mm thick	–	–	–	–	–	m²	**9.28**
black butyl; 1.50 mm thick	–	–	–	–	–	m²	**32.44**
Fine gravel; 100 mm; evenly spread over area							
of pool; by hand	3.33	0.13	2.41	–	3.33	m²	**5.74**
Selected topsoil from excavation; 100 mm;							
evenly spread over area of pool; by hand	–	0.13	2.41	–	–	m²	**2.41**
Extra over selected topsoil for spreading							
imported topsoil over area of pool; by hand	33.60	–	–	–	33.60	m³	**33.60**
Pool surrounds and ornament;							
Haddonstone Ltd; Portland Bath or							
terracotta cast stone							
Pool surrounds; installed to pools or water							
feature construction priced separately;							
surrounds and copings to 112.5 mm internal							
brickwork							
C4HSKVP half small pool surround; internal							
diameter 1780 mm; kerb features continuous							
moulding enriched with ovolvo and palmette							
designs; inclusive of plinth and integral							
conch shell vases flanked by dolphins	1250.00	16.00	308.00	–	1263.20	nr	**1571.20**
C4SKVP small pool surround as above but							
with full circular construction; internal							
diameter 1780 mm	2508.00	48.00	924.00	–	2524.76	nr	**3448.76**
C4MKVP medium pool surround; internal							
diameter 2705 mm; inclusive of plinth and							
integral vases	3762.00	48.00	924.00	–	3797.18	nr	**4721.18**
C4XLKVP extra large pool surround;							
internal diameter 5450 mm	5016.00	140.00	2695.00	–	5058.14	nr	**7753.14**
Pool centre pieces and fountains; inclusive of							
plumbing and pumps							
HC350 Lotus Bowl; 1830 mm wide with							
C1700 triple Dolphin Fountain; HD2900							
Doric Pedestal	3843.00	8.00	154.00	–	3846.11	nr	**4000.11**
C251 Gothic Fountain and Gothic Upper							
Base A350; free-standing fountain	1079.00	4.00	77.00	–	1082.11	nr	**1159.11**
HC521 Romanesque Fountain; free-							
standing bowl with self-circulating fountain;							
filled with cobbles; 815 mm diameter ×							
348 mm high	475.00	2.00	38.50	–	558.69	nr	**597.19**
C300 Lion Fountain; 610 mm high on							
fountain base C305; 280 mm high	233.00	2.00	38.50	–	236.11	nr	**274.61**

S15 FOUNTAINS/WATER FEATURES/SWIMMING POOLS

Item Excluding site overheads and profit	PC £	Labour hours	Labour £	Plant £	Material £	Unit	Total rate £
LAKES AND PONDS – cont							
Wall fountains, watertanks and fountains;							
inclusive of installation drainage,							
automatic top up, pump and balancing tank							
Capital Garden Products Ltd							
Lion Wall Fountain F010;							
970 × 940 × 520 mm	–	4.00	77.00	–	1697.50	nr	**1774.50**
Dolphin Wall Fountain F001; 740 × 510 mm	–	4.00	77.00	–	1538.00	nr	**1615.00**
Dutch Master Wall Fountain F012;							
740 × 410 mm	–	4.00	77.00	–	1564.00	nr	**1641.00**
James II Watertank 2801;							
730 × 730 × 760 mm high; 405 litres	–	4.00	77.00	–	1709.60	nr	**1786.60**
James II Fountain 4901 bp;							
710 × 1300 × 1440 mm high; 655 litres	–	4.00	77.00	–	1560.75	nr	**1637.75**
Marble wall fountains; Architectural							
Heritage Ltd; inclusive of installation,							
drainage, automatic top up, pump and							
balancing tank							
Breccia Pernice; marble wall fountain							
supported by two dolphins, 1770 mm high ×							
1000 mm wide × 640 mm deep	16400.00	4.00	77.00	–	17850.00	nr	**17927.00**
The River God; Verona marble wall mask							
fountain; 890 mm high × 810 mm wide ×							
230 mm deep	5400.00	4.00	77.00	–	6850.00	nr	**6927.00**
Fountain kits; typical prices of submersible							
units comprising fountain pumps, fountain							
nozzles, underwater spotlights, nozzle							
extension armatures, underwater terminal							
boxes and electrical control panels							
Single aerated white foamy water columns;							
ascending jet 70 mm diameter; descending							
water up to four times larger; jet height							
adjustable between 1.00–1.70 m	4095.00	–	–	–	4694.00	nr	**4694.00**
Single aerated white foamy water columns;							
ascending jet 110 mm diameter; descending							
water up to four times larger; jet height							
adjustable between 1.50–3.00 m	6825.00	–	–	–	7749.80	nr	**7749.80**
FOUNTAINS AND WATER FEATURES							
Waterfall construction; Fairwater Water							
Garden Design Consultants Ltd							
Stone placed on top of butyl liner; securing							
with concrete and dressing to form natural							
rock pools and edgings							
Portland stone; m³ rate	–	–	–	–	–	m³	**697.31**
Portland stone; tonne rate	–	–	–	–	–	tonne	**386.12**

S15 FOUNTAINS/WATER FEATURES/SWIMMING POOLS

Item Excluding site overheads and profit	PC £	Labour hours	Labour £	Plant £	Material £	Unit	Total rate £
Balancing tank; blockwork construction; inclusive of recirculation pump and pond level control; 110 mm balancing pipe to pond; waterproofed with preformed polypropylene membrane; pipework mains water top-up and overflow							
450 × 600 × 1000 mm	–	–	–	–	–	nr	1450.00
Extra for pump							
2000 gallons per hour; submersible	–	–	–	–	–	nr	215.00
Water features; Fairwater Water Garden **Design Consultants Ltd**							
Natural stream; butyl lined level changes of 1.00 m; water pumped from lower pond (not included) via balancing tank; level changes via Purbeck stone water falls							
20 m long stream	–	–	–	–	–	nr	20000.00
Bespoke free-standing water wall; steel or glass panel; self contained recirculation system and reservoir							
water wall; 2.00 m high × 850 mm wide	–	–	–	–	–	nr	9371.25
SWIMMING POOLS							
Natural swimming pools; Fairwater Water **Garden Design Consultants Ltd**							
Natural swimming pool of 50 m² area; shingle regeneration zone 50 m² planted with marginal and aquatic plants	–	–	–	–	–	nr	60637.50
Guncast Swimming Pools Ltd; excavation **and blinding of swimming pools; works by** **machine inclusive of allowances for** **working space and wall thicknesses;** **excavation of all trenches for pool** **pipeworks and electrical services**							
Regular shaped pools							
100 m² pool	–	–	–	–	–	nr	8640.00
72 m² pool	–	–	–	–	–	nr	6480.00
60 m² pool	–	–	–	–	–	nr	6480.00
Irregular shaped pools							
100 m² pool	–	–	–	–	–	nr	9072.00
72 m² pool	–	–	–	–	–	nr	6804.00
60 m² pool	–	–	–	–	–	nr	6804.00

S15 FOUNTAINS/WATER FEATURES/SWIMMING POOLS

Item Excluding site overheads and profit	PC £	Labour hours	Labour £	Plant £	Material £	Unit	Total rate £
SWIMMING POOLS – cont							
Guncast Swimming Pools Ltd;							
construction of a rectangular swimming							
pool in situ concrete corner steps,							
graduated shell depths of 1.1–2.5 m, water							
depth of 1.0–2.4 m as standard skimmer							
pool, tiled corner steps; excludes pump							
house construction							
Inclusive of drainage layer, formwork							
reinforcement and all Gunite construction,							
rendered internally and bespoke mosaic							
surface finish; bullnosed indian sandstone							
copings							
pool size 17.0 × 6.0 m; 172,500 litres	–	–	–	–	–	nr	117600.00
pool size 12.0 × 6.0 m; 125,750 litres	–	–	–	–	–	nr	94200.00
pool size 12.0 × 5.0 m; 102,150 litres	–	–	–	–	–	nr	69850.00
Guncast Swimming Pools Ltd; options and							
accessories for swimming pools							
Underwater lighting							
8 nr underwater floodlights 300w 12V;							
stainless steel	–	–	–	–	–	nr	5220.00
2 nr underwater floodlights 300w 12 V;							
stainless steel	–	–	–	–	–	nr	1305.00
Automatic pool cleaner	–	–	–	–	–	nr	1595.00
Automatic pool covers							
insulated automatic slatted cover; 12 × 6 m							
inclusive of additional surface skimmer and							
balance pipe in pit; housed in decking (not							
included)	–	–	–	–	–	nr	18990.00
solar cover; 17 × 6 m Ocea; housed in wall							
cave pit; logic motor; concealed with							
stainless steel frame and removable pvc							
panels; tiled or rendered to match pool	–	–	–	–	–	nr	32240.00
Driglide pool cover; 17 × 6 m selected							
colours on trackway	–	–	–	–	–	nr	33430.00
Ozone disinfection system	–	–	–	–	–	nr	5112.00
Guncast Swimming Pools Ltd;							
refurbishment of an existing swimming							
pool; with Roman bay end steps; no leaks							
present in existing pool; existing pipework							
and existing filtration systems and pump							
Empty pool, retain existing blockwork							
structure; retile pool; replace copings as indian							
sandstone bullnose; prepare shell; fix mosaic;							
pressure test							
pool size 8.50 × 4.25 m; 59,000 litres	–	–	–	–	–	nr	22900.00

S15 FOUNTAINS/WATER FEATURES/SWIMMING POOLS

Item Excluding site overheads and profit	PC £	Labour hours	Labour £	Plant £	Material £	Unit	Total rate £
Guncast Swimming Pools Ltd; options and accessories for swimming pools							
Pool heating options							
propane boiler 250,000 BTU	–	–	–	–	–	nr	3870.00
pool heating via propane boiler 400000 BTU	–	–	–	–	–	nr	4895.00
heat pump 100 × 1000 mm concrete base	–	–	–	–	–	nr	4275.00
pool heating via heat pump	–	–	–	–	–	nr	4275.00
Diving board	–	–	–	–	–	nr	2025.00
Pool maintenance kit							
telescopic pole, wall/floor brush; skimmer net; flexible vacuum hose; underwater vacuum cleaner; pool water test kit; thermometer; pool treatment chemicals	–	–	–	–	–	nr	865.00
INTERNAL WATER FEATURES							
Interanal Water features; Fairwater Water Garden Design Consultants Ltd; Stainless steel rill; powder coated black 3000 mm × 250 mm; 2 nr letterbox weirs to maintain water level in rill; WRAS compliant header tank; modified balancing tank with chemical dosing unit; circulation pump and pipework;							
3.00 m × 250 mm wide	–	–	–	–	–	nr	3842.00
Brimming bowl with catchment tray; Fairwater Water Garden Design Consultants Ltd; Marble brimming bowl. recessed Polypro catchment tray 1.30 × 1.30 × 120 mm; pipework; WRAS compliant header tank balancing tank with Chemical dosing unit; circulation pump							
Brimming bowl = 1.00 m diameter	–	–	–	–	–	nr	3190.00
Design mockup and commissioning on the above	–	–	–	–	–	nr	2640.00
Maintenance 12 months; monthly	–	–	–	–	–	nr	1104.00

V ELECTRICAL INSTALLATION

Item Excluding site overheads and profit	PC £	Labour hours	Labour £	Plant £	Material £	Unit	Total rate £
V ELECTRICAL INSTALLATION							
CLARIFICATION NOTES ON LABOUR COSTS IN THIS SECTION							
General groundworks team Generally a three man team is used in this section; The column 'Labour hours' reports team hours. The column 'Labour £' reports the total cost of the team for the unit of work shown							
3 man team	–	1.00	57.75	–	–	hr	**57.75**
CABLES AND SWITCHING							
Trenching for electrical services 3 tonne excavator (bucket volume 0.13 m³); arisings laid alongside							
600 mm deep	–	–	–	2.20	–	m	**2.20**
800 mm deep	–	–	–	2.57	–	m	**2.57**
1.00 m deep	–	–	–	3.09	–	m	**3.09**
Extra over any types of excavating irrespective of depth for breaking out existing materials; heavy duty 110 volt breaker tool							
hard rock	–	5.00	96.25	23.52	–	m³	**119.77**
concrete	–	3.00	57.75	14.11	–	m³	**71.86**
reinforced concrete	–	4.00	77.00	23.52	–	m³	**100.52**
brickwork, blockwork or stonework	–	1.50	28.88	7.06	–	m³	**35.94**
By hand							
600 mm deep	–	0.24	4.63	–	–	m	**4.63**
800 mm deep	–	0.43	8.23	–	–	m	**8.23**
1.00 m deep	–	0.67	12.83	–	–	m	**12.83**
Backfilling of trenches; including laying of electrical marker tape; compacting lightly as work proceeds To trenches containing armoured cable							
by machine	0.13	–	–	2.62	0.13	m	**2.75**
by hand	0.13	0.20	3.85	–	0.13	m	**3.98**
To trenches containing ducted cable; including 150 mm sharp sand over the duct							
by machine	2.13	–	–	3.28	2.13	m	**5.41**
by hand	2.13	0.25	4.81	–	2.13	m	**6.94**
Cable to trenches (trenching operations not included); laying only cable Twin core steel wire armoured cable; 50 m drums							
1.5 mm core	0.92	–	–	–	0.92	m	**0.92**
2.5 mm core	1.16	–	–	–	1.16	m	**1.16**
Twin core steel wire armoured cable; lengths less than 50 m runs							
1.5 mm core	0.99	–	–	–	0.99	m	**0.99**
2.5 mm core	1.24	–	–	–	1.24	m	**1.24**

V ELECTRICAL INSTALLATION

Item Excluding site overheads and profit	PC £	Labour hours	Labour £	Plant £	Material £	Unit	Total rate £
Three core steel wire armoured cable; 50 m drums							
1.5 mm core	1.04	–	–	–	1.04	m	**1.04**
2.5 mm core	1.46	–	–	–	1.46	m	**1.46**
4.0 mm core	1.73	–	–	–	1.73	m	**1.73**
6.0 mm core	2.35	–	–	–	2.35	m	**2.35**
10.0 mm core	4.24	–	–	–	4.24	m	**4.24**
Three core steel wire armoured cable; lengths less than 50 m runs							
1.5 mm core	1.07	–	–	–	1.07	m	**1.07**
2.5 mm core	1.58	–	–	–	1.58	m	**1.58**
Ducts to trenches; twin wall flexible cable ducts with drawstrings; laid on 150 mm clean sharp sand							
Twin wall duct; laying to trenches							
63 mm × 50 m coils	1.00	0.02	0.39	–	1.00	m	**1.39**
110 mm × 50 m coils	1.50	0.02	0.39	–	1.50	m	**1.89**
Cable drawing through ducts							
straight runs up to 50 m lengths	–	1.00	19.25	–	–	nr	**19.25**
Terminations to armoured cables; cutting coiling and taping length of cable to receive connection to light fitting, transformer and the like; fixing to temporary post							
Armoured cable	–	0.33	6.41	–	–	m	**6.41**
Plain ducted cable	–	0.13	2.41	–	–	m	**2.41**
Junction boxes; fixing to cables to receive connections to light fittings or transformers; to IP68							
Underground jointing box							
2 way	29.00	0.33	6.41	–	29.00	m	**35.41**
3 way	34.00	0.57	11.00	–	34.00	m	**45.00**
Underground jointing box							
2 way	3.50	0.33	6.41	–	3.50	m	**9.91**
3 way	4.90	0.40	7.70	–	4.90	m	**12.60**
Electrical connections							
Connections to existing distribution boards per switching circuit located within 1 m of the distribution board; chasing of cables to walls not included	–	–	–	–	–	nr	**46.68**
Connections to switches; external power cable circuits; connections to internal wall switches; drilling through walls and making good; chasing of cables to walls not included							
price for the first circuit	4.00	1.00	19.25	–	50.68	nr	**69.93**
additional circuits	4.00	–	–	–	15.20	nr	**15.20**

V41 STREET/AREA FLOODLIGHTING

Item Excluding site overheads and profit	PC £	Labour hours	Labour £	Plant £	Material £	Unit	Total rate £
V41 STREET/AREA FLOODLIGHTING							
GENERAL							
Street area floodlighting – General Preamble: There are an enormous number of luminaires available which are designed for small scale urban and garden projects. The designs are continually changing and the landscape designer is advised to consult the manufacturer's latest catalogue. Most manufacturers supply light fittings suitable for column, bracket, bulkhead, wall or soffit mounting. Highway lamps and columns for trafficked roads are not included in this section as the design of highway lighting is a very specialized subject outside the scope of most landscape contracts. The IP reference number refers to the waterproof properties of the fitting; the higher the number the more waterproof the fitting. Most items can be fitted with time clocks or PIR controls.							
MARKET PRICES OF LAMPS							
Market prices of lamps							
Lamps							
70 w HQIT-S	–	–	–	–	8.08	nr	**8.08**
70 w HQIT	–	–	–	–	17.50	nr	**17.50**
70 w SON	–	–	–	–	6.80	nr	**6.80**
70 w SONT	–	–	–	–	10.20	nr	**10.20**
100 w SONT	–	–	–	–	12.05	nr	**12.05**
150 w SONT	–	–	–	–	14.55	nr	**14.55**
28 w 2D	–	–	–	–	9.80	nr	**9.80**
100 w GLS\E27	–	–	–	–	1.90	nr	**1.90**
CLARIFICATION NOTES ON LABOUR COSTS IN THIS SECTION							
General groundworks team Generally a two man team is used in this section; The column 'Labour hours' reports team hours. The column 'Labour £' reports the total cost of the team for the unit of work shown							
2 man expert team	–	1.00	38.50	–	–	hr	**38.50**
craftman	–	1.00	19.25	–	–	hr	**19.25**
subcontract electrician	–	–	–	–	–	hr	**35.00**

V41 STREET/AREA FLOODLIGHTING

Item Excluding site overheads and profit	PC £	Labour hours	Labour £	Plant £	Material £	Unit	Total rate £
BULKHEAD LIGHTING							
Bulkhead and canopy fittings; including fixing to wall and light fitting (lamp, final painting, electric wiring, connections or related fixtures such as switch gear and time clock mechanisms not included unless otherwise indicated)							
Bulkhead and canopy fittings; Targetti Poulsen Nyhavn Wall small domed top conical shade with rings; copper wall lantern; finished untreated copper to achieve verdigris finish; also available in white aluminium; 310 mm diameter shade; with wall mounting arm; to IP 44	627.00	0.25	9.63	–	627.60	nr	**637.23**
Bulkhead and canopy fittings; Sugg Lighting Princess IP54 backlamp; 375 × 229 mm; copper frame; stove painted in black; with chimney and lampholder	380.00	0.25	9.63	–	380.60	nr	**390.23**
Victoria IP54 backlamp; 502 × 323 mm; copper frame; polished copper finish; with chimney, door and lampholder	478.00	0.25	9.63	–	478.60	nr	**488.23**
Palace IP54 backlamp; 457 × 321 mm; copper frame; stove painted black finish; with chimney, door and lampholder	425.00	0.25	9.63	–	425.60	nr	**435.23**
Windsor IP54 backlamp; 650 × 306 mm; copper frame; polished and lacquered finish; with door and lampholder	468.00	0.25	9.63	–	468.60	nr	**478.23**
Windsor IP54 gas backlamp; 650 × 306 mm; copper frame; polished and lacquered finish; hinged door; double inverted cluster mantle with permanent pilot and mains solenoid	836.00	0.25	9.63	–	836.60	nr	**846.23**
FLOODLIGHTS							
Floodlighting; ground, wall or pole mounted; including fixing (lamp, final painting, electric wiring, connections or related fixtures such as switch gear and time clock mechanisms not included unless otherwise indicated)							
Floodlighting; Targetti Poulsen – Landscape Division							
SPR-12; multi-purpose ground/spike mounted wide angle floodlight; 70 w HIT or SON; to IP65	319.50	0.50	19.25	–	322.89	nr	**342.14**

V41 STREET/AREA FLOODLIGHTING

Item Excluding site overheads and profit	PC £	Labour hours	Labour £	Plant £	Material £	Unit	Total rate £
FLOODLIGHTS – cont							
Floodlighting – cont							
Floodlight accessories; Targetti Poulsen – Landscape Division							
earth spike for SPR-12	35.50	–	–	–	35.50	nr	**35.50**
louvre for SPR-12	48.00	–	–	–	48.00	nr	**48.00**
cowl for SPR-12	48.00	–	–	–	48.00	nr	**48.00**
barn doors for SPR-12	112.00	–	–	–	112.00	nr	**112.00**
Large area/pitch floodlighting; CU Phosco							
FL444 1000 w SON-T; floodlights with lamp							
and loose gear; narrow asymmetric beam	266.27	–	–	–	301.27	nr	**301.27**
FL444 2.0 kw MBIOS; floodlight with lamp							
and loose gear; projector beam	508.09	–	–	–	543.09	nr	**543.09**
Large area floodlighting; CU Phosco							
FL345/G/250S; floodlight with lamp and							
integral gear	328.70	–	–	–	363.70	nr	**363.70**
FL345/G/400 MBI; floodlight with lamp and							
integral gear	253.16	–	–	–	288.16	nr	**288.16**
Small area floodlighting; Sugg Lighting; floodlight for feature lighting; clear or toughened glass							
Scenario lamp; 150 w HPS-T	698.75	–	–	–	733.75	nr	**733.75**
Scenario lamp; 150 w HQI-T	698.75	–	–	–	733.75	nr	**733.75**
SPOTLIGHTS							
Spotlights for uplighting and for **illuminating signs and notice boards,** **statuary and other features) ground, wall or** **pole mounted; including fixing, light fitting** **and priming (lamp, final painting, electric** **wiring, connections or related fixtures such** **as switch gear and time clock mechanisms** **not included); all mains voltage (240 v)** **unless otherwise stated**							
Spotlights; Targetti Poulsen – Landscape Division							
WeeBee Spot SP-05 for low voltage halogen reflector; 20/35/50 w; c/w integral transformer and wall mounting box; to IP65	160.00	0.50	9.63	–	198.39	nr	**208.02**
Spotlighters, uplighters and cowl lighting; Havells Sylvania; TECHNO – SHORT ARM; head adjustable 130°; rotation 350°; projection 215 mm on 210 mm base plate; 355 mm high; integral gear; PG16 cable gland; black or aluminium							
Ref S.3517.09/14; 70 w CDMT/HQIT	463.03	–	–	–	498.03	nr	**498.03**
Ref S.3518.09/14; 150 w CDMT/HQIT	486.72	–	–	–	521.72	nr	**521.72**

V41 STREET/AREA FLOODLIGHTING

Item Excluding site overheads and profit	PC £	Labour hours	Labour £	Plant £	Material £	Unit	Total rate £
RECESSED LIGHTING							
Recessed uplighting; including walk/drive over fully recessed uplighting; excavating, ground fixing, concreting in and making good surfaces (electric wiring, connections or related fixtures such as switch gear and time-clock mechanisms not included unless otherwise stated) (Note: transformers will power multiple lights dependent on the distance between the light units); all mains voltage (240 v) unless otherwise stated							
Recessed uplighting; Targetti Poulsen – Landscape Division; 266 mm diameter; diecast aluminium with stainless steel top plate 10 mm toughened safety glass; 2000 kg drive over; integral control gear; to IP67							
IPR-14 HIT metal halide; white light; spot or flood or wall wash distribution	275.00	0.50	9.63	–	313.39	nr	**323.02**
IRR-14 HIT compact fluorescent; white light; low power consumption; flood distribution	275.00	0.50	9.63	–	313.39	nr	**323.02**
Accessories for IPR-14 uplighters; Targetti Poulsen – Landscape Division							
rockguard	154.50	–	–	–	154.50	nr	**154.50**
stainless steel installation sleeve	49.50	–	–	–	49.50	nr	**49.50**
anti glare louvre (internal tilt)	98.50	–	–	–	98.50	nr	**98.50**
Accessories for Nimbus uplighters; Targetti Poulsen – Landscape Division							
IS installation sleeve	57.00	–	–	–	57.00	nr	**57.00**
Wall recessed; Havells Sylvania; EOS range; integral gear; asymmetric reflector; toughened reeded glass							
MINI EOS; 145 × 90 mm; black or aluminium; 20 W QT9 lamp	63.87	–	–	–	107.62	nr	**107.62**
RECTANGULAR EOS; 270 × 145 mm; black or aluminium; 20 W QT9 lamp	123.24	–	–	–	183.09	nr	**183.09**
LIGHTED BOLLARDS							
Lighted bollards; including excavating, ground fixing, concreting in and making good surfaces (lamp, final painting, electric wiring, connections or related fixtures such as switch gear and time-clock mechanisms not included) (Note: all illuminated bollards must be earthed); heights given are from ground level to top of bollards							

V41 STREET/AREA FLOODLIGHTING

Item Excluding site overheads and profit	PC £	Labour hours	Labour £	Plant £	Material £	Unit	Total rate £
LIGHTED BOLLARDS – cont							
Lighted bollards; Targetti Poulsen							
Orbiter; vandal-resistant; head of cast							
aluminium; domed top; anti-glare rings;							
pole extruded aluminium; diffuser clear UV							
stabilized polycarbonate; powder coated;							
1040 mm high × 255 mm diameter; with							
root or base plate; IP44	654.00	2.50	96.25	–	686.63	nr	**782.88**
Waterfront; solidly proportioned; head of							
cast silumin; domed top; symmetrical							
distribution; pole extruded aluminium							
sandblasted or painted white; internal							
diffuser clear UV stabilized polycarbonate;							
865 mm high × diameter 260 mm; IP55	736.00	2.50	96.25	–	768.63	nr	**864.88**
Bysted; concentric louvred bollard; head							
cast iron; post COR-TEN steel; externally							
untreated to provide natural aging effect of							
uniform oxidized red surface finish; internal							
painted white; lamp diffuser rings of clear							
polycarbonate; 1130 mm high × 280 mm							
diameter; to IP44	1133.00	2.50	96.25	–	1165.63	nr	**1261.88**
Lighted bollards; Woodscape Ltd							
Illuminated bollard; in 'very durable							
hardwood'; integral die cast aluminium							
lighting unit; 165 × 165 mm × 1.00 m high	258.65	2.50	96.25	–	297.36	nr	**393.61**
STREET AND PRECINCT LIGHTING							
Precinct lighting lanterns; ground, wall or							
pole mounted; including fixing and light							
fitting (lamps, poles, brackets, final							
painting, electric wiring, connections or							
related fixtures such as switch gear and							
time clock mechanisms not included)							
Precinct lighting lanterns; Targetti Poulsen							
Nyhavn Park side entry 90° curved arm							
mounted; steel rings over domed top;							
housing cast silumin sandblasted with							
integral gear; protected by UV stabilized							
clear polycarbonate diffuser; IP55	650.00	0.50	19.25	–	650.00	nr	**669.25**
Kipp; bottom entry pole-top; shot blasted or							
lacquered Hanover design award; hinged							
diffuser for simple maintenance; indirect							
lighting technique ensures low glare; IP55	638.00	0.50	19.25	–	638.00	nr	**657.25**
Precinct lighting lanterns; Sugg Lighting							
Juno Dome; IP66; molded GRP body;							
graphite finish	560.00	0.50	19.25	–	560.00	nr	**579.25**
Juno Cone; IP66; molded GRP body;							
graphite finish	560.00	0.50	19.25	–	560.00	nr	**579.25**
Sharkon; IP65; cast aluminium body; silver							
and blue finish	560.00	0.50	19.25	–	560.00	nr	**579.25**

V41 STREET/AREA FLOODLIGHTING

Item Excluding site overheads and profit	PC £	Labour hours	Labour £	Plant £	Material £	Unit	Total rate £
REPRODUCTION STREET LIGHTING							
Reproduction street lanterns and columns; including excavating, concreting in, backfilling and disposing of spoil, or fixing to ground or wall, making good surfaces, light fitting and priming (final painting, electric wiring, connections or related fixtures such as switch gear and time-clock mechanisms not included) (Note: lanterns up to 14 inches and columns up to 7 ft are suitable for residential lighting)							
Reproduction street lanterns; Sugg Lighting; reproduction lanterns handmade to original designs; all with ES lamp holder							
Westminster IP54 hexagonal hinged top lantern with door; two piece folded polycarbonate glazing; lamp 100 w HQI T; integral photo electric cell							
small; 1016 mm high × 356 mm wide	876.13	0.50	9.63	–	876.13	nr	**885.76**
large; 1124 mm high × 760 mm wide	1376.00	0.50	9.63	–	1376.00	nr	**1385.63**
Guildhall IP54; handcrafted copper frame; clear polycarbonate glazing circular tapered lantern with hemispherical top; integral photo electric cell; with door							
small; 711 mm high × 330 mm wide	749.28	0.50	9.63	–	749.88	nr	**759.51**
medium; 1150 mm high × 432 mm wide	790.13	0.50	9.63	–	790.73	nr	**800.36**
large; 1370 mm high × 550 mm wide	1096.50	0.50	9.63	–	1097.10	nr	**1106.73**
Grosvenor circular lantern with door; copper frame; polished copper finish; polycarbonate glazing							
small; 790 × 330 mm; IP54	763.25	0.50	9.63	–	763.85	nr	**773.48**
medium; 1080 × 435 mm; IP65	790.13	0.50	9.63	–	790.73	nr	**800.36**
Classic Globe IP54 lantern with hinged outer frame; cast aluminium frame; black polyester powder coating							
medium; 965 × 483 mm	698.75	0.50	9.63	–	699.35	nr	**708.98**
Windsor lantern; copper frame; polished copper finish							
small; 905 × 356 mm; IP54; with door	568.00	0.50	9.63	–	568.60	nr	**578.23**
small; 905 × 356 mm; IP65; without door	530.00	0.50	9.63	–	530.60	nr	**540.23**
medium; 1124 × 420 mm; IP54; with door	620.00	0.50	9.63	–	620.60	nr	**630.23**
large; 1124 × 470 mm; IP54; with door	720.00	0.50	9.63	–	720.60	nr	**730.23**
medium; gas lantern; 1124 × 420 mm; IP54; with door; integral solenoid and pilot	1101.88	0.50	9.63	–	1102.48	nr	**1112.11**

V41 STREET/AREA FLOODLIGHTING

Item Excluding site overheads and profit	PC £	Labour hours	Labour £	Plant £	Material £	Unit	Total rate £
REPRODUCTION STREET LIGHTING – cont							
Reproduction brackets and suspensions;							
Sugg Lighting							
Iron brackets							
Bow bracket; 6'0"	1071.56	2.00	38.50	–	1072.16	nr	**1110.66**
Ornate iron bracket; large	478.38	2.00	38.50	–	478.98	nr	**517.48**
Ornate iron bracket; medium	462.25	2.00	38.50	–	462.85	nr	**501.35**
Swan neck iron bracket; large	451.50	2.00	38.50	–	452.10	nr	**490.60**
Swan neck iron bracket; medium	255.85	2.00	38.50	–	256.45	nr	**294.95**
Cast brackets							
Universal cast bracket	341.85	2.00	38.50	–	342.45	nr	**380.95**
Abbey bracket	202.10	2.00	38.50	–	202.70	nr	**241.20**
Short Abbey bracket	199.95	2.00	38.50	–	200.55	nr	**239.05**
Plaza cast bracket	167.16	2.00	38.50	–	167.76	nr	**206.26**
Base mountings							
Universal pedestal	339.70	2.00	38.50	–	340.30	nr	**378.80**
Universal plinth	193.50	2.00	38.50	–	194.10	nr	**232.60**
Reproduction lighting columns							
Reproduction lighting columns; Sugg Lighting							
Harborne C11; fabricated iron/steel heavy duty post with integral cast root; 3–5 m	1483.50	8.00	154.00	–	1491.47	nr	**1645.47**
Aylesbury C12; base fabricated heavy gauge 89 mm aluminium post; 3–5 m	2115.60	8.00	154.00	–	2123.57	nr	**2277.57**
Cannonbury C13; fabricated heavy gauge 89 mm aluminium post; 3–5 m	1558.75	8.00	154.00	–	1566.72	nr	**1720.72**
standard column C14; rooted British Steel 168/89 mm embellished post; 5–8 m	702.10	8.00	154.00	–	710.07	nr	**864.07**
Large Constitution Hill C22X; cast aluminium, steel cored post with extended spigot; 4.3 m	5622.25	6.00	115.50	–	5630.22	nr	**5745.72**
Cardiff C29; cast aluminium post; 3.5 m	2759.53	6.00	115.50	–	2767.50	nr	**2883.00**
Seven Dials C36; cast aluminium rooted post; 3.8 m	2038.20	8.00	154.00	–	2046.17	nr	**2200.17**
Royal Exchange C42; traditional cast aluminium post; welded multi-arm construction; 2.1 m	3113.20	6.00	115.50	–	3121.17	nr	**3236.67**

PART 5

Approximate Estimating Rates – Minor Works

APPROXIMATE ESTIMATES

Prices in this section are based upon the Prices for Measured Works, but allow for incidentals which would normally be measured separately in a Bill of Quantities. They do not include for Preliminaries which are priced elsewhere in this book.

Items shown as subcontract or specialist rates would normally include the specialist's overhead and profit. All other items which could fall within the scope of works of general landscape and external works contractors would not include profit.

Based on current commercial rates, profits of 15% to 35% may be added to these rates to indicate the likely 'with profit' values of the tasks below. The variation quoted above is dependent on the sector in which the works are taking place – domestic, public or commercial.

DEMOLITION AND SITE CLEARANCE

Item Excluding site overheads and profit	Unit	Total rate £
PLANT COSTS		
Hired plant		
Figures in brackets reflect the actual working hours per week of the hired machinery; prices all exclude delivery and collection charges and fuel costs		
Dumper; 3 tonne Thwaites; self drive (28 hours/week)	hr	4.10
Dumper; 6 ton Thwaites; self drive (28 hours/week)	hr	5.35
Excavator; tracked; 5 ton; self drive (28 hours/week)	hr	10.20
Fork lift; telehandler; self drive (28 hours/week)	hr	11.90
Fuel charge; Red diesel	hr	1.15
JCB 3CX 4 × 4 Sitemaster; fueled; operated	hr	44.50
Mini excavator; 1.5 tonne; self drive	hr	5.35
Mini excavator; JCB 803; rubber tracks; self drive	hr	8.05
Mini excavator; JCB 803; steel tracks; self drive	hr	8.05
Skip loader; 1 tonne	hr	3.90
HSS		
Access tower; alloy; 5.2 m	day	34.50
Cultivator; 110 kg	hr	8.85
Diamond blade consumable; 450 mm; concrete	mm	30.00
Heavy-duty breaker; 110 v; 2200 w	hr	4.70
Mesh fence (Heras fence); 2.85 × 2.0 m high	1 m/wk	2.35
Oxy-acetylene cutting kit	hr	4.70
Petrol masonry saw bench; 350 mm	hr	5.70
Petrol poker vibrator + 50 mm head	hr	2.75
Post hole borer; 1 man; daily rate	hr	5.50
Vibrating plate compactor	hr	3.85
DEMOLITION AND SITE CLEARANCE		
Demolish existing surfaces by hand held electric breaker; removal by grab		
Break up plain concrete slab; remove to licensed tip		
150 mm thick	m²	10.60
200 mm thick	m²	13.00
Break up reinforced concrete slab and remove to licensed tip		
150 mm thick	m²	16.90
200 mm thick	m²	21.50
300 mm thick	m²	40.00
Break out existing surface and associated 150 mm thick granular base load to remove off site by grab		
macadam 70 mm thick	m²	11.90
block paving 50 mm thick	m²	16.10
block paving 80 mm thick	m²	18.20
Demolish existing free-standing walls; grub out foundations; remove arisings to skip; backfill with imported topsoil; works by excavator, dumper and diesel breaker		
Brick wall; 112 mm thick		
300 mm high	m	16.10
500 mm high	m	25.00

DEMOLITION AND SITE CLEARANCE

Item Excluding site overheads and profit	Unit	Total rate £
Brick wall; 225 mm thick		
300 mm high	m	24.00
500 mm high	m	26.50
1.00 m high	m	35.00
1.20 m high	m	40.00
1.50 m high	m	43.00
1.80 m high	m	50.00
Demolish existing free-standing walls; grub out foundations; remove arisings to skip; backfill with imported topsoil; works by hand and diesel breaker		
Brick wall; 112 mm thick		
300 mm high	m	17.50
500 mm high	m	20.00
Brick wall; 225 mm thick		
300 mm high	m	27.00
500 mm high	m	30.00
1.00 m high	m	42.00
1.20 m high	m	49.50
1.50 m high	m	53.00
1.80 m high	m	62.00
Break out existing free-standing building; break out plain concrete base 150 mm thick and remove to skip distance 50 m; backfill with imported topsoil; all works by hand		
Timber buildings		
shed 6.0 m^2	nr	270.00
shed 10.0 m^2	nr	610.00
shed 15.0 m^2	nr	850.00
Site clearance – General		
Clear away light fencing and gates (chain link, chestnut paling, light boarded fence or similar) and remove to licensed tip	100 m	315.00
Strip turf; strip topsoil 250 mm thick move to stockpile 25 m		
all by machine; disposal of turf of to skip	100 m	750.00
by machine; disposal of turf by grab	100 m	730.00
strip and stack turf for preservation by hand; strip soil by machine	100 m	920.00
by hand; disposal to skip	100 m	2225.00
Clear mixed shrub area; dig out roots		
Groundcovers and small shrubs 20%; shrubs 1.0–2.0 m high 40%; shrubs 2.0–3.0 m high 20%; shrubs over 3.0 m high 20%		
clearance only	m^2	31.00
disposal to skip; chipped	m^2	18.00
disposal to skip; unchipped	m^2	35.00
disposal on site; unchipped	m^2	4.80

GROUNDWORK

Item Excluding site overheads and profit	Unit	Total rate £
EXCAVATION AND FILLING		
Cut and strip by machine turves 50 mm thick		
Load to barrows and stack on site not exceeding 25 m travel to stack	100 m²	520.00
Load to barrows and disposal off site by skip; distance 25 m	100 m²	1150.00
Excavate to reduce levels; mechanical		
Removal to spoil heaps		
excavated directly to loading position	m³	8.05
transporting to loading position 25 m distance	m³	14.40
Excavate to reduce levels; mechanical		
Removal off site by grab		
excavated directly to loading position	m³	40.00
transporting to loading position 25 m distance	m³	46.00
Excavate to reduce levels; mechanical		
Removal off site by skip		
excavated directly to loading position	m³	56.00
transporting to loading position 25 m distance	m³	62.00
Excavate to reduce levels; hand		
Removal off site by skip		
excavated directly to loading position	m³	180.00
excavated directly to loading position	m³	230.00
filled to bags and transporting to loading position 25 m distant	m³	350.00
Excavation and filling; mechanical		
Excavate existing soil on proposed turf or planting area to reduce levels; grade to levels; fill excavated area with topsoil from spoil heaps; removal of and excavated material by skip		
100 mm deep	m²	7.30
200 mm deep	m²	14.50
300 mm deep	m²	22.00
Spread excavated material to levels in layers not exceeding 150 mm; grade to finished levels to receive surface treatments		
By machine		
average thickness 100 mm	m²	1.15
average thickness 100 mm but with imported topsoil	m²	5.00
average thickness 200 mm	m²	1.75
average thickness 200 mm but with imported topsoil	m²	9.40
average thickness 250 mm	m²	2.05
average thickness 250 mm but with imported topsoil	m²	11.60
extra for work to banks exceeding 30° slope	30%	–

GROUNDWORK

Item	Unit	Total rate £
Excluding site overheads and profit		
Filling by hand		
Excavate existing soil on proposed turf or planting area to reduce levels; grade to levels; fill excavated area with topsoil from spoil heaps; removal of excavated material by skip		
100 mm deep	m²	8.85
200 mm deep	m²	17.70
300 mm deep	m²	26.50
500 mm deep	m²	44.00
TRENCHES		
Excavate trenches; remove excavated material off site by grab 25 m distance; fill trench to ground level with site mixed concrete 1:3:6; allow for movement of concrete and excavated material		
By machine		
300 mm wide × 250 mm deep	m	14.70
500 mm wide × 250 mm deep	m	21.50
750 mm wide × 350 mm deep	m	43.00
1200 mm wide × 600 mm deep	m	120.00
By hand		
300 mm wide × 250 mm deep	m	24.00
500 mm wide × 250 mm deep	m	38.00
750 mm wide × 350 mm deep	m	84.00
1200 mm wide × 600 mm deep	m	230.00

IN SITU CONCRETE

Item Excluding site overheads and profit	Unit	Total rate £
IN SITU CONCRETE		
Mix concrete on site; aggregates delivered in 10 tonne loads; deliver mixed concrete to location by mechanical dumper distance 25 m		
1:3:6	m³	110.00
1:2:4	m³	130.00
As above but ready mixed concrete		
10 N/mm²	m³	120.00
15 N/mm²	m³	120.00
Mix concrete on site; aggregates delivered in 10 tonne loads; deliver mixed concrete to location by barrow distance 25 m		
1:3:6	m³	155.00
1:2:4	m³	170.00
As above but aggregates delivered in 850 kg bulk bags		
1:3:6	m³	200.00
1:2:4	m³	210.00
As above but concrete discharged directly from ready mix lorry to required location		
10 N/mm²	m³	110.00
15 N/mm²	m³	115.00
Excavate foundation trench mechanically; remove spoil off site by grab; lay 1:3:6 site mixed concrete foundations; distance from mixer 25 m; depth of trench to be 225 mm deeper than foundation to allow for three underground brick courses priced separately		
Foundation size		
200 mm deep × 400 mm wide	m	18.30
300 mm deep × 500 mm wide	m	31.50
400 mm deep × 400 mm wide	m	32.00
400 mm deep × 600 mm wide	m	48.00
600 mm deep × 600 mm wide	m	69.00
As above but hand excavation and disposal to spoil heap 25 m by barrow; disposal off site by grab		
200 mm deep × 400 mm wide	m	41.00
300 mm deep × 500 mm wide	m	48.50
400 mm deep × 400 mm wide	m	46.00
400 mm deep × 600 mm wide	m	98.00
600 mm deep × 600 mm wide	m	135.00

BRICK/BLOCK WALLING

Item Excluding site overheads and profit	Unit	Total rate £
BRICK WALLING		
Excavate foundation trench 500 mm deep; remove spoil to dump off site; (all by machine) lay site mixed concrete foundations 1:3:6 350 × 150 mm thick; construct half brick wall with one brick piers at 2.0 m centres; laid in cement: lime: sand (1:1:6) mortar with flush joints; fair face one side; DPC two courses underground; engineering brick in cement: sand (1:3) mortar; coping of headers on end		
Wall 900 mm high above DPC		
in engineering brick (class B) PC £300.00/1000	m	235.00
in sandfaced facings PC £500.00/1000	m	270.00
in reclaimed bricks PC £1,000.00/1000	m	380.00
Excavate foundation trench 400 mm deep; remove spoil to dump off site; lay GEN 1 concrete foundations 450 mm wide × 250 mm thick; construct one brick wall with one and a half brick piers at 3.0 m centres; all in English Garden Wall bond; laid in cement: lime: sand (1:1:6) mortar with flush joints, fair face one side; DPC two courses engineering brick in cement: sand (1:3) mortar; engineering brick coping		
Wall 900 mm high above DPC		
in engineering brick (class B) PC £300.00/1000	m	360.00
in sandfaced facings PC £500.00/1000	m	405.00
in reclaimed bricks PC £1,000.00/1000	m	460.00
Wall 1200 mm high above DPC		
in engineering brick (class B) PC £300.00/1000	m	420.00
in sandfaced facings PC £500.00/1000	m	470.00
in reclaimed bricks PC £1,000.00/1000	m	540.00
Wall 1800 mm high above DPC		
in engineering brick (class B) PC £300.00/1000	m	620.00
in sandfaced facings PC £500.00/1000	m	700.00
in reclaimed bricks PC £1,000.00/1000	m	800.00
BLOCK WALLING		
Excavate foundation trench 450 mm deep; remove spoil to dump off site; lay GEN 1 concrete foundations 600 × 300 mm thick; construct wall of concrete block; two courses below ground		
Solid blocks 7 N/mm^2; wall 1.00 m high		
100 mm thick	m	97.00
Hollow blocks filled with concrete; wall 1.00 m high		
215 mm thick	m	160.00
Hollow blocks but with steel bar cast into the foundation; wall 1.00 m high		
215 mm thick	m	160.00
Solid blocks 7 N/mm^2; wall 1.80 m high		
100 mm thick	m	140.00
Hollow blocks filled with concrete; wall 1.80 m high		
215 mm thick	m	245.00
Hollow blocks but with steel bar cast into the foundation wall 1.80 m high		
215 mm thick	m	245.00

ROADS AND PAVINGS

Item Excluding site overheads and profit	Unit	Total rate £
BASES FOR PAVING		
Excavate ground and reduce levels to receive 38 mm thick slab and 25 mm mortar bed; dispose of excavated material off site; treat substrate with total herbicide		
Lay granular fill Type 1 150 mm thick laid to falls and compacted		
all by machine	m²	18.00
all by hand except disposal by grab	m²	35.00
Excavate ground and reduce levels to receive 65 mm thick surface and bed (not included); dispose of excavated material off site; treat substrate with total herbicide		
Lay 100 mm compacted hardcore; lay 1:2:4 concrete base 150 mm thick laid to falls		
all by machine	m²	43.00
all by hand except disposal by grab	m²	84.00
As above but inclusive of reinforcement fabric A142		
all by machine	m²	47.50
all by hand except disposal by grab	m²	89.00
Lay 100 mm compacted hardcore; lay 1:3:6 concrete base 150 mm thick laid to falls		
all by machine	m²	36.00
all by hand except disposal by grab	m²	77.00
As above but inclusive of reinforcement fabric A142		
all by machine	m²	40.50
all by hand except disposal by grab	m²	82.00
KERBS AND EDGINGS		
Note: excavation is by machine unless otherwise mentioned		
Excavate trench and construct concrete foundation 150 mm wide × 150 mm deep; lay precast concrete kerb units bedded in semi-dry concrete; slump 35 mm maximum; haunching one side; disposal of arisings off site		
Edgings laid straight		
50 × 150 mm	m	30.00
125 mm high × 150 mm wide; bullnosed	m	30.50
50 × 200 mm	m	31.00
Second hand granite setts		
100 × 100 mm	m	63.00
Single course; brick or block edgings laid to stretcher		
concrete blocks 200 × 100 mm	m	24.00
engineering bricks	m	30.00
paving bricks; PC £500.00/1000	m	31.00
Double course; brick or block edgings laid to stretcher		
concrete blocks 200 × 100 mm	m	27.00
engineering bricks	m	35.50
paving bricks; PC £500.00/1000	m	37.00
Single course brick or block edgings laid to header course (soldier course)		
blocks 200 × 100 × 60 mm; PC £8.60/m²; butt jointed	m	27.00
bricks 200 × 100 × 50 mm; PC £300.00/1000; butt jointed	m	30.00
bricks 200 × 100 × 50 mm; PC £300.00/1000; mortar joints	m	34.50

ROADS AND PAVINGS

Item Excluding site overheads and profit	Unit	Total rate £
Sawn yorkstone edgings; excavate for groundbeam; lay concrete 1:2:4 150 mm deep × 33.3% wider than the edging; on 35 mm thick mortar bed; inclusive of haunching one side		
Yorkstone 50 mm thick		
100 mm wide × random lengths	m	35.50
100 × 100 mm	m	40.00
200 mm wide × 100 mm long	m	48.00
250 mm wide × random lengths	m	46.00
500 mm wide × random lengths	m	82.00
Timber edgings; softwood		
150 × 38 mm		
laid straight	m	4.60
laid to curves	m	6.70
INTERLOCKING BLOCK PAVING		
Excavate ground; supply and lay granular fill Type 1 150 mm thick laid to falls and compacted; supply and lay block pavers; laid on 50 mm compacted sharp sand; vibrated; joints filled with loose sand excluding edgings or kerbs measured separately		
Concrete blocks		
200 × 100 × 60 mm	m²	60.00
200 × 100 × 80 mm	m²	63.00
Reduce levels; lay 150 mm granular material Type 1; lay precast concrete edging 50 × 150 mm; on concrete foundation 1:2:4; lay 200 × 100 × 60 mm vehicular block paving to 90° herringbone pattern; on 50 mm compacted sand bed; vibrated; jointed in sand and vibrated		
1.0 m wide clear width between edgings	m	100.00
1.5 m wide clear width between edgings	m	130.00
2.0 m wide clear width between edgings	m	160.00
3.0 m wide clear width between edgings	m	220.00
Works by hand; reduce levels; lay 150 mm granular material Type 1; lay edge restraint of block paving 200 mm wide on 150 mm thick concrete foundation 1:2:4 haunched; lay 200 × 100 × 60 mm vehicular block paving to 90° herringbone pattern; on 50 mm compacted sand bed; vibrated; jointed in sand and vibrated		
1.0 m wide clear width between edgings	m	94.00
1.5 m wide clear width between edgings	m	120.00
2.0 m wide clear width between edgings	m	150.00
3.0 m wide clear width between edgings	m	210.00
Works by hand; reduce levels; lay 150 mm granular material Type 1; lay edge restraint of block paving 200 mm wide on 150 mm thick concrete foundation 1:2:4 haunched; lay 200 × 100 × 60 mm vehicular block paving to 90° herringbone pattern; on 50 mm compacted sand bed; vibrated; jointed in sand and vibrated		
1.0 m wide clear width between edgings	m	110.00
1.5 m wide clear width between edgings	m	140.00
2.0 m wide clear width between edgings	m	180.00
3.0 m wide clear width between edgings	m	255.00

ROADS AND PAVINGS

Item Excluding site overheads and profit	Unit	Total rate £
BRICK PAVING		
WORKS BY MACHINE		
Excavate and lay base Type 1 150 mm thick remove arisings; all by machine; lay clay brick paving		
200 × 100 × 50 mm thick; butt jointed on 50 mm sharp sand bed		
PC £300.00/1000	m²	70.00
PC £600.00/1000	m²	86.00
200 × 100 × 50 mm thick; 10 mm mortar joints on 35 mm mortar bed		
PC £300.00/1000	m²	86.00
PC £600.00/1000	m²	99.00
Excavate and lay 150 mm site mixed concrete base 1:3:6 reinforced with A142 mesh; all by machine; remove arisings; lay clay brick paving		
200 × 100 × 50 mm thick; 10 mm mortar joints on 35 mm mortar bed; running or stretcher bond		
PC £300.00/1000	m²	110.00
PC £600.00/1000	m²	120.00
200 × 100 × 50 mm thick; 10 mm mortar joints on 35 mm mortar bed; butt jointed; herringbone bond		
PC £300.00/1000	m²	93.00
PC £600.00/1000	m²	110.00
Excavate and lay readymix concrete base 150 mm thick reinforced with A393 mesh; all by machine; remove arisings; lay clay brick paving		
215 × 102.5 × 50 mm thick; 10 mm mortar joints on 35 mm mortar bed		
PC £300.00/1000	m²	84.00
PC £600.00/1000	m²	99.00
WORKS BY HAND		
Excavate and lay base Type 1 150 mm thick by hand; arisings barrowed to spoil heap maximum distance 25 m and removal off site by grab; lay clay brick paving		
200 × 100 × 50 mm thick; butt jointed on 50 mm sharp sand bed		
PC £300.00/1000	m²	83.00
PC £600.00/1000	m²	98.00
Excavate and lay 150 mm concrete base; 1:3:6: site mixed concrete reinforced with A142 mesh; remove arisings to stockpile and then off site by grab; lay clay brick paving		
215 × 102.5 × 50 mm thick; 10 mm mortar joints on 35 mm mortar bed		
PC £300.00/1000	m²	100.00
PC £600.00/1000	m²	120.00

ROADS AND PAVINGS

Item Excluding site overheads and profit	Unit	Total rate £
NATURAL STONE/SLAB PAVING		
WORKS BY MACHINE		
Yorkstone slabs; excavate ground by machine and reduce levels to receive 65 mm thick slab and 35 mm mortar bed; dispose of excavated material off site; treat substrate with total herbicide; lay granular fill Type 1 150 mm thick laid to falls and compacted; lay to random rectangular pattern on 35 mm mortar bed		
New riven slabs		
laid random rectangular	m²	170.00
New riven slabs; but to 150 mm plain concrete base		
laid random rectangular	m²	190.00
Reclaimed Cathedral grade riven slabs		
laid random rectangular	m²	195.00
Reclaimed Cathedral grade riven slabs; but to 150 mm plain concrete base		
laid random rectangular	m²	215.00
New slabs sawn 6 sides		
laid random rectangular	m²	125.00
3 sizes; laid to coursed pattern	m²	130.00
New slabs sawn 6 sides; but to 150 mm plain concrete base		
laid random rectangular	m²	140.00
3 sizes; laid to coursed pattern	m²	140.00
Indian sandstone slabs; excavate ground by machine and reduce levels to receive 65 mm thick slab and 35 mm mortar bed; dispose of excavated material off site; treat substrate with total herbicide; lay granular fill Type 1 150 mm thick laid to falls and compacted; lay to random rectangular pattern on 35 mm mortar bed		
New riven slabs		
laid random rectangular	m²	105.00
WORKS BY HAND		
Yorkstone slabs; excavate ground by hand and reduce levels to receive 65 mm thick slab and 35 mm mortar bed; barrow all materials and arisings 25 m; dispose of excavated material off site by grab; treat substrate with total herbicide; lay granular fill Type 1 150 mm thick laid to falls and compacted; lay to random rectangular pattern on 35 mm mortar bed		
New riven slabs		
laid random rectangular	m²	185.00
New riven slabs laid random rectangular; but to 150 mm plain concrete base		
laid random rectangular	m²	200.00
Reclaimed Cathedral grade riven slabs		
laid random rectangular	m²	210.00
Reclaimed Cathedral grade riven slabs; but to 150 mm plain concrete base		
laid random rectangular	m²	225.00
New slabs sawn 6 sides		
laid random rectangular	m²	140.00
3 sizes; laid to coursed pattern	m²	140.00

ROADS AND PAVINGS

Item Excluding site overheads and profit	Unit	Total rate £
NATURAL STONE/SLAB PAVING – cont		
Yorkstone slabs – cont		
New slabs sawn 6 sides; but to 150 mm plain concrete base		
laid random rectangular	m²	**150.00**
3 sizes; laid to coursed pattern	m²	**155.00**
Indian sandstone slabs; excavate ground by hand and reduce levels to receive 65 mm thick slab and 35 mm mortar bed; barrow all materials and arisings 25 m; dispose of excavated material off site by grab; treat substrate with total herbicide; lay granular fill Type 1 150 mm thick laid to falls and compacted; lay to random rectangular pattern on 35 mm mortar bed		
New riven slabs		
laid random rectangular	m²	**120.00**

PREPARATION FOR SEEDING/TURFING

Item Excluding site overheads and profit	Unit	Total rate £
SURFACE PREPARATIONS		
Cultivate existing ground by pedestrian operated rotavator; spread and lightly consolidate topsoil brought from spoil heap 25 m distance in layers not exceeding 150 mm; grade to specified levels; remove stones over 25 mm; rake and grade to a fine tilth		
By machine		
100 mm thick	100 m²	155.00
200 mm thick	100 m²	250.00
300 mm thick	100 m²	350.00
500 mm thick	100 m²	540.00
By hand		
100 mm thick	100 m²	470.00
150 mm thick	100 m²	680.00
300 mm thick	100 m²	1275.00
500 mm thick	100 m²	2350.00
Lift and remove existing turf to skip 25 m distance; cultivate surface to receive new turf; rake to a fine tilth		
Works by hand		
normal turfed area	100 m²	600.00
compacted turfed area	100 m²	620.00
Lift and remove existing turf to skip 25 m distance; cultivate surface to receive new turf; rake to a fine tilth; fill area to receive turf with 50 mm imported topsoil		
Works by hand		
normal turfed area	100 m²	870.00
compacted turfed area	100 m²	880.00
SEEDING AND TURFING		
Domestic lawn areas		
Cultivate recently filled topsoil area; grade to levels and falls and rake to remove stones and debris; add fertilizers; add surface treatment as specified		
Turf areas		
Rolawn medallion	m²	5.60
Seeded areas		
grass seed PC £4.50/kg; application rate 35 g/m²	m²	1.10
grass seed PC £4.50/kg; application rate 50 g/m²	m²	1.10
grass seed PC £6.00/kg; application rate 35 g/m²	m²	1.10
grass seed PC £6.00/kg; application rate 50 g/m²	m²	1.25

PREPARATION FOR SEEDING/TURFING

Item Excluding site overheads and profit	Unit	Total rate £
SEEDING AND TURFING – cont		
Cultivate recently filled topsoil area; grade to		
levels and falls and rake to remove stones and		
debris; add fertilizers; add surface treatment as		
specified; maintain for one year watering and		
cutting 26 times during the summer; pedestrian		
mower with grass box; arisings removed off site		
Turf areas		
Rolawn medallion	m^2	**7.90**
Seeded areas		
grass seed PC £4.50/kg; application rate 35 g/m^2	m^2	**3.30**
grass seed PC £4.50/kg; application rate 50 g/m^2	m^2	**3.35**
grass seed PC £6.00/kg; application rate 35 g/m^2	m^2	**3.40**
grass seed PC £6.00/kg; application rate 50 g/m^2	m^2	**3.50**

PLANTING

Item Excluding site overheads and profit	Unit	Total rate £
TREE PLANTING		
Excavate tree pit by hand; fork over bottom of pit; plant tree with roots well spread out; backfill with excavated material, incorporating treeplanting compost at 1 m³ per 3 m³ of soil, one tree stake and two ties; tree pits square in sizes shown		
Light standard bare root tree in pit; PC £9.45		
600 × 600 mm deep	each	38.00
900 × 900 mm deep	each	50.00
Standard tree bare root tree in pit; PC £12.60		
600 × 600 mm deep	each	42.50
900 × 600 mm deep	each	54.00
Standard root balled tree in pit; PC £24.00		
600 × 600 mm deep	each	55.00
900 × 600 mm deep	each	72.00
Selected standard bare root tree in pit; PC £20.48		
900 × 900 mm deep	each	71.00
1.00 m × 1.00 m × 600 mm deep	each	99.00
Selected standard root ball tree in pit; PC £37.00		
900 × 600 mm deep	each	87.00
1.00 m × 1.00 m × 600 mm deep	each	115.00
Heavy standard bare root tree in pit; PC £44.10		
900 × 900 mm deep	each	100.00
1.00 m × 1.00 m × 600 mm deep	each	110.00
Heavy standard root ball tree in pit; PC £63.00		
900 × 600 mm deep	each	120.00
1.00 m × 1.00 m × 600 mm deep	each	145.00
Extra heavy standard bare root tree in pit; PC £58.80		
1.00 × 1.00 m deep	each	145.00
Extra heavy standard root ball tree in pit; PC £80.50		
1.00 × 1.00 m deep	each	170.00
1.50 m × 750 mm deep	each	200.00
SHRUB PLANTING		
Treat recently filled ground with systemic weedkiller; cultivate ground and clear arisings; add mushroom composts and fertilizers		
Cultivation by rotavator; all other works by hand		
compost 50 mm; general purpose fertilizer 35 g/m²	m²	4.20
compost 100 mm; general purpose fertilizer 35 g/m²	m²	7.10
compost 100 mm; Enmag 35 g/m²	m²	7.15
All works by hand		
compacted ground; compost 50 mm; general purpose fertilizer 35 g/m²	m²	5.05
ground previously planted but cleared of vegetable matter; compost 50 mm; general purpose fertilizer 35 g/m²	m²	4.90
ground previously planted but cleared of vegetable matter; compost 50 mm; general purpose fertilizer 35 g/m²	m²	4.75

PLANTING

Item Excluding site overheads and profit	Unit	Total rate £
SHRUB PLANTING – cont		
Excavate planting holes on 300 × 300 × 300 mm deep to area previously prepared; **plant shrubs PC £3.00 each in groups of 3 to 5 inclusive of transport from** **holding area setting out and final mulching 50 mm thick**		
By hand		
300 mm centres (11.11 plants per m²)	m²	66.00
400 mm centres (6.26 plants per m²)	m²	41.00
500 mm centres (4 plants per m²)	m²	26.00
750 mm centres (1.78 plants per m²)	m²	12.70
Cultivate and grade shrub bed; bring top 300 mm of topsoil to a fine tilth, **incorporating mushroom compost at 50 mm and Enmag slow release fertilizer;** **rake and bring to given levels; remove all stones and debris over 50 mm; dig** **planting holes average 300 × 300 × 300 mm deep; supply and plant specified** **shrubs in quantities as shown below; backfill with excavated material as above;** **water to field capacity and mulch 50 mm bark chips 20–40 mm size; water and** **weed regularly for 12 months; shrubs 2 litre PC £2.80; ground covers 9 cm PC** **£1.50**		
Shrubs at centres shown below		
300 mm centres	m²	63.00
400 mm centres	m²	41.00
500 mm centres	m²	28.00
600 mm centres	m²	21.50
750 mm centres	m²	16.70
900 mm centres	m²	14.00
Groundcover 30%/shrubs 70% at the distances shown below		
200/300 mm	m²	67.00
300/400 mm	m²	57.00
300/500 mm	m²	33.50
Groundcover 50%/shrubs 50% at the distances shown below		
200/300 mm	m²	77.00
300/400 mm	m²	42.00
300/500 mm	m²	37.00
400/500 mm	m²	28.00
Cultivate ground by machine and rake to level; plant bulbs as shown; bulbs PC **£25.00/100**		
15 bulbs per m²	m²	6.80
25 bulbs per m²	m²	11.20
50 bulbs per m²	m²	22.00

PLANTING

Item Excluding site overheads and profit	Unit	Total rate £
BEDDING		
Spray surface with glyphosate; lift and dispose of turf when herbicide action is complete; cultivate new area for bedding plants to 400 mm deep; spread compost 100 mm deep and chemical fertilizer Enmag and rake to fine tilth to receive new bedding plants; remove all arisings to skip		
Existing turf area		
disposal to skip	100 m²	**860.00**
disposal to compost area on site; distance 25 m	100 m²	**520.00**
Plant bedding to existing planting area; bedding planting PC £0.25 each		
Clear existing bedding; cultivate soil to 230 mm deep; incorporate compost 75 mm and rake to fine tilth; collect bedding from nursery and plant at 100 mm ccs; irrigate on completion; maintain weekly for 12 weeks		
mass planted; 100 mm centres	m²	**32.50**
to patterns; 100 mm centres	m²	**35.00**
mass planted; 150 mm centres	m²	**18.70**
to patterns; 150 mm centres	m²	**21.50**
mass planted; 200 mm centres	m²	**12.80**
to patterns; 200 mm centres	m²	**15.20**
Extra for watering by hand held hose pipe		
Flow rate 25 litres/minute		
10 litres/m²	100 m²	**15.30**
15 litres/m²	100 m²	**23.00**
20 litres/m²	100 m²	**30.00**
25 litres/m²	100 m²	**38.00**
Flow rate 40 litres/minute		
10 litres/m²	100 m²	**9.60**
15 litres/m²	100 m²	**14.40**
20 litres/m²	100 m²	**18.90**
25 litres/m²	100 m²	**24.00**

MAINTENANCE

Item Excluding site overheads and profit	Unit	Total rate £
GRASS CUTTING		
Cut grass with pedestrian mower; strim edges of shrub beds and edges to pavings; length of edges is 10% of the grass area; remove arisings to compost heap on site		
Lawn cut 22 times per year; cost per occasion		
pedestrian mower 400 mm wide	100 m²	4.50
pedestrian mower 450 mm wide	100 m²	3.80
pedestrian mower 500 mm wide	100 m²	3.60
Lawn cut 18 times per year; cost per occasion		
pedestrian mower 400 mm wide	100 m²	5.00
pedestrian mower 450 mm wide	100 m²	4.30
pedestrian mower 500 mm wide	100 m²	4.10
SHRUB BED MAINTENANCE		
Prune shrubs; cut back to soft growth; shrubs which have previously been pruned in the last 3 years		
Shrubs 1.00–2.00 m high		
1.00 m centres	10 m²	11.50
750 mm centres	10 m²	20.50
Shrubs		
1.00 m centres	10 m²	8.30
750 mm centres	10 m²	14.80
500 mm centres	10 m²	33.00
400 mm centres	10 m²	52.00
Weed newly planted shrub bed; irrigate to field capacity by hand; fork over surface of plant bed every four visits		
Regular visits		
newly planted areas; per visit	m²	0.80
established areas; per visit	m²	0.50
HEDGE MAINTENANCE		
Cut hedge by hand operated machinery; remove arising to compost heap on site; 2 sides and top		
Hedge up to 2.00 m high; depth of cut 300 mm		
hedge up to 1.0 m high	m	0.95
hedge up to 1.5 m high	m	1.70
hedge up to 2.0 m high	m	3.45
Reduce height of ornamental hedge using hand tools or mechanical hand tools; thickness of branches average 20–50 mm diameter; collect arisings and dispose on site; hedge measured 2 sides and top; assumed thickness 600 mm		
Hedge up to 2.00 m high		
reduce by 600 mm	m	2.35
reduce by 900 mm	m	7.20
Hedge 2.00–4.00 m high		
reduce by 600 mm	m	3.40
reduce by 900 mm	m	8.70

DRAINAGE

Item Excluding site overheads and profit	Unit	Total rate £
PIPE LAYING		
Pipe laying		
Excavate trench 600 mm deep by excavator; lay bedding and backfill as per material specification below; lay non-woven geofabric and fill with topsoil to ground level		
100 mm vitrified clay; laid on earth with excavated backfill	m	19.30
110 mm PVC-u drainpipe; laid on sand bed with gravel backfill	m	26.00
Excavate trench 600 mm deep by hand; lay bedding and backfill as per material specification below; lay non-woven geofabric and fill with topsoil to ground level		
100 mm vitrified clay; laid on earth with excavated backfill	m	23.50
110 mm PVC-u drainpipe; laid on sand bed with gravel backfill	m	30.00
LINEAR DRAINAGE		
Linear drainage to design sensitive areas		
Excavate trench by machine; lay Aco Brickslot channel drain on concrete base and surround to falls; all to manufacturers specifications		
paving surround to both sides of channel	m	29.00
Linear drainage to vehicular area		
Excavate trench by machine; lay Aco MultiDrain MD Brickslot; offset galvanized slot drain grating for M100PPD; load class C250 channel drain on concrete base and surround to falls; all to manufacturers specifications; paving surround to channel with brick paving PC £300.00/1000		
brickslot galvanized grating; paving surround to one side of channel	m	105.00
slotted galvanized grating; paving surround to both sides of channel	m	115.00
stainless steel grating; paving surround to one side of channel	m	200.00
Linear drainage to pedestrian area		
Excavate trench by machine; lay Aco Hexdrain channel drain on concrete base and surround to falls; all to manufacturers specifications; paving surround to channel with brick paving PC £300.00/1000		
with black plastic; paving surround to one side of channel	m	80.00
slotted galvanized grating; paving surround to both sides of channel	m	95.00
with brickslot grating; paving surround to both side of channel	m	150.00
Heelguard ductile grating; paving surround to both sides of channel	m	9.00
Accessories for channel drain		
Sump unit with sediment bucket	nr	180.00
End cap; inlet/outlet	nr	3.75

DRAINAGE

Item Excluding site overheads and profit	Unit	Total rate £
MANHOLES		
Inspection chambers; brick manhole; excavate pit for inspection chamber including earthwork support and disposal of spoil to dump on site not exceeding 100 m; lay concrete (1:2:4) base 1500 mm diameter × 200 mm thick; 110 mm vitrified clay channels; benching in concrete (1:3:6) allowing one outlet and two inlets for 110 mm diameter pipe; construct inspection chamber one brick thick walls of engineering brick Class B; backfill with excavated material; complete with two cast iron step irons; cover 600 × 450 mm ductile iron; light vehicle loading		
1200 × 1200 × 1200 mm		
excavation by machine	each	1175.00
excavation by hand	each	2400.00
1200 × 1200 × 1200 mm; by machine but with recessed cover		
600 × 450 mm; 5 tonne loading; filled with block pavers	each	1275.00
1200 × 1200 × 1500 mm		
excavation by machine	each	1375.00
excavation by hand	each	2750.00
Inspection chambers; polypropylene; excavate pit for inspection chamber including earthwork support and disposal of spoil to dump on site not exceeding 100 m; lay concrete (1:2:4) base 700 mm diameter × 200 mm thick		
Polypropylene 600 mm deep		
excavation by machine	each	390.00
excavation by hand	each	350.00
Polypropylene 1200 mm		
excavation by machine	each	500.00
excavation by hand	each	570.00
GULLIES		
Clay gully		
Excavate hole; supply and set in concrete vitrified clay trapped mud (dirt) gully complete with galvanized bucket and cast iron hinged locking grate and frame; lay kerb to gully		
1 tonne loading; RGP5; 100 mm outlet; 100 mm diameter; 225 mm internal width; 585 mm internal depth	each	360.00
Gullies PVC-u		
Excavate hole and lay 100 mm concrete (C20P) base 150 × 150 mm to suit given invert level of drain; connect to drainage system; backfill with Type 1 granular fill; install gully; complete with grate and frame; brick kerb to gully surround		
PVC-u universal gully with vertical hopper; plastic grid and frame included	each	215.00
PVC-u gully 100 mm with P trap; grid and frame included	each	220.00
yard gully trapped; 300 mm diameter; 600 mm deep; sediment bucket and ductile iron cover	each	550.00

DRAINAGE

Item Excluding site overheads and profit	Unit	Total rate £
SOAKAWAYS		
Soakaway Aquacell Wavin Plastics Ltd; preformed polypropylene soakaway infiltration crate units		
Excavate pits or trenches for soakaway units; place Aquacell polypropylene soakaway units in recommended configurations laid on 100 mm shingle; cover with terram and backfill to sides and top with 100 mm shingle; backfill with 150 mm topsoil		
4 crates; 760 litres	nr	700.00
8 crates; 1520 litres	nr	870.00
12 crates; 2280 litres	nr	1350.00
16 crates; 3050 litres	nr	1550.00

GARDEN LIGHTING

Item Excluding site overheads and profit	Unit	Total rate £
TRENCHING AND CABLE INSTALLATION		
Small garden lighting; 12 × 12 m		
Lighting to domestic gardens; trenching to 600 mm deep and installation of cable; backfilling and terminating cable; connections to switches and to mains distribution board		
Works by machine; main supply of 50 m armoured cables to 4 lighting transformers; transformers at maximum 10 m distance from mains supply		
4 nr Hunza spike lights on one circuit	nr	1325.00
4 nr Hunza spike lights and 4 nr step lights on one circuit; 4 transformers	nr	2800.00
Works by hand; main supply of 50 m armoured cables to 4 lighting transformers; transformers at maximum 10 m distance from mains supply		
4 nr Hunza spike lights on one circuit	nr	1825.00
4 nr Hunza spike lights and 4 nr step lights on one circuit; 4 transformers	nr	3550.00
Medium garden lighting; 25 × 15 m		
Lighting to domestic gardens; trenching to 600 mm deep and installation of cable; backfilling and terminating cable; connections to switches and to mains distribution board		
Works by machine; main supply of 100 m armoured cables to 6 lighting transformers; transformers at maximum 10 m distance from mains supply		
8 nr Hunza spike lights on one circuit	nr	2200.00
8 nr Hunza spike lights and 4 nr step lights on one circuit; 6 transformers	nr	3700.00
Works by hand; main supply of 100 m armoured cables to 8 lighting transformers; transformers at maximum 10 m distance from mains supply		
8 nr Hunza spike lights on one circuit	nr	3200.00
8nr Hunza spike lights and 4 nr step lights on one circuit; 4 transformers	nr	4300.00
Large garden lighting; 60 m × 30 m		
Lighting to domestic gardens; trenching to 600 mm deep and installation of cable; backfilling and terminating cable; connections to switches and to mains distribution board		
Works by machine and by hand (70/30%); main supply of 180 m armoured cables to 10 lighting transformers; transformers at maximum 10 m distance from mains supply		
8 nr Hunza spike lights, 4 Hunza wall lights and 4 Hunza step lights on 1 circuit	nr	5600.00
Works by hand 70/30 %; main supply of 180 m armoured cables to 10 lighting transformers; transformers at maximum 10 m distance from mains supply		
8 nr Hunza spike lights, 4 Hunza wall lights and 4 Hunza step lights on 1 circuit	nr	6600.00

Construction Contracts Questions and Answers

2nd Edition

David Chappell

What they said about the first edition: "A fascinating concept, full of knowledgeable gems put in the most frank of styles... A book to sample when the time is right and to come back to when another time is right, maybe again and again."
– *David A Simmonds, Building Engineer magazine*

• Is there a difference between inspecting and supervizing?

• What does 'time-barred' mean?

• Is the contractor entitled to take possession of a section of the work even though it is the contractor's fault that possession is not practicable?

Construction law can be a minefield. Professionals need answers which are pithy and straightforward, as well as legally rigorous. The two hundred questions in the book are real questions, picked from the thousands of telephone enquiries David Chappell has received as a Specialist Adviser to the Royal Institute of British Architects. Although the enquiries were originally from architects, the answers to most of them are of interest to project managers, contractors, QSs, employers and others involved in construction.

The material is considerably updated from the first edition – weeded, extended and almost doubled in coverage. The questions range in content from extensions of time, liquidated damages and loss and/or expense to issues of warranties, bonds, novation, practical completion, defects, valuation, certificates and payment, architects' instructions, adjudication and fees. Brief footnotes and a table of cases will be retained for those who may wish to investigate further.

August 2010: 216x138: 344pp
Pb: 978-0-415-56650-6: **£34.99**

To Order: Tel: +44 (0) 1235 400524 **Fax:** +44 (0) 1235 400525
or Post: Taylor and Francis Customer Services,
Bookpoint Ltd, Unit T1, 200 Milton Park, Abingdon, Oxon, OX14 4TA UK
Email: book.orders@tandf.co.uk

For a complete listing of all our titles visit:
www.tandf.co.uk

PART 6

Prices for Measured Works – Minor Works

INTRODUCTION

Typical Project Profile

Contract value	£10,000.00–£100,000.00
Labour rate (see page 3)	£20.75 per hour
Labour rate for maintenance contracts	–
Number of site staff	6–9
Project area	1200 m^2
Project location	Outer London
Project components	20% hard landscape
	80% soft landscape and planting
Access to works areas	Very good
Contract	Main contract
Delivery of materials	Part loads
Profit and site overheads	Excluded

A PRELIMINARIES

Item Excluding site overheads and profit	PC £	Labour hours	Labour £	Plant £	Material £	Unit	Total rate £
A PRELIMINARIES							
Site accommodation							
Toilet							
weekly rate	–	–	–	27.27	–	week	27.27
delivery and collection charge	–	–	–	50.50	–	nr	50.50
Site office; 3.6 × 2.4 m							
weekly rate	–	–	–	37.12	–	week	37.12
delivery and collection charge	–	–	–	254.52	–	nr	254.52
Secure storage							
site storage; 3.0 × 2.4 m	–	–	–	12.12	–	week	12.12
delivery and collection charge	–	–	–	50.50	–	nr	50.50
Site compound; inclusive of small office, toilet, secure storage; all on 100 mm compacted crushed concrete hardcore with security fencing							
first week including setup costs	–	8.00	166.00	503.20	151.00	nr	835.20
weekly costs	–	–	–	147.01	–	nr	147.01
Site plant hire delivery costs							
Delivery and collection charges for small plant to 5 tonne							
per machine	–	–	–	80.00	–	nr	80.00
Non-productive time; unloading costs							
Costs for hand offloading on minor works items where materials are delivered in small loads and machine or crane offloads are impractical							
project value; £10,000	–	16.00	332.00	–	–	nr	332.00
project value; £20,000	–	20.00	415.00	–	–	nr	415.00
project value; £30,000	–	24.00	498.00	–	–	nr	498.00
project value; £40,000	–	28.00	581.00	–	–	nr	581.00
project value; £50,000	–	32.00	664.00	–	–	nr	664.00
Tendering costs							
Produce quotations for projects issued by landscape designer; Plans and sections issued to contractors on paper copies; includes for initial site visit							
project value; £10,000	–	6.00	124.50	–	–	nr	124.50
project value; £20,000	–	8.00	166.00	–	–	nr	166.00
project value; £30,000	–	12.00	249.00	–	–	nr	249.00
project value; £40,000	–	14.00	290.50	–	–	nr	290.50
project value; £50,000	–	17.00	352.75	–	–	nr	352.75

A PRELIMINARIES

Item Excluding site overheads and profit	PC £	Labour hours	Labour £	Plant £	Material £	Unit	Total rate £
DESIGN COSTS							
Garden design to small gardens; Paul Cowell; PC Landscapes Ltd; typical costs for the design of a small garden by a professional landscape architect; indicative stage prices and project size shown							
Project value £15,000 (17.5%)							
initial brief and site survey	–	–	–	–	–	nr	300.00
concept plan	–	–	–	–	–	nr	1000.00
detailed drawings and planting plans	–	–	–	–	–	nr	750.00
tender documentation and project management	–	–	–	–	–	nr	500.00
Project value £25,000 (16%)							
initial brief and site survey	–	–	–	–	–	nr	375.00
concept plan	–	–	–	–	–	nr	1500.00
detailed drawings and planting plans	–	–	–	–	–	nr	1500.00
tender documentation and project management	–	–	–	–	–	nr	650.00
Project value £40,000 (15%)							
initial brief and site survey	–	–	–	–	–	nr	750.00
concept plan	–	–	–	–	–	nr	1250.00
detailed drawings and planting plans	–	–	–	–	–	nr	2750.00
tender documentation and project management	–	–	–	–	–	nr	1250.00
Project value £60,000 (12.5%)							
initial brief and site survey	–	–	–	–	–	nr	750.00
concept plan	–	–	–	–	–	nr	2000.00
detailed drawings and planting plans	–	–	–	–	–	nr	3000.00
tender documentation and project management	–	–	–	–	–	nr	1250.00
Project value £100,000 (10%)							
initial brief and site survey	–	–	–	–	–	nr	1000.00
concept plan	–	–	–	–	–	nr	2750.00
detailed drawings and planting plans	–	–	–	–	–	nr	4500.00
tender documentation and project management	–	–	–	–	–	nr	1750.00
Project value over £100,000 (8.5%)							
initial brief and site survey	–	–	–	–	–	nr	1000.00
concept plan	–	–	–	–	–	nr	2750.00
detailed drawings and planting plans	–	–	–	–	–	nr	4750.00
tender documentation and project management	–	–	–	–	–	nr	2000.00

C20 DEMOLITION

Item Excluding site overheads and profit	PC £	Labour hours	Labour £	Plant £	Material £	Unit	Total rate £
C20 DEMOLITION							
DEMOLISH STRUCTURES							
Demolish existing structures; removal to skip maximum distance 50 m; mechanical demolition; with 3 tonne excavator and dumper							
Brick wall							
112.5 mm thick	–	0.13	2.77	6.28	–	m²	9.05
225 mm thick	–	0.17	3.46	11.95	–	m²	15.41
337.5 mm thick	–	0.20	4.15	18.88	–	m²	23.03
450 mm thick	–	0.27	5.53	22.68	–	m²	28.21
Demolish existing structures; removal to skip maximum distance 50 m; all works by hand							
Brick wall							
112.5 mm thick	–	0.33	6.92	5.47	–	m²	12.39
225 mm thick	–	0.50	10.38	10.94	–	m²	21.32
337.5 mm thick	–	0.67	13.83	17.67	–	m²	31.50
450 mm thick	–	1.00	20.75	21.06	–	m²	41.81
Demolish existing structures; removal to skip maximum distance 50 m; by diesel or electric breaker; all other works by hand							
Brick wall							
112.5 mm thick	–	0.17	3.46	5.89	–	m²	9.35
225 mm thick	–	0.20	4.15	11.44	–	m²	15.59
337.5 mm thick	–	0.25	5.19	18.30	–	m²	23.49
450 mm thick	–	0.33	6.92	21.90	–	m²	28.82
Break out concrete footings associated with free-standing walls; inclusive of all excavation, removal to skip and backfilling with excavated material							
By mechanical breaker; diesel or electric							
plain concrete	–	1.50	31.13	59.70	–	m³	90.83
as above but loading to skip by hand;							
backfilling by hand	–	5.50	114.13	60.91	–	m³	175.04
reinforced concrete	–	2.50	51.88	68.45	–	m³	120.33

C20 DEMOLITION

Item Excluding site overheads and profit	PC £	Labour hours	Labour £	Plant £	Material £	Unit	Total rate £
DEMOLITION OF FREE-STANDING BUILDINGS							
Remove existing free-standing buildings; demolition by hand							
Timber buildings with suspended timber floor; hardstanding or concrete base not included; loading to skip							
shed 6.0 m²	–	2.00	41.50	77.76	–	nr	**119.26**
shed 10.0 m²	–	3.00	62.25	109.35	–	nr	**171.60**
shed 15.0 m²	–	3.50	72.63	109.35	–	nr	**181.98**
Timber building; insulated; with timber or concrete posts set in concrete, felt covered timber or tiled roof; internal walls cladding with timber or plasterboard; load arisings to skip							
timber structure 6.0 m²	–	2.50	51.88	262.44	–	nr	**314.32**
timber structure 12.0 m²	–	3.50	72.63	367.42	–	nr	**440.05**
timber structure 20.0 m²	–	8.00	166.00	497.18	–	nr	**663.18**
Demolition of free-standing brick buildings with tiled or sheet roof; concrete foundations measured separately; mechanical demolition maximum distance to stockpile 25 m; including for all access scaffolding and the like; maximum height of roof 4.00 m; including all doors, windows, guttering and down pipes; including disposal by grab							
Half brick thick							
10 m²	–	8.00	166.00	257.45	176.80	nr	**600.25**
20 m²	–	16.00	332.00	480.44	227.41	nr	**1039.85**
1 brick thick							
10 m²	–	10.00	207.50	445.98	353.60	nr	**1007.08**
20 m²	–	16.00	332.00	668.97	454.83	nr	**1455.80**
Cavity wall with blockwork inner skin and brick outer skin; insulated							
10 m²	–	12.00	249.00	634.51	491.40	nr	**1374.91**
20 m²	–	20.00	415.00	857.49	631.80	nr	**1904.29**

C20 DEMOLITION

Item Excluding site overheads and profit	PC £	Labour hours	Labour £	Plant £	Material £	Unit	Total rate £
DISCONNECTION OF SERVICES							
Extra over to the above for disconnection of services							
Electrical							
disconnection	–	–	–	–	–	nr	**70.00**
grub out cables and dispose; backfilling; by machine	–	–	–	5.10	–	m	**5.10**
grub out cables and dispose; backfilling; by hand	–	0.50	10.38	–	–	m	**10.38**
Water supply; foul or surface water drainage							
disconnection; capping off	–	1.00	20.75	–	70.00	nr	**90.75**
grub out pipes and dispose; backfilling; by machine	–	–	–	5.82	–	m	**5.82**
grub out pipes and dispose; backfilling; by hand	–	0.50	10.38	0.72	–	m	**11.10**

D11 SOIL STABILIZATION

Item Excluding site overheads and profit	PC £	Labour hours	Labour £	Plant £	Material £	Unit	Total rate £
D11 SOIL STABILIZATION							
TIMBER LOG WALLS							
Timber log retaining walls; timber fencing;							
AVS Fencing Supplies Ltd; tanalized							
softwood posts; all timber posts are kiln							
dried redwood with 15 year guarantee;							
walls below which have cut posts would							
not be subject to the suppliers guarantee							
Machine rounded softwood logs to trenches							
priced separately; disposal of excavated							
material priced separately; inclusive of 75 mm							
hardcore blinding to trench and backfilling							
trench with site mixed concrete 1:3:6;							
geofabric pinned to rear of logs; heights of							
logs above ground							
500 mm (constructed from 1.80 m lengths)	33.50	1.50	31.13	–	44.00	m	**75.13**
1.20 m (constructed from 1.80 m lengths)	67.00	1.30	26.98	–	85.62	m	**112.60**
1.60 m (constructed from 2.40 m lengths)	82.30	2.50	51.88	–	107.07	m	**158.95**
2.00 m (constructed from 3.00 m lengths)	103.00	3.50	72.63	–	133.93	m	**206.56**
As above but with 150 mm machine rounded							
timbers							
500 mm	42.28	2.50	51.88	–	52.78	m	**104.66**
1.20 m	140.79	1.75	36.31	–	159.41	m	**195.72**
1.60 m	140.93	3.00	62.25	–	165.71	m	**227.96**
As above but with 200 mm machine rounded							
timbers							
1.80 m (constructed from 2.40 m lengths)	187.90	4.00	83.00	–	218.83	m	**301.83**
RAILWAY SLEEPER WALLS							
Railway sleeper walls; AVS Fencing							
Supplies Ltd; construct retaining wall from							
railway sleepers; fixed with steel							
galvanized pins 12 mm driven into the							
ground; sleepers laid flat							
Grade 1 softwood; 2590 × 250 × 125 mm							
150 mm; 1 sleeper high	8.86	0.50	10.38	–	13.82	m	**24.20**
300 mm; 2 sleepers high	17.71	1.00	20.75	–	18.70	m	**39.45**
450 mm; 3 sleepers high	26.36	1.50	31.13	–	27.66	m	**58.79**
600 mm; 4 sleepers high	35.51	2.00	41.50	–	36.81	m	**78.31**

D11 SOIL STABILIZATION

Item Excluding site overheads and profit	PC £	Labour hours	Labour £	Plant £	Material £	Unit	Total rate £
RAILWAY SLEEPER WALLS – cont							
Railway sleeper walls – cont							
Grade 1 softwood as above but with 2 nr galvanized angle iron stakes set into concrete internally and screwed to the inside face of the sleepers							
750 mm; 5 sleepers high	43.93	2.50	51.88	–	73.62	m²	**125.50**
900 mm; 6 sleepers high	52.58	2.75	57.06	–	85.59	m²	**142.65**
Grade 1 hardwood; 2590 × 250 × 150 mm							
150 mm; 1 sleeper high	8.49	0.50	10.38	–	8.98	m	**19.36**
300 mm; 2 sleepers high	16.97	1.00	20.75	–	17.96	m	**38.71**
450 mm; 3 sleepers high	25.26	1.50	31.13	–	26.56	m	**57.69**
600 mm; 4 sleepers high	34.02	2.00	41.50	–	35.33	m	**76.83**
Grade 1 hardwood as above but with 2 nr galvanized angle iron stakes set into concrete internally and screwed to the inside face of the sleepers							
750 mm; 5 sleepers high	42.09	2.50	51.88	–	71.79	m²	**123.67**
900 mm; 6 sleepers high	50.38	2.75	57.06	–	83.39	m²	**140.45**
New pine softwood; 2400 × 250 × 125 mm							
120 mm; 1 sleeper high	10.53	0.50	10.38	–	11.02	m	**21.40**
240 mm; 2 sleepers high	21.06	1.00	20.75	–	22.05	m	**42.80**
360 mm; 3 sleepers high	31.58	1.50	31.13	–	32.89	m	**64.02**
480 mm; 4 sleepers high	42.11	2.00	41.50	–	43.41	m	**84.91**
New pine softwood as above but with 2 nr galvanized angle iron stakes set into concrete internally and screwed to the inside face of the sleepers							
600 mm; 5 sleepers high	52.64	2.50	51.88	–	82.34	m²	**134.22**
720 mm; 6 sleepers high	63.17	2.75	57.06	–	96.18	m²	**153.24**
New oak hardwood; 2600 × 220 × 130 mm							
130 mm; 1 sleeper high	12.57	0.50	10.38	–	13.06	m	**23.44**
260 mm; 2 sleepers high	25.14	1.00	20.75	–	26.13	m	**46.88**
390 mm; 3 sleepers high	37.70	1.50	31.13	–	39.01	m	**70.14**
520 mm; 4 sleepers high	50.27	2.00	41.50	–	51.57	m	**93.07**
New oak hardwood as above but with 2 nr galvanized angle iron stakes set into concrete internally and screwed to the inside face of the sleepers							
640 mm; 5 sleepers high	62.84	2.50	51.88	–	92.54	m²	**144.42**
760 mm; 6 sleepers high	75.41	2.75	57.06	–	108.42	m²	**165.48**

D11 SOIL STABILIZATION

Item Excluding site overheads and profit	PC £	Labour hours	Labour £	Plant £	Material £	Unit	Total rate £
Excavate foundation trench; set railway sleepers vertically on end in concrete 1:3:6 continuous foundation to 33.3% of their length to form retaining wall							
Grade 1 hardwood; finished height above ground level							
300 mm	22.57	3.00	62.25	2.01	27.81	m	**92.07**
500 mm	32.72	3.00	62.25	2.01	37.96	m	**102.22**
600 mm	45.47	3.00	62.25	2.01	55.36	m	**119.62**
750 mm	49.07	3.50	72.63	2.01	58.96	m	**133.60**
1.00 m	55.89	3.75	77.81	2.01	66.70	m	**146.52**
Excavate and place vertical steel universal beams 165 wide in concrete base at 2.590 centres; fix railway sleepers set horizontally between beams to form horizontal fence or retaining wall							
Grade 1 hardwood; bay length 2.590 m							
0.50 m high; 2 sleepers	16.83	2.50	51.88	–	63.83	bay	**115.71**
0.75 m high; 3 sleepers	24.73	2.60	53.95	–	94.85	bay	**148.80**
1.00 m high; 4 sleepers	33.15	3.00	62.25	–	122.92	bay	**185.17**
1.25 m high; 5 sleepers	42.33	3.50	72.63	–	157.94	bay	**230.57**
1.50 m high; 6 sleepers	49.73	3.00	62.25	–	188.09	bay	**250.34**
1.75 m high; 7 sleepers	57.88	3.00	62.25	–	196.25	bay	**258.50**
GABIONS							
Retaining walls; Maccaferri Ltd							
Wire mesh gabions; galvanized mesh 80 mm × 100 mm; filling with broken stones 125–200 mm size; wire down securely to manufacturer's instructions; filling front face by hand							
2 × 1 × 0.50 m	20.09	2.00	41.50	5.60	89.39	nr	**136.49**
2 × 1 × 1.00 m	28.23	4.00	83.00	11.20	166.83	nr	**261.03**
PVC coated gabions							
2 × 1 × 0.5 m	26.37	2.00	41.50	5.60	95.67	nr	**142.77**
2 × 1 × 1.00 m	35.97	4.00	83.00	11.20	174.57	nr	**268.77**

D20 EXCAVATION AND FILLING

Item Excluding site overheads and profit	PC £	Labour hours	Labour £	Plant £	Material £	Unit	Total rate £
D20 EXCAVATION AND FILLING							
MACHINE SELECTION							
MACHINE SELECTION TABLE							
Road Equipment Ltd; machine volumes for excavating/filling only and placing excavated material alongside or to a dumper; no bulkages are allowed for in the material volumes; these rates should be increased by user-preferred percentages to suit prevailing site conditions; the figures in the next section for 'Excavation mechanical' and filling allow for the use of banksmen within the rates shown below							
1.5 tonne excavators; digging volume							
1 cycle/minute; 0.04 m³	–	0.42	8.65	2.94	–	m³	**11.59**
2 cycles/minute; 0.08 m³	–	0.21	4.32	1.97	–	m³	**6.29**
3 cycles/minute; 0.12 m³	–	0.14	2.88	1.66	–	m³	**4.54**
3 tonne excavators; digging volume							
1 cycle/minute; 0.13 m³	–	0.13	2.66	1.74	–	m³	**4.40**
2 cycles/minute; 0.26 m³	–	0.06	1.33	1.29	–	m³	**2.62**
3 cycles/minute; 0.39 m³	–	0.04	0.89	1.33	–	m³	**2.22**
5 tonne excavators; digging volume							
1 cycle/minute; 0.28 m³	–	0.06	1.23	1.01	–	m³	**2.24**
2 cycles/minute; 0.56 m³	–	0.03	0.62	0.69	–	m³	**1.31**
3 cycles/minute; 0.84 m³	–	0.02	0.41	0.65	–	m³	**1.06**
Dumpers; Road Equipment Ltd							
1 tonne high tip skip loader; volume 0.485 m³ (775 kg)							
5 loads per hour	–	0.41	8.56	1.81	–	m³	**10.37**
7 loads per hour	–	0.29	6.11	1.37	–	m³	**7.48**
10 loads per hour	–	0.21	4.28	1.04	–	m³	**5.32**
3 tonne dumper; maximum volume 2.40 m³ (3.38 t); available volume 1.9 m³							
4 loads per hour	–	0.14	2.88	0.62	–	m³	**3.50**
5 loads per hour	–	0.11	2.31	0.51	–	m³	**2.82**
7 loads per hour	–	0.08	1.65	0.38	–	m³	**2.03**
10 loads per hour	–	0.06	1.15	0.30	–	m³	**1.45**

D20 EXCAVATION AND FILLING

Item Excluding site overheads and profit	PC £	Labour hours	Labour £	Plant £	Material £	Unit	Total rate £
SITE CLEARANCE							
Lifting turf for preservation							
hand lift and stack	–	0.08	1.73	–	–	m²	1.73
Load to wheelbarrows and dispose off site by skip							
25 m distance	–	0.02	0.43	3.00	–	m²	3.43
Note: The figures in this section relate to the machine capacities shown in the Major Works section of this book. The figures below allow for dig efficiency based on depth, bulkages of 25% on loamy soils (adjustments should be made for different soil types) and for a banksman.							
Site clearance; by machine; clear site of mature shrubs from existing cultivated beds; dig out roots by machine							
Mixed shrubs in beds; planting centres 500 mm average							
height less than 1 m	–	0.03	0.52	0.89	–	m²	1.41
1.00–1.50 m	–	0.04	0.83	1.43	–	m²	2.26
1.50–2.00 m; pruning to ground level by hand	–	0.10	2.08	3.57	–	m²	5.65
2.00–3.00 m; pruning to ground level by hand	–	0.10	2.08	7.14	–	m²	9.22
3.00–4.00 m; pruning to ground level by hand	–	0.20	4.15	11.89	–	m²	16.04
Site clearance; by hand; clear site of mature shrubs from existing cultivated beds; dig out roots							
Mixed shrubs in beds; planting centres 500 mm average							
height less than 1 m	–	0.33	6.92	–	–	m²	6.92
1.00–1.50 m	–	0.50	10.38	–	–	m²	10.38
1.50–2.00 m	–	1.00	20.75	–	–	m²	20.75
2.00–3.00 m	–	2.00	41.50	–	–	m²	41.50
3.00–4.00 m	–	3.00	62.25	–	–	m²	62.25
Disposal of material from site clearance operations							
Shrubs and groundcovers less than 1.00 m height; disposal to skip							
deciduous shrubs; not chipped; winter	–	0.05	1.04	12.00	–	m²	13.04
deciduous shrubs; chipped; winter	–	0.05	1.04	2.88	–	m²	3.92
evergreen or deciduous shrubs; chipped; summer	–	0.08	1.56	8.28	–	m²	9.84

D20 EXCAVATION AND FILLING

| Item | PC | Labour | Labour | Plant | Material | Unit | Total |
Excluding site overheads and profit	£	hours	£	£	£		rate £
SITE CLEARANCE – cont							
Disposal of material from site clearance							
operations – cont							
Shrubs 1.00–2.00 m height; disposal to skip							
deciduous shrubs; not chipped; winter	–	0.17	3.46	36.00	–	m²	**39.46**
deciduous shrubs; chipped; winter	–	0.25	5.19	8.28	–	m²	**13.47**
evergreen or deciduous shrubs; chipped;							
summer	–	0.30	6.22	15.84	–	m²	**22.06**
Shrubs or hedges 2.00–3.00 m height;							
disposal to skip							
deciduous plants non-woody growth; not							
chipped; winter	–	0.25	5.19	36.00	–	m²	**41.19**
deciduous shrubs; chipped; winter	–	0.50	10.38	19.08	–	m²	**29.46**
evergreen or deciduous shrubs non-woody							
growth; chipped; summer	–	0.67	13.83	23.76	–	m²	**37.59**
evergreen or deciduous shrubs woody							
growth; chipped; summer	–	1.50	31.13	25.91	–	m²	**57.04**
Shrubs and groundcovers less than 1.00 m							
height; disposal to spoil heap							
deciduous shrubs; not chipped; winter	–	0.05	1.04	–	–	m²	**1.04**
deciduous shrubs; chipped; winter	–	0.05	1.04	1.08	–	m²	**2.12**
evergreen or deciduous shrubs; chipped;							
summer	–	0.08	1.56	1.08	–	m²	**2.64**
Shrubs 1.00–2.00 m height; disposal to spoil							
heap							
deciduous shrubs; chipped; winter	–	0.25	5.19	1.08	–	m²	**6.27**
deciduous shrubs; not chipped; winter	–	0.17	3.46	–	–	m²	**3.46**
evergreen or deciduous shrubs; chipped;							
summer	–	0.30	6.22	1.44	–	m²	**7.66**
Shrubs or hedges 2.00–3.00 m height;							
disposal to spoil heaps							
deciduous plants non-woody growth; not							
chipped; winter	–	0.25	5.19	–	–	m²	**5.19**
deciduous shrubs; chipped; winter	–	0.50	10.38	1.08	–	m²	**11.46**
evergreen or deciduous shrubs non-woody							
growth; chipped; summer	–	0.67	13.83	2.16	–	m²	**15.99**
evergreen or deciduous shrubs woody							
growth; chipped; summer	–	1.50	31.13	4.31	–	m²	**35.44**
EXCAVATION							
Excavating; mechanical; topsoil for							
preservation; depositing alongside							
3 tonne tracked excavator (bucket volume							
0.13 m³)							
average depth 100 mm	–	0.02	0.50	0.86	–	m²	**1.36**
average depth 150 mm	–	0.03	0.70	1.20	–	m²	**1.90**
average depth 200 mm	–	0.04	0.83	1.43	–	m²	**2.26**
average depth 250 mm	–	0.04	0.91	1.57	–	m²	**2.48**
average depth 300 mm	–	0.05	1.00	1.71	–	m²	**2.71**

D20 EXCAVATION AND FILLING

Item Excluding site overheads and profit	PC £	Labour hours	Labour £	Plant £	Material £	Unit	Total rate £
Excavating; mechanical; topsoil for preservation; depositing to spoil heaps by wheeled dumper; spoil heaps maximum 50 m distance							
3 tonne tracked excavator (bucket volume 0.13 m³)							
average depth 100 mm	–	0.02	0.50	1.29	–	m²	1.79
average depth 150 mm	–	0.03	0.70	1.86	–	m²	2.56
average depth 200 mm	–	0.04	0.83	2.30	–	m²	3.13
average depth 250 mm	–	0.04	0.91	2.66	–	m²	3.57
average depth 300 mm	–	0.05	1.00	3.02	–	m²	4.02
Excavating; mechanical; to reduce levels							
3 tonne excavator (bucket volume 0.13 m³)							
maximum depth not exceeding 0.25 m	–	0.13	2.66	5.39	–	m³	8.05
maximum depth not exceeding 1.00 m	–	0.13	2.66	5.12	–	m³	7.78
Pits; 3 tonne tracked excavator							
maximum depth not exceeding 0.25 m	–	0.58	12.04	1.08	–	m³	13.12
maximum depth not exceeding 1.00 m	–	0.50	10.38	4.82	–	m³	15.20
maximum depth not exceeding 2.00 m	–	0.60	12.45	5.79	–	m³	18.24
Trenches; width not exceeding 0.30 m; 3 tonne excavator							
maximum depth not exceeding 0.25 m	–	1.33	27.67	5.36	–	m³	33.03
maximum depth not exceeding 1.00 m	–	0.69	14.21	2.75	–	m³	16.96
maximum depth not exceeding 2.00 m	–	0.60	12.45	2.41	–	m³	14.86
Trenches; width exceeding 0.30 m; 3 tonne excavator							
maximum depth not exceeding 0.25 m	–	0.60	12.45	2.41	–	m³	14.86
maximum depth not exceeding 1.00 m	–	0.50	10.38	2.01	–	m³	12.39
maximum depth not exceeding 2.00 m	–	0.38	7.98	1.55	–	m³	9.53
Extra over any types of excavating irrespective of depth for breaking out existing materials; heavy duty 110 volt breaker tool							
hard rock	–	5.00	103.75	23.52	–	m³	127.27
concrete	–	3.00	62.25	14.12	–	m³	76.37
reinforced concrete	–	4.00	83.00	23.52	–	m³	106.52
brickwork, blockwork or stonework	–	1.50	31.13	7.06	–	m³	38.19
Extra over any types of excavating irrespective of depth for breaking out existing hard pavings; 1600 watt, 110 volt breaker							
concrete; 100 mm thick	–	0.20	4.15	0.50	–	m²	4.65
concrete; 150 mm thick	–	0.25	5.19	0.63	–	m²	5.82
concrete; 200 mm thick	–	0.29	5.93	0.72	–	m²	6.65
concrete; 300 mm thick	–	0.40	8.30	1.01	–	m²	9.31
reinforced concrete; 100 mm thick	–	0.33	6.92	3.19	–	m²	10.11
reinforced concrete; 150 mm thick	–	0.40	8.30	3.83	–	m²	12.13
reinforced concrete; 200 mm thick	–	0.50	10.38	4.79	–	m²	15.17
reinforced concrete; 300 mm thick	–	1.00	20.75	9.57	–	m²	30.32
tarmacadam; 75 mm thick	–	0.14	2.97	0.36	–	m²	3.33
tarmacadam and hardcore; 150 mm thick	–	0.20	4.15	0.50	–	m²	4.65

D20 EXCAVATION AND FILLING

Item Excluding site overheads and profit	PC £	Labour hours	Labour £	Plant £	Material £	Unit	Total rate £
EXCAVATION – cont							
Excavating – cont							
Extra over any types of excavating irrespective							
of depth for taking up							
precast concrete paving slabs	–	0.08	1.73	0.21	–	m²	1.94
natural stone paving	–	0.13	2.59	0.31	–	m²	2.90
cobbles	–	0.17	3.46	0.42	–	m²	3.88
brick paviors	–	0.17	3.46	0.42	–	m²	3.88
Excavating; hand							
Topsoil for preservation; loading to barrows							
average depth 100 mm	–	0.30	6.22	–	–	m²	6.22
average depth 200 mm	–	0.60	12.45	–	–	m²	12.45
average depth 300 mm	–	0.90	18.68	–	–	m²	18.68
over 300 mm	–	3.00	62.25	–	–	m³	62.25
Excavating; hand							
Topsoil to reduce levels							
maximum depth not exceeding 0.25 m	–	2.52	52.29	–	–	m³	52.29
maximum depth not exceeding 1.00 m	–	3.12	64.74	–	–	m³	64.74
Pits							
maximum depth not exceeding 0.25 m	–	2.67	55.33	–	–	m³	55.33
maximum depth not exceeding 1.00 m	–	3.47	71.93	–	–	m³	71.93
maximum depth not exceeding 2.00 m							
(includes earthwork support)	–	6.93	143.87	–	–	m³	242.01
Trenches; width not exceeding 0.30 m							
maximum depth not exceeding 0.25 m	–	2.86	59.28	–	–	m³	59.28
maximum depth not exceeding 1.00 m	–	3.72	77.21	–	–	m³	77.21
maximum depth not exceeding 1.00 m	–	3.72	77.21	–	–	m³	126.28
Trenches; width exceeding 0.30 m wide							
maximum depth not exceeding 0.25 m	–	2.86	59.28	–	–	m³	59.28
maximum depth not exceeding 1.00 m	–	4.00	83.00	–	–	m³	83.00
maximum depth not exceeding 2.00 m							
(includes earthwork support)	–	6.00	124.50	–	–	m³	222.64
Excavating; mechanical; trenches for							
electrical services							
3 tonne excavator (bucket volume 0.13 m³);							
arisings laid alongside							
600 mm deep	–	–	–	1.70	–	m	1.70
800 mm deep	–	–	–	1.98	–	m	1.98
1.00 m deep	–	–	–	2.38	–	m	2.38
By hand							
600 mm deep	–	0.24	4.99	–	–	m	4.99
800 mm deep	–	0.43	8.87	–	–	m	8.87
1.00 mm deep	–	0.67	13.83	–	–	m	13.83

D20 EXCAVATION AND FILLING

Item Excluding site overheads and profit	PC £	Labour hours	Labour £	Plant £	Material £	Unit	Total rate £
FILLING							
Filling to make up levels; mechanical							
(3 tonne tracked excavator); depositing in							
layers 150 mm maximum thickness							
Arising from the excavations							
maximum thickness less than 0.25 m	–	0.10	2.16	3.71	–	m^3	**5.87**
average thickness exceeding 0.25 m	–	0.08	1.66	2.85	–	m^3	**4.51**
Obtained from on site spoil heaps; average							
25 m distance; multiple handling							
maximum thickness less than 0.25 m	–	0.16	3.33	6.76	–	m^3	**10.09**
average thickness exceeding 0.25 m	–	0.13	2.70	5.69	–	m^3	**8.39**
Obtained off site; planting quality topsoil PC							
£28.00/m^3							
maximum thickness less than 0.25 m	35.00	0.14	2.97	6.15	35.00	m^3	**44.12**
average thickness exceeding 0.25 m	35.00	0.13	2.70	5.69	35.00	m^3	**43.39**
Obtained off site; crushed concrete hardcore;							
PC £16.20/ m^3							
maximum thickness less than 0.25 m	19.76	0.14	2.90	14.22	19.76	m^3	**36.88**
maximum thickness exceeding 0.25 m	19.76	0.13	2.64	13.66	19.76	m^3	**36.06**
Backfilling of trenches; including laying of							
electrical marker tape; compacting lightly							
as work proceeds; spreading any							
remaining material							
To trenches containing armoured cable							
by machine	0.13	–	–	1.78	0.13	m	**1.91**
by hand	0.13	0.20	4.15	–	0.13	m	**4.28**
To trenches containing ducted cable including							
150 mm sharp sand over the duct							
by machine	2.19	–	–	2.23	2.19	m	**4.42**
by hand	0.13	0.25	5.19	–	2.19	m	**7.38**
Filling to make up levels; hand; depositing							
in layers 150 mm maximum thickness							
Arising from the excavations							
average thickness exceeding 0.25 m	–	0.75	15.56	–	–	m^3	**15.56**
Obtained from on site spoil heaps; average							
25 m distance; multiple handling							
average thickness exceeding 0.25 m thick	–	1.25	25.94	–	–	m^3	**25.94**
Obtained off site; planting quality topsoil PC							
£32.25/m^3 (10 tonne loads)							
average thickness exceeding 0.25 m thick	66.52	1.25	25.94	–	66.52	m^3	**92.46**

D20 EXCAVATION AND FILLING

Item Excluding site overheads and profit	PC £	Labour hours	Labour £	Plant £	Material £	Unit	Total rate £
DISPOSAL							
Note: Bulkage factors have been applied to the figures below for disposal							
Disposal; mechanical							
Light soils and loams (bulking factor of 1.25)							
Excavated material; off site; to tip;							
mechanically loaded by grab; capacity of load							
10–12 m³ (18 tonne)							
inert material (mixed building waste and							
landfill	–	–	–	–	–	m³	31.68
soil (sandy and loam); dry	–	–	–	–	–	m³	31.68
soil (sandy and loam); wet	–	–	–	–	–	m³	33.60
broken out compacted materials such as							
road bases and the like	–	–	–	–	–	m³	35.20
soil (clay); dry	–	–	–	–	–	m³	44.55
soil (clay); wet	–	–	–	–	–	m³	53.40
Disposal by skip; 8 yd³ (6 m³)							
Excavated material; off site; to tip							
by machine	–	–	–	47.66	–	m³	47.66
by hand	–	4.00	83.00	43.20	–	m³	126.20
Disposal on site							
Excavated material to spoil heaps							
average 25 m distance	–	–	–	6.39	–	m³	6.39
average 50 m distance	–	–	–	7.03	–	m³	7.03
average 100 m distance	–	–	–	7.67	–	m³	7.67
average 200 m distance	–	–	–	9.59	–	m³	9.59
Excavated material; spreading on site							
average 25 m distance	–	–	–	7.18	–	m³	7.18
average 50 m distance	–	–	–	7.86	–	m³	7.86
average 100 m distance	–	–	–	8.99	–	m³	8.99
average 200 m distance	–	–	–	9.89	–	m³	9.89
Disposal; hand							
Excavated material; on site; in spoil heaps							
average 25 m distance	–	2.40	49.80	–	–	m³	49.80
average 50 m distance	–	2.64	54.78	–	–	m³	54.78
average 100 m distance	–	3.00	62.25	–	–	m³	62.25
average 200 m distance	–	3.60	74.70	–	–	m³	74.70
Disposal hand; excavated material in bags							
Filling bags 25 kg and barrowing through							
buildings							
average 25 m distance	–	8.40	174.30	–	–	m³	174.30
average 50 m distance	–	9.20	190.90	–	–	m³	190.90
average 100 m distance	–	10.10	209.57	–	–	m³	209.57
average 200 m distance	–	11.60	240.70	–	–	m³	240.70

D20 EXCAVATION AND FILLING

Item Excluding site overheads and profit	PC £	Labour hours	Labour £	Plant £	Material £	Unit	Total rate £
Excavated material; spreading on site							
average 25 m distance	–	2.64	54.78	–	–	m³	**54.78**
average 50 m distance	–	3.00	62.25	–	–	m³	**62.25**
average 100 m distance	–	3.60	74.70	–	–	m³	**74.70**
average 200 m distance	–	4.20	87.15	–	–	m³	**87.15**
SURFACE TREATMENTS							
Compacting							
Bottoms of excavations	–	0.01	0.10	0.05	–	m²	**0.15**
Grading operations; surface previously excavated to reduce levels to prepare to receive subsequent treatments; grading to accurate levels and falls 20 mm tolerances							
Clay or heavy soils or hardcore							
3 tonne excavator	–	0.02	0.41	0.71	–	m²	**1.12**
5 tonne excavator	–	0.04	0.83	0.20	–	m²	**1.03**
by hand	–	0.10	2.08	–	–	m²	**2.08**
Loamy topsoils							
3 tonne excavator	–	0.01	0.28	0.47	–	m²	**0.75**
5 tonne excavator	–	0.03	0.55	0.14	–	m²	**0.69**
by hand	–	0.05	1.04	–	–	m²	**1.04**
Sand or graded granular materials							
3 tonne excavator	–	0.01	0.21	0.36	–	m²	**0.57**
5 tonne excavator	–	0.02	0.41	0.10	–	m²	**0.51**
by hand	–	0.03	0.69	–	–	m²	**0.69**
Excavating; by hand							
Topsoil for preservation; loading to barrows							
average depth 100 mm	–	0.24	4.98	–	–	m²	**4.98**
average depth 150 mm	–	0.36	7.47	–	–	m²	**7.47**
average depth 200 mm	–	0.58	11.95	–	–	m²	**11.95**
average depth 250 mm	–	0.72	14.94	–	–	m²	**14.94**
average depth 300 mm	–	0.86	17.93	–	–	m²	**17.93**
Grading operations; surface recently filled to raise levels to prepare to receive subsequent treatments; grading to accurate levels and falls 20 mm tolerances							
Clay or heavy soils or hardcore							
3 tonne excavator	–	0.02	0.32	0.54	–	m²	**0.86**
by hand	–	0.08	1.73	–	–	m²	**1.73**
Loamy topsoils							
3 tonne excavator	–	0.01	0.21	0.37	–	m²	**0.58**
by hand	–	0.04	0.83	–	–	m²	**0.83**
Sand or graded granular materials							
3 tonne excavator	–	0.01	0.16	0.27	–	m²	**0.43**
by hand	–	0.03	0.59	–	–	m²	**0.59**

D20 EXCAVATION AND FILLING

Item Excluding site overheads and profit	PC £	Labour hours	Labour £	Plant £	Material £	Unit	Total rate £
SURFACE TREATMENTS – cont							
Grading operations – cont							
Surface preparation							
Trimming surfaces of cultivated ground to final levels, removing roots stones and debris exceeding 50 mm in any direction to tip off site; slopes less than 15°							
clean ground with minimal stone content	–	0.25	5.19	–	–	100 m²	**5.19**
slightly stony; 0.5 kg stones per m²	–	0.33	6.92	0.02	–	100 m²	**6.94**
very stony; 1.0–3.0 kg stones per m²	–	0.50	10.38	0.04	–	100 m²	**10.42**
clearing mixed slightly contaminated rubble inclusive of roots and vegetation	–	0.50	10.38	9.00	–	100 m²	**19.38**
clearing brick-bats, stones and clean rubble	–	0.60	12.45	0.16	–	100 m²	**12.61**
Hand cultivation; preparation for turfing/ seeding areas							
Cultivate surface to receive landscape surface treatments using hand fork or implement; 1 spit (300 mm deep), break up ground and bring to a fine tilth by rake							
compacted ground	–	0.07	1.38	–	–	m²	**1.38**
ground previously supporting turf; lifting of turf not included	–	0.06	1.30	–	–	m²	**1.30**
ground previously planted with small to medium shrubs; removal of planting not included	–	0.06	1.15	–	–	m²	**1.15**
soft ground	–	0.04	0.83	–	–	m²	**0.83**
Hand cultivation; preparation for planting							
Cultivate surface to receive landscape surface treatments using hand fork or implement; 1 spit (300 mm deep), break up ground and bring to a fine tilth by rake							
compacted ground	–	0.05	1.04	–	–	m²	**1.04**
ground previously supporting turf; lifting of turf not included	–	0.04	0.87	–	–	m²	**0.87**
ground previously planted with small to medium shrubs; removal of planting not included	–	0.04	0.74	–	–	m²	**0.74**
soft ground	–	0.03	0.59	–	–	m²	**0.59**

E10 IN SITU CONCRETE

Item Excluding site overheads and profit	PC £	Labour hours	Labour £	Plant £	Material £	Unit	Total rate £
E10 IN SITU CONCRETE							
MIXING OF CONCRETE							
General							
Please see the notes on concrete mix designations in the Major Works section of this book under E10 In Situ Concrete.							
Concrete mixes; mixed on site; costs for producing concrete; prices for commonly used mixes for various types of work; based on an 85 litre concrete mixer							
Roughest type mass concrete such as footings, road haunchings 300 mm thick; aggregates delivered in 10 tonne loads							
1:3:6	106.22	–	–	–	106.22	m³	**106.22**
As above but aggregates delivered in 850 kg bulk bags							
1:3:6	147.50	–	–	–	147.50	m³	**147.50**
Most ordinary use of concrete such as mass walls above ground, road slabs, etc. and general reinforced concrete work; aggregates delivered in 10 tonne loads							
1:2:4	90.16	1.51	31.40	–	90.16	m³	**121.56**
As above but aggregates delivered in 850 kg bulk bags							
1:2:4	51.45	1.51	31.40	–	130.53	m³	**161.93**
Watertight floors, pavements and walls, tanks, pits, steps, paths, surface of two course roads; reinforced concrete where extra strength is required (aggregates delivered in 10 tonne loads)							
1:1.5:3	104.54	1.51	31.40	–	104.54	m³	**135.94**
As above but aggregates delivered in 850 kg bulk bags							
1:1.5:3	65.84	1.51	31.40	–	144.92	m³	**176.32**
PLACING OF CONCRETE							
Plain in situ concrete; site mixed; **10 N/mm² – 40 mm aggregate (1:3:6)** **(aggregate delivery indicated)**							
Foundations							
ordinary Portland cement; 10 tonne ballast loads	93.77	0.33	20.54	–	93.77	m³	**114.31**
ordinary Portland cement; 850 kg bulk bags	147.50	0.33	20.54	–	147.50	m³	**168.04**

E10 IN SITU CONCRETE

Item Excluding site overheads and profit	PC £	Labour hours	Labour £	Plant £	Material £	Unit	Total rate £
PLACING OF CONCRETE – cont							
Plain in situ concrete – cont							
Foundations; poured on or against earth or unblinded hardcore							
ordinary Portland cement; 10 tonne ballast loads	98.45	0.33	20.54	–	98.45	m³	**118.99**
ordinary Portland cement; 850 kg bulk bags	154.88	0.33	20.54	–	154.88	m³	**175.42**
Isolated foundations							
ordinary Portland cement; 10 tonne ballast loads	96.11	0.66	41.09	–	96.11	m³	**137.20**
ordinary Portland cement; 850 kg bulk bags	151.19	0.66	41.09	–	151.19	m³	**192.28**
Plain in situ concrete; site mixed; 21 N/mm² – 20 mm aggregate (1:2:4)							
Foundations							
ordinary Portland cement; 10 tonne ballast loads	125.81	0.33	20.54	–	125.81	m³	**146.35**
ordinary Portland cement; 850 kg bulk bags	–	0.33	20.54	–	167.19	m³	**187.73**
Foundations; poured on or against earth or unblinded hardcore							
ordinary Portland cement; 10 tonne ballast loads	128.88	0.33	20.54	–	128.88	m³	**149.42**
ordinary Portland cement; 850 kg bulk bags	128.88	0.50	30.81	–	128.88	m³	**159.69**
Isolated foundations							
ordinary Portland cement	122.57	0.66	41.09	–	122.57	m³	**163.66**
Reinforced in situ concrete; site mixed; 21 N/mm² – 20 mm aggregate (1:2:4); aggregates delivered in 10 tonne loads							
Foundations							
ordinary Portland cement	125.81	0.36	22.60	–	125.81	m³	**148.41**
Foundations; poured on or against earth or unblinded hardcore							
ordinary Portland cement	128.69	0.36	22.60	–	128.69	m³	**151.29**
Isolated foundations							
ordinary Portland cement	125.81	0.73	45.19	–	125.81	m³	**171.00**
Plain in situ concrete; ready mixed; Tarmac Southern; 10 N/mm² mixes; suitable for mass concrete fill and blinding							
Foundations							
GEN1; designated mix	81.97	0.50	30.81	–	81.97	m³	**112.78**
ST2; standard mix	85.58	0.50	30.81	–	85.58	m³	**116.39**
Foundations; poured on or against earth or unblinded hardcore							
GEN1; designated mix	83.97	0.52	32.36	–	83.97	m³	**116.33**
ST2; standard mix	87.66	0.52	32.36	–	87.66	m³	**120.02**
Isolated foundations							
GEN1; designated mix	81.97	0.66	41.09	–	81.97	m³	**123.06**
ST2; standard mix	85.58	0.66	41.09	–	85.58	m³	**126.67**

E20 FORMWORK FOR IN SITU CONCRETE

Item Excluding site overheads and profit	PC £	Labour hours	Labour £	Plant £	Material £	Unit	Total rate £
Plain in situ concrete; ready mixed; Tarmac Southern; 15 N/mm^2 mixes; suitable for oversite below suspended slabs and strip footings in non-aggressive soils							
Foundations							
GEN 2; designated mix	84.00	0.50	30.81	–	84.00	m^3	**114.81**
ST3; standard mix	89.05	0.50	30.81	–	89.05	m^3	**119.86**
Foundations; poured on or against earth or unblinded hardcore							
GEN2; designated mix	84.00	0.52	32.36	–	84.00	m^3	**116.36**
ST3; standard mix	86.93	0.52	32.36	–	86.93	m^3	**119.29**
Isolated foundations							
GEN2; designated mix	84.00	0.66	41.09	–	84.00	m^3	**125.09**
ST3; standard mix	86.93	0.66	41.09	–	86.93	m^3	**128.02**
Concrete mixed on site; Easymix Concrete Ltd							
Concrete mixer on lorry; maximum standing time 3 m^3/hr							
C20 concrete	104.50	0.33	20.54	–	104.50	m^3	**125.04**
C25 concrete	112.20	0.33	20.54	–	112.20	m^3	**132.74**
C30 concrete	121.00	0.33	20.54	–	121.00	m^3	**141.54**
Concrete mixed on site; Easymix Concrete Ltd; concrete delivered by pump							
Concrete mixer on lorry; maximum standing time 3 m^3/hr							
C20 concrete	104.50	0.02	1.03	–	132.00	m^3	**133.03**
C25 concrete	112.20	0.33	20.54	–	139.70	m^3	**160.24**
C30 concrete	121.00	0.02	1.03	–	148.50	m^3	**149.53**
E20 FORMWORK FOR IN SITU CONCRETE							
FORMWORK							
Plain vertical formwork; basic finish							
Sides of foundations							
height exceeding 1.00 m	–	2.00	41.50	–	14.16	m^2	**55.66**
height not exceeding 250 mm	–	1.00	20.75	–	3.72	m	**24.47**
height 250–500 mm	–	1.00	20.75	–	7.08	m	**27.83**
height 500 mm–1.00 m	–	1.50	31.13	–	14.16	m	**45.29**
Sides of foundations; left in							
height over 1.00 m	–	2.00	41.50	–	50.37	m^2	**91.87**
height not exceeding 250 mm	–	1.00	20.75	–	14.42	m	**35.17**
height 250–500 mm	–	1.00	20.75	–	28.48	m	**49.23**
height 500 mm–1.00 m	–	1.50	31.13	–	56.97	m	**88.10**

E30 REINFORCEMENT FOR IN SITU CONCRETE

Item Excluding site overheads and profit	PC £	Labour hours	Labour £	Plant £	Material £	Unit	Total rate £
E30 REINFORCEMENT FOR IN SITU CONCRETE							
FABRIC REINFORCEMENT							
For reinforcement bar prices please see the Major Works section of this book.							
Reinforcement fabric; lapped; in beds or suspended slabs							
Fabric 3.6 × 2.0 m²							
ref A142 (2.22 kg/m²)	2.90	0.22	4.57	–	2.90	m²	**7.47**
ref A393 (6.16 kg/m²)	4.95	0.28	5.81	–	4.95	m²	**10.76**
Fabric 2.4 × 4.8 m²							
ref A98 (1.54 kg/m²)	1.08	0.22	4.57	–	1.08	m²	**5.65**
ref A142 (2.22 kg/m²)	1.61	0.22	4.57	–	1.61	m²	**6.18**
ref A193 (3.02 kg/m²)	1.94	0.22	4.57	–	1.94	m²	**6.51**
ref A252 (3.95 kg/m²)	2.62	0.24	4.98	–	2.62	m²	**7.60**
ref A393 (6.16 kg/m²)	3.90	0.28	5.81	–	3.90	m²	**9.71**

F10 BRICK/BLOCK WALLING

Item Excluding site overheads and profit	PC £	Labour hours	Labour £	Plant £	Material £	Unit	Total rate £
F10 BRICK/BLOCK WALLING							
MORTAR MIXES							
Batching quantities for these mortar mixes may be found in the Memoranda section of this book.							
Mortar mixes; common mixes for various types of work; mortar mixed on site; prices based on builders merchant rates for cement; mechanically mixed							
Aggregates delivered in 850 kg bulk bags							
1:3	–	0.75	15.56	–	173.85	m³	**189.41**
1:4	–	0.75	15.56	–	136.66	m³	**152.22**
1:1:6	–	0.75	15.56	–	169.80	m³	**185.36**
Aggregates delivered in 10 tonne loads							
1:2	–	0.75	15.56	–	164.01	m³	**179.57**
1:3	–	0.75	15.56	–	133.97	m³	**149.53**
1:4	–	0.75	15.56	–	110.27	m³	**125.83**
1:1:6	–	0.75	15.56	–	138.69	m³	**154.25**
Mortar mixes; common mixes for various types of work; mortar mixed on site; prices based on builders merchant rates for cement; hand mixed							
Aggregates delivered in 850 kg bulk bags							
1:3	–	2.00	41.50	–	175.91	m³	**217.41**
1:4	–	2.00	41.50	–	138.39	m³	**179.89**
1:1:6	–	2.00	41.50	–	171.53	m³	**213.03**
BRICK WALLING							
Variation in brick prices							
Add or subtract the following amounts for every £1.00/1000 difference in the PC price of the measured items below							
half brick thick	–	–	–	–	0.06	m²	**0.06**
one brick thick	–	–	–	–	0.13	m²	**0.13**
one and a half brick thick	–	–	–	–	0.19	m²	**0.19**
two brick thick	–	–	–	–	0.25	m²	**0.25**
Movement of materials							
Loading to wheelbarrows and transporting to location; per 215 mm thick walls; maximum 25 m distance	–	0.42	8.65	–	–	m²	**8.65**

F10 BRICK/BLOCK WALLING

Item Excluding site overheads and profit	PC £	Labour hours	Labour £	Plant £	Material £	Unit	Total rate £
BRICK WALLING – cont							
Class B engineering bricks; PC £290.00/ 1000; double Flemish bond in cement mortar (1:3)							
Walls							
half brick thick	–	1.05	65.57	–	19.63	m²	85.20
one brick thick	–	2.11	131.14	–	39.26	m²	170.40
one and a half brick thick	–	3.16	196.75	–	58.89	m²	255.64
two brick thick	–	4.21	262.28	–	78.52	m²	340.80
Walls; curved; mean radius 6 m							
half brick thick	–	1.59	98.72	–	20.38	m²	119.10
one brick thick	–	3.17	197.33	–	39.26	m²	236.59
Walls; curved; mean radius 1.5 m							
half brick thick	–	2.11	131.43	–	20.38	m²	151.81
one brick thick	–	4.23	263.03	–	39.26	m²	302.29
Walls; tapering; one face battering; average							
one and a half brick thick	–	3.41	212.51	–	58.89	m²	271.40
two brick thick	–	5.12	318.76	–	78.52	m²	397.28
Walls; battering (retaining)							
one and a half brick thick	–	3.79	236.12	–	58.89	m²	295.01
two brick thick	–	5.69	354.18	–	78.52	m²	432.70
Isolated piers							
one brick thick	–	3.07	191.25	–	43.75	m²	235.00
one and a half brick thick	–	3.95	245.89	–	65.62	m²	311.51
two brick thick	–	4.39	273.22	–	88.99	m²	362.21
three brick thick	–	5.44	338.79	–	131.24	m²	470.03
Projections; vertical							
one brick × half brick	–	0.23	14.38	–	4.38	m	18.76
one brick × one brick	–	0.46	28.76	–	8.77	m	37.53
one and a half brick × one brick	–	0.69	43.14	–	13.16	m	56.30
two brick × one brick	–	0.76	47.25	–	17.55	m	64.80
Walls; half brick thick							
in honeycomb bond	–	0.59	36.98	–	14.58	m²	51.56
in quarter bond	–	0.55	34.24	–	18.88	m²	53.12
Facing bricks; PC £500.00/1000; English garden wall bond; in gauged mortar (1:1:6); facework one side							
Walls							
half brick thick	–	1.05	65.57	–	33.81	m²	99.38
half brick thick (using site cut snap headers to form bond)	–	1.65	102.46	–	33.81	m²	136.27
one brick thick	–	2.11	131.14	–	65.31	m²	196.45
one and a half brick thick	–	3.16	196.72	–	101.44	m²	298.16
two brick thick	–	4.21	262.28	–	135.25	m²	397.53
Walls; curved; mean radius 6 m							
half brick thick	–	1.59	98.72	–	36.09	m²	134.81
one brick thick	–	3.17	197.33	–	70.63	m²	267.96

F10 BRICK/BLOCK WALLING

Item Excluding site overheads and profit	PC £	Labour hours	Labour £	Plant £	Material £	Unit	Total rate £
Walls; curved; mean radius 1.5 m							
half brick thick	–	2.11	131.43	–	35.34	m²	166.77
one brick thick	–	4.23	263.03	–	69.13	m²	332.16
Walls; tapering; one face battering; average							
one and a half brick thick	–	3.79	236.08	–	105.94	m²	342.02
two brick thick	–	5.69	354.18	–	141.25	m²	495.43
Walls; battering (retaining)							
one and a half brick thick	–	3.41	212.47	–	105.94	m²	318.41
two brick thick	–	5.12	318.76	–	141.25	m²	460.01
Isolated piers; English bond; facework all round							
one brick thick	–	3.07	191.25	–	72.25	m²	263.50
one and a half brick thick	–	3.95	245.89	–	108.38	m²	354.27
two brick thick	–	4.39	273.22	–	146.05	m²	419.27
three brick thick	–	5.44	338.79	–	216.76	m²	555.55
Projections; vertical							
one brick × half brick	–	0.31	19.12	–	7.54	m	26.66
one brick × one brick	–	0.61	38.25	–	15.08	m	53.33
one and a half brick × one brick	–	0.92	57.38	–	22.62	m	80.00
two brick × one brick	–	1.01	62.84	–	30.16	m	93.00
Brickwork fair faced both sides; facing bricks in gauged mortar (1:1:6)							
Extra for fair face both sides; flush, struck, weathered or bucket-handle pointing	–	0.29	18.21	–	–	m²	18.21
Reclaimed bricks; PC £1000.00/1000; English garden wall bond; in gauged mortar (1:1:6)							
Walls							
half brick thick (stretcher bond)	–	1.05	65.57	–	65.31	m²	130.88
half brick thick (using site cut snap headers to form bond)	–	1.65	102.46	–	65.31	m²	167.77
one brick thick	–	2.11	131.14	–	128.23	m²	259.37
one and a half brick thick	–	3.16	196.75	–	192.34	m²	389.09
two brick thick	–	4.21	262.28	–	256.45	m²	518.73
Walls; curved; mean radius 6 m							
half brick thick	–	1.59	98.72	–	69.08	m²	167.80
one brick thick	–	3.17	197.33	–	136.63	m²	333.96
Walls; curved; mean radius 1.50 m							
half brick thick	–	2.11	131.43	–	72.08	m²	203.51
one brick thick	–	4.21	262.28	–	142.63	m²	404.91
Walls; stretcher bond; wall ties at 450 centres vertically and horizontally							
one brick thick	–	2.30	143.07	–	130.01	m²	273.08
two brick thick	–	4.60	286.09	–	261.80	m²	547.89
Brickwork fair faced both sides; facing bricks in gauged mortar (1:1:6)							
extra for fair face both sides; flush, struck, weathered or bucket-handle pointing	–	0.22	13.70	–	–	m²	13.70

F10 BRICK/BLOCK WALLING

Item Excluding site overheads and profit	PC £	Labour hours	Labour £	Plant £	Material £	Unit	Total rate £
BRICK WALLING – cont							
Brick copings							
Copings; all brick headers-on-edge; two angles rounded 53 mm radius; flush pointing top and both sides as work proceeds; one brick wide; horizontal							
machine-made specials	51.49	0.23	14.60	–	52.38	m	**66.98**
handmade specials	65.23	0.23	14.60	–	66.13	m	**80.73**
Extra over copings for two courses machine-made tile creasings, projecting 25 mm each side; 260 mm wide copings; horizontal	6.42	0.17	10.27	–	6.87	m	**17.14**
Copings; all brick headers-on-edge; flush pointing top and both sides as work proceeds; one brick wide; horizontal							
bricks PC £500.00/1000	7.16	0.23	14.60	–	7.46	m	**22.06**
bricks PC £700.00/1000	9.56	0.23	14.60	–	9.88	m	**24.48**
BLOCK WALLING							
Dense aggregate concrete blocks; Tarmac Topblock or other equal and approved; in gauged mortar (1:2:9)							
Solid blocks 7 N/mm^2							
440 × 215 × 100 mm thick	8.93	0.40	24.65	–	10.12	m^2	**34.77**
440 × 215 × 140 mm thick	14.44	0.43	26.71	–	15.78	m^2	**42.49**
Solid blocks 7 N/mm^2 laid flat							
440 × 100 × 215 mm thick	18.21	1.06	66.15	–	21.42	m^2	**87.57**
Hollow concrete blocks							
440 × 215 × 215 mm thick	23.38	0.43	26.71	–	24.73	m^2	**51.44**
Filling of hollow concrete blocks with concrete as work proceeds; tamping and compacting							
440 × 215 × 215 mm thick	22.44	0.07	4.11	–	22.44	m^2	**26.55**
BRICK PIERS							
Isolated brick piers in English bond							
Engineering brick PC £290.00/1000 (not SMM)							
one brick thick (225 mm)	–	0.54	33.57	–	10.03	m	**43.60**
one and a half bricks thick (337.5 mm)	–	0.99	61.63	–	29.94	m	**91.57**
two bricks thick (450 mm)	–	1.49	92.44	–	47.75	m	**140.19**
Brick PC £300.00/1000 (not SMM)							
one brick thick (225 mm)	–	0.54	33.57	–	10.32	m	**43.89**
one and a half bricks thick (337.5 mm)	–	0.99	61.63	–	30.56	m	**92.19**
two bricks thick (450 mm)	–	1.49	92.44	–	48.85	m	**141.29**
three bricks thick (675 mm)	–	2.76	171.94	–	99.82	m	**271.76**
Brick PC £800.00/1000 (not SMM)							
one brick thick (225 mm)	–	0.54	33.57	–	24.32	m	**57.89**
one and a half bricks thick (337.5 mm)	–	0.99	61.63	–	60.94	m	**122.57**
two bricks thick (450 mm)	–	1.49	92.44	–	102.85	m	**195.29**
three bricks thick (675 mm)	–	2.76	171.94	–	221.32	m	**393.26**

J30 LIQUID APPLIED TANKING/DAMP-PROOFING

Item Excluding site overheads and profit	PC £	Labour hours	Labour £	Plant £	Material £	Unit	Total rate £
J30 LIQUID APPLIED TANKING/ DAMP-PROOFING							
Tanking and damp-proofing; Ruberoid Building Products; Synthaprufe cold applied bituminous emulsion waterproof coating							
Synthaprufe; to smooth finished concrete or screeded slabs; flat; blinding with sand							
two coats	8.13	0.22	4.61	–	8.30	m²	**12.91**
three coats	12.19	0.31	6.43	–	12.36	m²	**18.79**
Synthaprufe; to fair faced brickwork with flush joints, rendered brickwork or smooth finished concrete walls; vertical							
two coats	9.14	0.29	5.93	–	9.32	m²	**15.25**
three coats	13.41	0.40	8.30	–	13.58	m²	**21.88**
Tanking and damp-proofing; RIW Ltd							
Liquid asphaltic composition; to smooth finished concrete screeded slabs or screeded slabs; flat							
two coats	3.35	0.33	6.92	–	3.35	m²	**10.27**
Liquid asphaltic composition; fair faced brickwork with flush joints, rendered brickwork or smooth finished concrete walls; vertical							
two coats	3.35	0.50	10.38	–	3.35	m²	**13.73**
Heviseal; to smooth finished concrete or screeded slabs; to surfaces of ponds, tanks, planters; flat							
two coats	9.46	0.33	6.92	–	9.46	m²	**16.38**
Heviseal; to fair faced brickwork with flush joints, rendered brickwork or smooth finished concrete walls; to surfaces of retaining walls, ponds, tanks, planters; vertical							
two coats	9.46	0.50	10.38	–	9.46	m²	**19.84**

M60 PAINTING/CLEAR FINISHING

Item Excluding site overheads and profit	PC £	Labour hours	Labour £	Plant £	Material £	Unit	Total rate £
M60 PAINTING/CLEAR FINISHING							
METAL SURFACES							
Prepare; touch up primer; two undercoats and one finishing coat of gloss oil paint; on metal surfaces							
General surfaces							
girth exceeding 300 mm	1.38	0.67	13.83	–	1.38	m²	**15.21**
isolated surfaces; girth not exceeding 300 mm	0.46	0.40	8.30	–	0.46	m	**8.76**
isolated areas not exceeding 0.50 m² irrespective of girth	0.69	0.67	13.83	–	0.69	nr	**14.52**
Ornamental railings; each side measured overall							
girth exceeding 300 mm	1.38	0.75	15.56	–	1.38	m²	**16.94**
WOOD SURFACES							
Prepare; one coat primer; two undercoats and one finishing coat of gloss oil paint; on wood surfaces							
General surfaces							
girth exceeding 300 mm	1.29	0.40	8.30	–	1.29	m²	**9.59**
isolated surfaces; girth not exceeding 300 mm	0.39	0.20	4.15	–	0.39	m	**4.54**
isolated areas not exceeding 0.50 m² irrespective of girth	0.64	0.36	7.54	–	0.64	nr	**8.18**
Prepare; two coats of Protek wood preservative; on wood surfaces							
Planed surfaces							
girth exceeding 300 mm	0.33	0.07	1.42	–	0.33	m²	**1.75**
isolated surfaces; girth not exceeding 300 mm	1.10	0.17	3.55	–	1.10	m	**4.65**
Sawn surfaces							
girth exceeding 300 mm	0.22	0.10	2.13	–	0.22	m²	**2.35**
isolated surfaces; girth not exceeding 300 mm	0.07	0.17	3.55	–	0.07	m	**3.62**
Prepare; two coats of Sadolin wood preservative; on wood surfaces							
Planed surfaces							
girth exceeding 300 mm	2.32	0.07	1.42	–	2.32	m²	**3.74**
isolated surfaces; girth not exceeding 300 mm	0.77	0.14	2.83	–	0.77	m	**3.60**
Sawn surfaces							
girth exceeding 300 mm	3.72	0.10	2.13	–	3.72	m²	**5.85**
isolated surfaces; girth not exceeding 300 mm	0.62	0.17	3.55	–	0.62	m	**4.17**

M60 PAINTING/CLEAR FINISHING

Item Excluding site overheads and profit	PC £	Labour hours	Labour £	Plant £	Material £	Unit	Total rate £
Prepare; proprietary solution primer; two coats of dark stain; on wood surfaces							
General surfaces							
girth exceeding 300 mm	1.33	0.17	3.46	–	1.33	m²	**4.79**
isolated surfaces; girth not exceeding 300 mm	0.44	0.13	2.59	–	0.44	m	**3.03**
MASONRY SURFACES							
Two coats Weathershield exterior paint; to clean, dry surfaces; in accordance with manufacturer's instructions							
Brick or block walls							
girth exceeding 300 mm	–	0.33	6.92	–	1.09	m²	**8.01**
Cement render or concrete walls							
girth exceeding 300 mm	–	0.28	5.81	–	1.09	m²	**6.90**

Q10 KERBS/EDGINGS/CHANNELS/PAVING ACCESSORIES

Item Excluding site overheads and profit	PC £	Labour hours	Labour £	Plant £	Material £	Unit	Total rate £
Q10 KERBS/EDGINGS/CHANNELS/PAVING ACCESSORIES							
FOUNDATIONS TO KERBS							
Foundations to kerbs							
Excavating trenches; width 300 mm; 3 tonne excavator							
depth 300 mm	–	0.10	2.08	0.40	1.97	m	**4.45**
depth 400 mm	–	0.11	2.31	0.45	2.64	m	**5.40**
By hand	–	0.60	12.50	–	3.29	m	**15.79**
Excavating trenches; width 450 mm; 3 tonne excavator; disposal off site							
depth 300 mm	–	0.13	2.59	0.60	2.96	m	**6.15**
depth 400 mm	–	0.14	2.96	0.69	3.96	m	**7.61**
Disposal							
Excavated material; off site; to tip							
grab loaded 7.25 m³	–	–	–	–	–	m³	**1.97**
Disposal by skip; 8 yd³ (6 m³)							
by machine	–	–	–	41.39	–	m³	**41.39**
by hand	–	4.00	83.00	43.20	–	m³	**126.20**
Foundations; to kerbs, edgings or channels; in situ concrete; 21 N/mm² – 20 mm aggregate (1:2:4) site mixed; one side against earth face, other against formwork (not included); site mixed concrete							
Site mixed concrete							
150 wide × 100 mm deep	–	0.13	2.77	–	1.84	m	**4.61**
150 wide × 150 mm deep	–	0.17	3.46	–	2.76	m	**6.22**
200 wide × 150 mm deep	–	0.20	4.15	–	3.68	m	**7.83**
300 wide × 150 mm deep	–	0.23	4.84	–	5.52	m	**10.36**
600 wide × 200 mm deep	–	0.29	5.93	–	14.73	m	**20.66**
Ready mixed concrete							
150 wide × 100 mm deep	–	0.13	2.77	–	1.25	m	**4.02**
150 wide × 150 mm deep	–	0.17	3.46	–	1.88	m	**5.34**
200 wide × 150 mm deep	–	0.20	4.15	–	2.51	m	**6.66**
300 wide × 150 mm deep	–	0.23	4.84	–	3.76	m	**8.60**
600 wide × 200 mm deep	–	0.29	5.93	–	10.03	m	**15.96**
Formwork; sides of foundations (this will usually be required to one side of each kerb foundation adjacent to road subbases)							
100 mm deep	–	0.04	0.86	–	1.20	m	**2.06**
150 mm deep	–	0.04	0.86	–	1.80	m	**2.66**

Q10 KERBS/EDGINGS/CHANNELS/PAVING ACCESSORIES

Item Excluding site overheads and profit	PC £	Labour hours	Labour £	Plant £	Material £	Unit	Total rate £
KERBS							
Marshalls Mono; small element precast concrete kerb system; Keykerb Large (KL) upstand of 100–125 mm; on 150 mm concrete foundation including haunching with in situ concrete 1:3:6 1 side							
Bullnosed or half battered; 100 × 127 × 200 mm							
laid straight	13.57	0.80	16.60	–	14.55	m	**31.15**
radial blocks laid to curve; 8 blocks/1/4 circle – 500 mm radius	10.45	0.80	16.60	–	11.43	m	**28.03**
radial blocks laid to curve; 8 radial blocks; alternating 8 standard blocks/1/4 circle – 1000 mm radius	17.23	1.25	25.94	–	18.21	m	**44.15**
radial blocks; alternating 16 standard blocks/1/4 circle – 1500 mm radius	15.85	1.50	31.13	–	16.83	m	**47.96**
internal angle 90°	5.72	0.20	4.15	–	5.72	nr	**9.87**
external angle	5.72	0.20	4.15	–	5.72	nr	**9.87**
drop crossing kerbs; KL half battered to KL splay; LH and RH	11.98	1.00	20.75	–	12.96	pair	**33.71**
drop crossing kerbs; KL half battered to KS bullnosed	11.67	1.00	20.75	–	12.65	pair	**33.40**
Marshalls Mono; small element precast concrete kerb system; Keykerb Small (KS) upstand of 25–50 mm; on 150 mm concrete foundation including haunching with in situ concrete 1:3:6 1 side							
Half battered							
laid straight	9.53	0.80	16.60	–	10.51	m	**27.11**
radial blocks laid to curve; 8 blocks/1/4 circle – 500 mm radius	7.84	1.00	20.75	–	8.82	m	**29.57**
radial blocks laid to curve; 8 radial blocks; alternating 8 standard blocks/1/4 circle – 1000 mm radius	12.61	1.25	25.94	–	13.59	m	**39.53**
radial blocks; alternating 16 standard blocks/1/4 circle – 1500 mm radius	13.49	1.50	31.13	–	14.48	m	**45.61**
internal angle 90°	5.72	0.20	4.15	–	5.72	nr	**9.87**
external angle	5.72	0.20	4.15	–	5.72	nr	**9.87**
EDGINGS							
Precast concrete edging units; including haunching with in situ concrete; 11.50 N/mm² – 40 mm aggregate both sides							
Edgings; rectangular, bullnosed or chamfered							
50 × 150 mm	1.57	0.33	6.92	–	6.25	m	**13.17**
125 × 150 mm bullnosed	1.71	0.33	6.92	–	6.40	m	**13.32**
50 × 200 mm	2.34	0.33	6.92	–	7.03	m	**13.95**
50 × 250 mm	2.72	0.33	6.92	–	7.41	m	**14.33**
50 × 250 mm flat top	3.35	0.33	6.92	–	8.04	m	**14.96**

Q10 KERBS/EDGINGS/CHANNELS/PAVING ACCESSORIES

Item Excluding site overheads and profit	PC £	Labour hours	Labour £	Plant £	Material £	Unit	Total rate £
EDGINGS – cont							
Second-hand granite setts; 100 × 100 mm; **bedding in cement mortar (1:4); on 150 mm** **deep concrete foundations; including** **haunching with in situ concrete;** **11.50 N/mm² – 40 mm aggregate one side**							
Edgings							
300 mm wide	13.24	1.33	27.67	–	18.00	m	**45.67**
Brick or block stretchers; bedding in **cement mortar (1:4); on 150 mm deep** **concrete foundations, including haunching** **with in situ concrete; 11.50 N/mm² – 40 mm** **aggregate one side**							
Single course							
concrete paving blocks; PC £9.05/m²;							
200 × 100 × 60 mm	0.91	0.31	6.38	–	4.06	m	**10.44**
engineering bricks; PC £290.00/1000;							
215 × 102.5 × 65 mm	1.35	0.40	8.30	–	4.51	m	**12.81**
paving bricks; PC £500.00/1000;							
215 × 102.5 × 65 mm	2.22	0.40	8.30	–	5.38	m	**13.68**
Two courses							
concrete paving blocks; PC £9.05/m²;							
200 × 100 × 60 mm	1.81	0.40	8.30	–	5.54	m	**13.84**
engineering bricks; PC £290.00/1000;							
215 × 102.5 × 65 mm	2.70	0.57	11.86	–	6.43	m	**18.29**
paving bricks; PC £500.00/1000;							
215 × 102.5 × 65 mm	4.44	0.57	11.86	–	8.17	m	**20.03**
Three courses							
concrete paving blocks; PC £9.05/m²;							
200 × 100 × 60 mm	2.71	0.44	9.22	–	6.44	m	**15.66**
engineering bricks; PC £290.00/1000;							
215 × 102.5 × 65 mm	4.06	0.67	13.83	–	7.78	m	**21.61**
paving bricks; PC £500.00/1000;							
215 × 102.5 × 65 mm	6.67	0.67	13.83	–	10.39	m	**24.22**
Brick or block headers; brick on flat; **bedding in cement mortar (1:4); on 150 mm** **deep concrete foundations; including** **haunching with in situ concrete;** **11.50 N/mm² – 40 mm aggregate one side**							
Single course							
concrete paving blocks; PC £9.05/m²;							
200 × 100 × 60 mm	1.81	0.62	12.77	–	4.97	m	**17.74**
engineering bricks; PC £290.00/1000;							
215 × 102.5 × 65 mm	2.70	0.80	16.60	–	5.86	m	**22.46**
paving bricks; PC £500.00/1000;							
215 × 102.5 × 65 mm	4.44	0.80	16.60	–	7.60	m	**24.20**

Q10 KERBS/EDGINGS/CHANNELS/PAVING ACCESSORIES

Item Excluding site overheads and profit	PC £	Labour hours	Labour £	Plant £	Material £	Unit	Total rate £
Bricks on edge; bedding in cement mortar (1:4); on 150 mm deep concrete foundations; including haunching with in situ concrete; 11.50 N/mm^2 – 40 mm aggregate one side							
One brick wide							
engineering bricks; 215 × 102.5 × 65 mm	4.06	0.57	11.86	–	7.92	m	**19.78**
paving bricks; 215 × 102.5 × 65 mm	6.67	0.57	11.86	–	10.53	m	**22.39**
Two courses; stretchers laid on edge; 225 mm wide							
engineering bricks; 215 × 102.5 × 65 mm	8.11	1.20	24.90	–	13.24	m	**38.14**
paving bricks; 215 × 102.5 × 65 mm	13.33	1.20	24.90	–	18.45	m	**43.35**
Extra over bricks on edge for standard kerbs to one side; haunching in concrete							
125 × 255 mm; ref HB2; SP	4.01	0.44	9.22	–	6.83	m	**16.05**
New granite sett edgings; CED Ltd; 100 × 100 × 100 mm; bedding in cement mortar (1:4); including haunching with in situ concrete; 11.50 N/mm^2 – 40 mm aggregate one side; concrete foundations (not included)							
Edgings 100 × 100 × 100 mm; ref 1R 'better quality' with 20 mm pointing gaps							
100 mm wide	3.31	1.33	27.60	–	6.23	m	**33.83**
2 rows; 220 mm wide	6.61	2.22	46.06	–	10.39	m	**56.45**
3 rows; 340 mm wide	9.92	3.00	62.25	–	12.89	m	**75.14**
Edgings 100 × 100 × 200 mm; ref 1R 'better quality' with 20 mm pointing gaps							
100 mm wide	3.38	0.75	15.56	–	6.29	m	**21.85**
2 rows; 220 mm wide	6.75	1.50	31.13	–	10.53	m	**41.66**
3 rows; 340 mm wide	10.13	2.25	46.69	–	13.09	m	**59.78**
Edgings 100 × 100 × 100 mm; ref 8R 'standard quality' with 20 mm pointing gaps							
100 mm wide	2.89	1.33	27.60	–	5.81	m	**33.41**
2 rows; 220 mm wide	5.77	2.22	46.06	–	9.55	m	**55.61**
3 rows; 340 mm wide	8.66	3.00	62.25	–	11.63	m	**73.88**
Edgings 100 × 100 × 200 mm; ref 8R 'standard quality' with 20 mm pointing gaps							
100 mm wide	2.99	0.75	15.56	–	5.90	m	**21.46**
2 rows; 220 mm wide	5.97	1.50	31.13	–	9.75	m	**40.88**
3 rows; 340 mm wide	8.96	2.25	46.69	–	11.93	m	**58.62**

Q10 KERBS/EDGINGS/CHANNELS/PAVING ACCESSORIES

Item Excluding site overheads and profit	PC £	Labour hours	Labour £	Plant £	Material £	Unit	Total rate £
EDGINGS – cont							
Reclaimed granite setts edgings; CED Ltd;							
100 × 100 mm; bedding in cement mortar							
(1:4); including haunching with in situ							
concrete; 11.50 N/mm² – 40 mm aggregate							
one side; on concrete foundations (not							
included)							
Edgings 100 × 100 × 100 mm; with 20 mm							
pointing gaps							
100 mm wide	3.94	1.33	27.60	–	6.86	m	**34.46**
2 rows; 220 mm wide	7.88	2.22	46.06	–	11.66	m	**57.72**
3 rows; 340 mm wide	11.83	3.00	62.25	–	14.80	m	**77.05**
METAL EDGINGS							
Mild steel edgings; 6 mm galvanized steel;							
Gardenlink Ltd							
Bespoke edgings; complete with bolts for							
adjoining sections and mild steel pins at							
1.50 m centres							
100 mm straight	–	–	–	–	–	m	**17.71**
150 mm straight	–	–	–	–	–	m	**20.58**
300 mm straight	–	–	–	–	–	m	**30.36**
Permaloc AshphaltEdge; Kinley Systems;							
extruded aluminium alloy L shaped edging							
with 5.33 mm exposed upper lip; edging							
fixed to roadway base and edge profile							
with 250 mm steel fixing spike; laid to							
straight or curvilinear road edge;							
subsequently filled with macadam (not							
included)							
Depth of macadam							
38 mm	7.52	0.02	0.35	–	7.52	m	**7.87**
51 mm	8.48	0.02	0.36	–	8.48	m	**8.84**
64 mm	9.37	0.02	0.38	–	9.37	m	**9.75**
76 mm	11.33	0.02	0.41	–	11.33	m	**11.74**
102 mm	13.18	0.02	0.43	–	13.18	m	**13.61**
Permaloc Cleanline; Kinley Systems; heavy							
duty straight profile edging; for edgings to							
soft landscape beds or turf areas;							
3.2 mm × 102 high; 3.2 mm thick with							
4.75 mm exposed upper lip; fixed to form							
straight or curvilinear edge with 305 mm							
fixing spike							
Milled aluminium							
100 mm deep	8.62	0.02	0.38	–	8.62	m	**9.00**
Black							
100 mm deep	8.62	0.02	0.38	–	8.62	m	**9.00**

Q10 KERBS/EDGINGS/CHANNELS/PAVING ACCESSORIES

Item Excluding site overheads and profit	PC £	Labour hours	Labour £	Plant £	Material £	Unit	Total rate £
Permaloc Permastrip; Kinley Systems; **heavy duty L shaped profile maintenance** **strip; 3.2 mm × 89 mm high with 5.2 mm** **exposed top lip; for straight or gentle** **curves on paths or bed turf interfaces;** **fixed to form straight or curvilinear edge** **with standard 305 mm stake; other stake** **lengths available**							
Milled aluminium							
89 mm deep	8.62	0.02	0.38	–	8.62	m	**9.00**
Black							
89 mm deep	8.62	0.02	0.38	–	8.62	m	**9.00**
Permaloc Proline; Kinley Systems; medium **duty straight profiled maintenance strip;** **3.2 mm × 102 mm high with 3.18 mm** **exposed top lip; for straight or gentle** **curves on paths or bed turf interfaces;** **fixed to form straight or curvilinear edge** **with standard 305 mm stake; other stake** **lengths available**							
Milled aluminium							
89 mm deep	7.50	0.02	0.38	–	7.50	m	**7.88**
Black							
89 mm deep	7.50	0.02	0.38	–	7.50	m	**7.88**
Exceledge SuperEdge; recycled vinyl **edging; black; 3 × 150 mm high with 5 mm** **diameter bead on each edge; for edging to** **soft landscape turf or bedding areas;** **formed by heat to required angle, or fixed** **to form straight or curvilinear lines using** **350 mm galvanized fixing stakes**							
150 mm deep	5.83	0.02	0.38	–	5.83	m	**6.21**
Exceledge BrickEdge; recycled vinyl hard **landscape edging former; black; 50 mm** **high with 76 mm return; 3 mm thick wall for** **the retention of bricks, blocks and loose or** **bonded gravels; fixed with steel fixing** **spikes at 300 mm centres**							
Flexible; 50 mm deep	6.35	0.02	0.38	–	6.35	m	**6.73**
Rigid; 50 mm deep	6.35	0.02	0.38	–	6.35	m	**6.73**

Q10 KERBS/EDGINGS/CHANNELS/PAVING ACCESSORIES

Item Excluding site overheads and profit	PC £	Labour hours	Labour £	Plant £	Material £	Unit	Total rate £
METAL EDGINGS – cont							
Exceledge EdgeKing; polyethylene flexible edging; black; used for turfed areas and retention of loose fill materials; with 25 mm diameter rounded hose top for bold demarcation and safety in play areas; fixed with steel vee-stakes at 45° angle into adjacent material or to prefixed timber supports using screws and washers							
125 mm deep	4.37	0.02	0.38	–	4.37	m	**4.75**
Corner pieces	8.98	–	–	–	8.98	each	**8.98**
Joiner plugs	0.79	–	–	–	0.79	each	**0.79**
Exceledge Borderline; tree rings; preformed unit from 100 mm high rolled steel in galvanized or powder coated (black or brown) finish; for circular traditional lawn edging around trees, bushes, flag poles or part of a feature; fixed with flat steel matching stakes							
700 mm	20.50	0.50	10.38	–	20.50	each	**30.88**
900 mm	28.69	0.55	11.41	–	28.69	each	**40.10**
1200 mm	32.70	0.60	12.45	–	32.70	each	**45.15**
1500 mm	37.34	0.65	13.49	–	37.34	each	**50.83**
1800 mm	41.18	0.70	14.53	–	41.18	each	**55.71**
CHANNELS							
Channels; bedding in cement mortar (1:3); joints pointed flush; on concrete foundations (not included)							
Three courses stretchers; 350 mm wide; quarter bond to form dished channels							
engineering bricks; PC £300.00/1000;							
215 × 102.5 × 65 mm	4.06	1.00	20.75	–	5.33	m	**26.08**
paving bricks; PC £500.00/1000;							
215 × 102.5 × 65 mm	6.67	1.00	20.75	–	7.94	m	**28.69**
Three courses granite setts; 340 mm wide; to form dished channels							
340 mm wide	13.24	2.00	41.50	–	16.15	m	**57.65**

Q20 GRANULAR SUBBASES TO ROADS/PAVINGS

Item Excluding site overheads and profit	PC £	Labour hours	Labour £	Plant £	Material £	Unit	Total rate £
Q20 GRANULAR SUBBASES TO ROADS/ **PAVINGS**							
BASES FOR PAVINGS							
Hardcore bases; obtained off site; **PC £16.20/m³**							
By machine							
100 mm thick	1.62	0.05	1.04	2.02	1.62	m²	**4.68**
150 mm thick	2.43	0.07	1.38	2.61	2.43	m²	**6.42**
over 150 thick	16.20	0.50	10.38	20.17	16.20	m³	**46.75**
By hand							
100 mm thick	1.62	0.20	4.15	0.23	1.62	m²	**6.00**
150 mm thick	2.92	0.30	6.23	0.23	2.92	m²	**9.38**
Hardcore; difference for each £1.00 increase/ decrease in PC price per m³; price will vary with type and source of hardcore							
average 75 mm thick	–	–	–	–	0.08	m²	**0.08**
average 100 mm thick	–	–	–	–	0.10	m²	**0.10**
average 150 mm thick	–	–	–	–	0.15	m²	**0.15**
average 200 mm thick	–	–	–	–	0.20	m²	**0.20**
average 250 mm thick	–	–	–	–	0.25	m²	**0.25**
average 300 mm thick	–	–	–	–	0.30	m²	**0.30**
exceeding 300 mm thick	–	–	–	–	1.10	m³	**1.10**
Type 1 granular fill base; PC £21.50/tonne **(£47.30/m³ compacted)**							
By machine							
100 mm thick	4.73	0.03	0.58	0.52	4.73	m²	**5.83**
150 mm thick	7.09	0.03	0.52	0.71	7.09	m²	**8.32**
By hand (mechanical compaction)							
100 mm thick	4.73	0.17	3.46	0.08	4.73	m²	**8.27**
150 mm thick	7.09	0.25	5.19	0.12	7.09	m²	**12.40**
SURFACE TREATMENTS AND BLINDINGS							
Surface treatments							
Sand blinding; to hardcore base (not included); 25 mm thick	0.86	0.03	0.69	–	0.86	m²	**1.55**
Sand blinding; to hardcore base (not included); 50 mm thick	1.72	0.05	1.04	–	1.72	m²	**2.76**
Filter fabrics; to hardcore base (not included)	0.47	0.01	0.21	–	0.47	m²	**0.68**

Q21 IN SITU CONCRETE ROADS/PAVINGS

Item Excluding site overheads and profit	PC £	Labour hours	Labour £	Plant £	Material £	Unit	Total rate £
Q21 IN SITU CONCRETE ROADS/PAVINGS							
FORMWORK FOR CONCRETE ROADS AND PAVINGS							
Formwork for in situ concrete							
Sides of foundations							
height not exceeding 250 mm	3.00	0.05	1.04	–	3.29	m	**4.33**
Extra over formwork for curved work 6 m radius	–	0.25	5.19	–	–	m	**5.19**
Steel road forms; to edges of beds or faces of foundations							
150 mm wide	–	0.20	4.15	1.40	–	m	**5.55**
REINFORCEMENT							
Reinforcement; fabric; side laps 150 mm; head laps 300 mm; mesh 200 × 200 mm; in roads, footpaths or pavings							
Fabric							
A142 (2.22 kg/m^2)	2.90	0.08	1.72	–	2.90	m^2	**4.62**
A193 (3.02 kg/m^2)	0.38	0.08	1.72	–	0.38	m^2	**2.10**
IN SITU CONCRETE ROADS							
Unreinforced concrete; on prepared subbase (not included)							
Paths and roads; 21.00 N/mm^2 – 20 mm aggregate (1:2:4) mechanically mixed on site							
100 mm thick	12.89	0.13	2.59	–	12.89	m^2	**15.48**
150 mm thick	19.33	0.17	3.46	–	19.33	m^2	**22.79**
Reinforced in situ concrete; mechanically mixed on site; normal Portland cement; on hardcore base (not included); reinforcement (not included)							
Roads; 11.50 N/mm^2 – 40 mm aggregate (1:3:6)							
100 mm thick	9.61	0.40	8.30	–	9.61	m^2	**17.91**
150 mm thick	14.42	0.60	12.45	–	14.42	m^2	**26.87**
200 mm thick	19.22	0.80	16.60	–	19.22	m^2	**35.82**
Roads; 21.00 N/mm^2 – 20 mm aggregate (1:2:4)							
100 mm thick	12.58	0.40	8.30	–	12.58	m^2	**20.88**
150 mm thick	18.88	0.60	12.45	–	18.88	m^2	**31.33**
200 mm thick	25.16	0.80	16.60	–	25.16	m^2	**41.76**
Roads; 25.00 N/mm^2 – 20 mm aggregate GEN 4 ready mixed							
100 mm thick	8.75	0.40	8.30	–	8.75	m^2	**17.05**
150 mm thick	13.13	0.60	12.45	–	13.13	m^2	**25.58**
200 mm thick	17.50	0.80	16.60	–	17.50	m^2	**34.10**
250 mm thick	21.87	1.00	20.75	0.31	21.87	m^2	**42.93**

Q21 IN SITU CONCRETE ROADS/PAVINGS

Item Excluding site overheads and profit	PC £	Labour hours	Labour £	Plant £	Material £	Unit	Total rate £
Reinforced in situ concrete; ready mixed;							
discharged directly into location from							
supply lorry; normal Portland cement; on							
hardcore base (not included);							
reinforcement (not included)							
Roads; 11.50 N/mm² – 40 mm aggregate							
(1:3:6)							
100 mm thick	8.35	0.16	3.32	–	8.35	m²	**11.67**
150 mm thick	12.52	0.24	4.98	–	12.52	m²	**17.50**
200 mm thick	16.70	0.36	7.47	–	16.70	m²	**24.17**
250 mm thick	20.87	0.54	11.21	0.31	20.87	m²	**32.39**
300 mm thick	25.05	0.66	13.70	0.31	25.05	m²	**39.06**
Roads; 21.00 N/mm² – 20 mm aggregate							
(1:2:4)							
100 mm thick	8.62	0.16	3.32	–	8.62	m²	**11.94**
150 mm thick	12.94	0.24	4.98	–	12.94	m²	**17.92**
200 mm thick	17.25	0.36	7.47	–	17.25	m²	**24.72**
250 mm thick	21.56	0.54	11.21	0.31	21.56	m²	**33.08**
300 mm thick	25.87	0.66	13.70	0.31	25.87	m²	**39.88**
Roads; 26.00 N/mm² – 20 mm aggregate							
(1:1:5:3)							
100 mm thick	8.76	0.16	3.32	–	8.76	m²	**12.08**
150 mm thick	13.13	0.24	4.98	–	13.13	m²	**18.11**
200 mm thick	17.51	0.36	7.47	–	17.51	m²	**24.98**
250 mm thick	21.89	0.54	11.21	0.31	21.89	m²	**33.41**
300 mm thick	26.27	0.66	13.70	0.31	26.27	m²	**40.28**
Concrete sundries							
Treating surfaces of unset concrete; grading to							
cambers; tamping with 75 mm thick steel shod							
tamper or similar	–	0.13	2.77	–	–	m²	**2.77**
Expansion joints							
13 mm thick joint filler; formwork							
width or depth not exceeding 150 mm	2.29	0.20	4.15	–	4.79	m	**8.94**
width or depth 150–300 mm	2.11	0.25	5.19	–	7.11	m	**12.30**
width or depth 300–450 mm	0.95	0.30	6.22	–	8.45	m	**14.67**
25 mm thick joint filler; formwork							
width or depth not exceeding 150 mm	5.19	0.20	4.15	–	7.69	m	**11.84**
width or depth 150–300 mm	5.19	0.25	5.19	–	10.19	m	**15.38**
width or depth 300–450 mm	5.19	0.30	6.22	–	12.70	m	**18.92**
Sealants; sealing top 25 mm of joint with							
rubberized bituminous compound	1.28	0.25	5.19	–	1.28	m	**6.47**

Q22 COATED MACADAM/ASPHALT ROADS/PAVINGS

Item Excluding site overheads and profit	PC £	Labour hours	Labour £	Plant £	Material £	Unit	Total rate £
Q22 COATED MACADAM/ASPHALT ROADS/PAVINGS							
MACADAM SURFACING							
Coated macadam/asphalt roads/pavings – General							
Preamble: The prices for all in situ finishings to roads and footpaths include for work to falls, crossfalls or slopes not exceeding 15° from horizontal; for laying on prepared bases (not included) and for rolling with an appropriate roller.							
Users should note the new terminology for the surfaces described below which is to European standard descriptions. The now redundant descriptions for each course are shown in brackets.							
Note: The rates below are for smaller areas up to 200 m².							
Macadam surfacing; Spadeoak Construction Co Ltd; surface (wearing) course; 20 mm of 6 mm dense bitumen macadam							
Hand lay							
limestone aggregate	–	–	–	–	–	m²	**11.63**
granite aggregate	–	–	–	–	–	m²	**11.60**
red	–	–	–	–	–	m²	**14.30**
Macadam surfacing; Spadeoak Construction Co Ltd; surface (wearing) course; 30 mm of 10 mm dense bitumen macadam							
Hand lay							
limestone aggregate	–	–	–	–	–	m²	**13.47**
granite aggregate	–	–	–	–	–	m²	**13.64**
red	–	–	–	–	–	m²	**26.19**
Macadam surfacing; Spadeoak Construction Co Ltd; surface (wearing) course; 40 mm of 10 mm dense bitumen macadam							
Hand lay							
limestone aggregate	–	–	–	–	–	m²	**16.67**
granite aggregate	–	–	–	–	–	m²	**16.87**
red	–	–	–	–	–	m²	**29.45**

Q22 COATED MACADAM/ASPHALT ROADS/PAVINGS

Item Excluding site overheads and profit	PC £	Labour hours	Labour £	Plant £	Material £	Unit	Total rate £
Macadam surfacing; Spadeoak **Construction Co Ltd; binder (base) course;** **50 mm of 20 mm dense bitumen macadam** Hand lay							
limestone aggregate	–	–	–	–	–	m²	16.70
granite aggregate	–	–	–	–	–	m²	16.94
Macadam surfacing; Spadeoak **Construction Co Ltd; binder (base) course;** **60 mm of 20 mm dense bitumen macadam** Hand lay							
limestone aggregate	–	–	–	–	–	m²	18.17
granite aggregate	–	–	–	–	–	m²	18.44
Macadam surfacing; Spadeoak **Construction Co Ltd; base (roadbase)** **course; 75 mm of 28 mm dense bitumen** **macadam** Hand lay							
limestone aggregate	–	–	–	–	–	m²	21.89
granite aggregate	–	–	–	–	–	m²	22.26
Areas up to 300 m²							
Macadam surfacing; Spadeoak **Construction Co Ltd; base (roadbase)** **course; 100 mm of 28 mm dense bitumen** **macadam** Hand lay							
limestone aggregate	–	–	–	–	–	m²	52.24
granite aggregate	–	–	–	–	–	m²	27.15
Macadam surfacing; Spadeoak **Construction Co Ltd; base (roadbase)** **course; 150 mm of 28 mm dense bitumen** **macadam in two layers** Hand lay							
limestone aggregate	–	–	–	–	–	m²	42.84
granite aggregate	–	–	–	–	–	m²	43.54
Macadam surfacing; Spadeoak **Construction Co Ltd; Base (roadbase)** **course; 200 mm of 28 mm dense bitumen** **macadam in two layers** Hand lay							
limestone aggregate	–	–	–	–	–	m²	36.24
granite aggregate	–	–	–	–	–	m²	54.38

Q23 GRAVEL/HOGGIN/WOODCHIP ROADS/PAVINGS

Item Excluding site overheads and profit	PC £	Labour hours	Labour £	Plant £	Material £	Unit	Total rate £
RESIN BOUND MACADAM							
Resin bound macadam pavings; hand laid;							
Bituchem; to pedestrian or vehicular hard							
landscape areas; laid to base course (not							
included)							
Natratex clear resin bound macadam							
25 mm thick to pedestrian areas	22.50	0.13	2.59	–	22.50	m²	**25.09**
30 mm thick to vehicular areas	26.25	0.17	3.46	–	26.25	m²	**29.71**
Colourtex coloured resin bound macadam to							
pedestrian or vehicular hard landscape areas							
25 mm thick	22.50	0.13	2.59	–	22.50	m²	**25.09**
RESIN BOUND AGGREGATES							
Bound aggregates; Addagrip Surface							
Treatments UK Ltd; natural decorative							
resin bonded surface dressing laid to							
concrete, macadam or to plywood panels							
priced separately							
Primer coat to macadam or concrete base	–	–	–	–	–	m²	**5.00**
Golden pea gravel 1–3 mm							
buff adhesive	–	–	–	–	–	m²	**25.00**
red adhesive	–	–	–	–	–	m²	**25.00**
green adhesive	–	–	–	–	–	m²	**25.00**
Golden pea gravel 2–5 mm							
buff adhesive	–	–	–	–	–	m²	**28.00**
Chinese bauxite 1–3 mm							
buff adhesive	–	–	–	–	–	m²	**25.00**
Cobalt Blue Glass; 6 mm							
15 mm depth	–	–	–	–	–	m²	**85.00**
Midnight Grey, Chocolate, Derbyshire Cream,							
All Gold; 6 mm							
15 mm depth	–	–	–	–	–	m²	**50.00**
18 mm depth	–	–	–	–	–	m²	**55.00**
Q23 GRAVEL/HOGGIN/WOODCHIP ROADS/							
PAVINGS							
PATH PREPARATION							
Excavation and path preparation;							
excavating; 300 mm deep; to width of path;							
depositing excavated material at sides of							
excavation							
By machine							
width 1.00 m	–	–	–	4.76	–	m²	**4.76**
width 1.50 m	–	–	–	3.96	–	m²	**3.96**
width 2.00 m	–	–	–	3.40	–	m²	**3.40**
width 3.00 m	–	–	–	2.85	–	m²	**2.85**

Q23 GRAVEL/HOGGIN/WOODCHIP ROADS/PAVINGS

Item Excluding site overheads and profit	PC £	Labour hours	Labour £	Plant £	Material £	Unit	Total rate £
By hand							
width 1.00 m	–	0.70	14.61	–	–	m²	**14.61**
width 1.50 m	–	1.05	21.79	–	–	m²	**21.79**
width 2.00 m	–	1.40	29.05	–	–	m²	**29.05**
width 3.00 m	–	2.10	43.58	–	–	m²	**43.58**
Excavating trenches; in centre of							
pathways; 100 mm flexible drain pipes;							
filling with clean broken stone or gravel							
rejects							
300 × 450 mm deep							
by machine	7.28	0.10	2.08	1.78	7.28	m	**11.14**
by hand	7.28	0.40	8.30	–	7.28	m	**15.58**
Hand trimming and compacting reduced							
surface of pathway; by vibrating roller							
width 1.00 m	–	0.05	1.04	0.23	–	m	**1.27**
width 1.50 m	–	0.04	0.92	0.21	–	m	**1.13**
width 2.00 m	–	0.04	0.83	0.19	–	m	**1.02**
width 3.00 m	–	0.04	0.83	0.19	–	m	**1.02**
Permeable membranes; to trimmed and							
compacted surface of pathway							
Terram 1000	0.47	0.02	0.41	–	0.47	m²	**0.88**
Filling to make up levels							
Obtained off site; hardcore; PC £16.20/m³							
150 mm thick	2.43	0.07	1.38	2.61	2.43	m²	**6.42**
Obtained off site; granular fill type 1; PC							
£21.50/tonne (£47.3/m³ compacted)							
100 mm thick	4.73	0.03	0.58	0.52	4.73	m²	**5.83**
150 mm thick	7.09	0.03	0.52	0.71	7.09	m²	**8.32**
Surface treatments							
Sand blinding; to hardcore (not included)							
50 mm thick	1.89	0.04	0.83	–	1.89	m²	**2.72**
Filter fabric; to hardcore (not included)	0.47	0.01	0.21	–	0.47	m²	**0.68**
EDGINGS TO PATHWAYS							
Permaloc AshphaltEdge; Kinley Systems;							
extruded aluminium alloy L shaped edging							
with 5.33 mm exposed upper lip; edging							
fixed to roadway base and edge profile							
with 250 mm steel fixing spike; laid to							
straight or curvilinear road edge;							
subsequently filled with macadam (not							
included)							
Depth of macadam							
38 mm	7.52	0.02	0.35	–	7.52	m	**7.87**
51 mm	8.48	0.02	0.36	–	8.48	m	**8.84**
64 mm	9.37	0.02	0.38	–	9.37	m	**9.75**
76 mm	11.33	0.02	0.41	–	11.33	m	**11.74**
102 mm	13.18	0.02	0.43	–	13.18	m	**13.61**

Q23 GRAVEL/HOGGIN/WOODCHIP ROADS/PAVINGS

Item Excluding site overheads and profit	PC £	Labour hours	Labour £	Plant £	Material £	Unit	Total rate £
EDGINGS TO PATHWAYS – cont							
Permaloc Cleanline; Kinley Systems; heavy							
duty straight profile edging; for edgings to							
soft landscape beds or turf areas;							
3.2 mm × 102 high; 3.2 mm thick with							
4.75 mm exposed upper lip; fixed to form							
straight or curvilinear edge with 305 mm							
fixing spike							
Milled aluminium							
100 mm deep	8.62	0.02	0.38	–	8.62	m	**9.00**
Black							
100 mm deep	8.62	0.02	0.38	–	8.62	m	**9.00**
Permaloc Permastrip; Kinley Systems;							
heavy duty L shaped profile maintenance							
strip; 3.2 mm × 89 mm high with 5.2 mm							
exposed top lip; for straight or gentle							
curves on paths or bed turf interfaces;							
fixed to form straight or curvilinear edge							
with standard 305 mm stake; other stake							
lengths available							
Milled aluminium							
89 mm deep	8.62	0.02	0.38	–	8.62	m	**9.00**
Black							
89 mm deep	8.62	0.02	0.38	–	8.62	m	**9.00**
Permaloc Proline; Kinley Systems; medium							
duty straight profiled maintenance strip;							
3.2 mm × 102 mm high with 3.18 mm							
exposed top lip; for straight or gentle							
curves on paths or bed turf interfaces;							
fixed to form straight or curvilinear edge							
with standard 305 mm stake; other stake							
lengths available							
Milled aluminium							
89 mm deep	7.50	0.02	0.38	–	7.50	m	**7.88**
Black							
89 mm deep	7.50	0.02	0.38	–	7.50	m	**7.88**
Timber edging boards							
Boards; 50 × 50 × 750 mm timber pegs at							
1000 mm centres (excavations and hardcore							
under edgings not included)							
Straight							
38 × 150 mm treated softwood edge boards	2.54	0.10	2.08	–	2.54	m	**4.62**
50 × 150 mm treated softwood edge boards	2.75	0.10	2.08	–	2.75	m	**4.83**
Curved							
38 × 150 mm treated softwood edge boards	2.54	0.20	4.15	–	2.54	m	**6.69**
50 × 150 mm treated softwood edge boards	2.75	0.25	5.19	–	2.75	m	**7.94**

Q23 GRAVEL/HOGGIN/WOODCHIP ROADS/PAVINGS

Item Excluding site overheads and profit	PC £	Labour hours	Labour £	Plant £	Material £	Unit	Total rate £
GRANULAR PAVINGS							
Footpath gravels; porous self-binding gravel							
Breedon Special Aggregates; Golden Gravel or equivalent; rolling wet; on hardcore base (not included); for pavements; to falls and crossfalls and to slopes not exceeding 15° from horizontal; over 300 mm wide							
50 mm thick	11.82	0.05	1.04	0.61	11.82	m²	**13.47**
by hand; 50 mm thick	11.82	0.20	4.15	0.46	11.82	m²	**16.43**
Breedon Special Aggregates; Wayfarer specially formulated fine gravel for use on golf course pathways							
50 mm thick	10.36	0.03	0.59	0.41	10.36	m²	**11.36**
by hand; 50 mm thick	10.36	0.10	2.08	0.27	10.36	m²	**12.71**
Hoggin (stabilized); PC £54.58/m³ on hardcore base (not included); to falls and crossfalls and to slopes not exceeding 15° from horizontal; over 300 mm wide							
100 mm thick	11.91	0.02	0.41	0.35	11.91	m²	**12.67**
by hand; 100 mm thick	11.91	0.10	2.08	0.19	11.91	m²	**14.18**
150 mm thick	17.87	0.03	0.52	0.43	17.87	m²	**18.82**
Ballast; as dug; watering; rolling; on hardcore base (not included)							
100 mm thick	4.38	0.13	2.77	1.16	4.38	m²	**8.31**
150 mm thick	6.56	0.03	0.52	0.39	6.56	m²	**7.47**
CED Ltd; Cedec gravel; self-binding; laid to inert (non-limestone) base measured separately; compacting							
red, gold or silver; 50 mm thick	15.82	0.03	0.59	0.32	15.82	m²	**16.73**
by hand; red, gold or silver; 50 mm thick	15.82	0.07	1.38	0.09	15.82	m²	**17.29**
Grundon Ltd; Coxwell self-binding path gravels laid and compacted to excavation or base measured separately							
50 mm thick	5.69	0.08	1.73	0.19	5.69	m²	**7.61**
Footpath gravels; porous loose gravels							
Washed shingle; on prepared base (not included)							
25–50 mm size; 25 mm thick	1.02	0.02	0.37	0.07	1.02	m²	**1.46**
25–50 mm size; 75 mm thick	3.05	0.05	1.09	0.21	3.05	m²	**4.35**
50–75 mm size; 25 mm thick	1.02	0.02	0.41	0.08	1.02	m²	**1.51**
50–75 mm size; 75 mm thick	3.05	0.07	1.38	0.32	3.05	m²	**4.75**
Pea shingle; on prepared base (not included)							
10–15 mm size; 25 mm thick	1.01	0.02	0.37	0.07	1.01	m²	**1.45**
5–10 mm size; 75 mm thick	3.04	0.05	1.09	0.21	3.04	m²	**4.34**
Breedon Special Aggregates; Breedon Buff decorative limestone chippings							
50 mm thick	5.27	0.01	0.21	0.08	5.27	m²	**5.56**
by hand; 50 mm thick	5.27	0.03	0.69	–	5.27	m²	**5.96**

Q24 INTERLOCKING BRICK/BLOCK ROADS/PAVINGS

Item Excluding site overheads and profit	PC £	Labour hours	Labour £	Plant £	Material £	Unit	Total rate £
GRANULAR PAVINGS – cont							
Footpath gravels – cont							
Breedon Special Aggregates; Brindle or Moorland Black chippings							
50 mm thick	6.40	0.01	0.21	0.08	6.40	m²	**6.69**
by hand; 50 mm thick	6.40	0.03	0.69	–	6.40	m²	**7.09**
Breedon Special Aggregates; slate chippings; plum/blue							
50 mm thick	9.47	0.01	0.21	0.08	9.47	m²	**9.76**
by hand; 50 mm thick	9.47	0.03	0.69	–	9.47	m²	**10.16**
WOODCHIP PATHWAYS							
Wood chip surfaces; Melcourt Industries Ltd							
Wood chips; to surface of pathways by hand; levelling and spreading by hand (excavation and preparation not included) (items labelled FSC are Forest Stewardship Council certified)							
Walk Chips; FSC; 100 mm thick (25 m³ loads)	4.06	0.02	0.35	–	4.06	m²	**4.41**
Woodfibre; FSC; 100 mm thick	2.51	0.05	1.04	–	2.51	m²	**3.55**
Q24 INTERLOCKING BRICK/BLOCK ROADS/PAVINGS							
BLOCK PAVINGS							
Precast concrete block edgings; PC £8.60/m²; 200 × 100 × 60 mm; on prepared base (not included); haunching one side							
Edgings; butt joints							
stretcher course	0.86	0.17	3.46	–	3.47	m	**6.93**
header course	1.72	0.27	5.53	–	4.51	m	**10.04**
Precast concrete vehicular paving blocks; Marshalls Plc; on prepared base (not included); on 50 mm compacted sharp sand bed; blocks laid in 7 mm loose sand and vibrated; joints filled with sharp sand and vibrated; level and to falls only							
Trafica paving blocks; 450 × 450 × 70 mm							
Perfecta finish; colour natural	29.97	0.50	10.38	0.14	31.86	m²	**42.38**
Perfecta finish; colour buff	34.61	0.50	10.38	0.14	36.51	m²	**47.03**
Saxon finish; colour natural	26.28	0.50	10.38	0.14	28.17	m²	**38.69**
Saxon finish; colour buff	30.27	0.50	10.38	0.14	32.16	m²	**42.68**

Q24 INTERLOCKING BRICK/BLOCK ROADS/PAVINGS

Item Excluding site overheads and profit	PC £	Labour hours	Labour £	Plant £	Material £	Unit	Total rate £
Precast concrete vehicular paving blocks;							
Keyblock; Marshalls Plc; on prepared base							
(not included); on 50 mm compacted sharp							
sand bed; blocks laid in 7 mm loose sand							
and vibrated; joints filled with sharp sand							
and vibrated; level and to falls only							
Herringbone bond							
200 × 100 × 60 mm; natural grey	9.03	1.50	31.12	0.14	10.99	m²	**42.25**
200 × 100 × 60 mm; colours	9.50	1.50	31.12	0.14	11.47	m²	**42.73**
200 × 100 × 80 mm; natural grey	10.05	1.50	31.12	0.14	12.01	m²	**43.27**
200 × 100 × 80 mm; colours	11.35	1.50	31.12	0.14	13.31	m²	**44.57**
Basketweave bond							
200 × 100 × 60 mm; natural grey	9.03	1.20	24.90	0.14	10.99	m²	**36.03**
200 × 100 × 60 mm; colours	9.50	1.20	24.90	0.14	11.47	m²	**36.51**
200 × 100 × 80 mm; natural grey	10.05	1.20	24.90	0.14	12.01	m²	**37.05**
200 × 100 × 80 mm; colours	11.35	1.20	24.90	0.14	13.31	m²	**38.35**
Precast concrete vehicular paving blocks;							
Charcon Hard Landscaping; on prepared							
base (not included); on 50 mm compacted							
sharp sand bed; blocks laid in 7 mm loose							
sand and vibrated; joints filled with sharp							
sand and vibrated; level and to falls only							
Europa concrete blocks							
200 × 100 × 60 mm; natural grey	14.63	1.50	31.12	0.14	16.59	m²	**47.85**
200 × 100 × 60 mm; colours	15.66	1.50	31.12	0.14	17.62	m²	**48.88**
Parliament concrete blocks							
200 × 100 × 65 mm; natural grey	34.54	1.50	31.12	0.14	36.51	m²	**67.77**
200 × 100 × 65 mm; colours	34.54	1.50	31.12	0.14	36.51	m²	**67.77**
POLYETHYLENE GRASSBLOCKS							
Recycled polyethylene grassblocks;							
Boddingtons Grass Reinforcement							
Solutions; Boddingtons Ltd; interlocking							
units laid to prepared base or rootzone (not							
included)							
BodPave85; load bearing <400 tonnes per m²;							
500 × 500 × 50 mm deep; 35 mm ground spike							
1–50 m²	18.15	0.20	4.15	0.19	21.41	m²	**25.75**
51–500 m²	15.13	0.20	4.15	0.19	18.39	m²	**22.73**
GrassProtecta; extruded expanded							
polyethylene flexible mesh laid to existing							
grass surface or newly seeded areas to							
provide heavy surface protection from traffic							
and pedestrians							
Standard; 1.2 kg/m²; 20 × 2 m; up to 320 m²	8.25	0.01	0.17	–	8.25	m²	**8.42**
Standard; 1.2 kg/m²; 20 × 2 m; 321 m² to							
3000 m²	7.15	0.01	0.17	–	7.15	m²	**7.32**
Heavy; 2 kg/m²; 20 × 2 m; up to 320 m²	9.35	0.01	0.17	–	9.35	m²	**9.52**
Heavy; 2 kg/m²; 20 × 2 m; 321 m² to							
3000 m²	8.80	0.01	0.17	–	8.80	m²	**8.97**

Q25 SLAB/BRICK/SETT/COBBLE PAVINGS

Item Excluding site overheads and profit	PC £	Labour hours	Labour £	Plant £	Material £	Unit	Total rate £
POLYETHYLENE GRASSBLOCKS – cont							
Recycled polyethylene grassblocks – cont							
TurfProtecta; extruded polyethylene flexible mesh laid to existing grass surface or newly seeded areas to provide surface protection from traffic including vehicle or animal wear and tear							
Standard; 30 m × 2 m; up to 300 m^2	3.19	0.01	0.17	–	3.19	m^2	**3.36**
Standard; 30 m × 2 m; 301 m^2 to 600 m^2	2.65	0.01	0.17	–	2.65	m^2	**2.82**
Premium; 30 m × 2 m; up to 300 m^2	3.29	0.01	0.17	–	3.29	m^2	**3.46**
Premium; 30 m × 2 m; 301 m^2 to 600 m^2	2.83	0.01	0.17	–	2.83	m^2	**3.00**
Q25 SLAB/BRICK/SETT/COBBLE PAVINGS							
EDGINGS TO PAVINGS							
Bricks – General							
Preamble: Bricks shall be hard, well burnt, non-dusting, resistant to frost and sulphate attack and true to shape, size and sample.							
Movement of materials							
Mechanically offloading bricks; loading wheelbarrows; transporting maximum 25 m distance	–	0.20	4.15	–	–	m^2	**4.15**
Edge restraints; to brick paving; on prepared base (not included); 65 mm thick bricks; PC £300.00/1000; haunching one side							
Header course							
200 × 100 mm; butt joints	3.00	0.27	5.53	–	5.79	m	**11.32**
210 × 105 mm; mortar joints	2.67	0.50	10.38	–	5.78	m	**16.16**
Stretcher course							
200 × 100 mm; butt joints	1.50	0.17	3.46	–	4.11	m	**7.57**
210 × 105 mm; mortar joints	1.36	0.33	6.92	–	4.13	m	**11.05**
Variation in brick prices; add or subtract the following amounts for every £1.00/1000 difference in the PC price							
Edgings							
100 mm wide	–	–	–	–	0.05	10 m	**0.05**
200 mm wide	–	–	–	–	0.11	10 m	**0.11**
102.5 mm wide	–	–	–	–	0.05	10 m	**0.05**
215 mm wide	–	–	–	–	0.09	10 m	**0.09**

Q25 SLAB/BRICK/SETT/COBBLE PAVINGS

Item Excluding site overheads and profit	PC £	Labour hours	Labour £	Plant £	Material £	Unit	Total rate £
Edge restraints; to block paving; on prepared base (not included); 200 × 100 × 80 mm; PC £9.57/m²; haunching one side							
Header course							
200 × 100 mm; butt joints	1.91	0.27	5.53	–	4.70	m	**10.23**
Stretcher course							
200 × 100 mm; butt joints	0.96	0.17	3.46	–	3.57	m	**7.03**
BLOCK PAVING							
Concrete paviors; Marshalls Plc; on prepared base (not included); bedding on 50 mm sand; kiln dried sand joints swept in							
Keyblock paviors							
200 × 100 × 60 mm; grey	8.60	0.40	8.30	0.14	10.56	m²	**19.00**
200 × 100 × 60 mm; colours	9.05	0.40	8.30	0.14	11.01	m²	**19.45**
200 × 100 × 80 mm; grey	9.57	0.44	9.22	0.14	11.53	m²	**20.89**
200 × 100 × 80 mm; colours	10.81	0.44	9.22	0.14	12.77	m²	**22.13**
Concrete cobble paviors; Charcon Hard Landscaping; Concrete Products; on prepared base (not included); bedding on 50 mm sand; kiln dried sand joints swept in							
Paviors							
Woburn blocks; 100–201 × 134 × 80 mm; random sizes	36.31	0.67	13.83	0.14	38.27	m²	**52.24**
Woburn blocks; 100–201 × 134 × 80 mm; single size	36.31	0.50	10.38	0.14	38.27	m²	**48.79**
Woburn blocks; 100–201 × 134 × 60 mm; random sizes	29.40	0.67	13.83	0.14	31.36	m²	**45.33**
Woburn blocks; 100–201 × 134 × 60 mm; single size	29.37	0.50	10.38	0.14	31.33	m²	**41.85**
Concrete setts; on 25 mm sand; compacted; vibrated; joints filled with sand; natural or coloured; well rammed hardcore base (not included)							
Marshalls Plc; Tegula cobble paving							
60 mm thick; random sizes	23.14	0.57	11.86	0.14	25.11	m²	**37.11**
60 mm thick; single size	23.14	0.45	9.43	0.14	25.28	m²	**34.85**
80 mm thick; random sizes	26.66	0.57	11.86	0.14	28.80	m²	**40.80**
80 mm thick; single size	26.66	0.45	9.43	0.14	28.80	m²	**38.37**
cobbles 80 × 80 × 60 mm thick; traditional	41.42	0.56	11.53	0.14	43.56	m²	**55.23**
Cobbles							
Charcon Hard Landscaping; country setts							
100 mm thick; random sizes	43.74	1.00	20.75	0.14	45.88	m²	**66.77**
100 mm thick; single size	44.83	0.67	13.83	0.14	46.97	m²	**60.94**

Q25 SLAB/BRICK/SETT/COBBLE PAVINGS

Item Excluding site overheads and profit	PC £	Labour hours	Labour £	Plant £	Material £	Unit	Total rate £
BRICK PAVING							
Clay brick pavings; on prepared base (not included); bedding on 50 mm sharp sand; kiln dried sand joints							
Pavings; 200 × 100 × 65 mm wirecut chamfered paviors							
brick; PC £500.00/1000	25.63	1.97	40.86	0.24	27.76	m²	**68.86**
Clay brick pavings; 200 × 100 × 50 mm; laid to running stretcher or stack bond only; on prepared base (not included); bedding on cement: sand (1:4) pointing mortar as work proceeds							
PC £600.00/1000							
laid on edge	48.81	4.76	98.81	–	56.84	m²	**155.65**
laid on edge but pavior 65 mm thick	41.00	3.81	79.05	–	49.03	m²	**128.08**
laid flat	26.62	2.20	45.65	–	31.41	m²	**77.06**
PC £500.00/1000							
laid on edge	40.67	4.76	98.81	–	48.70	m²	**147.51**
laid on edge but pavior 65 mm thick	34.16	3.81	79.05	–	42.19	m²	**121.24**
laid flat	22.19	2.20	45.65	–	26.97	m²	**72.62**
PC £400.00/1000							
laid on edge	32.54	4.76	98.81	–	40.57	m²	**139.38**
laid on edge but pavior 65 mm thick	27.33	3.81	79.05	–	35.36	m²	**114.41**
laid flat	17.75	2.20	45.65	–	22.53	m²	**68.18**
PC £300.00/1000							
laid on edge	24.40	4.76	98.81	–	32.43	m²	**131.24**
laid on edge but pavior 65 mm thick	20.50	3.81	79.05	–	28.53	m²	**107.58**
laid flat	13.31	2.20	45.65	–	18.10	m²	**63.75**
Clay brick pavings; 200 × 100 × 50 mm; butt jointed laid herringbone or basketweave pattern only; on prepared base (not included); bedding on 50 mm sharp sand							
PC £600.00/1000							
laid on edge	61.50	2.26	46.86	0.34	63.90	m²	**111.10**
laid flat	30.75	1.44	29.95	0.34	32.91	m²	**63.20**
PC £500.00/1000							
laid flat	25.63	1.44	29.95	0.34	27.78	m²	**58.07**
PC £400.00/1000							
laid flat	20.50	1.44	29.95	0.34	22.66	m²	**52.95**
PC £300.00/1000							
laid flat	15.38	1.44	29.95	0.34	17.53	m²	**47.82**

Q25 SLAB/BRICK/SETT/COBBLE PAVINGS

Item Excluding site overheads and profit	PC £	Labour hours	Labour £	Plant £	Material £	Unit	Total rate £
Clay brick pavings; 215 × 102.5 × 65 mm; on prepared base (not included); bedding on cement: sand (1:4) pointing mortar as work proceeds							
Paving bricks; PC £600.00/1000; herringbone bond							
laid on edge	36.44	3.55	73.76	–	43.15	m²	**116.91**
laid flat	23.70	2.37	49.18	–	30.41	m²	**79.59**
Paving bricks; PC £600.00/1000; basketweave bond							
laid on edge	36.44	2.37	49.18	–	43.15	m²	**92.33**
laid flat	23.70	1.58	32.79	–	30.41	m²	**63.20**
Paving bricks; PC £600.00/1000; running or stack bond							
laid on edge	36.44	1.90	39.35	–	43.15	m²	**82.50**
laid flat	23.70	1.26	26.23	–	30.41	m²	**56.64**
Paving bricks; PC £500.00/1000; herringbone bond							
laid on edge	30.37	3.55	73.76	–	40.92	m²	**114.68**
laid flat	19.75	2.37	49.18	–	26.46	m²	**75.64**
Paving bricks; PC £500.00/1000; basketweave bond							
laid on edge	30.37	2.37	49.18	–	34.22	m²	**83.40**
laid flat	19.75	1.58	32.79	–	26.46	m²	**59.25**
Paving bricks; PC £500.00/1000; running or stack bond							
laid on edge	30.37	1.90	39.35	–	34.22	m²	**73.57**
laid flat	19.75	1.26	26.23	–	26.46	m²	**52.69**
Paving bricks; PC £400.00/1000; herringbone bond							
laid on edge	24.29	3.55	73.76	–	28.14	m²	**101.90**
laid flat	15.80	2.37	49.18	–	22.38	m²	**71.56**
Paving bricks; PC £400.00/1000; basketweave bond							
laid on edge	24.29	2.37	49.18	–	28.14	m²	**77.32**
laid flat	15.80	1.58	32.79	–	22.38	m²	**55.17**
Paving bricks; PC £400.00/1000; running or stack bond							
laid on edge	24.29	1.90	39.35	–	28.14	m²	**67.49**
laid flat	15.80	1.26	26.23	–	22.38	m²	**48.61**
Paving bricks; PC £300.00/1000; herringbone bond							
laid on edge	17.77	3.55	73.76	–	21.63	m²	**95.39**
laid flat	11.85	2.37	49.18	–	18.56	m²	**67.74**
Paving bricks; PC £300.00/1000; basketweave bond							
laid on edge	17.77	2.37	49.18	–	21.63	m²	**70.81**
laid flat	11.85	1.58	32.79	–	18.56	m²	**51.35**

Q25 SLAB/BRICK/SETT/COBBLE PAVINGS

Item Excluding site overheads and profit	PC £	Labour hours	Labour £	Plant £	Material £	Unit	Total rate £
BRICK PAVING – cont							
Clay brick pavings – cont							
Paving bricks; PC £300.00/1000; running or stack bond							
laid on edge	17.77	1.90	39.34	–	21.63	m²	**60.97**
laid flat	11.85	1.26	26.23	–	18.56	m²	**44.79**
Cutting							
curved cutting	–	0.44	9.22	6.29	–	m	**15.51**
raking cutting	–	0.33	6.92	5.02	–	m	**11.94**
Add or subtract the following amounts for every £10.00/1000 difference in the prime cost of bricks							
Butt joints							
200 × 100	–	–	–	–	0.53	m²	**0.53**
215 × 102.5	–	–	–	–	0.48	m²	**0.48**
10 mm mortar joints							
200 × 100	–	–	–	–	0.45	m²	**0.45**
215 × 102.5	–	–	–	–	0.41	m²	**0.41**
PRECAST CONCRETE SLAB PAVING							
Precast concrete pavings; Charcon Hard Landscaping; on prepared subbase (not included); bedding on 25 mm thick cement: sand mortar (1:4); butt joints; straight both ways; jointing in cement: sand (1:3) brushed in; on 50 mm thick sharp sand base							
Pavings; natural grey							
400 × 400 × 65 mm chamfered	31.44	0.40	8.30	–	36.30	m²	**44.60**
450 × 450 × 50 mm chamfered	16.94	0.44	9.22	–	21.80	m²	**31.02**
450 × 450 × 70 mm chamfered	19.51	0.44	9.22	–	24.37	m²	**33.59**
600 × 300 × 50 mm	15.33	0.44	9.22	–	20.20	m²	**29.42**
450 × 600 × 50 mm	13.52	0.44	9.22	–	18.38	m²	**27.60**
600 × 600 × 50 mm	10.89	0.40	8.30	–	15.75	m²	**24.05**
750 × 600 × 50 mm	10.29	0.40	8.30	–	15.15	m²	**23.45**
900 × 600 × 50 mm	9.41	0.40	8.30	–	14.27	m²	**22.57**
Pavings; coloured							
450 × 450 × 70 mm chamfered	22.86	0.44	9.22	–	27.73	m²	**36.95**
450 × 600 × 50 mm	20.48	0.44	9.22	–	25.35	m²	**34.57**
400 × 400 × 65 mm chamfered	31.44	0.40	8.30	–	36.30	m²	**44.60**
600 × 600 × 50 mm	17.03	0.40	8.30	–	21.89	m²	**30.19**
750 × 600 × 50 mm	15.71	0.40	8.30	–	20.58	m²	**28.88**
900 × 600 × 50 mm	13.97	0.40	8.30	–	18.83	m²	**27.13**

Q25 SLAB/BRICK/SETT/COBBLE PAVINGS

Item Excluding site overheads and profit	PC £	Labour hours	Labour £	Plant £	Material £	Unit	Total rate £
Precast concrete pavings; Charcon Hard Landscaping; on prepared subbase (not included); bedding on 25 mm thick cement: sand mortar (1:4); butt joints; straight both ways; jointing in cement: sand (1:3) brushed in; on 50 mm thick sharp sand base							
Appalacian rough textured exposed aggregate pebble paving							
600 × 600 × 65 mm	41.58	0.50	10.38	–	44.73	m²	**55.11**
Precast concrete pavings; Marshalls Plc; Heritage imitation riven yorkstone paving; on prepared subbase measured separately; bedding on 25 mm thick cement: sand mortar (1:4); pointed straight both ways cement: sand (1:3)							
Square and rectangular paving							
450 × 300 × 38 mm	40.77	1.00	20.75	–	45.01	m²	**65.76**
450 × 450 × 38 mm	24.55	0.75	15.56	–	28.80	m²	**44.36**
600 × 300 × 38 mm	26.82	0.80	16.60	–	31.06	m²	**47.66**
600 × 450 × 38 mm	27.20	0.75	15.56	–	31.44	m²	**47.00**
600 × 600 × 38 mm	26.07	0.50	10.38	–	30.39	m²	**40.77**
Extra labours for laying the a selection of the above sizes to random rectangular pattern	–	0.33	6.92	–	–	m²	**6.92**
Radial paving for circles							
circle with centre stone and first ring (8 slabs); 450 × 230/560 × 38 mm; diameter 1.54 m (total area 1.86 m²)	70.36	1.50	31.13	–	76.17	nr	**107.30**
circle with second ring (16 slabs); 450 × 300/460 × 38 mm; diameter 2.48 m (total area 4.83 m²)	181.40	4.00	83.00	–	196.34	nr	**279.34**
circle with third ring (16 slabs); 450 × 470/625 × 38 mm; diameter 3.42 m (total area 9.18 m²)	330.04	8.00	166.00	–	358.18	nr	**524.18**
Stepping stones							
380 diameter × 38 mm	5.05	0.20	4.15	–	11.34	nr	**15.49**
asymmetrical 560 × 420 × 38 mm	7.53	0.20	4.15	–	13.82	nr	**17.97**
Precast concrete pavings; Marshalls Plc; Chancery imitation reclaimed riven yorkstone paving; on prepared subbase measured separately; bedding on 25 mm thick cement: sand mortar (1:4); pointed straight both ways cement: sand (1:3)							
Square and rectangular paving							
300 × 300 × 45 mm	38.48	1.00	20.75	–	42.73	m²	**63.48**
450 × 300 × 45 mm	26.64	0.90	18.68	–	30.89	m²	**49.57**
600 × 300 × 45 mm	25.92	0.80	16.60	–	30.16	m²	**46.76**
600 × 450 × 45 mm	26.01	0.75	15.56	–	30.26	m²	**45.82**
450 × 450 × 45 mm	23.42	0.75	15.56	–	27.67	m²	**43.23**
600 × 600 × 45 mm	26.24	0.50	10.38	–	30.48	m²	**40.86**

Q25 SLAB/BRICK/SETT/COBBLE PAVINGS

Item Excluding site overheads and profit	PC £	Labour hours	Labour £	Plant £	Material £	Unit	Total rate £
PRECAST CONCRETE SLAB PAVING – cont							
Precast concrete pavings – cont							
Extra labours for laying the a selection of the above sizes to random rectangular pattern	–	0.33	6.92	–	–	m²	6.92
Radial paving for circles							
circle with centre stone and first ring							
(8 slabs); 450 × 230/560 × 38 mm; diameter							
1.54 m (total area 1.86 m²)	85.65	1.50	31.13	–	91.46	nr	122.59
circle with second ring (16 slabs);							
450 × 300/460 × 38 mm; diameter 2.48 m							
(total area 4.83 m²)	220.85	4.00	83.00	–	235.79	nr	318.79
circle with third ring (16 slabs); 450 × 470/							
625 × 38 mm; diameter 3.42 m (total area							
9.18 m²)	402.13	8.00	166.00	–	430.27	nr	596.27
Squaring off set for 2 ring circle							
16 slabs; 2.72 m²	235.25	1.00	20.75	–	235.25	nr	256.00
RECONSTITUTED STONE PAVINGS							
Reconstituted yorkstone aggregate pavings; Marshalls Plc; Saxon on prepared subbase measured separately; bedding on 25 mm thick cement: sand mortar (1:4) ;on 50 mm thick sharp sand base							
Square and rectangular paving in various colours; butt joints straight both ways							
300 × 300 × 35 mm	37.52	0.80	16.60	–	43.04	m²	59.64
600 × 300 × 35 mm	24.42	0.65	13.49	–	29.94	m²	43.43
450 × 450 × 50 mm	26.67	0.75	15.56	–	32.19	m²	47.75
600 × 600 × 35 mm	18.28	0.50	10.38	–	23.81	m²	34.19
600 × 600 × 50 mm	22.96	0.55	11.41	–	28.48	m²	39.89
Square and rectangular paving in natural; butt joints straight both ways							
300 × 300 × 35 mm	31.30	0.80	16.60	–	36.83	m²	53.43
450 × 450 × 50 mm	22.74	0.70	14.53	–	28.26	m²	42.79
600 × 300 × 35 mm	21.45	0.75	15.56	–	26.97	m²	42.53
600 × 600 × 35 mm	15.76	0.50	10.38	–	21.29	m²	31.67
600 × 600 × 50 mm	19.12	0.60	12.45	–	24.65	m²	37.10
Radial paving for circles; 20 mm joints							
circle with centre stone and first ring							
(8 slabs); 450 × 230/560 × 35 mm; diameter							
1.54 m (total area 1.86 m²)	72.67	1.50	31.13	–	78.81	nr	109.94
circle with second ring (16 slabs);							
450 × 300/460 × 35 mm; diameter 2.48 m							
(total area 4.83 m²)	177.15	4.00	83.00	–	193.01	nr	276.01
circle with third ring (24 slabs); 450 × 310/							
430 × 35 mm; diameter 3.42 m (total area							
9.18 m²)	333.87	8.00	166.00	–	364.20	nr	530.20

Q25 SLAB/BRICK/SETT/COBBLE PAVINGS

Item Excluding site overheads and profit	PC £	Labour hours	Labour £	Plant £	Material £	Unit	Total rate £
STONE PAVING							
Natural stone, slab or granite paving –							
General							
Preamble: Provide paving slabs of the							
specified thickness in random sizes but not							
less than 25 slabs per 10 m² of surface area,							
to be laid in parallel courses with joints							
alternately broken and laid to falls.							
Granite setts; bedding on 25 mm cement:							
sand (1:3)							
Natural granite setts; 100 × 100 mm to							
125 × 150 mm; × 150 to 250 mm length; riven							
surface							
new; standard grade	29.61	2.00	41.50	–	38.58	m²	**80.08**
new; high grade	32.81	2.00	41.50	–	41.78	m²	**83.28**
reclaimed; cleaned	39.88	2.00	41.50	–	48.85	m²	**90.35**
Natural stone, slate or granite flag pavings;							
CED Ltd; on prepared base (not included);							
bedding on 25 mm cement: sand (1:3);							
cement: sand (1:3) joints							
Yorkstone; riven laid random rectangular							
new slabs; 40–60 mm thick	107.29	1.71	35.57	–	111.23	m²	**146.80**
reclaimed slabs; Cathedral grade;							
50–75 mm thick	111.43	2.80	58.10	–	115.36	m²	**173.46**
Donegal quartzite slabs; standard tiles							
200 × random lengths × 15–25 mm	96.35	4.20	87.19	–	100.28	m²	**187.47**
250 × random lengths × 15–25 mm	96.35	3.82	79.23	–	100.28	m²	**179.51**
300 × random lengths × 15–25 mm	96.35	3.47	72.05	–	100.28	m²	**172.33**
350 × random lengths × 15–25 mm	96.35	3.16	65.48	–	100.28	m²	**165.76**
400 × random lengths × 15–25 mm	96.35	2.87	59.54	–	100.28	m²	**159.82**
450 × random lengths × 15–25 mm	101.17	2.61	54.12	–	105.10	m²	**159.22**
Natural yorkstone, pavings or edgings;							
Johnsons Wellfield Quarries; sawn 6 sides;							
50 mm thick; on prepared base measured							
separately; bedding on 25 mm cement:							
sand (1:3); cement: sand (1:3) joints							
Paving							
laid to random rectangular pattern	62.60	1.71	35.57	–	66.34	m²	**101.91**
laid to coursed laying pattern; 3 sizes	67.53	1.72	35.78	–	71.26	m²	**107.04**
Paving; single size							
600 × 600 mm	73.04	0.85	17.64	–	76.78	m²	**94.42**
600 × 400 mm	73.04	1.00	20.75	–	76.78	m²	**97.53**
300 × 200 mm	82.69	2.00	41.50	–	86.43	m²	**127.93**
215 × 102.5 mm	85.45	2.50	51.88	–	89.19	m²	**141.07**

Q25 SLAB/BRICK/SETT/COBBLE PAVINGS

Item Excluding site overheads and profit	PC £	Labour hours	Labour £	Plant £	Material £	Unit	Total rate £
STONE PAVING – cont							
Natural yorkstone, pavings or edgings –							
cont							
Paving; cut to template off site; 600 × 600 mm;							
radius							
1.00 m	178.50	3.33	69.17	–	182.24	m²	**251.41**
2.50 m	178.50	2.00	41.50	–	182.24	m²	**223.74**
5.00 m	178.50	2.00	41.50	–	182.24	m²	**223.74**
Edgings							
100 mm wide × random lengths	7.72	0.50	10.38	–	8.09	m	**18.47**
100 mm × 100 mm	8.54	0.50	10.38	–	12.28	m	**22.66**
100 mm × 200 mm	17.09	0.50	10.38	–	20.83	m	**31.21**
250 mm wide × random lengths	16.88	0.40	8.30	–	20.62	m	**28.92**
500 mm wide × random lengths	33.77	0.33	6.92	–	37.51	m	**44.43**
Yorkstone edgings; 600 mm long × 250 mm							
wide; cut to radius							
1.00 m to 3.00	53.75	0.50	10.38	–	54.12	m	**64.50**
3.00 m to 5.00 m	53.75	0.44	9.22	–	54.12	m	**63.34**
exceeding 5.00 m	53.75	0.40	8.30	–	54.12	m	**62.42**
Natural yorkstone, pavings or edgings;							
Johnsons Wellfield Quarries; sawn 6 sides;							
75 mm thick; on prepared base measured							
separately; bedding on 25 mm cement:							
sand (1:3); cement: sand (1:3) joints							
Paving							
laid to random rectangular pattern	75.14	0.95	19.71	–	78.88	m²	**98.59**
laid to coursed laying pattern; 3 sizes	80.65	0.95	19.71	–	84.39	m²	**104.10**
Paving; single size							
600 × 600 mm	86.16	0.95	19.71	–	89.90	m²	**109.61**
600 × 400 mm	86.16	0.95	19.71	–	89.90	m²	**109.61**
300 × 200 mm	97.02	0.75	15.56	–	100.76	m²	**116.32**
215 × 102.5 mm	99.22	2.50	51.88	–	102.96	m²	**154.84**
Paving; cut to template off site; 600 × 600 mm;							
radius							
1.00 m	210.00	4.00	83.00	–	213.74	m²	**296.74**
2.50 m	210.00	2.50	51.88	–	213.74	m²	**265.62**
5.00 m	210.00	2.50	51.88	–	213.74	m²	**265.62**
Edgings							
100 mm wide × random lengths	9.65	0.60	12.45	–	10.02	m	**22.47**
100 mm × 100 mm	9.92	0.60	12.45	–	13.66	m	**26.11**
100 mm × 200 mm	1.98	0.50	10.38	–	5.72	m	**16.10**
250 mm wide × random lengths	20.16	0.50	10.38	–	23.90	m	**34.28**
500 mm wide × random lengths	40.33	0.40	8.30	–	44.07	m	**52.37**
Edgings; 600 mm long × 250 mm wide; cut to							
radius							
1.00 m to 3.00	62.02	0.60	12.45	–	62.39	m	**74.84**
3.00 m to 5.00 m	62.02	0.50	10.38	–	62.39	m	**72.77**
exceeding 5.00 m	62.02	0.44	9.22	–	62.39	m	**71.61**

Q25 SLAB/BRICK/SETT/COBBLE PAVINGS

Item Excluding site overheads and profit	PC £	Labour hours	Labour £	Plant £	Material £	Unit	Total rate £
CED Ltd; Indian sandstone, riven pavings or edgings; calibrated +/-26 mm thick; on prepared base measured separately; bedding on 25 mm cement: sand (1:3); cement: sand (1:3) joints							
Paving							
laid to random rectangular pattern	30.66	2.40	49.81	–	34.40	m²	**84.21**
laid to coursed laying pattern; 3 sizes	30.66	2.00	41.50	–	34.40	m²	**75.90**
Paving; single size							
600 × 600 mm	30.66	1.00	20.75	–	34.40	m²	**55.15**
600 × 400 mm	30.66	1.25	25.94	–	34.40	m²	**60.34**
400 × 400 mm	30.66	1.67	34.58	–	34.40	m²	**68.98**
Natural stone, slate or granite flag pavings; CED Ltd; on prepared base (not included); bedding on 25 mm cement: sand (1:3); cement: sand (1:3) joints							
Granite paving; sawn 6 sides; textured top							
new slabs; silver grey, blue grey or yellow;							
50 mm thick	56.70	1.71	35.57	–	60.63	m²	**96.20**
new slabs; black; 50 mm thick	79.82	1.71	35.57	–	83.75	m²	**119.32**
Cobble pavings – General							
Cobbles should be embedded by hand, tight-butted, endwise to a depth of 60% of their length. A dry grout of rapid-hardening cement: sand (1:2) shall be brushed over the cobbles until the interstices are filled to the level of the adjoining paving. Surplus grout shall then be brushed off and a light, fine spray of water applied over the area.							
Cobble pavings							
Cobbles; to present a uniform colour in panels or varied in colour as required							
Scottish beach cobbles; 200–100 mm	41.80	2.00	41.50	–	49.82	m²	**91.32**
Scottish beach cobbles; 100–75 mm	40.62	2.50	51.88	–	48.64	m²	**100.52**
Scottish beach cobbles; 75–50 mm	28.52	3.33	69.17	–	36.55	m²	**105.72**
TIMBER DECKING							
Timber decking							
Supports for timber decking; softwood joists to receive decking boards; joists at 400 mm centres; Southern yellow pine							
38 × 88 mm	18.20	1.00	20.75	–	20.19	m²	**40.94**
50 × 150 mm	12.54	1.00	20.75	–	14.53	m²	**35.28**
50 × 125 mm	10.45	1.00	20.75	–	12.44	m²	**33.19**

Q26 SPECIAL SURFACINGS/PAVINGS FOR SPORT/GENERAL AMENITY

Item Excluding site overheads and profit	PC £	Labour hours	Labour £	Plant £	Material £	Unit	Total rate £
TIMBER DECKING – cont							
Timber decking – cont							
Hardwood decking; yellow Balau; grooved or smooth; 6 mm joints							
deck boards; 90 mm wide × 19 mm thick	19.99	1.00	20.75	–	22.26	m²	**43.01**
deck boards; 145 mm wide × 21 mm thick	32.26	1.00	20.75	–	34.54	m²	**55.29**
deck boards; 145 mm wide × 28 mm thick	33.21	1.00	20.75	–	35.48	m²	**56.23**
Hardwood decking; Ipe; smooth; 6 mm joints							
deck boards; 90 mm wide × 19 mm thick	45.18	1.00	20.75	–	47.46	m²	**68.21**
deck boards; 145 mm wide × 19 mm thick	46.34	1.00	20.75	–	48.62	m²	**69.37**
Western Red Cedar; 6 mm joints							
prime deck grade; 90 mm wide × 40 mm thick	38.13	1.00	20.75	–	40.41	m²	**61.16**
prime deck grade; 142 mm wide × 40 mm thick	40.54	1.00	20.75	–	42.82	m²	**63.57**
Handrails and base rail; fixed to posts at 2.00 m centres							
posts; 100 × 100 × 1370 mm high	9.19	1.00	20.75	–	9.73	m	**30.48**
posts; turned; 1220 mm high	17.83	1.00	20.75	–	18.36	m	**39.11**
Handrails; balusters							
square balusters at 100 mm centres	39.90	0.50	10.38	–	40.43	m	**50.81**
square balusters at 300 mm centres	13.29	0.33	6.92	–	13.71	m	**20.63**
turned balusters at 100 mm centres	63.00	0.50	10.38	–	63.53	m	**73.91**
turned balusters at 300 mm centres	20.98	0.33	6.92	–	21.41	m	**28.33**
Q26 SPECIAL SURFACINGS/PAVINGS FOR SPORT/GENERAL AMENITY							
PLAY AREAS							
Market prices of surfacing materials							
Surfacings; Melcourt Industries Ltd (items labelled FSC are Forest Stewardship Council certified)							
Playbark® 10/50; per 25 m³ load	–	–	–	–	63.60	m³	**63.60**
Playbark® 8/25; per 25 m³ load	–	–	–	–	62.10	m³	**62.10**
Playchips®; per 25 m³ load; FSC	–	–	–	–	41.75	m³	**41.75**
Kushyfall; per 25 m³ load; FSC	–	–	–	–	39.70	m³	**39.70**
Softfall; per 25 m³ load	–	–	–	–	31.40	m³	**31.40**
Playsand; per 10 t load	–	–	–	–	101.16	m³	**101.16**
Woodfibre; per 25 m³ load; FSC	–	–	–	–	36.95	m³	**36.95**
Playgrounds; SMP Playgrounds Ltd							
Tiles; on prepared base (not included)							
Premier 25; 1000 × 1000 × 25 mm; black; for general use; freefall height up to 0.8 m	48.00	0.20	4.15	–	57.00	m²	**61.15**
Premier 70; 1000 × 1000 × 70 mm; black; for higher equipment; freefall height up to 2.6 m	84.00	0.25	5.19	–	93.00	m²	**98.19**

Q30 SEEDING/TURFING

Item Excluding site overheads and profit	PC £	Labour hours	Labour £	Plant £	Material £	Unit	Total rate £
Playgrounds; Melcourt Industries Ltd							
Playbark® ; on drainage layer (not included);							
minimum 300 mm settled depth							
Playbark® 8/25; 8–25 mm particles; red/							
brown	18.63	0.35	7.28	–	18.63	m²	**25.91**
Playbark® 10/50; 10–50 mm particles; red/							
brown	21.20	0.55	11.43	–	21.20	m²	**32.63**
Playgrounds; timber edgings							
Timber edging boards; 50 × 50 × 750 mm							
timber pegs at 1000 mm centres; excavations							
and hardcore under edgings (not included)							
50 × 150 mm; treated softwood edge							
boards	2.75	0.13	2.59	–	2.75	m	**5.34**
38 × 150 mm; treated softwood edge							
boards	1.82	0.13	2.59	–	2.54	m	**5.13**
Q30 SEEDING/TURFING							
GENERAL							
Seeding/turfing – General							
Market prices of seeding materials							
Please see market prices in the Major Works							
section of this book.							
Market prices of chemicals and application							
rates							
Please see market prices in the Major Works							
section of this book.							
PREPARATION OF SEEDING AREAS							
Cultivation							
Breaking up existing ground; using pedestrian							
operated tine cultivator or rotavator							
100 mm deep	–	0.50	10.38	5.91	–	100 m²	**16.29**
150 mm deep	–	0.57	11.86	6.76	–	100 m²	**18.62**
200 mm deep	–	0.67	13.83	7.88	–	100 m²	**21.71**
As above but in heavy clay or wet soils							
100 mm deep	–	0.67	13.83	7.88	–	100 m²	**21.71**
150 mm deep	–	0.80	16.60	9.46	–	100 m²	**26.06**
200 mm deep	–	1.00	20.75	11.82	–	100 m²	**32.57**
Rolling cultivated ground lightly; using							
self-propelled agricultural roller	–	0.06	1.15	1.68	–	100 m²	**2.83**
Importing and storing selected and approved							
topsoil; inclusive of settlement							
small quantities; less than 15 m³	33.60	–	–	–	33.60	m³	**33.60**
over 15 m³	33.60	–	–	–	33.60	m³	**33.60**

Q30 SEEDING/TURFING

Item Excluding site overheads and profit	PC £	Labour hours	Labour £	Plant £	Material £	Unit	Total rate £
PREPARATION OF SEEDING AREAS – cont							
Spreading and lightly consolidating approved topsoil (imported or from spoil heaps); in layers not exceeding 150 mm; travel distance from spoil heaps not exceeding 25 m							
By machine (imported topsoil not included)							
minimum depth 100 mm	–	–	–	0.98	–	m²	**0.98**
minimum depth 200 mm	–	–	–	1.95	–	m²	**1.95**
minimum depth 300 mm	–	–	–	2.93	–	m²	**2.93**
minimum depth 500 mm	–	–	–	4.88	–	m²	**4.88**
over 500 mm	–	–	–	9.77	–	m³	**9.77**
By hand (imported topsoil not included)							
minimum depth 100 mm	–	0.20	4.15	–	–	m²	**4.15**
minimum depth 150 mm	–	0.30	6.23	–	–	m²	**6.23**
minimum depth 300 mm	–	0.60	12.43	–	–	m²	**12.43**
minimum depth 450 mm	–	0.90	18.69	–	–	m²	**18.69**
over 450 mm deep	–	2.22	46.06	–	–	m³	**46.06**
Extra over for spreading topsoil to slopes 15–30°; by machine or hand	–	–	–	–	–	10%	**–**
Extra over for spreading topsoil to slopes over 30°; by machine or hand	–	–	–	–	–	25%	**–**
Extra over for spreading topsoil from spoil heaps; travel exceeding 100 m; by machine							
100–150 m	–	0.01	0.26	0.08	–	m³	**0.34**
150–200 m	–	0.02	0.38	0.12	–	m³	**0.50**
200–300 m	–	0.03	0.58	0.18	–	m³	**0.76**
Extra over spreading topsoil for travel exceeding 100 m; by hand							
100 m	–	0.83	17.29	–	–	m³	**17.29**
200 m	–	1.67	34.59	–	–	m³	**34.59**
300 m	–	2.50	51.88	–	–	m³	**51.88**
Evenly grading; to general surfaces to bring to finished levels							
by pedestrian operated rotavator	–	–	0.08	0.05	–	m²	**0.13**
by hand	–	0.01	0.21	–	–	m²	**0.21**
Extra over grading for slopes 15–30°; by machine or hand	–	–	–	–	–	10%	**–**
Extra over grading for slopes over 30°; by machine or hand	–	–	–	–	–	25%	**–**
Apply screened topdressing to grass surfaces; spread using Tru-Lute							
sand soil mixes 90/10 to 50/50	0.12	–	0.04	0.06	0.12	m²	**0.22**
Spread only existing cultivated soil to final levels using Tru-Lute							
cultivated soil	–	–	0.04	0.06	–	m²	**0.10**

Q30 SEEDING/TURFING

Item Excluding site overheads and profit	PC £	Labour hours	Labour £	Plant £	Material £	Unit	Total rate £
Clearing stones; disposing off site; to distance not exceeding 13 km							
by hand; stones not exceeding 50 mm in any direction; loading to skip 4.6 m³	–	0.01	0.21	0.04	–	m²	**0.25**
Lightly cultivating; weeding; to fallow areas; disposing debris off site							
by hand	–	0.01	0.29	0.29	–	m²	**0.58**
SURFACE APPLICATIONS							
Surface applications and soil additives; pre-seeding; material delivered to a maximum of 25 m from area of application; applied by hand							
Soil conditioners; to cultivated ground; mushroom compost; delivered in 25 m³ loads; PC £23.70/m³; including turning in							
1 m³ per 40 m² = 25 mm thick	0.59	0.02	0.46	–	0.59	m²	**1.05**
1 m³ per 20 m² = 50 mm thick	1.19	0.04	0.92	–	1.19	m²	**2.11**
1 m³ per 13.33 m² = 75 mm thick	1.78	0.07	1.38	–	1.78	m²	**3.16**
1 m³ per 10 m² = 100 mm thick	2.37	0.08	1.66	–	2.37	m²	**4.03**
Soil conditioners; to cultivated ground; mushroom compost; delivered in 70 m³ loads; PC £12.00/m³; including turning in							
1 m³ per 40 m² = 25 mm thick	0.30	0.02	0.46	–	0.30	m²	**0.76**
1 m³ per 20 m² = 50 mm thick	0.53	0.04	0.92	–	0.53	m²	**1.45**
1 m³ per 13.33 m² = 75 mm thick	0.90	0.07	1.38	–	0.90	m²	**2.28**
1 m³ per 10 m² = 100 mm thick	1.20	0.08	1.66	–	1.20	m²	**2.86**
Preparation of seedbeds – General Preamble: for preliminary operations see 'Cultivation' section.							
Preparation of seedbeds; soil preparation							
Lifting selected and approved topsoil from spoil heaps; passing through 6 mm screen; removing debris	–	0.08	1.73	4.55	0.02	m³	**6.30**
Topsoil; supply only; PC £28.00/m³; allowing for 20% settlement; 20 tonne loads							
25 mm	0.84	–	–	–	0.84	m²	**0.84**
50 mm	1.68	–	–	–	1.68	m²	**1.68**
100 mm	3.36	–	–	–	3.36	m²	**3.36**
150 mm	5.04	–	–	–	5.04	m²	**5.04**
200 mm	6.72	–	–	–	6.72	m²	**6.72**
250 mm	8.40	–	–	–	8.40	m²	**8.40**
300 mm	10.08	–	–	–	10.08	m²	**10.08**
400 mm	13.44	–	–	–	13.44	m²	**13.44**
450 mm	15.12	–	–	–	15.12	m²	**15.12**

Q30 SEEDING/TURFING

Item Excluding site overheads and profit	PC £	Labour hours	Labour £	Plant £	Material £	Unit	Total rate £
SURFACE APPLICATIONS – cont							
Preparation of seedbeds – cont							
Topsoil; supply only; PC £32.25/m³; allowing							
for 20% settlement; 10 tonne loads							
25 mm	0.97	–	–	–	0.97	m²	0.97
50 mm	1.94	–	–	–	1.94	m²	1.94
100 mm	3.87	–	–	–	3.87	m²	3.87
150 mm	5.80	–	–	–	5.80	m²	5.80
200 mm	7.74	–	–	–	7.74	m²	7.74
250 mm	9.68	–	–	–	9.68	m²	9.68
300 mm	11.61	–	–	–	11.61	m²	11.61
400 mm	15.48	–	–	–	15.48	m²	15.48
450 mm	17.41	–	–	–	17.41	m²	17.41
Spreading topsoil to form seedbeds							
average thickness not exceeding 200 mm							
(topsoil not included); by machine; grading							
and cultivation not included							
Excavated material; from spoil heaps							
average 25 m distance	–	0.10	2.08	6.81	–	m³	8.89
average 50 m distance	–	0.10	2.08	7.63	–	m³	9.71
average 100 m distance	–	0.10	2.08	8.32	–	m³	10.40
average 200 m distance	–	0.10	2.08	10.40	–	m³	12.48
Excavated material; spreading on site							
25 mm deep	–	–	–	0.25	–	m²	0.25
50 mm deep	–	–	–	0.50	–	m²	0.50
75 mm deep	–	–	–	0.75	–	m²	0.75
100 mm deep	–	–	–	1.00	–	m²	1.00
150 mm deep	–	–	–	1.20	–	m²	1.20
Spreading only topsoil to form seedbeds							
(topsoil not included); by hand							
25 mm deep	–	0.03	0.52	–	–	m²	0.52
50 mm deep	–	0.03	0.69	–	–	m²	0.69
75 mm deep	–	0.04	0.89	–	–	m²	0.89
100 mm deep	–	0.05	1.04	–	–	m²	1.04
150 mm deep	–	0.08	1.56	–	–	m²	1.56
Bringing existing topsoil to a fine tilth for							
seeding; by raking or harrowing; stones not to							
exceed 6 mm; by machine	–	–	0.08	0.04	–	m²	0.12
Bringing existing topsoil to a fine tilth for							
seeding; by raking or harrowing; stones not to							
exceed 6 mm; by hand	–	0.01	0.18	–	–	m²	0.18

Q30 SEEDING/TURFING

Item Excluding site overheads and profit	PC £	Labour hours	Labour £	Plant £	Material £	Unit	Total rate £
Preparation of seedbeds; soil treatments							
For the following topsoil improvement and							
seeding operations add or subtract the							
following amounts for every £0.10 difference in							
the material cost price							
35 g/m²	–	–	–	–	0.35	100 m²	**0.35**
50 g/m²	–	–	–	–	0.50	100 m²	**0.50**
70 g/m²	–	–	–	–	0.70	100 m²	**0.70**
100 g/m²	–	–	–	–	1.00	100 m²	**1.00**
Pre-seeding fertilizers (12+00+09); PC £0.73/kg;							
to seedbeds; by hand							
35 g/m²	2.56	0.17	3.46	–	2.56	100 m²	**6.02**
50 g/m²	3.66	0.17	3.46	–	3.66	100 m²	**7.12**
70 g/m²	5.12	0.17	3.46	–	5.12	100 m²	**8.58**
100 g/m²	7.32	0.20	4.15	–	7.32	100 m²	**11.47**
125 g/m²	9.14	0.20	4.15	–	9.14	100 m²	**13.29**
SEEDING							
Seeding							
Grass seed; spreading in two operations;							
PC £4.50/kg (for changes in material prices							
please refer to table above); by hand							
35 g/m²	15.75	0.17	3.46	–	15.75	100 m²	**19.21**
50 g/m²	22.50	0.17	3.46	–	22.50	100 m²	**25.96**
70 g/m²	31.50	0.17	3.46	–	31.50	100 m²	**34.96**
100 g/m²	45.00	0.17	3.46	–	45.00	100 m²	**48.46**
125 g/m²	56.25	0.20	4.15	–	56.25	100 m²	**60.40**
Grass seed; spreading in two operations;							
PC £6.00/kg (for changes in material prices							
please refer to table above); by hand							
35 g/m²	21.00	0.17	3.46	–	21.00	100 m²	**24.46**
50 g/m²	30.00	0.17	3.46	–	30.00	100 m²	**33.46**
70 g/m²	42.00	0.17	3.46	–	42.00	100 m²	**45.46**
100 g/m²	60.00	0.17	3.46	–	60.00	100 m²	**63.46**
125 g/m²	75.00	0.20	4.15	–	75.00	100 m²	**79.15**
Extra over seeding by hand for slopes over							
30° (allowing for the actual area but measured							
in plan)							
35 g/m²	2.34	–	0.08	–	2.34	100 m²	**2.42**
50 g/m²	3.38	–	0.08	–	3.38	100 m²	**3.46**
70 g/m²	4.72	–	0.08	–	4.72	100 m²	**4.80**
100 g/m²	6.75	–	0.09	–	6.75	100 m²	**6.84**
125 g/m²	8.41	–	0.09	–	8.41	100 m²	**8.50**
Harrowing seeded areas; light chain harrow	–	–	–	0.25	–	100 m²	**0.25**
Raking over seeded areas							
by hand	–	0.80	16.60	–	–	100 m²	**16.60**

Q30 SEEDING/TURFING

Item Excluding site overheads and profit	PC £	Labour hours	Labour £	Plant £	Material £	Unit	Total rate £
SEEDING – cont							
Seeding – cont							
Rolling seeded areas; light roller							
by pedestrian operated mechanical roller	–	0.08	1.73	0.77	–	100 m²	**2.50**
by hand drawn roller	–	0.17	3.46	–	–	100 m²	**3.46**
Extra over harrowing, raking or rolling seeded							
areas for slopes over 30°; by machine or hand	–	–	–	–	–	25%	**–**
Turf edging; to seeded areas; 300 mm wide	1.99	0.13	2.77	–	1.99	m²	**4.76**
PREPARATION OF TURFING AREAS							
Preparation of turf beds							
Rolling turf to be lifted; lifting by hand or							
mechanical turf stripper; stacks to be not more							
than 1 m high							
cutting only preparing to lift; pedestrian turf							
cutter	–	0.75	15.56	9.91	–	100 m²	**25.47**
lifting and stacking; by hand	–	8.33	172.92	–	–	100 m²	**172.92**
Rolling up; moving to stacks							
distance not exceeding 100 m	–	2.50	51.88	–	–	100 m²	**51.88**
extra over rolling and moving turf to stacks							
to transport per additional 100 m	–	0.83	17.29	–	–	100 m²	**17.29**
Lifting selected and approved topsoil from							
spoil heaps							
passing through 6 mm screen; removing							
debris	–	0.17	3.46	9.11	–	m³	**12.57**
Extra over lifting topsoil and passing through							
screen for imported topsoil; plus 20%							
allowance for settlement	33.60	–	–	–	33.60	m³	**33.60**
Spreading topsoil to form turf beds (topsoil not							
included); by machine							
25 mm deep	–	–	–	0.24	–	m²	**0.24**
50 mm deep	–	–	–	0.49	–	m²	**0.49**
75 mm deep	–	–	–	0.73	–	m²	**0.73**
100 mm deep	–	–	–	0.98	–	m²	**0.98**
150 mm deep	–	–	–	1.46	–	m²	**1.46**
Bringing existing topsoil to a fine tilth for turfing							
by raking or harrowing; stones not to exceed							
6 mm; by hand	–	0.01	0.28	–	–	m²	**0.28**
Spreading topsoil to form turf beds (topsoil not							
included); by hand							
25 mm deep	–	0.04	0.83	–	–	m²	**0.83**
50 mm deep	–	0.08	1.66	–	–	m²	**1.66**
75 mm deep	–	0.12	2.49	–	–	m²	**2.49**
100 mm deep	–	0.15	3.12	–	–	m²	**3.12**
150 mm deep	–	0.23	4.67	–	–	m²	**4.67**

Q30 SEEDING/TURFING

Item Excluding site overheads and profit	PC £	Labour hours	Labour £	Plant £	Material £	Unit	Total rate £
Operations after spreading of topsoil; by hand							
Bringing existing topsoil to a fine tilth for turfing by raking or harrowing; stones not to exceed 6 mm							
topsoil spread by machine (soil compacted)	–	0.03	0.52	–	–	m²	**0.52**
topsoil spread by hand	–	0.02	0.35	–	–	m²	**0.35**
Turfing							
Turfing; laying only; to stretcher bond; butt joints; including providing and working from barrow plank runs where necessary to surfaces not exceeding 30° from horizontal							
specially selected lawn turves from previously lifted stockpile	–	0.20	4.15	–	–	m²	**4.15**
cultivated lawn turves; to larger open areas	–	0.11	2.21	–	–	m²	**2.21**
cultivated lawn turves; to domestic or garden areas	–	0.13	2.77	–	–	m²	**2.77**
TURFING							
Industrially grown turf; PC prices listed represent the general range of industrial turf prices for sportsfields and amenity purposes; prices will vary with quantity and site location							
Rolawn							
RB Medallion; sports fields, domestic lawns, general landscape	1.99	0.13	2.77	–	1.99	m²	**4.76**
Inturf							
Inturf Masters; formal lawns, golf greens, bowling greens and low maintenance areas	2.68	0.13	2.77	–	2.68	m²	**5.45**
Inturf Classic; golf tees, surrounds, lawns, parks, general purpose landscaping areas and winter sports	2.08	0.13	2.77	–	2.08	m²	**4.85**
Inturf Ornamental; medium fine turf for ornamental lawns, fairways and green surrounds	2.18	0.13	2.77	–	2.18	m²	**4.95**
Firming turves with wooden beater	–	0.01	0.21	–	–	m²	**0.21**
Rolling turfed areas; light roller							
by pedestrian operated mechanical roller	–	0.08	1.73	0.77	–	100 m²	**2.50**
by hand drawn roller	–	0.17	3.46	–	–	100 m²	**3.46**
Dressing with finely sifted topsoil; brushing into joints	0.04	0.05	1.04	–	0.04	m²	**1.08**
Turfing; laying only							
to slopes over 30°; to diagonal bond (measured as plan area – add 15% to these rates for the incline area of 30° slopes)	–	0.12	2.49	–	–	m²	**2.49**

Q30 SEEDING/TURFING

Item Excluding site overheads and profit	PC £	Labour hours	Labour £	Plant £	Material £	Unit	Total rate £
TURFING – cont							
Industrially grown turf – cont							
Extra over laying turfing for pegging down turves							
wooden or galvanized wire pegs; 200 mm long; 2 pegs per 0.50 m²	1.58	0.01	0.28	–	1.58	m²	1.86
ARTIFICIAL GRASS							
Artificial grass; Trulawn Ltd; laid to Type 1 50 mm and 25 mm sharp sand bed							
Please see the Major Works section for full descriptions of these products							
Trulawn Value	–	–	–	–	–	m²	37.23
Trulawn Continental	–	–	–	–	–	m²	39.55
Trulawn Play	–	–	–	–	–	m²	41.74
Trulawn Optimum	–	–	–	–	–	m²	50.05
Trulawn Luxury	–	–	–	–	–	m²	52.25
MAINTENANCE OF GRASS AREAS							
Maintenance operations (Note: the following rates apply to aftercare maintenance executed as part of a landscaping contract only)							
Initial cutting; to turfed areas							
20 mm high; using pedestrian guided power driven cylinder mower; including boxing off cuttings (stone picking and rolling not included)	–	0.18	3.73	0.31	–	100 m²	4.04
Repairing damaged grass areas							
scraping out; removing slurry; from ruts and holes; average 100 mm deep	–	0.13	2.77	–	–	m²	2.77
100 mm topsoil	–	0.13	2.77	–	3.36	m²	6.13
Repairing damaged grass areas; sowing grass seed to match existing or as specified; to individually prepared worn patches							
35 g/m²	0.18	0.01	0.21	–	0.18	m²	0.39
50 g/m²	0.26	0.01	0.21	–	0.26	m²	0.47
Using pedestrian operated mechanical equipment and blowers							
grassed areas with perimeters of mature trees such as sports fields and amenity areas	–	0.04	0.83	0.05	–	100 m²	0.88
grassed areas containing ornamental trees and shrub beds	–	0.10	2.08	0.13	–	100 m²	2.21
verges	–	0.07	1.38	0.09	–	100 m²	1.47

Q30 SEEDING/TURFING

Item Excluding site overheads and profit	PC £	Labour hours	Labour £	Plant £	Material £	Unit	Total rate £
By hand							
grassed areas with perimeters of mature trees such as sports fields and amenity areas	–	0.05	1.04	0.07	–	100 m²	**1.11**
grassed areas containing ornamental trees and shrub beds	–	0.08	1.73	0.11	–	100 m²	**1.84**
verges	–	1.00	20.75	1.30	–	100 m²	**22.05**
Removal of arisings							
areas with perimeters of mature trees	–	0.01	0.14	0.12	1.32	100 m²	**1.58**
areas containing ornamental trees and shrub beds	–	0.02	0.41	0.45	3.30	100 m²	**4.16**
Cutting grass to specified height; per cut							
ride-on triple cylinder mower	–	0.01	0.29	0.14	–	100 m²	**0.43**
ride-on triple rotary mower	–	0.01	0.29	–	–	100 m²	**0.29**
pedestrian mower (open areas)	–	0.18	3.73	0.39	–	100 m²	**4.12**
pedestrian mower (small areas or areas with obstacles)	–	0.33	6.92	0.47	–	100 m²	**7.39**
Cutting rough grass; per cut							
power flail or scythe cutter	–	0.04	0.73	–	–	100 m²	**0.73**
Extra over cutting grass for slopes not exceeding 30°	–	–	–	–	–	10%	**–**
Extra over cutting grass for slopes exceeding 30°	–	–	–	–	–	40%	**–**
Cutting fine sward							
pedestrian operated seven-blade cylinder lawn mower	–	0.14	2.90	0.23	–	100 m²	**3.13**
Extra over cutting fine sward for boxing off cuttings							
pedestrian mower	–	0.03	0.58	0.05	–	100 m²	**0.63**
Cutting areas of rough grass							
scythe	–	1.00	20.75	–	–	100 m²	**20.75**
sickle	–	2.00	41.50	–	–	100 m²	**41.50**
petrol operated strimmer	–	0.30	6.23	0.44	–	100 m²	**6.67**
Cutting areas of rough grass which contain trees or whips							
petrol operated strimmer	–	0.40	8.30	0.59	–	100 m²	**8.89**
Extra over cutting rough grass for on site raking up and dumping	–	0.33	6.92	–	–	100 m²	**6.92**
Trimming edge of grass areas; edging tool							
with petrol powered strimmer	–	0.13	2.77	0.20	–	100 m	**2.97**
by hand	–	0.67	13.83	–	–	100 m	**13.83**
Rolling grass areas; light roller							
by pedestrian operated mechanical roller	–	0.08	1.73	0.77	–	100 m²	**2.50**
by hand drawn roller	–	0.17	3.46	–	–	100 m²	**3.46**
Aerating grass areas; to a depth of 100 mm							
using pedestrian-guided motor powered solid or slitting tine turf aerator	–	0.18	3.63	2.95	–	100 m²	**6.58**
using hollow tine aerator; including sweeping up and dumping corings	–	0.50	10.38	5.89	–	100 m²	**16.27**
using hand aerator or fork	–	1.67	34.58	–	–	100 m²	**34.58**

Q31 PLANTING

Item Excluding site overheads and profit	PC £	Labour hours	Labour £	Plant £	Material £	Unit	Total rate £
MAINTENANCE OF GRASS AREAS – cont							
Maintenance operations (Note: the							
following rates apply to aftercare							
maintenance executed as part of a							
landscaping contract only) – cont							
Extra over aerating grass areas for on site							
sweeping up and dumping corings	–	0.17	3.46	–	–	100 m²	**3.46**
Switching off dew; from fine turf areas	–	0.20	4.15	–	–	100 m²	**4.15**
Scarifying grass areas to break up thatch;							
removing dead grass							
using self-propelled scarifier; including							
removing and disposing of grass on site	–	0.33	6.92	0.15	–	100 m²	**7.07**
by hand	–	3.03	62.88	–	–	100 m²	**62.88**
For the following topsoil improvement and							
seeding operations add or subtract the							
following amounts for every £0.10 difference in							
the material cost price							
35 g/m²	–	–	–	–	0.35	100 m²	**0.35**
50 g/m²	–	–	–	–	0.50	100 m²	**0.50**
70 g/m²	–	–	–	–	0.70	100 m²	**0.70**
100 g/m²	–	–	–	–	1.00	100 m²	**1.00**
Top dressing fertilizers (7+7+7); PC £0.81/kg;							
to seedbeds; by hand							
35 g/m²	3.75	0.17	3.46	–	3.75	100 m²	**7.21**
50 g/m²	5.36	0.17	3.46	–	5.36	100 m²	**8.82**
70 g/m²	7.51	0.17	3.46	–	7.51	100 m²	**10.97**
Watering turf; evenly; at a rate of 5 litres/m²							
using sprinkler equipment and with							
sufficient water pressure to run 1 nr 15 m							
radius sprinkler	–	0.02	0.41	–	–	100 m²	**0.41**
using hand-held watering equipment	–	0.25	5.19	–	–	100 m²	**5.19**
Q31 PLANTING							
GENERALLY							
Site protection; temporary protective							
fencing							
Cleft chestnut rolled fencing; to 100 mm							
diameter chestnut posts; driving into firm							
ground at 3 m centres; pales at 50 mm centres							
900 mm high	4.42	0.11	2.21	–	9.68	m	**11.89**
1100 mm high	4.90	0.11	2.21	–	10.16	m	**12.37**
1500 mm high	7.66	0.11	2.21	–	12.93	m	**15.14**
Extra over temporary protective fencing for							
removing and making good (no allowance for							
reuse of material)	–	0.07	1.38	0.27	–	m	**1.65**

Q31 PLANTING

Item Excluding site overheads and profit	PC £	Labour hours	Labour £	Plant £	Material £	Unit	Total rate £
Root decompaction; Terravent; Goroots Ltd							
200 mm thick	474.00	12.90	267.77	1.97	474.00	100 m²	**743.74**
Decompaction and aeration of soil in the root zone of a tree or plant; followed by an infusion of beneficial mychorrhizal fungi and other liquid amendments into the freshly aerated soil; mulching the root zone	–	–	–	–	1.00	m²	**1.00**
CULTIVATION AND SURFACE PREPARATION							
Cultivation							
Breaking up existing ground; using pedestrian operated tine cultivator or rotavator							
100 mm deep	–	0.55	11.41	5.91	–	100 m²	**17.32**
150 mm deep	–	0.63	13.04	6.76	–	100 m²	**19.80**
200 mm deep	–	0.73	15.22	7.88	–	100 m²	**23.10**
As above but in heavy clay or wet soils							
100 mm deep	–	0.73	15.22	7.88	–	100 m²	**23.10**
150 mm deep	–	0.88	18.26	9.46	–	100 m²	**27.72**
200 mm deep	–	1.10	22.82	11.82	–	100 m²	**34.64**
Importing only selected and approved topsoil							
1–14 m³	33.60	–	–	–	33.60	m³	**33.60**
over 15 m³	33.60	–	–	–	33.60	m³	**33.60**
Imported topsoil; spreading and lightly consolidating approved topsoil in layers not exceeding 150 mm; travel distance from offloading point not exceeding 25 m							
By machine							
minimum depth 100 mm	–	–	–	0.98	3.50	m²	**4.48**
minimum depth 200 mm	–	–	–	1.95	7.00	m²	**8.95**
minimum depth 300 mm	–	–	–	2.93	10.50	m²	**13.43**
minimum depth 500 mm	–	–	–	4.88	17.50	m²	**22.38**
over 500 mm	–	–	–	9.77	35.00	m³	**44.77**
By hand							
minimum depth 100 mm	–	0.20	4.15	–	3.50	m²	**7.65**
minimum depth 150 mm	–	0.30	6.23	–	5.25	m²	**11.48**
minimum depth 300 mm	–	0.60	12.43	–	10.50	m²	**22.93**
minimum depth 500 mm	–	0.90	18.69	–	17.50	m²	**36.19**
over 500 mm deep	–	2.22	46.06	–	35.00	m³	**81.06**

Q31 PLANTING

Item Excluding site overheads and profit	PC £	Labour hours	Labour £	Plant £	Material £	Unit	Total rate £
CULTIVATION AND SURFACE PREPARATION – cont							
Spreading and lightly consolidating approved topsoil (imported or from spoil heaps); in layers not exceeding 150 mm; travel distance from spoil heaps not exceeding 25 m							
By machine (imported topsoil not included)							
minimum depth 100 mm	–	–	–	0.98	–	m²	**0.98**
minimum depth 200 mm	–	–	–	1.95	–	m²	**1.95**
minimum depth 300 mm	–	–	–	2.93	–	m²	**2.93**
minimum depth 500 mm	–	–	–	4.88	–	m²	**4.88**
over 500 mm deep	–	–	–	9.77	–	m³	**9.77**
By hand (imported topsoil not included)							
minimum depth 100 mm	–	0.20	4.15	–	–	m²	**4.15**
minimum depth 150 mm	–	0.30	6.23	–	–	m²	**6.23**
minimum depth 300 mm	–	0.60	12.43	–	–	m²	**12.43**
minimum depth 500 mm	–	0.90	18.69	–	–	m²	**18.69**
over 500 mm deep	–	2.22	46.06	–	–	m³	**46.06**
Extra over for spreading topsoil to slopes 15–30° by machine or hand	–	–	–	–	–	10%	**–**
Extra over for spreading topsoil to slopes over 30° by machine or hand	–	–	–	–	–	25%	**–**
Extra over spreading topsoil for travel exceeding 100 m; by machine							
100–150 m	–	–	–	0.65	–	m³	**0.65**
150–200 m	–	–	–	0.87	–	m³	**0.87**
200–300 m	–	–	–	1.16	–	m³	**1.16**
Extra over spreading topsoil for travel exceeding 100 m; by hand							
100 m	–	2.50	51.88	–	–	m³	**51.88**
200 m	–	3.50	72.63	–	–	m³	**72.63**
300 m	–	4.50	93.38	–	–	m³	**93.38**
Evenly grading; to general surfaces to bring to finished levels							
by pedestrian operated rotavator	–	–	0.08	0.05	–	m²	**0.13**
by hand	–	0.01	0.21	–	–	m²	**0.21**
Extra over grading for slopes 15–30° by machine or hand	–	–	–	–	–	10%	**–**
Extra over grading for slopes over 30° by machine or hand	–	–	–	–	–	25%	**–**
Clearing stones; disposing off site							
by hand; stones not exceeding 50 mm in any direction; loading to skip 4.6 m³	–	0.01	0.21	0.04	–	m²	**0.25**
Lightly cultivating; weeding; to fallow areas; disposing debris off site							
by hand	–	0.01	0.29	0.29	–	m²	**0.58**

Q31 PLANTING

Item Excluding site overheads and profit	PC £	Labour hours	Labour £	Plant £	Material £	Unit	Total rate £
Preparation of planting operations; spreading only; movement of material to planting beds not included							
Mushroom compost (20 m³ loads); from not further than 25 m from location; cultivating into topsoil by pedestrian operated cultivator							
50 mm thick	118.50	2.86	59.29	1.97	118.50	100 m²	**179.76**
100 mm thick	237.00	6.05	125.52	1.97	237.00	100 m²	**364.49**
150 mm thick	355.50	8.90	184.77	1.97	355.50	100 m²	**542.24**
LANDSCAPE CHEMICAL APPLICATIONS							
Preparation of planting operations; herbicides and granular additives							
For the following topsoil improvement and planting operations add or subtract the following amounts for every £0.10 difference in the material cost price							
35 g/m²	–	–	–	–	0.35	100 m²	**0.35**
50 g/m²	–	–	–	–	0.50	100 m²	**0.50**
70 g/m²	–	–	–	–	0.70	100 m²	**0.70**
100 g/m²	–	–	–	–	1.00	100 m²	**1.00**
General herbicides; in accordance with manufacturer's instructions; knapsack spray application							
Roundup Pro Biactive 360 at 50 ml/100 m²	0.37	0.33	6.92	–	0.37	100 m²	**7.29**
Fertilizers; in top 150 mm of topsoil; at 35 g/m²							
Mascot Microfine; 8+0+6 + 2% Mg + 4% Fe	5.64	0.12	2.54	–	5.64	100 m²	**8.18**
Enmag; controlled release fertilizer (8–9 months); 11+22+09	9.77	0.12	2.54	–	9.77	100 m²	**12.31**
Mascot Outfield; 9+5+5	3.67	0.12	2.54	–	3.67	100 m²	**6.21**
Mascot Outfield; 4+10+10	3.75	0.12	2.54	–	3.75	100 m²	**6.29**
Fertilizers; in top 150 mm of topsoil at 70 g/m²							
Mascot Microfine; 8+0+6 + 2% Mg + 4% Fe	11.29	0.12	2.54	–	11.29	100 m²	**13.83**
Enmag; controlled release fertilizer (8–9 months); 11+22+09	19.54	0.12	2.54	–	19.54	100 m²	**22.08**
Mascot Oufield; 9+5+5	7.33	0.12	2.54	–	7.33	100 m²	**9.87**
Mascot Outfield; 4+10+10	7.51	0.12	2.54	–	7.51	100 m²	**10.05**
Preparation of planting areas; movement of materials to location maximum 25 m from offload location							
By machine	–	–	–	7.99	–	m³	**7.99**
By hand	–	1.00	20.75	–	–	m³	**20.75**

Q31 PLANTING

Item Excluding site overheads and profit	PC £	Labour hours	Labour £	Plant £	Material £	Unit	Total rate £
TREE PLANTING							
Tree planting; pre-planting operations							
Excavating tree pits; depositing soil alongside pits; by machine							
600 mm × 600 mm × 600 mm deep	–	0.15	3.05	0.59	–	nr	3.64
900 mm × 900 mm × 600 mm deep	–	0.33	6.85	1.33	–	nr	8.18
1.00 m × 1.00 m × 600 mm deep	–	0.61	12.73	1.64	–	nr	14.37
1.25 m × 1.25 m × 600 mm deep	–	0.96	19.93	2.57	–	nr	22.50
1.00 m × 1.00 m × 1.00 m deep	–	1.02	21.22	2.74	–	nr	23.96
1.50 m × 1.50 m × 750 mm deep	–	1.73	35.81	4.62	–	nr	40.43
1.50 m × 1.50 m × 1.00 m deep	–	2.30	47.63	6.15	–	nr	53.78
1.75 m × 1.75 m × 1.00 m deep	–	3.13	64.99	8.39	–	nr	73.38
2.00 m × 2.00 m × 1.00 m deep	–	4.09	84.88	10.96	–	nr	95.84
Excavating tree pits; depositing soil alongside pits; by hand							
600 mm × 600 mm × 600 mm deep	–	0.44	9.13	–	–	nr	9.13
900 mm × 900 mm × 600 mm deep	–	1.00	20.75	–	–	nr	20.75
1.00 m × 1.00 m × 600 mm deep	–	1.13	23.34	–	–	nr	23.34
1.25 m × 1.25 m × 600 mm deep	–	1.93	40.05	–	–	nr	40.05
1.00 m × 1.00 m × 1.00 m deep	–	2.06	42.74	–	–	nr	42.74
1.50 m × 1.50 m × 750 mm deep	–	3.47	72.00	–	–	nr	72.00
1.75 m × 1.50 m × 750 mm deep	–	4.05	84.04	–	–	nr	84.04
1.50 m × 1.50 m × 1.00 m deep	–	4.63	96.07	–	–	nr	96.07
2.00 m × 2.00 m × 750 mm deep	–	6.17	128.03	–	–	nr	128.03
2.00 m × 2.00 m × 1.00 m deep	–	8.23	170.78	–	–	nr	170.78
Breaking up subsoil in tree pits; to a depth of 200 mm	–	0.03	0.69	–	–	m²	0.69
Spreading and lightly consolidating approved topsoil (imported or from spoil heaps); in layers not exceeding 150 mm; distance from spoil heaps not exceeding 50 m (imported topsoil not included); by machine							
minimum depth 100 mm	–	–	–	1.56	–	m²	1.56
minimum depth 150 mm	–	–	–	2.34	–	m²	2.34
minimum depth 300 mm	–	–	–	4.28	–	m²	4.28
minimum depth 450 mm	–	–	–	6.18	–	m²	6.18
Spreading and lightly consolidating approved topsoil (imported or from spoil heaps); in layers not exceeding 150 mm; distance from spoil heaps not exceeding 100 m (imported topsoil not included); by hand							
minimum depth 100 mm	–	0.13	2.59	–	–	m²	2.59
minimum depth 150 mm	–	0.19	3.89	–	–	m²	3.89
minimum depth 300 mm	–	0.38	7.80	–	–	m²	7.80
minimum depth 450 mm	–	0.56	11.72	–	–	m²	11.72
Extra for filling tree pits with imported topsoil; PC £28.00/m³; plus allowance for 20% settlement							
depth 100 mm	–	–	–	–	3.36	m²	3.36
depth 150 mm	–	–	–	–	5.04	m²	5.04
depth 200 mm	–	–	–	–	6.72	m²	6.72

Q31 PLANTING

Item Excluding site overheads and profit	PC £	Labour hours	Labour £	Plant £	Material £	Unit	Total rate £
depth 300 mm	–	–	–	–	10.08	m²	**10.08**
depth 400 mm	–	–	–	–	13.44	m²	**13.44**
depth 450 mm	–	–	–	–	15.12	m²	**15.12**
depth 500 mm	–	–	–	–	16.80	m²	**16.80**
depth 600 mm	–	–	–	–	20.16	m²	**20.16**
Add or deduct the following amounts for every £0.50 change in the material price of topsoil							
depth 100 mm	–	–	–	–	0.06	m²	**0.06**
depth 150 mm	–	–	–	–	0.09	m²	**0.09**
depth 200 mm	–	–	–	–	0.12	m²	**0.12**
depth 300 mm	–	–	–	–	0.18	m²	**0.18**
depth 400 mm	–	–	–	–	0.24	m²	**0.24**
depth 450 mm	–	–	–	–	0.27	m²	**0.27**
depth 500 mm	–	–	–	–	0.30	m²	**0.30**
depth 600 mm	–	–	–	–	0.36	m²	**0.36**
Tree staking							
J Toms Ltd; extra over trees for tree stake(s); driving 500 mm into firm ground; trimming to approved height; including two tree ties to approved pattern							
one stake; 1.52 m long × 32 mm × 32 mm	1.50	0.20	4.15	–	1.50	nr	**5.65**
two stakes; 1.21 m long × 25 mm × 25 mm	1.62	0.30	6.22	–	1.62	nr	**7.84**
two stakes; 1.52 m long × 32 mm × 32 mm	2.22	0.30	6.22	–	2.22	nr	**8.44**
three stakes; 1.52 m long × 32 mm × 32 mm	3.33	0.36	7.47	–	3.33	nr	**10.80**
Tree anchors							
Platipus Anchors Ltd; extra over trees for tree anchors							
RF1 rootball kit; for 75 to 220 mm girth; 2 to 4.5 m high; inclusive of Plati-Mat PM1	34.99	1.00	20.75	–	34.99	nr	**55.74**
RF2; rootball kit; for 220 to 450 mm girth; 4.5 to 7.5 m high; inclusive of Plati-Mat PM2	58.48	1.33	27.60	–	58.48	nr	**86.08**
RF3; rootball kit; for 450 to 750 mm girth; 7.5 to 12 m high; inclusive of Plati-Mat PM3	116.90	1.50	31.13	–	116.90	nr	**148.03**
CG1; guy fixing kit; 75 to 220 mm girth; 2 to 4.5 m high	19.08	1.67	34.58	–	19.08	nr	**53.66**
CG2; guy fixing kit; 220 to 450 mm girth; 4.5 to 7.5 m high	38.59	2.00	41.50	–	38.59	nr	**80.09**
installation tools; drive rod for RF1/CG1 kits	–	–	–	–	62.79	nr	**62.79**
installation tools; drive rod for RF2/CG2 kits	–	–	–	–	93.20	nr	**93.20**
Extra over trees for land drain to tree pits; 100 mm diameter perforated flexible agricultural drain; including excavating drain trench; laying pipe; backfilling	3.19	1.00	20.75	–	3.19	m	**23.94**

Q31 PLANTING

Item Excluding site overheads and profit	PC £	Labour hours	Labour £	Plant £	Material £	Unit	Total rate £
TREE PLANTING – cont							
Tree planting; tree pit additives							
Melcourt Industries Ltd; Topgrow;							
incorporating into topsoil at 1 part Topgrow to							
3 parts excavated topsoil; supplied in 75 L							
bags; pit size							
600 × 600 × 600 mm	1.76	0.02	0.41	–	1.76	nr	**2.17**
900 × 900 × 900 mm	5.95	0.06	1.25	–	5.95	nr	**7.20**
1.00 × 1.00 × 1.00 m	8.16	0.24	4.98	–	8.16	nr	**13.14**
1.25 × 1.25 × 1.25 m	15.95	0.40	8.30	–	15.95	nr	**24.25**
1.50 × 1.50 × 1.50 m	27.56	0.90	18.68	–	27.56	nr	**46.24**
Melcourt Industries Ltd; Topgrow;							
incorporating into topsoil at 1 part Topgrow to							
3 parts excavated topsoil; supplied in 70 m³							
loose loads; pit size							
600 × 600 × 600 mm	1.51	0.02	0.35	–	1.51	nr	**1.86**
900 × 900 × 900 mm	5.11	0.05	1.04	–	5.11	nr	**6.15**
1.00 × 1.00 × 1.00 m	7.01	0.20	4.15	–	7.01	nr	**11.16**
1.25 × 1.25 × 1.25 m	13.70	0.33	6.92	–	13.70	nr	**20.62**
1.50 × 1.50 × 1.50 m	23.67	0.75	15.56	–	23.67	nr	**39.23**
Mulching of tree pits; Melcourt Industries							
Ltd (items labelled FSC are Forest							
Stewardship Council certified)							
Spreading mulch; to individual trees;							
maximum distance 25 m (mulch not included)							
50 mm thick	–	0.05	1.01	–	–	m²	**1.01**
75 mm thick	–	0.07	1.51	–	–	m²	**1.51**
100 mm thick	–	0.10	2.02	–	–	m²	**2.02**
Mulch; Bark Nuggets®; to individual trees;							
delivered in 25 m³ loads; maximum distance							
25 m							
50 mm thick	2.87	0.05	1.01	–	2.87	m²	**3.88**
75 mm thick	4.31	0.05	1.04	–	4.31	m²	**5.35**
100 mm thick	5.74	0.07	1.38	–	5.74	m²	**7.12**
Mulch; Amenity Bark Mulch; FSC; to individual							
trees; delivered in 25 m³ loads; maximum							
distance 25 m							
50 mm thick	2.08	0.05	1.01	–	2.08	m²	**3.09**
75 mm thick	3.12	0.07	1.51	–	3.12	m²	**4.63**
100 mm thick	4.16	0.07	1.38	–	4.16	m²	**5.54**
Trees; planting labours only							
Bare root trees; including backfilling with							
previously excavated material (all other							
operations and materials not included)							
light standard; 6–8 cm girth	–	0.35	7.26	–	–	nr	**7.26**
standard; 8–10 cm girth	–	0.40	8.30	–	–	nr	**8.30**
selected standard; 10–12 cm girth	–	0.58	12.04	–	–	nr	**12.04**
heavy standard; 12–14 cm girth	–	0.83	17.29	–	–	nr	**17.29**
extra heavy standard; 14–16 cm girth	–	1.00	20.75	–	–	nr	**20.75**

Q31 PLANTING

Item Excluding site overheads and profit	PC £	Labour hours	Labour £	Plant £	Material £	Unit	Total rate £
Root balled trees; including backfilling with previously excavated material (all other operations and materials not included)							
standard; 8–10 cm girth	–	0.50	10.38	–	–	nr	**10.38**
selected standard; 10–12 cm girth	–	0.60	12.45	–	–	nr	**12.45**
heavy standard; 12–14 cm girth	–	0.80	16.60	–	–	nr	**16.60**
extra heavy standard; 14–16 cm girth	–	1.50	31.13	–	–	nr	**31.13**
16–18 cm girth	–	1.30	26.89	31.82	–	nr	**58.71**
18–20 cm girth	–	1.60	33.20	39.29	–	nr	**72.49**
20–25 cm girth	–	4.50	93.38	121.37	–	nr	**214.75**
25–30 cm girth	–	6.00	124.50	160.37	–	nr	**284.87**
30–35 cm girth	–	11.00	228.25	286.40	–	nr	**514.65**
Tree planting; root balled trees; advanced nursery stock and semi-mature – General Preamble: the cost of planting semi-mature trees will depend on the size and species, and on the access to the site for tree handling machines. Prices should be obtained for individual trees and planting.							
Tree planting; bare root trees; nursery stock; James Coles & Sons (Nurseries) Ltd Acer platanoides; including backfillling with excavated material (other operations not included)							
light standard; 6–8 cm girth	9.45	0.35	7.26	–	9.45	nr	**16.71**
standard; 8–10 cm girth	12.60	0.40	8.30	–	12.60	nr	**20.90**
selected standard; 10–12 cm girth	20.48	0.58	12.04	–	20.48	nr	**32.52**
heavy standard; 12–14 cm girth	44.10	0.83	17.29	–	44.10	nr	**61.39**
extra heavy standard; 14–16 cm girth	58.80	1.00	20.75	–	58.80	nr	**79.55**
Carpinus betulus; including backfillling with excavated material (other operations not included)							
light standard; 6–8 cm girth	13.91	0.35	7.26	–	13.91	nr	**21.17**
standard; 8–10 cm girth	27.82	0.40	8.30	–	27.82	nr	**36.12**
selected standard; 10–12 cm girth	39.64	0.58	12.04	–	39.64	nr	**51.68**
heavy standard; 12–14 cm girth	41.21	0.83	17.30	–	41.21	nr	**58.51**
extra heavy standard; 14–16 cm girth	48.56	1.00	20.75	–	48.56	nr	**69.31**
Fraxinus excelsior; including backfillling with excavated material (other operations not included)							
light standard; 6–8 cm girth	10.24	0.35	7.26	–	10.24	nr	**17.50**
standard; 8–10 cm girth	16.80	0.40	8.30	–	16.80	nr	**25.10**
selected standard; 10–12 cm girth	23.63	0.58	12.04	–	23.63	nr	**35.67**
heavy standard; 12–14 cm girth	41.21	0.83	17.22	–	41.21	nr	**58.43**
extra heavy standard; 14–16 cm girth	49.88	1.00	20.75	–	49.88	nr	**70.63**

Q31 PLANTING

Item Excluding site overheads and profit	PC £	Labour hours	Labour £	Plant £	Material £	Unit	Total rate £
TREE PLANTING – cont							
Tree planting – cont							
Prunus avium 'Plena'; including backfillling with excavated material (other operations not included)							
light standard; 6–8 cm girth	10.24	0.36	7.55	–	10.24	nr	**17.79**
standard; 8–10 cm girth	16.80	0.40	8.30	–	16.80	nr	**25.10**
selected standard; 10–12 cm girth	30.98	0.58	12.04	–	30.98	nr	**43.02**
heavy standard; 12–14 cm girth	49.88	0.83	17.29	–	49.88	nr	**67.17**
extra heavy standard; 14–16 cm girth	66.15	1.00	20.75	–	66.15	nr	**86.90**
Quercus robur; including backfillling with excavated material (other operations not included)							
light standard; 6–8 cm girth	22.05	0.35	7.26	–	22.05	nr	**29.31**
standard; 8–10 cm girth	32.29	0.40	8.30	–	32.29	nr	**40.59**
selected standard; 10–12 cm girth	44.10	0.58	12.04	–	44.10	nr	**56.14**
heavy standard; 12–14 cm girth	61.69	0.83	17.29	–	61.69	nr	**78.98**
Robinia pseudoacacia 'Frisia'; including backfillling with excavated material (other operations not included)							
light standard; 6–8 cm girth	22.05	0.35	7.26	–	22.05	nr	**29.31**
standard; 8–10 cm girth	29.40	0.40	8.30	–	29.40	nr	**37.70**
selected standard; 10–12 cm girth	46.99	0.58	12.04	–	46.99	nr	**59.03**
Tree planting; root balled trees; nursery stock; James Coles & Sons (Nurseries) Ltd							
Acer platanoides; including backfillling with excavated material (other operations not included)							
standard; 8–10 cm girth	24.00	0.48	9.96	–	24.00	nr	**33.96**
selected standard; 10–12 cm girth	37.00	0.56	11.63	–	37.00	nr	**48.63**
heavy standard; 12–14 cm girth	63.00	0.76	15.86	–	63.00	nr	**78.86**
extra heavy standard; 14–16 cm girth	80.50	1.20	24.90	–	80.50	nr	**105.40**
Carpinus betulus; including backfillling with excavated material (other operations not included)							
standard; 8–10 cm girth	38.50	0.48	9.96	–	38.50	nr	**48.46**
selected standard; 10–12 cm girth	55.25	0.56	11.63	–	55.25	nr	**66.88**
heavy standard; 12–14 cm girth	85.50	0.76	15.86	–	85.50	nr	**101.36**
extra heavy standard; 14–16 cm girth	150.50	1.20	24.90	–	150.50	nr	**175.40**
Tree planting; containerized trees; nursery stock; James Coles & Sons (Nurseries) Ltd							
Acer platanoides 'Emerald Queen'; including backfillling with excavated material (other operations not included)							
standard; 8–10 cm girth	44.75	0.48	9.96	–	44.75	each	**54.71**
selected standard; 10–12 cm girth	56.00	0.56	11.63	–	56.00	each	**67.63**
heavy standard; 12–14 cm girth	63.00	0.76	15.86	–	63.00	each	**78.86**
extra heavy standard; 14–16 cm girth	105.00	1.20	24.90	–	105.00	each	**129.90**

Q31 PLANTING

Item Excluding site overheads and profit	PC £	Labour hours	Labour £	Plant £	Material £	Unit	Total rate £
Carpinus betulus; including backfillling with excavated material (other operations not included)							
standard; 8–10 cm girth	44.75	0.48	9.96	–	44.75	each	**54.71**
selected standard; 10–12 cm girth	56.00	0.56	11.63	–	56.00	each	**67.63**
heavy standard; 12–14 cm girth	91.00	0.76	15.86	–	91.00	each	**106.86**
extra heavy standard; 14–16 cm girth	112.00	1.20	24.90	–	112.00	each	**136.90**
Fraxinus excelsior 'Altena'; including backfillling with excavated material (other operations not included)							
standard; 8–10 cm girth	42.00	0.48	9.96	–	42.00	each	**51.96**
selected standard; 10–12 cm girth	67.25	0.56	11.63	–	67.25	each	**78.88**
heavy standard; 12–14 cm girth	84.00	0.76	15.86	–	84.00	each	**99.86**
extra heavy standard; 14–16 cm girth	112.00	1.20	24.90	–	112.00	each	**136.90**
Prunus avium 'Plena'; including backfillling with excavated material (other operations not included)							
selected standard; 10–12 cm girth	70.00	0.56	11.63	–	70.00	each	**81.63**
heavy standard; 12–14 cm girth	84.00	0.76	15.86	–	84.00	each	**99.86**
extra heavy standard; 14–16 cm girth	98.00	1.20	24.90	–	98.00	each	**122.90**
Quercus robur; including backfillling with excavated material (other operations not included)							
standard; 8–10 cm girth	49.00	0.48	9.96	–	49.00	each	**58.96**
selected standard; 10–12 cm girth	77.00	0.56	11.63	–	77.00	each	**88.63**
heavy standard; 12–14 cm girth	105.00	0.76	15.86	–	105.00	each	**120.86**
extra heavy standard; 14–16 cm girth	119.00	1.20	24.90	–	119.00	each	**143.90**
Betula utilis jaquemontii; multi-stemmed; including backfillling with excavated material (other operations not included)							
175/200 mm high	84.00	0.48	9.96	–	84.00	each	**93.96**
200/250 mm high	98.00	0.56	11.63	–	98.00	each	**109.63**
250/300 mm high	120.00	0.76	15.86	–	120.00	each	**135.86**
300/350 mm high	196.00	1.20	24.90	–	196.00	each	**220.90**
Tree planting; Airpot container grown trees; advanced nursery stock and semi-mature; Deepdale Trees Ltd Acer platanoides 'Emerald Queen'; including backfilling with excavated material (other operations not included)							
16–18 cm girth	95.00	1.98	41.09	35.33	95.00	nr	**171.42**
18–20 cm girth	130.00	2.18	45.19	38.86	130.00	nr	**214.05**
20–25 cm girth	190.00	2.38	49.30	42.39	190.00	nr	**281.69**
25–30 cm girth	250.00	2.97	61.63	70.66	250.00	nr	**382.29**
30–35 cm girth	450.00	3.96	82.17	92.41	450.00	nr	**624.58**

Q31 PLANTING

Item Excluding site overheads and profit	PC £	Labour hours	Labour £	Plant £	Material £	Unit	Total rate £
TREE PLANTING – cont							
Tree planting – cont							
Aesculus briotti; including backfilling with excavated material (other operations not included)							
16–18 cm girth	110.00	1.98	41.09	35.33	110.00	nr	**186.42**
18–20 cm girth	130.00	1.60	33.20	38.86	130.00	nr	**202.06**
20–25 cm girth	200.00	2.38	49.30	42.39	200.00	nr	**291.69**
25–30 cm girth	300.00	2.97	61.63	85.49	300.00	nr	**447.12**
30–35 cm girth	450.00	3.96	82.17	92.41	450.00	nr	**624.58**
Tree planting; Airpot container grown trees; semi-mature and mature trees; Deepdale Trees Ltd; planting and back filling; planted by telehandler or by crane; delivery included; all other operations priced separately							
Semi-mature trees indicative prices							
40–45 cm girth	550.00	4.00	83.00	55.69	550.00	nr	**688.69**
45–50 cm girth	750.00	4.00	83.00	55.69	750.00	nr	**888.69**
55–60 cm girth	1350.00	6.00	124.50	55.69	1350.00	nr	**1530.19**
60–70 cm girth	2500.00	7.00	145.25	74.61	2500.00	nr	**2719.86**
70–80 cm girth	3500.00	7.50	155.63	93.53	3500.00	nr	**3749.16**
80–90 cm girth	4500.00	8.00	166.00	111.37	4500.00	nr	**4777.37**
Tree planting; root balled trees; advanced nursery stock and semi-mature; Lorenz von Ehren							
Acer platanoides 'Emerald Queen'; including backfilling with excavated material (other operations not included)							
16–18 cm girth	93.50	1.30	26.89	4.02	93.50	nr	**124.41**
18–20 cm girth	115.50	1.60	33.20	4.02	115.50	nr	**152.72**
20–25 cm girth	143.00	4.50	93.38	16.90	143.00	nr	**253.28**
25–30 cm girth	187.00	6.00	124.50	21.09	187.00	nr	**332.59**
30–35 cm girth	335.50	11.00	228.25	28.37	335.50	nr	**592.12**
Quercus palustris; including backfilling with excavated material (other operations not included)							
16–18 cm girth	137.50	1.30	26.89	4.02	137.50	nr	**168.41**
18–20 cm girth	165.00	1.60	33.20	4.02	165.00	nr	**202.22**
20–25 cm girth	209.00	4.50	93.38	16.90	209.00	nr	**319.28**
25–30 cm girth	247.50	6.00	124.50	21.09	247.50	nr	**393.09**
30–35 cm girth	374.00	11.00	228.25	28.37	374.00	nr	**630.62**

Q31 PLANTING

Item Excluding site overheads and profit	PC £	Labour hours	Labour £	Plant £	Material £	Unit	Total rate £
Tilia cordata 'Greenspire'; including backfilling with excavated material (other operations not included)							
16–18 cm girth	104.50	1.30	26.89	4.02	104.50	nr	**135.41**
18–20 cm girth	126.50	1.60	33.20	4.02	126.50	nr	**163.72**
20–25 cm girth	154.00	4.50	93.38	16.90	154.00	nr	**264.28**
25–30 cm girth; 5 × transplanted 4.0–5.0 m tall	165.00	6.00	124.50	21.09	165.00	nr	**310.59**
30–35 cm girth; 5 × transplanted 5.0–7.0 m tall	242.00	11.00	228.25	28.37	242.00	nr	**498.62**
Betula pendula; 3 stems; including backfilling with excavated material (other operations not included)							
3.0–3.5 m high	66.00	1.98	41.09	35.33	66.00	nr	**142.42**
3.5–4.0 m high	99.00	1.60	33.20	38.86	99.00	nr	**171.06**
4.0–4.5 m high	143.00	2.38	49.30	55.44	143.00	nr	**247.74**
4.5–5.0 m high	165.00	2.97	61.63	85.49	165.00	nr	**312.12**
5.0–6.0 m high	231.00	3.96	82.17	92.41	231.00	nr	**405.58**
6.0–7.0 m high	385.00	4.50	93.38	107.56	385.00	nr	**585.94**
Pinus sylvestris; including backfilling with excavated material (other operations not included)							
3.0–3.5 m high	440.00	1.98	41.09	35.33	440.00	nr	**516.42**
3.5–4.0 m high	550.00	1.60	33.20	38.86	550.00	nr	**622.06**
4.0–4.5 m high	759.01	2.38	49.30	42.39	759.01	nr	**850.70**
4.5–5.0 m high	1045.01	2.97	61.63	85.49	1045.01	nr	**1192.13**
5.0–6.0 m high	1650.02	3.96	82.17	92.41	1650.02	nr	**1824.60**
6.0–7.0 m high	2750.03	4.50	93.38	111.13	2750.03	nr	**2954.54**
Tree planting; containerized trees; Lorenz von Ehren; to the tree prices above add for Airpot containerization only							
Tree size							
20–25 cm	–	–	–	–	45.00	nr	**45.00**
25–30 cm	–	–	–	–	75.00	nr	**75.00**
30–35 cm	–	–	–	–	95.00	nr	**95.00**
35–40 cm	–	–	–	–	130.00	nr	**130.00**
40–45 cm	–	–	–	–	150.00	nr	**150.00**
45–50 cm	–	–	–	–	190.00	nr	**190.00**
50–60 cm	–	–	–	–	250.00	nr	**250.00**
60–70 cm	–	–	–	–	320.00	nr	**320.00**
70–80 cm	–	–	–	–	390.00	nr	**390.00**
80–90 cm	–	–	–	–	450.00	nr	**450.00**

Q31 PLANTING

Item Excluding site overheads and profit	PC £	Labour hours	Labour £	Plant £	Material £	Unit	Total rate £
SHRUB PLANTING							
Shrub planting – General Preamble: for preparation of planting areas see 'Cultivation' at the beginning of the planting section.							
Shrub planting Setting out; selecting planting from holding area; loading to wheelbarrows; planting as plan or as directed; distance from holding area maximum 50 m; plants 2–3 litre containers							
plants in groups of 100 nr minimum	–	0.01	0.24	–	–	nr	**0.24**
plants in groups of 10–100 nr	–	0.02	0.34	–	–	nr	**0.34**
plants in groups of 3–5 nr	–	0.03	0.52	–	–	nr	**0.52**
single plants not grouped	–	0.04	0.83	–	–	nr	**0.83**
Forming planting holes; in cultivated ground (cultivating not included); by mechanical auger; trimming holes by hand; depositing excavated material alongside holes							
250 mm diameter	–	0.03	0.69	0.06	–	nr	**0.75**
250 × 250 mm	–	0.04	0.83	0.11	–	nr	**0.94**
300 × 300 mm	–	0.08	1.56	0.14	–	nr	**1.70**
Hand excavation; forming planting holes; in cultivated ground (cultivating not included); depositing excavated material alongside holes							
100 × 100 × 100 mm deep; with mattock or hoe	–	0.01	0.14	–	–	nr	**0.14**
250 × 250 × 300 mm deep	–	0.04	0.83	–	–	nr	**0.83**
300 × 300 × 300 mm deep	–	0.06	1.15	–	–	nr	**1.15**
400 × 400 × 400 mm deep	–	0.13	2.59	–	–	nr	**2.59**
500 × 500 × 500 mm deep	–	0.25	5.19	–	–	nr	**5.19**
600 × 600 × 600 mm deep	–	0.43	8.98	–	–	nr	**8.98**
900 × 900 × 600 mm deep	–	1.00	20.75	–	–	nr	**20.75**
1.00 m × 1.00 m × 600 mm deep	–	1.23	25.52	–	–	nr	**25.52**
1.25 m × 1.25 m × 600 mm deep	–	1.93	40.05	–	–	nr	**40.05**
Hand excavation; forming planting holes; in uncultivated ground; depositing excavated material alongside holes							
100 × 100 × 100 mm deep; with mattock or hoe	–	0.03	0.52	–	–	nr	**0.52**
250 × 250 × 300 mm deep	–	0.06	1.15	–	–	nr	**1.15**
300 × 300 × 300 mm deep	–	0.06	1.30	–	–	nr	**1.30**
400 × 400 × 400 mm deep	–	0.25	5.19	–	–	nr	**5.19**
500 × 500 × 500 mm deep	–	0.33	6.75	–	–	nr	**6.75**
600 × 600 × 600 mm deep	–	0.55	11.41	–	–	nr	**11.41**
900 × 900 × 600 mm deep	–	1.25	25.94	–	–	nr	**25.94**
1.00 m × 1.00 m × 600 mm deep	–	1.54	31.90	–	–	nr	**31.90**
1.25 m × 1.25 m × 600 mm deep	–	2.41	50.06	–	–	nr	**50.06**

Q31 PLANTING

Item Excluding site overheads and profit	PC £	Labour hours	Labour £	Plant £	Material £	Unit	Total rate £
Bare root planting; to planting holes (forming holes not included); including backfilling with excavated material (bare root plants not included)							
bare root 1+1; 30–90 mm high	–	0.02	0.35	–	–	nr	0.35
bare root 1+2; 90–120 mm high	–	0.02	0.35	–	–	nr	0.35
Containerized planting; to planting holes (forming holes not included); including backfilling with excavated material (shrub or ground cover not included)							
9 cm pot	–	0.01	0.21	–	–	nr	0.21
2 litre container	–	0.02	0.41	–	–	nr	0.41
3 litre container	–	0.02	0.46	–	–	nr	0.46
5 litre container	–	0.03	0.69	–	–	nr	0.69
10 litre container	–	0.05	1.04	–	–	nr	1.04
15 litre container	–	0.07	1.38	–	–	nr	1.38
20 litre container	–	0.08	1.73	–	–	nr	1.73
Shrub planting; 2 litre containerized plants; in cultivated ground (cultivating not included); PC £3.00/nr							
average 2 plants per m²	–	0.06	1.16	–	6.00	m²	7.16
average 3 plants per m²	–	0.08	1.74	–	9.00	m²	10.74
average 4 plants per m²	–	0.11	2.33	–	12.00	m²	14.33
average 6 plants per m²	–	0.17	3.49	–	18.00	m²	21.49
Extra over shrubs for stakes	0.70	0.02	0.35	–	0.70	nr	1.05
Mushroom compost; 25 m³ loads; delivered not further than 25 m from location; cultivating into topsoil by pedestrian operated machine							
50 mm thick	118.50	2.86	59.29	1.97	118.50	100 m²	179.76
100 mm thick	237.00	6.05	125.52	1.97	237.00	100 m²	364.49
150 mm thick	355.50	8.90	184.77	1.97	355.50	100 m²	542.24
200 mm thick	474.00	12.90	267.77	1.97	474.00	100 m²	743.74
Manure; 20 m³ loads; delivered not further than 25 m from location; cultivating into topsoil by pedestrian operated machine							
50 mm thick	236.25	2.86	59.29	1.97	236.25	100 m²	297.51
100 mm thick	496.13	6.05	125.52	1.97	496.13	100 m²	623.62
150 mm thick	744.19	8.90	184.77	1.97	744.19	100 m²	930.93
200 mm thick	992.25	12.90	267.77	1.97	992.25	100 m²	1261.99
Fertilizers (7+7+7); PC £0.81/kg; to beds; by hand							
35 g/m²	3.75	0.17	3.46	–	3.75	100 m²	7.21
50 g/m²	5.36	0.17	3.46	–	5.36	100 m²	8.82
70 g/m²	7.51	0.17	3.46	–	7.51	100 m²	10.97
Fertilizers; Enmag; PC £2.66/kg; controlled release fertilizer; to beds; by hand							
35 g/m²	9.30	0.17	3.46	–	9.30	100 m²	12.76
50 g/m²	13.29	0.17	3.46	–	13.29	100 m²	16.75
70 g/m²	19.93	0.17	3.46	–	19.93	100 m²	23.39
Note: for machine incorporation of fertilizers and soil conditioners see 'Cultivation'.							

Q31 PLANTING

Item Excluding site overheads and profit	PC £	Labour hours	Labour £	Plant £	Material £	Unit	Total rate £
HERBACEOUS/GROUNDCOVER/BULB/ AQUATIC PLANTING							
Herbaceous and groundcover planting							
Herbaceous plants; PC £1.50/nr; including forming planting holes in cultivated ground (cultivating not included); backfilling with excavated material; 1 litre containers							
average 4 plants per m² – 500 mm centres	6.00	0.09	1.94	–	6.00	m²	**7.94**
average 6 plants per m² – 408 mm centres	–	0.14	2.91	–	9.00	m²	**11.91**
average 8 plants per m² – 354 mm centres	–	0.19	3.88	–	12.00	m²	**15.88**
Note: for machine incorporation of fertilizers and soil conditioners see 'Cultivation'.							
Bulb planting							
Bulbs; including forming planting holes in cultivated area (cultivating not included); backfilling with excavated material							
small	13.00	0.83	17.29	–	13.00	100 nr	**30.29**
medium	22.00	0.83	17.29	–	22.00	100 nr	**39.29**
large	25.00	0.91	18.86	–	25.00	100 nr	**43.86**
Bulbs; in grassed area; using bulb planter; including backfilling with screened topsoil or peat and cut turf plug							
small	13.00	1.67	34.58	–	13.00	100 nr	**47.58**
medium	22.00	1.67	34.58	–	22.00	100 nr	**56.58**
large	25.00	2.00	41.50	–	25.00	100 nr	**66.50**
Aquatic planting							
Aquatic plants; in prepared growing medium in pool; plant size 2–3 litre containerized (plants not included)	–	0.04	0.83	–	–	nr	**0.83**
MAINTENANCE OF PLANTING AREAS							
Operations after planting							
Initial cutting back to shrubs and hedge plants; including disposal of all cuttings	–	1.00	20.75	–	–	100 m²	**20.75**
Mulch; Melcourt Industries Ltd; Bark Nuggets®; to plant beds; delivered in 25 m³ loads; maximum distance 25 m							
50 mm thick	2.55	0.03	0.61	–	2.55	m²	**3.16**
75 mm thick	4.31	0.07	1.38	–	4.31	m²	**5.69**
100 mm thick	5.74	0.09	1.84	–	5.74	m²	**7.58**
Mulch; Melcourt Industries Ltd; Amenity Bark Mulch; FSC; to plant beds; delivered in 25 m³ loads; maximum distance 25 m							
50 mm thick	2.08	0.04	0.92	–	2.08	m²	**3.00**
75 mm thick	3.12	0.07	1.38	–	3.12	m²	**4.50**
100 mm thick	4.16	0.09	1.84	–	4.16	m²	**6.00**

Q35 LANDSCAPE MAINTENANCE

Item Excluding site overheads and profit	PC £	Labour hours	Labour £	Plant £	Material £	Unit	Total rate £
Maintenance operations (Note: the following rates apply to aftercare maintenance executed as part of a landscaping contract only)							
Weeding and hand forking planted areas; including disposing weeds and debris on site; areas maintained weekly	–	–	0.08	–	–	m^2	**0.08**
Weeding and hand forking planted areas; including disposing weeds and debris on site; areas maintained monthly	–	0.01	0.21	–	–	m^2	**0.21**
Mulch; Melcourt Industries Ltd; Bark Nuggets®; to plant beds; delivered in 25 m^3 loads; maximum distance 25 m							
50 mm thick	2.87	0.04	0.92	–	2.87	m^2	**3.79**
75 mm thick	4.31	0.07	1.38	–	4.31	m^2	**5.69**
Mulch; Melcourt Industries Ltd; Amenity Bark Mulch; FSC; to plant beds; delivered in 25 m^3 loads; maximum distance 25 m							
50 mm thick	2.08	0.04	0.92	–	2.08	m^2	**3.00**
75 mm thick	3.12	0.07	1.38	–	3.12	m^2	**4.50**
Watering planting; evenly; at a rate of 5 litres/m^2							
using hand-held watering equipment	–	0.25	5.19	–	–	100 m^2	**5.19**
using sprinkler equipment and with sufficient water pressure to run 1 nr 15 m radius sprinkler	–	0.14	2.89	–	–	100 m^2	**2.89**
Work to existing planting							
Cutting and trimming ornamental hedges; to specified profiles; including cleaning out hedge bottoms; hedge cut 2 occasions per annum; by hand							
up to 2.00 m high	–	0.03	0.69	–	0.33	m	**1.02**
2.00–4.00 m high	–	0.05	1.04	1.97	0.67	m	**3.68**
Q35 LANDSCAPE MAINTENANCE							
GRASS CUTTING							
Grass cutting – pedestrian operated equipment							
Using cylinder lawn mower fitted with not less than five cutting blades, front and rear rollers; on surface not exceeding 30° from horizontal; arisings let fly; width of cut							
51 cm	–	0.06	1.26	0.20	–	100 m	**1.46**
61 cm	–	0.05	1.06	0.18	–	100 m	**1.24**

Q35 LANDSCAPE MAINTENANCE

Item Excluding site overheads and profit	PC £	Labour hours	Labour £	Plant £	Material £	Unit	Total rate £
GRASS CUTTING – cont							
Grass cutting – pedestrian operated equipment – cont							
Using rotary self-propelled mower; width of cut							
45 cm	–	0.08	1.60	0.06	–	100 m²	**1.66**
40 cm	–	0.11	2.31	0.06	–	100 m²	**2.37**
Add for using grass box for collecting and depositing arisings							
removing and depositing arisings	–	0.05	1.04	–	–	100 m²	**1.04**
Add for 30–50° from horizontal	–	–	–	–	–	33%	**–**
Add for slopes exceeding 50°	–	–	–	–	–	100%	**–**
Cutting grass or light woody undergrowth; using trimmer with nylon cord or metal disc cutter; on surface							
not exceeding 30° from horizontal	–	0.20	4.15	0.29	–	100 m²	**4.44**
30–50° from horizontal	–	0.40	8.30	0.59	–	100 m²	**8.89**
exceeding 50° from horizontal	–	0.50	10.38	0.73	–	100 m²	**11.11**
Grass cutting – collecting arisings							
Extra over for tractor drawn and self-propelled machinery using attached grass boxes; depositing arisings							
22 cuts per year	–	0.05	1.04	–	–	100 m²	**1.04**
18 cuts per year	–	0.08	1.56	–	–	100 m²	**1.56**
12 cuts per year	–	0.10	2.08	–	–	100 m²	**2.08**
4 cuts per year	–	0.25	5.19	–	–	100 m²	**5.19**
Disposing arisings							
22 cuts per year	–	0.01	0.16	0.03	0.05	100 m²	**0.24**
18 cuts per year	–	0.01	0.20	0.04	0.06	100 m²	**0.30**
12 cuts per year	–	0.01	0.30	0.05	0.09	100 m²	**0.44**
4 cuts per year	–	0.04	0.90	0.17	0.26	100 m²	**1.33**
Scarifying by hand							
hand implement	–	0.50	10.38	–	–	100 m²	**10.38**
add for disposal of arisings	–	0.03	0.52	1.35	16.50	100 m²	**18.37**
TURF MAINTENANCE							
Rolling							
Rolling grassed area; equipment towed by tractor; once over; using							
smooth roller	–	0.01	0.30	0.43	–	100 m²	**0.73**
Turf aeration							
By hand							
hand fork; to effect a minimum penetration of 100 mm and spaced 150 mm apart	–	1.33	27.67	–	–	100 m²	**27.67**
hollow tine hand implement; to effect a minimum penetration of 100 mm and spaced 150 mm apart	–	2.00	41.50	–	–	100 m²	**41.50**
collection of arisings by hand	–	3.00	62.25	–	–	100 m²	**62.25**

Q35 LANDSCAPE MAINTENANCE

Item Excluding site overheads and profit	PC £	Labour hours	Labour £	Plant £	Material £	Unit	Total rate £
Turf areas; surface treatments and top dressing; Boughton Loam Ltd Apply screened topdressing to grass surfaces; spread using Tru-Lute							
sand soil mixes 90/10 to 50/50	0.12	–	0.04	0.06	0.12	m²	**0.22**
CHEMICAL APPLICATIONS							
Spraying; labour rates only; for chemical rates please use the tables at the beginning of the section Q35 Landscape Maintenance in the Major Works section of this book Herbicide applications; standard backpack spray applicators; application to maintain 1.00 m diameter clear circles (0.79 m²) around new planting							
plants at 1.50 m centres; 4444 nr/ha	–	0.25	5.12	–	–	100 m²	**5.12**
plants at 1.75 m centres; 3265 nr/ha	–	0.18	3.76	–	–	100 m²	**3.76**
plants at 2.00 m centres; 2500 nr/ha	–	0.14	2.88	–	–	100 m²	**2.88**
understory of mature planting or trees	–	0.13	2.59	–	–	100 m²	**2.59**
mass spraying low vegetation or hard surfaces	–	0.10	2.08	–	–	100 m²	**2.08**
GROUNDS MAINTENANCE							
Leaf clearance Using pedestrian operated mechanical equipment and blowers							
grassed areas with perimeters of mature trees such as sports fields and amenity areas	–	0.04	0.83	0.05	–	100 m²	**0.88**
grassed areas containing ornamental trees and shrub beds	–	0.10	2.08	0.13	–	100 m²	**2.21**
verges	–	0.07	1.38	0.09	–	100 m²	**1.47**
By hand							
grassed areas with perimeters of mature trees such as sports fields and amenity areas	–	0.05	1.04	0.07	–	100 m²	**1.11**
grassed areas containing ornamental trees and shrub beds	–	0.08	1.73	0.11	–	100 m²	**1.84**
verges	–	1.00	20.75	1.30	–	100 m²	**22.05**
Removal of arisings							
areas with perimeters of mature trees	–	0.01	0.14	0.12	1.32	100 m²	**1.58**
areas containing ornamental trees and shrub beds	–	0.02	0.41	0.45	3.30	100 m²	**4.16**
Litter clearance Collection and disposal of litter from grassed area							
areas exceeding 1000 m²	–	0.01	0.21	0.29	–	100 m²	**0.50**
area not exceeding 1000 m²	–	0.04	0.83	0.29	–	100 m²	**1.12**

Q35 LANDSCAPE MAINTENANCE

Item Excluding site overheads and profit	PC £	Labour hours	Labour £	Plant £	Material £	Unit	Total rate £
GROUNDS MAINTENANCE – cont							
Edge maintenance							
Maintain edges where lawn abuts pathway or							
hard surface using							
strimmer	–	0.01	0.10	0.01	–	m	**0.11**
shears	–	0.02	0.35	–	–	m	**0.35**
Maintain edges where lawn abuts plant bed							
using							
mechanical edging tool	–	0.01	0.14	0.03	–	m	**0.17**
shears	–	0.01	0.23	–	–	m	**0.23**
half moon edging tool	–	0.02	0.41	–	–	m	**0.41**
Tree guards, stakes and ties							
Adjusting existing tree tie	–	0.03	0.69	–	–	nr	**0.69**
Taking up single or double tree stake and ties;							
removing and disposing	–	0.05	1.04	–	–	nr	**1.04**
Pruning shrubs							
Trimming ground cover planting							
soft groundcover; vinca ivy and the like	–	1.00	20.75	–	–	100 m²	**20.75**
woody groundcover; cotoneaster and the like	–	1.50	31.13	–	–	100 m²	**31.13**
Pruning massed shrub border (measure							
ground area)							
shrub beds pruned annually	–	0.01	0.21	–	–	m²	**0.21**
shrub beds pruned hard every 3 years	–	0.03	0.58	–	–	m²	**0.58**
Cutting off dead heads							
bush or standard rose	–	0.05	1.04	–	–	nr	**1.04**
climbing rose	–	0.08	1.73	–	–	nr	**1.73**
Pruning roses							
bush or standard rose	–	0.05	1.04	–	–	nr	**1.04**
climbing or rambling rose; tying in as							
required	–	0.07	1.38	–	–	nr	**1.38**
Pruning ornamental shrubs; height before							
pruning (increase these rates by 50% if							
pruning work has not been executed during							
the previous two years)							
not exceeding 1 m	–	0.04	0.83	–	–	nr	**0.83**
1–2 m	–	0.06	1.15	–	–	nr	**1.15**
exceeding 2 m	–	0.13	2.59	–	–	nr	**2.59**
Removing excess growth etc. from face of							
building etc.; height before pruning							
not exceeding 2 m	–	0.03	0.59	–	–	nr	**0.59**
2–4 m	–	0.05	1.04	–	–	nr	**1.04**
4–6 m	–	0.08	1.73	–	–	nr	**1.73**
6–8 m	–	0.13	2.59	–	–	nr	**2.59**
8–10 m	–	0.14	2.97	–	–	nr	**2.97**
Removing epicormic growth from base of							
shrub or trunk and base of tree; any height;							
any diameter; number of growths							
not exceeding 10	–	0.05	1.04	–	–	nr	**1.04**
10–20	–	0.07	1.38	–	–	nr	**1.38**

Q35 LANDSCAPE MAINTENANCE

Item Excluding site overheads and profit	PC £	Labour hours	Labour £	Plant £	Material £	Unit	Total rate £
Beds, borders and planters							
Lifting							
bulbs	–	0.50	10.38	–	–	100 nr	**10.38**
tubers or corms	–	0.40	8.30	–	–	100 nr	**8.30**
established herbaceous plants; hoeing and depositing for replanting	–	2.00	41.50	–	–	100 nr	**41.50**
Temporary staking and tying in herbaceous plant	–	0.03	0.69	–	0.24	nr	**0.93**
Cutting down spent growth of herbaceous plant; clearing arisings							
unstaked	–	0.02	0.41	–	–	nr	**0.41**
staked; not exceeding 4 stakes per plant; removing stakes and putting into store	–	0.03	0.52	–	–	nr	**0.52**
Hand weeding							
newly planted areas	–	2.00	41.50	–	–	100 m²	**41.50**
established areas	–	0.50	10.38	–	–	100 m²	**10.38**
Removing grasses from groundcover areas	–	3.00	62.31	–	–	100 m²	**62.31**
Hand digging with fork; not exceeding 150 mm deep; breaking down lumps; leaving surface with a medium tilth	–	1.33	27.66	–	–	100 m²	**27.66**
Hand digging with fork or spade to an average depth of 230 mm; breaking down lumps; leaving surface with a medium tilth	–	2.00	41.50	–	–	100 m²	**41.50**
Hand hoeing; not exceeding 50 mm deep; leaving surface with a medium tilth		0.40	8.30	–	–	100 m²	**8.30**
Hand raking to remove stones etc.; breaking down lumps; leaving surface with a fine tilth prior to planting	–	0.67	13.83	–	–	100 m²	**13.83**
Hand weeding; planter or window box; not exceeding 1.00 m²							
ground level box	–	0.05	1.04	–	–	nr	**1.04**
box accessed by stepladder	–	0.08	1.73	–	–	nr	**1.73**
Spreading only compost, mulch or processed bark to a depth of 75 mm							
on shrub bed with existing mature planting	–	0.09	1.89	–	–	m²	**1.89**
recently planted areas	–	0.07	1.38	–	–	m²	**1.38**
groundcover and herbaceous areas	–	0.08	1.56	–	–	m²	**1.56**
Clearing cultivated area of leaves, litter and other extraneous debris; using hand implement							
weekly maintenance	–	0.13	2.59	–	–	100 m²	**2.59**
daily maintenance	–	0.02	0.35	–	–	100 m²	**0.35**
BEDDING							
Bedding							
Lifting							
bedding plants; hoeing and depositing for disposal	–	3.00	62.25	–	–	100 m²	**62.25**
Hand digging with fork; not exceeding 150 mm deep; breaking down lumps; leaving surface with a medium tilth	–	0.75	15.56	–	–	100 m²	**15.56**

Q35 LANDSCAPE MAINTENANCE

Item Excluding site overheads and profit	PC £	Labour hours	Labour £	Plant £	Material £	Unit	Total rate £
BEDDING – cont							
Bedding – cont							
Hand weeding							
newly planted areas	–	2.00	41.50	–	–	100 m²	**41.50**
established areas	–	0.50	10.38	–	–	100 m²	**10.38**
Hand digging with fork or spade to an average depth of 230 mm; breaking down lumps; leaving surface with a medium tilth	–	0.50	10.38	–	–	100 m²	**10.38**
Hand hoeing: not exceeding 50 mm deep; leaving surface with a medium tilth	–	0.40	8.30	–	–	100 m²	**8.30**
Hand raking to remove stones etc.; breaking down lumps; leaving surface with a fine tilth prior to planting	–	0.67	13.83	–	–	100 m²	**13.83**
Hand weeding; planter or window box; not exceeding 1.00 m²							
ground level box	–	0.05	1.04	–	–	nr	**1.04**
box accessed by stepladder	–	0.08	1.73	–	–	nr	**1.73**
Spreading only; compost, mulch or processed bark to a depth of 75 mm							
on shrub bed with existing mature planting	–	0.09	1.89	–	–	m²	**1.89**
recently planted areas	–	0.07	1.38	–	–	m²	**1.38**
groundcover and herbaceous areas	–	0.08	1.56	–	–	m²	**1.56**
Collecting bedding from nursery	–	3.00	62.25	18.00	–	100 m²	**80.25**
Setting out							
mass planting single variety	–	0.13	2.59	–	–	m²	**2.59**
pattern	–	0.33	6.92	–	–	m²	**6.92**
Planting only							
massed bedding plants	–	0.20	4.15	–	–	m²	**4.15**
Clearing cultivated area of leaves, litter and other extraneous debris; using hand implement							
weekly maintenance	–	0.13	2.59	–	–	100 m²	**2.59**
daily maintenance	–	0.02	0.35	–	–	100 m²	**0.35**
HEDGE CUTTING							
Hedge cutting; field hedges cut once or twice annually							
Trimming sides and top using hand tool or hand held mechanical tools							
not exceeding 2 m high	–	0.10	2.08	0.16	–	10 m²	**2.24**
2 to 4 m high	–	0.33	6.92	0.52	–	10 m²	**7.44**
Hedge cutting; ornamental							
Trimming sides and top using hand tool or hand held mechanical tools							
not exceeding 2 m high	–	0.13	2.59	0.20	–	10 m²	**2.79**
2 to 4 m high	–	0.50	10.38	0.78	–	10 m²	**11.16**

Q35 LANDSCAPE MAINTENANCE

Item Excluding site overheads and profit	PC £	Labour hours	Labour £	Plant £	Material £	Unit	Total rate £
Hedge cutting; reducing width; hand tool or hand held mechanical tools							
Not exceeding 2 m high							
average depth of cut not exceeding							
300 mm	–	0.10	2.08	0.16	–	10 m²	**2.24**
average depth of cut 300 to 600 mm	–	0.83	17.29	1.31	–	10 m²	**18.60**
average depth of cut 600 to 900 mm	–	1.25	25.94	1.96	–	10 m²	**27.90**
2 to 4 m high							
average depth of cut not exceeding							
300 mm	–	0.03	0.52	0.04	–	10 m²	**0.56**
average depth of cut 300 to 600 mm	–	0.13	2.59	0.20	–	10 m²	**2.79**
average depth of cut 600 to 900 mm	–	2.50	51.88	3.92	–	10 m²	**55.80**
4 to 6 m high							
average depth of cut not exceeding							
300 mm	–	0.10	2.08	0.16	–	m²	**2.24**
average depth of cut 300 to 600 mm	–	0.17	3.46	0.26	–	m²	**3.72**
average depth of cut 600 to 900 mm	–	0.50	10.38	0.78	–	m²	**11.16**
Hedge cutting; reducing height; hand tool or hand held mechanical tools							
Not exceeding 2 m high							
average depth of cut not exceeding							
300 mm	–	0.03	0.52	0.04	–	m²	**0.56**
average depth of cut 300 to 600 mm	–	0.04	0.83	0.06	–	m²	**0.89**
average depth of cut 600 to 900 mm	–	0.10	2.08	0.16	–	m²	**2.24**
2 to 4 m high							
average depth of cut not exceeding							
300 mm	–	0.05	1.04	0.08	–	m²	**1.12**
average depth of cut 300 to 600 mm	–	0.07	1.38	0.10	–	m²	**1.48**
average depth of cut 600 to 900 mm	–	0.13	2.59	0.20	–	m²	**2.79**
4 to 6 m high							
average depth of cut not exceeding							
300 mm	–	0.11	2.31	0.17	–	m²	**2.48**
average depth of cut 300 to 600 mm	–	0.20	4.15	0.31	–	m²	**4.46**
average depth of cut 600 to 900 mm	–	0.50	10.38	0.78	–	m²	**11.16**
Hedge cutting; removal and disposal of arisings							
Sweeping up and depositing arisings							
300 mm cut	–	0.05	1.04	–	–	10 m²	**1.04**
600 mm cut	–	0.20	4.15	–	–	10 m²	**4.15**
900 mm cut	–	0.40	8.30	–	–	10 m²	**8.30**
Chipping arisings							
300 mm cut	–	0.02	0.41	0.43	–	10 m²	**0.84**
600 mm cut	–	0.08	1.73	1.80	–	10 m²	**3.53**
900 mm cut	–	0.20	4.15	4.31	–	10 m²	**8.46**
Disposal of unchipped arisings							
300 mm cut	–	0.02	0.35	0.68	2.20	10 m²	**3.23**
600 mm cut	–	0.03	0.69	1.35	3.30	10 m²	**5.34**
900 mm cut	–	0.08	1.73	3.38	8.25	10 m²	**13.36**

Q40 FENCING

Item Excluding site overheads and profit	PC £	Labour hours	Labour £	Plant £	Material £	Unit	Total rate £
HEDGE CUTTING – cont							
Hedge cutting – cont							
Disposal of chipped arisings							
300 mm cut	–	–	0.07	0.23	3.30	10 m²	**3.60**
600 mm cut	–	0.02	0.35	0.23	6.60	10 m²	**7.18**
900 mm cut	–	0.03	0.69	0.45	16.50	10 m²	**17.64**
IRRIGATION AND WATERING							
Irrigation and watering							
Hand held hosepipe; flow rate 25 litres per							
minute; irrigation requirement							
10 litres/m²	–	0.74	15.29	–	–	100 m²	**15.29**
15 litres/m²	–	1.10	22.82	–	–	100 m²	**22.82**
20 litres/m²	–	1.46	30.36	–	–	100 m²	**30.36**
25 litres/m²	–	1.84	38.12	–	–	100 m²	**38.12**
Hand held hosepipe; flow rate 40 litres per							
minute; irrigation requirement							
10 litres/m²	–	0.46	9.59	–	–	100 m²	**9.59**
15 litres/m²	–	0.69	14.38	–	–	100 m²	**14.38**
20 litres/m²	–	0.91	18.94	–	–	100 m²	**18.94**
25 litres/m²	–	1.15	23.78	–	–	100 m²	**23.78**
Q40 FENCING							
TEMPORARY/PROTECTIVE FENCING							
Temporary fencing; HSS Hire; mesh framed							
mesh unclimbable fencing; including							
precast concrete supports and couplings							
Weekly hire; 2.85 × 2 m high							
weekly hire rate	–	–	–	2.35	–	m	**2.35**
erection of fencing; labour only	–	0.13	2.59	–	–	m	**2.59**
removal of fencing; loading to collection							
vehicle	–	0.08	1.73	–	–	m	**1.73**
delivery charge	–	–	–	0.80	–	m	**0.80**
return haulage charge	–	–	–	0.60	–	m	**0.60**
Protective fencing							
Cleft chestnut rolled fencing; fixing to 100 mm							
diameter chestnut posts; driving into firm							
ground at 3 m centres							
900 mm high	4.42	0.11	2.21	–	9.68	m	**11.89**
1200 mm high	5.14	0.16	3.32	–	10.40	m	**13.72**
1500 mm high; 3 strand	7.66	0.21	4.43	–	12.93	m	**17.36**

Q40 FENCING

Item Excluding site overheads and profit	PC £	Labour hours	Labour £	Plant £	Material £	Unit	Total rate £
TIMBER FENCING							
Timber fencing; AVS Fencing Supplies Ltd; **tanalized softwood fencing**							
Timber lap panels; pressure treated; fixed to timber posts 75 × 75 mm in 1:3:6 concrete; at 1.90 centres							
900 mm high	8.63	0.67	13.83	–	19.80	m	**33.63**
1200 mm high	8.81	0.75	15.56	–	20.31	m	**35.87**
1500 mm high	8.97	0.80	16.60	–	20.27	m	**36.87**
1800 mm high	9.78	0.90	18.68	–	22.64	m	**41.32**
Timber lap panels; fixed to slotted concrete posts 100 × 100 mm in 1:3:6 concrete at 1.88 m centres							
900 mm high	8.72	0.75	15.56	–	24.02	m	**39.58**
1200 mm high	8.90	0.80	16.60	–	24.20	m	**40.80**
1500 mm high	14.57	0.85	17.64	–	28.66	m	**46.30**
1800 mm high	21.97	0.90	18.68	–	36.06	m	**54.74**
extra for corner posts	22.73	0.90	18.68	–	35.78	nr	**54.46**
Extra over panel fencing for 300 mm high trellis tops; slats at 100 mm centres; including additional length of posts	4.70	0.10	2.08	–	4.70	m	**6.78**
Closeboarded fencing; to concrete posts 100 × 100 mm; 2 nr oak Arris rails; 100 × 22 mm softwood pales lapped 13 mm; including excavating and backfilling into firm ground at 3.00 m centres; setting in concrete 1:3:6; timber gravel board 150 × 50 mm							
900 mm high	16.18	1.00	20.75	–	22.24	m	**42.99**
1050 mm high	17.83	1.10	22.82	–	22.72	m	**45.54**
1500 mm high	19.87	1.15	23.86	–	24.76	m	**48.62**
Closeboarded fencing; to concrete posts 100 × 100 mm; 3 nr oak Arris rails; 100 × 22 mm softwood pales lapped 13 mm; including excavating and backfilling into firm ground at 3.00 m centres; setting in concrete; timber gravel board 150 × 50 mm							
1650 mm high	24.97	1.25	25.94	–	29.86	m	**55.80**
1800 mm high	25.12	1.30	26.98	–	30.02	m	**57.00**
extra over to the above for concrete gravel board 150 × 50 mm in lieu of timber gravel board	4.52	0.11	2.31	–	4.59	m	**6.90**
Closeboarded fencing; 2 nr oak Arris rails; to timber posts 100 × 100 mm; 100 × 22 mm softwood pales lapped 13 mm; including excavating and backfilling into firm ground at 3.00 m centres; setting in concrete 1:3:6							
900 mm high	12.76	1.00	20.75	–	22.56	m	**43.31**
1050 mm high	18.15	1.10	22.82	–	23.04	m	**45.86**
1200 mm high	20.19	1.15	23.86	–	25.08	m	**48.94**

Q40 FENCING

Item Excluding site overheads and profit	PC £	Labour hours	Labour £	Plant £	Material £	Unit	Total rate £
TIMBER FENCING – cont							
Closeboarded fencing; 3 nr oak Arris rails; to							
timber posts 100 × 100 mm; 100 × 22 mm							
softwood pales lapped 13 mm; including							
excavating and backfilling into firm ground at							
3.00 m centres; setting in concrete 1:3:6							
1350 mm high	20.07	0.95	19.76	–	28.70	m	**48.46**
1650 mm high	24.89	1.30	26.98	–	29.78	m	**56.76**
1800 mm high	25.01	1.35	28.01	–	29.90	m	**57.91**
extra over for post 125 × 100 mm; for							
1800 mm high fencing	9.26	–	–	–	9.26	nr	**9.26**
extra over to the above for counter rail	0.68	0.10	2.08	–	0.68	m	**2.76**
extra over to the above for capping rail	1.35	0.10	2.08	–	1.35	m	**3.43**
Closeboarded fencing; 2 nr cant rails; nailed to							
timber posts 100 × 100 mm; 125 × 22 mm							
softwood pales lapped 13 mm; including							
excavating and backfilling into firm ground at							
3.00 m centres; setting in concrete 1:3:6							
900 mm high	7.49	1.00	20.75	–	14.69	m	**35.44**
1050 mm high	10.28	1.10	22.82	–	15.17	m	**37.99**
1200 mm high	11.60	1.10	22.82	–	16.49	m	**39.31**
Closeboarded fencing; 3 nr cant rails; nailed to							
timber posts 100 × 100 mm; 125 × 22 mm							
softwood pales lapped 13 mm; including							
excavating and backfilling into firm ground at							
3.00 m centres; setting in concrete 1:3:6							
1350 mm high	10.41	1.15	23.86	–	17.61	m	**41.47**
1800 mm high	14.70	1.20	24.90	–	19.59	m	**44.49**
extra over for post 125 × 100 mm; for							
1800 mm high fencing	–	–	–	–	–	nr	**–**
Palisade fencing; 19 × 75 mm softwood vertical							
palings with pointed tops at 150 mm centres;							
nailing to 2 nr horizontal softwood Arris rails;							
morticed to 100 × 100 mm softwood posts with							
weathered tops at 3.00 m centres; setting in							
concrete							
900 mm high	18.54	0.90	18.68	–	23.53	m	**42.21**
1050 mm high	19.75	0.90	18.68	–	24.64	m	**43.32**
1200 mm high	20.08	0.95	19.71	–	25.07	m	**44.78**
extra over for rounded tops	0.80	–	–	–	0.80	m	**0.80**
Post-and-rail fencing; 90 × 38 mm softwood							
horizontal rails; fixing with galvanized nails to							
150 × 75 mm softwood posts; including							
excavating and backfilling into firm ground at							
1.80 m centres; all treated timber							
1200 mm high; 3 horizontal rails	9.08	0.35	7.27	–	9.17	m	**16.44**
1200 mm high; 4 horizontal rails	10.42	0.35	7.27	–	10.51	m	**17.78**

Q40 FENCING

Item Excluding site overheads and profit	PC £	Labour hours	Labour £	Plant £	Material £	Unit	Total rate £
Cleft rail fencing; oak or chestnut tapered rails 2.80 m long; morticed into joints; to 125 × 100 mm softwood posts 1.95 m long; including excavating and backfilling into firm ground at 2.80 m centres							
two rails	20.37	0.25	5.19	–	20.37	m	**25.56**
three rails	23.21	0.28	5.81	–	23.21	m	**29.02**
four rails	26.05	0.35	7.27	–	26.05	m	**33.32**
Hit and Miss horizontal rail fencing; treated softwood; 87 × 38 mm top and bottom rails; 100 × 22 mm vertical boards arranged alternately on opposite side of rails; to 100 × 100 mm posts; including excavating and backfilling into firm ground; setting in concrete at 1.8 m centres							
1600 mm high	30.26	1.20	24.90	–	33.22	m	**58.12**
1800 mm high	33.31	1.33	27.67	–	36.27	m	**63.94**
2000 mm high	36.37	1.40	29.05	–	39.33	m	**68.38**
Palisade fencing; 100 × 22 mm softwood vertical palings with flat tops; nailing to 3 nr 50 × 100 mm horizontal softwood rails; housing into 100 × 100 mm softwood posts with weathered tops at 3.00 m centres; setting in concrete; all treated timber							
1800 mm high	23.61	0.47	9.69	–	28.67	m	**38.36**
1800 mm high	32.06	0.47	9.68	–	37.12	m	**46.80**
Post-and-rail fencing; 2 nr 90 × 38 mm softwood horizontal rails; fixing with galvanized nails to 150 × 75 mm softwood posts; including excavating and backfilling into firm ground at 1.80 m centres; all treated timber							
1200 mm high	7.74	0.30	6.22	–	7.83	m	**14.05**
Post-and-rail fencing; 3 nr 90 × 38 mm softwood horizontal rails; fixing with galvanized nails to 150 × 75 mm softwood posts; including excavating and backfilling into firm ground at 1.80 m centres; all treated timber							
1200 mm high	9.08	0.35	7.27	–	9.17	m	**16.44**
Morticed post-and-rail fencing; 3 nr horizontal 90 × 38 mm softwood rails; fixing with galvanized nails; 90 × 38 mm softwood centre prick posts; to 150 × 75 mm softwood posts; including excavating and backfilling into firm ground at 2.85 m centres; all treated timber							
1200 mm high	8.56	0.35	7.27	–	8.65	m	**15.92**
1350 mm high five rails	10.91	0.40	8.30	–	11.00	m	**19.30**

Q40 FENCING

Item Excluding site overheads and profit	PC £	Labour hours	Labour £	Plant £	Material £	Unit	Total rate £
TIMBER FENCING – cont							
Cleft rail fencing; chestnut tapered rails 2.85 m long; morticed into joints; to 125 × 100 mm softwood posts 2.1 m long; including excavating and backfilling into firm ground at 2.75 m centres							
two rails	30.88	0.25	5.19	–	30.88	m	**36.07**
three rails	38.12	0.28	5.81	–	38.12	m	**43.93**
four rails	45.93	0.35	7.27	–	45.93	m	**53.20**
STRAINED WIRE FENCING							
Boundary fencing; strained wire and wire mesh; AVS Fencing Supplies Ltd							
Strained wire fencing; concrete posts only at 2750 mm centres; 610 mm below ground; excavating holes; filling with concrete; replacing topsoil; disposing surplus soil off site							
900 mm high	3.36	0.56	11.53	0.86	5.02	m	**17.41**
1200 mm high	4.21	0.56	11.53	0.86	5.87	m	**18.26**
1800 mm high	5.76	0.78	16.14	2.14	7.41	m	**25.69**
Extra over strained wire fencing for concrete straining posts with one strut; posts and struts 610 mm below ground; struts, cleats, stretchers, winders, bolts and eye bolts; excavating holes; filling to within 150 mm of ground level with concrete (1:12) – 40 mm aggregate; replacing topsoil; disposing surplus soil off site							
900 mm high	21.28	0.67	13.90	1.92	55.02	nr	**70.84**
1200 mm high	24.87	0.67	13.84	1.93	59.10	nr	**74.87**
1800 mm high	34.93	0.67	13.84	1.93	71.85	nr	**87.62**
Extra over strained wire fencing for concrete straining posts with two struts; posts and struts 610 mm below ground; excavating holes; filling to within 150 mm of ground level with concrete (1:12) – 40 mm aggregate; replacing topsoil; disposing surplus soil off site							
900 mm high	32.21	0.91	18.88	2.47	82.91	nr	**104.26**
1200 mm high	37.11	0.85	17.64	1.93	93.26	nr	**112.83**
1800 mm high	52.59	0.85	17.64	1.93	112.89	nr	**132.46**
Strained wire fencing; painted steel angle posts only at 2750 mm centres; 610 mm below ground; driving in							
900 mm high; 40 × 40 × 5 mm	3.04	0.03	0.63	–	3.04	m	**3.67**
1200 mm high; 40 × 40 × 5 mm	3.46	0.03	0.69	–	3.46	m	**4.15**
1400 mm high; 40 × 40 × 5 mm	4.54	0.06	1.26	–	4.54	m	**5.80**
1800 mm high; 40 × 40 × 5 mm	4.95	0.07	1.51	–	4.95	m	**6.46**
1800 mm high; 45 × 45 × 5 mm; with extension for three rows barbed wire	6.35	0.12	2.51	–	6.35	m	**8.86**

Q40 FENCING

Item Excluding site overheads and profit	PC £	Labour hours	Labour £	Plant £	Material £	Unit	Total rate £
Painted steel straining posts with two struts for strained wire fencing; setting in concrete							
900 mm high; 50 × 50 × 6 mm	54.14	1.00	20.75	–	54.14	nr	**74.89**
1200 mm high; 50 × 50 × 6 mm	65.40	1.00	20.75	–	65.40	nr	**86.15**
1500 mm high; 50 × 50 × 6 mm	81.71	1.00	20.75	–	81.71	nr	**102.46**
1800 mm high; 50 × 50 × 6 mm	84.69	1.00	20.75	–	84.69	nr	**105.44**
Strained wire; to posts (posts not included); 3 mm galvanized wire; fixing with galvanized stirrups							
900 mm high; 2 wire	0.18	0.03	0.69	–	0.35	m	**1.04**
1200 mm high; 3 wire	0.28	0.05	0.97	–	0.45	m	**1.42**
1400 mm high; 3 wire	0.28	0.05	0.97	–	0.45	m	**1.42**
1800 mm high; 3 wire	0.28	0.05	0.97	–	0.45	m	**1.42**
Barbed wire; to posts (posts not included); 3 mm galvanized wire; fixing with galvanized stirrups							
900 mm high; 2 wire	0.27	0.07	1.38	–	0.44	m	**1.82**
1200 mm high; 3 wire	0.41	0.09	1.93	–	0.58	m	**2.51**
1400 mm high; 3 wire	0.41	0.09	1.93	–	0.58	m	**2.51**
1800 mm high; 3 wire	0.41	0.09	1.93	–	0.58	m	**2.51**
Chain link fencing; AVS Fencing Supplies Ltd; to strained wire and posts priced separately; 3 mm galvanized wire; 50 mm mesh; galvanized steel components; fixing to line wires threaded through posts and strained with eye-bolts; posts (not included)							
900 mm high	3.79	0.07	1.38	–	3.96	m	**5.34**
1200 mm high	5.30	0.07	1.38	–	5.47	m	**6.85**
1800 mm high	7.35	0.10	2.08	–	7.52	m	**9.60**
2400 mm high	9.95	0.10	2.08	–	10.12	m	**12.20**
Chain link fencing; to strained wire and posts priced separately; 3.15 mm plastic coated galvanized wire (wire only 2.50 mm); 50 mm mesh; galvanized steel components; fencing with line wires threaded through posts and strained with eye-bolts; posts (not included) (Note: plastic coated fencing can be cheaper than galvanized finish as wire of a smaller cross-sectional area can be used)							
900 mm high	2.68	0.07	1.38	–	3.02	m	**4.40**
1200 mm high	3.70	0.07	1.38	–	4.04	m	**5.42**
1800 mm high	5.26	0.13	2.59	–	6.62	m	**9.21**
Extra over strained wire fencing for cranked arms and galvanized barbed wire							
1 row	3.98	0.02	0.35	–	3.98	m	**4.33**
2 rows	4.12	0.05	1.04	–	4.12	m	**5.16**
3 rows	4.25	0.05	1.04	–	4.25	m	**5.29**

Q40 FENCING

Item Excluding site overheads and profit	PC £	Labour hours	Labour £	Plant £	Material £	Unit	Total rate £
STRAINED WIRE FENCING – cont							
Boundary fencing – cont							
Field fencing; Jacksons Fencing; welded wire mesh; fixed to posts and straining wires measured separately							
cattle fence; 1100 m high; 114 × 300 mm at bottom to 230 × 300 mm at top	1.43	0.10	2.08	–	1.74	m	3.82
sheep fence; 900 mm high; 140 × 300 mm at bottom to 230 × 300 mm at top	0.95	0.10	2.08	–	1.26	m	3.34
deer fence; 1900 mm high; 89 × 150 mm at bottom to 267 × 300 mm at top	2.92	0.13	2.59	–	3.23	m	5.82
Extra for concreting in posts	–	0.50	10.38	–	4.50	nr	14.88
Extra for straining post	10.76	0.75	15.56	–	10.76	nr	26.32
Boundary fencing; strained wire and wire mesh; Jacksons Fencing							
Tubular chain link fencing; galvanized; plastic coated; 60.3 mm diameter posts at 3.0 m centres; setting 700 mm into ground; choice of ten mesh colours; including excavating holes; backfilling and removing surplus soil; with top rail only							
900 mm high	–	–	–	–	–	m	33.68
1200 mm high	–	–	–	–	–	m	35.78
1800 mm high	–	–	–	–	–	m	39.89
2000 mm high	–	–	–	–	–	m	42.10
Tubular chain link fencing; galvanized; plastic coated; 60.3 mm diameter posts at 3.0 m centres; cranked arms and 3 lines barbed wire; setting 700 mm into ground; including excavating holes; backfilling and removing surplus soil; with top rail only							
1800 mm high	–	–	–	–	–	m	41.47
2000 mm high	–	–	–	–	–	m	42.68
RABBIT NETTING							
Rabbit netting; AVS Fencing Supplies Ltd; timber stakes; peeled kiln dried pressure treated; pointed; 1.8 m posts driven 900 mm into ground at 3 m centres (line wires and netting priced separately)							
75–100 mm stakes	1.98	0.25	5.19	–	1.98	m	7.17
Corner posts or straining posts 150 mm diameter 2.3 m high set in concrete; centres to suit local conditions or changes of direction							
1 strut	13.10	1.00	20.75	5.49	20.95	each	47.19
2 strut	15.90	1.00	20.75	5.49	23.76	each	50.00

Q40 FENCING

Item Excluding site overheads and profit	PC £	Labour hours	Labour £	Plant £	Material £	Unit	Total rate £
Strained wire; to posts (posts not included); 3 mm galvanized wire; fixing with galvanized stirrups							
900 mm high; 2 wire	0.18	0.03	0.69	–	0.35	m	**1.04**
1200 mm high; 3 wire	0.28	0.05	0.97	–	0.45	m	**1.42**
Rabbit netting; 31 mm; 19 gauge; 1050 mm high netting fixed to posts; line wires and straining posts or corner posts all priced separately							
900 high turned in	1.29	0.04	0.83	–	1.29	m	**2.12**
900 high buried 150 mm in trench	1.29	0.08	1.73	–	1.29	m	**3.02**
TRELLIS							
Traditional trellis panels; The Garden Trellis Company; bespoke trellis panels for decorative, screening or security applications; timber planed all round; height of trellis 1800 mm							
Free-standing panels; posts 70 × 70 mm set in concrete; timber frames mitred and grooved 45 × 34 mm; heights and widths to suit; slats 32 × 10 mm joinery quality tanalized timber at 100 mm ccs; elements fixed by galvanized staples; capping rail 70 × 34 mm							
HV68; horizontal and vertical slats; softwood	97.20	1.00	20.75	–	103.20	m	**123.95**
D68; diagonal slats; softwood	111.60	1.00	20.75	–	117.60	m	**138.35**
HV68; horizontal and vertical slats; hardwood iroko	234.00	1.00	20.75	–	240.00	m	**260.75**
D68; diagonal slats; hardwood iroko or Western Red Cedar	255.60	1.00	20.75	–	261.60	m	**282.35**
Trellis panels fixed to face of existing wall or railings; timber frames mitred and grooved 45 × 34 mm; heights and widths to suit; slats 32 × 10 mm joinery quality tanalized timber at 100 mm ccs; elements fixed by galvanized staples; capping rail 70 × 34 mm							
HV68; horizontal and vertical slats; softwood	93.60	0.50	10.38	–	95.87	m	**106.25**
D68; diagonal slats; softwood	104.40	0.50	10.38	–	106.67	m	**117.05**
HV68; horizontal and vertical slats; iroko or Western Red Cedar	234.00	0.50	10.38	–	236.27	m	**246.65**
D68; diagonal slats; iroko or Western Red Cedar	237.60	0.50	10.38	–	239.87	m	**250.25**

Q40 FENCING

Item Excluding site overheads and profit	PC £	Labour hours	Labour £	Plant £	Material £	Unit	Total rate £
TRELLIS – cont							
Contemporary style trellis panels; The Garden Trellis Company; bespoke trellis panels for decorative, screening or security applications; timber planed all round							
Free-standing panels 30/15; posts 70 × 70 mm set in concrete with 90 × 30 mm top capping; slats 30 × 14 mm with 15 mm gaps; vertical support at 450 mm ccs							
joinery treated softwood	144.00	1.00	20.75	–	150.00	m	**170.75**
hardwood iroko or Western Red Cedar	252.00	1.00	20.75	–	258.00	m	**278.75**
Panels 30/15 face fixed to existing wall or fence; posts 70 × 70 mm set in concrete with 90 × 30 mm top capping; slats 30 × 14 mm with 15 mm gaps; vertical support at 450 mm ccs							
joinery treated softwood	144.00	0.50	10.38	–	146.27	m	**156.65**
hardwood iroko or Western Red Cedar	230.40	0.50	10.38	–	232.67	m	**243.05**
Integral arches to ornamental trellis panels; The Garden Trellis Company							
Arches to trellis panels in 45 × 34 mm grooved timbers to match framing; fixed to posts							
R450; ¼ circle; joinery treated softwood	45.00	1.50	31.13	–	51.00	nr	**82.13**
R450; ¼ circle; hardwood iroko or Western Red Cedar	60.00	1.50	31.13	–	66.00	nr	**97.13**
Arches to span 1800 mm wide							
joinery treated softwood	60.00	1.50	31.13	–	66.00	nr	**97.13**
hardwood iroko or Western Red Cedar	80.00	1.50	31.13	–	86.00	nr	**117.13**
Painting or staining of trellis panels; high quality coatings							
microporous opaque paint or spirit based stain	–	–	–	–	28.00	m²	**28.00**
CONCRETE FENCING							
Concrete fencing							
Panel fencing; to precast concrete posts; in 2 m bays; setting posts 600 mm into ground; sandfaced finish							
900 mm high	11.74	0.25	5.19	–	13.52	m	**18.71**
1200 mm high	15.68	0.25	5.19	–	17.45	m	**22.64**
Panel fencing; to precast concrete posts; in 2 m bays; setting posts 750 mm into ground; sandfaced finish							
1500 mm high	20.46	0.33	6.92	–	22.23	m	**29.15**
1800 mm high	24.52	0.36	7.54	–	26.29	m	**33.83**
2100 mm high	28.39	0.40	8.30	–	30.16	m	**38.46**
2400 mm high	32.63	0.40	8.30	–	34.40	m	**42.70**
extra over capping panel to all of the above	6.72	–	–	–	6.72	m	**6.72**

Q40 FENCING

Item Excluding site overheads and profit	PC £	Labour hours	Labour £	Plant £	Material £	Unit	Total rate £
WINDBREAK FENCING							
Windbreak fencing							
Fencing; English Woodlands; Shade and							
Shelter Netting windbreak fencing; green; to							
100 mm diameter treated softwood posts;							
setting 450 mm into ground; fixing with							
50 × 25 mm treated softwood battens nailed to							
posts; including excavating and backfilling into							
firm ground; setting in concrete at 3 m centres							
1200 mm high	1.25	0.16	3.23	–	5.56	m	**8.79**
1800 mm high	1.87	0.16	3.23	–	6.18	m	**9.41**
GATES							
Gates – General							
Preamble: gates in fences; see specification							
for fencing, as gates in traditional or							
proprietary fencing systems are usually							
constructed of the same materials and finished							
as the fencing itself.							
Gates; hardwood; AVS Fencing Supplies							
Ltd							
Hardwood entrance gate; five bar diamond							
braced; curved hanging stile; planed iroko;							
fixed to 150 × 150 mm softwood posts;							
inclusive of hinges and furniture							
ref 1100 040; 0.9 m wide	220.58	5.00	103.75	–	285.97	nr	**389.72**
ref 1100 041; 1.2 m wide	235.19	5.00	103.75	–	300.58	nr	**404.33**
ref 1100 042; 1.5 m wide	306.56	5.00	103.75	–	371.95	nr	**475.70**
ref 1100 043; 1.8 m wide	324.61	5.00	103.75	–	390.00	nr	**493.75**
ref 1100 044; 2.1 m wide	355.42	5.00	103.75	–	420.81	nr	**524.56**
ref 1100 045; 2.4 m wide	334.50	5.00	103.75	–	399.89	nr	**503.64**
ref 1100 047; 3.0 m wide	408.55	5.00	103.75	–	473.94	nr	**577.69**
ref 1100 048; 3.3 m wide	426.71	5.00	103.75	–	492.10	nr	**595.85**
ref 1100 049; 3.6 m wide	441.28	5.00	103.75	–	506.67	nr	**610.42**
Hardwood field gate; five bar diamond braced;							
planed iroko; fixed to 150 × 150 mm softwood							
posts; inclusive of hinges and furniture							
ref 1100 100; 0.9 m wide	157.78	4.00	83.00	–	223.17	nr	**306.17**
ref 1100 101; 1.2 m wide	171.42	4.00	83.00	–	236.81	nr	**319.81**
ref 1100 102; 1.5 m wide	204.38	4.00	83.00	–	269.77	nr	**352.77**
ref 1100 103; 1.8 m wide	219.68	4.00	83.00	–	285.07	nr	**368.07**
ref 1100 104; 2.1 m wide	270.50	4.00	83.00	–	335.89	nr	**418.89**
ref 1100 105; 2.4 m wide	286.59	4.00	83.00	–	351.98	nr	**434.98**
ref 1100 106; 2.7 m wide	302.73	4.00	83.00	–	368.12	nr	**451.12**
ref 1100 108; 3.3 m wide	334.96	4.00	83.00	–	400.35	nr	**483.35**
ref 1100 109; 3.6 m wide	351.90	4.00	83.00	–	417.29	nr	**500.29**

Q40 FENCING

Item Excluding site overheads and profit	PC £	Labour hours	Labour £	Plant £	Material £	Unit	Total rate £
GATES – cont							
Gates; softwood; Jacksons Fencing							
Timber field gates; treated softwood; including							
wrought iron ironmongery; five bar type;							
diamond braced; 1.80 m high; to 200 × 200 mm							
posts; setting 750 mm into firm ground							
width 2400 mm	98.10	10.00	207.50	–	220.82	nr	**428.32**
width 2700 mm	100.48	10.00	207.50	–	223.20	nr	**430.70**
width 3000 mm	107.73	10.00	207.50	–	230.45	nr	**437.95**
width 3300 mm	115.42	10.00	207.50	–	238.14	nr	**445.64**
Featherboard garden gates; treated softwood;							
including ironmongery; to 100 × 120 mm posts;							
1 nr diagonal brace							
1.0 × 1.2 m high	26.64	3.00	62.25	–	113.22	nr	**175.47**
1.0 × 1.5 m high	35.10	3.00	62.25	–	128.92	nr	**191.17**
1.0 × 1.8 m high	39.24	3.00	62.25	–	140.31	nr	**202.56**
Picket garden gates; treated softwood;							
including ironmongery; to match picket fence;							
width 1000 mm; to 100 × 120 mm posts; 1 nr							
diagonal brace							
950 mm high	68.22	3.00	62.25	–	104.80	nr	**167.05**
1200 mm high	71.91	3.00	62.25	–	108.49	nr	**170.74**
1800 mm high	84.38	3.00	62.25	–	133.56	nr	**195.81**
Gates; tubular steel; Jacksons Fencing							
Tubular mild steel field gates; galvanized;							
including ironmongery; diamond braced;							
1.80 m high; to tubular steel posts; setting in							
concrete							
width 3000 mm	126.27	5.00	103.75	–	235.40	nr	**339.15**
width 3300 mm	133.88	5.00	103.75	–	243.00	nr	**346.75**
width 3600 mm	141.66	5.00	103.75	–	250.79	nr	**354.54**
width 4200 mm	157.23	5.00	103.75	–	266.36	nr	**370.11**
Kissing gates							
Kissing gates; Jacksons Fencing; in							
galvanized metal bar; fixing to fencing posts							
(posts not included); 1.65 × 1.30 × 1.00 m high	243.00	5.00	103.75	–	243.00	nr	**346.75**
Stiles							
Stiles; 2 nr posts; setting into firm ground; 3 nr							
rails; 2 nr treads	72.00	3.00	62.25	–	84.00	nr	**146.25**

R12 DRAINAGE BELOW GROUND

Item Excluding site overheads and profit	PC £	Labour hours	Labour £	Plant £	Material £	Unit	Total rate £
R12 DRAINAGE BELOW GROUND							
EXCAVATING FOR DRAINAGE SERVICES							
Excavating trenches; using 3 tonne tracked excavator; to receive pipes; grading bottoms; earthwork support; filling with excavated material to within 150 mm of finished surfaces and compacting; completing fill with topsoil; disposal of surplus soil							
Services not exceeding 200 mm nominal size							
average depth of run not exceeding 0.50 m	–	0.02	0.41	3.75	1.74	m	**5.90**
average depth of run not exceeding 0.75 m	–	0.03	0.69	5.29	1.74	m	**7.72**
average depth of run not exceeding 1.00 m	–	0.05	1.04	9.11	1.74	m	**11.89**
average depth of run not exceeding 1.25 m	–	0.10	2.08	12.08	1.74	m	**15.90**
Granular beds to trenches; lay granular material; to trenches excavated separately; to receive pipes (not included)							
300 mm wide × 100 mm thick							
reject sand	–	0.03	0.52	0.89	0.96	m	**2.37**
shingle 20 mm aggregate	–	0.03	0.52	0.89	1.16	m	**2.57**
sharp sand	–	0.03	0.52	0.89	1.03	m	**2.44**
300 mm wide × 150 mm thick							
reject sand	–	0.04	0.78	1.34	0.48	m	**2.60**
shingle 20 mm aggregate	–	0.04	0.78	1.34	0.58	m	**2.70**
sharp sand	–	0.04	0.78	1.34	0.52	m	**2.64**
Excavating trenches; using 3 tonne tracked excavator; to receive pipes; grading bottoms; earthwork support; filling with imported granular material type 2 and compacting; disposal of surplus soil							
Services not exceeding 200 mm nominal size							
average depth of run not exceeding 0.50 m	1.62	0.02	0.41	2.57	4.02	m	**7.00**
average depth of run not exceeding 0.75 m	2.43	0.03	0.52	3.04	6.03	m	**9.59**
average depth of run not exceeding 1.00 m	3.24	0.03	0.59	4.23	8.04	m	**12.86**
average depth of run not exceeding 1.25 m	4.05	0.03	0.59	7.40	10.05	m	**18.04**
Excavating trenches; using 3 tonne tracked excavator; to receive pipes; grading bottoms; earthwork support; filling with concrete ready mixed ST2; disposal of surplus soil							
Services not exceeding 200 mm nominal size							
average depth of run not exceeding 0.50 m	12.52	0.02	0.41	2.38	14.92	m	**17.71**
average depth of run not exceeding 0.75 m	18.79	0.03	0.52	3.04	22.39	m	**25.95**
average depth of run not exceeding 1.00 m	25.05	0.03	0.59	4.23	29.85	m	**34.67**
average depth of run not exceeding 1.25 m	31.31	0.03	0.59	7.40	37.31	m	**45.30**

R12 DRAINAGE BELOW GROUND

Item Excluding site overheads and profit	PC £	Labour hours	Labour £	Plant £	Material £	Unit	Total rate £
INSPECTION CHAMBERS							
Inspection chambers							
Excavating pits; starting from ground level; by machine							
maximum depth not exceeding 1.00 m	–	–	–	4.49	–	m³	**4.49**
maximum depth not exceeding 2.00 m	–	–	–	4.31	–	m³	**4.31**
Disposal of excavated material; depositing on site in permanent spoil heaps; average 50 m	–	–	–	7.99	–	m³	**7.99**
Surface treatments; compacting; bottoms of excavations	–	0.05	1.04	–	–	m²	**1.04**
Earthwork support; distance between opposing faces not exceeding 2.00 m							
maximum depth not exceeding 1.00 m	–	0.20	4.15	–	24.93	m³	**29.08**
maximum depth not exceeding 2.00 m	–	0.30	6.22	–	24.93	m³	**31.15**
maximum depth not exceeding 4.00 m	–	0.67	13.82	–	12.14	m³	**25.96**
Inspection chambers; in situ concrete							
Beds; plain in situ concrete; 11.50 N/mm² – 40 mm aggregate							
thickness not exceeding 150 mm	–	1.00	20.75	–	122.74	m³	**143.49**
Benchings in bottoms; plain in situ concrete; 25.50 N/mm² – 20 mm aggregate							
thickness 150–450 mm	–	2.00	41.50	–	122.74	m³	**164.24**
Isolated cover slabs; reinforced in situ concrete; 21.00 N/mm² – 20 mm aggregate							
thickness not exceeding 150 mm	–	4.00	83.00	–	122.74	m³	**205.74**
Fabric reinforcement; A193 (3.02 kg/m²) in cover slabs	0.38	0.06	1.30	–	0.38	m²	**1.68**
Formwork to reinforced in situ concrete; isolated cover slabs							
soffits; horizontal	–	3.28	68.06	–	19.67	m²	**87.73**
height not exceeding 250 mm	–	0.97	20.13	–	12.39	m	**32.52**
Brickwork							
Walls to manholes; bricks; PC £300.00/1000; in cement mortar (1:3)							
one brick thick	63.00	3.00	62.25	–	71.97	m²	**134.22**
one and a half brick thick	94.50	4.00	83.00	–	107.96	m²	**190.96**
two brick thick projection of footing or the like	126.00	4.80	99.60	–	143.94	m²	**243.54**
Walls to manholes; engineering bricks; PC £290.00/1000; in cement mortar (1:3)							
one brick thick	36.51	3.00	62.25	–	45.49	m²	**107.74**
one and a half brick thick	54.77	3.00	62.25	–	63.74	m²	**125.99**
two brick thick projection of footing or the like	73.03	3.00	62.25	–	82.00	m²	**144.25**

R12 DRAINAGE BELOW GROUND

Item Excluding site overheads and profit	PC £	Labour hours	Labour £	Plant £	Material £	Unit	Total rate £
Extra over common or engineering bricks in any mortar for fair face; flush pointing as work proceeds; English bond walls or the like	–	0.13	2.77	–	–	m²	2.77
In situ finishings; cement: sand mortar (1:3); steel trowelled; 13 mm one coat work to manhole walls; to brickwork or blockwork base; over 300 mm wide	–	0.80	16.60	–	2.99	m²	19.59
Building into brickwork; ends of pipes; making good facings or renderings							
small	–	0.20	4.15	–	–	nr	4.15
large	–	0.30	6.22	–	–	nr	6.22
extra large	–	0.40	8.30	–	–	nr	8.30
extra large; including forming ring arch cover	–	0.50	10.38	–	–	nr	10.38
Inspection chambers; polypropylene; Hepworth Plc							
Mini access chamber; up to 600 mm deep; including cover and frame							
300 mm diameter × 600 mm deep; three 100/110 mm inlets	201.76	3.00	62.25	–	204.57	nr	266.82
Up to 1200 mm deep; including polymer cover and frame with screw down lid							
475 mm diameter × 940 mm deep; five 100/110 mm inlets; supplied with four stoppers in inlets	298.28	4.00	83.00	–	397.53	nr	480.53
Extra over for square ductile iron cover and frame to 1 tonne load; screw down lid	–	–	–	–	41.93	nr	41.93
Step irons; Ashworth Ltd; drainage systems; malleable cast iron; galvanized; building into joints							
General purpose pattern; for one brick walls	4.50	0.17	3.53	–	4.50	nr	8.03
Best quality vitrified clay half section channels; Hepworth Plc; bedding and jointing in cement: mortar (1:2)							
Channels; straight							
100 mm	8.50	0.80	16.60	–	12.09	m	28.69
150 mm	14.14	1.00	20.75	–	17.73	m	38.48
225 mm	31.74	1.35	28.01	–	37.13	m	65.14
300 mm	65.16	1.80	37.35	–	68.75	m	106.10
Bends; 15, 30, 45 or 90°							
100 mm bends	7.65	0.75	15.56	–	9.44	nr	25.00
150 mm bends	13.22	0.90	18.68	–	16.81	nr	35.49
225 mm bends	51.22	1.20	24.90	–	54.82	nr	79.72
300 mm bends	104.43	1.10	22.82	–	108.92	nr	131.74

R12 DRAINAGE BELOW GROUND

Item Excluding site overheads and profit	PC £	Labour hours	Labour £	Plant £	Material £	Unit	Total rate £
INSPECTION CHAMBERS – cont							
Intercepting traps; Hepworth Plc							
Vitrified clay; inspection arms; brass stoppers;							
iron levers; chains and staples; galvanized;							
staples cut and pinned to brickwork; cement:							
mortar (1:2) joints to vitrified clay pipes and							
channels; bedding and surrounding in							
concrete; 11.50 N/mm^2 – 40 mm aggregate;							
cutting and fitting brickwork; making good							
facings							
100 mm inlet; 100 mm outlet	111.11	3.00	62.25	–	127.61	nr	**189.86**
150 mm inlet; 150 mm outlet	160.21	2.00	41.50	–	178.66	nr	**220.16**
PIPES AND FITTINGS							
Clay pipes and fittings; Hepworth Plc;							
Supersleve							
100 mm clay pipes; polypropylene slip							
coupling; in trenches (trenches not included)							
laid straight	6.71	0.25	5.19	–	9.81	m	**15.00**
short runs under 3.00 m	6.71	0.31	6.48	–	9.81	m	**16.29**
Extra over 100 mm clay pipes for							
bends; 15–90°; single socket	13.98	0.25	5.19	–	13.98	nr	**19.17**
junction; 45 or 90°; double socket	29.45	0.25	5.19	–	29.45	nr	**34.64**
slip couplings; polypropylene	4.95	0.08	1.73	–	4.95	nr	**6.68**
gully with P trap; 100 mm; 154 × 154 mm							
plastic grating	45.63	1.00	20.75	–	67.08	nr	**87.83**
PVC-u pipes and fittings; Wavin Plastics							
Ltd; OsmaDrain system							
110 mm PVC-u pipes; in trenches (trenches							
not included)							
laid straight	15.29	0.08	1.66	–	15.29	m	**16.95**
short runs under 3.00 m	15.29	0.12	2.49	–	15.29	m	**17.78**
Extra over 110 mm PVC-u pipes for							
bends; short radius	29.87	0.25	5.19	–	29.87	nr	**35.06**
bends; long radius	55.98	0.25	5.19	–	55.98	nr	**61.17**
junctions; equal; double socket	35.63	0.25	5.19	–	35.63	nr	**40.82**
slip couplings	17.28	0.10	2.08	–	17.28	nr	**19.36**
adaptors to clay	33.67	0.50	10.38	–	33.85	nr	**44.23**

R12 DRAINAGE BELOW GROUND

Item Excluding site overheads and profit	PC £	Labour hours	Labour £	Plant £	Material £	Unit	Total rate £
GULLIES							
Kerbs to gullies							
Kerbs to gullies; in one course Class B engineering bricks; to 4 nr sides; rendering in cement: mortar (1:3); dished to gully gratings	2.32	1.00	20.75	–	3.22	nr	23.97
Gullies; vitrified clay; Hepworth Plc; bedding in concrete; 11.50 N/mm² – 40 mm aggregate							
Yard gullies (mud); trapped; domestic duty (up to 1 tonne)							
100 mm outlet; 100 mm diameter; 225 mm internal width; 585 mm internal depth	144.49	3.50	72.63	–	145.14	nr	217.77
Yard gullies (mud); trapped; medium duty (up to 5 tonnes)							
100 mm outlet; 100 mm diameter; 225 mm internal width; 585 mm internal depth	144.49	3.50	72.63	–	145.14	nr	217.77
150 mm outlet; 100 mm diameter; 225 mm internal width; 585 mm internal depth	223.80	3.50	72.63	–	224.45	nr	297.08
Combined filter and silt bucket for yard gullies							
225 mm diameter	52.25	–	–	–	52.25	nr	52.25
Road gullies; trapped with rodding eye							
100 mm outlet; 300 mm internal diameter; 600 mm internal depth	138.97	3.50	72.63	–	139.62	nr	212.25
150 mm outlet; 300 mm internal diameter; 600 mm internal depth	142.31	3.50	72.63	–	142.96	nr	215.59
150 mm outlet; 400 mm internal diameter; 750 mm internal depth	165.04	3.50	72.63	–	165.69	nr	238.32
150 mm outlet; 450 mm internal diameter; 900 mm internal depth	223.30	3.50	72.63	–	223.94	nr	296.57
Hinged gratings and frames for gullies; alloy							
193 mm for 150 mm diameter gully	–	–	–	–	34.62	nr	34.62
120 × 120 mm	–	–	–	–	12.31	nr	12.31
150 × 150 mm	–	–	–	–	22.28	nr	22.28
230 × 230 mm	–	–	–	–	40.73	nr	40.73
316 × 316 mm	–	–	–	–	107.97	nr	107.97
Hinged gratings and frames for gullies; cast iron							
265 mm for 225 mm diameter gully	–	–	–	–	69.18	nr	69.18
150 × 150 mm	–	–	–	–	22.28	nr	22.28
230 × 230 mm	–	–	–	–	40.73	nr	40.73
316 × 316 mm	–	–	–	–	107.97	nr	107.97
Universal gully trap; PVC-u; Wavin Plastics Ltd; OsmaDrain system; bedding in concrete; 11.50 N/mm² – 40 mm aggregate							
Universal gully fitting; comprising gully trap only							
110 mm outlet; 110 mm diameter; 205 mm internal depth	25.34	3.50	72.63	–	25.98	nr	98.61

R12 DRAINAGE BELOW GROUND

Item Excluding site overheads and profit	PC £	Labour hours	Labour £	Plant £	Material £	Unit	Total rate £
GULLIES – cont							
Universal gully trap – cont							
Vertical inlet hopper c\w plastic grate							
272 × 183 mm	34.05	0.25	5.19	–	34.05	nr	**39.24**
Sealed access hopper							
110 × 110 mm	80.05	0.25	5.19	–	80.05	nr	**85.24**
Universal gully; PVC-u; Wavin Plastics Ltd;							
OsmaDrain system; accessories to							
universal gully trap							
Hoppers; backfilling with clean granular							
material; tamping; surrounding in lean mix							
concrete							
plain hopper; with 110 spigot 150 mm long	27.83	0.40	8.30	–	29.48	nr	**37.78**
vertical inlet hopper; with 110 spigot							
150 mm long	34.05	0.40	8.30	–	35.70	nr	**44.00**
sealed access hopper; with 110 spigot							
150 mm long	80.05	0.40	8.30	–	81.70	nr	**90.00**
plain hopper; solvent weld to trap	19.01	0.40	8.30	–	20.66	nr	**28.96**
vertical inlet hopper; solvent wed to trap	32.66	0.40	8.30	–	34.32	nr	**42.62**
sealed access cover; PVC-u	40.77	0.10	2.08	–	40.77	nr	**42.85**
Gullies PVC-u; Wavin Plastics Ltd;							
OsmaDrain system; bedding in concrete;							
11.50 N/mm² – 40 mm aggregate							
Bottle gully; providing access to the drainage							
system for cleaning							
bottle gully; 228 × 228 × 317 mm deep	67.58	0.50	10.38	–	68.15	nr	**78.53**
sealed access cover; PVC-u; 217 × 217 mm	52.54	0.10	2.08	–	52.54	nr	**54.62**
grating; ductile iron; 215 × 215 mm	40.50	0.10	2.08	–	40.50	nr	**42.58**
bottle gully riser; 325 mm	8.71	0.50	10.38	–	9.93	nr	**20.31**
Yard gully; trapped; 300 mm diameter; 600 mm							
deep; including catchment bucket and ductile							
iron cover and frame; medium duty loading							
305 mm diameter; 600 mm deep	409.80	2.50	51.88	–	413.45	nr	**465.33**
LINEAR DRAINAGE							
Linear Drainage							
Loadings for slot drains							
A15 – 1.5 tonne; pedestrian use							
B125 – 12.5 tonne; domestic use							
C250 – 25 tonne; car parks, supermarkets,							
industrial units							
D400 – 40 tonne; highways							
E600 – 60 tonne; forklifts							

R12 DRAINAGE BELOW GROUND

Item Excluding site overheads and profit	PC £	Labour hours	Labour £	Plant £	Material £	Unit	Total rate £
Linear drainage; channels; Aco Building Products; laid to concrete bed C25 on compacted granular base on 200 mm deep concrete bed; haunched with 200 mm concrete surround; all in 750 × 430 mm wide trench with compacted 200 mm granular base surround (excavation and subbase not included)							
HexDrain channel drain; recycled polypropylene; load class A15; 1000 × 125 × 80 mm							
with black plastic grating	20.25	1.20	24.90	–	30.45	each	**55.35**
with galvanized steel grating	25.50	1.20	24.90	–	35.70	each	**60.60**
with black plastic brickslot grating	20.25	1.20	24.90	–	30.45	each	**55.35**
Accessories for HexDrain							
endcap; black plastic	3.75	–	–	–	3.75	each	**3.75**
corner unit; with metallic effect plastic grating	22.20	0.50	10.38	–	26.89	each	**37.27**
sump unit; black plastic; 250 mm depth	35.70	1.00	20.75	–	40.39	each	**61.14**
corner unit; with black plastic grating; 125 mm vertical outlet	23.25	0.75	15.56	–	23.80	each	**39.36**
corner unit; brickslot; black plastic; 125 mm vertical outlet	23.25	0.50	10.38	–	27.94	each	**38.32**
Universal 820 drain union; PVC-u; 110 mm diameter	4.10	–	–	–	4.10	each	**4.10**
RainDrain lightweight polymer concrete channel; load class A15; 1000 × 118 × 97 mm							
with galvanized steel grating	28.95	1.20	24.90	–	33.64	each	**58.54**
Accessories for RainDrain							
foul air trap; horizontal; 110 mm diameter	19.05	–	–	–	19.05	each	**19.05**
closing endcap	4.70	–	–	–	4.70	each	**4.70**
outlet endcap; 110 mm diameter	9.26	–	–	–	9.26	each	**9.26**
sump; with galvanized steel grating and sediment bucket; 500 mm	72.35	0.50	10.38	–	77.04	each	**87.42**
Linear drainage; channels; Aco Building Products; laid to concrete bed C25 on compacted granular base on 200 mm deep concrete bed; haunched with 200 mm concrete surround; all in 750 × 430 mm wide trench with compacted 200 mm granular base surround (excavation and subbase not included)							
Aco MultiDrain M100PPD; recycled polypropylene drainage channel; range of gratings to complement installations which require discreet slot drainage							
142 mm wide × 150 mm deep	71.50	1.20	24.90	–	81.70	m	**106.60**

R12 DRAINAGE BELOW GROUND

Item Excluding site overheads and profit	PC £	Labour hours	Labour £	Plant £	Material £	Unit	Total rate £
LINEAR DRAINAGE – cont							
Linear drainage – cont							
Accessories for M100PPD							
connectors; vertical outlet 110 mm	14.65	–	–	–	14.65	nr	**14.65**
connectors; vertical outlet 160 mm	16.45	–	–	–	16.45	nr	**16.45**
sump units; black plastic with plastic silt							
bucket; 110/160 mm inlet/outlets	130.45	1.50	31.13	–	149.20	nr	**180.33**
universal gully and bucket;							
440 × 440 × 1315 mm deep	762.55	3.00	62.25	–	797.71	nr	**859.96**
Aco drainlock gratings for M100PPD system to loading A15							
slotted galvanized steel	22.15	–	–	–	22.15	m	**22.15**
perforated galvanized steel	30.45	–	–	–	30.45	m	**30.45**
Aco drainlock gratings for M100PPD system to loading C250							
Heelguard; composite black; 500 mm long							
with security locking	26.25	–	–	–	26.25	m	**26.25**
intercept; ductile iron; 500 mm long	34.35	–	–	–	34.35	m	**34.35**
slotted galvanized steel; 1.00 m long	57.30	–	–	–	57.30	m	**57.30**
perforated galvanized steel; 1.00 m long	63.10	–	–	–	63.10	m	**63.10**
mesh galvanized steel; 1.00 m long	42.90	–	–	–	42.90	m	**42.90**
Aco MultiDrain MD brickslot; offset galvanized slot drain grating for M100PPD; load class C250							
brickslot galvanized steel; 1.00 m	81.75	–	–	–	81.75	m	**81.75**
Aco MultiDrain MD brickslot; offset galvanized slot drain grating for M100PPD; load class C250 – 400							
Aco gratings for M100PPD brickslot; galvanized steel; load class C250/D400; 1.00 m							
brickslot; galvanized steel; 1.00 m	81.75	–	–	–	81.75	m	**81.75**
brickslot; stainless steel; 1.00 m	167.10	–	–	–	167.10	m	**167.10**
Channel footpath drain							
Aco channel drain to pedestrian footpaths; connection to existing kerbdrain; laying on 150 mm concrete base and surround; excavation subbase not included	27.09	1.20	24.90	–	37.30	m	**62.20**
Channel drains; Aco Technologies Ltd; Aco MultiDrain MD polymer concrete channel drainage system; traditional channel and grate drainage solution							
Shallow depth channels M100D system; 135 mm wide							
100 mm deep	70.95	1.20	24.90	–	81.15	m	**106.05**
Accessories for MultiDrain MD							
sump unit; complete with sediment bucket	127.75	1.50	31.13	–	127.75	nr	**158.88**

R12 DRAINAGE BELOW GROUND

Item Excluding site overheads and profit	PC £	Labour hours	Labour £	Plant £	Material £	Unit	Total rate £
ACCESS COVERS AND FRAMES							
Access Covers and Frames							
FACTA (Fabricated Access Cover Trade Association) class:							
A – 0.5 tonne maximum slow moving wheel load							
AA – 1.5 tonne maximum slow moving wheel load							
AAA – 2.5 tonne maximum slow moving wheel load							
B – 5 tonne maximum slow moving wheel load							
C – 6.5 tonne maximum slow moving wheel load							
D – 11 tonne maximum slow moving wheel load							
Access covers and frames; solid top; galvanized; Steelway Brickhouse; Bristeel; bedding frame in cement mortar (1:3); cover in grease and sand; clear opening sizes; base size shown in brackets (50 mm depth)							
FACTA AA; single seal							
450 × 450 mm (520 × 520 mm)	71.40	1.50	31.13	–	82.18	nr	**113.31**
600 × 450 mm (670 × 520 mm)	79.24	1.80	37.35	–	91.21	nr	**128.56**
600 × 600 mm (670 × 670 mm)	84.73	2.00	41.50	–	99.10	nr	**140.60**
FACTA AA; double seal							
450 × 450 mm (560 × 560 mm)	115.99	1.50	31.13	–	126.77	nr	**157.90**
60 × 450 mm (710 × 560 mm)	127.45	1.80	37.35	–	139.42	nr	**176.77**
600 × 600 mm (710 × 710 mm)	139.05	2.00	41.50	–	153.42	nr	**194.92**
FACTA B; single seal							
450 × 450 mm (520 × 520 mm)	92.30	1.50	31.13	–	103.08	nr	**134.21**
600 × 450 mm (670 × 520 mm)	115.23	1.80	37.35	–	127.20	nr	**164.55**
600 × 600 mm (670 × 670 mm)	133.95	2.00	41.50	–	148.32	nr	**189.82**
FACTA B; double seal							
450 × 450 mm (560 × 560 mm)	133.19	1.50	31.13	–	143.97	nr	**175.10**
600 × 450 mm (710 × 560 mm)	148.28	1.80	37.35	–	160.25	nr	**197.60**
600 × 600 mm (710 × 710 mm)	175.40	2.00	41.50	–	189.77	nr	**231.27**

R13 LAND DRAINAGE

Item Excluding site overheads and profit	PC £	Labour hours	Labour £	Plant £	Material £	Unit	Total rate £
ACCESS COVERS AND FRAMES – cont							
Access covers and frames; recessed; galvanized; Steelway Brickhouse; Bripave; bedding frame in cement mortar (1:3); cover in grease and sand; clear opening sizes; base size shown in brackets (for block depths 50 mm, 65 mm, 80 mm or 100 mm)							
FACTA AA							
450 × 450 mm (562 × 562 mm)	206.95	1.50	31.13	–	217.73	nr	**248.86**
600 × 450 mm (712 × 562 mm)	227.00	1.80	37.35	–	238.97	nr	**276.32**
600 × 600 mm (712 × 712 mm)	235.30	2.00	41.50	–	505.37	nr	**546.87**
750 × 600 mm (862 × 712 mm)	262.56	2.40	49.80	–	555.14	nr	**604.94**
FACTA B							
450 × 450 mm (562 × 562 mm)	207.31	1.50	31.13	–	218.09	nr	**249.22**
600 × 450 mm (712 × 562 mm)	241.00	1.80	37.35	–	252.97	nr	**290.32**
600 × 600 mm (712 × 712 mm)	244.00	2.00	41.50	–	258.37	nr	**299.87**
750 × 600 mm (862 × 712 mm)	244.79	2.40	49.80	–	260.36	nr	**310.16**
FACTA D							
450 × 450 mm (562 × 562 mm)	232.35	1.50	31.13	–	243.13	nr	**274.26**
600 × 450 mm (712 × 562 mm)	246.86	1.80	37.35	–	258.83	nr	**296.18**
600 × 600 mm (712 × 712 mm)	255.05	2.00	41.50	–	269.42	nr	**310.92**
750 × 600 mm (862 × 712 mm)	310.40	2.40	49.80	–	325.97	nr	**375.77**
Extra over manhole frames and covers for							
filling recessed manhole covers with brick paviors; PC £305.00/1000	13.42	1.00	20.75	–	14.13	m²	**34.88**
filling recessed manhole covers with vehicular paving blocks; PC £9.05/m²	9.05	0.75	15.56	–	9.77	m²	**25.33**
filling recessed manhole covers with concrete paving flags; PC £14.41/m²	14.42	0.35	7.26	–	14.85	m²	**22.11**
R13 LAND DRAINAGE							
EXCAVATION FOR LAND DRAINS							
Land drainage; excavating for drains; by 3 tonne tracked excavator; including disposing spoil to spoil heaps not exceeding 100 m							
Width 225 mm							
depth 450 mm	–	–	–	377.06	–	100 m	**377.06**
depth 600 mm	–	–	–	387.53	–	100 m	**387.53**
depth 700 mm	–	–	–	398.91	–	100 m	**398.91**
depth 900 mm	–	–	–	424.94	–	100 m	**424.94**
depth 1000 mm	–	–	–	439.90	–	100 m	**439.90**
depth 1200 mm	–	–	–	456.44	–	100 m	**456.44**

R13 LAND DRAINAGE

Item Excluding site overheads and profit	PC £	Labour hours	Labour £	Plant £	Material £	Unit	Total rate £
Land drainage; excavating for drains; by hand; including disposing spoil to spoil heaps not exceeding 100 m							
Width 150 mm							
depth 450 mm	–	22.03	457.12	–	–	100 m	**457.12**
depth 600 mm	–	29.38	609.63	–	–	100 m	**609.63**
depth 700 mm	–	34.27	711.10	–	–	100 m	**711.10**
depth 900 mm	–	44.06	914.25	–	–	100 m	**914.25**
Width 225 mm							
depth 450 mm	–	33.05	685.79	–	–	100 m	**685.79**
depth 600 mm	–	44.06	914.25	–	–	100 m	**914.25**
depth 700 mm	–	51.41	1066.76	–	–	100 m	**1066.76**
depth 900 mm	–	66.10	1371.58	–	–	100 m	**1371.58**
depth 1000 mm	–	146.88	3047.76	–	–	100 m	**3047.76**
PIPE LAYING							
Wavin Plastics Ltd; flexible plastic perforated pipes in trenches (not included); to a minimum depth of 450 mm (couplings not included)							
OsmaDrain; flexible plastic perforated pipes in trenches (not included); to a minimum depth of 450 mm (couplings not included)							
80 mm diameter; available in 100 m coil	179.22	2.00	41.50	–	179.22	100 m	**220.72**
100 mm diameter; available in 100 m coil	290.22	2.00	41.50	–	290.22	100 m	**331.72**
160 mm diameter; available in 35 m coil	719.40	2.00	41.50	–	719.40	100 m	**760.90**
WavinCoil; plastic pipe junctions							
80 × 80 mm	8.09	0.05	1.04	–	8.09	nr	**9.13**
100 × 100 mm	9.04	0.05	1.04	–	9.04	nr	**10.08**
100 × 60 mm	8.62	0.05	1.04	–	8.62	nr	**9.66**
100 × 80 mm	8.62	0.05	1.04	–	8.62	nr	**9.66**
160 × 160 mm	21.51	0.05	1.04	–	21.51	nr	**22.55**
WavinCoil; couplings for flexible pipes							
80 mm diameter	2.85	0.03	0.69	–	2.85	nr	**3.54**
100 mm diameter	3.16	0.03	0.69	–	3.16	nr	**3.85**
160 mm diameter	4.26	0.03	0.69	–	4.26	nr	**4.95**

R13 LAND DRAINAGE

Item Excluding site overheads and profit	PC £	Labour hours	Labour £	Plant £	Material £	Unit	Total rate £
BACKFILLING OF TRENCHES							
Market prices of backfilling materials							
Sand	38.40	–	–	–	38.40	m³	**38.40**
Gravel rejects	36.96	–	–	–	36.96	m³	**36.96**
Topsoil; allowing for 20% settlement	33.60	–	–	–	33.60	m³	**33.60**
Land drainage; backfilling trench after laying pipes with gravel rejects or similar; blind filling with ash or sand; topping with 150 mm topsoil from dumps not exceeding 100 m; by machine							
Width 150 mm							
depth 450 mm	–	3.20	66.40	53.95	235.44	100 m	**355.79**
depth 600 mm	–	4.20	87.15	71.93	316.80	100 m	**475.88**
depth 750 mm	–	4.86	100.84	89.92	361.44	100 m	**552.20**
depth 900 mm	–	6.20	128.65	107.90	462.24	100 m	**698.79**
Width 225 mm							
depth 450 mm	–	4.80	99.60	82.72	319.12	100 m	**501.44**
depth 600 mm	–	6.30	130.72	107.90	475.36	100 m	**713.98**
depth 750 mm	–	7.80	161.85	134.84	319.12	100 m	**615.81**
depth 900 mm	–	9.30	192.97	165.45	319.12	100 m	**677.54**
Land drainage; backfilling trench after laying pipes with gravel rejects or similar; blind filling with ash or sand; topping with 150 mm topsoil from dumps not exceeding 100 m; by hand							
Width 150 mm							
depth 450 mm	–	18.63	386.57	–	207.72	100 m	**594.29**
depth 600 mm	–	24.84	515.43	–	301.68	100 m	**817.11**
depth 750 mm	–	31.05	644.29	–	283.32	100 m	**927.61**
depth 900 mm	–	37.26	773.14	–	447.12	100 m	**1220.26**

V ELECTRICAL

Item Excluding site overheads and profit	PC £	Labour hours	Labour £	Plant £	Material £	Unit	Total rate £
V ELECTRICAL							
TRENCHING FOR ELECTRICAL SERVICES							
Excavating; mechanical; trenches for electrical services							
3 tonne excavator (bucket volume 0.13 m³); arisings laid alongside							
600 mm deep	–	–	–	1.70	–	m	**1.70**
800 mm deep	–	–	–	1.98	–	m	**1.98**
1.00 m deep	–	–	–	2.38	–	m	**2.38**
Extra over any types of excavating irrespective of depth for breaking out existing materials; heavy duty 110 volt breaker tool							
hard rock	–	5.00	103.75	23.52	–	m³	**127.27**
concrete	–	3.00	62.25	14.12	–	m³	**76.37**
reinforced concrete	–	4.00	83.00	23.52	–	m³	**106.52**
brickwork, blockwork or stonework	–	1.50	31.13	7.06	–	m³	**38.19**
By hand							
600 mm deep	–	0.24	4.99	–	–	m	**4.99**
800 mm deep	–	0.43	8.87	–	–	m	**8.87**
1.00 mm deep	–	0.67	13.83	–	–	m	**13.83**
Backfilling of trenches; including laying of electrical marker tape; compacting lightly as work proceeds; spreading any remaining material							
To trenches containing armoured cable							
by machine	0.13	–	–	1.78	0.13	m	**1.91**
by hand	0.13	0.20	4.15	–	0.13	m	**4.28**
To trenches containing ducted cable including 150 mm sharp sand over the duct							
by machine	2.19	–	–	2.23	2.19	m	**4.42**
by hand	0.13	0.25	5.19	–	2.19	m	**7.38**
V41 EXTERIOR LIGHTS – LOW VOLTAGE							
Low voltage lights; connecting to transformers (transformers shown separately); fixing as described							
Hunza Spike spotlight; 63.5 mm diameter × 75 mm long; c/w 20/35/50 w lamp							
stainless steel	110.96	–	–	–	122.63	nr	**122.63**
copper	69.56	–	–	–	81.23	nr	**81.23**
black	44.53	–	–	–	56.20	nr	**56.20**
Hunza Spike spotlight; adjustable; 63.5 mm diameter × 75 mm long; c/w 20/35/50 w lamp							
stainless steel	110.96	–	–	–	122.63	nr	**122.63**
copper	79.92	–	–	–	91.59	nr	**91.59**
black	52.56	–	–	–	64.23	nr	**64.23**

V ELECTRICAL

Item Excluding site overheads and profit	PC £	Labour hours	Labour £	Plant £	Material £	Unit	Total rate £
V41 EXTERIOR LIGHTS – LOW VOLTAGE – **cont**							
Low voltage lights – cont							
Hunza pole lights; single; fixing to concrete							
base 150 × 150 × 150 mm							
stainless steel	173.51	1.00	20.75	–	185.18	nr	**205.93**
copper	136.60	1.00	20.75	–	148.77	nr	**169.52**
black	75.75	1.00	20.75	–	87.92	nr	**108.67**
Hunza pole lights; double; fixing to concrete							
base 150 × 150 × 150 mm							
stainless steel	277.40	1.00	20.75	–	289.57	nr	**310.32**
copper	136.60	1.00	20.75	–	148.77	nr	**169.52**
black	75.75	1.00	20.75	–	87.92	nr	**108.67**
Recessed wall lights; eyelid steplights; fixing							
to brick or stone walls; inclusive of core drilling							
stainless steel	277.40	1.50	31.13	–	289.07	nr	**320.20**
copper	170.94	1.50	31.13	–	182.61	nr	**213.74**
black	356.33	1.50	31.13	–	368.00	nr	**399.13**
Hunza; recessed deck path or lawn lights;							
inclusive of core drilling excavation and all							
making good							
stainless steel light installed to deck	196.00	0.66	13.70	–	196.00	nr	**209.70**
stainless steel light installed to path	196.00	1.08	22.41	–	196.00	nr	**218.41**
stainless steel light installed to lawn	154.76	0.20	4.15	–	166.31	nr	**170.46**
stainless steel driveway light	196.00	1.08	22.41	–	196.00	nr	**218.41**
V90 ELECTRICAL CABLES							
Cable to trenches (trenching operations							
not included); laying only cable							
Twin core steel wire armoured cable; 50 m							
drums							
1.5 mm core	0.92	0.02	0.41	–	0.92	m	**1.33**
2.5 mm core	1.16	0.02	0.41	–	1.16	m	**1.57**
Twin core steel wire armoured cable; lengths							
less than 50 m runs							
1.5 mm core	1.01	0.02	0.41	–	1.01	m	**1.42**
2.5 mm core	1.24	0.02	0.41	–	1.24	m	**1.65**
Three core steel wire armoured cable; 50 m							
drums							
1.5 mm core	1.04	0.02	0.41	–	1.04	m	**1.45**
2.5 mm core	1.46	0.02	0.41	–	1.46	m	**1.87**
4.0 mm core	1.73	0.02	0.41	–	1.73	m	**2.14**
6.0 mm core	2.34	0.02	0.46	–	2.34	m	**2.80**
10.0 mm core	4.24	0.03	0.52	–	4.24	m	**4.76**
Three core steel wire armoured cable; lengths							
less than 50 m runs							
1.5 mm core	1.07	0.02	0.41	–	1.07	m	**1.48**
2.5 mm core	1.58	0.02	0.41	–	1.58	m	**1.99**

V ELECTRICAL

Item Excluding site overheads and profit	PC £	Labour hours	Labour £	Plant £	Material £	Unit	Total rate £
Ducts to trenches; twin wall flexible cable ducts with drawstrings; laid on 150 mm clean sharp sand							
Twin wall duct; laying to trenches							
63 mm × 50 m coils	1.00	0.02	0.41	–	1.00	m	**1.41**
110 mm × 50 m coils	1.50	0.02	0.41	–	1.50	m	**1.91**
Cable drawing through ducts							
straight runs up to 50 m lengths	–	1.00	20.75	–	–	nr	**20.75**
Terminations to armoured cables; cutting, coiling and taping length of cable to receive connection to light fitting, transformer and the like; (all not included) fixing to temporary posts							
Armoured cable	–	0.33	6.92	–	–	nr	**6.92**
Plain ducted cable	–	0.13	2.59	–	–	nr	**2.59**
Junction boxes; fixing to cables to receive connections to light fittings or transformers; to IP68							
Underground jointing box							
2 way	29.00	–	–	–	46.50	nr	**46.50**
3 way	34.00	–	–	–	54.00	nr	**54.00**
Above ground jointing box							
2 way	3.50	0.33	6.92	–	3.50	nr	**10.42**
3 way	4.90	–	–	–	18.90	nr	**18.90**
Power connections							
Internal connections and switches for control of external lighting; connections of switches or switch bank to distribution board and internal cables to switches							
per circuit located within 1.00 m of the distribution board; chasing of cables to walls not included	–	–	–	–	–	nr	**35.00**
Power connections to switches							
power cable to switches including builders work; drilling through walls and making good; chasing of cables to walls not included; price per circuit for the first circuit	–	1.00	20.75	–	74.00	nr	**94.75**
additional circuits located in the same location	–	–	–	–	12.75	nr	**12.75**

V ELECTRICAL

Item Excluding site overheads and profit	PC £	Labour hours	Labour £	Plant £	Material £	Unit	Total rate £
V90 EXTERIOR LIGHTING **TRANSFORMERS**							
Outdoor transformers for 12 volt exterior **lighting; connecting to junction box (not** **included); IP67**							
Single light transformers; 100 × 68 × 72 mm; 50 vA							
fused	24.50	–	–	–	71.17	nr	**71.17**
unfused	22.50	–	–	–	22.50	nr	**22.50**
Two light transformer; 130 × 85 × 85 mm; 100 vA							
fused	38.50	–	–	–	38.50	nr	**38.50**
unfused	35.50	–	–	–	35.50	nr	**35.50**
Three light transformer; 130 × 85 × 85 mm; 150 vA							
fused	40.38	–	–	–	40.38	nr	**40.38**
unfused	39.90	–	–	–	39.90	nr	**39.90**
Four light transformer; 130 × 85 × 100 mm; 200 vA							
fused	51.13	–	–	–	51.13	nr	**51.13**
unfused	45.20	–	–	–	45.20	nr	**45.20**

Design for Outdoor Recreation

2nd Edition

S. Bell

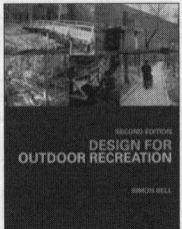

Design for Outdoor Recreation takes a detailed look at all aspects of design of facilities needed by visitors to outdoor recreation destinations. The book is a comprehensive manual for planners, designers and managers of recreation taking them through the processes of design and enabling them to find the most appropriate balance between visitor needs and the capacity of the landscape. A range of different aspects are covered including car parking, information signing, hiking, waterside activities, wildlife watching and camping.

This second edition incorporates new examples from overseas, including Australia, New Zealand, Japan and Eastern Europe as well as focusing on more current issues such as accessibility and the changing demands for recreational use.

July 2008: 276x219: 240pp
Pb: 978-0-415-44172-8: **£55.00**

To Order: Tel: +44 (0) 1235 400524 **Fax:** +44 (0) 1235 400525
or Post: Taylor and Francis Customer Services,
Bookpoint Ltd, Unit T1, 200 Milton Park, Abingdon, Oxon, OX14 4TA UK
Email: book.orders@tandf.co.uk

For a complete listing of all our titles visit:
www.tandf.co.uk

PART 7

Common Arrangement of Work Sections

The main work sections relevant to landscape work and their grouping:

A   Preliminaries/general conditions
     A10   Project particulars
     A11   Tender and contract documents
     A12   The site/existing buildings
     A13   Description of the work
     A20   The contract/subcontract
     A30   Employer's requirements: Tendering/
              Subletting/Supply
     A31   Employer's requirements: Provision,
              content and use of documents
     A32   Employer's requirements: Management
              of the Works
     A33   Employer's requirements: Quality stan-
              dards/control
     A34   Employer's requirements: Security/
              Safety/Protection
     A35   Employer's requirements: Specific
              limitations on method/sequence/timing/use
              of site
     A36   Employer's requirements: facilities/
              temporary works/services
     A37   Employer's requirements: operation/
              maintenance of the finished building
     A40   Contractor's general cost items:
              Management and staff
     A41   Contractor's general cost items: Site
              accommodation
     A42   Contractor's general cost items: Services
              and facilities
     A43   Contractor's general cost items:
              Mechanical plant
     A44   Contractor's general cost items:
              Temporary works
     A50   Works/Products by/on behalf of the
              Employer
     A51   Nominated subcontractors
     A52   Nominated suppliers
     A53   Work by statutory authorities/undertakers
     A54   Provisional work

     A55   Dayworks
     A60   Preliminaries/General conditions for
              demolition contract
     A61   Preliminaries/General conditions for
              investigation/survey contract
     A62   Preliminaries/General conditions for
              piling/embedded retaining wall contract
     A63   Preliminaries/General conditions for
              landscape contract
     A70   General specification requirements for
              work package
B   Complete buildings/structures/units
     B10   Prefabricated buildings/structures
     B11   Prefabricated building units
C   Existing site/buildings/services
     C10   Site survey
     C11   Ground investigation
     C12   Underground services survey
     C13   Building fabric survey
     C20   Demolition
     C50   Repairing/Renovating/Conserving metal
D   Groundwork
     D11   Soil stabilization
     D20   Excavating and filling
     D41   Crib walls/gabions/reinforced earth
E   In situ concrete/large precast concrete
     E05   In situ concrete construction generally
     E10   Mixing/Casting/Curing in situ concrete
     E20   Formwork for in situ concrete
     E30   Reinforcement for in situ concrete
F   Masonry
     F10   Brick/Block walling
     F20   Natural stone rubble walling
     F21   Natural stone ashlar walling/dressings
     F22   Cast stone ashlar walling/dressings
     F30   Accessories/Sundry items for brick/
              block/stone walling
     F31   Precast concrete sills/lintels/copings/
              features

G   Structural/carcassing metal/timber
     G31 Prefabricated timber unit decking
H   Cladding/covering
     H51 Natural stone slab cladding/features
     H52 Cast stone slab cladding/features
J   Waterproofing
     J10 Specialist waterproof rendering
     J20 Mastic asphalt tanking/damp-proofing
     J21 Mastic asphalt roofing/insulation/finishes
     J22 Proprietary roof decking with asphalt finish
     J30 Liquid applied tanking/damp-proofing
     J31 Liquid applied waterproof roof coatings
     J40 Flexible sheet tanking/damp-proofing
     J44 Sheet linings for pools/lakes/waterways
     J50 Green Roof systems
M   Surface finishes
     M10 Cement: sand/Concrete screeds
     M20 Plastered/Rendered/Roughcast coatings
     M40 Stone/Concrete/Quarry/Ceramic tiling/
          Mosaic
     M60 Painting/clear finishing
P   Building fabric sundries
     P30 Trenches/Pipeways/Pits for buried
          engineering services
Q   Paving/planting/fencing/site furniture

Q10 Kerbs/Edgings/Channels/paving
     accessories
Q20 Granular subbases to roads/pavings
Q21 In situ concrete roads/pavings
Q22 Coated macadam/Asphalt roads/pavings
Q23 Gravel/Hoggin/Woodchip roads/pavings
Q24 Interlocking brick/block roads/pavings
Q25 Slab/Brick/Sett/Cobble pavings
Q26 Special surfacings/pavings for sport/
     general amenity
Q30 Seeding/Turfing
Q31 Planting
Q32 Planting in Special environments
Q35 Landscape maintenance
Q40 Fencing
Q50 Site/Street furniture/equipment
R   Disposal systems
R12 Drainage below ground
R13 Land drainage
S   Piped supply systems
S10 Cold water
S14 Irrigation
S15 Fountains/Water features
V   Electrical supply/power/lighting systems
V41 Street/Area/Flood lighting

PART 8

Fees for Professional Services

LANDSCAPE ARCHITECTS' FEES

The Landscape Institute no longer sets any fee scales for its members. A publication which offers guidance in determining fees for different types of project is available from:

The Landscape Institute
33 Great Portland Street
London
W1W 8QG

Telephone: 020 7299 4500

The publication is entitled 'Engaging a Landscape Consultant – Guidance for Clients on Fees 2002'

The document refers to suggested fee systems on the following basis:

- Time Charged Fee Basis
- Lump Sum Fee Basis
- Percentage Fee Basis
- Retainer Fee Basis

In regard to the 'Percentage Fee Basis' the publication lists various project types and relates them to complexity ratings for works valued at £22,500 and above. The suggested fee scales are dependent on the complexity rating of the project.

The following are samples of the Complexity Categories:

Category 1: Golf Courses, Country Parks and Estates, Planting Schemes

Category 2: Coastal, River and Agricultural Works, Roads, Rural Recreation Schemes

Category 3: Hospital Grounds, Sport Stadia, Urban Offices and Commercial Properties, Housing

Category 4: Domestic/Historic Garden Design, Urban Rehabilitation and Environmental Improvements

Table Showing Suggested Percentage Fees for Various Categories of Complexity of Landscape Design or Consultancy

Project Value £	Complexity			
	1	2	3	4
22,500	15.00%	16.50%	18.00%	21.00%
30,000	13.75	15.00	16.50	19.25
50,000	10.50	12.75	14.00	16.25
100,000	9.50	10.50	11.50	13.50
150,000	8.75	9.50	10.50	12.50
200,000	8.00	8.75	9.75	11.25
300,000	7.50	8.25	9.00	10.50
500,000	6.75	7.50	8.25	9.75
750,000	6.50	7.25	7.75	9.25
1,000,000	6.25	7.00	7.50	8.75
10,000,000	6.25	6.75	7.25	8.25

(extrapolated from the graph/curve chart in the aforementioned publication)

Guide to Stage Payments of Fees, Relevant Fee Basis and Proportion of Fee Applicable to Lump Sum and Percentage Fee Basis. *Details of Preliminary, Standard and Other Services are set out in detail in the Landscape Consultant's Appointment.*

Work Stage	Relevant Fee Basis			Proportion of Fee	
	Time	Lump	%	*Proportion of fee*	Total
Preliminary Services					
A Inception	✓	✓	n/a	n/a	n/a
B Feasibility	✓	✓	n/a	n/a	n/a
Standard Services					
C Outline Proposals	✓	✓	✓	15%	15%
D Sketch Scheme					
Proposals	✓	✓	✓	15%	30%
E Detailed Proposals	✓	✓	✓	15%	45%
FG Production Information	✓	✓	✓	20%	65%
HJ Tender Action &					
Contract Preparation	✓	✓	✓	5%	70%
K Operations on Site	✓	✓	✓	25%	95%
L Completion	✓	✓	✓	5%	100%
Other Services	✓	✓	n/a		

Timing of Fee Payments

Percentage fees are normally paid at the end of each work stage. Time based fees are normally paid at monthly intervals. Lump sum fees are normally paid at intervals by agreement. Retainer or term commission fees are normally paid in advance, for predetermined periods of service.

WORKED EXAMPLES OF PERCENTAGE FEE CALCULATIONS

Worked Example 1

Project Type   Caravan Site
Services Required To Detailed Proposals – Work Stages C to E
Budget   £120,000

Step 1 Decide on Work Type and therefore Complexity Rating – Complexity Rating 2
Step 2 Decide on Services required and Proportion of Fee – To Detailed Proposals, 45%
Step 3 Read off Graph, Complexity Rating 2, the % fee of £120,000 – Graph Fee 9.9%
Step 4 Multiply the Proportion of Fee (45%) by the Graph Fee (9.9%) – Adjusted Fee – 4.46%
Step 5 Calculate the Guide Fee (4.46% of £120,000) – Guide Fee – £5,352
Step 6 Agree fee with Client, complete Memorandum of Agreement & Schedule of Services & Fees

Worked Example 2

Project Type   New Housing
Services Required Full Standard Services – Work Stages C to L
Budget   £350,000

Step 1 Decide on Work Type and therefore Complexity Rating – Complexity Rating 3
Step 2 Decide on Services required and Proportion of Fee – To Completion, 100%
Step 3 Read off Graph, Complexity Rating 3, the % fee of £350,000 – Graph Fee 8.8%
Step 4 Multiply the Proportion of Fee (100%) by the Graph Fee (8.8%) – Adjusted Fee – 8.8%
Step 5 Calculate the Guide Fee (8.8% of £350,000) – Guide Fee – £30,800
Step 6 Agree fee with Client, complete Memorandum of Agreement & Schedule of Services & Fees

Worked Example 3

Project Type   Urban Environmental Improvements
Services Required To Production Information – Work Stages C to G
Budget   £1,250,000

Step 1 Decide on Work Type and therefore Complexity Rating – Complexity Rating 4
Step 2 Decide on Services required and Proportion of Fee – To Production Information, 65%
Step 3 Read off Graph, Complexity Rating 4, the % fee of £1,250,000 – Graph Fee 8.6%
Step 4 Multiply the Proportion of Fee (65%) by the Graph Fee (8.6%) – Adjusted Fee – 5.59%
Step 5 Calculate the Guide Fee (5.59% of £1,250,000) – Guide Fee – £69,875
Step 6 Agree fee with Client, complete Memorandum of Agreement & Schedule of Services & Fees

Spon's Asia-Pacific Construction Costs Handbook

4th Edition

Edited by **Davis Langdon & Seah**

Spon's Asia Pacific Construction Costs Handbook includes construction cost data for 20 countries. This new edition has been extended to include Pakistan and Cambodia. Australia, UK and America are also included, to facilitate comparison with construction costs elsewhere. Information is presented for each country in the same way, as follows:

• key data on the main economic and construction indicators.

• an outline of the national construction industry, covering structure, tendering and contract procedures, materials cost data, regulations and standards

• labour and materials cost data

• Measured rates for a range of standard construction work items

• Approximate estimating costs per unit area for a range of building types

• price index data and exchange rate movements against £ sterling, $US and Japanese Yen

The book also includes a Comparative Data section to facilitate country-to-country comparisons. Figures from the national sections are grouped in tables according to national indicators, construction output, input costs and costs per square metre for factories, offices, warehouses, hospitals, schools, theatres, sports halls, hotels and housing.

This unique handbook will be an essential reference for all construction professionals involved in work outside their own country and for all developers or multinational companies assessing comparative development costs.

April 2010: 234x156: 480pp
Hb: 978-0-415-46565-6: **£120.00**

To Order: Tel: +44 (0) 1235 400524 **Fax:** +44 (0) 1235 400525
or Post: Taylor and Francis Customer Services,
Bookpoint Ltd, Unit T1, 200 Milton Park, Abingdon, Oxon, OX14 4TA UK
Email: book.orders@tandf.co.uk

For a complete listing of all our titles visit:
www.tandf.co.uk

PART 9

Daywork and Prime Cost Notes

When work is carried out which cannot be valued in any other way it is customary to assess the value on a cost basis with an allowance to cover overheads and profit. The basis of costing is a matter for agreement between the parties concerned, but definitions of prime cost for the building industry have been prepared and published jointly by the Royal Institution of Chartered Surveyors and the National Federation of Building Trades Employers (now the Construction Confederation) for the convenience of those who wish to use them. These documents are reproduced with the permission of the Royal Institution of Chartered Surveyors, which owns the copyright.

The daywork schedule published by the Civil Engineering Contractors Association is included in the External Works and Landscape Price Book's companion title, *Spon's Civil Engineering and Highway Works Price Book*.

For larger Prime Cost contracts the reader is referred to the form of contract issued by the Royal Institute of British Architects.

BUILDING INDUSTRY

DEFINITION OF PRIME COST OF DAYWORK CARRIED OUT UNDER A BUILDING CONTRACT (JUNE 2007 – THIRD EDITION)

This definition of Prime Cost is published by the Royal Institution of Chartered Surveyors and the Construction Confederation, for convenience and for use by people who choose to use it. Members of the Construction Confederation are not in any way debarred from defining Prime Cost and rendering their accounts for work carried out on that basis in any way they choose. Building owners are advised to reach agreement with contractors on the Definition of Prime Cost to be used prior to issuing instructions.

INTRODUCTION

This new edition of the Definition includes two options for dealing with the prime cost of labour:

Option 'A' – Percentage Addition, is based upon the traditional method of pricing labour in daywork, and allows for a percentage addition to be made for incidental costs, overheads and profit, to the prime cost of labour applicable at the time the daywork is carried out.

Option 'B' – All-inclusive Rates, includes not only the prime cost of labour but also includes an allowance for incidental costs, overheads and profit. The all-inclusive rates are deemed to be fixed for the period of the contract. However, where a fluctuating price contract is used, or where the rates in the contract are to be index-linked, the all-inclusive rates shall be adjusted by a suitable index in accordance with the contract conditions.

Model documentation, intended for inclusion in a building contract, is included in Appendix A, which illustrates how the Definition of Prime Cost may be applied in practice.

Example calculations of the Prime Cost of Labour in Daywork are given in Appendix B.

SECTION 1 – APPLICATION

1.1   This Definition provides a basis for the valuation of daywork executed under such building contracts as provide for its use.
1.2   It is not applicable in the case of daywork executed after the date of practical completion.
1.3   It is applicable to works carried out incidental to contract work but may not be deemed appropriate for use in 'daywork only' work or work carried out on an 'hourly' basis only, for which the 'Definition of Prime Cost of Building Works of a Jobbing or Maintenance Character' may be more suitable.
1.4   The terms 'contract' and 'contractor' herein shall be read as 'subcontract' and 'subcontractor' as applicable.
1.5   Dayworks are to be calculated by reference to the rate(s) current and prevailing on the day the work is carried out, except where Option 'B' for labour is used which may be adjusted by a suitable index in accordance with the contract conditions.

SECTION 2 – COMPOSITION OF TOTAL CHARGES

2.1   The prime cost of daywork comprises the sum of the following costs:
       2.1.1 Labour as defined in Section 3.
       2.1.2 Material and goods as defined in Section 4.
       2.1.3 Plant as defined in Section 5.
2.2   Incidental costs, overheads and profit as defined in Section 6, as provided in the building contract and expressed therein as percentage adjustments are applicable to each of 2.1.1, (Option A for Labour – Section 3) – 2.1.3. NB: If using Option 'B' for the labour element of prime cost in Section 3, incidental costs, overheads and profit are deemed included.

SECTION 3 – LABOUR

Option A – Percentage Addition

3.1   The prime cost of labour is defined in 3.5. Incidental costs, overheads and profit should be added as defined in Section 6.
3.2   The standard wage rates, payments and expenses referred to below and the standard working hours referred to in 3.3 are those laid down for the time being in the rules or decisions of the Construction Industry Joint Council (CIJC) and the terms of the Building and Civil Engineering Benefits Scheme (managed by the Building and Civil Engineering Holidays Scheme Management Ltd) applicable to the works, or the rules or decisions or agreements of such body, other than the CIJC, as may be applicable relating to the grade and type of operative concerned at the time when and in the area where the daywork is executed.
3.3   Hourly base rates for labour are computed by dividing the annual prime cost of labour, based upon standard working hours and as defined in 3.5, by the number of standard working hours per annum (see Example 1 on page 29)
3.4   The hourly rates computed in accordance with 3.3 shall be applied in respect of the time spent by operatives directly engaged on daywork, including those operating mechanical plant and transport and erecting and dismantling other plant (unless otherwise expressly provided in the building contract) and handling and distributing the materials and goods used in the daywork.
3.5   The annual prime cost of labour comprises the following:
       (a)   Standard or guaranteed minimum weekly earnings.*
       (b)   All other guaranteed minimum payments (unless included in Section 6). *
       (c)   Differentials or extra payments in respect of skill, responsibility, discomfort, inconvenience or risk (excluding those in respect of supervisory responsibility – see 3.6). *
       (d)   Payments in respect of public holidays.
       (e)   Any amounts which may become payable by the Contractor to or in respect of operatives arising from the operation of the rules or decisions referred to in 3.2 which are not provided for in 3.5 (a) to (d) or in Section 6. *
       (f)   Employer's contributions to industry's annual holiday with pay scheme or payment in lieu thereof.
       (g)   Employer's contributions to industry's welfare benefits scheme or payment in lieu thereof.
       (h)   Employer's National Insurance contributions applicable to 3.5 (a) to (g).

(i)   Any contribution, levy or tax imposed by statute, payable by the contractor in his capacity as an employer, or compliance with any legislation which has a direct effect on the cost of labour. *

3.6   Differentials or extra payments in respect of supervisory responsibility are excluded from the annual prime cost (see Section 6). The time of supervisory staff such as principals, foremen, gangers, leading hands and the like, when working manually, is admissible under this Section only at the appropriate standard/normal rates for the grade of operative suitable for the operation concerned.

3.7   An example calculation of a typical standard hourly base rate is provided in Example 1 on page 29.

Non-productive Overtime

3.8   * The prime cost for non-productive overtime should be based only on the hourly payments for items marked with an asterisk in 3.5 #.

3.9   Not used.

Option B – All-inclusive Rates

3.10   The prime cost of labour is based on the all-inclusive rates for labour provided for in the building contract. The all-inclusive rates are to include all costs associated with employing the labour including all items listed in 3.5.

3.11   The all-inclusive hourly rates are also to include all costs, fixed and time-related charges, overheads and profit (as defined in Section 6) in connection with labour.

3.12   The all-inclusive hourly rates shall be applied in respect of the time actually spent by the operatives directly engaged on daywork, including those operating mechanical plant and transport and erecting and dismantling other plant (unless otherwise expressly provided in the building contract) and handling and distributing the materials and goods used in the daywork.

3.13   The time of supervisory staff, such as principals, foremen, gangers, leading hands and the like, when working manually, is admissible under this Section only at the appropriate all-inclusive hourly rates for the grade of operative suitable for the operations concerned. Any extra payment in respect of supervisory responsibility is not allowable.

3.14   The all-inclusive rates are deemed to be fixed for the period of the contract. However, where a fluctuating price contract is used, or where the rates in the contract are to be index-linked, the all-inclusive rates shall be adjusted by a suitable index in accordance with the contract conditions.

Non-productive Overtime

3.15   Allowance for non-productive overtime should be made in accordance with the Model Documentation included in Appendix A #.

SECTION 4 – MATERIALS AND GOODS

4.1   The prime cost of materials and goods obtained specifically for the daywork is the invoice cost after deducting all trade discounts and any portion of cash discounts in excess of 5%, plus any appropriate handling and delivery charges.

4.2   The prime cost of materials and goods supplied from the Contractor's stock is based upon the current market prices after deducting all trade discounts and any portion of cash discounts in excess of 5%, plus any appropriate handling charges.

4.3   Any Value Added Tax which is treated, or is capable of being treated, as input tax (as defined in the Finance Act, 1972, or any re-enactment or amendment thereof or substitution thereof) by the Contractor is excluded, for the purpose of calculations.

SECTION 5 – PLANT

5.1   Unless otherwise stated in the building contract, the prime cost of plant comprises the cost of the following:
       (a)   Use or hire of mechanical operated plant and transport for the time employed/engaged for the day-
              work.
       (b)   Use of non-mechanical plant (excluding non-mechanical hand tools) for the time employed/engaged
              for the daywork.
       (c)   Transport/delivery to and from site and erection and dismantling where applicable.
       (d)   Qualified professional operators (e.g. crane drivers) not employed by the contractor (see 5.5 below).
5.2   Where plant is hired, the prime cost of plant shall be the invoice cost after deducting all trade discounts and
       any portion of cash discount in excess of 5%.
5.3   Where plant is not hired, the prime cost of plant shall be calculated in accordance with the latest edition of
       the Royal Institution of Chartered Surveyor's (RICS) Schedule of Basic Plant Charges for Use in Connection
       with Daywork Under a Building Contract.
5.4   The use of non-mechanical hand tools and of erected scaffolding, staging, trestles or the like is excluded
       (see Section 6).
5.5   Where hired or other plant is operated by the Contractor's operatives, the operative's time is to be included
       under Section 3 unless otherwise provided in the contract.
5.6   Any Value Added Tax which is treated, or is capable of being treated, as input tax (as defined by the
       Finance Act, 1972, or any re-enactment or amendment thereof or substitution therefor) by the Contractor is
       excluded, for the purposes of calculation.

SECTION 6 – INCIDENTAL COSTS, OVERHEADS AND PROFIT

6.1   The percentage adjustments provided in the building contract, which are applicable to each of the totals of
       Sections 3 (Option A), 4 and 5, include the following: #
       (a)   Head Office charges.
       (b)   Site staff, including site supervision.
       (c)   The additional cost of overtime (other than that referred to in #).
       (d)   Time lost due to inclement weather.
       (e)   The additional cost of bonuses and all other incentive payments in excess of any guaranteed mini-
              mum included in 3.5 (a).
       (f)   Apprentices' study time.
       (g)   Subsistence, lodging and periodic allowances.
       (h)   Fares and travelling allowances.
       (i)   Sick pay or insurance in respect thereof.
       (j)   Third-party and employers' liability insurance.
       (k)   Liability in respect of redundancy payments to employees.
       (l)   Employers' National Insurance contributions not included in Section 3.5.
       (m)   Tool allowances.
       (n)   Use and maintenance of non-mechanical hand tools.
       (o)   Use of erected scaffolding, staging, trestles or the like.
       (p)   Use of tarpaulins, plastic sheeting or the like, all necessary protective clothing, artificial lighting,
              safety and welfare facilities, storage and the like that may be available on the site.
       (q)   Any variation to basic rates required by the Contractor in cases where the building contract provides
              for the use of a specified schedule of basic plant charges (to the extent that no other provision is
              made for such variation – see Section 5).
       (r)   All other liabilities and obligations whatsoever not specifically referred to in this Section nor charge-
              able under any other Section.
       (s)   Any variation in welfare/pension payments from industry standard.
       (t)   Profit (including main contractor's profit as appropriate).

Non-productive Overtime

6.2   When calculating the percentage adjustment for incidental costs, overheads and profit, if the Option A calculation of price cost of labour is prescribed in the contract, it should be borne in mind that not all items listed in 6.1 are necessarily applicable to non-productive overtime. When Option B is prescribed, non-productive overtime should be shown separately in the contract documents as detailed in the Model Documentation in Appendix A

\# The additional cost of non-productive overtime, where specifically ordered by the Architect/Supervising Officer/ Contract Administrator/Employer's Agent, shall only be chargeable on the terms of prior written agreement between the parties to the building contract.

APPENDIX A

Model Documentation for Inclusion in a Building Contract

This model document is included to illustrate how the Definition of Prime Cost may be applied in practice. It does not form part of the Definition. It is, however, in a form agreed between the RICS and the Construction Confederation and its use in this form amended only as required to suit the specific building contract is encouraged.

Where Using Option A for Labour

Dayworks

The Contractor will be paid as defined below for the cost of works carried out as daywork in accordance with the building contract.

For building works, the prime cost of daywork will be calculated in accordance with the latest *Definition of Prime Cost of Daywork carried out under a Building Contract,* (*State edition*_____), published by the Royal Institution of Chartered Surveyors and the Construction Confederation.

For electrical works, the prime cost of daywork will be calculated in accordance with the latest *Definition of Prime Cost of Daywork carried out under an Electrical Contract,* (*State edition*_____), published by the Royal Institution of Chartered Surveyors, the Electrical Contractors' Association and 'SELECT' the Electrical Contractors' Association of Scotland.

For heating and ventilating work etc., the prime cost of daywork will be calculated in accordance with the latest *Definition of Prime Cost of Daywork carried out under a Heating, Ventilating, Air Conditioning, Refrigeration, Pipework and/or Domestic Engineering Contract,* (*State edition*_____), published by the Royal Institution of Chartered Surveyors and the Heating and Ventilating Contractors' Association.

For plumbing work, the prime cost of daywork will be calculated in accordance with the latest *Definition of Prime Cost of Daywork carried out under a Plumbing Contract,* (*State edition*_____), published by the Royal Institution of Chartered Surveyors, the Association of Plumbing and Heating Contractors and the Scottish and Northern Ireland Plumbing Employers' Confederation.

Labour

Building Operatives	Provisional Sum	£
Add for Incidental Costs, Overheads and Profit	............%	£
Electrical Operatives	Provisional Sum	£
Add for Incidental Costs, Overheads and Profit	............%	£
Heating and Ventilating Operatives	Provisional Sum	£
Add for Incidental Costs, Overheads and Profit	............%	£
Plumbing Operatives	Provisional Sum	£
Add for Incidental Costs, Overheads and Profit	............%	£

Non-productive Overtime

Building Operatives	Provisional Sum	£
Add for Incidental Costs, Overheads and Profit	............%	£
Electrical Operatives	Provisional Sum	£
Add for Incidental Costs, Overheads and Profit	............%	£
Heating and Ventilating Operatives	Provisional Sum	£
Add for Incidental Costs, Overheads and Profit	............%	£
Plumbing Operatives	Provisional Sum	£
Add for Incidental Costs, Overheads and Profit	............%	£

Where using Option B for Labour

Dayworks

The Contractor will be paid as defined below for the cost of works carried out as daywork in accordance with the building contract.

For building works, the prime cost of daywork will be calculated in accordance with the latest *Definition of Prime Cost of Daywork carried out under a Building Contract, (State edition_____)*, published by the Royal Institution of Chartered Surveyors and the Construction Confederation.

For electrical works, the prime cost of daywork will be calculated in accordance with the latest *Definition of Prime Cost of Daywork carried out under an Electrical Contract, (State edition_____)*, published by the Royal Institution of Chartered Surveyors, the Electrical Contractors' Association and 'SELECT' the Electrical Contractors' Association of Scotland.

For heating and ventilating work etc., the prime cost of daywork will be calculated in accordance with the latest *Definition of Prime Cost of Daywork carried out under a Heating, Ventilating, Air Conditioning, Refrigeration, Pipework and/or Domestic Engineering Contract, (State edition_____)*, published by the Royal Institution of Chartered Surveyors and the Heating and Ventilating Contractors' Association.

For plumbing work, the prime cost of daywork will be calculated in accordance with the latest *Definition of Prime Cost of Daywork carried out under a Plumbing Contract, (State edition_____)*, published by the Royal Institution of Chartered Surveyors, the Association of Plumbing and Heating Contractors and the Scottish and Northern Ireland Plumbing Employers' Confederation.

Labour

The Contractor must state below the all-inclusive prime cost hourly rates required for labour as defined in Section 3 (Option B) and the core working ours to which they apply.

Core Hours

General Operatives	£............ per hour
Skilled Operatives (all grades)	£............ per hour
Craft Operatives	£............ per hour

Other Grades/Trades:

....	£............ per hour
....	£............ per hour
....	£............ per hour
....	£............ per hour
....	£............ per hour

Core hours are ____am to ____pm Monday to Friday (excluding statutory holidays)

Overtime specifically ordered by the Architect/Supervising Officer/Contract Administrator/ Employer's Agent

The non-productive element of overtime should be as defined in the relevant Working Rule Agreement. However, if different, please state below.

Trade	Day	Time	Non-Productive Element (hours)
....	....	..........to.........	....
....	....	..........to.........	....
....	....	..........to.........	....
....	....	..........to.........	....
....	....	..........to.........	....
....	....	..........to.........	....
....	....	..........to.........	....

Provide the all-inclusive prime cost of labour as defined in Section 3 (Option B)

Productive Hours
[ ] hours (Provisional) General Operatives @ £.........per hour £
[ ] hours (Provisional) General Operatives @ £.........per hour £
[ ] hours (Provisional) General Operatives @ £.........per hour £

Other Grades/Trades:
[ ] hours (Provisional) General Operatives @ £.........per hour £
[ ] hours (Provisional) General Operatives @ £.........per hour £
[ ] hours (Provisional) General Operatives @ £.........per hour £

Non-productive Hours
[ ] hours (Provisional) General Operatives @ £.........per hour £
[ ] hours (Provisional) General Operatives @ £.........per hour £
[ ] hours (Provisional) General Operatives @ £.........per hour £

Other Grades/Trades:

[   ] hours (Provisional) General Operatives   @ £.........per hour

[   ] hours (Provisional) General Operatives   @ £.........per hour

[   ] hours (Provisional) General Operatives   @ £.........per hour

Materials and Goods

Provide for the prime cost of materials and goods
as defined in Section 4 (Provisional)   £ [ ]

Add the percentage addition for incidental costs,
overheads and profit as defined in Section 6   _____%

Plant

Provide for the prime cost of plant hired by the
Contractor as defined in Section 5 (Provisional)   £ [ ]

Add the percentage addition for incidental costs,
overheads and profit as defined in Section 6   _____%

Rates for plant not hired by the Contractor shall be as set out in *The Schedule of Basic Plant Charges for Use in Connection with Daywork Under a Building Contract* published by the Royal Institution of Chartered Surveyors (_____ Edition dated _____)

Provide for the prime cost of plant not hired by the
Contractor, as defined in Section 5 (Provisional)   £ [ ]

Add the percentage addition for incidental costs,
overheads and profit as defined in Section 6   _____%

PART 10

Tables and Memoranda

This part of the book contains the following sections:

CONVERSION TABLES

CONVERSION TABLES

Length	Unit	Conversion factors			
Millimetre	mm	1 in	= 25.4 mm	1 mm	= 0.0394 in
Centimetre	cm	1 in	= 2.54 cm	1 cm	= 0.3937 in
Metre	m	1 ft	= 0.3048 m	1 m	= 3.2808 ft
		1 yd	= 0.9144 m		= 1.0936 yd
Kilometre	km	1 mile	= 1.6093 km	1 km	= 0.6214 mile

Note:
1 cm = 10 mm	1 ft = 12 in
1 m = 1 000 mm	1 yd = 3 ft
1 km = 1 000 m	1 mile = 1 760 yd

Area	Unit	Conversion factors			
Square Millimetre	mm^2	$1\ in^2$	$= 645.2\ mm^2$	$1\ mm^2$	$= 0.0016\ in^2$
Square Centimetre	cm^2	$1\ in^2$	$= 6.4516\ cm^2$	$1\ cm^2$	$= 1.1550\ in^2$
Square Metre	m^2	$1\ ft^2$	$= 0.0929\ m^2$	$1\ m^2$	$= 10.764\ ft^2$
		$1\ yd^2$	$= 0.8361\ m^2$	$1\ m^2$	$= 1.1960\ yd^2$
Square Kilometre	km^2	$1\ mile^2$	$= 2.590\ km^2$	$1\ km^2$	$= 0.3861\ mile^2$

Note:
$1\ cm^2 = 100\ mm^2$	$1\ ft^2 = 144\ in^2$
$1\ m^2 = 10\ 000\ cm^2$	$1\ yd^2 = 9\ ft^2$
$1\ km^2 = 100$ hectares	$1\ acre = 4\ 840\ yd^2$
	$1\ mile^2 = 640$ acres

Volume	Unit	Conversion factors			
Cubic Centimetre	cm^3	$1\ cm^3$	$= 0.0610\ in^3$	$1\ in^3$	$= 16.387\ cm^3$
Cubic Decimetre	dm^3	$1\ dm^3$	$= 0.0353\ ft^3$	$1\ ft^3$	$= 28.329\ dm^3$
Cubic Metre	m^3	$1\ m^3$	$= 35.3147\ ft^3$	$1\ ft^3$	$= 0.0283\ m^3$
		$1\ m^3$	$= 1.3080\ yd^3$	$1\ yd^3$	$= 0.7646\ m^3$
Litre	l	1 l	= 1.76 pint	1 pint	= 0.5683 l
			= 2.113 US pt		= 0.4733 US l

Note:
$1\ dm^3 = 1\ 000\ cm^3$	$1\ ft^3 = 1\ 728\ in^3$	1 pint = 20 fl oz
$1\ m^3 = 1\ 000\ dm^3$	$1\ yd^3 = 27\ ft^3$	1 gal = 8 pints
1 l = $1\ dm^3$		

Neither the Centimetre nor Decimetre are SI units, and as such their use, particularly that of the Decimetre, is not widespread outside educational circles.

Mass	Unit	Conversion factors			
Milligram	mg	1 mg	= 0.0154 grain	1 grain	= 64.935 mg
Gram	g	1 g	= 0.0353 oz	1 oz	= 28.35 g
Kilogram	kg	1 kg	= 2.2046 lb	1 lb	= 0.4536 kg
Tonne	t	1 t	= 0.9842 ton	1 ton	= 1.016 t

Note:
1 g = 1000 mg	1 oz = 437.5 grains	1 cwt = 112 lb
1 kg = 1000 g	1 lb = 16 oz	1 ton = 20 cwt
1 t = 1000 kg	1 stone = 14 lb	

CONVERSION TABLES

Force	Unit	Conversion factors			
Newton	N	1 lbf	= 4.448 N	1 kgf	= 9.807 N
Kilonewton	kN	1 lbf	= 0.004448 kN	1 ton f	= 9.964 kN
Meganewton	MN	100 tonf	= 0.9964 MN		

Pressure and stress	Unit	Conversion factors	
Kilonewton per square metre	kN/m^2	1 lbf/in^2	= 6.895 kN/m^2
		1 bar	= 100 kN/m^2
Meganewton per square metre	MN/m^2	1 tonf/ft^2	= 107.3 kN/m^2 = 0.1073 MN/m^2
		1 kgf/cm^2	= 98.07 kN/m^2
		1 lbf/ft^2	= 0.04788 kN/m^2

Coefficient of consolidation (Cv) or swelling	Unit	Conversion factors	
Square metre per year	m^2/year	1 cm^2/s	= 3 154 m^2/year
		1 ft^2/year	= 0.0929 m^2/year

Coefficient of permeability	Unit	Conversion factors	
Metre per second	m/s	1 cm/s	= 0.01 m/s
Metre per year	m/year	1 ft/year	= 0.3048 m/year
			= 0.9651 × (10)^8 m/s

Temperature	Unit	Conversion factors	
Degree Celsius	°C	°C = 5/9 × (°F − 32)	°F = (9 × °C)/ 5 + 32

CONVERSION TABLES

SPEED CONVERSION

km/h	m/min	mph	fpm
1	16.7	0.6	54.7
2	33.3	1.2	109.4
3	50.0	1.9	164.0
4	66.7	2.5	218.7
5	83.3	3.1	273.4
6	100.0	3.7	328.1
7	116.7	4.3	382.8
8	133.3	5.0	437.4
9	150.0	5.6	492.1
10	166.7	6.2	546.8
11	183.3	6.8	601.5
12	200.0	7.5	656.2
13	216.7	8.1	710.8
14	233.3	8.7	765.5
15	250.0	9.3	820.2
16	266.7	9.9	874.9
17	283.3	10.6	929.6
18	300.0	11.2	984.3
19	316.7	11.8	1038.9
20	333.3	12.4	1093.6
21	350.0	13.0	1148.3
22	366.7	13.7	1203.0
23	383.3	14.3	1257.7
24	400.0	14.9	1312.3
25	416.7	15.5	1367.0
26	433.3	16.2	1421.7
27	450.0	16.8	1476.4
28	466.7	17.4	1531.1
29	483.3	18.0	1585.7
30	500.0	18.6	1640.4
31	516.7	19.3	1695.1
32	533.3	19.9	1749.8
33	550.0	20.5	1804.5
34	566.7	21.1	1859.1
35	583.3	21.7	1913.8
36	600.0	22.4	1968.5
37	616.7	23.0	2023.2
38	633.3	23.6	2077.9
39	650.0	24.2	2132.5
40	666.7	24.9	2187.2

CONVERSION TABLES

km/h	m/min	mph	fpm
41	683.3	25.5	2241.9
42	700.0	26.1	2296.6
43	716.7	26.7	2351.3
44	733.3	27.3	2405.9
45	750.0	28.0	2460.6
46	766.7	28.6	2515.3
47	783.3	29.2	2570.0
48	800.0	29.8	2624.7
49	816.7	30.4	2679.4
50	833.3	31.1	2734.0

GEOMETRY

Two dimensional figures

Figure	Diagram of figure	Surface area	Perimeter
Square		a^2	$4a$
Rectangle		ab	$2(a+b)$
Triangle		$\frac{1}{2}ch$	$a+b+c$
Circle		πr^2 $\frac{1}{4}\pi d^2$ where $2r=d$	$2\pi r$ πd
Parallelogram		ah	$2(a+b)$
Trapezium		$\frac{1}{2}h(a+b)$	$a+b+c+d$
Ellipse		Approximately πab	$\pi(a+b)$
Hexagon		$2.6 \times a^2$	

GEOMETRY

Figure	Diagram of figure	Surface area	Perimeter
Octagon		$4.83 \times a^2$	6a
Sector of a circle		$\frac{1}{2}$rb or $\frac{q}{360}\pi r^2$ note b = angle $\frac{q}{360} \times \pi 2r$	
Segment of a circle		S − T where S = area of sector, T = area of triangle	
Bellmouth		$\frac{3}{14} \times r^2$	

GEOMETRY

Three dimensional figures

Figure	Diagram of figure	Surface area	Volume
Cube		$6a^2$	a^3
Cuboid/ rectangular block		$2(ab + ac + bc)$	abc
Prism/ triangular block		$bd + hc + dc + ad$	$\frac{1}{2}hcd$
Cylinder		$2\pi r^2 + 2\pi h$	$\pi r^2 h$ $\frac{1}{4}\pi d^2 h$
Sphere		$4\pi r^2$	$\frac{4}{3}\pi r^3$
Segment of sphere		$2\pi Rh$	$\frac{1}{6}\pi h(3r^2 + h^2)$ $\frac{1}{3}\pi h^2(3R - H)$
Pyramid		$(a + b)l + ab$	$\frac{1}{3}abh$

GEOMETRY

Figure	Diagram of figure	Surface area	Volume
Frustum of a pyramid		$l(a + b + c + d) + \sqrt{(ab + cd)}$ [rectangular figure only]	$\frac{h}{3}(ab + cd + \sqrt{abcd})$
Cone		$\pi r l$ (excluding base) $\pi r l + \pi r^2$ (including base)	$\frac{1}{3}\pi r^2 h$ $\frac{1}{12}\pi d^2 h$
Frustum of a cone		$\pi r^2 + \pi R^2 + \pi l(R + r)$	$\frac{1}{3}\pi(R^2 + Rr + r^2)$

FORMULAE

Formulae

Formula	Description
Pythagoras Theorem	$A^2 = B^2 + C^2$ where A is the hypotenuse of a right-angled triangle and B and C are the two adjacent sides

Simpsons Rule	The Area is divided into an even number of strips of equal width, and therefore has an odd number of ordinates at the division points
	
	$$\text{area} = \frac{S(A + 2B + 4C)}{3}$$
	
	where S = common interval (strip width)
	           A = sum of first and last ordinates
	           B = sum of remaining odd ordinates
	           C = sum of the even ordinates
	
	The Volume can be calculated by the same formula, but by substituting the area of each coordinate rather than its length

Trapezoidal Rule	A given trench is divided into two equal sections, giving three ordinates, the first, the middle and the last
	
	$$\text{volume} = \frac{S \times (A + B + 2C)}{2}$$
	
	where S = width of the strips
	           A = area of the first section
	           B = area of the last section
	           C = area of the rest of the sections

Prismoidal Rule	A given trench is divided into two equal sections, giving three ordinates, the first, the middle and the last
	
	$$\text{volume} = \frac{L \times (A + 4B + C)}{6}$$
	
	where L = total length of trench
	           A = area of the first section
	           B = area of the middle section
	           C = area of the last section

TYPICAL THERMAL CONDUCTIVITY OF BUILDING MATERIALS

(Always check manufacturer's details – variation will occur depending on product and nature of materials)

	Thermal conductivity (W/mK)		Thermal conductivity (W/mK)
Acoustic plasterboard	0.25	Oriented strand board	0.13
Aerated concrete slab (500 kg/m^3)	0.16	Outer leaf brick	0.77
Aluminium	237	Plasterboard	0.22
Asphalt (1700 kg/m^3)	0.5	Plaster dense (1300 kg/m^3)	0.5
Bitumen-impregnated fibreboard	0.05	Plaster lightweight (600 kg/m^3)	0.16
Blocks (standard grade 600 kg/m^3)	0.15	Plywood (950 kg/m^3)	0.16
Blocks (solar grade 460 kg/m^3)	0.11	Prefabricated timber wall panels (check manufacturer)	0.12
Brickwork (outer leaf 1700 kg/m^3)	0.84	Screed (1200 kg/m^3)	0.41
Brickwork (inner leaf 1700 kg/m^3)	0.62	Stone chippings (1800 kg/m^3)	0.96
Dense aggregate concrete block 1800 kg/m^3 (exposed)	1.21	Tile hanging (1900 kg/m^3)	0.84
Dense aggregate concrete block 1800 kg/m^3 (protected)	1.13	Timber (650 kg/m^3)	0.14
Calcium silicate board (600 kg/m^3)	0.17	Timber flooring (650 kg/m^3)	0.14
Concrete general	1.28	Timber rafters	0.13
Concrete (heavyweight 2300 kg/m^3)	1.63	Timber roof or floor joists	0.13
Concrete (dense 2100 kg/m^3 typical floor)	1.4	Roof tile (1900 kg/m^3)	0.84
Concrete (dense 2000 kg/m^3 typical floor)	1.13	Timber blocks (650 kg/m^3)	0.14
Concrete (medium 1400 kg/m^3)	0.51	Cellular glass	0.045
Concrete (lightweight 1200 kg/m^3)	0.38	Expanded polystyrene	0.034
Concrete (lightweight 600 kg/m^3)	0.19	Expanded polystyrene slab (25 kg/m^3)	0.035
Concrete slab (aerated 500 kg/m^3)	0.16	Extruded polystyrene	0.035
Copper	390	Glass mineral wool	0.04
External render sand/cement finish	1	Mineral quilt (12 kg/m^3)	0.04
External render (1300 kg/m^3)	0.5	Mineral wool slab (25 kg/m^3)	0.035
Felt – Bitumen layers (1700 kg/m^3)	0.5	Phenolic foam	0.022
Fibreboard (300 kg/m^3)	0.06	Polyisocyanurate	0.025
Glass	0.93	Polyurethane	0.025
Marble	3	Rigid polyurethane	0.025
Metal tray used in wriggly tin concrete floors (7800 kg/m^3)	50	Rock mineral wool	0.038
Mortar (1750 kg/m^3)	0.8		

EARTHWORK

Weights of Typical Materials Handled by Excavators

The weight of the material is that of the state in its natural bed and includes moisture
Adjustments should be made to allow for loose or compacted states

Material	Mass (kg/m^3)	Mass (lb/cu yd)
Ashes, dry	610	1028
Ashes, wet	810	1365
Basalt, broken	1954	3293
Basalt, solid	2933	4943
Bauxite, crushed	1281	2159
Borax, fine	849	1431
Caliche	1440	2427
Cement, clinker	1415	2385
Chalk, fine	1221	2058
Chalk, solid	2406	4055
Cinders, coal, ash	641	1080
Cinders, furnace	913	1538
Clay, compacted	1746	2942
Clay, dry	1073	1808
Clay, wet	1602	2700
Coal, anthracite, solid	1506	2538
Coal, bituminous	1351	2277
Coke	610	1028
Dolomite, lumpy	1522	2565
Dolomite, solid	2886	4864
Earth, dense	2002	3374
Earth, dry, loam	1249	2105
Earth, Fullers, raw	673	1134
Earth, moist	1442	2430
Earth, wet	1602	2700
Felsite	2495	4205
Fieldspar, solid	2613	4404
Fluorite	3093	5213
Gabbro	3093	5213
Gneiss	2696	4544
Granite	2690	4534
Gravel, dry 1/4 to 2 inch	1682	2835
Gravel, dry, loose	1522	2565
Gravel, wet 1/4 to 2 inch	2002	3374
Gypsum, broken	1450	2444
Gypsum, solid	2787	4697
Hardcore (consolidated)	1928	3249
Lignite, dry	801	1350
Limestone, broken	1554	2619
Limestone, solid	2596	4375
Magnesite, magnesium ore	2993	5044
Marble	2679	4515
Marl, wet	2216	3735
Mica, broken	1602	2700
Mica, solid	2883	4859
Peat, dry	400	674
Peat, moist	700	1179

Tables and Memoranda

EARTHWORK

Material	Mass (kg/m³)	Mass (lb/cu yd)
Peat, wet	1121	1889
Potash	1281	2159
Pumice, stone	640	1078
Quarry waste	1438	2423
Quartz sand	1201	2024
Quartz, solid	2584	4355
Rhyolite	2400	4045
Sand and gravel, dry	1650	2781
Sand and gravel, wet	2020	3404
Sand, dry	1602	2700
Sand, wet	1831	3086
Sandstone, solid	2412	4065
Shale, solid	2637	4444
Slag, broken	2114	3563
Slag, furnace granulated	961	1619
Slate, broken	1370	2309
Slate, solid	2667	4495
Snow, compacted	481	810
Snow, freshly fallen	160	269
Taconite	2803	4724
Trachyte	2400	4045
Trap rock, solid	2791	4704
Turf	400	674
Water	1000	1685

Transport Capacities

Type of vehicle	Capacity of vehicle	
	Payload	Heaped capacity
Wheelbarrow	150	0.10
1 tonne dumper	1250	1.00
2.5 tonne dumper	4000	2.50
Articulated dump truck (Volvo A20 6 × 4)	18500	11.00
Articulated dump truck (Volvo A35 6 × 6)	32000	19.00
Large capacity rear dumper (Euclid R35)	35000	22.00
Large capacity rear dumper (Euclid R85)	85000	50.00

EARTHWORK

Machine Volumes for Excavating and Filling

Machine type	Cycles per minute	Volume per minute (m³)
1.5 tonne excavator	1	0.04
	2	0.08
	3	0.12
3 tonne excavator	1	0.13
	2	0.26
	3	0.39
5 tonne excavator	1	0.28
	2	0.56
	3	0.84
7 tonne excavator	1	0.28
	2	0.56
	3	0.84
21 tonne excavator	1	1.21
	2	2.42
	3	3.63
Backhoe loader JCB3CX excavator	1	0.28
Rear bucket capacity 0.28 m³	2	0.56
	3	0.84
Backhoe loader JCB3CX loading	1	1.00
Front bucket capacity 1.00 m³	2	2.00

Machine Volumes for Excavating and Filling

Machine type	Loads per hour	Volume per hour (m³)
1 tonne high tip skip loader	5	2.43
Volume 0.485 m³	7	3.40
	10	4.85
3 tonne dumper	4	7.60
Max volume 2.40 m³	5	9.50
Available volume 1.9 m³	7	13.30
	10	19.00
6 tonne dumper	4	15.08
Max volume 3.40 m³	5	18.85
Available volume 3.77 m³	7	26.39
	10	37.70

EARTHWORK

Bulkage of Soils (after excavation)

Type of soil	Approximate bulking of 1 m³ after excavation
Vegetable soil and loam	25–30%
Soft clay	30–40%
Stiff clay	10–20%
Gravel	20–25%
Sand	40–50%
Chalk	40–50%
Rock, weathered	30–40%
Rock, unweathered	50–60%

Shrinkage of Materials (on being deposited)

Type of soil	Approximate bulking of 1 m³ after excavation
Clay	10%
Gravel	8%
Gravel and sand	9%
Loam and light sandy soils	12%
Loose vegetable soils	15%

Voids in Material Used as Subbases or Beddings

Material	m³ of voids/m³
Alluvium	0.37
River grit	0.29
Quarry sand	0.24
Shingle	0.37
Gravel	0.39
Broken stone	0.45
Broken bricks	0.42

Angles of Repose

Type of soil		Degrees
Clay	– dry	30
	– damp, well drained	45
	– wet	15–20
Earth	– dry	30
	– damp	45
Gravel	– moist	48
Sand	– dry or moist	35
	– wet	25
Loam		40

EARTHWORK

Slopes and Angles

Ratio of base to height	Angle in degrees
5:1	11
4:1	14
3:1	18
2:1	27
1½:1	34
1:1	45
1:1½	56
1:2	63
1:3	72
1:4	76
1:5	79

Grades (in Degrees and Percents)

Degrees	Percent	Degrees	Percent
1	1.8	24	44.5
2	3.5	25	46.6
3	5.2	26	48.8
4	7.0	27	51.0
5	8.8	28	53.2
6	10.5	29	55.4
7	12.3	30	57.7
8	14.0	31	60.0
9	15.8	32	62.5
10	17.6	33	64.9
11	19.4	34	67.4
12	21.3	35	70.0
13	23.1	36	72.7
14	24.9	37	75.4
15	26.8	38	78.1
16	28.7	39	81.0
17	30.6	40	83.9
18	32.5	41	86.9
19	34.4	42	90.0
20	36.4	43	93.3
21	38.4	44	96.6
22	40.4	45	100.0

EARTHWORK

Bearing Powers

Ground conditions		Bearing power		
		kg/m^2	lb/in^2	Metric t/m^2
Rock,	broken	483	70	50
	solid	2415	350	240
Clay,	dry or hard	380	55	40
	medium dry	190	27	20
	soft or wet	100	14	10
Gravel,	cemented	760	110	80
Sand,	compacted	380	55	40
	clean dry	190	27	20
Swamp and alluvial soils		48	7	5

Earthwork Support

Maximum depth of excavation in various soils without the use of earthwork support

Ground conditions	Feet (ft)	Metres (m)
Compact soil	12	3.66
Drained loam	6	1.83
Dry sand	1	0.3
Gravelly earth	2	0.61
Ordinary earth	3	0.91
Stiff clay	10	3.05

It is important to note that the above table should only be used as a guide. Each case must be taken on its merits and, as the limited distances given above are approached, careful watch must be kept for the slightest signs of caving in

CONCRETE WORK

Weights of Concrete and Concrete Elements

Type of material		kg/m³	lb/cu ft
Ordinary concrete (dense aggregates)			
Non-reinforced plain or mass concrete			
Nominal weight		2305	144
Aggregate	– limestone	2162 to 2407	135 to 150
	– gravel	2244 to 2407	140 to 150
	– broken brick	2000 (av)	125 (av)
	– other crushed stone	2326 to 2489	145 to 155
Reinforced concrete			
Nominal weight		2407	150
Reinforcement	– 1%	2305 to 2468	144 to 154
	– 2%	2356 to 2519	147 to 157
	– 4%	2448 to 2703	153 to 163
Special concretes			
Heavy concrete			
Aggregates	– barytes, magnetite	3210 (min)	200 (min)
	– steel shot, punchings	5280	330
Lean mixes			
Dry-lean (gravel aggregate)		2244	140
Soil-cement (normal mix)		1601	100

CONCRETE WORK

Type of material		kg/m² per mm thick	lb/sq ft per inch thick
Ordinary concrete (dense aggregates)			
Solid slabs (floors, walls etc.)			
Thickness:	75 mm or 3 in	184	37.5
	100 mm or 4 in	245	50
	150 mm or 6 in	378	75
	250 mm or 10 in	612	125
	300 mm or 12 in	734	150
Ribbed slabs			
Thickness:	125 mm or 5 in	204	42
	150 mm or 6 in	219	45
	225 mm or 9 in	281	57
	300 mm or 12 in	342	70
Special concretes			
Finishes etc.			
	Rendering, screed etc.		
	Granolithic, terrazzo	1928 to 2401	10 to 12.5
	Glass-block (hollow) concrete	1734 (approx)	9 (approx)
Prestressed concrete		Weights as for reinforced concrete (upper limits)	
Air-entrained concrete		Weights as for plain or reinforced concrete	

CONCRETE WORK

Average Weight of Aggregates

Materials	Voids %	Weight kg/m³
Sand	39	1660
Gravel 10–20 mm	45	1440
Gravel 35–75 mm	42	1555
Crushed stone	50	1330
Crushed granite (over 15 mm)	50	1345
(n.e. 15 mm)	47	1440
'All-in' ballast	32	1800–2000

Material	kg/m³	lb/cu yd
Vermiculite (aggregate)	64–80	108–135
All-in aggregate	1999	125

Applications and Mix Design

Site mixed concrete

Recommended mix	Class of work suitable for	Cement (kg)	Sand (kg)	Coarse aggregate (kg)	Nr 25 kg bags cement per m³ of combined aggregate
1:3:6	Roughest type of mass concrete such as footings, road haunching over 300 mm thick	208	905	1509	8.30
1:2.5:5	Mass concrete of better class than 1:3:6 such as bases for machinery, walls below ground etc..	249	881	1474	10.00
1:2:4	Most ordinary uses of concrete, such as mass walls above ground, road slabs etc. and general reinforced concrete work	304	889	1431	12.20
1:1.5:3	Watertight floors, pavements and walls, tanks, pits, steps, paths, surface of 2 course roads, reinforced concrete where extra strength is required	371	801	1336	14.90
1:1:2	Works of thin section such as fence posts and small precast work	511	720	1206	20.40

CONCRETE WORK

Ready mixed concrete

Application	Designated concrete	Standardized prescribed concrete	Recommended consistence (nominal slump class)
Foundations			
Mass concrete fill or blinding	GEN 1	ST2	S3
Strip footings	GEN 1	ST2	S3
Mass concrete foundations			
Single storey buildings	GEN 1	ST2	S3
Double storey buildings	GEN 3	ST4	S3
Trench fill foundations			
Single storey buildings	GEN 1	ST2	S4
Double storey buildings	GEN 3	ST4	S4
General applications			
Kerb bedding and haunching	GEN 0	ST1	S1
Drainage works – immediate support	GEN 1	ST2	S1
Other drainage works	GEN 1	ST2	S3
Oversite below suspended slabs	GEN 1	ST2	S3
Floors			
Garage and house floors with no embedded steel	GEN 3	ST4	S2
Wearing surface: Light foot and trolley traffic	RC30	ST4	S2
Wearing surface: General industrial	RC40	N/A	S2
Wearing surface: Heavy industrial	RC50	N/A	S2
Paving			
House drives, domestic parking and external parking	PAV 1	N/A	S2
Heavy-duty external paving	PAV 2	N/A	S2

CONCRETE WORK

Prescribed Mixes for Ordinary Structural Concrete

Weights of cement and total dry aggregates in kg to produce approximately one cubic metre of fully compacted concrete together with the percentages by weight of fine aggregate in total dry aggregates

Conc. grade	Nominal max size of aggregate (mm)	40		20		14		10	
	Workability	Med.	High	Med.	High	Med.	High	Med.	High
	Limits to slump that may be expected (mm)	50–100	100–150	25–75	75–125	10–50	50–100	10–25	25–50
7	Cement (kg)	180	200	210	230	–	–	–	–
	Total aggregate (kg)	1950	1850	1900	1800	–	–	–	–
	Fine aggregate (%)	30–45	30–45	35–50	35–50	–	–	–	–
10	Cement (kg)	210	230	240	260	–	–	–	–
	Total aggregate (kg)	1900	1850	1850	1800	–	–	–	–
	Fine aggregate (%)	30–45	30–45	35–50	35–50	–	–	–	–
15	Cement (kg)	250	270	280	310	–	–	–	–
	Total aggregate (kg)	1850	1800	1800	1750	–	–	–	–
	Fine aggregate (%)	30–45	30–45	35–50	35–50	–	–	–	–
20	Cement (kg)	300	320	320	350	340	380	360	410
	Total aggregate (kg)	1850	1750	1800	1750	1750	1700	1750	1650
	Sand								
	Zone 1 (%)	35	40	40	45	45	50	50	55
	Zone 2 (%)	30	35	35	40	40	45	45	50
	Zone 3 (%)	30	30	30	35	35	40	40	45
25	Cement (kg)	340	360	360	390	380	420	400	450
	Total aggregate (kg)	1800	1750	1750	1700	1700	1650	1700	1600
	Sand								
	Zone 1 (%)	35	40	40	45	45	50	50	55
	Zone 2 (%)	30	35	35	40	40	45	45	50
	Zone 3 (%)	30	30	30	35	35	40	40	45
30	Cement (kg)	370	390	400	430	430	470	460	510
	Total aggregate (kg)	1750	1700	1700	1650	1700	1600	1650	1550
	Sand								
	Zone 1 (%)	35	40	40	45	45	50	50	55
	Zone 2 (%)	30	35	35	40	40	45	45	50
	Zone 3 (%)	30	30	30	35	35	40	40	45

REINFORCEMENT

REINFORCEMENT

Weights of Bar Reinforcement

Nominal sizes (mm)	Cross-sectional area (mm²)	Mass (kg/m)	Length of bar (m/tonne)
6	28.27	0.222	4505
8	50.27	0.395	2534
10	78.54	0.617	1622
12	113.10	0.888	1126
16	201.06	1.578	634
20	314.16	2.466	405
25	490.87	3.853	260
32	804.25	6.313	158
40	1265.64	9.865	101
50	1963.50	15.413	65

Weights of Bars (at specific spacings)

Weights of metric bars in kilogrammes per square metre

Size (mm)	Spacing of bars in millimetres									
	75	100	125	150	175	200	225	250	275	300
6	2.96	2.220	1.776	1.480	1.27	1.110	0.99	0.89	0.81	0.74
8	5.26	3.95	3.16	2.63	2.26	1.97	1.75	1.58	1.44	1.32
10	8.22	6.17	4.93	4.11	3.52	3.08	2.74	2.47	2.24	2.06
12	11.84	8.88	7.10	5.92	5.07	4.44	3.95	3.55	3.23	2.96
16	21.04	15.78	12.63	10.52	9.02	7.89	7.02	6.31	5.74	5.26
20	32.88	24.66	19.73	16.44	14.09	12.33	10.96	9.87	8.97	8.22
25	51.38	38.53	30.83	25.69	22.02	19.27	17.13	15.41	14.01	12.84
32	84.18	63.13	50.51	42.09	36.08	31.57	28.06	25.25	22.96	21.04
40	131.53	98.65	78.92	65.76	56.37	49.32	43.84	39.46	35.87	32.88
50	205.51	154.13	123.31	102.76	88.08	77.07	68.50	61.65	56.05	51.38

Basic weight of steelwork taken as 7850 kg/m³
Basic weight of bar reinforcement per metre run = 0.00785 kg/mm²
The value of π has been taken as 3.141592654

REINFORCEMENT

Fabric Reinforcement

Preferred range of designated fabric types and stock sheet sizes

Fabric reference	Longitudinal wires			Cross wires			
	Nominal wire size (mm)	Pitch (mm)	Area (mm²/m)	Nominal wire size (mm)	Pitch (mm)	Area (mm²/m)	Mass (kg/m²)
Square mesh							
A393	10	200	393	10	200	393	6.16
A252	8	200	252	8	200	252	3.95
A193	7	200	193	7	200	193	3.02
A142	6	200	142	6	200	142	2.22
A98	5	200	98	5	200	98	1.54
Structural mesh							
B1131	12	100	1131	8	200	252	10.90
B785	10	100	785	8	200	252	8.14
B503	8	100	503	8	200	252	5.93
B385	7	100	385	7	200	193	4.53
B283	6	100	283	7	200	193	3.73
B196	5	100	196	7	200	193	3.05
Long mesh							
C785	10	100	785	6	400	70.8	6.72
C636	9	100	636	6	400	70.8	5.55
C503	8	100	503	5	400	49.0	4.34
C385	7	100	385	5	400	49.0	3.41
C283	6	100	283	5	400	49.0	2.61
Wrapping mesh							
D98	5	200	98	5	200	98	1.54
D49	2.5	100	49	2.5	100	49	0.77

Stock sheet size 4.8 m × 2.4 m, Area 11.52 m²

Average weight kg/m³ of steelwork reinforcement in concrete for various building elements

Substructure	kg/m³ concrete	Substructure	kg/m³ concrete
Pile caps	110–150	Plate slab	150–220
Tie beams	130–170	Cant slab	145–210
Ground beams	230–330	Ribbed floors	130–200
Bases	125–180	Topping to block floor	30–40
Footings	100–150	Columns	210–310
Retaining walls	150–210	Beams	250–350
Raft	60–70	Stairs	130–170
Slabs – one way	120–200	Walls – normal	40–100
Slabs – two way	110–220	Walls – wind	70–125

Note: For exposed elements add the following %:
Walls 50%, Beams 100%, Columns 15%

FORMWORK

Formwork Stripping Times – Normal Curing Periods

Conditions under which concrete is maturing	Minimum periods of protection for different types of cement					
	Number of days (where the average surface temperature of the concrete exceeds 10°C during the whole period)			Equivalent maturity (degree hours) calculated as the age of the concrete in hours multiplied by the number of degrees Celsius by which the average surface temperature of the concrete exceeds 10°C		
	Other	SRPC	OPC or RHPC	Other	SRPC	OPC or RHPC
1. Hot weather or drying winds	7	4	3	3500	2000	1500
2. Conditions not covered by 1	4	3	2	2000	1500	1000

KEY
OPC – Ordinary Portland Cement
RHPC – Rapid-hardening Portland Cement
SRPC – Sulphate-resisting Portland Cement

Minimum Period before Striking Formwork

	Minimum period before striking		
	Surface temperature of concrete		
	16°C	17°C	t°C (0–25)
Vertical formwork to columns, walls and large beams	12 hours	18 hours	300 hours t+10
Soffit formwork to slabs	4 days	6 days	100 days t+10
Props to slabs	10 days	15 days	250 days t+10
Soffit formwork to beams	9 days	14 days	230 days t+10
Props to beams	14 days	21 days	360 days t+10

MASONRY

Number of Bricks required for Various Types of Work per m² of Walling

Description	Brick size	
	215 × 102.5 × 50 mm	215 × 102.5 × 65 mm
Half brick thick		
Stretcher bond	74	59
English bond	108	86
English garden wall bond	90	72
Flemish bond	96	79
Flemish garden wall bond	83	66
One brick thick and cavity wall of two half brick skins		
Stretcher bond	148	119

Quantities of Bricks and Mortar required per m² of Walling

	Unit	No of bricks required	Mortar required (cubic metres)		
Standard bricks			**No frogs**	**Single frogs**	**Double frogs**
Brick size 215 × 102.5 × 50 mm					
half brick wall (103 mm)	m²	72	0.022	0.027	0.032
2 × half brick cavity wall (270 mm)	m²	144	0.044	0.054	0.064
one brick wall (215 mm)	m²	144	0.052	0.064	0.076
one and a half brick wall (322 mm)	m²	216	0.073	0.091	0.108
Mass brickwork	m³	576	0.347	0.413	0.480
Brick size 215 × 102.5 × 65 mm					
half brick wall (103 mm)	m²	58	0.019	0.022	0.026
2 × half brick cavity wall (270 mm)	m²	116	0.038	0.045	0.055
one brick wall (215 mm)	m²	116	0.046	0.055	0.064
one and a half brick wall (322 mm)	m²	174	0.063	0.074	0.088
Mass brickwork	m³	464	0.307	0.360	0.413
Metric modular bricks			**Perforated**		
Brick size 200 × 100 × 75 mm					
90 mm thick	m²	67	0.016	0.019	
190 mm thick	m²	133	0.042	0.048	
290 mm thick	m²	200	0.068	0.078	
Brick size 200 × 100 × 100 mm					
90 mm thick	m²	50	0.013	0.016	
190 mm thick	m²	100	0.036	0.041	
290 mm thick	m²	150	0.059	0.067	
Brick size 300 × 100 × 75 mm					
90 mm thick	m²	33	–	0.015	
Brick size 300 × 100 × 100 mm					
90 mm thick	m²	44	0.015	0.018	

Note: Assuming 10 mm thick joints

MASONRY

Mortar required per m² Blockwork (9.88 blocks/m²)

Wall thickness	75	90	100	125	140	190	215
Mortar m³/m²	0.005	0.006	0.007	0.008	0.009	0.013	0.014

Mortar Group	Cement: lime: sand	Masonry cement: sand	Cement: sand with plasticizer
1	1:0–0.25:3		
2	1:0.5:4–4.5	1:2.5-3.5	1:3–4
3	1:1:5–6	1:4–5	1:5–6
4	1:2:8–9	1:5.5–6.5	1:7–8
5	1:3:10–12	1:6.5–7	1:8

Group 1: strong inflexible mortar
Group 5: weak but flexible

All mixes within a group are of approximately similar strength
Frost resistance increases with the use of plasticizers
Cement: lime: sand mixes give the strongest bond and greatest resistance to rain penetration
Masonry cement equals ordinary Portland cement plus a fine neutral mineral filler and an air entraining agent

Calcium Silicate Bricks

Type	Strength	Location
Class 2 crushing strength	14.0 N/mm²	not suitable for walls
Class 3	20.5 N/mm²	walls above dpc
Class 4	27.5 N/mm²	cappings and copings
Class 5	34.5 N/mm²	retaining walls
Class 6	41.5 N/mm²	walls below ground
Class 7	48.5 N/mm²	walls below ground

The Class 7 calcium silicate bricks are therefore equal in strength to Class B bricks
Calcium silicate bricks are not suitable for DPCs

Durability of Bricks	
FL	Frost resistant with low salt content
FN	Frost resistant with normal salt content
ML	Moderately frost resistant with low salt content
MN	Moderately frost resistant with normal salt content

MASONRY

Brickwork Dimensions

No. of horizontal bricks	Dimensions (mm)	No. of vertical courses	Height of vertical courses (mm)
½	112.5	1	75
1	225.0	2	150
1½	337.5	3	225
2	450.0	4	300
2½	562.5	5	375
3	675.0	6	450
3½	787.5	7	525
4	900.0	8	600
4½	1012.5	9	675
5	1125.0	10	750
5½	1237.5	11	825
6	1350.0	12	900
6½	1462.5	13	975
7	1575.0	14	1050
7½	1687.5	15	1125
8	1800.0	16	1200
8½	1912.5	17	1275
9	2025.0	18	1350
9½	2137.5	19	1425
10	2250.0	20	1500
20	4500.0	24	1575
40	9000.0	28	2100
50	11250.0	32	2400
60	13500.0	36	2700
75	16875.0	40	3000

TIMBER

Weights of Timber

Material	kg/m³	lb/cu ft
General	806 (avg)	50 (avg)
Douglas fir	479	30
Yellow pine, spruce	479	30
Pitch pine	673	42
Larch, elm	561	35
Oak (English)	724 to 959	45 to 60
Teak	643 to 877	40 to 55
Jarrah	959	60
Greenheart	1040 to 1204	65 to 75
Quebracho	1285	80
Material	**kg/m² per mm thickness**	**lb/sq ft per inch thickness**
Wooden boarding and blocks		
Softwood	0.48	2.5
Hardwood	0.76	4
Hardboard	1.06	5.5
Chipboard	0.76	4
Plywood	0.62	3.25
Blockboard	0.48	2.5
Fibreboard	0.29	1.5
Wood-wool	0.58	3
Plasterboard	0.96	5
Weather boarding	0.35	1.8

TIMBER

Conversion Tables (for timber only)

Inches	Millimetres	Feet	Metres
1	25	1	0.300
2	50	2	0.600
3	75	3	0.900
4	100	4	1.200
5	125	5	1.500
6	150	6	1.800
7	175	7	2.100
8	200	8	2.400
9	225	9	2.700
10	250	10	3.000
11	275	11	3.300
12	300	12	3.600
13	325	13	3.900
14	350	14	4.200
15	375	15	4.500
16	400	16	4.800
17	425	17	5.100
18	450	18	5.400
19	475	19	5.700
20	500	20	6.000
21	525	21	6.300
22	550	22	6.600
23	575	23	6.900
24	600	24	7.200

Planed Softwood
The finished end section size of planed timber is usually 3/16" less than the original size from which it is produced. This however varies slightly depending upon availability of material and origin of the species used.

Standards (timber) to cubic metres and cubic metres to standards (timber)

Cubic metres	Cubic metres standards	Standards
4.672	1	0.214
9.344	2	0.428
14.017	3	0.642
18.689	4	0.856
23.361	5	1.070
28.033	6	1.284
32.706	7	1.498
37.378	8	1.712
42.050	9	1.926
46.722	10	2.140
93.445	20	4.281
140.167	30	6.421
186.890	40	8.561
233.612	50	10.702
280.335	60	12.842
327.057	70	14.982
373.779	80	17.122

TIMBER

1 cu metre = 35.3148 cu ft = 0.21403 std

1 cu ft = 0.028317 cu metres

1 std = 4.67227 cu metres

Basic sizes of sawn softwood available (cross-sectional areas)

Thickness (mm)	Width (mm)								
	75	100	125	150	175	200	225	250	300
16	X	X	X	X					
19	X	X	X	X					
22	X	X	X	X					
25	X	X	X	X	X	X	X	X	X
32	X	X	X	X	X	X	X	X	X
36	X	X	X	X					
38	X	X	X	X	X	X	X		
44	X	X	X	X	X	X	X	X	X
47*	X	X	X	X	X	X	X	X	X
50	X	X	X	X	X	X	X	X	X
63	X	X	X	X	X	X	X		
75	X	X	X	X	X	X	X	X	
100		X		X		X		X	X
150				X		X			X
200						X			
250								X	
300									X

* This range of widths for 47 mm thickness will usually be found to be available in construction quality only

Note: The smaller sizes below 100 mm thick and 250 mm width are normally but not exclusively of European origin. Sizes beyond this are usually of North and South American origin

Basic lengths of sawn softwood available (metres)

1.80	2.10	3.00	4.20	5.10	6.00	7.20
	2.40	3.30	4.50	5.40	6.30	
	2.70	3.60	4.80	5.70	6.60	
		3.90			6.90	

Note: Lengths of 6.00 m and over will generally only be available from North American species and may have to be recut from larger sizes

TIMBER

Reductions from basic size to finished size by planning of two opposed faces

Purpose	Reductions from basic sizes for timber			
	15–35 mm	36–100 mm	101–150 mm	over 150 mm
a) Constructional timber	3 mm	3 mm	5 mm	6 mm
b) Matching interlocking boards	4 mm	4 mm	6 mm	6 mm
c) Wood trim not specified in BS 584	5 mm	7 mm	7 mm	9 mm
d) Joinery and cabinet work	7 mm	9 mm	11 mm	13 mm

Note: The reduction of width or depth is overall the extreme size and is exclusive of any reduction of the face by the machining of a tongue or lap joints

Maximum Spans for Various Roof Trusses

Maximum permissible spans for rafters for Fink trussed rafters

Basic size (mm)	Actual size (mm)	Pitch (degrees)								
		15 (m)	17.5 (m)	20 (m)	22.5 (m)	25 (m)	27.5 (m)	30 (m)	32.5 (m)	35 (m)
38 × 75	35 × 72	6.03	6.16	6.29	6.41	6.51	6.60	6.70	6.80	6.90
38 × 100	35 × 97	7.48	7.67	7.83	7.97	8.10	8.22	8.34	8.47	8.61
38 × 125	35 × 120	8.80	9.00	9.20	9.37	9.54	9.68	9.82	9.98	10.16
44 × 75	41 × 72	6.45	6.59	6.71	6.83	6.93	7.03	7.14	7.24	7.35
44 × 100	41 × 97	8.05	8.23	8.40	8.55	8.68	8.81	8.93	9.09	9.22
44 × 125	41 × 120	9.38	9.60	9.81	9.99	10.15	10.31	10.45	10.64	10.81
50 × 75	47 × 72	6.87	7.01	7.13	7.25	7.35	7.45	7.53	7.67	7.78
50 × 100	47 × 97	8.62	8.80	8.97	9.12	9.25	9.38	9.50	9.66	9.80
50 × 125	47 × 120	10.01	10.24	10.44	10.62	10.77	10.94	11.00	11.00	11.00

Tables and Memoranda

TIMBER

Sizes of Internal and External Doorsets

Description	Internal size (mm)	Permissible deviation	External size (mm)	Permissible deviation
Coordinating dimension: height of door leaf height sets	2100		2100	
Coordinating dimension: height of ceiling height set	2300 2350 2400 2700 3000		2300 2350 2400 2700 3000	
Coordinating dimension: width of all doorsets S = Single leaf set D = Double leaf set	600 S 700 S 800 S&D 900 S&D 1000 S&D 1200 D 1500 D 1800 D 2100 D		900 S 1000 S 1200 D 1800 D 2100 D	
Work size: height of door leaf height set	2090	± 2.0	2095	± 2.0
Work size: height of ceiling height set	2285 2335 2385 2685 2985	± 2.0	2295 2345 2395 2695 2995	± 2.0
Work size: width of all doorsets S = Single leaf set D = Double leaf set	590 S 690 S 790 S&D 890 S&D 990 S&D 1190 D 1490 D 1790 D 2090 D	± 2.0	895 S 995 S 1195 D 1495 D 1795 D 2095 D	± 2.0
Width of door leaf in single leaf sets F = Flush leaf P = Panel leaf	526 F 626 F 726 F&P 826 F&P 926 F&P	± 1.5	806 F&P 906 F&P	± 1.5
Width of door leaf in double leaf sets F = Flush leaf P = Panel leaf	362 F 412 F 426 F 562 F&P 712 F&P 826 F&P 1012 F&P	± 1.5	552 F&P 702 F&P 852 F&P 1002 F&P	± 1.5
Door leaf height for all doorsets	2040	± 1.5	1994	± 1.5

ROOFING

Total Roof Loadings for Various Types of Tiles/Slates

		Roof load (slope) kg/m^2		
		Slate/Tile	Roofing underlay and battens^2	Total dead load kg/m
Asbestos cement slate (600 × 300)		21.50	3.14	24.64
Clay tile	interlocking	67.00	5.50	72.50
	plain	43.50	2.87	46.37
Concrete tile	interlocking	47.20	2.69	49.89
	plain	78.20	5.50	83.70
Natural slate (18" × 10")		35.40	3.40	38.80
		Roof load (plan) kg/m^2		
Asbestos cement slate (600 × 300)		28.45	76.50	104.95
Clay tile	interlocking	53.54	76.50	130.04
	plain	83.71	76.50	60.21
Concrete tile	interlocking	57.60	76.50	134.10
	plain	96.64	76.50	173.14

ROOFING

Tiling Data

Product		Lap (mm)	Gauge of battens	No. slates per m²	Battens (m/m²)	Weight as laid (kg/m²)
CEMENT SLATES						
Eternit slates	600 × 300 mm	100	250	13.4	4.00	19.50
(Duracem)		90	255	13.1	3.92	19.20
		80	260	12.9	3.85	19.00
		70	265	12.7	3.77	18.60
	600 × 350 mm	100	250	11.5	4.00	19.50
		90	255	11.2	3.92	19.20
	500 × 250 mm	100	200	20.0	5.00	20.00
		90	205	19.5	4.88	19.50
		80	210	19.1	4.76	19.00
		70	215	18.6	4.65	18.60
	400 × 200 mm	90	155	32.3	6.45	20.80
		80	160	31.3	6.25	20.20
		70	165	30.3	6.06	19.60
CONCRETE TILES/SLATES						
Redland Roofing						
Stonewold slate	430 × 380 mm	75	355	8.2	2.82	51.20
Double Roman tile	418 × 330 mm	75	355	8.2	2.91	45.50
Grovebury pantile	418 × 332 mm	75	343	9.7	2.91	47.90
Norfolk pantile	381 × 227 mm	75	306	16.3	3.26	44.01
		100	281	17.8	3.56	48.06
Renown interlocking tile	418 × 330 mm	75	343	9.7	2.91	46.40
'49' tile	381 × 227 mm	75	306	16.3	3.26	44.80
		100	281	17.8	3.56	48.95
Plain, vertical tiling	265 × 165 mm	35	115	52.7	8.70	62.20
Marley Roofing						
Bold roll tile	420 × 330 mm	75	344	9.7	2.90	47.00
		100	–	10.5	3.20	51.00
Modern roof tile	420 × 330 mm	75	338	10.2	3.00	54.00
		100	–	11.0	3.20	58.00
Ludlow major	420 × 330 mm	75	338	10.2	3.00	45.00
		100	–	11.0	3.20	49.00
Ludlow plus	387 × 229 mm	75	305	16.1	3.30	47.00
		100	–	17.5	3.60	51.00
Mendip tile	420 × 330 mm	75	338	10.2	3.00	47.00
		100	–	11.0	3.20	51.00
Wessex	413 × 330 mm	75	338	10.2	3.00	54.00
		100	–	11.0	3.20	58.00
Plain tile	267 × 165 mm	65	100	60.0	10.00	76.00
		75	95	64.0	10.50	81.00
		85	90	68.0	11.30	86.00
Plain vertical tiles (feature)	267 × 165 mm	35	110	53.0	8.70	67.00
		34	115	56.0	9.10	71.00

ROOFING

Slate Nails, Quantity per Kilogram

Length	Type			
	Plain wire	**Galvanized wire**	**Copper nail**	**Zinc nail**
28.5 mm	325	305	325	415
34.4 mm	286	256	254	292
50.8 mm	242	224	194	200

Metal Sheet Coverings

Thicknesses and weights of sheet metal coverings								
Lead to BS 1178								
BS Code No	3	4	5	6	7	8		
Colour code	Green	Blue	Red	Black	White	Orange		
Thickness (mm)	1.25	1.80	2.24	2.50	3.15	3.55		
Density kg/m²	14.18	20.41	25.40	30.05	35.72	40.26		
Copper to BS 2870								
Thickness (mm)		0.60	0.70					
Bay width								
Roll (mm)		500	650					
Seam (mm)		525	600					
Standard width to form bay	600	750						
Normal length of sheet	1.80	1.80						
Zinc to BS 849								
Zinc Gauge (Nr)	9	10	11	12	13	14	15	16
Thickness (mm)	0.43	0.48	0.56	0.64	0.71	0.79	0.91	1.04
Density (kg/m²)	3.1	3.2	3.8	4.3	4.8	5.3	6.2	7.0
Aluminium to BS 4868								
Thickness (mm)	0.5	0.6	0.7	0.8	0.9	1.0	1.2	
Density (kg/m²)	12.8	15.4	17.9	20.5	23.0	25.6	30.7	

ROOFING

Type of felt	Nominal mass per unit area (kg/10 m)	Nominal mass per unit area of fibre base (g/m²)	Nominal length of roll (m)
Class 1			
1B fine granule	14	220	10 or 20
surfaced bitumen	18	330	10 or 20
	25	470	10
1E mineral surfaced bitumen	38	470	10
1F reinforced bitumen	15	160 (fibre)	15
		110 (hessian)	
1F reinforced bitumen, aluminium faced	13	160 (fibre)	15
		110 (hessian)	
Class 2			
2B fine granule surfaced bitumen asbestos	18	500	10 or 20
2E mineral surfaced bitumen asbestos	38	600	10
Class 3			
3B fine granule surfaced bitumen glass fibre	18	60	20
3E mineral surfaced bitumen glass fibre	28	60	10
3E venting base layer bitumen glass fibre	32	60*	10
3H venting base layer bitumen glass fibre	17	60*	20

* Excluding effect of perforations

GLAZING

GLAZING

Nominal thickness (mm)	Tolerance on thickness (mm)	Approximate weight (kg/m²)	Normal maximum size (mm)
Float and polished plate glass			
3	+ 0.2	7.50	2140 × 1220
4	+ 0.2	10.00	2760 × 1220
5	+ 0.2	12.50	3180 × 2100
6	+ 0.2	15.00	4600 × 3180
10	+ 0.3	25.00)	6000 × 3300
12	+ 0.3	30.00)	
15	+ 0.5	37.50	3050 × 3000
19	+ 1.0	47.50)	3000 × 2900
25	+ 1.0	63.50)	
Clear sheet glass			
2 *	+ 0.2	5.00	1920 × 1220
3	+ 0.3	7.50	2130 × 1320
4	+ 0.3	10.00	2760 × 1220
5 *	+ 0.3	12.50)	2130 × 2400
6 *	+ 0.3	15.00)	
Cast glass			
3	+ 0.4		
	− 0.2	6.00)	2140 × 1280
4	+ 0.5	7.50)	
5	+ 0.5	9.50	2140 × 1320
6	+ 0.5	11.50)	3700 × 1280
10	+ 0.8	21.50)	
Wired glass (Cast wired glass)			
6	+ 0.3	−)	
	− 0.7	)	3700 × 1840
7	+ 0.7	−)	
(Polished wire glass)			
6	+ 1.0	−	330 × 1830

* The 5 mm and 6 mm thickness are known as *thick drawn sheet*. Although 2 mm sheet glass is available it is not recommended for general glazing purposes

METAL

Weights of Metals

Material	kg/m³	lb/cu ft
Metals, steel construction, etc.		
Iron		
– cast	7207	450
– wrought	7687	480
– ore – general	2407	150
– (crushed) Swedish	3682	230
Steel	7854	490
Copper		
– cast	8731	545
– wrought	8945	558
Brass	8497	530
Bronze	8945	558
Aluminium	2774	173
Lead	11322	707
Zinc (rolled)	7140	446
	g/mm² per metre	**lb/sq ft per foot**
Steel bars	7.85	3.4
Structural steelwork	Net weight of member @ 7854 kg/m³	
riveted	+ 10% for cleats, rivets, bolts, etc.	
welded	+ 1.25% to 2.5% for welds, etc.	
Rolled sections		
beams	+ 2.5%	
stanchions	+ 5% (extra for caps and bases)	
Plate		
web girders	+ 10% for rivets or welds, stiffeners, etc.	
	kg/m	**lb/ft**
Steel stairs: industrial type		
1 m or 3 ft wide	84	56
Steel tubes		
50 mm or 2 in bore	5 to 6	3 to 4
Gas piping		
20 mm or ¾ in	2	1¼

METAL

Universal Beams BS 4: Part 1: 2005

Designation	Mass (kg/m)	Depth of section (mm)	Width of section (mm)	Thickness		Surface area (m²/m)
				Web (mm)	Flange (mm)	
1016 × 305 × 487	487.0	1036.1	308.5	30.0	54.1	3.20
1016 × 305 × 438	438.0	1025.9	305.4	26.9	49.0	3.17
1016 × 305 × 393	393.0	1016.0	303.0	24.4	43.9	3.15
1016 × 305 × 349	349.0	1008.1	302.0	21.1	40.0	3.13
1016 × 305 × 314	314.0	1000.0	300.0	19.1	35.9	3.11
1016 × 305 × 272	272.0	990.1	300.0	16.5	31.0	3.10
1016 × 305 × 249	249.0	980.2	300.0	16.5	26.0	3.08
1016 × 305 × 222	222.0	970.3	300.0	16.0	21.1	3.06
914 × 419 × 388	388.0	921.0	420.5	21.4	36.6	3.44
914 × 419 × 343	343.3	911.8	418.5	19.4	32.0	3.42
914 × 305 × 289	289.1	926.6	307.7	19.5	32.0	3.01
914 × 305 × 253	253.4	918.4	305.5	17.3	27.9	2.99
914 × 305 × 224	224.2	910.4	304.1	15.9	23.9	2.97
914 × 305 × 201	200.9	903.0	303.3	15.1	20.2	2.96
838 × 292 × 226	226.5	850.9	293.8	16.1	26.8	2.81
838 × 292 × 194	193.8	840.7	292.4	14.7	21.7	2.79
838 × 292 × 176	175.9	834.9	291.7	14.0	18.8	2.78
762 × 267 × 197	196.8	769.8	268.0	15.6	25.4	2.55
762 × 267 × 173	173.0	762.2	266.7	14.3	21.6	2.53
762 × 267 × 147	146.9	754.0	265.2	12.8	17.5	2.51
762 × 267 × 134	133.9	750.0	264.4	12.0	15.5	2.51
686 × 254 × 170	170.2	692.9	255.8	14.5	23.7	2.35
686 × 254 × 152	152.4	687.5	254.5	13.2	21.0	2.34
686 × 254 × 140	140.1	383.5	253.7	12.4	19.0	2.33
686 × 254 × 125	125.2	677.9	253.0	11.7	16.2	2.32
610 × 305 × 238	238.1	635.8	311.4	18.4	31.4	2.45
610 × 305 × 179	179.0	620.2	307.1	14.1	23.6	2.41
610 × 305 × 149	149.1	612.4	304.8	11.8	19.7	2.39
610 × 229 × 140	139.9	617.2	230.2	13.1	22.1	2.11
610 × 229 × 125	125.1	612.2	229.0	11.9	19.6	2.09
610 × 229 × 113	113.0	607.6	228.2	11.1	17.3	2.08
610 × 229 × 101	101.2	602.6	227.6	10.5	14.8	2.07
533 × 210 × 122	122.0	544.5	211.9	12.7	21.3	1.89
533 × 210 × 109	109.0	539.5	210.8	11.6	18.8	1.88
533 × 210 × 101	101.0	536.7	210.0	10.8	17.4	1.87
533 × 210 × 92	92.1	533.1	209.3	10.1	15.6	1.86
533 × 210 × 82	82.2	528.3	208.8	9.6	13.2	1.85
457 × 191 × 98	98.3	467.2	192.8	11.4	19.6	1.67
457 × 191 × 89	89.3	463.4	191.9	10.5	17.7	1.66
457 × 191 × 82	82.0	460.0	191.3	9.9	16.0	1.65
457 × 191 × 74	74.3	457.0	190.4	9.0	14.5	1.64
457 × 191 × 67	67.1	453.4	189.9	8.5	12.7	1.63
457 × 152 × 82	82.1	465.8	155.3	10.5	18.9	1.51
457 × 152 × 74	74.2	462.0	154.4	9.6	17.0	1.50
457 × 152 × 67	67.2	458.0	153.8	9.0	15.0	1.50
457 × 152 × 60	59.8	454.6	152.9	8.1	13.3	1.50
457 × 152 × 52	52.3	449.8	152.4	7.6	10.9	1.48
406 × 178 × 74	74.2	412.8	179.5	9.5	16.0	1.51
406 × 178 × 67	67.1	409.4	178.8	8.8	14.3	1.50
406 × 178 × 60	60.1	406.4	177.9	7.9	12.8	1.49

Tables and Memoranda

METAL

Designation	Mass (kg/m)	Depth of section (mm)	Width of section (mm)	Thickness Web (mm)	Flange (mm)	Surface area (m²/m)
406 × 178 × 50	54.1	402.6	177.7	7.7	10.9	1.48
406 × 140 × 46	46.0	403.2	142.2	6.8	11.2	1.34
406 × 140 × 39	39.0	398.0	141.8	6.4	8.6	1.33
356 × 171 × 67	67.1	363.4	173.2	9.1	15.7	1.38
356 × 171 × 57	57.0	358.0	172.2	8.1	13.0	1.37
356 × 171 × 51	51.0	355.0	171.5	7.4	11.5	1.36
356 × 171 × 45	45.0	351.4	171.1	7.0	9.7	1.36
356 × 127 × 39	39.1	353.4	126.0	6.6	10.7	1.18
356 × 127 × 33	33.1	349.0	125.4	6.0	8.5	1.17
305 × 165 × 54	54.0	310.4	166.9	7.9	13.7	1.26
305 × 165 × 46	46.1	306.6	165.7	6.7	11.8	1.25
305 × 165 × 40	40.3	303.4	165.0	6.0	10.2	1.24
305 × 127 × 48	48.1	311.0	125.3	9.0	14.0	1.09
305 × 127 × 42	41.9	307.2	124.3	8.0	12.1	1.08
305 × 127 × 37	37.0	304.4	123.3	7.1	10.7	1.07
305 × 102 × 33	32.8	312.7	102.4	6.6	10.8	1.01
305 × 102 × 28	28.2	308.7	101.8	6.0	8.8	1.00
305 × 102 × 25	24.8	305.1	101.6	5.8	7.0	0.992
254 × 146 × 43	43.0	259.6	147.3	7.2	12.7	1.08
254 × 146 × 37	37.0	256.0	146.4	6.3	10.9	1.07
254 × 146 × 31	31.1	251.4	146.1	6.0	8.6	1.06
254 × 102 × 28	28.3	260.4	102.2	6.3	10.0	0.904
254 × 102 × 25	25.2	257.2	101.9	6.0	8.4	0.897
254 × 102 × 22	22.0	254.0	101.6	5.7	6.8	0.890
203 × 133 × 30	30.0	206.8	133.9	6.4	9.6	0.923
203 × 133 × 25	25.1	203.2	133.2	5.7	7.8	0.915
203 × 102 × 23	23.1	203.2	101.8	5.4	9.3	0.790
178 × 102 × 19	19.0	177.8	101.2	4.8	7.9	0.738
152 × 89 × 16	16.0	152.4	88.7	4.5	7.7	0.638
127 × 76 × 13	13.0	127.0	76.0	4.0	7.6	0.537

METAL

Universal Columns BS 4: Part 1: 2005

Designation	Mass (kg/m)	Depth of section (mm)	Width of section (mm)	Thickness Web (mm)	Flange (mm)	Surface area (m²/m)
356 × 406 × 634	633.9	474.7	424.0	47.6	77.0	2.52
356 × 406 × 551	551.0	455.6	418.5	42.1	67.5	2.47
356 × 406 × 467	467.0	436.6	412.2	35.8	58.0	2.42
356 × 406 × 393	393.0	419.0	407.0	30.6	49.2	2.38
356 × 406 × 340	339.9	406.4	403.0	26.6	42.9	2.35
356 × 406 × 287	287.1	393.6	399.0	22.6	36.5	2.31
356 × 406 × 235	235.1	381.0	384.8	18.4	30.2	2.28
356 × 368 × 202	201.9	374.6	374.7	16.5	27.0	2.19
356 × 368 × 177	177.0	368.2	372.6	14.4	23.8	2.17
356 × 368 × 153	152.9	362.0	370.5	12.3	20.7	2.16
356 × 368 × 129	129.0	355.6	368.6	10.4	17.5	2.14
305 × 305 × 283	282.9	365.3	322.2	26.8	44.1	1.94
305 × 305 × 240	240.0	352.5	318.4	23.0	37.7	1.91
305 × 305 × 198	198.1	339.9	314.5	19.1	31.4	1.87
305 × 305 × 158	158.1	327.1	311.2	15.8	25.0	1.84
305 × 305 × 137	136.9	320.5	309.2	13.8	21.7	1.82
305 × 305 × 118	117.9	314.5	307.4	12.0	18.7	1.81
305 × 305 × 97	96.9	307.9	305.3	9.9	15.4	1.79
254 × 254 × 167	167.1	289.1	265.2	19.2	31.7	1.58
254 × 254 × 132	132.0	276.3	261.3	15.3	25.3	1.55
254 × 254 × 107	107.1	266.7	258.8	12.8	20.5	1.52
254 × 254 × 89	88.9	260.3	256.3	10.3	17.3	1.50
254 × 254 × 73	73.1	254.1	254.6	8.6	14.2	1.49
203 × 203 × 86	86.1	222.2	209.1	12.7	20.5	1.24
203 × 203 × 71	71.0	215.8	206.4	10.0	17.3	1.22
203 × 203 × 60	60.0	209.6	205.8	9.4	14.2	1.21
203 × 203 × 52	52.0	206.2	204.3	7.9	12.5	1.20
203 × 203 × 46	46.1	203.2	203.6	7.2	11.0	1.19
152 × 152 × 37	37.0	161.8	154.4	8.0	11.5	0.912
152 × 152 × 30	30.0	157.6	152.9	6.5	9.4	0.901
152 × 152 × 23	23.0	152.4	152.2	5.8	6.8	0.889

METAL

Joists BS 4: Part 1: 2005 (retained for reference, Corus have ceased manufacture in UK)

Designation	Mass (kg/m)	Depth of section (mm)	Width of section (mm)	Thickness Web (mm)	Flange (mm)	Surface area (m²/m)
254 × 203 × 82	82.0	254.0	203.2	10.2	19.9	1.210
203 × 152 × 52	52.3	203.2	152.4	8.9	16.5	0.932
152 × 127 × 37	37.3	152.4	127.0	10.4	13.2	0.737
127 × 114 × 29	29.3	127.0	114.3	10.2	11.5	0.646
127 × 114 × 27	26.9	127.0	114.3	7.4	11.4	0.650
102 × 102 × 23	23.0	101.6	101.6	9.5	10.3	0.549
102 × 44 × 7	7.5	101.6	44.5	4.3	6.1	0.350
89 × 89 × 19	19.5	88.9	88.9	9.5	9.9	0.476
76 × 76 × 13	12.8	76.2	76.2	5.1	8.4	0.411

Parallel Flange Channels

Designation	Mass (kg/m)	Depth of section (mm)	Width of section (mm)	Thickness Web (mm)	Flange (mm)	Surface area (m²/m)
430 × 100 × 64	64.4	430	100	11.0	19.0	1.23
380 × 100 × 54	54.0	380	100	9.5	17.5	1.13
300 × 100 × 46	45.5	300	100	9.0	16.5	0.969
300 × 90 × 41	41.4	300	90	9.0	15.5	0.932
260 × 90 × 35	34.8	260	90	8.0	14.0	0.854
260 × 75 × 28	27.6	260	75	7.0	12.0	0.79
230 × 90 × 32	32.2	230	90	7.5	14.0	0.795
230 × 75 × 26	25.7	230	75	6.5	12.5	0.737
200 × 90 × 30	29.7	200	90	7.0	14.0	0.736
200 × 75 × 23	23.4	200	75	6.0	12.5	0.678
180 × 90 × 26	26.1	180	90	6.5	12.5	0.697
180 × 75 × 20	20.3	180	75	6.0	10.5	0.638
150 × 90 × 24	23.9	150	90	6.5	12.0	0.637
150 × 75 × 18	17.9	150	75	5.5	10.0	0.579
125 × 65 × 15	14.8	125	65	5.5	9.5	0.489
100 × 50 × 10	10.2	100	50	5.0	8.5	0.382

METAL

Equal Angles BS EN 10056-1

Designation	Mass (kg/m)	Surface area (m²/m)
200 × 200 × 24	71.1	0.790
200 × 200 × 20	59.9	0.790
200 × 200 × 18	54.2	0.790
200 × 200 × 16	48.5	0.790
150 × 150 × 18	40.1	0.59
150 × 150 × 15	33.8	0.59
150 × 150 × 12	27.3	0.59
150 × 150 × 10	23.0	0.59
120 × 120 × 15	26.6	0.47
120 × 120 × 12	21.6	0.47
120 × 120 × 10	18.2	0.47
120 × 120 × 8	14.7	0.47
100 × 100 × 15	21.9	0.39
100 × 100 × 12	17.8	0.39
100 × 100 × 10	15.0	0.39
100 × 100 × 8	12.2	0.39
90 × 90 × 12	15.9	0.35
90 × 90 × 10	13.4	0.35
90 × 90 × 8	10.9	0.35
90 × 90 × 7	9.61	0.35
90 × 90 × 6	8.30	0.35

Unequal Angles BS EN 10056-1

Designation	Mass (kg/m)	Surface area (m²/m)
200 × 150 × 18	47.1	0.69
200 × 150 × 15	39.6	0.69
200 × 150 × 12	32.0	0.69
200 × 100 × 15	33.7	0.59
200 × 100 × 12	27.3	0.59
200 × 100 × 10	23.0	0.59
150 × 90 × 15	26.6	0.47
150 × 90 × 12	21.6	0.47
150 × 90 × 10	18.2	0.47
150 × 75 × 15	24.8	0.44
150 × 75 × 12	20.2	0.44
150 × 75 × 10	17.0	0.44
125 × 75 × 12	17.8	0.40
125 × 75 × 10	15.0	0.40
125 × 75 × 8	12.2	0.40
100 × 75 × 12	15.4	0.34
100 × 75 × 10	13.0	0.34
100 × 75 × 8	10.6	0.34
100 × 65 × 10	12.3	0.32
100 × 65 × 8	9.94	0.32
100 × 65 × 7	8.77	0.32

METAL

Structural Tees Split from Universal Beams BS 4: Part 1: 2005

Designation	Mass (kg/m)	Surface area (m²/m)
305 × 305 × 90	89.5	1.22
305 × 305 × 75	74.6	1.22
254 × 343 × 63	62.6	1.19
229 × 305 × 70	69.9	1.07
229 × 305 × 63	62.5	1.07
229 × 305 × 57	56.5	1.07
229 × 305 × 51	50.6	1.07
210 × 267 × 61	61.0	0.95
210 × 267 × 55	54.5	0.95
210 × 267 × 51	50.5	0.95
210 × 267 × 46	46.1	0.95
210 × 267 × 41	41.1	0.95
191 × 229 × 49	49.2	0.84
191 × 229 × 45	44.6	0.84
191 × 229 × 41	41.0	0.84
191 × 229 × 37	37.1	0.84
191 × 229 × 34	33.6	0.84
152 × 229 × 41	41.0	0.76
152 × 229 × 37	37.1	0.76
152 × 229 × 34	33.6	0.76
152 × 229 × 30	29.9	0.76
152 × 229 × 26	26.2	0.76

Universal Bearing Piles BS 4: Part 1: 2005

Designation	Mass (kg/m)	Depth of Section (mm)	Width of Section (mm)	Thickness Web (mm)	Thickness Flange (mm)
356 × 368 × 174	173.9	361.4	378.5	20.3	20.4
356 × 368 × 152	152.0	356.4	376.0	17.8	17.9
356 × 368 × 133	133.0	352.0	373.8	15.6	15.7
356 × 368 × 109	108.9	346.4	371.0	12.8	12.9
305 × 305 × 223	222.9	337.9	325.7	30.3	30.4
305 × 305 × 186	186.0	328.3	320.9	25.5	25.6
305 × 305 × 149	149.1	318.5	316.0	20.6	20.7
305 × 305 × 126	126.1	312.3	312.9	17.5	17.6
305 × 305 × 110	110.0	307.9	310.7	15.3	15.4
305 × 305 × 95	94.9	303.7	308.7	13.3	13.3
305 × 305 × 88	88.0	301.7	307.8	12.4	12.3
305 × 305 × 79	78.9	299.3	306.4	11.0	11.1
254 × 254 × 85	85.1	254.3	260.4	14.4	14.3
254 × 254 × 71	71.0	249.7	258.0	12.0	12.0
254 × 254 × 63	63.0	247.1	256.6	10.6	10.7
203 × 203 × 54	53.9	204.0	207.7	11.3	11.4
203 × 203 × 45	44.9	200.2	205.9	9.5	9.5

METAL

Hot Formed Square Hollow Sections EN 10210 S275J2H & S355J2H

Size (mm)	Wall thickness (mm)	Mass (kg/m)	Superficial area (m²/m)
40 × 40	2.5	2.89	0.154
	3.0	3.41	0.152
	3.2	3.61	0.152
	3.6	4.01	0.151
	4.0	4.39	0.150
	5.0	5.28	0.147
50 × 50	2.5	3.68	0.194
	3.0	4.35	0.192
	3.2	4.62	0.192
	3.6	5.14	0.191
	4.0	5.64	0.190
	5.0	6.85	0.187
	6.0	7.99	0.185
	6.3	8.31	0.184
60 × 60	3.0	5.29	0.232
	3.2	5.62	0.232
	3.6	6.27	0.231
	4.0	6.90	0.230
	5.0	8.42	0.227
	6.0	9.87	0.225
	6.3	10.30	0.224
	8.0	12.50	0.219
70 × 70	3.0	6.24	0.272
	3.2	6.63	0.272
	3.6	7.40	0.271
	4.0	8.15	0.270
	5.0	9.99	0.267
	6.0	11.80	0.265
	6.3	12.30	0.264
	8.0	15.00	0.259
80 × 80	3.2	7.63	0.312
	3.6	8.53	0.311
	4.0	9.41	0.310
	5.0	11.60	0.307
	6.0	13.60	0.305
	6.3	14.20	0.304
	8.0	17.50	0.299
90 × 90	3.6	9.66	0.351
	4.0	10.70	0.350
	5.0	13.10	0.347
	6.0	15.50	0.345
	6.3	16.20	0.344
	8.0	20.10	0.339
100 × 100	3.6	10.80	0.391
	4.0	11.90	0.390
	5.0	14.70	0.387
	6.0	17.40	0.385
	6.3	18.20	0.384
	8.0	22.60	0.379
	10.0	27.40	0.374
120 × 120	4.0	14.40	0.470
	5.0	17.80	0.467

METAL

Size (mm)	Wall thickness (mm)	Mass (kg/m)	Superficial area (m²/m)
	6.0	21.20	0.465
	6.3	22.20	0.464
	8.0	27.60	0.459
	10.0	33.70	0.454
	12.0	39.50	0.449
	12.5	40.90	0.448
140 × 140	5.0	21.00	0.547
	6.0	24.90	0.545
	6.3	26.10	0.544
	8.0	32.60	0.539
	10.0	40.00	0.534
	12.0	47.00	0.529
	12.5	48.70	0.528
150 × 150	5.0	22.60	0.587
	6.0	26.80	0.585
	6.3	28.10	0.584
	8.0	35.10	0.579
	10.0	43.10	0.574
	12.0	50.80	0.569
	12.5	52.70	0.568
Hot formed from seamless hollow	16.0	65.2	0.559
160 × 160	5.0	24.10	0.627
	6.0	28.70	0.625
	6.3	30.10	0.624
	8.0	37.60	0.619
	10.0	46.30	0.614
	12.0	54.60	0.609
	12.5	56.60	0.608
	16.0	70.20	0.599
180 × 180	5.0	27.30	0.707
	6.0	32.50	0.705
	6.3	34.00	0.704
	8.0	42.70	0.699
	10.0	52.50	0.694
	12.0	62.10	0.689
	12.5	64.40	0.688
	16.0	80.20	0.679
200 × 200	5.0	30.40	0.787
	6.0	36.20	0.785
	6.3	38.00	0.784
	8.0	47.70	0.779
	10.0	58.80	0.774
	12.0	69.60	0.769
	12.5	72.30	0.768
	16.0	90.30	0.759
250 × 250	5.0	38.30	0.987
	6.0	45.70	0.985
	6.3	47.90	0.984
	8.0	60.30	0.979
	10.0	74.50	0.974
	12.0	88.50	0.969
	12.5	91.90	0.968

METAL

Size (mm)	Wall thickness (mm)	Mass (kg/m)	Superficial area (m²/m)
	16.0	115.00	0.959
300 × 300	6.0	55.10	1.18
	6.3	57.80	1.18
	8.0	72.80	1.18
	10.0	90.20	1.17
	12.0	107.00	1.17
	12.5	112.00	1.17
	16.0	141.00	1.16
350 × 350	8.0	85.40	1.38
	10.0	106.00	1.37
	12.0	126.00	1.37
	12.5	131.00	1.37
	16.0	166.00	1.36
400× 400	8.0	97.90	1.58
	10.0	122.00	1.57
	12.0	145.00	1.57
	12.5	151.00	1.57
	16.0	191.00	1.56
(Grade S355J2H only)	20.00*	235.00	1.55

Note: * SAW process

METAL

Hot Formed Square Hollow Sections JUMBO RHS: JIS G3136

Size (mm)	Wall thickness (mm)	Mass (kg/m)	Superficial area (m²/m)
350 × 350	19.0	190.00	1.33
	22.0	217.00	1.32
	25.0	242.00	1.31
400 × 400	22.0	251.00	1.52
	25.0	282.00	1.51
450 × 450	12.0	162.00	1.76
	16.0	213.00	1.75
	19.0	250.00	1.73
	22.0	286.00	1.72
	25.0	321.00	1.71
	28.0 *	355.00	1.70
	32.0 *	399.00	1.69
500 × 500	12.0	181.00	1.96
	16.0	238.00	1.95
	19.0	280.00	1.93
	22.0	320.00	1.92
	25.0	360.00	1.91
	28.0 *	399.00	1.90
	32.0 *	450.00	1.89
	36.0 *	498.00	1.88
550 × 550	16.0	263.00	2.15
	19.0	309.00	2.13
	22.0	355.00	2.12
	25.0	399.00	2.11
	28.0 *	443.00	2.10
	32.0 *	500.00	2.09
	36.0 *	555.00	2.08
	40.0 *	608.00	2.06
600 × 600	25.0 *	439.00	2.31
	28.0 *	487.00	2.30
	32.0 *	550.00	2.29
	36.0 *	611.00	2.28
	40.0 *	671.00	2.26
700 × 700	25.0 *	517.00	2.71
	28.0 *	575.00	2.70
	32.0 *	651.00	2.69
	36.0 *	724.00	2.68
	40.0 *	797.00	2.68

Note: * SAW process

METAL

Hot Formed Rectangular Hollow Sections: EN10210 S275J2h & S355J2H

Size (mm)	Wall thickness (mm)	Mass (kg/m)	Superficial area (m²/m)
50 × 30	2.5	2.89	0.154
	3.0	3.41	0.152
	3.2	3.61	0.152
	3.6	4.01	0.151
	4.0	4.39	0.150
	5.0	5.28	0.147
60 × 40	2.5	3.68	0.194
	3.0	4.35	0.192
	3.2	4.62	0.192
	3.6	5.14	0.191
	4.0	5.64	0.190
	5.0	6.85	0.187
	6.0	7.99	0.185
	6.3	8.31	0.184
80 × 40	3.0	5.29	0.232
	3.2	5.62	0.232
	3.6	6.27	0.231
	4.0	6.90	0.230
	5.0	8.42	0.227
	6.0	9.87	0.225
	6.3	10.30	0.224
	8.0	12.50	0.219
76.2 × 50.8	3.0	5.62	0.246
	3.2	5.97	0.246
	3.6	6.66	0.245
	4.0	7.34	0.244
	5.0	8.97	0.241
	6.0	10.50	0.239
	6.3	11.00	0.238
	8.0	13.40	0.233
90 × 50	3.0	6.24	0.272
	3.2	6.63	0.272
	3.6	7.40	0.271
	4.0	8.15	0.270
	5.0	9.99	0.267
	6.0	11.80	0.265
	6.3	12.30	0.264
	8.0	15.00	0.259
100 × 50	3.0	6.71	0.292
	3.2	7.13	0.292
	3.6	7.96	0.291
	4.0	8.78	0.290
	5.0	10.80	0.287
	6.0	12.70	0.285
	6.3	13.30	0.284
	8.0	16.30	0.279

METAL

Size (mm)	Wall thickness (mm)	Mass (kg/m)	Superficial area (m²/m)
100 × 60	3.0	7.18	0.312
	3.2	7.63	0.312
	3.6	8.53	0.311
	4.0	9.41	0.310
	5.0	11.60	0.307
	6.0	13.60	0.305
	6.3	14.20	0.304
	8.0	17.50	0.299
120 × 60	3.6	9.70	0.351
	4.0	10.70	0.350
	5.0	13.10	0.347
	6.0	15.50	0.345
	6.3	16.20	0.344
	8.0	20.10	0.339
120 × 80	3.6	10.80	0.391
	4.0	11.90	0.390
	5.0	14.70	0.387
	6.0	17.40	0.385
	6.3	18.20	0.384
	8.0	22.60	0.379
	10.0	27.40	0.374
150 × 100	4.0	15.10	0.490
	5.0	18.60	0.487
	6.0	22.10	0.485
	6.3	23.10	0.484
	8.0	28.90	0.479
	10.0	35.30	0.474
	12.0	41.40	0.469
	12.5	42.80	0.468
160 × 80	4.0	14.40	0.470
	5.0	17.80	0.467
	6.0	21.20	0.465
	6.3	22.20	0.464
	8.0	27.60	0.459
	10.0	33.70	0.454
	12.0	39.50	0.449
	12.5	40.90	0.448
200 × 100	5.0	22.60	0.587
	6.0	26.80	0.585
	6.3	28.10	0.584
	8.0	35.10	0.579
	10.0	43.10	0.574
	12.0	50.80	0.569
	12.5	52.70	0.568
	16.0	65.20	0.559
250 × 150	5.0	30.40	0.787
	6.0	36.20	0.785
	6.3	38.00	0.784
	8.0	47.70	0.779
	10.0	58.80	0.774
	12.0	69.60	0.769
	12.5	72.30	0.768
	16.0	90.30	0.759

METAL

Size (mm)	Wall thickness (mm)	Mass (kg/m)	Superficial area (m²/m)
300 × 200	5.0	38.30	0.987
	6.0	45.70	0.985
	6.3	47.90	0.984
	8.0	60.30	0.979
	10.0	74.50	0.974
	12.0	88.50	0.969
	12.5	91.90	0.968
	16.0	115.00	0.959
400 × 200	6.0	55.10	1.18
	6.3	57.80	1.18
	8.0	72.80	1.18
	10.0	90.20	1.17
	12.0	107.00	1.17
	12.5	112.00	1.17
	16.0	141.00	1.16
450 × 250	8.0	85.40	1.38
	10.0	106.00	1.37
	12.0	126.00	1.37
	12.5	131.00	1.37
	16.0	166.00	1.36
500 × 300	8.0	98.00	1.58
	10.0	122.00	1.57
	12.0	145.00	1.57
	12.5	151.00	1.57
	16.0	191.00	1.56
	20.0	235.00	1.55

METAL

Hot Formed Circular Hollow Sections EN 10210 S275J2H & S355J2H

Outside diameter (mm)	Wall thickness (mm)	Mass (kg/m)	Superficial area (m²/m)
21.3	3.2	1.43	0.067
26.9	3.2	1.87	0.085
33.7	3.0	2.27	0.106
	3.2	2.41	0.106
	3.6	2.67	0.106
	4.0	2.93	0.106
42.4	3.0	2.91	0.133
	3.2	3.09	0.133
	3.6	3.44	0.133
	4.0	3.79	0.133
48.3	2.5	2.82	0.152
	3.0	3.35	0.152
	3.2	3.56	0.152
	3.6	3.97	0.152
	4.0	4.37	0.152
	5.0	5.34	0.152
60.3	2.5	3.56	0.189
	3.0	4.24	0.189
	3.2	4.51	0.189
	3.6	5.03	0.189
	4.0	5.55	0.189
	5.0	6.82	0.189
76.1	2.5	4.54	0.239
	3.0	5.41	0.239
	3.2	5.75	0.239
	3.6	6.44	0.239
	4.0	7.11	0.239
	5.0	8.77	0.239
	6.0	10.40	0.239
	6.3	10.80	0.239
88.9	2.5	5.33	0.279
	3.0	6.36	0.279
	3.2	6.76	0.27
	3.6	7.57	0.279
88.9	4.0	8.38	0.279
	5.0	10.30	0.279
	6.0	12.30	0.279
	6.3	12.80	0.279
114.3	3.0	8.23	0.359
	3.2	8.77	0.359
	3.6	9.83	0.359
	4.0	10.09	0.359
	5.0	13.50	0.359
	6.0	16.00	0.359
	6.3	16.80	0.359

METAL

Outside diameter (mm)	Wall thickness (mm)	Mass (kg/m)	Superficial area (m²/m)
139.7	3.2	10.80	0.439
	3.6	12.10	0.439
	4.0	13.40	0.439
	5.0	16.60	0.439
	6.0	19.80	0.439
	6.3	20.70	0.439
	8.0	26.00	0.439
	10.0	32.00	0.439
168.3	3.2	13.00	0.529
	3.6	14.60	0.529
	4.0	16.20	0.529
	5.0	20.10	0.529
	6.0	24.00	0.529
	6.3	25.20	0.529
	8.0	31.60	0.529
	10.0	39.00	0.529
	12.0	46.30	0.529
	12.5	48.00	0.529
193.7	5.0	23.30	0.609
	6.0	27.80	0.609
	6.3	29.10	0.609
	8.0	36.60	0.609
	10.0	45.30	0.609
193.7	12.0	53.80	0.609
	12.5	55.90	0.609
219.1	5.0	26.40	0.688
	6.0	31.50	0.688
	6.3	33.10	0.688
	8.0	41.60	0.688
	10.0	51.60	0.688
	12.0	61.30	0.688
	12.5	63.70	0.688
	16.0	80.10	0.688
244.5	5.0	29.50	0.768
	6.0	35.30	0.768
	6.3	37.00	0.768
	8.0	46.70	0.768
	10.0	57.80	0.768
	12.0	68.80	0.768
	12.5	71.50	0.768
	16.0	90.20	0.768
273.0	5.0	33.00	0.858
	6.0	39.50	0.858
	6.3	41.40	0.858
	8.0	52.30	0.858
	10.0	64.90	0.858
	12.0	77.20	0.858
	12.5	80.30	0.858
	16.0	101.00	0.858

Tables and Memoranda

METAL

Outside diameter (mm)	Wall thickness (mm)	Mass (kg/m)	Superficial area (m²/m)
323.9	5.0	39.30	1.02
	6.0	47.00	1.02
	6.3	49.30	1.02
	8.0	62.30	1.02
	10.0	77.40	1.02
	12.0	92.30	1.02
	12.5	96.00	1.02
	16.0	121.00	1.02
355.6	6.3	54.30	1.12
	8.0	68.60	1.12
	10.0	85.30	1.12
	12.0	102.00	1.12
	12.5	106.00	1.12
	16.0	134.00	1.12
406.4	6.3	62.20	1.28
	8.0	79.60	1.28
	10.0	97.80	1.28
	12.0	117.00	1.28
	12.5	121.00	1.28
	16.0	154.00	1.28
457.0	6.3	70.00	1.44
	8.0	88.60	1.44
	10.0	110.00	1.44
	12.0	132.00	1.44
	12.5	137.00	1.44
	16.0	174.00	1.44
508.0	6.3	77.90	1.60
	8.0	98.60	1.60
	10.0	123.00	1.60
	12.0	147.00	1.60
	12.5	153.00	1.60
	16.0	194.00	1.60

METAL

Spacing of Holes in Angles

Nominal leg length (mm)	Spacing of holes						Maximum diameter of bolt or rivet		
	A	B	C	D	E	F	A	B and C	D, E and F
200		75	75	55	55	55		30	20
150		55	55					20	
125		45	60					20	
120									
100	55						24		
90	50						24		
80	45						20		
75	45						20		
70	40						20		
65	35						20		
60	35						16		
50	28						12		
45	25								
40	23								
30	20								
25	15								

KERBS, PAVING, ETC.

KERBS/EDGINGS/CHANNELS

Precast Concrete Kerbs to BS 7263

Straight kerb units: length from 450 to 915 mm

150 mm high × 125 mm thick		
bullnosed	type BN	
half battered	type HB3	
255 mm high × 125 mm thick		
45° splayed	type SP	
half battered	type HB2	
305 mm high × 150 mm thick		
half battered	type HB1	
Quadrant kerb units		
150 mm high × 305 and 455 mm radius to match	type BN	type QBN
150 mm high × 305 and 455 mm radius to match	type HB2, HB3	type QHB
150 mm high × 305 and 455 mm radius to match	type SP	type QSP
255 mm high × 305 and 455 mm radius to match	type BN	type QBN
255 mm high × 305 and 455 mm radius to match	type HB2, HB3	type QHB
225 mm high × 305 and 455 mm radius to match	type SP	type QSP
Angle kerb units		
305 × 305 × 225 mm high × 125 mm thick		
bullnosed external angle	type XA	
splayed external angle to match type SP	type XA	
bullnosed internal angle	type IA	
splayed internal angle to match type SP	type IA	
Channels		
255 mm wide × 125 mm high flat	type CS1	
150 mm wide × 125 mm high flat type	CS2	
255 mm wide × 125 mm high dished	type CD	

KERBS, PAVING, ETC.

Transition kerb units			
from kerb type SP to HB	left handed	type TL	
	right handed	type TR	
from kerb type BN to HB	left handed	type DL1	
	right handed	type DR1	
from kerb type BN to SP	left handed	type DL2	
	right handed	type DR2	

Number of kerbs required per quarter circle (780 mm kerb lengths)

Radius (m)	Number in quarter circle
12	24
10	20
8	16
6	12
5	10
4	8
3	6
2	4
1	2

Precast Concrete Edgings

Round top type ER	Flat top type EF	Bullnosed top type EBN
150 × 50 mm	150 × 50 mm	150 × 50 mm
200 × 50 mm	200 × 50 mm	200 × 50 mm
250 × 50 mm	250 × 50 mm	250 × 50 mm

KERBS, PAVING, ETC.

BASES

Cement Bound Material for Bases and Subbases

CBM1:	very carefully graded aggregate from 37.5–75 ym, with a 7-day strength of 4.5 N/mm^2
CBM2:	same range of aggregate as CBM1 but with more tolerance in each size of aggregate with a 7-day strength of 7.0 N/mm^2
CBM3:	crushed natural aggregate or blast furnace slag, graded from 37.5 mm – 150 ym for 40 mm aggregate, and from 20–75 ym for 20 mm aggregate, with a 7-day strength of 10 N/mm^2
CBM4:	crushed natural aggregate or blast furnace slag, graded from 37.5 mm – 150 ym for 40 mm aggregate, and from 20–75 ym for 20 mm aggregate, with a 7-day strength of 15 N/mm^2

INTERLOCKING BRICK/BLOCK ROADS/PAVINGS

Sizes of Precast Concrete Paving Blocks

Type R blocks
200 × 100 × 60 mm
200 × 100 × 65 mm
200 × 100 × 80 mm
200 × 100 × 100 mm

Type S
Any shape within a 295 mm space

Sizes of clay brick pavers
200 × 100 × 50 mm
200 × 100 × 65 mm
210 × 105 × 50 mm
210 × 105 × 65 mm
215 × 102.5 × 50 mm
215 × 102.5 × 65 mm

Type PA: 3 kN
Footpaths and pedestrian areas, private driveways, car parks, light vehicle traffic and over-run

Type PB: 7 kN
Residential roads, lorry parks, factory yards, docks, petrol station forecourts, hardstandings, bus stations

PAVING AND SURFACING

Weights and Sizes of Paving and Surfacing

Description of item	Size	Quantity per tonne
Paving 50 mm thick	900 × 600 mm	15
Paving 50 mm thick	750 × 600 mm	18
Paving 50 mm thick	600 × 600 mm	23
Paving 50 mm thick	450 × 600 mm	30
Paving 38 mm thick	600 × 600 mm	30
Path edging	914 × 50 × 150 mm	60
Kerb (including radius and tapers)	125 × 254 × 914 mm	15
Kerb (including radius and tapers)	125 × 150 × 914 mm	25
Square channel	125 × 254 × 914 mm	15
Dished channel	125 × 254 × 914 mm	15
Quadrants	300 × 300 × 254 mm	19
Quadrants	450 × 450 × 254 mm	12
Quadrants	300 × 300 × 150 mm	30
Internal angles	300 × 300 × 254 mm	30
Fluted pavement channel	255 × 75 × 914 mm	25
Corner stones	300 × 300 mm	80
Corner stones	360 × 360 mm	60
Cable covers	914 × 175 mm	55
Gulley kerbs	220 × 220 × 150 mm	60
Gulley kerbs	220 × 200 × 75 mm	120

KERBS, PAVING, ETC.

Weights and Sizes of Paving and Surfacing

Material	kg/m^3	lb/cu yd
Tarmacadam	2306	3891
Macadam (waterbound)	2563	4325
Vermiculite (aggregate)	64–80	108–135
Terracotta	2114	3568
Cork – compressed	388	24
	kg/m^2	lb/sq ft
Clay floor tiles, 12.7 mm	27.3	5.6
Pavement lights	122	25
Damp-proof course	5	1
	kg/m^2 per mm thickness	lb/sq ft per inch thickness
Paving Slabs (stone)	2.3	12
Granite setts	2.88	15
Asphalt	2.30	12
Rubber flooring	1.68	9
Polyvinyl chloride	1.94 (avg)	10 (avg)

Coverage (m^2) Per Cubic Metre of Materials Used as Subbases or Capping Layers

Consolidated thickness laid in (mm)	Square metre coverage		
	Gravel	Sand	Hardcore
50	15.80	16.50	–
75	10.50	11.00	–
100	7.92	8.20	7.42
125	6.34	6.60	5.90
150	5.28	5.50	4.95
175	–	–	4.23
200	–	–	3.71
225	–	–	3.30
300	–	–	2.47

KERBS, PAVING, ETC.

Approximate Rate of Spreads

Average thickness of course (mm)	Description	Approximate rate of spread			
		Open Textured		Dense, Medium & Fine Textured	
		(kg/m^2)	(m^2/t)	(kg/m^2)	(m^2/t)
35	14 mm open textured or dense wearing course	60–75	13–17	70–85	12–14
40	20 mm open textured or dense base course	70–85	12–14	80–100	10–12
45	20 mm open textured or dense base course	80–100	10–12	95–100	9–10
50	20 mm open textured or dense, or 28 mm dense base course	85–110	9–12	110–120	8–9
60	28 mm dense base course, 40 mm open textured of dense base course or 40 mm single course as base course		8–10	130–150	7–8
65	28 mm dense base course, 40 mm open textured or dense base course or 40 mm single course	100–135	7–10	140–160	6–7
75	40 mm single course, 40 mm open textured or dense base course, 40 mm dense roadbase	120–150	7–8	165–185	5–6
100	40 mm dense base course or roadbase	–	–	220–240	4–4.5

KERBS, PAVING, ETC.

Surface Dressing Roads: Coverage (m²) per Tonne of Material

Size in mm	Sand	Granite chips	Gravel	Limestone chips
Sand	168	–	–	–
3	–	148	152	165
6	–	130	133	144
9	–	111	114	123
13	–	85	87	95
19	–	68	71	78

Sizes of Flags

Reference	Nominal size (mm)	Thickness (mm)
A	600 × 450	50 and 63
B	600 × 600	50 and 63
C	600 × 750	50 and 63
D	600 × 900	50 and 63
E	450 × 450	50 and 70 chamfered top surface
F	400 × 400	50 and 65 chamfered top surface
G	300 × 300	50 and 60 chamfered top surface

Sizes of Natural Stone Setts

Width (mm)		Length (mm)		Depth (mm)
100	×	100	×	100
75	×	150 to 250	×	125
75	×	150 to 250	×	150
100	×	150 to 250	×	100
100	×	150 to 250	×	150

SEEDING/TURFING AND PLANTING

Topsoil Quality

Topsoil grade	Properties
Premium	Natural topsoil, high fertility, loamy texture, good soil structure, suitable for intensive cultivation.
General purpose	Natural or manufactured topsoil of lesser quality than Premium, suitable for agriculture or amenity landscape, may need fertilizer or soil structure improvement.
Economy	Selected subsoil, natural mineral deposit such as river silt or greensand. The grade comprises two subgrades; 'Low clay' and 'High clay' which is more liable to compaction in handling. This grade is suitable for low-production agricultural land and amenity woodland or conservation planting areas.

Forms of Trees

Standards:	Shall be clear with substantially straight stems. Grafted and budded trees shall have no more than a slight bend at the union. Standards shall be designated as Half, Extra light, Light, Standard, Selected standard, Heavy, and Extra heavy.
Sizes of Standards	
Heavy standard	12–14 cm girth × 3.50 to 5.00 m high
Extra Heavy standard	14–16 cm girth × 4.25 to 5.00 m high
Extra Heavy standard	16–18 cm girth × 4.25 to 6.00 m high
Extra Heavy standard	18–20 cm girth × 5.00 to 6.00 m high
Semi-mature trees:	Between 6.0 m and 12.0 m tall with a girth of 20 to 75 cm at 1.0 m above ground.
Feathered trees:	Shall have a defined upright central leader, with stem furnished with evenly spread and balanced lateral shoots down to or near the ground.
Whips:	Shall be without significant feather growth as determined by visual inspection.
Multi-stemmed trees:	Shall have two or more main stems at, near, above or below ground.

Seedlings grown from seed and not transplanted shall be specified when ordered for sale as:

1+0	one year old seedling
2+0	two year old seedling
1+1	one year seed bed, one year transplanted = two year old seedling
1+2	one year seed bed, two years transplanted = three year old seedling
2+1	two year seed bed, one year transplanted = three year old seedling
1u1	two years seed bed, undercut after 1 year = two year old seedling
2u2	four years seed bed, undercut after 2 years = four year old seedling

SEEDING/TURFING AND PLANTING

Cuttings

The age of cuttings (plants grown from shoots, stems, or roots of the mother plant) shall be specified when ordered for sale. The height of transplants and undercut seedlings/cuttings (which have been transplanted or undercut at least once) shall be stated in centimetres. The number of growing seasons before and after transplanting or undercutting shall be stated.

0 + 1	one year cutting
0 + 2	two year cutting
0 + 1 + 1	one year cutting bed, one year transplanted = two year old seedling
0 + 1 + 2	one year cutting bed, two years transplanted = three year old seedling

Grass Cutting Capacities in m² per hour

Speed mph	Width of cut in metres												
	0.5	0.7	1.0	1.2	1.5	1.7	2.0	2.0	2.1	2.5	2.8	3.0	3.4
1.0	724	1127	1529	1931	2334	2736	3138	3219	3380	4023	4506	4828	5472
1.5	1086	1690	2293	2897	3500	4104	4707	4828	5069	6035	6759	7242	8208
2.0	1448	2253	3058	3862	4667	5472	6276	6437	6759	8047	9012	9656	10944
2.5	1811	2816	3822	4828	5834	6840	7846	8047	8449	10058	11265	12070	13679
3.0	2173	3380	4587	5794	7001	8208	9415	9656	10139	12070	13518	14484	16415
3.5	2535	3943	5351	6759	8167	9576	10984	11265	11829	14082	15772	16898	19151
4.0	2897	4506	6115	7725	9334	10944	12553	12875	13518	16093	18025	19312	21887
4.5	3259	5069	6880	8690	10501	12311	14122	14484	15208	18105	20278	21726	24623
5.0	3621	5633	7644	9656	11668	13679	15691	16093	16898	20117	22531	24140	27359
5.5	3983	6196	8409	10622	12834	15047	17260	17703	18588	22128	24784	26554	30095
6.0	4345	6759	9173	11587	14001	16415	18829	19312	20278	24140	27037	28968	32831
6.5	4707	7322	9938	12553	15168	17783	20398	20921	21967	26152	29290	31382	35566
7.0	5069	7886	10702	13518	16335	19151	21967	22531	23657	28163	31543	33796	38302

Number of Plants per m²: For Plants Planted on an Evenly Spaced Grid

Planting distances

mm	0.10	0.15	0.20	0.25	0.35	0.40	0.45	0.50	0.60	0.75	0.90	1.00	1.20	1.50
0.10	100.00	66.67	50.00	40.00	28.57	25.00	22.22	20.00	16.67	13.33	11.11	10.00	8.33	6.67
0.15	66.67	44.44	33.33	26.67	19.05	16.67	14.81	13.33	11.11	8.89	7.41	6.67	5.56	4.44
0.20	50.00	33.33	25.00	20.00	14.29	12.50	11.11	10.00	8.33	6.67	5.56	5.00	4.17	3.33
0.25	40.00	26.67	20.00	16.00	11.43	10.00	8.89	8.00	6.67	5.33	4.44	4.00	3.33	2.67
0.35	28.57	19.05	14.29	11.43	8.16	7.14	6.35	5.71	4.76	3.81	3.17	2.86	2.38	1.90
0.40	25.00	16.67	12.50	10.00	7.14	6.25	5.56	5.00	4.17	3.33	2.78	2.50	2.08	1.67
0.45	22.22	14.81	11.11	8.89	6.35	5.56	4.94	4.44	3.70	2.96	2.47	2.22	1.85	1.48
0.50	20.00	13.33	10.00	8.00	5.71	5.00	4.44	4.00	3.33	2.67	2.22	2.00	1.67	1.33
0.60	16.67	11.11	8.33	6.67	4.76	4.17	3.70	3.33	2.78	2.22	1.85	1.67	1.39	1.11
0.75	13.33	8.89	6.67	5.33	3.81	3.33	2.96	2.67	2.22	1.78	1.48	1.33	1.11	0.89
0.90	11.11	7.41	5.56	4.44	3.17	2.78	2.47	2.22	1.85	1.48	1.23	1.11	0.93	0.74
1.00	10.00	6.67	5.00	4.00	2.86	2.50	2.22	2.00	1.67	1.33	1.11	1.00	0.83	0.67
1.20	8.33	5.56	4.17	3.33	2.38	2.08	1.85	1.67	1.39	1.11	0.93	0.83	0.69	0.56
1.50	6.67	4.44	3.33	2.67	1.90	1.67	1.48	1.33	1.11	0.89	0.74	0.67	0.56	0.44

SEEDING/TURFING AND PLANTING

Grass Clippings Wet: Based on 3.5 m³/tonne

Annual kg/100 m²	Average 20 cuts kg/100 m²	m²/tonne	m²/m³
32.0	1.6	61162.1	214067.3

Nr of cuts	22	20	18	16	12	4
kg/cut	1.45	1.60	1.78	2.00	2.67	8.00
	Area capacity of 3 tonne vehicle per load					
m²	206250	187500	168750	150000	112500	37500
Load m³	100 m² units/m³ of vehicle space					
1	196.4	178.6	160.7	142.9	107.1	35.7
2	392.9	357.1	321.4	285.7	214.3	71.4
3	589.3	535.7	482.1	428.6	321.4	107.1
4	785.7	714.3	642.9	571.4	428.6	142.9
5	982.1	892.9	803.6	714.3	535.7	178.6

Transportation of Trees

To unload large trees a machine with the necessary lifting strength is required. The weight of the trees must therefore be known in advance. The following table gives a rough overview. The additional columns with root ball dimensions and the number of plants per trailer provide additional information for example about preparing planting holes and calculating unloading times.

Girth in cm	Rootball diameter in cm	Ball height in cm	Weight in kg	Numbers of trees per trailer
16–18	50–60	40	150	100–120
18–20	60–70	40–50	200	80–100
20–25	60–70	40–50	270	50–70
25–30	80	50–60	350	50
30–35	90–100	60–70	500	12–18
35–40	100–110	60–70	650	10–15
40–45	110–120	60–70	850	8–12
45–50	110–120	60–70	1100	5–7
50–60	130–140	60–70	1600	1–3
60–70	150–160	60–70	2500	1
70–80	180–200	70	4000	1
80–90	200–220	70–80	5500	1
90–100	230–250	80–90	7500	1
100–120	250–270	80–90	9500	1

Data supplied by Lorenz von Ehren GmbH
The information in the table is approximate; deviations depend on soil type, genus and weather

FENCING AND GATES

Types of Preservative

Creosote (tar oil) can be 'factory' applied	by pressure to BS 144: pts 1&2 by immersion to BS 144: pt 1 by hot and cold open tank to BS 144: pts 1&2
Copper/chromium/arsenic (CCA)	by full cell process to BS 4072 pts 1&2
Organic solvent (OS)	by double vacuum (vacvac) to BS 5707 pts 1&3 by immersion to BS 5057 pts 1&3
Pentachlorophenol (PCP)	by heavy oil double vacuum to BS 5705 pts 2&3
Boron diffusion process (treated with disodium octaborate to BWPA Manual 1986)	

Note: Boron is used on green timber at source and the timber is supplied dry

Cleft Chestnut Pale Fences

Pales	Pale spacing	Wire lines	
900 mm	75 mm	2	temporary protection
1050 mm	75 or 100 mm	2	light protective fences
1200 mm	75 mm	3	perimeter fences
1350 mm	75 mm	3	perimeter fences
1500 mm	50 mm	3	narrow perimeter fences
1800 mm	50 mm	3	light security fences

Close-Boarded Fences

Close-boarded fences 1.05 to 1.8 m high
Type BCR (recessed) or BCM (morticed) with concrete posts 140 × 115 mm tapered and Type BW with timber posts

Palisade Fences

Wooden palisade fences
Type WPC with concrete posts 140 × 115 mm tapered and Type WPW with timber posts

For both types of fence:
Height of fence 1050 mm: two rails
Height of fence 1200 mm: two rails
Height of fence 1500 mm: three rails
Height of fence 1650 mm: three rails
Height of fence 1800 mm: three rails

FENCING AND GATES

Post and Rail Fences

Wooden post and rail fences
Type MPR 11/3 morticed rails and Type SPR 11/3 nailed rails
Height to top of rail 1100 mm
Rails: three rails 87 mm, 38 mm

Type MPR 11/4 morticed rails and Type SPR 11/4 nailed rails
Height to top of rail 1100 mm
Rails: four rails 87 mm, 38 mm

Type MPR 13/4 morticed rails and Type SPR 13/4 nailed rails
Height to top of rail 1300 mm
Rail spacing 250 mm, 250 mm, and 225 mm from top
Rails: four rails 87 mm, 38 mm

Steel Posts

Rolled steel angle iron posts for chain link fencing

Posts	Fence height	Strut	Straining post
1500 × 40 × 40 × 5 mm	900 mm	1500 × 40 × 40 × 5 mm	1500 × 50 × 50 × 6 mm
1800 × 40 × 40 × 5 mm	1200 mm	1800 × 40 × 40 × 5 mm	1800 × 50 × 50 × 6 mm
2000 × 45 × 45 × 5 mm	1400 mm	2000 × 45 × 45 × 5 mm	2000 × 60 × 60 × 6 mm
2600 × 45 × 45 × 5 mm	1800 mm	2600 × 45 × 45 × 5 mm	2600 × 60 × 60 × 6 mm
3000 × 50 × 50 × 6 mm	1800 mm	2600 × 45 × 45 × 5 mm	3000 × 60 × 60 × 6 mm
with arms			

Concrete Posts

Concrete posts for chain link fencing

Posts and straining posts	Fence height	Strut
1570 mm 100 × 100 mm	900 mm	1500 mm × 75 × 75 mm
1870 mm 125 × 125 mm	1200 mm	1830 mm × 100 × 75 mm
2070 mm 125 × 125 mm	1400 mm	1980 mm × 100 × 75 mm
2620 mm 125 × 125 mm	1800 mm	2590 mm × 100 × 85 mm
3040 mm 125 × 125 mm	1800 mm	2590 mm × 100 × 85 mm (with arms)

FENCING AND GATES

Rolled Steel Angle Posts

Rolled steel angle posts for rectangular wire mesh (field) fencing

Posts	Fence height	Strut	Straining post
1200 × 40 × 40 × 5 mm	600 mm	1200 × 75 × 75 mm	1350 × 100 × 100 mm
1400 × 40 × 40 × 5 mm	800 mm	1400 × 75 × 75 mm	1550 × 100 × 100 mm
1500 × 40 × 40 × 5 mm	900 mm	1500 × 75 × 75 mm	1650 × 100 × 100 mm
1600 × 40 × 40 × 5 mm	1000 mm	1600 × 75 × 75 mm	1750 × 100 × 100 mm
1750 × 40 × 40 × 5 mm	1150 mm	1750 × 75 × 100 mm	1900 × 125 × 125 mm

Concrete Posts

Concrete posts for rectangular wire mesh (field) fencing

Posts	Fence height	Strut	Straining post
1270 × 100 × 100 mm	600 mm	1200 × 75 × 75 mm	1420 × 100 × 100 mm
1470 × 100 × 100 mm	800 mm	1350 × 75 × 75 mm	1620 × 100 × 100 mm
1570 × 100 × 100 mm	900 mm	1500 × 75 × 75 mm	1720 × 100 × 100 mm
1670 × 100 × 100 mm	600 mm	1650 × 75 × 75 mm	1820 × 100 × 100 mm
1820 × 125 × 125 mm	1150 mm	1830 × 75 × 100 mm	1970 × 125 × 125 mm

Cleft Chestnut Pale Fences

Timber Posts

Timber posts for wire mesh and hexagonal wire netting fences

Round timber for general fences

Posts	Fence height	Strut	Straining post
1300 × 65 mm dia.	600 mm	1200 × 80 mm dia.	1450 × 100 mm dia.
1500 × 65 mm dia.	800 mm	1400 × 80 mm dia.	1650 × 100 mm dia.
1600 × 65 mm dia.	900 mm	1500 × 80 mm dia.	1750 × 100 mm dia.
1700 × 65 mm dia.	1050 mm	1600 × 80 mm dia.	1850 × 100 mm dia.
1800 × 65 mm dia.	1150 mm	1750 × 80 mm dia.	2000 × 120 mm dia.

Squared timber for general fences

Posts	Fence height	Strut	Straining post
1300 × 75 × 75 mm	600 mm	1200 × 75 × 75 mm	1450 × 100 × 100 mm
1500 × 75 × 75 mm	800 mm	1400 × 75 × 75 mm	1650 × 100 × 100 mm
1600 × 75 × 75 mm	900 mm	1500 × 75 × 75 mm	1750 × 100 × 100 mm
1700 × 75 × 75 mm	1050 mm	1600 × 75 × 75 mm	1850 × 100 × 100 mm
1800 × 75 × 75 mm	1150 mm	1750 × 75 × 75 mm	2000 × 125 × 100 mm

FENCING AND GATES

Steel Fences to BS 1722: Part 9: 1992

	Fence height	Top/bottom rails and flat posts	Vertical bars
Light	1000 mm	40 × 10 mm 450 mm in ground	12 mm dia. at 115 mm cs
	1200 mm	40 × 10 mm 550 mm in ground	12 mm dia. at 115 mm cs
	1400 mm	40 × 10 mm 550 mm in ground	12 mm dia. at 115 mm cs
Light	1000 mm	40 × 10 mm 450 mm in ground	16 mm dia. at 120 mm cs
	1200 mm	40 × 10 mm 550 mm in ground	16 mm dia. at 120 mm cs
	1400 mm	40 × 10 mm 550 mm in ground	16 mm dia. at 120 mm cs
Medium	1200 mm	50 × 10 mm 550 mm in ground	20 mm dia. at 125 mm cs
	1400 mm	50 × 10 mm 550 mm in ground	20 mm dia. at 125 mm cs
	1600 mm	50 × 10 mm 600 mm in ground	22 mm dia. at 145 mm cs
	1800 mm	50 × 10 mm 600 mm in ground	22 mm dia. at 145 mm cs
Heavy	1600 mm	50 × 10 mm 600 mm in ground	22 mm dia. at 145 mm cs
	1800 mm	50 × 10 mm 600 mm in ground	22 mm dia. at 145 mm cs
	2000 mm	50 × 10 mm 600 mm in ground	22 mm dia. at 145 mm cs
	2200 mm	50 × 10 mm 600 mm in ground	22 mm dia. at 145 mm cs

Notes: Mild steel fences: round or square verticals; flat standards and horizontals. Tops of vertical bars may be bow-top, blunt, or pointed. Round or square bar railings

Timber Field Gates to BS 3470: 1975

Gates made to this standard are designed to open one way only
All timber gates are 1100 mm high
Width over stiles 2400, 2700, 3000, 3300, 3600, and 4200 mm
Gates over 4200 mm should be made in two leaves

Steel Field Gates to BS 3470: 1975

All steel gates are 1100 mm high	
Heavy duty:	width over stiles 2400, 3000, 3600 and 4500 mm
Light duty:	width over stiles 2400, 3000, and 3600 mm

FENCING AND GATES

Domestic Front Entrance Gates to BS 4092: Part 1: 1966

Metal gates:	Single gates are 900 mm high minimum, 900 mm, 1000 mm and 1100 mm wide

Domestic Front Entrance Gates to BS 4092: Part 2: 1966

Wooden gates:	All rails shall be tenoned into the stiles
	Single gates are 840 mm high minimum, 801 mm and 1020 mm wide
	Double gates are 840 mm high minimum, 2130, 2340 and 2640 mm wide

Timber Bridle Gates to BS 5709:1979 (Horse or Hunting Gates)

Gates open one way only	
Minimum width between posts	1525 mm
Minimum height	1100 mm

Timber Kissing Gates to BS 5709:1979

Minimum width	700 mm
Minimum height	1000 mm
Minimum distance between shutting posts	600 mm
Minimum clearance at mid-point	600 mm

Metal Kissing Gates to BS 5709:1979

Sizes are the same as those for timber kissing gates
Maximum gaps between rails 120 mm

Categories of Pedestrian Guard Rail to BS 3049:1976

Class A for normal use
Class B where vandalism is expected
Class C where crowd pressure is likely

DRAINAGE

Width required for Trenches for Various Diameters of Pipes

Pipe diameter (mm)	Trench n.e. 1.50 m deep	Trench over 1.50 m deep
n.e. 100 mm	450 mm	600 mm
100–150 mm	500 mm	650 mm
150–225 mm	600 mm	750 mm
225–300 mm	650 mm	800 mm
300–400 mm	750 mm	900 mm
400–450 mm	900 mm	1050 mm
450–600 mm	1100 mm	1300 mm

Weights and Dimensions – Vitrified Clay Pipes

Product	Nominal diameter (mm)	Effective length (mm)	BS 65 limits of tolerance min (mm)	max (mm)	Crushing strength (kN/m)	Weight (kg/pipe)	(kg/m)
Supersleve	100	1600	96	105	35.00	14.71	9.19
	150	1750	146	158	35.00	29.24	16.71
Hepsleve	225	1850	221	236	28.00	84.03	45.42
	300	2500	295	313	34.00	193.05	77.22
	150	1500	146	158	22.00	37.04	24.69
Hepseal	225	1750	221	236	28.00	85.47	48.84
	300	2500	295	313	34.00	204.08	81.63
	400	2500	394	414	44.00	357.14	142.86
	450	2500	444	464	44.00	454.55	181.63
	500	2500	494	514	48.00	555.56	222.22
	600	2500	591	615	57.00	796.23	307.69
	700	3000	689	719	67.00	1111.11	370.45
	800	3000	788	822	72.00	1351.35	450.45
Hepline	100	1600	95	107	22.00	14.71	9.19
	150	1750	145	160	22.00	29.24	16.71
	225	1850	219	239	28.00	84.03	45.42
	300	1850	292	317	34.00	142.86	77.22
Hepduct (conduit)	90	1500	–	–	28.00	12.05	8.03
	100	1600	–	–	28.00	14.71	9.19
	125	1750	–	–	28.00	20.73	11.84
	150	1750	–	–	28.00	29.24	16.71
	225	1850	–	–	28.00	84.03	45.42
	300	1850	–	–	34.00	142.86	77.22

DRAINAGE

Weights and Dimensions – Vitrified Clay Pipes

Nominal internal diameter (mm)	Nominal wall thickness (mm)	Approximate weight (kg/m)
150	25	45
225	29	71
300	32	122
375	35	162
450	38	191
600	48	317
750	54	454
900	60	616
1200	76	912
1500	89	1458
1800	102	1884
2100	127	2619

Wall thickness, weights and pipe lengths vary, depending on type of pipe required

The particulars shown above represent a selection of available diameters and are applicable to strength class 1 pipes with flexible rubber ring joints

Tubes with Ogee joints are also available

DRAINAGE

Weights and Dimensions – PVC-u Pipes

	Nominal size	Mean outside diameter (mm)		Wall thickness	Weight
		min	max	(mm)	(kg/m)
Standard pipes	82.4	82.4	82.7	3.2	1.2
	110.0	110.0	110.4	3.2	1.6
	160.0	160.0	160.6	4.1	3.0
	200.0	200.0	200.6	4.9	4.6
	250.0	250.0	250.7	6.1	7.2
Perforated pipes heavy grade	As above	As above	As above	As above	As above
thin wall	82.4	82.4	82.7	1.7	–
	110.0	110.0	110.4	2.2	–
	160.0	160.0	160.6	3.2	–

Width of Trenches Required for Various Diameters of Pipes

Pipe diameter (mm)	Trench n.e. 1.5 m deep (mm)	Trench over 1.5 m deep (mm)
n.e. 100	450	600
100–150	500	650
150–225	600	750
225–300	650	800
300–400	750	900
400–450	900	1050
450–600	1100	1300

DRAINAGE

DRAINAGE BELOW GROUND AND LAND DRAINAGE

Flow of Water Which Can Be Carried by Various Sizes of Pipe

Clay or concrete pipes

	Gradient of pipeline							
	1:10	1:20	1:30	1:40	1:50	1:60	1:80	1:100
Pipe size	Flow in litres per second							
DN 100 15.0	8.5	6.8	5.8	5.2	4.7	4.0	3.5	
DN 150 28.0	19.0	16.0	14.0	12.0	11.0	9.1	8.0	
DN 225 140.0	95.0	76.0	66.0	58.0	53.0	46.0	40.0	

Plastic pipes

	Gradient of pipeline							
	1:10	1:20	1:30	1:40	1:50	1:60	1:80	1:100
Pipe size	Flow in litres per second							
82.4 mm i/dia.	12.0	8.5	6.8	5.8	5.2	4.7	4.0	3.5
110 mm i/dia.	28.0	19.0	16.0	14.0	12.0	11.0	9.1	8.0
160 mm i/dia.	76.0	53.0	43.0	37.0	33.0	29.0	25.0	22.0
200 mm i/dia.	140.0	95.0	76.0	66.0	58.0	53.0	46.0	40.0

Vitrified (Perforated) Clay Pipes and Fittings to BS En 295-5 1994

Length not specified		
75 mm bore	**250 mm bore**	**600 mm bore**
100	300	700
125	350	800
150	400	1000
200	450	1200
225	500	

Precast Concrete Pipes: Prestressed Non-pressure Pipes and Fittings: Flexible Joints to BS 5911: Pt. 103: 1994

Rationalized metric nominal sizes: 450, 500	
Length:	500–1000 by 100 increments
	1000–2200 by 200 increments
	2200–2800 by 300 increments
Angles: length:	450–600 angles 45, 22.5,11.25°
	600 or more angles 22.5, 11.25°

DRAINAGE

Precast Concrete Pipes: Un-reinforced and Circular Manholes and Soakaways to BS 5911: Pt. 200: 1994

Nominal sizes:	
Shafts:	675, 900 mm
Chambers:	900, 1050, 1200, 1350, 1500, 1800, 2100, 2400, 2700, 3000 mm
Large chambers:	To have either tapered reducing rings or a flat reducing slab in order to accept the standard cover
Ring depths:	1. 300–1200 mm by 300 mm increments except for bottom slab and rings below cover slab, these are by 150 mm increments
	2. 250–1000 mm by 250 mm increments except for bottom slab and rings below cover slab, these are by 125 mm increments
Access hole:	750 × 750 mm for DN 1050 chamber
	1200 × 675 mm for DN 1350 chamber

Calculation of Soakaway Depth

The following formula determines the depth of concrete ring soakaway that would be required for draining given amounts of water.

$$h = \frac{4ar}{3\pi D^2}$$

h = depth of the chamber below the invert pipe
a = the area to be drained
r = the hourly rate of rainfall (50 mm per hour)
π = pi
D = internal diameter of the soakaway

This table shows the depth of chambers in each ring size which would be required to contain the volume of water specified. These allow a recommended storage capacity of $\frac{1}{3}$ (one third of the hourly rainfall figure).

Table Showing Required Depth of Concrete Ring Chambers in Metres

Area m²	50	100	150	200	300	400	500
Ring size							
0.9	1.31	2.62	3.93	5.24	7.86	10.48	13.10
1.1	0.96	1.92	2.89	3.85	5.77	7.70	9.62
1.2	0.74	1.47	2.21	2.95	4.42	5.89	7.37
1.4	0.58	1.16	1.75	2.33	3.49	4.66	5.82
1.5	0.47	0.94	1.41	1.89	2.83	3.77	4.72
1.8	0.33	0.65	0.98	1.31	1.96	2.62	3.27
2.1	0.24	0.48	0.72	0.96	1.44	1.92	2.41
2.4	0.18	0.37	0.55	0.74	1.11	1.47	1.84
2.7	0.15	0.29	0.44	0.58	0.87	1.16	1.46
3.0	0.12	0.24	0.35	0.47	0.71	0.94	1.18

DRAINAGE

Precast Concrete Inspection Chambers and Gullies to BS 5911: Part 230: 1994

Nominal sizes:	375 diameter, 750, 900 mm deep
	450 diameter, 750, 900, 1050, 1200 mm deep
Depths:	from the top for trapped or un-trapped units:
	centre of outlet 300 mm
	invert (bottom) of the outlet pipe 400 mm
Depth of water seal for trapped gullies:	
	85 mm, rodding eye int. dia. 100 mm
Cover slab:	65 mm min

Bedding Flexible Pipes: PVC-u Or Ductile Iron

Type 1 =	100 mm fill below pipe, 300 mm above pipe: single size material
Type 2 =	100 mm fill below pipe, 300 mm above pipe: single size or graded material
Type 3 =	100 mm fill below pipe, 75 mm above pipe with concrete protective slab over
Type 4 =	100 mm fill below pipe, fill laid level with top of pipe
Type 5 =	200 mm fill below pipe, fill laid level with top of pipe
Concrete =	25 mm sand blinding to bottom of trench, pipe supported on chocks, 100 mm concrete under the pipe, 150 mm concrete over the pipe

DRAINAGE

Bedding Rigid Pipes: Clay or Concrete
(for vitrified clay pipes the manufacturer should be consulted)

Class D:	Pipe laid on natural ground with cut-outs for joints, soil screened to remove stones over 40 mm and returned over pipe to 150 m min depth. Suitable for firm ground with trenches trimmed by hand.
Class N:	Pipe laid on 50 mm granular material of graded aggregate to Table 4 of BS 882, or 10 mm aggregate to Table 6 of BS 882, or as dug light soil (not clay) screened to remove stones over 10 mm. Suitable for machine dug trenches.
Class B:	As Class N, but with granular bedding extending half way up the pipe diameter.
Class F:	Pipe laid on 100 mm granular fill to BS 882 below pipe, minimum 150 mm granular fill above pipe: single size material. Suitable for machine dug trenches.
Class A:	Concrete 100 mm thick under the pipe extending half way up the pipe, backfilled with the appropriate class of fill. Used where there is only a very shallow fall to the drain. Class A bedding allows the pipes to be laid to an exact gradient.
Concrete surround:	25 mm sand blinding to bottom of trench, pipe supported on chocks, 100 mm concrete under the pipe, 150 mm concrete over the pipe. It is preferable to bed pipes under slabs or wall in granular material.

PIPED SUPPLY SYSTEMS

Identification of Service Tubes From Utility to Dwellings

Utility	Colour	Size	Depth
British Telecom	grey	54 mm od	450 mm
Electricity	black	38 mm od	450 mm
Gas	yellow	42 mm od rigid 60 mm od convoluted	450 mm
Water	may be blue	(normally untubed)	750 mm

ELECTRICAL SUPPLY/POWER/LIGHTING SYSTEMS

Electrical Insulation Class En 60.598 BS 4533

Class 1:	luminaires comply with class 1 (I) earthed electrical requirements
Class 2:	luminaires comply with class 2 (II) double insulated electrical requirements
Class 3:	luminaires comply with class 3 (III) electrical requirements

Protection to Light Fittings

BS EN 60529:1992 Classification for degrees of protection provided by enclosures.
(IP Code – International or ingress Protection)

1st characteristic: against ingress of solid foreign objects

The figure	2	indicates that fingers cannot enter
	3	that a 2.5 mm diameter probe cannot enter
	4	that a 1.0 mm diameter probe cannot enter
	5	the fitting is dust proof (no dust around live parts)
	6	the fitting is dust tight (no dust entry)

2nd characteristic: ingress of water with harmful effects

The figure	0	indicates unprotected
	1	vertically dripping water cannot enter
	2	water dripping 15° (tilt) cannot enter
	3	spraying water cannot enter
	4	splashing water cannot enter
	5	jetting water cannot enter
	6	powerful jetting water cannot enter
	7	proof against temporary immersion
	8	proof against continuous immersion

Optional additional codes:		A–D protects against access to hazardous parts
	H	High voltage apparatus
	M	fitting was in motion during water test
	S	fitting was static during water test
	W	protects against weather

Marking code arrangement:	(example) IPX5S = IP (International or Ingress Protection)
	X (denotes omission of first characteristic)
	5 = jetting
	S = static during water test

RAIL TRACKS

	kg/m of track	lb/ft of track
Standard gauge Bull-head rails, chairs, transverse timber (softwood) sleepers etc.	245	165
Main lines Flat-bottom rails, transverse prestressed concrete sleepers, etc.	418	280
Add for electric third rail	51	35
Add for crushed stone ballast	2600	1750
	kg/m²	**lb/sq ft**
Overall average weight – rails connections, sleepers, ballast, etc.	733	150
	kg/m of track	**lb/ft of track**
Bridge rails, longitudinal timber sleepers, etc.	112	75

RAIL TRACKS

Heavy Rails

British Standard Section No.	Rail height (mm)	Foot width (mm)	Head width (mm)	Min web thickness (mm)	Section weight (kg/m)
Flat Bottom Rails					
60 A	114.30	109.54	57.15	11.11	30.62
70 A	123.82	111.12	60.32	12.30	34.81
75 A	128.59	114.30	61.91	12.70	37.45
80 A	133.35	117.47	63.50	13.10	39.76
90 A	142.88	127.00	66.67	13.89	45.10
95 A	147.64	130.17	69.85	14.68	47.31
100 A	152.40	133.35	69.85	15.08	50.18
110 A	158.75	139.70	69.85	15.87	54.52
113 A	158.75	139.70	69.85	20.00	56.22
50 'O'	100.01	100.01	52.39	10.32	24.82
80 'O'	127.00	127.00	63.50	13.89	39.74
60R	114.30	109.54	57.15	11.11	29.85
75R	128.59	122.24	61.91	13.10	37.09
80R	133.35	127.00	63.50	13.49	39.72
90R	142.88	136.53	66.67	13.89	44.58
95R	147.64	141.29	68.26	14.29	47.21
100R	152.40	146.05	69.85	14.29	49.60
95N	147.64	139.70	69.85	13.89	47.27
Bull Head Rails					
95R BH	145.26	69.85	69.85	19.05	47.07

Light Rails

British Standard Section No.	Rail height (mm)	Foot width (mm)	Head width (mm)	Min web thickness (mm)	Section weight (kg/m)
Flat Bottom Rails					
20M	65.09	55.56	30.96	6.75	9.88
30M	75.41	69.85	38.10	9.13	14.79
35M	80.96	76.20	42.86	9.13	17.39
35R	85.73	82.55	44.45	8.33	17.40
40	88.11	80.57	45.64	12.3	19.89
Bridge Rails					
13	48.00	92	36.00	18.0	13.31
16	54.00	108	44.50	16.0	16.06
20	55.50	127	50.00	20.5	19.86
28	67.00	152	50.00	31.0	28.62
35	76.00	160	58.00	34.5	35.38
50	76.00	165	58.50	–	50.18
Crane Rails					
A65	75.00	175.00	65.00	38.0	43.10
A75	85.00	200.00	75.00	45.0	56.20
A100	95.00	200.00	100.00	60.0	74.30
A120	105.00	220.00	120.00	72.0	100.00
175CR	152.40	152.40	107.95	38.1	86.92

RAIL TRACKS

Fish Plates

British Standard Section No.	Overall plate length		Hole diameter	Finished weight per pair	
	4 Hole (mm)	6 Hole (mm)	(mm)	4 Hole (kg/pair)	6 Hole (kg/pair)
For British Standard Heavy Rails: Flat Bottom Rails					
60 A	406.40	609.60	20.64	9.87	14.76
70 A	406.40	609.60	22.22	11.15	16.65
75 A	406.40	–	23.81	11.82	17.73
80 A	406.40	609.60	23.81	13.15	19.72
90 A	457.20	685.80	25.40	17.49	26.23
100 A	508.00	–	pear	25.02	–
110 A (shallow)	507.00	–	27.00	30.11	54.64
113 A (heavy)	507.00	–	27.00	30.11	54.64
50 'O' (shallow)	406.40	–	–	6.68	10.14
80 'O' (shallow)	495.30	–	23.81	14.72	22.69
60R (shallow)	406.40	609.60	20.64	8.76	13.13
60R (angled)	406.40	609.60	20.64	11.27	16.90
75R (shallow)	406.40	–	23.81	10.94	16.42
75R (angled)	406.40	–	23.81	13.67	–
80R (shallow)	406.40	609.60	23.81	11.93	17.89
80R (angled)	406.40	609.60	23.81	14.90	22.33
For British Standard Heavy Rails: Bull head rails					
95R BH (shallow)	–	457.20	27.00	14.59	14.61
For British Standard Light Rails: Flat Bottom Rails					
30M	355.6	–	–	–	2.72
35M	355.6	–	–	–	2.83
40	355.6	–	–	3.76	–

FRACTIONS, DECIMALS AND MILLIMETRE EQUIVALENTS

FRACTIONS, DECIMALS AND MILLIMETRE EQUIVALENTS

Fractions	Decimals	(mm)		Fractions	Decimals	(mm)
1/64	0.015625	0.396875		33/64	0.515625	13.096875
1/32	0.03125	0.79375		17/32	0.53125	13.49375
3/64	0.046875	1.190625		35/64	0.546875	13.890625
1/16	0.0625	1.5875		9/16	0.5625	14.2875
5/64	0.078125	1.984375		37/64	0.578125	14.684375
3/32	0.09375	2.38125		19/32	0.59375	15.08125
7/64	0.109375	2.778125		39/64	0.609375	15.478125
1/8	0.125	3.175		5/8	0.625	15.875
9/64	0.140625	3.571875		41/64	0.640625	16.271875
5/32	0.15625	3.96875		21/32	0.65625	16.66875
11/64	0.171875	4.365625		43/64	0.671875	17.065625
3/16	0.1875	4.7625		11/16	0.6875	17.4625
13/64	0.203125	5.159375		45/64	0.703125	17.859375
7/32	0.21875	5.55625		23/32	0.71875	18.25625
15/64	0.234375	5.953125		47/64	0.734375	18.653125
1/4	0.25	6.35		3/4	0.75	19.05
17/64	0.265625	6.746875		49/64	0.765625	19.446875
9/32	0.28125	7.14375		25/32	0.78125	19.84375
19/64	0.296875	7.540625		51/64	0.796875	20.240625
5/16	0.3125	7.9375		13/16	0.8125	20.6375
21/64	0.328125	8.334375		53/64	0.828125	21.034375
11/32	0.34375	8.73125		27/32	0.84375	21.43125
23/64	0.359375	9.128125		55/64	0.859375	21.828125
3/8	0.375	9.525		7/8	0.875	22.225
25/64	0.390625	9.921875		57/64	0.890625	22.621875
13/32	0.40625	10.31875		29/32	0.90625	23.01875
27/64	0.421875	10.71563		59/64	0.921875	23.415625
7/16	0.4375	11.1125		15/16	0.9375	23.8125
29/64	0.453125	11.50938		61/64	0.953125	24.209375
15/32	0.46875	11.90625		31/32	0.96875	24.60625
31/64	0.484375	12.30313		63/64	0.984375	25.003125
1/2	0.5	12.7		1.0	1	25.4

IMPERIAL STANDARD WIRE GAUGE (SWG)

IMPERIAL STANDARD WIRE GAUGE (SWG)

SWG No.	Diameter (inches)	Diameter (mm)	SWG No.	Diameter (inches)	Diameter (mm)
7/0	0.5	12.7	23	0.024	0.61
6/0	0.464	11.79	24	0.022	0.559
5/0	0.432	10.97	25	0.02	0.508
4/0	0.4	10.16	26	0.018	0.457
3/0	0.372	9.45	27	0.0164	0.417
2/0	0.348	8.84	28	0.0148	0.376
1/0	0.324	8.23	29	0.0136	0.345
1	0.3	7.62	30	0.0124	0.315
2	0.276	7.01	31	0.0116	0.295
3	0.252	6.4	32	0.0108	0.274
4	0.232	5.89	33	0.01	0.254
5	0.212	5.38	34	0.009	0.234
6	0.192	4.88	35	0.008	0.213
7	0.176	4.47	36	0.008	0.193
8	0.16	4.06	37	0.007	0.173
9	0.144	3.66	38	0.006	0.152
10	0.128	3.25	39	0.005	0.132
11	0.116	2.95	40	0.005	0.122
12	0.104	2.64	41	0.004	0.112
13	0.092	2.34	42	0.004	0.102
14	0.08	2.03	43	0.004	0.091
15	0.072	1.83	44	0.003	0.081
16	0.064	1.63	45	0.003	0.071
17	0.056	1.42	46	0.002	0.061
18	0.048	1.22	47	0.002	0.051
19	0.04	1.016	48	0.002	0.041
20	0.036	0.914	49	0.001	0.031
21	0.032	0.813	50	0.001	0.025
22	0.028	0.711			

PIPES, WATER, STORAGE, INSULATION

WATER PRESSURE DUE TO HEIGHT

Imperial

Head (Feet)	Pressure (lb/in^2)		Head (Feet)	Pressure (lb/in^2)
1	0.43		70	30.35
5	2.17		75	32.51
10	4.34		80	34.68
15	6.5		85	36.85
20	8.67		90	39.02
25	10.84		95	41.18
30	13.01		100	43.35
35	15.17		105	45.52
40	17.34		110	47.69
45	19.51		120	52.02
50	21.68		130	56.36
55	23.84		140	60.69
60	26.01		150	65.03
65	28.18			

Metric

Head (m)	Pressure (bar)		Head (m)	Pressure (bar)
0.5	0.049		18.0	1.766
1.0	0.098		19.0	1.864
1.5	0.147		20.0	1.962
2.0	0.196		21.0	2.06
3.0	0.294		22.0	2.158
4.0	0.392		23.0	2.256
5.0	0.491		24.0	2.354
6.0	0.589		25.0	2.453
7.0	0.687		26.0	2.551
8.0	0.785		27.0	2.649
9.0	0.883		28.0	2.747
10.0	0.981		29.0	2.845
11.0	1.079		30.0	2.943
12.0	1.177		32.5	3.188
13.0	1.275		35.0	3.434
14.0	1.373		37.5	3.679
15.0	1.472		40.0	3.924
16.0	1.57		42.5	4.169
17.0	1.668		45.0	4.415

1 bar	=	14.5038 lbf/in^2
1 lbf/in^2	=	0.06895 bar
1 metre	=	3.2808 ft or 39.3701 in
1 foot	=	0.3048 metres
1 in wg	=	2.5 mbar (249.1 N/m^2)

PIPES, WATER, STORAGE, INSULATION

Dimensions and Weights of Copper Pipes to BSEN 1057, BSEN 12499, BSEN 14251

Outside Diameter (mm)	Internal Diameter (mm)	Weight per Metre (kg)	Internal Diameter (mm)	Weight per Metre (kg)	Internal Diameter (mm)	Weight per Metre (kg)
	Formerly Table X		Formerly Table Y		Formerly Table Z	
6	4.80	0.0911	4.40	0.1170	5.00	0.0774
8	6.80	0.1246	6.40	0.1617	7.00	0.1054
10	8.80	0.1580	8.40	0.2064	9.00	0.1334
12	10.80	0.1914	10.40	0.2511	11.00	0.1612
15	13.60	0.2796	13.00	0.3923	14.00	0.2031
18	16.40	0.3852	16.00	0.4760	16.80	0.2918
22	20.22	0.5308	19.62	0.6974	20.82	0.3589
28	26.22	0.6814	25.62	0.8985	26.82	0.4594
35	32.63	1.1334	32.03	1.4085	33.63	0.6701
42	39.63	1.3675	39.03	1.6996	40.43	0.9216
54	51.63	1.7691	50.03	2.9052	52.23	1.3343
76.1	73.22	3.1287	72.22	4.1437	73.82	2.5131
108	105.12	4.4666	103.12	7.3745	105.72	3.5834
133	130.38	5.5151	–	–	130.38	5.5151
159	155.38	8.7795	–	–	156.38	6.6056

Dimensions of Stainless Steel Pipes to BS 4127

Outside Diameter (mm)	Maximum Outside Diameter (mm)	Minimum Outside Diameter (mm)	Wall Thickness (mm)	Working Pressure (bar)
6	6.045	5.940	0.6	330
8	8.045	7.940	0.6	260
10	10.045	9.940	0.6	210
12	12.045	11.940	0.6	170
15	15.045	14.940	0.6	140
18	18.045	17.940	0.7	135
22	22.055	21.950	0.7	110
28	28.055	27.950	0.8	121
35	35.070	34.965	1.0	100
42	42.070	41.965	1.1	91
54	54.090	53.940	1.2	77

PIPES, WATER, STORAGE, INSULATION

Dimensions of Steel Pipes to BS 1387

Nominal Size	Approx. Outside Diameter	Outside Diameter				Thickness		
		Light		Medium & Heavy		Light	Medium	Heavy
		Max	Min	Max	Min			
(mm)	(mm)	(mm)	(mm)	(mm)	(mm)	(mm)	(mm)	(mm)
6	10.20	10.10	9.70	10.40	9.80	1.80	2.00	2.65
8	13.50	13.60	13.20	13.90	13.30	1.80	2.35	2.90
10	17.20	17.10	16.70	17.40	16.80	1.80	2.35	2.90
15	21.30	21.40	21.00	21.70	21.10	2.00	2.65	3.25
20	26.90	26.90	26.40	27.20	26.60	2.35	2.65	3.25
25	33.70	33.80	33.20	34.20	33.40	2.65	3.25	4.05
32	42.40	42.50	41.90	42.90	42.10	2.65	3.25	4.05
40	48.30	48.40	47.80	48.80	48.00	2.90	3.25	4.05
50	60.30	60.20	59.60	60.80	59.80	2.90	3.65	4.50
65	76.10	76.00	75.20	76.60	75.40	3.25	3.65	4.50
80	88.90	88.70	87.90	89.50	88.10	3.25	4.05	4.85
100	114.30	113.90	113.00	114.90	113.30	3.65	4.50	5.40
125	139.70	–	–	140.60	138.70	–	4.85	5.40
150	165.1*	–	–	166.10	164.10	–	4.85	5.40

* 165.1 mm (6.5in) outside diameter is not generally recommended except where screwing to BS 21 is necessary
All dimensions are in accordance with ISO R65 except approximate outside diameters which are in accordance with ISO R64
Light quality is equivalent to ISO R65 Light Series II

Approximate Metres Per Tonne of Tubes to BS 1387

Nom. Size	BLACK						GALVANIZED					
	Plain/screwed ends			Screwed & socketed			Plain/screwed ends			Screwed & socketed		
	L	M	H	L	M	H	L	M	H	L	M	H
(mm)	(m)	(m)	(m)	(m)	(m)	(m)	(m)	(m)	(m)	(m)	(m)	(m)
6	2765	2461	2030	2743	2443	2018	2604	2333	1948	2584	2317	1937
8	1936	1538	1300	1920	1527	1292	1826	1467	1254	1811	1458	1247
10	1483	1173	979	1471	1165	974	1400	1120	944	1386	1113	939
15	1050	817	688	1040	811	684	996	785	665	987	779	661
20	712	634	529	704	628	525	679	609	512	673	603	508
25	498	410	336	494	407	334	478	396	327	474	394	325
32	388	319	260	384	316	259	373	308	254	369	305	252
40	307	277	226	303	273	223	296	268	220	292	264	217
50	244	196	162	239	194	160	235	191	158	231	188	157
65	172	153	127	169	151	125	167	149	124	163	146	122
80	147	118	99	143	116	98	142	115	97	139	113	96
100	101	82	69	98	81	68	98	81	68	95	79	67
125	–	62	56	–	60	55	–	60	55	–	59	54
150	–	52	47	–	50	46	–	51	46	–	49	45

The figures for 'plain or screwed ends' apply also to tubes to BS 1775 of equivalent size and thickness
Key:
L – Light
M – Medium
H – Heavy

PIPES, WATER, STORAGE, INSULATION

Flange Dimension Chart to BS 4504 & BS 10

Normal Pressure Rating (PN 6) 6 Bar

Nom. Size	Flange Outside Dia.	Table 6/2 Forged Welding Neck	Table 6/3 Plate Slip on	Table 6/4 Forged Bossed Screwed	Table 6/5 Forged Bossed Slip on	Table 6/8 Plate Blank	Raised Face Dia.	Raised Face T'ness	Nr. Bolt Hole	Size of Bolt
15	80	12	12	12	12	12	40	2	4	M10 × 40
20	90	14	14	14	14	14	50	2	4	M10 × 45
25	100	14	14	14	14	14	60	2	4	M10 × 45
32	120	14	16	14	14	14	70	2	4	M12 × 45
40	130	14	16	14	14	14	80	3	4	M12 × 45
50	140	14	16	14	14	14	90	3	4	M12 × 45
65	160	14	16	14	14	14	110	3	4	M12 × 45
80	190	16	18	16	16	16	128	3	4	M16 × 55
100	210	16	18	16	16	16	148	3	4	M16 × 55
125	240	18	20	18	18	18	178	3	8	M16 × 60
150	265	18	20	18	18	18	202	3	8	M16 × 60
200	320	20	22	–	20	20	258	3	8	M16 × 60
250	375	22	24	–	22	22	312	3	12	M16 × 65
300	440	22	24	–	22	22	365	4	12	M20 × 70

Normal Pressure Rating (PN 16) 16 Bar

Nom. Size	Flange Outside Dia.	Table 6/2 Forged Welding Neck	Table 6/3 Plate Slip on	Table 6/4 Forged Bossed Screwed	Table 6/5 Forged Bossed Slip on	Table 6/8 Plate Blank	Raised Face Dia.	Raised Face T'ness	Nr. Bolt Hole	Size of Bolt
15	95	14	14	14	14	14	45	2	4	M12 × 45
20	105	16	16	16	16	16	58	2	4	M12 × 50
25	115	16	16	16	16	16	68	2	4	M12 × 50
32	140	16	16	16	16	16	78	2	4	M16 × 55
40	150	16	16	16	16	16	88	3	4	M16 × 55
50	165	18	18	18	18	18	102	3	4	M16 × 60
65	185	18	18	18	18	18	122	3	4	M16 × 60
80	200	20	20	20	20	20	138	3	8	M16 × 60
100	220	20	20	20	20	20	158	3	8	M16 × 65
125	250	22	22	22	22	22	188	3	8	M16 × 70
150	285	22	22	22	22	22	212	3	8	M20 × 70
200	340	24	24	–	24	24	268	3	12	M20 × 75
250	405	26	26	–	26	26	320	3	12	M24 × 90
300	460	28	28	–	28	28	378	4	12	M24 × 90

PIPES, WATER, STORAGE, INSULATION

Minimum Distances Between Supports/Fixings

Material	BS Nominal Pipe Size		Pipes – Vertical	Pipes – Horizontal on to low gradients
	(inch)	(mm)	Support distance in metres	Support distance in metres
Copper	0.50	15.00	1.90	1.30
	0.75	22.00	2.50	1.90
	1.00	28.00	2.50	1.90
	1.25	35.00	2.80	2.50
	1.50	42.00	2.80	2.50
	2.00	54.00	3.90	2.50
	2.50	67.00	3.90	2.80
	3.00	76.10	3.90	2.80
	4.00	108.00	3.90	2.80
	5.00	133.00	3.90	2.80
	6.00	159.00	3.90	2.80
muPVC	1.25	32.00	1.20	0.50
	1.50	40.00	1.20	0.50
	2.00	50.00	1.20	0.60
Polypropylene	1.25	32.00	1.20	0.50
	1.50	40.00	1.20	0.50
uPVC	–	82.40	1.20	0.50
	–	110.00	1.80	0.90
	–	160.00	1.80	1.20
Steel	0.50	15.00	2.40	1.80
	0.75	20.00	3.00	2.40
	1.00	25.00	3.00	2.40
	1.25	32.00	3.00	2.40
	1.50	40.00	3.70	2.40
	2.00	50.00	3.70	2.40
	2.50	65.00	4.60	3.00
	3.00	80.40	4.60	3.00
	4.00	100.00	4.60	3.00
	5.00	125.00	5.50	3.70
	6.00	150.00	5.50	4.50
	8.00	200.00	8.50	6.00
	10.00	250.00	9.00	6.50
	12.00	300.00	10.00	7.00
	16.00	400.00	10.00	8.25

PIPES, WATER, STORAGE, INSULATION

Litres of Water Storage Required Per Person Per Building Type

Type of Building	Storage (litres)
Houses and flats (up to 4 bedrooms)	120/bedroom
Houses and flats (more than 4 bedrooms)	100/bedroom
Hostels	90/bed
Hotels	200/bed
Nurses homes and medical quarters	120/bed
Offices with canteen	45/person
Offices without canteen	40/person
Restaurants	7/meal
Boarding schools	90/person
Day schools – Primary	15/person
Day schools – Secondary	20/person

Recommended Air Conditioning Design Loads

Building Type	Design Loading
Computer rooms	500 W/m² of floor area
Restaurants	150 W/m² of floor area
Banks (main area)	100 W/m² of floor area
Supermarkets	25 W/m² of floor area
Large office block (exterior zone)	100 W/m² of floor area
Large office block (interior zone)	80 W/m² of floor area
Small office block (interior zone)	80 W/m² of floor area

PIPES, WATER, STORAGE, INSULATION

Capacity and Dimensions of Galvanized Mild Steel Cisterns – BS 417

Capacity (litres)	BS type (SCM)	Length (mm)	Dimensions Width (mm)	Depth (mm)
18	45	457	305	305
36	70	610	305	371
54	90	610	406	371
68	110	610	432	432
86	135	610	457	482
114	180	686	508	508
159	230	736	559	559
191	270	762	584	610
227	320	914	610	584
264	360	914	660	610
327	450/1	1220	610	610
336	450/2	965	686	686
423	570	965	762	787
491	680	1090	864	736
709	910	1070	889	889

Capacity of Cold Water Polypropylene Storage Cisterns – BS 4213

Capacity (litres)	BS type (PC)	Maximum height (mm)
18	4	310
36	8	380
68	15	430
91	20	510
114	25	530
182	40	610
227	50	660
273	60	660
318	70	660
455	100	760

PIPES, WATER, STORAGE, INSULATION

Minimum Insulation Thickness to Protect Against Freezing for Domestic Cold Water Systems (8 Hour Evaluation Period)

Pipe size (mm)	Insulation thickness (mm)					
	Condition 1			Condition 2		
	λ = 0.020	λ = 0.030	λ = 0.040	λ = 0.020	λ = 0.030	λ = 0.040
Copper pipes						
15	11	20	34	12	23	41
22	6	9	13	6	10	15
28	4	6	9	4	7	10
35	3	5	7	4	5	7
42	3	4	5	8	4	6
54	2	3	4	2	3	4
76	2	2	3	2	2	3
Steel pipes						
15	9	15	24	10	18	29
20	6	9	13	6	10	15
25	4	7	9	5	7	10
32	3	5	6	3	5	7
40	3	4	5	3	4	6
50	2	3	4	2	3	4
65	2	2	3	2	3	3

Condition 1: water temperature 7°C; ambient temperature –6°C; evaluation period 8 h; permitted ice formation 50%; normal installation, i.e. inside the building and inside the envelope of the structural insulation
Condition 2: water temperature 2°C; ambient temperature –6°C; evaluation period 8 h; permitted ice formation 50%; extreme installation, i.e. inside the building but outside the envelope of the structural insulation
λ = thermal conductivity [W/(mK)]

Insulation Thickness for Chilled And Cold Water Supplies to Prevent Condensation

On a Low Emissivity Outer Surface (0.05, i.e. Bright Reinforced Aluminium Foil) with an Ambient Temperature of +25°C and a Relative Humidity of 80%

Steel pipe size (mm)	t = +10			t = +5			t = 0		
	Insulation thickness (mm)			Insulation thickness (mm)			Insulation thickness (mm)		
	λ = 0.030	λ = 0.040	λ = 0.050	λ = 0.030	λ = 0.040	λ = 0.050	λ = 0.030	λ = 0.040	λ = 0.050
15	16	20	25	22	28	34	28	36	43
25	18	24	29	25	32	39	32	41	50
50	22	28	34	30	39	47	38	49	60
100	26	34	41	36	47	57	46	60	73
150	29	38	46	40	52	64	51	67	82
250	33	43	53	46	60	74	59	77	94
Flat surfaces	39	52	65	56	75	93	73	97	122

t = temperature of contents (°C)
λ = thermal conductivity at mean temperature of insulation [W/(mK)]

PIPES, WATER, STORAGE, INSULATION

Insulation Thickness for Non-domestic Heating Installations to Control Heat Loss

Steel pipe size (mm)	t = 75			t = 100			t = 150		
	Insulation thickness (mm)			Insulation thickness (mm)			Insulation thickness (mm)		
	λ = 0.030	λ = 0.040	λ = 0.050	λ = 0.030	λ = 0.040	λ = 0.050	λ = 0.030	λ = 0.040	λ = 0.050
10	18	32	55	20	36	62	23	44	77
15	19	34	56	21	38	64	26	47	80
20	21	36	57	23	40	65	28	50	83
25	23	38	58	26	43	68	31	53	85
32	24	39	59	28	45	69	33	55	87
40	25	40	60	29	47	70	35	57	88
50	27	42	61	31	49	72	37	59	90
65	29	43	62	33	51	74	40	63	92
80	30	44	62	35	52	75	42	65	94
100	31	46	63	37	54	76	45	68	96
150	33	48	64	40	57	77	50	73	100
200	35	49	65	42	59	79	53	76	103
250	36	50	66	43	61	80	55	78	105

t = hot face temperature (°C)
λ = thermal conductivity at mean temperature of insulation [W/(mK)]

Index

European Gardens
History, Philosophy and Design

Tom Turner

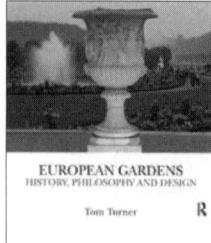

Garden design and usage has been a feature of human civilisation as far back as Neolithic times, when the first gardens began to be used for residential, horticultural and sacred tasks. Tom Turner follows the entire history of the European garden from its prehistoric rootsright up to the present day in this beautifully illustrated book.

European Gardens is divided into ten periods of history and garden development, detailing the advancement of land usage for over 10,000 years. Some of the topics covered in this comprehensive book include the Egyptian gardens of the Pharaohs, the castle gardens of medieval times, eclectic gardens of the nineteenth century and abstract gardens of the last 100 years. The geographical scope of this book covers the whole of the European continent, and touches the garden designs of North Africa and the Middle East.

Tom Turner is a skilled landscape architect and garden historian, who supports his engaging writing with his own detailed plans and diagrams. *European Gardens* also features almost 1,000 colour photographs from across the continent allowing the reader to see for themselves how the design and structure of gardens has developed over time.

A companion to the *Asian Gardens* book, published by Routledge in 2010, *European Gardens* is a development of the original Garden History book from 2004

December 2010: 250x250: 424pp
Hb: 978-0-415-49684-1: **£35.00**

To Order: Tel: +44 (0) 1235 400524 **Fax:** +44 (0) 1235 400525
or Post: Taylor and Francis Customer Services,
Bookpoint Ltd, Unit T1, 200 Milton Park, Abingdon, Oxon, OX14 4TA UK
Email: book.orders@tandf.co.uk

For a complete listing of all our titles visit:
www.tandf.co.uk

Electronic Version Single-User Licence Agreement

We welcome you as a user of this Spon Price Book Software and hope that you find it a useful and valuable tool. Please read this document carefully. **This is a legal agreement** between you (hereinafter referred to as the "Licensee") and Taylor and Francis Books Ltd. (the "Publisher"), which defines the terms under which you may use the Product. **By accessing and retrieving the access code on the label inside the back cover of this book you agree to these terms and conditions outlined herein. If you do not agree to these terms you must return the Product to your supplier intact, with the seal on the label unbroken and with the access code not accessed.**

1. **Definition of the Product**
   The product which is the subject of this Agreement, (the "Product") consists of internet access for searching and viewing the price book data from *Spon's External Works and Landscape Price Book 2013* online.

2. **Commencement and licence**
   2.1 This Agreement commences upon the breaking open of the document containing the access code by the Licensee (the "Commencement Date").
   2.2 This is a licence agreement (the "Agreement") for the use of the Product by the Licensee, and not an agreement for sale.
   2.3 The Publisher licenses the Licensee on a non-exclusive and non-transferable basis to use the Product on condition that the Licensee complies with this Agreement. The Licensee acknowledges that it is only permitted to use the Product in accordance with this Agreement.

3. **Multiple use**
   For more than one user or for a wide area network or consortium, use of the Product is only permissible with the purchase from the Publisher of a multiple-user licence and adherence to the terms and conditions of that licence.

4. **Installation and Use**
   4.1 The Licensee may provide access to the Product for individual study in the following manner: The Licensee may install the Product on a secure local area network on a single site for use by one user.
   4.2 The Licensee shall be responsible for installing the Product and for the effectiveness of such installation.
   4.3 Text from the Product may be incorporated in a coursepack. Such use is only permissible with the express permission of the Publisher in writing and requires the payment of the appropriate fee as specified by the Publisher and signature of a separate licence agreement.
   4.4 The Product is a free addition to the book and the Publisher is under no obligation to provide any technical support.

5. **Permitted Activities**
   5.1 The Licensee shall be entitled to use the Product for its own internal purposes;
   5.2 The Licensee acknowledges that its rights to use the Product are strictly set out in this Agreement, and all other uses (whether expressly mentioned in Clause 6 below or not) are prohibited.

6. **Prohibited Activities**
   The following are prohibited without the express permission of the Publisher:
   6.1 The commercial exploitation of any part of the Product.
   6.2 The rental, loan, (free or for money or money's worth) or hire purchase of this product, save with the express consent of the Publisher.
   6.3 Any activity which raises the reasonable prospect of impeding the Publisher's ability or opportunities to market the Product.
   6.4 Any networking, physical or electronic distribution or dissemination of the product save as expressly permitted by this Agreement.
   6.5 Any reverse engineering, decompilation, disassembly or other alteration of the Product save in accordance with applicable national laws.
   6.6 The right to create any derivative product or service from the Product save as expressly provided for in this Agreement.
   6.7 Any alteration, amendment, modification or deletion from the Product, whether for the purposes of error correction or otherwise.

7. General Responsibilities of the License

7.1 The Licensee will take all reasonable steps to ensure that the Product is used in accordance with the terms and conditions of this Agreement.

7.2 The Licensee acknowledges that damages may not be a sufficient remedy for the Publisher in the event of breach of this Agreement by the Licensee, and that an injunction may be appropriate.

7.3 The Licensee undertakes to keep the Product safe and to use its best endeavours to ensure that the product does not fall into the hands of third parties, whether as a result of theft or otherwise.

7.4 Where information of a confidential nature relating to the product of the business affairs of the Publisher comes into the possession of the Licensee pursuant to this Agreement (or otherwise), the Licensee agrees to use such information solely for the purposes of this Agreement, and under no circumstances to disclose any element of the information to any third party save strictly as permitted under this Agreement. For the avoidance of doubt, the Licensee's obligations under this sub-clause 7.4 shall survive the termination of this Agreement.

8. Warrant and Liability

8.1 The Publisher warrants that it has the authority to enter into this agreement and that it has secured all rights and permissions necessary to enable the Licensee to use the Product in accordance with this Agreement.

8.2 The Publisher warrants that the Product as supplied on the Commencement Date shall be free of defects in materials and workmanship, and undertakes to replace any defective Product within 28 days of notice of such defect being received provided such notice is received within 30 days of such supply. As an alternative to replacement, the Publisher agrees fully to refund the Licensee in such circumstances, if the Licensee so requests, provided that the Licensee returns this copy of *Spon's External Works and Landscape Price Book 2013* to the Publisher. The provisions of this sub-clause 8.2 do not apply where the defect results from an accident or from misuse of the product by the Licensee.

8.3 Sub-clause 8.2 sets out the sole and exclusive remedy of the Licensee in relation to defects in the Product.

8.4 The Publisher and the Licensee acknowledge that the Publisher supplies the Product on an "as is" basis. The Publisher gives no warranties:

8.4.1 that the Product satisfies the individual requirements of the Licensee; or

8.4.2 that the Product is otherwise fit for the Licensee's purpose; or

8.4.3 that the Product is compatible with the Licensee's hardware equipment and software operating environment.

8.5 The Publisher hereby disclaims all warranties and conditions, express or implied, which are not stated above.

8.6 Nothing in this Clause 8 limits the Publisher's liability to the Licensee in the event of death or personal injury resulting from the Publisher's negligence.

8.7 The Publisher hereby excludes liability for loss of revenue, reputation, business, profits, or for indirect or consequential losses, irrespective of whether the Publisher was advised by the Licensee of the potential of such losses.

8.8 The Licensee acknowledges the merit of independently verifying the price book data prior to taking any decisions of material significance (commercial or otherwise) based on such data. It is agreed that the Publisher shall not be liable for any losses which result from the Licensee placing reliance on the data under any circumstances.

8.9 Subject to sub-clause 8.6 above, the Publisher's liability under this Agreement shall be limited to the purchase price.

9. Intellectual Property Rights

9.1 Nothing in this Agreement affects the ownership of copyright or other intellectual property rights in the Product.

9.2 The Licensee agrees to display the Publishers' copyright notice in the manner described in the Product.

9.3 The Licensee hereby agrees to abide by copyright and similar notice requirements required by the Publisher, details of which are as follows:

"© 2013 Taylor & Francis. All rights reserved. All materials in *Spon's External Works and Landscape Price Book 2013* are copyright protected. All rights reserved. No such materials may be used, displayed, modified, adapted, distributed, transmitted, transferred, published or otherwise reproduced in any form or by any means now or hereafter developed other than strictly in accordance with the terms of the licence agreement enclosed with *Spon's External Works and Landscape Price Book 2013*. However, text and images may be printed and copied for research and private study within the preset program limitations. Please note the copyright notice above, and that any text or images printed or copied must credit the source."

9.4 This Product contains material proprietary to and copyedited by the Publisher and others. Except for the licence granted herein, all rights, title and interest in the Product, in all languages, formats and media throughout the world, including copyrights therein, are and remain the property of the Publisher or other copyright holders identified in the Product.

10. Non-assignment

This Agreement and the licence contained within it may not be assigned to any other person or entity without the written consent of the Publisher.

11. Termination and Consequences of Termination.

11.1 The Publisher shall have the right to terminate this Agreement if:

11.1.1 the Licensee is in material breach of this Agreement and fails to remedy such breach (where capable of remedy) within 14 days of a written notice from the Publisher requiring it to do so; or

11.1.2 the Licensee becomes insolvent, becomes subject to receivership, liquidation or similar external administration; or

11.1.3 the Licensee ceases to operate in business.

11.2 The Licensee shall have the right to terminate this Agreement for any reason upon two month's written notice. The Licensee shall not be entitled to any refund for payments made under this Agreement prior to termination under this sub-clause 11.2.

11.3 Termination by either of the parties is without prejudice to any other rights or remedies under the general law to which they may be entitled, or which survive such termination (including rights of the Publisher under sub-clause 7.4 above).

11.4 Upon termination of this Agreement, or expiry of its terms, the Licensee must destroy all copies and any back up copies of the product or part thereof.

12. General

12.1 *Compliance with export provisions*

The Publisher hereby agrees to comply fully with all relevant export laws and regulations of the United Kingdom to ensure that the Product is not exported, directly or indirectly, in violation of English law.

12.2 *Force majeure*

The parties accept no responsibility for breaches of this Agreement occurring as a result of circumstances beyond their control.

12.3 *No waiver*

Any failure or delay by either party to exercise or enforce any right conferred by this Agreement shall not be deemed to be a waiver of such right.

12.4 *Entire agreement*

This Agreement represents the entire agreement between the Publisher and the Licensee concerning the Product. The terms of this Agreement supersede all prior purchase orders, written terms and conditions, written or verbal representations, advertising or statements relating in any way to the Product.

12.5 *Severability*

If any provision of this Agreement is found to be invalid or unenforceable by a court of law of competent jurisdiction, such a finding shall not affect the other provisions of this Agreement and all provisions of this Agreement unaffected by such a finding shall remain in full force and effect.

12.6 *Variations*

This agreement may only be varied in writing by means of variation signed in writing by both parties.

12.7 *Notices*

All notices to be delivered to: Spon's Price Books, Taylor & Francis Books Ltd., 3 Park Square, Milton Park, Abingdon, Oxfordshire, OX14 4RN, UK.

12.8 *Governing law*

This Agreement is governed by English law and the parties hereby agree that any dispute arising under this Agreement shall be subject to the jurisdiction of the English courts.

If you have any queries about the terms of this licence, please contact:

Spon's Price Books
Taylor & Francis Books Ltd.
3 Park Square, Milton Park, Abingdon, Oxfordshire, OX14 4RN

Multiple-user use of the Spon Press Software and eBook

To buy a licence to install *Spon's Online* on a secure local area network or a wide area network, and for the supply of network key files, for an agreed number of users please contact:

Spon's Price Books
Taylor & Francis Books Ltd.
2 Park Square, Milton Park, Abingdon, Oxfordshire, OX14 4RN
Tel: +44 (0) 207 017 6000
www.pricebooks.co.uk

Number of users	Licence cost
2–5	£450
6–10	£915
11–20	£1400
21–30	£2150
31–50	£4200
51–75	£5900
76–100	£7100
Over 100	Please contact Spon for details

Spon's Online Installation and Use Instructions

General

Spon's Online is available via the web using the given access code as outlined below.

No CD-ROM is supplied. For set up and use you will need internet access.

Setting up

Open your internet browser, type in www.pricebooks.co.uk/downloads and follow the instructions on screen.

Retrieve your unique access code by scratching off the security-protected label inside the back cover of this book.

Type in your access code as directed.

When your access code is validated successfully, please give your email address and choose a memorable password to register yourself.

You will be sent an activation email instantly and can begin to use *Spon's Online*.

Please note: you will only be allowed one use of the access code and one opportunity to register, although you will be able to change your password later if you wish.

Use

The set-up process will give you access via www.pricebooks.co.uk/SponsOnline through your regular internet browser.

Technical Support

Spon's Online is a free addition to the book.

For help with setting up please visit
www.pricebooks.co.uk/esupport

Guidance on use is provided on
www.pricebooks.co.uk/SponsOnline

Free Updates

with three easy steps…

1. Register today on www.pricebooks.co.uk/updates

2. We'll alert you by email when new updates are posted on our website

3. Then go to www.pricebooks.co.uk/updates
   and download the update.

All four Spon Price Books – *Architects' and Builders'*, *Civil Engineering and Highway Works*, *External Works and Landscape* and *Mechanical and Electrical Services* – are supported by an updating service. Two or three updates are loaded on our website during the year, typically in November, February and May. Each gives details of changes in prices of materials, wage rates and other significant items, with regional price level adjustments for Northern Ireland, Scotland and Wales and regions of England. The updates terminate with the publication of the next annual edition.

As a purchaser of a Spon Price Book you are entitled to this updating service for this 2013 edition – free of charge. Simply register via the website www.pricebooks.co.uk/updates and we will send you an email when each update becomes available.

If you haven't got internet access or if you've some questions about the updatees please write to us at Spon Price Book Updates, Spon Press Marketing Department, 3 Park Square, Milton Park, Abingdon, Oxfordshire, OX14 4RN.

Find out more about Spon b**
Visit www.sponpress.com for r